The commentary text in this handbook is written to assist users in understanding and applying the provisions of the Code. The commentary explains the reasoning behind the Code's requirements and provides numerous examples, tables, photographs, and illustrations. The commentary text is printed in blue type to distinguish it from the text of the Code. Please note that the commentary is not part of the Code and therefore is not enforceable.

A conversion burn
heating oil. This in
ing the entire appli

MW01716769

**EXHIBIT 10.16** *Gas Conversion Burner. (Courtesy of Wayne Combustion Systems)*

## 10.6 Decorative Appliances for Installation in Vented Fireplaces

Decorative appliances (sometimes called "gas logs") for installation in vented fireplaces, which are covered in Section 10.6 and shown in Exhibit 10.17, are self-contained, freestanding, fuel-gas-burning appliances designed for installation only in vented fireplaces. These decorative appliances rely on fireplace venting to vent their products of combustion.

**EXHIBIT 10.17** *Decorative Gas Appliance Installed in a Wood-Burning Fireplace. (Courtesy of DESA International)*

The primary function of a decorative appliance lies in the aesthetic effect of its flames. The four different types of decorative appliances are described as follows:

1. Coal basket: An open-flame appliance with a metal basket filled with simulated coals.
2. Fireplace insert: An open-flame, radiant-type appliance mounted in a decorative metal panel that covers the fireplace opening. This appliance has a provision for venting to the fireplace chimney.
3. Gas log: An open-flame appliance consisting of a metal frame or base supporting simulated logs.
4. Radiant appliance: An open-flame appliance designed primarily to convert the energy in fuel gas to radiant heat by means of refractory or similarly radiating materials. A radiant heater typically has no external jacket.

### 10.6.1* Prohibited Installations.

Decorative appliances for installation in vented fireplaces shall not be installed in bathrooms or bedrooms unless the appliance is listed and the bedroom or bathroom has the required vol-



interconnected piping and other portions of the air-conditioning unit. Provisions for expansion of the water in the piping system and remote chiller will protect the chilled water coil.

(3) Refrigeration coils shall be installed in parallel with or on the downstream side of central furnaces to avoid condensation in the heating element, unless the furnace has been specifically listed for downstream installation. With a parallel flow arrangement, the dampers or other means used to control flow of air shall be sufficiently tight to prevent any circulation of cooled air through the furnace.

Subsection 10.3.8(3) requires that only a furnace specifically constructed and designed for installation downstream from a cooling coil be so installed, and that the heat exchanger and burners be of corrosion-resistant material. If the furnace is listed for downstream installation, the tests conducted on the furnace will show that there is no danger of condensate dripping on the burners and pilots during the cooling cycle or that the heat exchanger will corrode. When a parallel flow system is designed, there should be face and bypass dampers in front of both the cooling coil and the heating unit, to ensure that there is no recirculation of chilled air back over the heat exchanger.

Exhibit 10.12 illustrates the permissible location of refrigeration coils in forced air furnaces. Exhibit 10.12(a) shows the refrigeration coil upstream to the furnace fan, which is not permitted by the code. Exhibit 10.12(b) shows the refrigeration coil downstream from the furnace fan, which is permitted by the code. Note that the installation in Exhibit 10.12(b) would not require a listed furnace and would meet the code requirements if blower capacity is sufficient.

(4) Means shall be provided for disposal of condensate and to prevent dripping of condensate on the heating element.

A typical coil for installation in a plenum above an upflow furnace, complying with the condensate drain requirement of 10.3.8(4), is shown in Exhibit 10.12(b). The "A-type" coil is constructed so that the condensate forming on the coil will travel down the edge of the coil fins to the drain pan connection. Note also that this particular coil has a secondary drain

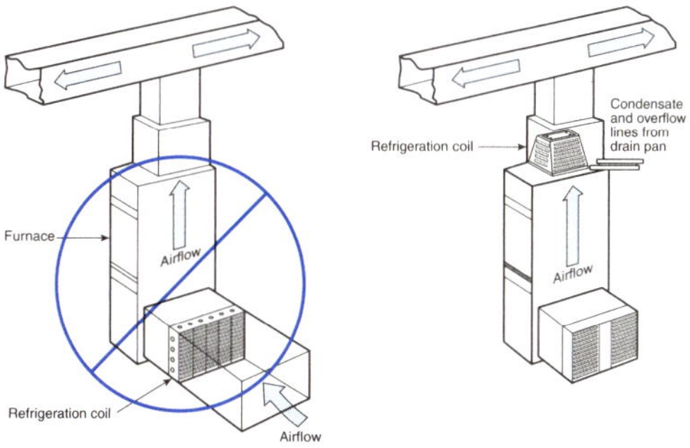


(a) Refrigeration coil upstream to furnace fan

(b) Refrigeration coil downstream of furnace fan

**EXHIBIT 10.12** *Refrigeration Coils in Forced Air Furnaces: (a) Location of Coils Not Permitted and (b) Location of Coils Permitted.*



Commentary art is set within blue lines and labeled "Exhibit." The caption is printed in blue ink. The commentary exhibits, including both drawings and photographs, provide detailed views of the Code's concepts and are numbered sequentially throughout each chapter.

# National Fuel Gas Code Handbook

## SEVENTH EDITION

EDITED BY

Theodore C. Lemoff, P.E.
Principal Gases Engineer, NFPA

With the complete text of the 2009 edition of NFPA® 54, *National Fuel Gas Code*

NFPA® National Fire Protection Association®
Quincy, Massachusetts

AGA American Gas Association
Washington, DC

Product Manager: Debra Rose
Developmental Editor: Khela Thorne
Project Editor: Irene Herlihy
Permissions Editor: Josiane Domenici
Copy Editor: Kathleen Allain
Composition: Modern Graphics, Inc.

Art Coordinator: Cheryl Langway
Illustrations: George Nichols
Cover Design: McCusker Design
Manufacturing Manager: Ellen Glisker
Printer: R.R. Donnelley/Willard


One Batterymarch Park
Quincy, Massachusetts 02169-7471

American Gas Association
400 N. Capitol St. NW
Suite 450
Washington, DC 20001



**Notice Concerning Liability:** Publication of this handbook is for the purpose of circulating information and opinion among those concerned for fire and electrical safety and related subjects. While every effort has been made to achieve a work of high quality, neither NFPA®, AGA, nor the contributors to this handbook guarantee the accuracy or completeness of or assume any liability in connection with the information and opinions contained in this handbook. NFPA, AGA, and the contributors shall in no event be liable for any personal injury, property, or other damages of any nature whatsoever, whether special, indirect, consequential, or compensatory, directly or indirectly resulting from the publication, use of, or reliance upon this handbook.

This handbook is published with the understanding that NFPA, AGA, and the contributors to this handbook are supplying information and opinion but are not attempting to render engineering or other professional services. If such services are required, the assistance of an appropriate professional should be sought.

NFPA codes and standards are made available for use subject to Important Notices and Legal Disclaimers, which appear at the end of this handbook and can also be viewed at *www.nfpa.org/disclaimers.*

**Notice Concerning Code Interpretations:** This seventh edition of *National Fuel Gas Code Handbook* is based on the 2009 edition of NFPA 54, *National Fuel Gas Code.* All NFPA codes, standards, recommended practices, and guides are developed in accordance with the published procedures of the NFPA by technical committees comprised of volunteers drawn from a broad array of relevant interests. The handbook contains the complete text of NFPA 54 and any applicable Formal Interpretations issued by the Association. These documents are accompanied by explanatory commentary and other supplementary materials.

The commentary and supplementary materials in this handbook are not a part of the Code and do not constitute Formal Interpretations of the NFPA (which can be obtained only through requests processed by the responsible technical committees in accordance with the published procedures of the NFPA). The commentary and supplementary materials, therefore, solely reflect the personal opinions of the editor or other contributors and do not necessarily represent the official position of the NFPA or its technical committees.



NFPA No.: 54HB09
ISBN-13: 978-0-87765-827-6
ISBN-10: 0-87765-827-7
Library of Congress Control No.: 2008933729

Printed in the United States of America
09 10 11 12 13 5 4 3 2 1

# Contents

# Preface

This seventh edition of the *National Fuel Gas Code Handbook* continues the work begun in 1988 to help users of the *National Fuel Gas Code* to understand and use the code more easily. The code must be written using a technically clear, usable, and enforceable style suitable for adoption into law. This handbook provides commentary that brings the legalistic language of the code to life.

This handbook has grown considerably since its first edition, published in 1988. New material in the code accounts for part of this growth, but most of the increase comes from added commentary, revised supplements, and calculation worksheets that are intended to be reproduced and used.

The purpose of this handbook is to assist enforcers or users of the *National Fuel Gas Code* in installing gas piping, gas appliances, and venting for gas appliances.

The 2009 edition of the code contains several important changes. The reader is referred to Supplement 8, Technical/Substantive Changes from the 2006 Edition to the 2009 Edition of NFPA 54, where the significant changes and the reasons thereof are explained.


## Acknowledgments

The contributions of all those who participated in the development of this seventh edition of the handbook are very much appreciated. First and foremost, I express gratitude to the National Fuel Gas Code Committee, the authors of the code. The committee's attention to technical detail and hard work within the consensus code-making process make NFPA 54 a document that commands respect.

For each edition, the assistance of experts in a field covered by the code is obtained to assist in keeping the commentary up to date, to the degree possible. The development of this seventh edition of the *National Fuel Gas Code Handbook* was assisted by three contributors, Thomas Crane, Scott Sollars, and Matt Wilber of Crane Engineering in Minneapolis. Crane Engineering is an engineering firm that concentrates in forensic engineering. Their work in engineering and scientific research, development, and forensic analysis provided a solid background for their recommendations for the handbook commentary. The process of review involved many back and forth discussions of what the code meant and what was reality, and both parties agree that the handbook benefits from this process.

In addition, I thank Paul Cabot, committee secretary and staff liaison to the ANSI Z223 committee, for his review of the commentary and recommendations to improve the book.

Many readers of the handbook volunteer improvements, and for this edition one stands out. Mr. Hall Virgil of Indianapolis, Indiana, retired from Carrier Corporation, has provided his expertise in the areas of venting and furnaces. His contributions are appreciated.

Of course, any edition of a book builds on the editions that have preceded it, and the following people are acknowledged for their work on previous editions of this handbook:

- Richard White, Orrin Burwell, J. Herbert Witte, and Forrest G. Hammaker, Jr., assisted with the editing of the inaugural edition of the *National Fuel Gas Code Handbook* in 1988.

- Joseph Drechsler contributed significantly to the second edition, published in 1992.
- Alan Callahan assisted with the editing of the second edition. Committee members Ross Burnside and Mike Gorham were contributing authors to the third edition.
- Robert Borgeson and Bradford Wong were contributing authors to the fourth edition.
- Bradley Wong and Jay Mullowney were contributing authors to the fifth edition.
- Jim Brewer and Roger Rotundo were contributing authors to the sixth edition.

I also appreciate the efforts of the NFPA staff who made this book possible. My sincere thanks go to Product Manager Debra Rose for her assistance and gentle reminders of the practical side of producing a handbook. Thanks also go to Khela Thorne, Developmental Editor; Irene Herlihy, Project Editor; Josiane Domenici, Permissions Editor; and Kathleen Allain, Copy Editor. Finally, I thank my wife, Sharon, for her continued understanding of my time away from home and for my time at home spent on this book and other NFPA projects.

Theodore C. Lemoff, P.E.
Editor

## History of the *National Fuel Gas Code*

The *National Fuel Gas Code* covers the installation of gas piping and gas appliances in buildings. The history of the code dates back approximately 75 years and involves a number of organizations, all with the common goal of providing for the safe use of gas-burning equipment.

The following highlights significant steps in the development of the *National Fuel Gas Code*.

**1913** The National Fire Protection Association's Committee on Explosives and Combustibles, whose scope covered flammable gases, appointed a subcommittee to prepare a fire protection code for city gas installation.

**1920** After several years of cooperative work with the U.S. Bureau of Standards and other interested agencies, the NFPA Committee on Explosives and Combustibles presented the NFPA with a preliminary code, which was adopted in 1920.

The American Gas Association (AGA) asked the American Engineering Standards Committee to draft a *Gas Safety Code* on installation practices under its customary procedure for standards development.

**1927–1928** On March 10, 1927, the American Standards Association approved the *American Standard Gas Safety Code for Installation and Work in Buildings,* K2-1927. The first edition of AGA's *Requirements for House Piping and Appliance Installation* was published in June 1928.

**1933** The NFPA's code for city gas installation, revised in 1932 with the active cooperation of the AGA and the U.S. Bureau of Standards, was approved by the American Standards Association as the *American Recommended Practice for the Installation, Maintenance and Use of Piping and Fittings for City Gas*, Z27-1933. It was printed by NFPA in 1938 in the *National Fire Codes*, Vol. I, "Flammable Liquids, Gases, Chemicals and Explosives." (It was not until 1950 that the NFPA established a general policy of identifying its standards by a number, as well as a title.)

**1940** AGA published the second edition of the *Requirements and Recommended Practice for House Piping and Appliance Installation.*

**1950** *Installation of Gas Appliances and Gas Piping* was published under two separate covers, as ASA Z21.30-1950 and NFPA 54-1950.

**1959** The 1959 edition of the standard incorporated the essential features of NFPA 52, *Liquefied Petroleum Gas Piping and Appliance Installation in Buildings.*

**1971** A Conference Group on Piping and Installation Standards, comprised of representatives of AGA, ASME, and NFPA, was held to consider the development of a national fuel gas code. The work of this group led to the combination of American National Standards Z21.30 and NFPA 54, Z83.1 and NFPA 54A, and ASME B31.2, *Fuel Gas Piping,* into a national fuel gas code. ANSI approved the formation and the scope of activities of an American National Standards Committee on National Fuel Gas Code, Z223, cosponsored by AGA, ASME, and NFPA. ASME left the project after withdrawing ASME B31.4, *Building Services Piping.*

**1974** ANSI and NFPA approved the first edition of the *National Fuel Gas Code* in 1974. Subsequent editions of the *National Fuel Gas Code* were issued in 1980, 1984, 1988, 1992, 1996, 1999, 2002, 2006, and 2009.

# About the Contributors

### Paul Cabot

Paul started his career in the gas utility industry after college by joining the Brooklyn Union Gas Company in 1981. He held a number of engineering positions while at BUG before joining AGA in 1991. He assists the association's efforts to help ensure that uniform and safe installation practices for natural gas are adopted into the nation's building codes. He currently serves as the Secretary to the National Fuel Gas Code Committees, ANSI Z223 and NFPA 54, and is responsible for coordinating efforts between the International Code Council and AGA on the International Fuel Gas Code. Paul is also the Secretary to ANSI Accredited Standards Committee Z380, Gas Piping Technology Committee, which writes guidance material on the federal safety regulations that govern the operation of natural gas pipeline and distribution systems. Paul is a Certified Gas Engineer by the American Society of Gas Engineers.

### Thomas R. Crane

Thomas R. Crane, P.E., has more than 35 years of experience providing engineering expertise to a wide range of designers, manufacturers, testing and investigative clients. He has gained national recognition for his innovative and effective test programs and systems. Tom founded Crane Engineering in 1978 and continues to provide managerial and technical direction for the firm. Tom has provided consulting engineering services to the fuel gas industry for more than 30 years. His industry knowledge continues to be sustained by his presence as a member of the National Fuel Gas Code (NFPA 54) technical committee for more than 12 years.

### Scott A. Sollars

Scott A. Sollars, P.E., has more than 16 years of experience as a mechanical engineer and has worked at Crane Engineering for 13 of those years providing engineering evaluation, performing failure analyses of mechanical systems and equipment, and mechanical testing and modeling. Scott is nationally recognized for his work in the area of fuel gas combustion and fire and explosion analysis. His extensive consultation experience with the fuel gas industry includes clients in component manufacturing, appliance fabrication, and system installation. He is active in his memberships in the American Society of Gas Engineers and ASHRAE.

### Matt Wilber

Matt Wilber, a Gas Application Specialist at Crane Engineering, has more than 30 years of hands-on experience in the gas industry as a technician and manager of codes and standards for CenterPoint Energy. He has served on numerous technical standards writing committees and assisted in the development of national and state legislation relating to standards, codes, and safety issues impacting fuel gas customers. Matt Wilber is currently chairing the ANSI Z21/83 Committee's Central Furnace Technical Advisory Group and has served on numerous state and ANSI Z21/83 sub-committees. At Crane Engineering, Matt performs analysis of mechanical systems and design, mechanical systems failure, explosions and gas-related fires, and carbon monoxide for Crane Engineering. He consults on natural gas and LP systems and controls.

# About the Editor

After graduating from the City College of New York with a Bachelor of Engineering (Chemical) degree, Theodore C. Lemoff was employed in the chemical and petrochemical industry by Proctor and Gamble Company, Sun Chemical Corporation, and Badger Engineers, Inc., in various engineering positions. He joined NFPA in 1985 and currently serves as NFPA's principal gases engineer. He serves as NFPA staff liaison to the National Fuel Gas Code Committee, and he is also responsible for other NFPA standards on flammable gases.

In addition to being the editor of the *National Fuel Gas Code Handbook*, Ted is the editor of the *LP-Gas Code Handbook* and co-author of the *NFPA Pocket Guide to Fuel Gas Storage and Use*.

Ted represents NFPA on the ASME Code for Pressure Piping Committee (National Interest Review Group) (B31), the pipeline safety advisory committee of the U.S. Department of Transportation, and the IAPMO Mechanical and Plumbing Code Committees.

Ted is a registered professional engineer in the Commonwealth of Massachusetts and is a member of the American Institute of Chemical Engineers and the Society of Fire Protection Engineers, the Society of Gas Engineers, and the International Association of Plumbing and Mechanical Officials.

## PART ONE

# NFPA® 54, *National Fuel Gas Code*, with Commentary

Part One of this *National Fuel Gas Code Handbook* includes the complete text and figures of the 2009 edition of NFPA 54, *National Fuel Gas Code*. The text, tables, and figures from the code are printed in black and are the official requirements of NFPA 54. Line drawings and photographs from the code are labeled as "Figures."

An asterisk (*) following a code paragraph number indicates that advisory annex material pertaining to that paragraph appears in Annex A. Paragraphs that begin with the letter A are extracted from Annex A of the code. Although printed in black ink, this nonmandatory material is purely explanatory in nature. For ease of use, this handbook places Annex A material immediately after the code paragraph to which it refers and identifies it with an icon in the margin.

In addition to code text and annexes, Part One includes explanatory commentary that provides the history and other background information for specific paragraphs in the code. This insightful commentary takes the reader behind the scenes, into the reasons underlying the requirements.

Commentary text, captions, and tables are printed in blue, to clarify identification of commentary material. So that the reader can easily distinguish between line drawings and photographs of the code and those of the commentary, line drawings, graphs, and photographs in the commentary are labeled as "Exhibits."

This edition of the handbook includes a frequently asked questions feature. The marginal FAQs are based on the questions most commonly asked of the NFPA 54 staff. The handbook also features a tool designed to help users easily identify important new or revised elements in the standard. New or revised material is identified with a "new" icon in the margin of the book next to that material.

# Administration

## CHAPTER 1

Chapter 1 provides the administrative requirements in the areas of the scope, purpose, retroactivity, equivalency, and enforcement for NFPA 54/ANSI Z223.1, *National Fuel Gas Code*. These areas are required by the *Manual of Style for NFPA Technical Committee Documents*. Chapter 1 covers the following:

- Scope and application of the code: This section includes both installations that are covered by the code and those that are not. (See Section 1.1.)
- Purpose: Although NFPA 54/ANSI Z223.1 does not have an official purpose statement, this section explains why the committee has reserved this section for future use. (See Section 1.2.)
- Retroactivity: NFPA 54/ANSI Z223.1 is retroactive only when this is specifically stated in the code. There are no retroactive requirements in the 2009 edition. (See Section 1.3.)
- Equivalent materials, methods of construction, or installation procedures not specifically addressed in the code: This section of Chapter 1 explains the basis for the authority having jurisdiction to permit alternatives. (See Section 1.4.)
- Enforcement of the code: This section of Chapter 1 states which authorities administer and enforce the code. (See Section 1.5.)

## 1.1 Scope

The scope defines the application (and nonapplication) of NFPA 54/ANSI Z223.1. The scope is important because users of this or any code must know whether or not the right document is being used.

### 1.1.1 Applicability.

The *National Fuel Gas Code* is the American National Standard that applies to the installation of fuel gas piping systems and equipment and appliances that are supplied with natural gas; manufactured gas; liquefied petroleum gas (LP-Gas), in the vapor phase only; LP-Gas–air mixtures; mixtures of these gases; and gas–air mixtures in the flammable range.

◀**FAQ**
What does NPFA 54 cover?

Natural gas, as its name implies, is a naturally occurring product found in many parts of the world. It is recovered by drilling wells into underground pockets of natural gas. The recovered gas is piped to homes and businesses via collection, cleanup, transmission, and utility distribution piping. Another source of natural gas is synthetic natural gas, which is produced by cracking naphtha or other chemical feedstocks to supplement natural gas supplies. Under suitable conditions of elevated pressure and low temperature, natural gas can be liquefied to reduce its volume for purposes of storage or transportation. Liquefied natural gas (LNG) is the source of a larger amount of the natural gas used in North America over the last 5 years, and it appears it will have a larger share in the foreseeable future. The increased use of LNG will have minimal impact on this code, as the gas being delivered to homes and businesses remains the same.

There are no standards or conventions that specify a composition of natural gas. Natural gas consists principally of methane, but it also contains ethane and small amounts of propane, butane, and higher hydrocarbons. Also, natural gas can contain small amounts of nitrogen, carbon dioxide, hydrogen sulfide, and helium. Normally, natural gas contracts specify a heating value [usually 940 Btu/ft$^3$ to 1080 Btu/ft$^3$ (35.0 kJ/m$^3$ to 40.2 kJ/m$^3$)] and a maximum amount of hydrogen sulfide [typically 0.3 grains per 100 ft$^3$ (6.8 mg/m$^3$)]. Hydrogen sulfide is a natural gas contaminant that is corrosive to copper and brass. Some contractual arrangements also limit the carbon dioxide content and water content of natural gas.

Manufactured gas was the first fuel gas distributed in piping systems in cities. It was used in most communities in North America until the early 1950s, when the natural gas transmission system was significantly expanded in the United States. Manufactured gas, a low Btu fuel gas produced by one of several processes, was used primarily for lighting and cooking. Although no longer used in North America (except where produced as a byproduct of the manufacture of metallurgical coke), manufactured gas is still used in some other parts of the world. Refer to Commentary Table 1.1 for the typical composition of some manufactured gases.

LP-Gases include propane and butane, which are used as fuel gases. Pure propane, due to its low boiling point of approximately −44°F (−42°C), vaporizes rapidly and is used extensively as a fuel gas in colder climates. Pure normal butane, due to its higher boiling point of approximately 31°F (−0.6°C), requires a vaporizer for most applications. It is not widely used in the United States as a fuel gas. It is used in warmer countries as a fuel gas and for other applications not covered by this code.

Propane–air mixtures are used by some gas utility companies to supplement natural gas supplies during peak demand periods, such as in extremely cold weather, and by some large users of natural gas as a standby fuel during periods when natural gas supplies are curtailed or are economically unattractive.

Propane has a significantly higher heating value per unit of vapor volume than natural gas [approximately 2500 Btu/ft$^3$ (93 kJ/m$^3$) versus approximately 1000 Btu/ft$^3$ (37 kJ/m$^3$) for natural gas], and mixing propane with air provides a fuel with burning characteristics similar to those of natural gas. Propane–air mixtures are always well above the upper flammable limit of approximately 10 percent for propane in air.

Commentary Table 1.1 gives typical properties of several fuel gases. The compositions of these fuels can vary widely, and the table should not be used for design purposes. Instead, the gas supplier should be contacted for data on the fuel gas to be used.

The storage, handling, and use of liquid LP-Gas are not included under the scope of NFPA 54/ANSI Z223.1. NFPA 58, *Liquefied Petroleum Gas Code*, covers liquid LP-Gas. LP-Gas has two sources: it is a byproduct of natural gas production that is removed during the cleanup of natural gas prior to its entering the gas transmission system and it is a product of petroleum refineries.

**1.1.1.1** This code is a safety code that shall apply to the installation of fuel gas piping systems, appliances, equipment, and related accessories as shown in 1.1.1.1(A) through 1.1.1.1(D).

**(A)** Coverage of piping systems shall extend from the point of delivery to the appliance connections. For other than undiluted liquefied petroleum gas (LP-Gas) systems, the point of delivery shall be considered to be the outlet of the service meter assembly or the outlet of the service regulator or service shutoff valve where no meter is provided. For undiluted LP-Gas, the point of delivery shall be considered to be the outlet of the final pressure regulator, exclusive of line gas regulators, in the system.

As addressed in 1.1.1.1(A), the point of delivery is the outlet of the gas meter where a meter is used, or it is the outlet of the service regulator or service shutoff valve when no meter is installed. The installation of transmission and distribution piping and the gas meter in the United States is covered by the U.S. Department of Transportation (49 CFR 192). These

***COMMENTARY TABLE 1.1** Typical Properties of Typical Fuel Gases*

| Typical Fuel Gases % Constituents and Properties | | Natural Gas | Commercial Propane | Propane HD 5 | Commercial Butane | Water Gas | Producer Gas | Coke Oven Gas |
|---|---|---|---|---|---|---|---|---|
| Butane(s) | | 0.6 | | | 95 min. combined | | | |
| Butylene | | | | | | | | |
| Carbon dioxide | | 0.4 | | | | 5.1 | 4.3 | 2.0 |
| Carbon monoxide | | | | | | 40.2 | 26.7 | 6.2 |
| Ethane | | 6.0 | | | | | | |
| Hydrogen | | | | | | 50.0 | 13.9 | 53.2 |
| Methane | | 81.1 | | | | 0.7 | 3.1 | 26.7 |
| Nitrogen | | 9.2 | | | | 4.0 | 51.5 | 7.0 |
| Pentane(s) | | 0.2 | | | | | | |
| Propane | | 2.1 | 95 min. combined | 95 | | | | |
| Propylene | | | | | | | | |
| Others, unknown, misc. | | 0.4 | 5 max. | 5 max. | 5 max. | Some | 0.5 | 4.9 |
| Heat of combustion (Btu per $ft^3$) | Gross | 1000 | 2500 | 2500 | 3200 | 260 | 165 | 570 |
| | Net | 900 | 2300 | 2300 | 2950 | 240 | 155 | 510 |
| Specific gravity (air = 1.00) | | 0.65 | 1.52 | 1.45 | 1.95 | 0.70 | 0.86 | 0.4 |
| Flammability limits (% volume in air) | Lower | 3.9 | 2.4 | 2.1 | 1.9 | 6.9 | 16.8 | 5.0 |
| | Upper | 15.0 | 9.6 | 10.1 | 8.6 | 69.5 | 73.7 | 28.0 |
| Combustion air ($ft^3$ approx.) | Per $ft^3$ gas | 10 | 25 | 25 | 32 | 2.6 | 1.7 | 5.7 |
| | Per 100 Btu | 1 | 1 | 1 | 1 | 1 | 1 | 1 |

Note: For SI units, 1 Btu/$ft^3$ = 37 kJ/$m^3$; 1 ft/sec = 0.305 m/sec; 1 $ft^3$ = 0.028 $m^3$; 1 Btu = 1055 J.

regulations can be found in ANSI Z380, *Guide for Gas Transmission and Distribution Piping*.

The point of delivery where LP-Gas is used is the outlet of the final pressure regulator. Prior to the 1996 edition of this code, the point of delivery was the outlet of the first-stage pressure regulator. This modification recognized a change in NFPA 58 that mandated two-stage pressure regulation for most LP-Gas systems covered by the *National Fuel Gas Code*. Prior to this change in NFPA 58, buildings could be supplied using a single-stage regulator, or a two-stage regulator system. Therefore, to prevent confusion on the part of Code users, the split was at the discharge of the first-stage pressure regulator, which could be the only propane pressure regulator in the system.

Exhibit 1.1 shows a typical propane supply (illustrating the split between the *National Fuel Gas Code* and NFPA 58) and a typical natural gas supply (illustrating the split between the *National Fuel Gas Code* and Federal Pipeline Regulations [49 CFR 192]). Normally, only one gas supply will serve a building; however, this code does not prohibit more than one fuel gas being used in a building. In many industrial buildings, it is normal to have a fuel and a standby fuel (typically natural gas and propane, respectively) for economic reasons.

***EXHIBIT 1.1*** *Code Jurisdictions for Typical Propane Supply and Typical Natural Gas Supply.*

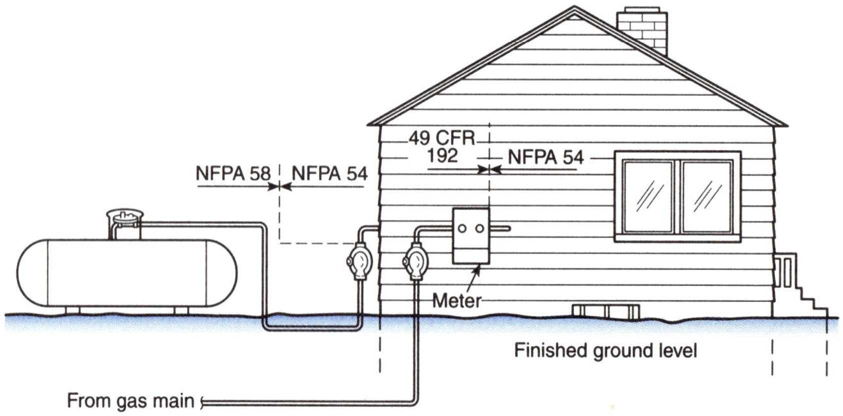


**(B)** The maximum operating pressure shall be 125 psi (862 kPa).

Since the 1988 edition, the maximum operating gas pressure covered by the *National Fuel Gas Code* has been 125 psi (862 kPa), as required by 1.1.1.1(B). Prior to the 1988 edition, the limit was 60 psi (414 kPa), and ASME B31.2, *Fuel Gas Piping*, published by the American Society of Mechanical Engineers, covered gas piping for pressures of 60 psi to 125 psi (410 kPa to 862 kPa). Originally, ASME had planned to incorporate the ASME B31.2 standard into a new standard (ASME B31.9), to cover all building service piping systems, including gas piping above 60 psi (410 kPa). Instead, the ASME B31.9 committee chose to delete all coverage for fuel gas piping systems from proposed ASME B31.9, *Building Services Piping Code*, and to withdraw ASME B31.2. Because there was no longer any standard covering fuel gas piping systems operating at pressures in excess of 60 psi (410 kPa) downstream of the point of delivery, the *National Fuel Gas Code* was revised to address this area.

The *National Fuel Gas Code* does not prohibit the use of gas at pressures over 125 psi (862 kPa), but users of fuel gases at pressures higher than 125 psi (862 kPa) must find other codes for piping requirements.

Piping systems operated at pressures in excess of 125 psi (862 kPa) should be designed and constructed for specific equipment that requires these inlet pressures. The piping installation should be constructed in accordance with comprehensive specifications that are consistent with piping design and installation practice for pressures being used. Additional information regarding the design of gas piping systems above 125 psi (862 kPa) can be found in ASME B31.1, *Power Piping*, or in ASME B31.3, *Process Piping*.

*Exception No. 1: Piping systems for gas–air mixtures within the flammable range are limited to a maximum pressure of 10 psi (69 kPa).*

As stated in 1.1.1.1(B), Exception No. 1, the pressure of flammable gas–air mixtures (5 percent to 15 percent natural gas in air and 2.15 percent to 9.6 percent propane in air) is limited to 10 psi (69 kPa), to minimize any hazard. If ignited, flammable gas–air mixtures can propagate flame back through a piping system. Flame propagation can cause the pipe to explode. The hazard is increased with increased gas pressure.

*Exception No. 2: LP-Gas piping systems are limited to 20 psi (140 kPa), except as provided in 5.5.2.*

The pressure of LP-Gas piping systems is limited to 20 psi (140 kPa) in 1.1.1.1(B), Exception No. 2, to ensure that propane remains in the vapor phase. Below this pressure, propane will liquefy at temperatures below approximately −5°F (−21°C). Systems designed to operate below −5°F (−21°C) and butane systems are permitted by 5.5.2, but they must be designed either to accommodate liquid LP-Gas or to prevent LP-Gas vapor from condensing. LP-Gas systems operating above 20 psi are not likely to be found in residential or commercial installations, but they are occasionally used in industrial processes requiring high gas flows. Such systems must be designed to prevent condensation of liquid.

In the 2008 edition of NFPA 58, *Liquefied Petroleum Gas Code*, coverage was added for LP-Gas vapor systems with pressures of 20 psig through 50 psig (140 kPag through 345 kPag) in supplying appliances in industrial occupancies. These requirements can be used as the basis for a variance under Section 1.4, Equivalency, if needed.

**(C)** Requirements for piping systems shall include design, materials, components, fabrication, assembly, installation, testing, inspection, operation, and maintenance.

**(D)** Requirements for appliances, equipment, and related accessories shall include installation, combustion, and ventilation air and venting.

The safe operation of gas appliances occurs when the following conditions are met:

- The necessary amount of gas is provided at the intended pressure.
- Sufficient air is provided to completely burn the gas.
- The installation is such that heat produced by the appliance does not cause a building fire.
- The products of combustion are removed from the building, either by a flue or vent, or, for smaller appliances, by the normal air exchange in a building.

**1.1.1.2** This code shall not apply to the following items (reference standards for some of which appear in Annex M):

(1) Portable LP-Gas appliances and equipment of all types that are not connected to a fixed fuel piping system

◀ **FAQ**
What does NFPA 54 *not* cover?

This code does not cover or address the use of any type of LP-Gas equipment, such as outdoor gas grills, handheld soldering torches, and butane lighters, which is not connected to fixed fuel piping systems. Some of these appliances are covered by other codes and standards. For example, NFPA 58 covers portable outdoor gas grill cylinders connected to self-contained LP-Gas supply systems, and ANSI Z21.58, *Outdoor Cooking Gas Appliances*, covers gas grill construction and testing. NFPA 58 also covers handheld soldering torches. The use of butane lighters is not covered by any national standard.

(2) Installation of farm appliances and equipment such as brooders, dehydrators, dryers, and irrigation equipment

Natural gas distribution mains do not serve the vast majority of farms. Most gas-fired farm equipment operates on LP-Gas directly from a container. Examples of such appliances include LP-Gas engine–driven pumps, crop dryers, and poultry building heaters. NFPA 58 covers these and similar appliances.

(3) Raw material (feedstock) applications except for piping to special atmosphere generators

Because this code limits its scope to fuel gases, excluding other applications is appropriate. Natural gas or LP-Gas, when used as a raw material or feedstock for a chemical, petrochemical, or other application, is outside the scope of this code and, therefore, is not covered, as stated in 1.1.1.2(3). Such uses may fall under the scope of ASME B31.3.

The code, however, does cover piping to special atmosphere generators and special-use ovens that employ a reducing atmosphere. Special atmosphere generators include gas-burning devices in greenhouses. They are usually small units, and the gas supply is similar to piping serving other appliances covered under this code.

(4) Oxygen–fuel gas cutting and welding systems

NFPA 51, *Standard for the Design and Installation of Oxygen-Fuel Gas Systems for Cutting, Welding, and Allied Processes*, covers fuel gas cutting and welding systems, as addressed in 1.1.1.2(4). As both natural gas and propane are used with oxygen in buildings, the use of NFPA 51 is important to prevent fires and explosions. Fires and explosions can occur if fuel gas and oxygen mix in piping systems that are not designed for such components. Examples of the use of oxygen and fuel gas are found in jewelry manufacturing and repair shops and automobile muffler repair shops, where natural gas or LP-Gas is used with oxygen to cut off mufflers.

(5) Industrial gas applications using such gases as acetylene and acetylenic compounds, hydrogen, ammonia, carbon monoxide, oxygen, and nitrogen

As noted in 1.1.1.2(5), acetylene and acetylenic compounds, hydrogen, ammonia, carbon monoxide, oxygen, and nitrogen are excluded specifically from coverage in the *National Fuel Gas Code* because their properties are inherently different from methane, propane, and similar hydrocarbons. This code covers only fuel gases that are distributed commercially. The use of hydrogen as a fuel gas is not covered by the code, per 1.1.1.2(19). Where hydrogen is used as a fuel gas (for example, for a fuel cell), other standards, such as NFPA 853, *Standard for the Installation of Stationary Fuel Cell Power Systems*, must be used for the installation of the hydrogen piping, installation of the appliance, and venting.

(6) Petroleum refineries, pipeline compressor or pumping stations, loading terminals, compounding plants, refinery tank farms, and natural gas processing plants

Petroleum refineries and natural gas processing plants have tanks and vessels that hold flammable liquids and gases (including butane and propane) with differing compositions, as well as some toxic compounds. The piping requirements for these materials are beyond the scope of this code.

(7) Large integrated chemical plants or portions of such plants where flammable or combustible liquids or gases are produced by chemical reactions or used in chemical reactions

(8) LP-Gas installations at utility gas plants

NFPA 59, *Utility LP-Gas Plant Code*, covers LP-Gas installations at utility gas plants. These plants mix propane and air to make a heating fuel used by gas utility companies to supplement natural gas supplies during periods of peak demand, such as during extreme cold weather. While these plants have fallen out of favor as the most efficient way to provide supplemental gas, there are still many of them in service, and some new ones planned. The heating value of the propane–air mixture is selected to provide a performance equivalent to that of natural gas in appliances.

(9) Liquefied natural gas (LNG) installations

NFPA 59A, *Standard for the Production, Storage, and Handling of Liquefied Natural Gas (LNG)*, covers liquefied natural gas installations, as noted in 1.1.1.2(9). NFPA 52, *Vehicular Fuel Systems Code*, covers the use of liquefied natural gas as a vehicle fuel. The 2006 edition of NFPA 52 has been revised to incorporate NFPA 57, *Liquefied Natural Gas (LNG) Vehicular Fuel Systems Code*, which was withdrawn. NFPA 52 continues to cover the use of compressed natural gas (CNG) and also covers hydrogen for use as a vehicle fuel.

NEW (10) Fuel gas piping in electric utility power plants

ASME B31.1, *Power Piping*, addresses fuel gas piping in power plants.

(11) Proprietary items of equipment, apparatus, or instruments such as gas generating sets, compressors, and calorimeters

Proprietary items of equipment, such as gas-fueled generator sets, compressors, and calorimeters, represent specialized equipment that is therefore excluded from the code, as stated in 1.1.1.2(11). However, NFPA 37, *Standard for the Installation and Use of Stationary Combustion Engines and Gas Turbines*, covers gas-fueled generator sets, and NFPA 853, *Standard for the Installation of Stationary Fuel Cell Power Systems*, covers the installation of fuel cells.

(12) LP-Gas equipment for vaporization, gas mixing, and gas manufacturing

NFPA 58 covers LP-Gas vaporization and mixing, addressed in 1.1.1.2(12).

(13) LP-Gas piping for buildings under construction or renovations that is not to become part of the permanent building piping system — that is, temporary fixed piping for building heat

NFPA 58 covers temporary LP-Gas piping, addressed in 1.1.1.2(13).

(14) Installation of LP-Gas systems for railroad switch heating

NFPA 58 covers railroad switch heating, addressed in 1.1.1.2(14).

(15) Installation of LP-Gas and compressed natural gas (CNG) systems on vehicles

The *National Fuel Gas Code* does not cover vehicular gaseous fuel systems, as stated in 1.1.1.2(15). The following standards should be referenced for these installation requirements:

- NFPA 58 covers LP-Gas vehicle fuel systems.
- NFPA 52 covers compressed and liquefied natural gas fuel systems for vehicles.

(16) Gas piping, meters, gas pressure regulators, and other appurtenances used by the serving gas supplier in distribution of gas, other than undiluted LP-Gas

The gas piping, meters, gas pressure regulators, and other appurtenances addressed in 1.1.1.2(16) may be installed by the gas supplier, the mechanical contractor, or a plumber. U.S. Department of Transportation regulations cover these items. Where LP-Gas is supplied, NFPA 58 covers first-stage and second-stage regulators, as well as any meters located upstream of the second-stage pressure regulator.

(17) Building design and construction, except as specified herein

The *National Fuel Gas Code* is not intended to cover building design and construction, including means for limiting the spread of heat and smoke, which are the responsibility of NFPA *101®*, *Life Safety Code®*; *NFPA 5000®*, *Building Construction and Safety Code®*; or the locally adopted building code. Installers of fuel gas system components should be aware of the possible impact of the gas system on the spread of heat and smoke, especially in multistory buildings, and coordinate their activities with building safety authorities.

(18) Fuel gas systems on recreational vehicles manufactured in accordance with NFPA 1192, *Standard on Recreational Vehicles*
(19) Fuel gas systems using hydrogen as a fuel

Prior to the 2002 edition, the code was silent on the subject of hydrogen used as a fuel gas, as hydrogen piping systems were installed only in certain industrial buildings for non-fuel uses. Now hydrogen is excluded from the code in 1.1.1.2(19), because there is the possibility that hydrogen may be used as a fuel for fuel cells, and the piping requirements do not address the unique properties of hydrogen that require special considerations when installing a piping system. Specifically, hydrogen is much lighter than both natural gas and propane, and it has a much higher flame speed and lower minimum ignition energy than other fuel gases. In addition, the flammable range of hydrogen (4 percent to 75 percent) is much broader than that of natural gas (5 percent to 15 percent) and propane (2 percent to 10 percent). Finally, hydrogen is a very small molecule and can diffuse through metal piping and valves. These properties make hydrogen much easier to ignite than other fuel gases, and they have a much higher burning rate, which results in a higher probability of an explosion, as compared to other fuel gases. The significantly lower density allows hydrogen to mix with air much faster than other fuel gases. If the use of fuel cells in buildings that run on hydrogen becomes more popular, the committee may readdress this subject.

NFPA 52, *Vehicular Fuel Systems Code*, covers hydrogen as a vehicular fuel. Coverage includes the storage and dispensing of hydrogen as well as limited requirements onboard the vehicle that would apply primarily to vehicles converted to use hydrogen as a fuel (non-OEM vehicles).

NFPA 853, *Standard for the Installation of Stationary Fuel Cell Power Systems*, covers hydrogen as a fuel for stationary fuel cells. NFPA 853 applies to any size stationary fuel cell. NFPA 55, *Standard for the Storage, Use, and Handling of Compressed Gases and Cryogenic Fluids in Portable and Stationary Containers, Cylinders, and Tanks*, covers other uses of hydrogen as a fuel and it covers all hydrogen storage and piping systems.

As hydrogen usage increases, it is likely that more specific requirements will be developed for the applications that NFPA 52 and NFPA 853 do not cover.

This code does cover natural gas piping and connection to a fuel cell assembly that uses natural gas either directly or as a feedstock for conversion into hydrogen.

(20) Construction of appliances

The exclusion in 1.1.1.2(20) clarifies that the construction of appliances is not covered in the code. There has been some confusion because many appliances use aluminum piping and regulators, which are covered by the code. Standards exist for the construction of appliances, and these should be used rather than the code.

### 1.1.2 Other Standards.

**In applying this code, reference shall also be made to the manufacturers' instructions and the serving gas supplier regulations.**

For a list of standards specifically referenced in applying this code and other standards that can be used, see Chapter 2 and Annex M of the code.

This code lists the rules that govern the fuel gas industry. The code recognizes that appliances vary considerably and that it is impractical for the code to include all variations. Therefore, 1.1.2 requires compliance with the manufacturer's installation instructions for appliances, equipment, and some piping and venting materials. Manufacturer's instructions contain the specific instructions and rules for the proper installation of a manufacturer's appliance. The relevant listing laboratories have reviewed the installation instructions for listed products. These instructions include items specific to the product and reflect the testing of that product.

The fuel gas supplier may also add rules for installations, based on conditions in the area. Such issues as climate, geographical conditions, and the composition of the fuel can affect how appliances and piping should be installed in order to ensure a safe installation. When a state, city, or other political entity adopts the code, it can also amend the code based on local experience. The local rule should specify which part of the code is being altered. A local authority can require that the local municipality license the installer, or the local authority can alter the code if there is a local rule that is different from that specified in the code. For instance, the local authority can adopt a correction to the code that might be stated as follows: "In applying this code, reference shall also be made to the manufacturers' instructions, the serving gas supplier regulations, and State Fuel Gas Code Modifications." Such a requirement would indicate that the state has a set of modifications that must be followed in addition to the *National Fuel Gas Code*. In cases where local rules differ from the code, the local rules would be in effect.

## 1.2 Purpose

**(Reserved)**

The *Manual of Style for NFPA Technical Committee Documents* reserves Section 1.2 for a purpose statement. NFPA 54/ANSI Z223.1 does not have a statement of purpose, but this section has been reserved so that if a statement is added in the future, Chapter 1 will not need to be renumbered.

## 1.3 Retroactivity

Unless otherwise stated, the provisions of this code shall not be applied retroactively to existing systems that were in compliance with the provisions of the code in effect at the time of installation.

◀ **FAQ**
Can the requirements of NFPA 54 be applied retroactively?

In a standard that is widely adopted by public safety regulatory agencies and used in litigation, as is the *National Fuel Gas Code*, the subject of retroactivity is important. Although the National Fuel Gas Code Committee has the authority to make a provision retroactive, it has never done so. If a matter was a hazard to life and property, it could induce the committee to take such an action.

The code, therefore, is not intended to be retroactive. Any equipment or system complying with the code at the time of installation can be maintained in use as long as the equipment is not changed significantly.

A state, city, or other government body adopting the code could require retroactivity. If a question exists on this subject, the authority having jurisdiction should be queried or local legislation reviewed.

## 1.4 Equivalency

The provisions of this code are not intended to prevent the use of any material, method of construction, or installation procedure not specifically prescribed by this code, provided any such alternative is acceptable to the authority having jurisdiction *(see 3.2.2)*. The authority having jurisdiction shall require that sufficient evidence be submitted to substantiate any claims made regarding the safety of such alternatives.

It is not the intent of the code to prohibit safe practices in the installation of gas piping and equipment that have not yet been developed, or to prohibit new technology. Section 1.4 allows the authority having jurisdiction to require evidence to substantiate any claims and, with that evidence, to permit an installation using an alternate or new method or procedure. [See the definition of authority having jurisdiction (3.2.2 and A.3.2.2) for additional information on who may be an authority having jurisdiction.]

## 1.5 Enforcement

This code shall be administered and enforced by the authority having jurisdiction designated by the governing authority. *(See Annex L for sample wording for enabling legislation.)*

Section 1.5 provides a statement on who enforces the code. Refer to the definition of authority having jurisdiction (3.2.2 and A.3.2.2) for more information on who may be an authority having jurisdiction.

## REFERENCES CITED IN COMMENTARY

The following publications are available from the National Fire Protection Association, 1 Batterymarch Park, Quincy, MA 02169-7471, 1-800-344-3555, www.nfpa.org.

*Manual of Style for NFPA Technical Committee Documents*, 2004 edition.
NFPA 37, *Standard for the Installation and Use of Stationary Combustion Engines and Gas Turbines*, 2006 edition.
NFPA 51, *Standard for the Design and Installation of Oxygen–Fuel Gas Systems for Welding, Cutting, and Allied Processes*, 2007 edition.
NFPA 52, *Vehicular Fuel Systems Code*, 2006 edition.
NFPA 55, *Standard for the Storage, Use, and Handling of Compressed Gases and Cryogenic Fluids in Portable and Stationary Containers, Cylinders, and Tanks*, 2005 edition.
NFPA 58, *Liquefied Petroleum Gas Code*, 2008 edition.
NFPA 59, *Utility LP-Gas Plant Code*, 2008 edition.
NFPA 59A, *Standard for the Production, Storage, and Handling of Liquefied Natural Gas (LNG)*, 2009 edition.
NFPA *101*®, *Life Safety Code*®, 2009 edition.
NFPA 853, *Standard for the Installation of Stationary Fuel Cell Power Systems*, 2007 edition.
*NFPA 5000*®, *Building Construction and Safety Code*®, 2009 edition.

The following publications are available from the American Society of Mechanical Engineers, Three Park Avenue, New York, NY 10016-5990, 1-800-843-2763, www.asme.org.

ANSI/ASME B31.1, *Power Piping*, 2007.
ASME B31.2, *Fuel Gas Piping*, 1968.
ANSI/ASME B31.3, *Process Piping*, 2006.

The following publication is available from CSA America, Inc., 8501 East Pleasant Valley Road, Cleveland, OH 44131-5575, 216-524-4990, www.csa-america.org.

ANSI Z21.58/CGA 1.6, *Outdoor Cooking Gas Appliances*, 2007.

The following publication is available from the American Gas Association, 400 North Capitol Street, NW, Washington, DC 20001, 202-824-7000, www.aga.org.

ANSI Z380, *Guide for Gas Transmission and Distribution Piping*, 2003.

The following publication is available from the U.S. Government Printing Office, Washington, DC 20402, www.gpo.gov.

Title 49, Code of Federal Regulations, Part 192, "Transportation of Natural and Other Gas by Pipeline: Minimum Federal Safety Standard."

# Referenced Publications

# CHAPTER 2

Chapter 2, Referenced Publications, is a list of the codes and standards that are referenced in the *National Fuel Gas Code.*

This chapter lists the mandatory referenced publications. Annex M lists nonmandatory referenced publications. By locating the information immediately after Chapter 1, Administration, the user is presented with the complete list of publications needed for effective use of the code before reading the specific requirements. The provisions of the publications that are mandated by NFPA 54 are also requirements. Regardless of whether an actual requirement resides within NFPA 54 or is mandatorily referenced and appears only in the referenced publication, it is a requirement that must be met to achieve compliance with NFPA 54.

◀ **FAQ**
**What is the difference between the documents listed in Chapter 2 and those listed in Annex M?**

The reasons for locating all mandatory references in a single chapter are, first, to simplify use of the code and, second, to make it easier for adopting jurisdictions to update the references in only one location rather than throughout the code. The editions of the referenced publications listed in Chapter 2 are legally referenced editions unless the jurisdiction, when adopting the code, has updated the list of codes and standards.

NFPA 54/ANSI Z223.1 can mandate the use of a referenced document only if that document is a consensus-based code or standard.

## 2.1 General

The documents or portions thereof listed in this chapter are referenced within this code and shall be considered part of the requirements of this document.

## 2.2 NFPA Publications

National Fire Protection Association, 1 Batterymarch Park, Quincy, MA 02169-7471.

NFPA 30A, *Code for Motor Fuel Dispensing Facilities and Repair Garages,* 2008 edition.

NFPA 37, *Standard for the Installation and Use of Stationary Combustion Engines and Gas Turbines,* 2006 edition.

NFPA 51, *Standard for the Design and Installation of Oxygen–Fuel Gas Systems for Welding, Cutting, and Allied Processes,* 2007 edition.

NFPA 52, *Vehicular Fuel Systems Code,* 2006 edition.

NFPA 58, *Liquefied Petroleum Gas Code,* 2008 edition.

*NFPA 70®, National Electrical Code®*, 2008 edition.

NFPA 82, *Standard on Incinerators and Waste and Linen Handling Systems and Equipment,* 2004 edition.

NFPA 88A, *Standard for Parking Structures,* 2007 edition.

NFPA 90A, *Standard for the Installation of Air-Conditioning and Ventilating Systems,* 2009 edition.

NFPA 90B, *Standard for the Installation of Warm Air Heating and Air-Conditioning Systems,* 2009 edition.

NFPA 96, *Standard for Ventilation Control and Fire Protection of Commercial Cooking Operations,* 2008 edition.

NFPA 211, *Standard for Chimneys, Fireplaces, Vents, and Solid Fuel–Burning Appliances,* 2006 edition.

NFPA 409, *Standard on Aircraft Hangars,* 2004 edition.

NEW NFPA 780, *Standard for the Installation of Lightning Protection Systems,* 2008 edition.

NFPA 853, *Standard for the Installation of Stationary Fuel Cell Power Systems,* 2007 edition.

NFPA 1192, *Standard on Recreational Vehicles,* 2008 edition.

## 2.3 Other Publications

### 2.3.1 ASME Publications.

American Society of Mechanical Engineers, Three Park Avenue, New York, NY 10016-5990, (800)843-2763, www.asme.org.

ANSI/ASME B1.20.1, *Pipe Threads, General Purpose, Inch,* 1983 (Reaffirmed 2001).

ANSI/ASME B16.1, *Cast Iron Pipe Flanges and Flanged Fittings, Class 25, 125, 250, and 800,* 1998.

ANSI/ASME B16.20, *Metal Gaskets for Pipe Flanges, Ring Joint Spiral Wound and Jacketed,* 2000.

ANSI/ASME B36.10M, *Welded and Seamless Wrought Steel Pipe,* 2001.

### 2.3.2 ASTM Publications.

ASTM International, 100 Barr Harbor Drive, P.O. Box C700, West Conshohocken, PA 19428-2959, (610)832-9585, www.astm.org.

ASTM A 53, *Standard Specification for Pipe, Steel, Black and Hot-Dipped, Zinc-Coated Welded and Seamless,* 2001.

ASTM A 106, *Standard Specification for Seamless Carbon Steel Pipe for High-Temperature Service,* 1999.

ASTM A 254, *Standard Specification for Copper Brazed Steel Tubing,* 2002.

ASTM B 88, *Specification for Seamless Copper Water Tube,* 1999.

ASTM B 210, *Specification for Aluminum-Alloy Drawn Seamless Tubes,* 2000.

ASTM B 241, *Specification for Aluminum-Alloy Seamless Pipe and Seamless Extruded Tube,* 2000.

ASTM B 280, *Specification for Seamless Copper Tube for Air Conditioning and Refrigeration Field Service,* 1999.

ASTM D 2513, *Standard Specification for Thermoplastic Gas Pressure Pipe, Tubing, and Fittings,* 2001.

NEW ASTM E 136, *Standard Test Method for Behavior of Materials in a Vertical Tube Furnace at 750 Degrees C,* 2004.

ASTM F 1973, *Standard Specification for Factory Assembled Anodeless Risers and Transition Fittings in Polyethylene (PE) and Polyamide 11 (PA11) Fuel Gas Distribution Systems,* 2002.

NEW ASTM F 2509, *Standard Specification for Field-Assembled Anodeless Riser Kits for Use on Outside Diameter Controlled Polyethylene Gas Distribution Pipe and Tubing,* 2006.

### 2.3.3 CSA America Publications.

Canadian Standards Association, 8501 East Pleasant Valley Road, Cleveland, OH 44131-5575, (216)524-4990, www.csa-america.org.

ANSI Z21.8, *Installation of Domestic Gas Conversion Burners,* 2000.

NEW ANSI Z21.11.2, *Gas-Fired Room Heaters-Volume II, Unvented Room Heaters,* 2007.

ANSI Z21.24/CSA 6.10, *Standard for Connectors for Gas Appliances*, 2007.

ANSI Z21.41/CSA 6.9, *Quick-Disconnect Devices for Use with Gas Fuel Appliances,* 2003.

ANSI Z21.69/CSA 6.22, *Connectors for Movable Gas Appliances,* 2002.

ANSI Z21.75/CSA 6.27, *Connectors for Outdoor Gas Appliances and Manufactured Homes,* 2007.

ANSI Z21.80/CSA 3.7, *Line Pressure Regulators,* 2003.

ANSI Z21.90, *Gas Convenience Outlets and Optional Enclosures,* 2001.

ANSI Z83.4/CSA 3.7, *Non-Recirculating Direct Gas-Fired Industrial Air Heaters*, 1999.

ANSI Z83.18, *Recirculating Direct Gas-Fired Industrial Air Heaters,* 1990 (2000).

ANSI LC 1/CSA 6.26, *Fuel Gas Piping Systems Using Corrugated Stainless Steel Tubing,* 2008.

ANSI LC-4, *Press-Connect Copper and Copper Alloy Fittings for Use in Fuel Gas Distribution Systems,* 2007.

### 2.3.4 MSS Publications.

Manufacturers Standardization Society of the Valve and Fittings Industry, 127 Park Street, NE, Vienna, VA 22180-6671, (703)281-6613, www.mss-hq.com.

MSS SP-6, *Standard Finishes for Contact Faces of Pipe Flanges and Connecting-End Flanges of Valves and Fittings,* 2001.

ANSI/MSS SP-58, *Pipe Hangers and Supports — Materials, Design and Manufacture,* 1993.

### 2.3.5 UL Publications.

Underwriters Laboratories Inc., 333 Pfingsten Road, Northbrook, IL 60062-2096, www.ul.com.

ANSI/UL 651, *Schedule 40 and 80 Rigid PVC Conduit and Fittings,* 2005, Revised 2007.

### 2.3.6 U.S. Government Publications.

U.S. Government Printing Office, Washington, DC 20402, www.access.gpo.gov.

Title 49, Code of Federal Regulations, Part 192.

### 2.3.7 Other Publications.

*Merriam-Webster's Collegiate Dictionary*, 11th edition, Merriam-Webster, Inc., Springfield, MA, 2003.

## 2.4 References for Extracts in Mandatory Sections

*NFPA 70®*, *National Electrical Code®*, 2008 edition.

NFPA 211, *Standard for Chimneys, Fireplaces, Vents, and Solid Fuel–Burning Appliances,* 2006 edition.

NFPA 501, *Standard on Manufactured Housing,* 2005 edition.

*NFPA 5000®*, *Building Construction and Safety Code®*, 2009 edition.

# Definitions

## CHAPTER 3

Many of the definitions used throughout the *National Fuel Gas Code* are unique to the fuel gas industry, and the terms are defined for the reader in Chapter 3, Definitions. The definitions are organized into one of the following categories:

- General — Use of terms and general dictionary reference (See Section 3.1.)
- NFPA Official Definitions — Specific administrative terms as defined by the NFPA Standards Council (See Section 3.2.)
- General Definitions — General technical terms (See Section 3.3.)

Many terms not used in the *National Fuel Gas Code* are nonetheless useful to code users. These terms are defined in Annex K, Other Useful Definitions.

## 3.1 General

The definitions contained in this chapter shall apply to the terms used in this code. Where terms are not defined in this chapter or within another chapter, they shall be defined using their ordinarily accepted meanings within the context in which they are used. *Merriam-Webster's Collegiate Dictionary*, 11th edition, shall be the source for the ordinarily accepted meaning.

The code recognizes that many of the technical words used in the code are defined in the dictionary, and their definition is not needed in the code. Examples of this are the words *pipe* and *valve*. A specific dictionary, *Merriam-Webster's Collegiate Dictionary,* is to be used for terms not defined in Chapter 3.

## 3.2 NFPA Official Definitions

Section 3.2 provides the NFPA official definitions of terms that are used in all NFPA codes and standards.

**3.2.1* Approved.** Acceptable to the authority having jurisdiction.

**A.3.2.1 Approved.** The American Gas Association, American National Standards Institute, and the National Fire Protection Association do not approve, inspect, or certify any installations, procedures, appliances, equipment, or materials; nor do they approve or evaluate testing laboratories. In determining the acceptability of installations, procedures, appliances, equipment, or materials, the authority having jurisdiction may base acceptance on compliance with NFPA or other appropriate standards. In the absence of such standards, said authority may require evidence of proper installation, procedure, or use. The authority having jurisdiction may also refer to the listings or labeling practices *(see 3.2.2)* of an organization that is concerned with product evaluations and is thus in a position to determine compliance with AGA, ANSI, CSA, NFPA, or appropriate standards for the current production of listed items. Additional information regarding the coordination of appliance design, construction, and maintenance can be found in Annex B.

**3.2.2* Authority Having Jurisdiction (AHJ).** An organization, office, or individual responsible for enforcing the requirements of a code or standard, or for approving equipment, materials, an installation, or a procedure.

**A.3.2.2 Authority Having Jurisdiction (AHJ).** The phrase "authority having jurisdiction," or its acronym AHJ, is used in NFPA documents in a broad manner, since jurisdictions and approval agencies vary, as do their responsibilities. Where public safety is primary, the authority having jurisdiction may be a federal, state, local, or other regional department or individual such as a fire chief; fire marshal; chief of a fire prevention bureau, labor department, or health department; building official; electrical inspector; or others having statutory authority. For insurance purposes, an insurance inspection department, rating bureau, or other insurance company representative may be the authority having jurisdiction. In many circumstances, the property owner or his or her designated agent assumes the role of the authority having jurisdiction; at government installations, the commanding officer or departmental official may be the authority having jurisdiction.

As used in the definition of Authority Having Jurisdiction, equipment includes appliances and materials.

**3.2.3 Labeled.** Equipment or materials to which has been attached a label, symbol, or other identifying mark of an organization that is acceptable to the authority having jurisdiction and concerned with product evaluation, that maintains periodic inspection of production of labeled equipment or materials, and by whose labeling the manufacturer indicates compliance with appropriate standards or performance in a specified manner.

**3.2.4* Listed.** Equipment, materials, or services included in a list published by an organization that is acceptable to the authority having jurisdiction and concerned with evaluation of products or services, that maintains periodic inspection of production of listed equipment or materials or periodic evaluation of services, and whose listing states that either the equipment, material, or service meets appropriate designated standards or has been tested and found suitable for a specified purpose.

**A.3.2.4 Listed.** The means for identifying listed appliances and equipment may vary for each organization concerned with product evaluation; some organizations do not recognize appliances and equipment as listed unless it is also labeled. The authority having jurisdiction should utilize the system employed by the listing organization to identify a listed product.

As used in the definition of Listed, equipment includes appliances and materials.

**3.2.5 Shall.** Indicates a mandatory requirement.

The term *shall,* which is defined in 3.2.5, indicates a requirement of this code and mandates that a specific provision of NFPA 54 be followed. When the term *shall* is attached to a specific provision of the code, compliance with that provision is not optional.

## 3.3 General Definitions

Section 3.3 contains all the technical definitions of terms used in the code.

**3.3.1 Accessible.** Having access to but which first requires the removal of a panel, door, or similar covering of the item described.

What is the difference between the terms *accessible* and *readily accessible*?

*3.3.1.1* ***Readily Accessible.*** Having direct access without the need of removing or moving any panel, door, or similar covering of the item described.

FAQ ►

Some jurisdictions make reference to *accessible* and *readily accessible.* The term *readily accessible* is usually used where there is direct access to appliance controls and equipment shutoff valves without the removal of a panel, door, or similar covering. For example, the term applies in cases where immediate access to an appliance shutoff is required. The gas shutoff valve to an appliance may be a case where the local authority requires the valve be readily accessible so that it can be shut off in case of emergency, where the time it would take to remove a panel is not acceptable.

**3.3.2 Air.**

***3.3.2.1 Circulating Air.*** Air for cooling, heating, or ventilation distributed to habitable spaces.

***3.3.2.2 Dilution Air.*** Air that enters a draft hood or draft regulator and mixes with the flue gases.

Dilution air is air that is drawn from the room into the venting system by draft. In gas venting systems, dilution air is beneficial because it mixes dry room air with the relatively moist flue gases, thereby reducing flue gas dew point and the likelihood for condensation. Also, dilution air allows additional air into the system, which increases flow through the venting system, promoting faster warm-up of the venting system, which helps reduce condensation in the venting system.

***3.3.2.3 Excess Air.*** Air that passes through the combustion chamber and the appliance flues in excess of that which is theoretically required for complete combustion.

***3.3.2.4 Primary Air.*** The air introduced into a burner that mixes with the gas before it reaches the port or ports.

**3.3.3 Air Conditioning.** The treatment of air so as to control simultaneously its temperature, humidity, cleanness, and distribution to meet the requirements of a conditioned space.

**3.3.4 Air Shutter.** An adjustable device for varying the size of the primary air inlet(s).

**3.3.5 Anodeless Riser.** An assembly of steel-cased plastic pipe used to make the transition between plastic piping installed underground and metallic piping installed aboveground.

**3.3.6 Appliance.** Any device that utilizes gas as a fuel or raw material to produce light, heat, power, refrigeration, or air conditioning.

***3.3.6.1 Automatically Controlled Appliance.*** Appliance equipped with an automatic burner ignition and safety shutoff device and other automatic devices.

***3.3.6.2 Decorative Appliance for Installation in a Vented Fireplace.*** A self-contained, freestanding, fuel gas burning appliance designed for installation only in a vented fireplace and whose primary function lies in the aesthetic effect of the flame.

***3.3.6.3 Direct Vent Appliances.*** Appliances that are constructed and installed so that all air for combustion is derived directly from the outdoors and all flue gases are discharged to the outdoors.

***3.3.6.4 Fan-Assisted Combustion Appliance.*** An appliance equipped with an integral mechanical means to either draw or force products of combustion through the combustion chamber or heat exchanger.

***3.3.6.5 Food Service Appliance.***

***3.3.6.5.1 Baking and Roasting Gas Oven.*** An oven primarily intended for volume food preparation that may be composed of one or more sections or units of the following types: (1) cabinet oven, an oven having one or more cavities heated by a single burner or group of burners; (2) reel-type oven, an oven employing trays that are moved by mechanical means; or (3) sectional oven, an oven composed of one or more independently heated cavities.

***3.3.6.5.2 Gas Counter Appliance.*** An appliance such as a gas coffee brewer and coffee urn and any appurtenant water heating appliance, food and dish warmer, hot plate, and griddle.

***3.3.6.5.3 Gas Deep Fat Fryer.*** An appliance, including a cooking vessel in which oils or fats are placed to such a depth that the cooking food is essentially supported by displacement of the cooking fluid or a perforated container immersed in the cooking fluid rather than by the bottom of the vessel, designed primarily for use in hotels, restaurants, clubs, and similar institutions.

***3.3.6.5.4 Gas Range.*** A self-contained gas range providing for cooking, roasting, baking, or broiling, or any combination of these functions, and not designed specifically for domestic use.

***3.3.6.5.5 Gas Steam Cooker.*** An appliance that cooks, defrosts, or reconstitutes food by direct contact with steam.

***3.3.6.5.6 Gas Steam Generator.*** A separate appliance primarily intended to supply steam for use with food service appliances.

***3.3.6.5.7 Kettle.*** An appliance with a cooking chamber that is heated either by a steam jacket in which steam is generated by gas heat or by direct gas heat applied to the cooking chamber.

***3.3.6.6 Gas Counter Appliances.*** See 3.3.6.5.2.

***3.3.6.7 Household Cooking Appliance.*** An appliance for domestic food preparation, providing at least one function of (1) top or surface cooking, (2) oven cooking, or (3) broiling.

***3.3.6.7.1 Household Broiler Cooking Appliance.*** A unit that cooks primarily by radiated heat.

***3.3.6.7.2 Household Built-In Unit Cooking Appliance.*** A unit designed to be recessed into, placed upon, or attached to the construction of a building, but not for installation on the floor.

***3.3.6.8 Nonresidential, Low-Heat Appliance.*** A commercial, industrial, or institutional appliance needing a chimney capable of withstanding a continuous flue gas temperature not exceeding 1000°F (538°C). [**211,** 2006]

This definition of *nonresidential low-heat appliance* is extracted from NFPA 211, *Standard for Chimneys, Fireplaces, Vents, and Solid Fuel-Burning Appliances*, and coordinates definitions used in previous editions of both NFPA 211 and the *National Fuel Gas Code.*

***3.3.6.9 Nonresidential, Medium-Heat Appliance.*** A commercial, industrial, or institutional appliance needing a chimney capable of withstanding a continuous flue gas temperature not exceeding 1800°F (982°C). [**211,** 2006]

This definition of *nonresidential medium-heat appliance* is extracted from NFPA 211, *Standard for Chimneys, Fireplaces, Vents, and Solid Fuel-Burning Appliances*, and coordinates definitions used in previous editions of both NFPA 211 and the *National Fuel Gas Code*.

***3.3.6.10 Outdoor Cooking Appliance.*** A gas-fired cooking appliance for outdoor use only that is provided with a means of support by the manufacturer and is connected to a fixed gas piping system.

The definition of *outdoor cooking appliance* is intended to reflect the variety of outdoor cooking appliances currently being produced. Prior to the 2006 edition of the code, the definition only applied to outdoor appliances mounted on a post, which left other types of outdoor appliances, such as gas grills with wheels, undefined.

***3.3.6.11 Vented Appliance.***

***3.3.6.11.1* Category I Vented Appliance.*** An appliance that operates with a nonpositive vent static pressure and with a vent gas temperature that avoids excessive condensate production in the vent.

*A.3.3.6.11.1 Category I Vented Appliance.* For additional information on appliance categorization as shown in 3.3.6.11.1 through 3.3.6.11.4, see the appropriate Z21 and Z83 American National Standards.

*3.3.6.11.2 Category II Vented Appliance.* An appliance that operates with a nonpositive vent static pressure and with a vent gas temperature that may cause excessive condensate production in the vent.

*3.3.6.11.3 Category III Vented Appliance.* An appliance that operates with a positive vent static pressure and with a vent gas temperature that avoids excessive condensate production in the vent.

*3.3.6.11.4 Category IV Vented Appliance.* An appliance that operates with a positive vent static pressure and with a vent gas temperature that may cause excessive condensate production in the vent.

**3.3.7 Appliance Categorized Vent Diameter/Area.** The minimum vent diameter/area permissible for Category I appliances to maintain a nonpositive vent static pressure when tested in accordance with nationally recognized standards.

**3.3.8 Automatic Firecheck.** A device for stopping the progress of a flame front in burner mixture lines (flashback) and for automatically shutting off the fuel–air mixture.

**3.3.9 Backfire Preventer.** See 3.3.91, Safety Blowout.

**3.3.10 Baffle.** An object placed in an appliance to change the direction of or retard the flow of air, air–gas mixtures, or flue gases.

**3.3.11 Boiler.**

*3.3.11.1 Hot Water Heating Boiler.* A boiler designed to heat water for circulation through an external space heating system.

*3.3.11.2 Hot Water Supply Boiler.* A boiler used to heat water for purposes other than space heating.

*3.3.11.3 Low-Pressure Boiler.* A boiler that supplies steam at a pressure not exceeding 15 psi (100 kPa), or hot water at a pressure not exceeding 160 psi (1100 kPa) at a temperature not exceeding 250°F (121°C).

*3.3.11.4 Steam Boiler.* A boiler designed to convert water into steam that is supplied to an external system.

**3.3.12 Bonding Jumper.** A reliable conductor to ensure the required electrical conductivity between metal parts required to be electrically connected. [**70,** 2008] NEW

**3.3.13 Branch Line.** Gas piping that conveys gas from a supply line to the appliance.

**3.3.14 Breeching.** See 3.3.108, Vent Connector.

**3.3.15 Broiler.** A general term including broilers, salamanders, barbecues, and other devices cooking primarily by radiated heat, excepting toasters.

*3.3.15.1 Unit Broiler.* A broiler constructed as a separate appliance.

**3.3.16 Btu.** Abbreviation for British thermal unit, which is the quantity of heat required to raise the temperature of 1 pound of water 1 degree Fahrenheit (equivalent to 1055 joules).

**3.3.17 Burner.** A device for the final conveyance of gas, or a mixture of gas and air, to the combustion zone.

*3.3.17.1 Forced-Draft Burner.* See 3.3.17.5, Power Burner.

*3.3.17.2 Gas Conversion Burner.* A unit consisting of a burner and its controls utilizing gaseous fuel for installation in an appliance originally utilizing another fuel.

*3.3.17.3* ***Injection (Bunsen) Type Burner.*** A burner employing the energy of a jet of gas to inject air for combustion into the burner and mix it with the gas.

*3.3.17.4* ***Main Burner.*** A device or group of devices essentially forming an integral unit for the final conveyance of gas or a mixture of gas and air to the combustion zone and on which combustion takes place to accomplish the function for which the appliance is designed.

*3.3.17.5* ***Power Burner.*** A burner in which either gas or air, or both, are supplied at a pressure exceeding, for gas, the line pressure, and for air, atmospheric pressure; this added pressure being applied at the burner. A burner for which air for combustion is supplied by a fan ahead of the appliance is commonly designated as a forced-draft burner.

*3.3.17.5.1* ***Fan-Assisted Power Burner.*** A burner that uses either induced or forced draft.

**3.3.18 Chimney.** One or more passageways, vertical or nearly so, for conveying flue or vent gases to the outdoors. *(See also 3.3.107.2, Gas Vent; 3.3.107, Vent; and 3.3.100.7, Venting System.)*

*3.3.18.1* ***Exterior Masonry Chimneys.*** Masonry chimneys exposed to the outdoors on one or more sides below the roof line.

*3.3.18.2* ***Factory-Built Chimney.*** A chimney composed of listed factory-built components assembled in accordance with the manufacturer's installation instructions to form the completed chimney.

*3.3.18.3* ***Masonry Chimney.*** A field-constructed chimney of solid masonry units, bricks, stones, listed masonry chimney units, or reinforced Portland cement concrete, lined with suitable chimney flue liners.

*3.3.18.4* ***Metal Chimney.*** A field-constructed chimney of metal.

**3.3.19 Clothes Dryer.** An appliance used to dry wet laundry by means of heat derived from the combustion of fuel gases.

*3.3.19.1* ***Type 1 Clothes Dryer.*** Primarily used in family living environment. May or may not be coin-operated for public use.

*3.3.19.2* ***Type 2 Clothes Dryer.*** Used in business with direct intercourse of the function with the public. May or may not be operated by public or hired attendant. May or may not be coin-operated.

There are two definitions to differentiate dryers by their usage. A Type 1 Clothes Dryer has much less use than a Type 2 Clothes Dryer. The higher usage increases the probability of lint blockage and potential for a lint fire.

**3.3.20 Combustion.** As used herein, the rapid oxidation of fuel gases accompanied by the production of heat or heat and light. Complete combustion of a fuel is possible only in the presence of an adequate supply of oxygen.

**3.3.21 Combustion Chamber.** The portion of an appliance within which combustion occurs.

**3.3.22 Combustion Products.** Constituents resulting from the combustion of a fuel with the oxygen of the air, including the inert but excluding excess air.

Note that excess air is not a combustion product, as addressed in 3.3.22, but it is air that passes through the combustion chamber and the appliance flue in excess of that which, theoretically, is required for complete combustion.

**3.3.23 Condensate (Condensation).** The liquid that separates from a gas (including flue gas) due to a reduction in temperature or an increase in pressure.

**3.3.24 Consumption.** The maximum amount of gas per unit of time, usually expressed in cubic feet per hour, or Btu per hour, required for the operation of the appliance or appliances supplied.

**3.3.25 Controls.** Devices designed to regulate the gas, air, water, or electrical supply to an appliance. These may be manual or automatic.

*3.3.25.1 Limit Control.* A device responsive to changes in pressure, temperature, or liquid level for turning on, shutting off, or throttling the gas supply to an appliance.

**3.3.26 Cubic Foot ($ft^3$) of Gas.** The amount of gas that would occupy 1 $ft^3$ (0.03 $m^3$) when at a temperature of 60°F (16°C), saturated with water vapor and under a pressure equivalent to that of 30 in. w.c. (7.5 kPa).

**3.3.27 Deep Fat Fryer.** See 3.3.6.5.3, Gas Deep Fat Fryer.

**3.3.28 Design Certification.** The process by which a product is evaluated and tested by an independent laboratory to affirm that the product design complies with specific requirements.

**3.3.29 Device.**

*3.3.29.1 Automatic Gas Shutoff Device.* A device constructed so that the attainment of a water temperature in a hot water supply system in excess of some predetermined limit acts in such a way as to cause the gas to the system to be shut off.

*3.3.29.2 Pressure Limiting Device.* Equipment that under abnormal conditions will act to reduce, restrict, or shut off the supply of gas flowing into a system in order to prevent the gas pressure in that system from exceeding a predetermined value.

*3.3.29.3 Quick-Disconnect Device.* A hand-operated device that provides a means for connecting and disconnecting an appliance or an appliance connector to a gas supply and that is equipped with an automatic means to shut off the gas supply when the device is disconnected.

*3.3.29.4 Safety Shutoff Device.* A device that will shut off the gas supply to the controlled burner(s) in the event the source of ignition fails. This device can interrupt the flow of gas to main burner(s) only or to pilot(s) and main burner(s) under its supervision.

*3.3.29.5 Vent Damper Device.*

*3.3.29.5.1 Automatic Vent Damper Device.* A device that is intended for installation in the venting system, in the outlet of or downstream of the appliance draft hood, of an individual automatically operated appliance and that is designed to automatically open the venting system when the appliance is in operation and to automatically close off the venting system when the appliance is in a standby or shutdown condition.

*3.3.29.5.2 Electrically Operated, Automatic Vent Damper Device.* An automatic vent damper device that employs electrical energy to control the device.

*3.3.29.5.3 Mechanically Actuated, Automatic Vent Damper Device.* An automatic vent damper device dependent for operation on the direct application or transmission of mechanical energy without employing any type of energy conversion.

*3.3.29.5.4 Thermally Actuated, Automatic Vent Damper Device.* An automatic vent damper device dependent for operation exclusively on the direct conversion of the thermal energy of the vent gases into mechanical energy.

**3.3.30 Diversity Factor.** Ratio of the maximum probable demand to the maximum possible demand.

**3.3.31 Domestic Laundry Stove.** A fuel gas burning appliance consisting of one or more open-top-type burners mounted on high legs or having a cabinet base.

**3.3.32 Draft.** A pressure difference that causes gases or air to flow through a chimney, vent, flue, or appliance.

***3.3.32.1 Mechanical Draft.*** Draft produced by a fan or an air or steam jet. When a fan is located so as to push the flue gases through the chimney or vent, the draft is forced. When the fan is located so as to pull the flue gases through the chimney or vent, the draft is induced. [**211,** 2006]

***3.3.32.2 Natural Draft.*** Draft produced by the difference in the weight of a column of flue gases within a chimney or vent and a corresponding column of air of equal dimension outside the chimney or vent. [**211,** 2006]

The phrase *difference in the weight* referred to in the definition of natural draft in 3.3.32.2 is a characteristic created by the heat of the vent gases. Hot gas has a lower weight per unit of volume than cold gas. This comparison is a fancy way of saying that hot gas rises, a fact well known to those who operate hot air balloons. The difference in weight is the driving force for natural draft venting.

**3.3.33 Draft Hood.** A nonadjustable device built into an appliance, or made a part of the vent connector from an appliance, that is designed to (1) provide for the ready escape of the flue gases from the appliance in the event of no draft, backdraft, or stoppage beyond the draft hood, (2) prevent a backdraft from entering the appliance, and (3) neutralize the effect of stack action of the chimney or gas vent upon the operation of the appliance.

**3.3.34 Drip.** The container placed at a low point in a system of piping to collect condensate and from which it may be removed.

**3.3.35 Dry Gas.** A gas having a moisture and hydrocarbon dew point below any normal temperature to which the gas piping is exposed.

**3.3.36 Effective Ground-Fault Current Path.** An intentionally constructed, permanent, low impedance electrically conductive path designed and intended to carry electric fault current from the point of a ground fault on a wiring system to the electrical supply source.

The definition of *effective ground-fault current path* in 3.3.36 is an unusual one for this code because it is an electrical definition. The term is used in Section 7.13, Electrical Bonding and Grounding. Refer to the commentary following Section 7.13 for an explanation of the subject.

**3.3.37 Equipment.** Devices other than appliances.

**3.3.38 Explosion Heads (Soft Heads or Rupture Discs).** A protective device for relieving excessive pressure in a premix system by bursting of a rupturable disc.

**3.3.39 FAN Max.** The maximum input rating of a Category I, fan-assisted appliance attached to a vent or connector.

**3.3.40 FAN Min.** The minimum input rating of a Category I, fan-assisted appliance attached to a vent or connector.

The terms *FAN Max* and *FAN Min* are venting terms that are used only in Chapter 13, Sizing of Category I Venting Systems, and in Annex G, Sizing of Venting Systems Serving Appliances Equipped with Draft Hoods, Category I Appliances, and Appliances Listed for Use with Type B Vents. The terms are shorthand used in the vent sizing Tables 13.1(a) through 13.1(d) to determine chimney and vent capacity, in Tables 13.2(a) through 13.2(d) to determine common vent capacity (when more than one appliance is connected to a vent), and in Annex G, in examples of the use of the tables.

Commentary Table 3.1 is an excerpt from Table 13.1(a), showing the use of FAN Min and FAN Max for a Type B vent with a diameter of 3 in. (80 mm). A fan-assisted combustion appliance installed with a 6 ft (1.8 m) vent height and a 2 ft (0.6 m) lateral has a single ap-

*COMMENTARY TABLE 3.1 FAN Min and FAN Max*

| Vent Height H (ft) | Lateral L (ft) | FAN Min | FAN Max | NAT Max |
|---|---|---|---|---|
| 6 | 0 | 0 | 78 | 46 |
| | 2 | 13 | 51 | 36 |
| | 4 | 21 | 49 | 34 |
| | 6 | 25 | 46 | 32 |

For SI units: 1 ft = 0.305 m; 1000 Btu/hr = 0.293 kW.

pliance capacity from 13,000 to 51,000 Btu/hr. The shaded column, NAT Max, is used for Category I appliances that are not fan assisted.

**3.3.41 FAN+FAN.** The maximum combined appliance input rating of two or more Category I, fan-assisted appliances attached to the common vent.

**3.3.42 FAN+NAT.** The maximum combined appliance input rating of one or more Category I, fan-assisted appliances and one or more Category I, draft hood–equipped appliances attached to the common vent.

The terms *FAN+FAN*, *FAN+NAT*, and *NAT+NAT* (3.3.72) are used only in Chapter 13, Sizing of Category I Venting Systems, and in Annex G, Sizing of Venting Systems Serving Appliances Equipped with Draft Hoods, Category I Appliances, and Appliances Listed for Use with Type B Vents. The terms are shorthand used in vent sizing Tables 13.2(a) through 13.2(d) to determine vent connector capacity for vent connectors connecting one appliance to a chimney, common vent, or vent manifold.

Commentary Table 3.2 is an excerpt from the Common Vent Capacity part of Table 13.2(a) showing the use of the terms with a 4 in. (100 mm) diameter common vent. Note that a 4 in. (100 mm) diameter common vent that is 6 ft (1.8 m) high and serves two or more Category I appliances has the following capacities:

- A maximum capacity of 92,000 Btu/hr (27 kW) if both appliances are fan assisted
- A maximum capacity of 81,000 Btu/hr (24 kW) if both fan-assisted and non-fan-assisted appliances are connected
- A capacity of 65,000 Btu/hr (19 kW) if only non-fan-assisted appliances are connected

The shaded column, NAT+NAT, is used for Category I appliances that are not fan assisted.

*COMMENTARY TABLE 3.2 FAN+FAN, FAN+NAT, and NAT+NAT*

| Vent Height H (ft) | FAN+FAN | FAN+NAT | NAT+NAT |
|---|---|---|---|
| 6 | 92 | 81 | 65 |
| 8 | 101 | 90 | 73 |
| 10 | 110 | 97 | 79 |
| 15 | 125 | 112 | 91 |
| 20 | 136 | 123 | 102 |

**3.3.43 Fireplace.** A fire chamber and hearth constructed of noncombustible material for use with solid fuels and provided with a chimney.

***3.3.43.1 Gas Fireplace.***

***3.3.43.1.1 Direct Vent Gas Fireplace.*** A system consisting of (1) an appliance for indoor installation that allows the view of flames and provides the simulation of a solid fuel fireplace, (2) combustion air connections between the appliance and the vent air intake terminal, (3) flue-gas connections between the appliance and the vent-air intake terminal, and (4) a vent air intake terminal for installation outdoors, constructed such that all air for combustion is obtained from the outdoor atmosphere and all flue gases are discharged to the outdoor atmosphere.

***3.3.43.1.2 Vented Gas Fireplace.*** A vented appliance that allows the view of flames and provides the simulation of a solid fuel fireplace.

**3.3.44 Flame Arrester.** A nonvalve device for use in a gas–air mixture line containing a means for temporarily stopping the progress of a flame front (flashback).

Flame arresters are used with central premix systems (see 7.12.1), with some regulator vents [see 9.1.19(5)], and for bleed lines for diaphragm-type valves [see 9.1.20(4)].

**3.3.45 Flue.**

***3.3.45.1 Appliance Flue.*** The passage(s) within an appliance through which combustion products pass from the combustion chamber of the appliance to the draft hood inlet opening on an appliance equipped with a draft hood or to the outlet of the appliance on an appliance not equipped with a draft hood.

***3.3.45.2 Chimney Flue.*** The passage(s) in a chimney for conveying the flue or vent gases to the outdoors.

**3.3.46 Flue Collar.** That portion of an appliance designed for the attachment of a draft hood, vent connector, or venting system.

**3.3.47 Furnace.**

***3.3.47.1 Central Furnace.*** A self-contained appliance for heating air by transfer of heat of combustion through metal to the air and designed to supply heated air through ducts to spaces remote from or adjacent to the appliance location.

***3.3.47.2 Direct Vent Wall Furnace.*** A system consisting of an appliance, combustion air, and flue gas connections between the appliance and the outdoor atmosphere, and a vent cap supplied by the manufacturer and constructed so that all air for combustion is obtained from the outdoor atmosphere and all flue gases are discharged to the outdoor atmosphere.

***3.3.47.3 Duct Furnace.*** A furnace normally installed in distribution ducts of air conditioning systems to supply warm air for heating. This definition applies only to an appliance that depends for air circulation on a blower not furnished as part of the furnace.

***3.3.47.4 Enclosed Furnace.*** A specific heating, or heating and ventilating, furnace incorporating an integral total enclosure and using only outdoor air for combustion.

***3.3.47.5 Floor Furnace.*** A completely self-contained unit furnace suspended from the floor of the space being heated, taking air for combustion from outside this space.

***3.3.47.6 Forced-Air Furnace.*** A furnace equipped with a fan or blower that provides the primary means for circulation of air.

***3.3.47.7 Vented Wall Furnace.*** A self-contained, vented, fuel gas burning appliance complete with grilles or equivalent, designed for incorporation in or permanent attachment to the structure of a building and furnishing heated air, circulated by gravity or by a fan, directly into the space to be heated through openings in the casing.

**3.3.48 Furnace Plenum.** A compartment or chamber that is supplied with the furnace or constructed of ductwork that is attached to the inlet or outlet of a furnace or air handling unit and has one or more circulating air ducts connected to it.

**3.3.49 Garage.**

***3.3.49.1 Repair Garage.*** A building, structure, or portions thereof wherein major repair, painting, or body and fender work is performed on motorized vehicles or automobiles, and includes associated floor space used for offices, parking, and showrooms.

***3.3.49.2 Residential Garage.*** A building or room in which self-propelled passenger vehicles are or can be stored and that will not normally be used for other than minor service or repair operations on such stored vehicles.

**3.3.50 Gas Convenience Outlet.** A permanently mounted, hand-operated device providing a means for connecting and disconnecting an appliance or an appliance connector to the gas supply piping. The device includes an integral, manually operated gas valve with a nondisplaceable valve member so that disconnection can be accomplished only when the manually operated gas valve is in the closed position.

**3.3.51 Gases.** Include natural gas, manufactured gas, liquefied petroleum (LP) gas in the vapor phase only, liquefied petroleum gas–air mixtures, and mixtures of these gases, plus gas–air mixtures within the flammable range, with the fuel gas or the flammable component of a mixture being a commercially distributed product.

***3.3.51.1 Flue Gases.*** Products of combustion plus excess air in appliance flues or heat exchangers.

Note that flue gases include combustion products and excess air. (See commentary concerning *combustion products* following 3.3.22.) Vent gases include dilution air also. (See the definitions of *vent gases* in 3.3.51.3 and *dilution air* in 3.3.2.2.)

***3.3.51.2 Utility Gases.*** Natural gas, manufactured gas, liquefied petroleum gas–air mixtures, or mixtures of any of these gases.

***3.3.51.3 Vent Gases.*** Products of combustion from appliances plus excess air, plus dilution air in the venting system above the draft hood or draft regulator.

**3.3.52 Gas-Fired Air Conditioner.** An automatically operated appliance for supplying cooled and/or dehumidified air or chilled liquid.

**3.3.53 Gas-Fired Heat Pump.** An automatically operated appliance utilizing a refrigeration system for supplying either heated air or liquid or heated and/or cooled air or liquid.

**3.3.54 Gas-Mixing Machine.** Any combination of automatic proportioning control devices, blowers, or compressors that supply mixtures of gas and air to multiple burner installations where control devices or other accessories are installed between the mixing device and burner.

**3.3.55 Gravity.** See 3.3.96, Specific Gravity.

**3.3.56 Grounding Electrode.** A conducting object through which a direct connection to earth is established. [**70,** 2008] NEW

**3.3.57 Heater.**

***3.3.57.1 Direct Gas-Fired Nonrecirculating Industrial Air Heater.*** A nonrecirculating industrial air heater in which all the products of combustion generated by the appliance are released into the outdoor airstream being heated.

***3.3.57.2 Direct Gas-Fired Recirculating Industrial Air Heater.*** An air recirculating heater in which all of the products of combustion generated by the appliance are released into the airstream being heated.

*3.3.57.3 Infrared Heater.* A heater that directs a substantial amount of its energy output in the form of infrared energy into the area to be heated. Such heaters may be of either the vented or unvented type.

*3.3.57.4 Pool Heater.* An appliance designed for heating nonpotable water stored at atmospheric pressure, such as water in swimming pools, therapeutic pools, and similar applications.

*3.3.57.5 Unit Heater.*

*3.3.57.5.1 High-Static Pressure-Type Unit Heater.* A self-contained, automatically controlled, vented appliance having integral means for circulation of air against 0.2 in. (15 mm) $H_2O$ or greater static pressure.

*3.3.57.5.2 Low-Static Pressure-Type Unit Heater.* A self-contained, automatically controlled, vented, fuel gas burning appliance, intended for installation in the space to be heated without the use of ducts, having integral means for circulation of air, normally by a propeller fan(s), and may be equipped with louvers or face extensions made in accordance with the manufacturer's specifications.

*3.3.57.6 Unvented Room Heater.* An unvented, self-contained, freestanding, nonrecessed, fuel gas burning appliance for furnishing warm air by gravity or fan circulation to the space in which installed, directly from the heater without duct connection.

*3.3.57.7 Water Heater.* An appliance for supplying hot water for domestic or commercial purposes.

**3.3.58 Heating Value (Total).** The number of British thermal units produced by the combustion, at constant pressure, of 1 $ft^3$ (0.03 $m^3$) of gas when the products of combustion are cooled to the initial temperature of the gas and air, when the water vapor formed during combustion is condensed, and when all the necessary corrections have been applied.

**3.3.59 Hot Plate.** See 3.3.6.5.2, Gas Counter Appliance.

*3.3.59.1 Domestic Hot Plate.* A fuel -gas burning appliance consisting of one or more open-top-type burners mounted on short legs or a base.

**3.3.60 Hot Taps.** Piping connections made to operating pipelines or mains or other facilities while they are in operation. The connection of the branch piping to the operating line and the tapping of the operating line are done while it is under gas pressure.

**3.3.61 Ignition.**

*3.3.61.1 Automatic Ignition.* Ignition of gas at the burner(s) when the gas controlling device is turned on, including re-ignition if the flames on the burner(s) have been extinguished by means other than by the closing of the gas controlling device.

*3.3.61.2 Sources of Ignition.* Appliances or equipment that, because of their intended modes of use or operation, are capable of providing sufficient thermal energy to ignite flammable gas–air mixtures.

**3.3.62 Insulating Millboard.** A factory fabricated board formed with noncombustible materials, normally fibers, and having a thermal conductivity in the range of 1 Btu/in./$ft^2$/°F/hr (0.14 W/m/°K).

**3.3.63 Kettle.** See 3.3.6.5.7.

NEW **3.3.64 Leak Check.** An operation performed on a gas piping system to verify that the system does not leak.

**3.3.65 Manifold.**

*3.3.65.1 Common Vent Manifold.* A horizontal extension of the common vent within the room in which the appliances are installed.

*3.3.65.2* ***Gas Manifold.*** The conduit of an appliance that supplies gas to the individual burners.

**3.3.66 Manufactured Home.** A structure, transportable in one or more sections, which, in the traveling mode, is 8 body-ft (2.4 m) or more in width or 40 body-ft (12.2 m) or more in length or, when erected on site, is 320 $ft^2$ (29.7 $m^2$) or more and which is built on a permanent chassis and designed to be used as a dwelling, with or without a permanent foundation, when connected to the required utilities, and includes plumbing, heating, air-conditioning, and electrical systems contained therein; except that such terms shall include any structure which meets all the requirements of this paragraph except the size requirements and with respect to which the manufacturer voluntarily files a certification required by the regulatory agency. Calculations used to determine the number of square feet in a structure are based on the structure's exterior dimensions, measured at the largest horizontal projections when erected on site. These dimensions include all expandable rooms, cabinets, and other projections containing interior space, but do not include bay windows. [**501,** 2005]

**3.3.67 Material.**

*3.3.67.1* ***Combustible Material.*** As pertaining to materials adjacent to or in contact with heat-producing appliances, vent connectors, gas vents, chimneys, steam and hot water pipes, and warm air ducts, materials made of or surfaced with wood, compressed paper, plant fibers, or other materials that are capable of being ignited and burned. Such material shall be considered combustible even though flame-proofed, fire-retardant treated, or plastered.

*3.3.67.2* ***Noncombustible Material.*** A material that, in the form in which it is used and under the conditions anticipated, will not ignite, burn, support combustion, or release flammable vapors when subjected to fire or heat. Materials that are reported as passing ASTM E 136, *Standard Test Method for Behavior of Materials in a Vertical Tube Furnace at 750 Degrees C*, are considered noncombustible materials.

**3.3.68 Meter.** An instrument installed to measure the volume of gas delivered through it.

**3.3.69 Mixing Blower.** A motor-driven blower to produce gas–air mixtures for combustion through one or more gas burners or nozzles on a single-zone industrial heating appliance or on each control zone of a multizone industrial appliance or on each control zone of a multizone installation.

**3.3.70 NA.** Vent configuration that is not allowed due to potential for condensate formation or pressurization of the venting system or that is not applicable due to physical or geometric restraints.

**3.3.71 NAT Max.** The maximum input rating of a Category I, draft hood-equipped appliance attached to a vent or connector.

**3.3.72 NAT+NAT.** The maximum combined appliance input rating of two or more Category I, draft hood-equipped appliances attached to the common vent.

**3.3.73 Occupancy.**

*3.3.73.1* ***Health Care Occupancy.*** An occupancy used for purposes of medical or other treatment or care of four or more persons where such occupants are mostly incapable of self-preservation due to age, physical or mental disability, or because of security measures not under the occupants' control. [***5000,*** 2009]

*3.3.73.2* ***Residential Board and Care Occupancy.*** A building or portion thereof that is used for lodging and boarding of four or more residents, not related by blood or marriage to the owners or operators, for the purpose of providing personal care services. [***5000,*** 2009]

**3.3.74 Orifice.** The opening in a cap, spud, or other device whereby the flow of gas is limited and through which the gas is discharged to the burner.

NEW **3.3.75 Oven, Gas Baking and Roasting.** See 3.3.6.5.1, Baking and Roasting Gas Oven.

**3.3.76 Parking Structure.** A building, structure, or portion thereof used for the parking of motor vehicles.

NEW ***3.3.76.1 Basement or Underground Parking Structure.*** A parking structure or portion thereof located below finished ground level.

***3.3.76.2 Enclosed Parking Structure.*** Having exterior enclosing walls that have less than 25 percent of the total wall area open to atmosphere at each level using at least two sides of the structure.

**3.3.77 Pilot.** A small flame that is utilized to ignite the gas at the main burner or burners.

**3.3.78 Pipe.** Rigid conduit of iron, steel, copper, brass, aluminum, or plastic.

***3.3.78.1 Equivalent Length Pipe.*** The resistance of valves, controls, and fittings to gas flow expressed as equivalent length of straight pipe for convenience in calculating pipe sizes.

**3.3.79 Piping.** As used in this code, either pipe, tubing, or both. *(See also 3.3.78, Pipe, and 3.3.104, Tubing.)*

***3.3.79.1 Concealed Gas Piping.*** Gas piping that, when in place in a finished building, would require removal of permanent construction to gain access to the piping.

***3.3.79.2 Control Piping.*** All piping, valves, and fittings used to interconnect air, gas, or hydraulically operated control apparatus or instrument transmitters and receivers.

**3.3.80 Plenum.** A compartment or chamber to which one or more ducts are connected and that forms part of the air distribution system.

**3.3.81 Pressure.** Unless otherwise stated, a measurement expressed in pounds per square inch above atmospheric pressure.

***3.3.81.1 Atmospheric Pressure.*** The pressure of the weight of air on the surface of the earth, approximately 14.7 pounds per square inch (psia) (101 kPa absolute) at sea level.

***3.3.81.2 Back Pressure.*** Pressure against which a fluid is flowing, resulting from friction in lines, restrictions in pipes or valves, pressure in vessel to which fluid is flowing, hydrostatic head, or other impediment that causes resistance to fluid flow.

***3.3.81.3 Design Pressure.*** The maximum operating pressure permitted by this code, as determined by the design procedures applicable to the materials involved.

The term *design pressure* in 3.3.81.3 defines the maximum gas pressure allowed for a given installation, and takes into account the nature of the equipment, type of occupancy, piping location, and so forth.

***3.3.81.4 Maximum Working Pressure.*** The maximum pressure at which a piping system may be operated in accordance with the provisions of this code.

FAQ ▶ How is the term *maximum working pressure* different from the term *design pressure*?

*Maximum working pressure* can be determined from allowable stress considerations for piping material or from the nature of the installation itself and the code limitations thereon. The term *design pressure*, which is defined in 3.3.81.3, limits the pressure based on system considerations or appliance limitations. As defined in 3.3.81.4, the term *maximum working pressure* is nearly interchangeable with the term *design pressure*.

For example, a gas piping system in a building has a maximum working pressure, which is determined by the materials used. For a steel piping system, it might be 150 psi. The design pressure of the piping system is lower and is determined by the lowest inlet pressure of all appliances connected to the piping system, which could be 11 in. w.c. or 2 psi.

**3.3.82 Pressure Drop.** The loss in pressure due to friction or obstruction in pipes, valves, fittings, regulators, and burners.

**3.3.83 Pressure Test.** An operation performed to verify the gastight integrity of gas piping following its installation or modification.

**3.3.84 Purge.** To free a gas conduit of air or gas, or a mixture of gas and air.

**3.3.85 Qualified Agency.** Any individual, firm, corporation, or company that either in person or through a representative is engaged in and is responsible for (1) the installation, testing, or replacement of gas piping or (2) the connection, installation, testing, repair, or servicing of appliances and equipment; that is experienced in such work; that is familiar with all precautions required; and that has complied with all the requirements of the authority having jurisdiction.

**3.3.86 Range.** See 3.3.6.5.4, Gas Range.

**3.3.87 Refrigerator (Using Gas Fuel).** An appliance that is designed to extract heat from a suitable chamber.

**3.3.88 Regulator.**

***3.3.88.1 Appliance Regulator.*** A pressure regulator for controlling pressure to the appliance manifold.

***3.3.88.2 Draft Regulator.*** A device that functions to maintain a desired draft in the appliance by automatically reducing the draft to the desired value.

***3.3.88.2.1 Barometric Draft Regulator.*** A balanced damper device attached to a chimney, vent connector, breeching, or flue gas manifold to control chimney draft.

A barometric draft regulator, such as the one shown in Exhibit 3.1, controls draft by opening and closing a gate that allows room air to enter the venting system. The gate responds to changes in chimney draft, opening to allow more air into the venting system as draft increases, and closing to reduce air entering the venting system as draft decreases.

It should be noted that a barometric draft regulator can only decrease draft through the appliance and venting system. It cannot increase draft. Also see commentary following 12.13.4 for more information on the use of barometric draft regulators with gas appliances.

**EXHIBIT 3.1** *Barometric Draft Regulator. (Courtesy of Field Controls, LLC)*

***3.3.88.3 Line Gas Regulator.*** A pressure regulator placed in a gas line between the service regulator and the appliance regulator.

***3.3.88.4 Monitoring Regulator.*** A pressure regulator set in series with another pressure regulator for the purpose of automatically taking over in an emergency the control of the pressure downstream of the regulator in case that pressure tends to exceed a set maximum.

***3.3.88.5 Pressure Regulator.*** Equipment placed in a gas line for reducing, controlling, and maintaining the pressure in that portion of the piping system downstream of the equipment.

***3.3.88.6 Series Regulator.*** A pressure regulator in series with one or more other pressure regulators.

***3.3.88.7 Service Regulator.*** A pressure regulator installed by the serving gas supplier to reduce and limit the service line gas pressure to delivery pressure.

**3.3.89 Relief Opening.** The opening provided in a draft hood to permit the ready escape to the atmosphere of the flue products from the draft hood in the event of no draft, backdraft, or stoppage beyond the draft hood and to permit inspiration of air into the draft hood in the event of a strong chimney updraft.

**3.3.90 Room Large in Comparison with Size of Appliance.** Rooms having a volume equal to at least 12 times the total volume of a furnace or air-conditioning appliance and at least 16 times the total volume of a boiler.

The term *room large in comparison with size of appliance* is used in the clearance requirements covering the installation of central heating boilers, furnaces, and air-conditioning equipment, with regard to the size of the room in which a central heating boiler or furnace is installed. (See Chapter 10 of the code for these clearance requirements.) The definition ***does not apply*** to air for combustion and ventilation for these appliances.

**3.3.91 Safety Blowout (Backfire Preventer).** A protective device located in the discharge piping of large mixing machines, incorporating a bursting disc for excessive pressure release, means for stopping a flame front, and an electric switch or other release mechanism for actuating a built-in or separate safety shutoff.

**3.3.92 Service Head Adapter.** A transition fitting for use with plastic piping (which is encased in non-pressure-carrying metal pipe) that connects the metal pipe casing and plastic pipe and tubing to the remainder of the piping system.

**3.3.93 Service Meter Assembly.** The piping and fittings installed by the serving gas supplier to connect the inlet side of the meter to the gas service and to connect the outlet side of the meter to the customer's house or yard piping.

FAQ ►
Does the *National Fuel Gas Code* cover the service meter assembly?

The code does not cover the service meter assembly, which is defined in 3.3.93, because it precedes the gas meter. Coverage of the *National Fuel Gas Code* starts at the discharge of the gas meter. (See Exhibit 1.1.)

**3.3.94 Service Regulator.** See 3.3.88.5, Pressure Regulator; and 3.3.88.7, Service Regulator.

**3.3.95 Shutoff.** See 3.3.105, Valve.

**3.3.96 Specific Gravity.** As applied to gas, the ratio of the weight of a given volume to that of the same volume of air, both measured under the same conditions.

**3.3.97 Steam Cooker.** See 3.3.6.5.5, Gas Steam Cooker.

**3.3.98 Steam Generator.** See 3.3.6.5.6, Gas Steam Generator.

**3.3.99 Stress.** The resultant internal force that resists change in the size or shape of a body acted on by external forces. In this code, the term *stress* is often used as being synonymous with unit stress, which is the stress per unit area (psi).

***3.3.99.1 Hoop Stress.*** The stress in a pipe wall, acting circumferentially in a plane perpendicular to the longitudinal axis of the pipe and produced by the pressure of the fluid in the pipe.

**3.3.100 System.**

***3.3.100.1 Central Premix System.*** A system that distributes flammable gas–air mixtures to two or more remote stations.

***3.3.100.2 Fan-Assisted Combustion System.*** An appliance equipped with an integral mechanical means to either draw or force products of combustion through the combustion chamber or heat exchanger.

***3.3.100.3 Hybrid Pressure System.*** A piping system in which the pressure at the point of delivery is reduced by one or more line pressure regulators prior to the appliance connection.

*3.3.100.4 Mechanical Exhaust System.* Equipment installed in and made a part of the vent to provide the required flow of gases through the vent.

Gas flow can be accomplished through either forced draft (positive pressure in the vent) or induced draft (negative pressure in the vent). See also 3.3.32.1, Mechanical Draft.

*3.3.100.5 Natural Draft Venting System.* A venting system that relies on natural draft to convey the products of combustion.

*3.3.100.6 Piping System.* All piping, valves, and fittings from the outlet of the point of delivery from the supplier to the outlets of the equipment shutoff valves.

*3.3.100.7* Venting System.* A continuous open passageway from the flue collar or draft hood of an appliance to the outdoors for the purpose of removing flue or vent gases.

*A.3.3.100.7 Venting System.* A venting system is usually composed of a vent or a chimney and vent connector(s), if used, assembled to form the open passageway.

*3.3.100.7.1 Forced Mechanical Draft Venting System.* A venting system in which a fan or other mechanical device is used to cause the flow of flue or vent gases under positive vent pressure.

*3.3.100.7.2 Mechanical Draft Venting System.* A venting system designed to remove flue or vent gases by mechanical means, which may consist of an induced draft portion under nonpositive static pressure or a forced draft portion under positive static pressure.

**3.3.101 Tensile Strength.** The highest unit tensile stress (referred to the original cross section) a material can sustain before failure (psi).

**3.3.102 Thermostat.**

*3.3.102.1 Electric Switch Type Thermostat.* A device that senses changes in temperature and controls electrically, by means of separate components, the flow of gas to the burner(s) to maintain selected temperatures.

*3.3.102.2 Integral Gas Valve Type Thermostat.* An automatic device, actuated by temperature changes, designed to control the gas supply to the burner(s) in order to maintain temperatures between predetermined limits and in which the thermal actuating element is an integral part of the device: (1) graduating thermostat, a thermostat in which the motion of the valve is approximately in direct proportion to the effective motion of the thermal element induced by temperature change; (2) snap-acting thermostat, a thermostat in which the thermostatic valve travels instantly from the closed to the open position, and vice versa.

**3.3.103 Thread Joint Compounds.** Nonhardening materials used on pipe threads to ensure a seal.

**3.3.104 Tubing.** Semirigid conduit of copper, steel, aluminum, corrugated stainless steel tubing (CSST), or plastic.

This definition specifically lists CSST as a tubing because some code users have considered it to be a different material and did not apply the requirements for tubing to CSST.

**3.3.105 Valve.** A device used in piping to control the gas supply to any section of a system of piping or to an appliance.

*3.3.105.1 Appliance Shutoff Valve.* A valve located in the piping system used to shut off individual equipment.

*3.3.105.2 Automatic Valve.* An automatic or semiautomatic device consisting essentially of a valve and operator that control the gas supply to the burner(s) during operation of an appliance.

NEW *3.3.105.3 **Excess Flow Valve (EFV).*** A valve designed to activate when the fuel gas passing through it exceeds a prescribed flow rate.

*3.3.105.4 **Manual Reset Valve.*** An automatic shutoff valve installed in the gas supply piping and set to shut off when unsafe conditions occur. The device remains closed until manually reopened.

*3.3.105.5 **Relief Valve.*** A safety valve designed to forestall the development of a dangerous condition by relieving either pressure, temperature, or vacuum in a hot water supply system.

*3.3.105.5.1 **Pressure Relief Valve.*** A valve that automatically opens and closes a relief vent, depending on whether the pressure is above or below a predetermined value.

*3.3.105.5.2 **Temperature Relief Valve.*** A valve that automatically opens and automatically closes a relief vent, depending on whether the temperature is above or below a predetermined value.

*3.3.105.5.3 **Vacuum Relief Valve.*** A valve that automatically opens and closes a vent for relieving a vacuum within the hot water supply system, depending on whether the vacuum is above or below a predetermined value.

*3.3.105.6 **Service Shutoff Valve.*** A valve, installed by the serving gas supplier between the service meter or source of supply and the customer piping system, to shut off the entire piping system.

**3.3.106 Valve Member.** That part of a gas valve rotating within or in respect to the valve body that, by its position with respect to the valve body, controls the flow of gas.

*3.3.106.1 **Nondisplaceable Valve Member.*** A valve member that cannot be moved from its seat by a force applied to the handle or to any exterior portion of the valve.

**3.3.107 Vent.** A passageway used to convey flue gases from appliances or their vent connectors to the outdoors.

*3.3.107.1 **Common Vent.*** That portion of a vent or chimney system that conveys products of combustion from more than one appliance.

*3.3.107.2 **Gas Vent.*** A passageway composed of listed factory-built components assembled in accordance with the manufacturer's installation instructions for conveying vent gases from appliances or their vent connectors to the outdoors.

For an item to be called a *gas vent,* as defined in 3.3.107.2, it must be factory-made and listed for conveying vent gases, which are defined as products of combustion plus excess air plus dilution air. Therefore, installers cannot fabricate gas vents. In order to be called a gas vent and in order to be installed as required in Section 12.7, the materials used must be listed as a gas vent.

*3.3.107.2.1 **Special-Type Gas Vent.*** Gas vents for venting listed Category II, III, and IV appliances.

*3.3.107.2.2 **Type B Gas Vent.*** A vent for venting listed gas appliances with draft hoods and other Category I appliances listed for use with Type B gas vents.

The Type B gas vent, which is defined in 3.3.107.2.2, is a double-wall type with a relatively low permissible temperature of vent gases due to its construction, which incorporates an aluminum inner tube. An air space between the inner, aluminum tube and the outer, galvanized steel tube has insulating properties, which minimize heat loss. Where higher vent temperatures are anticipated, a Type L vent can be used. Type B gas vents (and all other gas vents) must be factory made and listed.

*3.3.107.2.3* ***Type B-W Gas Vent.*** A vent for venting listed wall furnaces.

The Type B-W gas vent, which is defined in 3.3.107.2.3, is usually an oval-shaped, double-wall type designed to fit within a conventional stud wall. The most common use for this type of vent is to vent wall furnaces. See Exhibit 10.14 for examples of Type B and Type B-W vent fittings. Type B-W gas vents (and all other gas vents) must be factory made and listed.

*3.3.107.2.4* ***Type L Gas Vent.*** A vent for venting appliances listed for use with Type L vents and appliances listed for use with Type B gas vents.

The Type L gas vent in 3.3.107.2.4 is rated for higher temperature vent gases than the Type B gas vent. It differs from Type B gas vents in that the inner tube is stainless steel, rather than aluminum. Type L gas vents can be used with appliances that are listed for Type B, but not vice versa. Type L gas vents (and all other gas vents) must be factory made and listed.

*3.3.107.3* ***Regulator Vent.*** The opening in the atmospheric side of the regulator housing permitting the in and out movement of air to compensate for the movement of the regulator diaphragm.

**3.3.108 Vent Connector.** The pipe or duct that connects a fuel gas burning appliance to a vent or chimney.

**3.3.109 Vent Offset.** An arrangement of two or more fittings and pipe installed for the purpose of locating a vertical section of vent pipe in a different but parallel plane with respect to an adjacent section of vertical vent pipe.

**3.3.110 Venting.** Removal of combustion products as well as process fumes to the outer air.

**3.3.111 Wall Head Adapter.** A transition fitting for terminating plastic pipe inside of buildings at the building wall.

**3.3.112 Zero Governor.** A regulating device that is normally adjusted to deliver gas at atmospheric pressure within its flow rating.

## REFERENCE CITED IN COMMENTARY

The following publication is available from the National Fire Protection Association, 1 Batterymarch Park, Quincy, MA 02169-7471, 1-800-344-3555, www.nfpa.org.

NFPA 211, *Standard for Chimneys, Fireplaces, Vents, and Solid Fuel-Burning Appliances*, 2006 edition.

# General

# CHAPTER 4

Chapter 4 covers the general requirements for fuel gas. Other general requirements concerning scope, purpose, retroactivity, equivalency, and enforcement are covered in Chapter 1. Chapter 4 general requirements cover the following:

- Limitation of installations, modifications, and repairs performed under the code to a qualified agency (See Section 4.1.)
- What must be done prior to and following an interruption of gas service (See Section 4.2.)
- Prevention of accidental ignition while repairs are made to gas piping (See Section 4.3.)

## 4.1 Qualified Agency

Installation, testing, and replacement of gas piping, appliances, or accessories, and repair and servicing of equipment, shall be performed only by a qualified agency.

Section 4.1 reflects that fuel gases are potentially dangerous if they are not handled properly. For this reason, the code specifically states (and defines) that only qualified persons are permitted to install and maintain gas utilization equipment.

*Qualified agency* is defined in 3.3.85. A representative of the serving gas supplier is an example of a qualified agency. Many states, counties, cities, and municipalities permit only licensed plumbers or gas fitters to install and maintain gas distribution systems and equipment. These entities may conduct examinations or recognize certificates of national testing organizations to establish a list of qualified agencies and individuals.

◄ **FAQ**
How can the consumer identify a qualified agency?

## 4.2 Interruption of Service

Following an interruption of service, the requirements of Chapter 8 must be followed prior to putting the gas system back into service.

### 4.2.1 Notification of Interrupted Service.

When the gas supply is to be turned off, it shall be the duty of the qualified agency to notify all affected users. Where two or more users are served from the same supply system, precautions shall be exercised to ensure that service only to the proper user is turned off.

*Exception: In cases of emergency, affected users shall be notified as soon as possible of the actions taken by the qualified agency.*

When gas service must be interrupted for other than emergency situations, users must be notified. This requirement in 4.2.1 clearly defines the responsibility of the qualified agency shutting off the gas supply to notify all affected users and to be aware of potential cross connections.

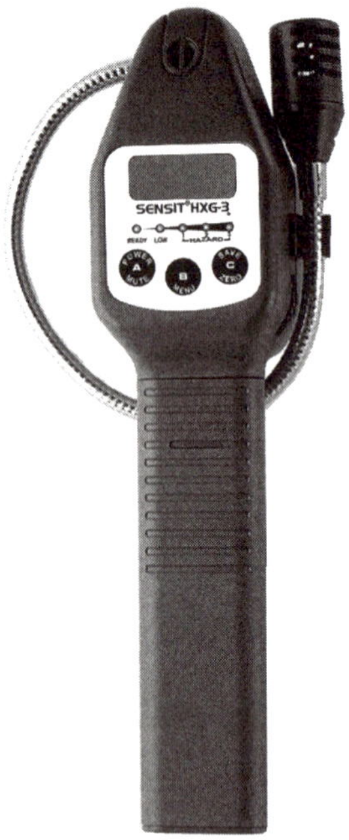


*EXHIBIT 4.1* *Combustible Gas Detector. (Courtesy of SENSIT)*

### 4.2.2 Work Interruptions.

When interruptions in work occur while repairs or alterations are being made to an existing piping system, the system shall be left in a safe condition.

If interruptions in work occur, as addressed in 4.2.2, all open-ended piping should be capped or plugged to prevent leakage if valves are opened inadvertently. Supply valves should be tagged "out of service" and locked, if practical.

## 4.3 Prevention of Accidental Ignition

### 4.3.1 Potential Ignition Sources.

Where work is being performed on piping that contains or has contained gas, the following shall apply:

(1) Provisions for electrical continuity shall be made before alterations are made in a metallic piping system.
(2) Smoking, open flames, lanterns, welding, or other sources of ignition shall not be permitted.
(3) A metallic electrical bond shall be installed around the location of cuts in metallic gas pipes made by other than cutting torches. Where cutting torches, welding, or other sources of ignition are unavoidable, it shall be determined that all sources of gas or gas–air mixtures have been secured and that all flammable gas or liquids have been cleared from the area. Piping shall be purged as required in Section 8.3 before welding or cutting with a torch is attempted.
(4) Artificial illumination shall be restricted to listed safety-type flashlights and safety lamps. Electric switches shall not be operated, on or off.

Subsection 4.3.1 provides rules for preventing accidental ignition when working on piping that contains, or has contained, gas. Electrical continuity must be maintained to prevent sparks, which are a source of ignition. Smoking, open flames, lanterns, welding, and other sources of ignition are not permitted. Artificial illumination is restricted to electrical devices listed for Class I, Division 1, Group D locations as defined in *NFPA 70®*, *National Electrical Code®*, and to locations where operation of electrical switches is prohibited. A combustible gas detector can be used to determine if flammable gas–air mixtures are present. See Exhibit 4.1 for a picture of a combustible gas detector.

For additional information on safety in cutting and welding, refer to NFPA 51B, *Standard for Fire Prevention During Welding, Cutting, and Other Hot Work.*

### 4.3.2 Handling of Flammable Liquids.

Flammable liquids are ignited easily and are of great concern in the workplace. Extreme caution must be used when storing and transferring flammable liquids. For additional information, refer to NFPA 30, *Flammable and Combustible Liquids Code.*

Gasoline is an extremely flammable liquid that usually is present as a fuel for portable welding machines, generators, or other internal combustion engines. Extreme caution must be observed when handling or storing this fuel.

Why would liquids captured in the drip leg of a gas system be of concern to the serving gas supplier?
FAQ ▶

**4.3.2.1 Drip Liquids.** Liquid that is removed from a drip in existing gas piping shall be handled to avoid spillage or ignition. The gas supplier shall be notified when drip liquids are removed.

Drip liquids, which are liquids that condense from natural gas or LP-Gas in piping systems, can be highly flammable and can cause operational problems for valves and other piping components. Extreme caution must be observed when drip liquids are found in piping systems that are open for work.

Although natural gas is mostly "dry," the lubrication used at compressor stations on the pipeline facilities results in oil in the system. This oil may contain PCBs, which are highly toxic. PCBs have not been used intentionally for at least 30 years. However, traces can still be present in pipeline and distribution systems. When oil is found in piping systems, care must be taken because the type and source of the liquid cannot be determined by sight. If it is suspected that the oil is from the transmission line, the local utility should be contacted as soon as possible, and the oil should be treated as a potential hazard until it is identified.

**4.3.2.2 Other Flammable Liquids.** Flammable liquids used by the installer shall be handled with precaution and shall not be left within the premises from the end of one working day to the beginning of the next.

## REFERENCES CITED IN COMMENTARY

The following publications are available from the National Fire Protection Association, 1 Batterymarch Park, Quincy, MA 02169-7471, 1-800-344-3555, www.nfpa.org.

NFPA 30, *Flammable and Combustible Liquids Code*, 2008 edition.

NFPA 51B, *Standard for Fire Prevention During Welding, Cutting, and Other Hot Work*, 2009 edition.

*NFPA 70®, National Electrical Code®*, 2008 edition.

# Gas Piping System Design, Materials, and Components

CHAPTER 5

Chapter 5 covers the requirements for piping design, materials, and components. Chapter 5 includes detailed requirements for the following:

- A piping plan, which can be required by the authority having jurisdiction for new and modified gas piping systems (See Section 5.1.)
- Provisions for location of the point of delivery (See Section 5.2.)
- Interconnection between gas piping systems to prevent contamination and inaccurate metering of gas (See Section 5.3.)
- Sizing of gas piping systems to ensure that sufficient gas can be provided to each appliance (Tables are included in Chapter 6 to size piping systems; other engineering methods may be used as well.) (See Section 5.4.)
- Piping system operating pressure limitations, which limit gas pressure in piping systems in residential and most commercial buildings (See Section 5.5.)
- Acceptable piping materials and joining methods (This section contains comprehensive requirements for all aspects of pipe joining, threads, flanges, and gaskets. This section also contains requirements for both metallic and plastic piping.) (See Section 5.6.)
- Gas meters (See Section 5.7.)
- Gas pressure regulators, including requirements for their location, protection, ventilation, and identification (See Section 5.8.)
- Overpressure protection devices to prevent the high pressures of gas distribution systems and propane tanks from entering building piping systems (High pressure can result in gas control failure and leakage of gas into a building.) (See Section 5.9.)
- Back pressure protection to prevent contamination of piped air, oxygen, or alternative fuel gas (See Section 5.10.)
- Low-pressure protection to prevent air from entering pipes (See Section 5.11.)
- Shutoff valves (See Section 5.12.)
- Sizing excess flow valves, where installed (See Section 5.13)
- Expansion and flexibility of gas piping systems, including thermal and seismic stresses (See Section 5.14.)

## 5.1 Piping Plan

### 5.1.1 Installation of Piping System.

Where required by the authority having jurisdiction, a piping sketch or plan shall be prepared before proceeding with the installation. This plan shall show the proposed location of piping, the size of different branches, the various load demands, and the location of the point of delivery.

Subsection 5.1.1 allows the authority having jurisdiction to require that a piping plan be submitted for review of the piping and sizing prior to an appliance installation.

If there are unusual appliances or piping schemes, they can be reviewed prior to issuing a permit. Review also enables the authority having jurisdiction to know the type and scope of the job before any work is done. The authority having jurisdiction should review the piping plan and ensure it contains the following details:

- Total length
- Length of the branches
- Total demand of the appliances, in cubic feet per hour for natural gas or Btu/hr for propane
- Type of piping material (e.g., Schedule 40 metallic, copper, or corrugated stainless steel)
- Pressure of gas serving the system (after the service regulator)
- Type of fuel gas (natural gas, propane, etc.)
- Pipe sizing table used

Examples of piping plans are shown in Exhibit 5.1 and in Annex C of the code.

### 5.1.2 Addition to Existing System.

**5.1.2.1** When additional appliances are being connected to a gas piping system, the existing piping shall be checked to determine whether it has adequate capacity *(see 5.4.3).*

**5.1.2.2** If inadequate, the existing system shall be enlarged as required, or separate gas piping of adequate capacity shall be provided.

The provisions in 5.1.2 recognize that preparing a plan or sketch of the proposed piping system is necessary, whether it is new or an extension of an existing system. By showing the loads and lengths of runs, the sizing can be determined by methods described in Section 5.4 and can help to prevent added gas utilization equipment from overloading the existing gas piping system.

## 5.2 Provision for Location of Point of Delivery

The location of the point of delivery shall be acceptable to the serving gas supplier.

To avoid unnecessary changes and reworking, the gas supplier should be consulted to determine the best point of delivery. Reworking can be a very important issue if the supplier and installer do not agree on the location of the point of delivery prior to the installation of piping. A natural gas supplier determines the point of delivery based on company policies or the regulations of the U.S. Department of Transportation or state public utility commission. State law may also govern the point of delivery for LP-Gas systems. The installer must know the

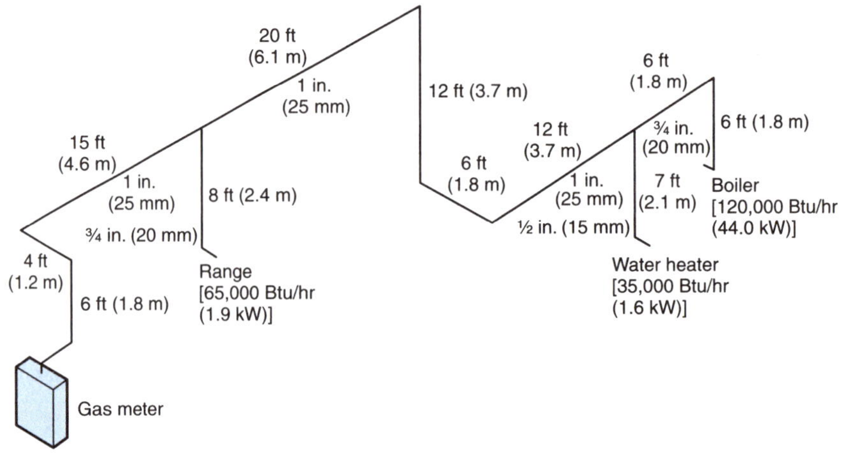


***EXHIBIT 5.1*** *Typical Piping Layout. (Courtesy of Richard E. White)*

point of delivery so that the length of the piping system can be properly determined in order to provide sufficient gas for proper appliance operation. Failure to do so may result in an inadequate piping system or one that does not coincide with the point of delivery. This situation may require re-piping of the system.

## 5.3 Interconnections Between Gas Piping Systems

### 5.3.1 Interconnections Supplying Separate Users.

Where two or more meters, or two or more service regulators where meters are not provided, are located on the same premises and supply separate users, the gas piping systems shall not be interconnected on the outlet side of the meters or service regulators.

Interconnected gas systems can be a safety issue. A problem can occur in apartment buildings and condominiums where each separate living unit is piped separately to a meter for proper billing. If a cross connection exists, maintenance, emergency, and utility personnel have no way of knowing that all services must be turned off to take the system safely out of service.

Interconnected gas systems are not to be confused with a natural gas supplier providing multiple meters or service regulators to provide additional capacity to a user. Such an interconnection falls under the meter or service regulator assembly, which is governed by the Department of Transportation in the United States.

### 5.3.2 Interconnections for Standby Fuels.

**5.3.2.1** Where a supplementary gas for standby use is connected downstream from a meter or a service regulator where a meter is not provided, equipment to prevent backflow shall be installed.

**5.3.2.2** A three-way valve installed to admit the standby supply and at the same time shut off the regular supply shall be permitted to be used for this purpose.

Large industrial users may use standby fuels for economic reasons. Where more than one fuel gas is used, the different gas supplies must not be mixed. If a building is served by a natural gas pipeline and uses a propane system for standby, it is most important that propane is not fed into the natural gas main and that natural gas is not fed into the propane storage tanks. The former will lead to an increased heating value of the natural gas, with resulting different combustion properties. Appliances adjusted for natural gas will not operate properly on a mixture of natural gas and propane. The latter will lead to operation of the pressure relief valves on propane tanks when the tanks are filled. The use of a three-way valve, such as the one shown in Exhibit 5.2, is one way to prevent the accidental mixing of two fuel supplies.

## 5.4 Sizing of Gas Piping Systems

The code requires that gas piping be sized to provide sufficient gas to meet the maximum load. Any of the three sizing methods specified in 5.4.3 can be used. The graphical sizing method shown in Annex C also may be used if approved by the authority having jurisdiction, as it is an engineering method. The sizing methods specified include the following:

- The pipe sizing tables or equations in Chapter 6 (Three methods are provided in Chapter 6 to use the tables — the longest length method, the branch length method, and the hybrid pressure method.)
- Other approved engineering methods, such as the graphical method in Annex C
- Sizing tables included in the manufacturer's installation instructions for listed piping systems

***EXHIBIT 5.2*** *Three-Way L-Type Valve Showing Different Operating Modes.*

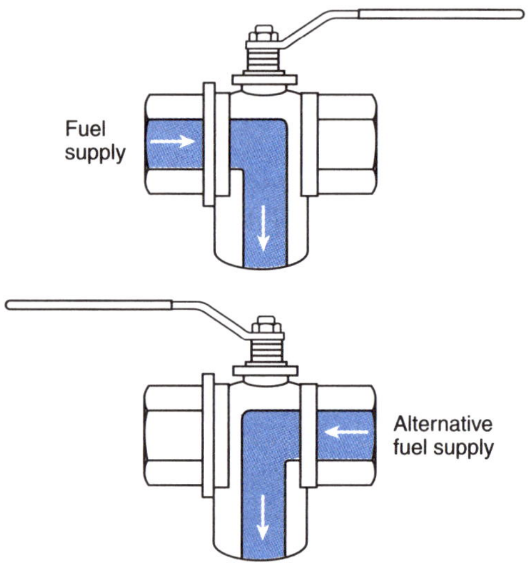


No matter which method is used, the piping system must provide sufficient gas for all appliances that can operate properly at any time.

### 5.4.1* General Considerations.

Gas piping systems shall be of such size and so installed as to provide a supply of gas sufficient to meet the maximum demand and supply gas to each appliance inlet at not less than the minimum supply pressure required by the appliance.

The requirement in 5.4.1 was revised in the 2009 edition to be more specific. Previously the requirement was that the piping system meet the maximum demand without "undue loss." This was changed to require that the minimum supply pressure of the appliances be provided, a much more specific requirement. Gas piping systems must supply the volume of gas required by each gas appliance such that:

- The pressure required by each appliance is at least the minimum in order for all appliances to perform properly under all operating conditions
- The maximum pressure specified by each appliance manufacturer is not exceeded

Providing this volume of gas, at the pressure required by each appliance, is the essence of the piping sizing requirements of the *National Fuel Gas Code.* Although the code provides many gas pipe sizing tables, they are not the code requirement. The code requirement, simply stated, is that if the appliances operate properly, the piping is properly sized.

In addition, ensuring that gas pressure to the connected equipment remains within the manufacturer's design limits when the equipment cycles "off" is important. A gas piping system's operating pressure has an effect on capacity, with higher initial pressures and their corresponding greater pressure drops enabling greater capacities for a given pipe or tube size. Gas piping systems that operate at pressures exceeding the connected gas utilization equipment's rated inlet pressure or that deliver gas at a pressure that varies outside the equipment manufacturer's design pressure limits must incorporate pressure regulators so that gas delivered to the equipment is within the design pressure range. See Section 5.8 for line gas pressure regulator requirements.

Regardless of which method is used to size the gas piping system, the volume of gas required by each appliance, and the pressure at which it operates, must be known. Knowing the amount of gas each appliance requires is not as straightforward as many users believe. In new construction the appliances may not be selected until after the piping is installed, and reasonable assumptions must be made. Replacement appliances may not be the same input as the appliances they replace, and can use more or less gas. If an older appliance is used, the nameplate may be missing or illegible, and Table 5.4.2.1 can provide an estimate.

Of course, the person sizing a gas piping system is not expected to be able to peer into the future and foresee what appliances might be installed in the future, but reasonable assumptions must be made.

**A.5.4.1** The size of gas piping depends on the following factors:

(1) Allowable loss in pressure *(see 5.4.4)* from point of delivery to appliance
(2) Maximum gas demand
(3) Length of piping and number of fittings
(4) Specific gravity of the gas
(5) Diversity factor
(6) Foreseeable future demand

Paragraph A.5.4.1 lists the general considerations for any fluid distribution system. Paragraphs 5.4.2 through 5.4.4 describe the topics in greater detail. The code also makes reference to the Chapter 6 sizing tables and to the Annex C material, which provides additional information to code users.

## 5.4.2* Maximum Gas Demand.

**A.5.4.2** To obtain the cubic feet per hour of gas required, divide the Btu per hour rating by the Btu per cubic foot heating value of the gas supplied. The heating value of the gas can be obtained from the local gas supplier.

Where the ratings of the appliances to be installed are not known, Table 5.4.2.1 shows the approximate demand of typical appliances by types.

***TABLE 5.4.2.1*** *Approximate Gas Input for Typical Appliances*

| *Appliance* | *Input Btu/hr (Approx.)* | *Appliance* | *Input Btu/hr (Approx.)* |
|---|---|---|---|
| **Space Heating Units** | | Water heater, automatic instantaneous | |
| Warm air furnace | | Capacity at 2 gal/min | 142,800 |
| Single family | 100,000 | Capacity at 4 gal/min | 285,000 |
| Multifamily, per unit | 60,000 | Capacity at 6 gal/min | 428,400 |
| Hydronic boiler | | Water heater, domestic, circulating or side-arm | 35,000 |
| Single family | 100,000 | | |
| Multifamily, per unit | 60,000 | **Cooking Appliances** | |
| **Space and Water Heating Units** | | Range, freestanding, domestic | 65,000 |
| Hydronic boiler | | Built-in oven or broiler unit, domestic | 25,000 |
| Single family | 120,000 | Built-in top unit, domestic | 40,000 |
| Multifamily, per unit | 75,000 | **Other Appliances** | |
| **Water Heating Appliances** | | Refrigerator | 3,000 |
| Water heater, automatic storage 30 gal to 40 gal tank | 35,000 | Clothes dryer, Type 1 (domestic) | 35,000 |
| | | Gas fireplace direct vent | 40,000 |
| Water heater, automatic storage 50 gal tank | 50,000 | Gas log | 80,000 |
| | | Barbecue | 40,000 |
| | | Gas light | 2,500 |

For SI units, 1000 Btu/hr = 0.293 W.

**5.4.2.1** The volume of gas to be provided (in cubic feet per hour) shall be determined directly from the manufacturers' input ratings of the appliances served. Where the input rating is not indicated, the gas supplier, appliance manufacturer, or a qualified agency shall be contacted, or the rating from Table 5.4.2.1 shall be used for estimating the volume of gas to be supplied.

Natural gas has a nominal heating value of 1000 Btu/ft$^3$ (37 kJ/m$^3$), and Table 6.2 (a) through Table 6.2 (v) for sizing natural gas systems provide capacity in cubic feet per hour. Propane gas has a heating value of about 2500 Btu/ft$^3$ (93 kJ/m$^3$), and Tables 6.3 (a) through Table 6.3 (m) for sizing propane systems provide capacity in thousands of Btu per hour. There is no technical reason for this difference, but it represents a system that many people have become comfortable with.

Table 5.4.2.1, Approximate Gas Input for Typical Appliances, provides gas input values for typical appliances, but the designer is cautioned because the input ratings of such domestic appliances as ranges, gas logs, and ovens vary widely.

In some commercial and industrial applications, equipment requires pressure greater than the normal 7 in. water column (w.c.) (1.7 kPa) for natural gas or 11 in. w.c. (2.7 kPa) for propane. In this case for natural gas systems, the gas supplier should be contacted to determine the availability of gas at a higher pressure, and the piping system should be designed according to that supply pressure. For propane systems, the propane regulator must be selected appropriately, and the propane storage containers must provide sufficient vaporization capacity over the range of temperatures anticipated. For more information about propane vaporization capacity, see the *NFPA Pocket Guide to Fuel Gas Storage and Use*.

The piping provisions of this code can be used for piping systems up to 125 psi (862 kPa). Pressure higher than 125 psi (862 kPa) is not prohibited, but other codes must be used for piping design and installation. [See the commentary following 1.1.1.1(B).]

**5.4.2.2** The total connected hourly load shall be used as the basis for piping sizing, assuming all appliances are operating at full capacity simultaneously.

*Exception: Sizing shall be permitted to be based upon established load diversity factors.*

Diversity of load, addressed in the exception to 5.4.2.2, is the reduction of load per unit in piping serving multiple units. Diversity of load is important in buildings such as apartment buildings, where many apartments are served by common supply piping and there are many gas ranges, water heaters, and furnaces or boilers on a single gas supply system. The reduction of pipe sizing of the supply system where a diversity of load exists assumes that appliances will not operate simultaneously. It is reasonable to assume that all occupants will not take showers at the same time, that they all will not operate cooking appliances at the same time, and that all furnaces will not operate simultaneously because of the cyclical nature of their operation. Gas suppliers may provide load diversity factors applicable to their service areas. Load diversity factors may vary by region or climate. Check with the local gas utility or propane supplier for more information.

### 5.4.3* Sizing Methods.

Gas piping shall be sized in accordance with one of the following:

(1) Pipe sizing tables or sizing equations in Chapter 6
(2) Other approved engineering methods acceptable to the authority having jurisdiction
(3) Sizing tables included in a listed piping system manufacturer's installation instructions

**A.5.4.3** The gas-carrying capacities for different sizes and lengths of iron pipe, or equivalent rigid pipe, and semirigid tubing are shown in the capacity tables in Chapter 6.

Table 6.2(a) through Table 6.2(t) indicate approximate capacities for single runs of piping. If the specific gravity of the gas is other than 0.60, correction factors should be applied. Correction factors for use with these tables are given in Table C.3.4.

For any gas piping system, for special appliances, or for conditions other than those covered by the capacity tables in Chapter 6, such as longer runs, greater gas demands, or greater pressure drops, the size of each gas piping system should be determined by the pipe sizing equations in Section 6.4 or by standard engineering methods acceptable to the authority having jurisdiction.

A suggested procedure for using the Chapter 6 tables to size a gas piping system is illustrated in Annex C.

Several pipe sizing methods are recognized in the code. These include the pipe sizing equations in Chapter 6, the longest length and branch length sizing methods, sizing tables provided by manufacturers of listed piping materials, such as CSST, and other approved sizing methods.

Although the code never mandated the use of the longest length method, most code users choose this method because it is easy to use and because it provides adequate pipe size with an ample margin for future expansion. The branch length method can also be selected, as well as the pipe sizing tables provided for listed piping systems. The pipe sizing equations in Chapter 6 can also be used. Because they have been revised to solve for diameter, they will be more useful than the former equations, which solved for the flow given a specific diameter. Other engineering methods, such as the graphical method provided in Annex C, are also permitted, but only when approved by the authority having jurisdiction.

**Longest Length Method.** The longest length method, using the tables or sizing equations in Chapter 6, is the most widely used method of sizing gas piping. The tables contain gas-carrying capacities for different sizes and lengths of iron pipe (or equivalently dimensioned rigid pipe), semi-rigid tubing, and CSST. The tables for natural gas are located in Section 6.2, and those for propane are located in Section 6.3. Both sections are arranged in order by piping material (pipe, tube, or CSST), pressure, and pressure drop allowed.

The longest length method is straightforward and can be used by any gas piping installer. The tables for pipe and tubing do not attempt to optimize pipe size to minimize material costs, because the potential savings to be had on a house or a small multifamily dwelling or commercial building will be much less than the cost of the time it would take an engineer to precisely calculate the system. In larger installations, an architect or engineer is usually involved, and it can be economical to use another sizing method because significant savings can result.

**Branch Length Method.** The branch length method may provide smaller pipe diameters than the longest length method in systems of widely varying lengths from the point of delivery to the appliances.

◄ **FAQ**
Why might the branch length method of calculating pipe size be useful?

Exhibits 5.3 and 5.4 show the pipe reduction of a small system using the branch length method versus the longest length method. Refer to the commentary in Chapter 6 for more information on the branch length method.

### 5.4.4 Allowable Pressure Drop.

NEW

The design pressure loss in any piping system under maximum probable flow conditions, from the point of delivery to the inlet connection of the appliance, shall be such that the supply pressure at the appliance is greater than or equal to the minimum pressure required by the appliance.

In the 2009 edition, 5.4.4 was clarified to allow a system to be designed for the minimum pressure drop required in the system. Previously, the system had to be designed for a pressure greater than the minimum pressure drop required in the system. Many code users do not understand the term *pressure drop* as used in a gas piping system. Pressure drop is an important concept, because it affects gas pipe sizing. In any gas piping system that is flowing gas, there is a certain friction, which slows the flow of gas and can be measured by the difference in pressure measurements along the piping system. Of interest is the difference, or pressure

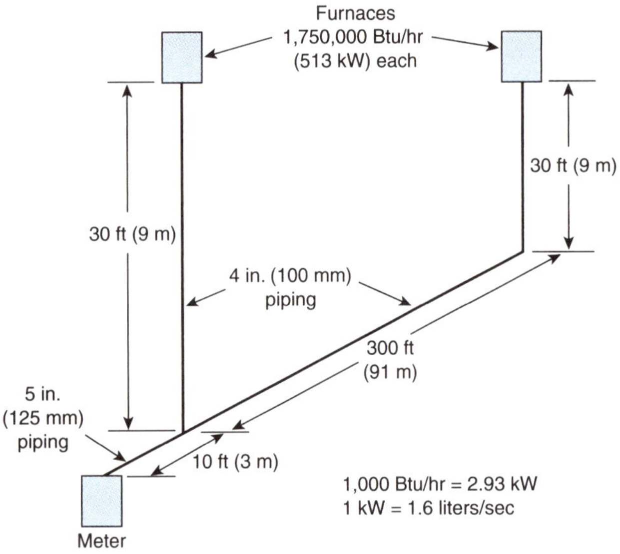


***EXHIBIT 5.3*** *Piping System Sized Using the Longest Length Method.*

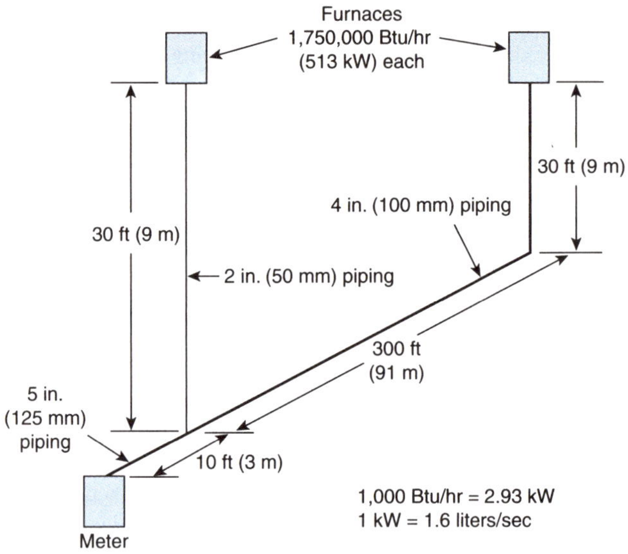


***EXHIBIT 5.4*** *Piping System Sized Using the Branch Length Method.*

drop, between the inlet pressure and the pressure at the most remote appliance. Pressure drop is important because an appliance will not operate properly if the gas pressure at the inlet to the appliance is less than that required by the manufacturer.

The appliance nameplate usually identifies the minimum gas inlet pressure. The piping designer should specify a pressure drop. The tables in Chapter 6 identify the pressure drops used to calculate the flows for the table in the identification block at the top right. When selecting a table, the user should check the pressure drop given for that table. If an excessive pressure drop is used, some appliances may not operate properly when the system is under full load. It is possible to use a compressor to boost gas pressure to make up for the pressure drop due to friction. This is rarely if ever done, though, because the cost of a compressor and the energy to operate it will almost always exceed the cost of larger pipe.

## 5.5 Piping System Operating Pressure Limitations

### 5.5.1 Maximum Design Operating Pressure.

The maximum design operating pressure for piping systems located inside buildings shall not exceed 5 psi (34 kPa) unless one or more of the following conditions are met:

(1)* The piping system is welded.

**A.5.5.1(1)** For welding specifications and procedures that can be used, see the API 1104, *Welding of Pipelines and Related Facilities*; AWS B2.1, *Standard for Welding Procedure and Performance Qualification*; or *ASME Boiler and Pressure Vessel Code*, Section IX.

(2) The piping is located in a ventilated chase or otherwise enclosed for protection against accidental gas accumulation.
(3) The piping is located inside buildings or separate areas of buildings used exclusively for one of the following:
    (a) Industrial processing or heating
    (b) Research

(c) Warehousing
(d) Boiler or mechanical rooms

Subsection 5.5.1 allows pressure above 5 psi (34 kPa) to be used without approval for the uses covered in 5.5.1(2), 5.5.1(3), and 5.5.1(4). Approval of piping systems pressures above 5 psi (34 kPa) for these uses was a code requirement in the 1999 and prior editions. It was deleted because higher pressures were not a problem, because the code provided no criteria for the authority having jurisdiction to use in approving or not approving higher pressures, and because approval was nearly always granted for the higher pressure.

When a leak occurs in a gas piping system, the flow of gas is proportional to the pressure. Most gas appliances are designed to require substantially less than 5 psi (34 kPa) for proper operation. Leakage through an opening in a piping system that operates at 7 in. w.c. (1.7 kPa) (natural gas) or 11 in. w.c. (2.7 kPa) (propane) would be less than 1/10th the leakage of a system operating at 5 psi (34 kPa), and less than 1/20th the leakage of a system operating at 20 psi (140 kPa). The 5 psi (34 kPa) general pressure limit for threaded gas piping systems in buildings recognizes these facts. For piping above 5 psi (34 kPa), welding is required because welded pipe joints have a lower possibility of leakage compared to threaded pipe joints.

◀ **FAQ**
**Why is gas pressure normally limited to 5 psi (34 kPa)?**

(4) The piping is a temporary installation for buildings under construction.

Temporary piping systems used during construction are exempt from the pressure limit in 5.5.1 because there is no exposure to the public.

(5) The piping serves appliances or equipment used for agricultural purposes.

NEW

The new requirement in 5.5.1(5) exempts fuel gas piping in buildings devoted to agricultural uses from the requirements in 5.5.1 to provide additional protection from gas leakage. This exemption recognizes that higher pressures are typically used on farms for livestock facilities and other purposes. Examples of such buildings are greenhouses, poultry breeding facilities, tobacco drying barns, and others. Note that 1.1.1.2 (2) exempts farm appliances and equipment from the scope of NFPA 54.

(6) The piping system is an LP-Gas piping system with a design operating pressure greater than 20 psi and complies with NFPA 58, *Liquefied Petroleum Gas Code.*

NEW

The new exception in 5.5.1(6) recognizes coverage that was added to NFPA 58 for propane piping systems operating above 20 psig (140 kPag) [to 50 psig (340 kPag)] and references NFPA 58 for systems above 20 psig (140 kPag).

### 5.5.2 LP-Gas Systems.

NEW

LP-Gas systems designed to operate below −5°F (−21°C) or with butane or a propane–butane mix shall be designed to either accommodate liquid LP-Gas or to prevent LP-Gas vapor from condensing back into a liquid.

The 20 psi (140 kPa) limit required by 5.5.1 is a practical one because, at higher pressures, propane droplets will begin to form at temperatures below −5°F (−21°C), a temperature that can be expected in many parts of the United States and Canada at some time during the year. LP-Gas in liquid form entering a burner designed for vapor will produce a large increase in flame size, a significant fire hazard. When vaporized, LP-Gas expands by a factor of about 270 to 1 and, although a lesser volume of liquid than vapor will pass through an orifice, a much greater quantity of Btu will be available to be burned.

It should be noted that while 5.5.2 is specifically applicable to LP-Gas systems, 5.5.1 is also applicable to LP-Gas systems. This has been somewhat clarified with the new 5.5.1(6), which covers LP-Gas.

Commentary Table 5.1 shows the pressure at which liquefaction occurs for the minimum expected system temperature for propane systems. Commentary Table 5.2 presents similar data for butane systems.

**COMMENTARY TABLE 5.1** *Liquefaction of Propane*

| Minimum Anticipated Temperature (°F) | Liquefaction Pressure (psi) |
|---|---|
| −5 | 20.0 |
| −10 | 16.7 |
| −15 | 13.6 |
| −20 | 10.7 |
| −25 | 8.0 |
| −30 | 5.6 |
| −35 | 3.4 |
| −40 | 1.5 |

For SI units: °C = 5/9 (°F − 32); 1 psi = 6.894 kPa.

**COMMENTARY TABLE 5.2** *Liquefaction of Butane*

| Minimum Anticipated Temperature (°F) | Liquefaction Pressure (psi) |
|---|---|
| 80 | 22.9 |
| 75 | 19.8 |
| 70 | 16.9 |
| 65 | 14.2 |
| 60 | 11.6 |
| 55 | 9.1 |
| 50 | 6.9 |
| 45 | 4.9 |
| 40 | 3.0 |
| 35 | 1.3 |

For SI units: °C = 5/9 (°F −32); 1 psi = 6.894 kPa.

NEW

## 5.6* Acceptable Piping Materials and Joining Methods

NEW

**A.5.6** Table A.5.6 is a list of piping materials and fittings that are allowed in the code.

The text of Table A.5.6 has been revised by a tentative interim amendment (TIA). See page 1.

Table A.5.6 is new and provides a list of allowed pipe, tubing, and fittings.

NEW

***TABLE A.5.6*** *Approved Pipe, Tube, Fittings, and Joints for Natural Gas and Liquefied Petroleum Fuel Applications*

| *Pipe Material* | *Standard* | *Fitting Types* | *Joint Types* | *Other Requirements* |
|---|---|---|---|---|
| ***Metallic Pipe*** | | | | |
| Black steel (minimum Schedule 40) | ASTM A 106* | Steel<br>Malleable iron<br>Steel cast iron<br>ASME B16.1*<br>Brass<br>Bronze<br>Special | Threaded<br>Flanged | Threads per ASME B1.20.1*.<br>Special fittings shall be appropriate for the application and acceptable to AHJ. |
| Galvanized steel (minimum Schedule 40) | ASTM A 53* | | | |
| Wrought iron (minimum Schedule 40) (Also known as low iron or wrought steel) | ASME B36.10M* | | | |
| Copper | None specified | Cast copper alloy<br>Bronze<br>Brass<br>Special | None specified | Prohibited where the gas contains more than an average of 0.3 grains of hydrogen sulfide per 100 scf of gas (0.7 mg/100 L).<br>Threads cannot form the joint seal.<br>Special fittings shall be appropriate for the application and acceptable to AHJ. |
| Copper alloy (brass) | None specified | | | |

*TABLE A.5.6 Continued*

| *Pipe Material* | *Standard* | *Fitting Types* | *Joint Types* | *Other Requirements* |
|---|---|---|---|---|
| Aluminum | ASTM B 241* | Aluminum<br>Special | None specified | Alloy 5456 is prohibited.<br>Threads cannot form the joint seal.<br>Coated to protect against external corrosion where it is in contact with masonry, plaster, or insulation, or is subject to repeated wettings by such liquids as water, detergents, or sewage.<br>Aluminum alloy pipe shall not be used in exterior locations or underground.<br>Special fiitings shall be appropriate for the application and acceptable to AHJ. |
| ***Metallic Tubing*** | | | | |
| Copper | ASTM B 88*<br>ASTM B 280* | Cast copper alloy<br>Wrought copper<br>Press fittings meeting ANSI LC4*<br>Forged copper alloy<br>Special | Brazed<br>Flanged/brazed<br>Brazed mechanically<br>Pressed (crimped)<br>Flared | Prohibited where the gas contains more than an average of 0.3 grains of hydrogen sulfide per 100 scf of gas (0.7 mg/100 L).<br>Brazed joints fabricated with alloys having a melting temperature greater than 1000°F.<br>Brazing alloys contain less than 0.05% phosphorus.<br>Minimum melting temperature greater than 1000°F.<br>Brazing alloys contain less than 0.05% phosphorus.<br>Flares, SAE J533 for single 45° flares.<br>Special fittings shall be appropriate for the application and acceptable to AHJ. |
| CSST | ANSI LC 1 / CSA 6.26* | ANSI LC 1 / CSA 6.26* | Manufacturer's installation instructions | Installation in accordance with the manufacturer's installation instructions. |
| Aluminum | ASTM B 210*<br>ASTM B 241* | Copper alloy (brass)<br>Special | Compression | Coated to protect against external corrosion where it is in contact with masonry, plaster, or insulation, or is subject to repeated wettings by such liquids as water, detergent, or sewage.<br>Aluminum-alloy tubing shall be used in exterior locations or underground.<br>Special fittings shall be appropriate for the application and acceptable to AHJ. |
| Steel | ASTM A 539*<br>ASTM A 254* | Special | | Special fittings shall be appropriate for the application and acceptable to AHJ. |
| ***Non-Metallic Pipe and Tube*** | | | | |
| Polyethylene (PE) | ASTM D 2513*<br>ASTM D 2513* (heat fusion) | Polyethylene (PE)<br>Service head adapters meeting Category I of ASTM D 2513*<br>Connections to metallic pipe meeting ASTM D 2513*, ASTM F 1973*, or ASTM F 2509* | Manufacturer's instructions<br>Compression-type mechanical joints<br>Heat fusion | Pipe and tubing installed outdoors underground only.<br>Pipe shall be marked "gas" and "ASTM D 2513."<br>Plastic pipe, tubing, fittings, and joints in undiluted liquefied petroleum gas piping systems shall be in accordance with NFPA 58*. |
| PVC | UL 651*<br>Schedule 40 or 80 | PVC UL 651*<br>Schedule 40 or 80 | Manufacturer's instructions | Regulator vents only<br>Outdoor installation only |

*Required standard. See Annex M for standard title.

### 5.6.1 General.

**5.6.1.1 Acceptable Materials.** Materials used for piping systems shall comply with the requirements of this chapter or shall be acceptable to the authority having jurisdiction.

Paragraph 5.6.1.1 clearly states that any piping material listed in this chapter is allowed for piping systems. Materials other than those listed in this chapter may be used with the approval of the authority having jurisdiction. The most widely used piping materials are listed in the code to assist users, but they are not intended to restrict the use of other materials that may be preferred in other cases. One example is stainless steel pipe, which is not widely used due to its higher cost, but in chemical plants with corrosive atmospheres it may be preferred.

Specific criteria for approval of piping materials are not provided in the code; however, the authority having jurisdiction should require proof that the material can hold the anticipated pressure, is compatible with the gas being used, has the mechanical strength needed for the installation, and has resistance to any corrosive environment in which the piping system may be placed.

**5.6.1.2 Used Materials.** Pipe, fittings, valves, or other materials shall not be used again unless they are free of foreign materials and have been ascertained to be adequate for the service intended.

While materials are not often reused in gas piping systems, there is no reason to prevent them from being used again if they retain their original strength and thickness. Approval should be obtained prior to installation because an inspection of the used material may be required.

**5.6.1.3 Other Materials.** Material not covered by the standards specifications listed herein shall meet the following criteria:

The code's intent is not to prohibit the use of new or different materials or technology. The intent of 5.6.1.3 is to provide the authority having jurisdiction with a basis on which to judge the acceptability of alternate piping materials that are not recognized specifically in the code (i.e., requiring evidence to substantiate the safety and suitability of the new piping materials or practices).

(1) Be investigated and tested to determine that it is safe and suitable for the proposed service
(2) Be recommended for that service by the manufacturer
(3) Be acceptable to the authority having jurisdiction

### 5.6.2 Metallic Pipe.

**5.6.2.1 Cast Iron.** Cast-iron pipe shall not be used.

**5.6.2.2 Steel and Wrought Iron.** Steel and wrought-iron pipe shall be at least of standard weight (Schedule 40) and shall comply with one of the following standards:

(1) ANSI/ASME B36.10, *Welded and Seamless Wrought-Steel Pipe*
(2) ASTM A 53, *Standard Specification for Pipe, Steel, Black and Hot-Dipped, Zinc-Coated Welded and Seamless*
(3) ASTM A 106, *Standard Specification for Seamless Carbon Steel Pipe for High-Temperature Service*

**5.6.2.3* Copper and Brass.** Copper and brass pipe shall not be used if the gas contains more than an average of 0.3 grains of hydrogen sulfide per 100 scf of gas (0.7 mg/100 L).

**A.5.6.2.3** An average of 0.3 grains of hydrogen sulfide per 100 scf of gas (0.7 mg/100 L) is equivalent to a trace as determined by ANSI/ASTM D 2385, *Method of Test for Hydrogen*

*Sulfide and Mercaptan Sulfur in Natural Gas (Cadmium Sulfate — Iodometric Titration Method),* or ANSI/ASTM D 2420, *Method of Test for Hydrogen Sulfide in Liquefied Petroleum (LP) Gases (Lead Acetate Method).*

Hydrogen sulfide is a naturally occurring impurity in natural gas and crude oil. LP-Gas is produced from two sources. Propane is one of the by-products of natural gas processing and is also produced by refining crude oil. Hydrogen sulfide is removed at the natural gas processing plant and in the crude oil–refining process. The hydrogen sulfide concentration stated in 5.6.2.3 is equivalent to a trace and should not have any significant corrosive effect on copper or brass pipe.

The natural gas is tested for the presence of hydrogen sulfide. Information on hydrogen sulfide levels should be available from the natural gas supplier. Propane is usually tested in the distribution system, and information may be available from the propane supplier.

◀ **FAQ**
Where could information on the hydrogen sulfide content of the gas supplied be found locally?

Hydrogen sulfide is highly corrosive to copper, brass, and other copper alloys. Copper and brass are used as a residential gas piping material and in valves, and corrosion by the gas is not acceptable.

◀ **FAQ**
Why is hydrogen sulfide limited in fuel gases?

If inspection of the inside of the pipe or tubing reveals accumulation of a fine black powder or black flakes, which may have a shiny finish, then hydrogen sulfide should be suspected. Hydrogen sulfide reacts with copper (or the copper in brass) to form copper sulfide. In such cases, replacement of all brass and copper components should be considered. The gas supplier should be notified, because other systems may also be similarly affected.

◀ **FAQ**
How can I tell if hydrogen sulfide has caused corrosion in a piping system?

**5.6.2.4 Threaded Copper, Brass, and Aluminum.** Threaded copper, brass, or aluminum alloy pipe shall not be used with gases corrosive to such material.

**5.6.2.5 Aluminum Alloy.** Aluminum alloy pipe shall comply with ASTM B 241, *Specification for Aluminum-Alloy Seamless Pipe and Seamless Extruded Tube* (except that the use of alloy 5456 is prohibited), and shall be marked at each end of each length indicating compliance. Aluminum alloy pipe shall be coated to protect against external corrosion where it is in contact with masonry, plaster, or insulation or is subject to repeated wettings by such liquids as water, detergents, or sewage.

Paragraph 5.6.2.5 requires that aluminum tubing be marked so that technicians and inspectors can see readily that installed tubing meets the proper specification. The coating requirement and prohibition of underground and exterior applications minimize corrosion.

Strong acids and strong bases readily corrode aluminum and its alloys. The assistance of a metallurgist or qualified corrosion specialist is recommended strongly in selecting an aluminum alloy, because the corrosion behavior of aluminum alloys varies significantly. Although they are used in some other countries, aluminum pipe and tubing are not generally used for gas piping in the United States. Aluminum tubing is used for piping that is part of appliances; however, this code does not cover such piping.

**5.6.2.6 Aluminum Installation.** Aluminum alloy pipe shall not be used in exterior locations or underground.

### 5.6.3 Metallic Tubing.

Seamless copper, aluminum alloy, or steel tubing shall not be used with gases corrosive to such material.

Copper, aluminum alloy, or steel tubing, addressed in 5.6.3, is permitted for the types and grades listed in the following subparagraphs provided that the gas is not corrosive to the tubing material. Note that aluminum alloy tubing is subject to the same restrictions as aluminum alloy pipe.

NEW **5.6.3.1 Steel.** Steel tubing shall comply with ASTM A 254, *Standard Specification for Copper Brazed Steel Tubing.*

Note that in the 2009 edition, reference to ASTM A 539, *Standard Specification for Electric Resistance-Welded Coiled Steel Tubing for Gas and Fuel Oil Lines*, was deleted, as the standard was withdrawn by ASTM, with no replacement.

**5.6.3.2* Copper and Brass.** Copper and brass tubing shall not be used if the gas contains more than an average of 0.3 grains of hydrogen sulfide per 100 scf of gas (0.7 mg/100 L). Copper tubing shall comply with standard Type K or Type L of ASTM B 88, *Specification for Seamless Copper Water Tube,* or ASTM B 280, *Specification for Seamless Copper Tube for Air Conditioning and Refrigeration Field Service.*

**A.5.6.3.2** See A.5.6.2.3.

Copper and brass tubing and fittings (except tin-lined copper tubing) should not be used if the gas contains more than an average of 0.3 grains of hydrogen sulfide per 100 scf of gas (0.7 mg/100 L).

Prohibiting copper and brass tubing when gas contains more than a trace of hydrogen sulfide is consistent with the requirement in 5.6.3.2 for copper and brass pipe. All natural gas that is distributed in pipelines, and all LP-Gas in the United States today, is treated to remove hydrogen sulfide. If natural gas is obtained directly from a gas well, this prohibition is important because there is no removal of hydrogen sulfide when gas is used directly from a well.

**5.6.3.3 Aluminum.** Aluminum alloy tubing shall comply with ASTM B 210, *Specification for Aluminum-Alloy Drawn Seamless Tubes,* or ASTM B 241, *Specification for Aluminum-Alloy Seamless Pipe and Seamless Extruded Tube.* Aluminum alloy tubing shall be coated to protect against external corrosion where it is in contact with masonry, plaster, or insulation or is subject to repeated wettings by such liquids as water, detergent, or sewage. Aluminum alloy tubing shall not be used in exterior locations or underground.

**5.6.3.4 Corrugated Stainless Steel.** Corrugated stainless steel tubing shall be listed in accordance with ANSI LC 1/CSA 6.26, *Fuel Gas Piping Systems Using Corrugated Stainless Steel Tubing.*

***EXHIBIT 5.5*** *Corrugated Stainless Steel Tubing. (Courtesy of Foster-Miller, Inc.)*

Corrugated stainless steel tubing (CSST), addressed in 5.6.3.4, was introduced to the North American market in the mid-1980s as a new gas piping material. It has been accepted and is being used throughout most areas. Supplement 2, Update on the Design and Installation Requirements for CSST Gas Piping Systems, provides more detail. (See also Exhibit 5.5.)

The *National Fuel Gas Code* permits the installation of systems using CSST listed in accordance with the requirements of ANSI LC 1/CSA 6.26, *Fuel Gas Piping Systems Using Corrugated Stainless Steel Tubing*. Note that manufacturer's installation instructions include any installation requirements in ANSI LC 1. There should not be any conflicts between the code and the manufacturer's installation instructions.

### 5.6.4 Plastic Pipe, Tubing, and Fittings.

NEW **5.6.4.1 Standard and Marking.** Plastic pipe, tubing, and fittings used to supply fuel gas shall conform to ASTM D 2513, *Standard Specification for Thermoplastic Gas Pressure Pipe, Tubing, and Fittings.* Pipe to be used shall be marked "gas" and "ASTM D 2513."

NEW **5.6.4.2* Regulator Vent Piping.** Plastic pipe, tubing, and fittings used to connect regulator vents to remote vent terminations shall be PVC conforming to ANSI/UL 651, *Schedule 40 and 80 Rigid PVC Conduit and Fittings.* PVC vent piping shall not be installed indoors.

NEW **A.5.6.4.2** The reference to ANSI/UL 651, *Schedule 40 and 80 Rigid PVC Conduit and Fittings*, is to require that PVC be a minimum of Schedule 40 and that it be resistant to the effects of ultraviolet light because it is likely to be exposed to the outdoors when used for regulator vents.

In the 2009 edition, 5.6.4.2 was revised to delete a restriction that plastic gas piping be installed outdoors, underground only. This change was made because installation is not within the scope of Chapter 5, and such restriction is in Chapter 7, Gas Piping Installation, in 7.1.7.1.

**5.6.4.3 Anodeless Risers.** Anodeless risers shall comply with the following:

(1) Factory-assembled anodeless risers shall be recommended by the manufacturer for the gas used and shall be leak tested by the manufacturer in accordance with written procedures.
(2) Service head adapters and field-assembled anodeless risers incorporating service head adapters shall be recommended by the manufacturer for the gas used and shall be design-certified to meet the requirements of Category I of ASTM D 2513, *Standard Specification for Thermoplastic Gas Pressure Pipe, Tubing, and Fittings,* and 49 CFR 192.281(e). The manufacturer shall provide the user qualified installation instructions as prescribed by 49 CFR 192.283(b).
(3) The use of plastic pipe, tubing, and fittings in undiluted LP-Gas piping systems shall be in accordance with NFPA 58, *Liquefied Petroleum Gas Code.*

A riser, such as the one shown in Exhibit 5.6, is an assembly that makes the transition from underground plastic pipe or tubing to aboveground metal pipe or tubing. Examples of such uses include serving a gaslight or providing gas to a detached garage for heating.

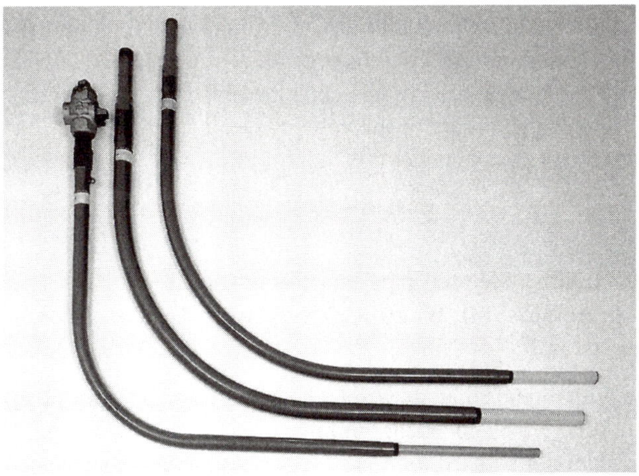

***EXHIBIT 5.6*** *Risers. (Courtesy of R. W. Lyall & Company, Inc.)*

Risers are available as one-piece, factory-assembled units and as field-assembled units. Factory-assembled risers have a baked, protective coating that makes the use of a sacrificial anode for corrosion protection unnecessary. Hence they are called "anodeless."

## 5.6.5 Workmanship and Defects.

Gas pipe, tubing, and fittings shall be clear and free from cutting burrs and defects in structure or threading and shall be thoroughly brushed and chip and scale blown. Defects in pipe, tubing, and fittings shall not be repaired. Defective pipe, tubing, and fittings shall be replaced. *(See 8.1.1.3.)*

Installers must inspect and clean piping materials before assembly to ensure compliance with the code. Small pieces of metal can clog the small clearances between moving parts and in orifices.

FAQ ► Can defective pipe or fittings be repaired?

Defects in pipe, tubing, or fittings must not be repaired. When defective pipe, tubing, or fittings are located in a system, the defective material must be replaced. Repairing defective components is not an appropriate remedy. Subsection 5.6.5 requires each installation to be made with defect-free materials.

Paragraph 8.1.1.3 covers testing repairs and additions to systems, including those in which defective materials have been replaced.

### 5.6.6 Protective Coating.

Where in contact with material or atmosphere exerting a corrosive action, metallic piping and fittings coated with a corrosion-resistant material shall be used. External or internal coatings or linings used on piping or components shall not be considered as adding strength.

Materials commonly used for protective coating include paint, polyethylene jacketing over a mastic primer, and fusion-bonded epoxy. Paint is often used on abovegrade applications, and the other coatings on underground pipe. Galvanization is not considered to be an adequate protective coating on underground installations. Galvanizing, or coating with zinc, provides a sacrificial coating, because the zinc will be corroded before steel. When the zinc is gone, the steel will corrode. It is not possible to inspect buried pipe to determine when all the zinc is gone.

Using a corrosion-resistant material for these applications is often more satisfactory than providing a corrosion protection system that inherently has a limited life. When using coated steel pipe underground, consideration generally is given to using cathodic protection as the means of protection from corrosion.

There are two types of cathodic protection systems: impressed current systems, which require a dc power source, and passive systems, which rely on a sacrificial material (zinc and aluminum are commonly used). Impressed current systems may be costly for the relatively short lengths of underground piping that are covered by the code. Passive systems should be monitored for voltage using a half-cell, and a record of the voltage must be maintained. Specialized equipment and training are required to conduct this monitoring.

### 5.6.7 Metallic Pipe Threads.

**5.6.7.1 Specifications for Pipe Threads.** Metallic pipe and fitting threads shall be taper pipe threads and shall comply with ANSI/ASME B1.20.1, *Pipe Threads, General Purpose, Inch.*

**5.6.7.2 Damaged Threads.** Pipe with threads that are stripped, chipped, corroded, or otherwise damaged shall not be used. Where a weld opens during the operation of cutting or threading, that portion of the pipe shall not be used.

**5.6.7.3 Number of Threads.** Field threading of metallic pipe shall be in accordance with Table 5.6.7.3.

Paragraph 5.6.7.3 establishes requirements for the number of threads for a given pipe size by reference to Table 5.6.7.3.

FAQ ► Why is the correct number of threads on metallic pipe so critical?

Properly cut pipe threads are tapered (the threads nearest the end of the pipe have a smaller diameter than the threads farthest away). If common threading dies (e.g., fixed cutting heads) are run too far onto the pipe during the threading process, the resulting threads will be straight (un-tapered) for a portion of the threads, then will taper. These straight threads are known as running threads, because the thread has been "run" down the pipe. An improperly threaded pipe may bottom out in the fitting before the tapered threads are firmly seated. Because pipe fittings depend on a tapered thread on the pipe to seal properly, an improperly threaded pipe can easily result in a leak at the joint. Proper use of a fixed die requires the threading operation to stop as soon as the end of the pipe exits the die.

*TABLE 5.6.7.3 Specifications for Threading Metallic Pipe*

| *Iron Pipe Size (in.)* | *Approximate Length of Threaded Portion (in.)* | *Approximate No. of Threads to Be Cut* |
|---|---|---|
| 1/2 | 3/4 | 10 |
| 3/4 | 3/4 | 10 |
| 1 | 7/8 | 10 |
| 1 1/4 | 1 | 11 |
| 1 1/2 | 1 | 11 |
| 2 | 1 | 11 |
| 2 1/2 | 1 1/2 | 12 |
| 3 | 1 1/2 | 12 |
| 4 | 1 5/8 | 13 |

For SI units, 1 in. = 25.4 mm.

**5.6.7.4 Thread Compounds.** Thread (joint) compounds (pipe dope) shall be resistant to the action of LP-Gas or to any other chemical constituents of the gases to be conducted through the piping.

When threaded pipe and fittings are assembled, the meshed threads form a mechanical seal that holds the pipe together but does not provide a gastight seal. Gastightness is provided by a thread compound, which must be used for a gastight piping system. Thread compounds should be applied only to the male pipe connection to avoid having the compound enter the piping system during assembly.

Teflon® tape is a type of thread compound. If Teflon® tape is used, it must be applied properly to prevent pieces of Teflon® from entering the gas piping system and causing problems downstream. The use of Teflon® tape is not prohibited by the code, but the component manufacturer's instructions (e.g., for equipment such as gas valves) should be consulted regarding restrictions on its use.

## 5.6.8 Metallic Piping Joints and Fittings.

The type of piping joint used shall be suitable for the pressure and temperature conditions and shall be selected giving consideration to joint tightness and mechanical strength under the service conditions. The joint shall be able to sustain the maximum end force due to the internal pressure and any additional forces due to temperature expansion or contraction, vibration, fatigue, or the weight of the pipe and its contents.

Subsection 5.6.8 lists considerations that influence the selection of a joining method. A threaded joint is not desirable for a system that is exposed to significant vibration or external mechanical loads.

**5.6.8.1* Pipe Joints.** Pipe joints shall be threaded, flanged, brazed, or welded. Where nonferrous pipe is brazed, the brazing materials shall have a melting point in excess of 1000°F (538°C). Brazing alloys shall not contain more than 0.05 percent phosphorus.

**A.5.6.8.1** For welding and brazing specifications and procedures that can be used, see API 1104, *Standard for Welding Pipelines and Related Facilities*; AWS B2.1, *Standard for Welding Procedure and Performance Qualification*; AWS B2.2, *Standard for Brazing Procedure and Performance Qualification*; or ASME *Boiler and Pressure Vessel Code*, Section IX.

Because some fuel gases can contain trace amounts of sulfur, brazing alloys containing more than 0.05 percent phosphorus are not permitted, as noted in 5.6.8.1. If both are present, sulfur and phosphorus will combine to destroy the joint. This prohibition applies even if a current gas supply contains no sulfur, because the gas supply can shift to a source containing a trace of sulfur. Most commercially available brazing alloys contain less than 0.05 percent phosphorus.

**5.6.8.2 Tubing Joints.** Tubing joints shall either be made with approved gas tubing fittings, be brazed with a material having a melting point in excess of 1000°F (538°C), or made by press-connect fittings complying with ANSI LC-4, *Press-Connect Copper and Copper Alloy Fittings for Use in Fuel Gas Distribution Systems*. Brazing alloys shall not contain more than 0.05 percent phosphorus.

In the 2009 edition, a new system of fittings for copper tubing was added, Press Connected Fittings. These are tubing fittings that are placed on the tube, and a tool is used to seal the fitting to the tube (see Exhibit 5.7). This tubing connection system has been in use for a number of years in plumbing applications, and it has been found to be reliable. CSA America has developed a standard for the press connect fitting, cited in the code. ANSI LC-4 is a standard for the fittings and the integrity of the connections. It does not cover the tool used to assemble the fittings.

*EXHIBIT 5.7 A Press Connect Tool. (Courtesy of Viega, LLC)*

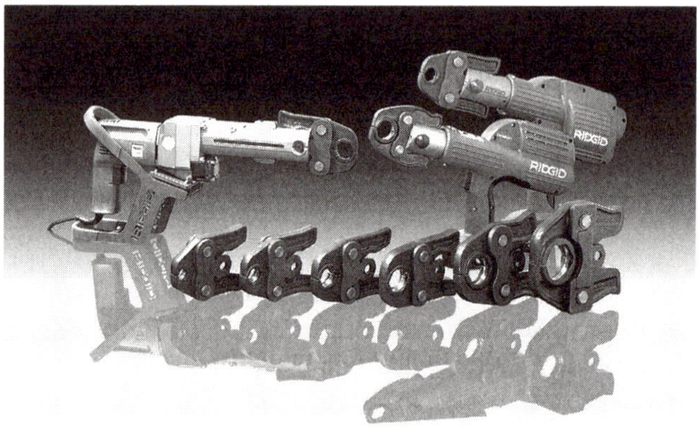


The word *approved* as used in the phrase *approved gas tubing fittings* in 5.6.8.2 is defined in 3.2.1 as follows:

**3.2.1* Approved.** Acceptable to the authority having jurisdiction.

**5.6.8.3 Flared Joints.** Flared joints shall be used only in systems constructed from nonferrous pipe and tubing where experience or tests have demonstrated that the joint is suitable for the conditions and where provisions are made in the design to prevent separation of the joints.

**5.6.8.4 Metallic Fittings (Including Valves, Strainers, Filters).** Metallic fittings shall comply with the following:

(1) Threaded fittings in sizes larger than 4 in. (100 mm) shall not be used unless acceptable to the authority having jurisdiction.

Paragraph 5.6.8.4(1) limits metallic threaded fittings larger than 4 in. (100 mm) to approved applications. They must be approved because of the difficulty in developing and maintaining tight threaded joints in these sizes. *Approved* is defined in 3.2.1, and appears above. Alternate fittings include welded and flanged fittings.

(2) Fittings used with steel or wrought-iron pipe shall be steel, brass, bronze, malleable iron, or cast iron.
(3) Fittings used with copper or brass pipe shall be copper, brass, or bronze.

Prior to the 2002 edition, the code required a minimum of 80 percent copper in fittings exposed to soil. This requirement is no longer included in the code.

(4) Fittings used with aluminum alloy pipe shall be of aluminum alloy.
(5) *Cast-Iron Fittings.* Cast-iron fittings shall comply with the following:

  (a) Flanges shall be permitted.
  (b) Bushings shall not be used.

Cast iron bushings are prohibited in 5.6.8.4(5)(b) because of the brittle nature of cast iron. Brittleness makes bushings less able to withstand overtightening during assembly or external mechanical abuse in service. Flammable gas–air mixtures in piping systems require more shock resistance than cast iron fittings can provide. Large cast iron fittings [flanged, because 5.6.8.4(1) prohibits large threaded fittings] cannot resist external loads adequately and are only allowed where approved.

  (c) Fittings shall not be used in systems containing flammable gas–air mixtures.
  (d) Fittings in sizes 4 in. (100 mm) and larger shall not be used indoors unless approved by the authority having jurisdiction.
  (e) Fittings in sizes 6 in. (150 mm) and larger shall not be used unless approved by the authority having jurisdiction.

(6) *Aluminum Alloy Fittings.* Threads shall not form the joint seal.

Paragraph 5.6.8.4(6) requires flanges, welded fittings, flared fittings, or metallic ball sleeve compression fittings to be used instead of direct threaded aluminum pipe-to-pipe connections. It is very difficult to form a gastight connection using aluminum threads because of the inherent low strength of the metal.

(7) *Zinc–Aluminum Alloy Fittings.* Fittings shall not be used in systems containing flammable gas–air mixtures.
(8) *Special Fittings.* Fittings such as couplings, proprietary-type joints, saddle tees, gland-type compression fittings, and flared, flareless, or compression-type tubing fittings shall be as follows:

  (a) Used within the fitting manufacturer's pressure–temperature recommendations
  (b) Used within the service conditions anticipated with respect to vibration, fatigue, thermal expansion, or contraction
  (c) Installed or braced to prevent separation of the joint by gas pressure or external physical damage
  (d) Acceptable to the authority having jurisdiction

A summary of the requirements for metal fittings in 5.6.8.4 is provided in Commentary Table 5.3.

### 5.6.9 Plastic Piping, Joints, and Fittings.

Plastic pipe, tubing, and fittings shall be joined in accordance with the manufacturers' instructions. The following shall be observed when making such joints:

(1) The joint shall be designed and installed so that the longitudinal pullout resistance of the joint will be at least equal to the tensile strength of the plastic piping material.
(2) Heat fusion joints shall be made in accordance with qualified procedures that have been established and proven by test to produce gastight joints at least as strong as the pipe or tubing being joined. Joints shall be made with the joining method recommended by the pipe manufacturer. Heat fusion fittings shall be marked "ASTM D 2513."
(3) Where compression-type mechanical joints are used, the gasket material in the fitting shall be compatible with the plastic piping and with the gas distributed by the system.

***COMMENTARY TABLE 5.3** Metal Fitting Requirements*

| *Fitting Material/Application* | *Requirement* |
|---|---|
| Threaded >4 in. | Be approved |
| Used with steel or wrought-iron pipe | Steel, brass, bronze, malleable iron, or cast iron only |
| Used with copper or brass pipe | Copper, brass, or bronze only |
| Used with aluminum alloy pipe | Aluminum alloy only<br>Threads shall not form the joint seal |
| Cast iron | No flanges or bushings<br>Do not use in systems containing flammable gas–air mixtures |
| Cast iron fittings, >4 in. indoors | Be approved |
| Cast iron fittings, 6 in. and larger | Be approved |
| Zinc–aluminum alloy fittings | Do not use in systems containing flammable gas–air mixtures |
| *Special fittings* (i.e., couplings, proprietary-type joints, saddle tees, gland-type compression fittings, and flared, flareless, or compression-type tubing fittings) | 1. Use within the fitting manufacturer's pressure–temperature recommendations<br>2. Use within the service conditions anticipated with respect to vibration, fatigue, thermal expansion, or contraction<br>3. Install or brace to prevent separation of the joint by gas pressure or external physical damage<br>4. Be approved |

An internal tubular rigid stiffener shall be used in conjunction with the fitting. The stiffener shall be flush with the end of the pipe or tubing and shall extend at least to the outside end of the compression fitting when installed. The stiffener shall be free of rough or sharp edges and shall not be a force fit in the plastic. Split tubular stiffeners shall not be used.

(4) Plastic piping joints and fittings for use in LP-Gas piping systems shall be in accordance with NFPA 58, *Liquefied Petroleum Gas Code.*

Because the manufacturer's instructions must be followed for any joining procedures, their recommendations become the basis of all installation procedures. Systems that are in common use and are recommended by most manufacturers for the joining of plastic gas pipe are addressed in 5.6.9(4) and shown in Exhibit 5.8. These include butt fusion, saddle fusion, socket fusion, electrofusion, and mechanical fittings or joints. One type of mechanical fitting is shown in Exhibit 5.9.

The code places additional requirements in 5.6.9(1), 5.6.9(2), and 5.6.9(3) to ensure a safe system. Note that, as with everything else involving the use of plastic pipe in LP-Gas systems, NFPA 58 is the applicable standard.

## 5.6.10 Flanges.

All flanges shall comply with ANSI/ASME B16.1, *Cast Iron Pipe Flanges and Flanged Fittings, Class 25, 125, 250, and 800*; ANSI/ASME B16.20, *Metal Gaskets for Pipe Flanges, Ring Joint Spiral Wound and Jacketed*; or MSS SP-6, *Standard Finishes for Contact Faces of Pipe Flanges and Connecting-End Flanges of Valves and Fittings.* The pressure–temperature ratings shall equal or exceed that required by the application.

**5.6.10.1 Flange Facings.** Standard facings shall be permitted for use under this code. Where 150 psi (1034 kPa) steel flanges are bolted to Class 125 cast-iron flanges, the raised face on the steel flange shall be removed.

*EXHIBIT 5.8* *Plastic Coiled Pipe. (Courtesy of Gastite)*

*EXHIBIT 5.9* *Mechanical Fittings. (Courtesy of R.W. Lyall & Company, Inc.)*

The requirement in 5.6.10.1 recognizes that if the face were left on the steel flange, the resulting structural forces on the cast iron flange could cause it to break.

◀ **FAQ**
**Why must the raised face of a steel flange be ground off when mated with a cast iron flange?**

**5.6.10.2 Lapped Flanges.** Lapped flanges shall be used only above ground or in exposed locations accessible for inspection.

### 5.6.11 Flange Gaskets.

The material for gaskets shall be capable of withstanding the design temperature and pressure of the piping system and the chemical constituents of the gas being conducted without change to its chemical and physical properties. The effects of fire exposure to the joint shall be considered in choosing the material.

**5.6.11.1** Acceptable materials shall include the following:

(1) Metal or metal-jacketed asbestos (plain or corrugated)
(2) Asbestos
(3) Aluminum "O" rings and spiral-wound metal gaskets

**5.6.11.2** When a flanged joint is opened, the gasket shall be replaced.

**5.6.11.3** Full-face gaskets shall be used with all bronze and cast-iron flanges.

"O" rings, spiral-wound metal gaskets (5.6.11.3), and some other types of flange gaskets either do not extend to the outside edge of the flange face or are not of full thickness all the way to the edge of the flange face. Either condition could place unacceptable mechanical strain on bronze or cast iron flanges.

## 5.7* Gas Meters

Section 5.7 does not cover gas meters supplied or specified by a natural gas supplier. The scope of NFPA 54 specifically states:

> **1.1.1.1(A)** Coverage of piping systems shall extend from the point of delivery to the appliance connections. For other than undiluted liquefied petroleum gas (LP-Gas) systems, the point of delivery shall be considered to be the outlet of the service meter assembly or the outlet of the service regulator or service shutoff valve where no meter is provided. . . .

Gas meters installed upstream of the point of delivery are covered by federal law in 49 CFR Part 192. Section 5.7 applies to any meters installed downstream of the point of delivery or utility meter in natural gas systems and downstream of the second-stage regulator in propane systems. Premises-owned meters are typically submeters that are downstream of the gas supplier's meter (sometimes called a master meter) in applications such as manufactured home courts, apartment buildings, or industrial properties where measurement of the gas flow downstream of the gas supplier's meter is necessary. The provisions of Section 5.7 are intended to ensure safe, trouble-free service from the meters and to protect them from physical damage. (See Exhibit 5.10.)

**A.5.7** This section applies to premises-owned meters *[see 1.1.1.2(16)]*.

### 5.7.1 Capacity.

Gas meters shall be selected for the maximum expected pressure and permissible pressure drop.

Small, residential-size meters, such as the one shown in Exhibit 5.11, measure their rated flow with a $^1/_2$ in. w.c. (0.1 kPa) pressure drop. Larger meters are dual-rated, and the designer can choose between the rated flow at a $^1/_2$ in. w.c. (0.1 kPa) pressure drop or a larger flow at a 2 in. w.c. (0.5 kPa) pressure drop. These pressure drops usually meet the requirements of 5.7.1, and the pressure drop can be accommodated in the piping system design.

*EXHIBIT 5.10* Gas Meter Protection. (Courtesy of Richard E. White)

### 5.7.2 Location.

**5.7.2.1** Gas meters shall be located in ventilated spaces readily accessible for examination, reading, replacement, or necessary maintenance.

The ventilated space requirement in 5.7.2.1 anticipates the release of a small amount of gas when a meter is replaced, and the possibility of a small amount of leakage from the meter itself.

**5.7.2.2** Gas meters shall not be placed where they will be subjected to damage, such as adjacent to a driveway, under a fire escape, in public passages, halls, or coal bins, or where they will be subject to excessive corrosion or vibration.

**5.7.2.3** Gas meters shall not be located where they will be subjected to extreme temperatures or sudden extreme changes in temperature. Meters shall not be located in areas where they are subjected to temperatures beyond those recommended by the manufacturer.

In the 2009 edition, the former paragraph 5.7.2.3 was deleted. It established a minimum distance of 3 ft (0.9 m) from a source of ignition to gas meters. The requirement was deleted because gas meters are not a source of gas discharge and therefore should not have location restrictions from sources of ignition.

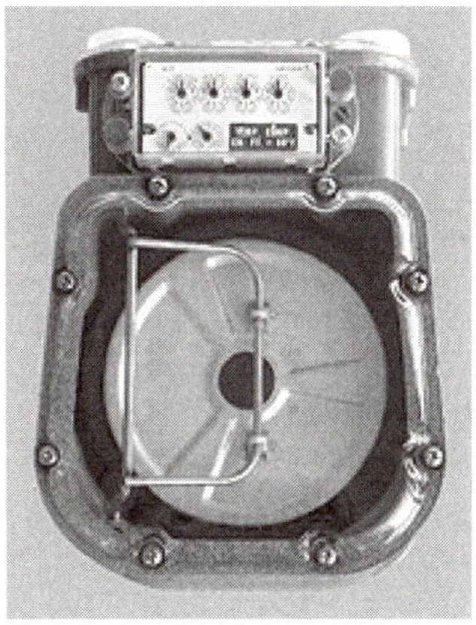

***EXHIBIT 5.11*** *Gas Meter with a Plastic Cover, Showing the Diaphragm.*

### 5.7.3 Supports.

Gas meters shall be supported or connected to rigid piping so as not to exert a strain on the meters. Where flexible connectors are used to connect a gas meter to downstream piping at mobile homes in mobile home parks, the meter shall be supported by a post or bracket placed in a firm footing or by other means providing equivalent support.

When flexible connectors are used to connect the gas meter to the downstream piping, the meter must be supported. (See Exhibit 5.12.)

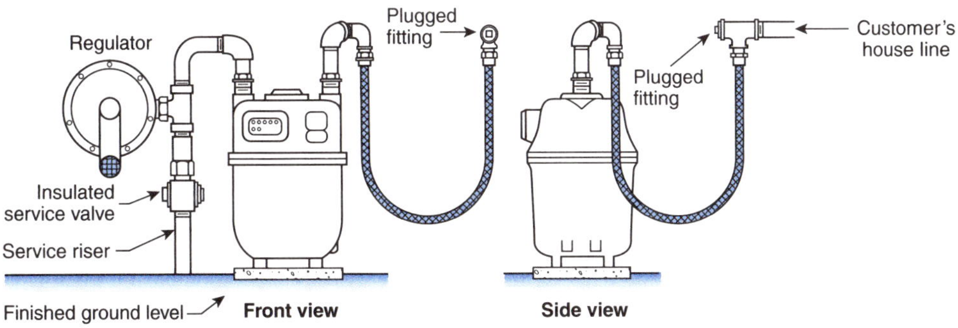


***EXHIBIT 5.12*** *Gas Meter Supported on a Concrete Pre-Cast Slab.*

### 5.7.4 Meter Protection.

Meters shall be protected against overpressure, back pressure, and vacuum, where such conditions are anticipated.

Gas meters have a maximum allowable operating pressure (MAOP) that must not be exceeded. Residential meters typically have MAOPs of 5 psi (34 kPa). Larger or special-application meters can carry MAOPs of 10 psi (70 kPa), 25 psi (170 kPa), or higher. Meters can leak air into the gas under vacuum conditions, an undesirable condition because the gas will be diluted with air. The amount of dilution will be relatively small, and it is highly unlikely to be large enough to create a flammable gas–air mixture. The amount may be sufficient to

*EXHIBIT 5.13 ID Tag. (Courtesy of Marking Services, Inc.)*

cause improper appliance operation, and it may result in the meter reading the gas–air mixture, and the user is billed for air.

### 5.7.5 Identification.

Gas piping at multiple meter installations shall be marked by a metal tag or other permanent means designating the building or the part of the building being supplied and attached by the installing agency.

When installing gas piping systems at a multiple meter location, the installer is the person who knows which piping system belongs to which appliances and to which designated part of the building. Therefore, the installer is in an ideal position to identify and attach to the piping system the permanent marking in the form of an ID tag, as shown in Exhibit 5.13, and as required by 5.7.5. It is important that the designations used on the tag be the same as those used for the location. For example, if a building is being divided into units 1 and 2, the piping system markings should be the same, "unit 1" and "unit 2," not "left" and "right" or "101" and "102." If a meter is not provided, the marking should be done at multiple piping locations, where the point of delivery may be the service shutoff valve or service regulator. [See 1.1.1.1(A).]

## 5.8* Gas Pressure Regulators

**A.5.8** This section applies to premises-owned regulators *[see 1.1.1.2(16)]*.

With increasing uses for gas appliances and with the economy of using higher pressures that allow the use of smaller-diameter gas piping systems (hybrid pressure systems), a need exists for line gas pressure regulators to deliver proper gas pressures to equipment or groups of equipment. This system parallels the use of higher electrical voltage (pressure) and smaller wires (pipes) with transformers (regulators) to supply the loads properly.

Note that Section 5.8 refers to line pressure regulators, not appliance regulators provided as part of an appliance. Appliance regulators are covered in 9.1.18.

### 5.8.1 Where Required.

A line gas pressure regulator or gas equipment pressure regulator, as applicable, shall be installed where the gas supply pressure is higher than that at which the branch supply line or appliances are designed to operate or vary beyond design pressure limits.

Where natural gas is used, the gas supplier should be consulted to determine the supply pressure being delivered. Where propane is used, the second-stage regulator determines the system supply pressure. If the piping system pressure at the inlet of any appliance on the piping system is greater than its maximum allowable inlet pressure, a line regulator must be installed. The piping system pressure determines the type and design of the line pressure regulator. The line pressure regulator addressed in 5.8.1 must be selected to meet the requirements of the system, including the inlet and outlet pressure rating and the volume of gas needed.

Where natural gas is supplied, the gas utility usually installs a pressure regulator upstream of the gas meter. This placement ensures constant pressure at the meter and correct metering and billing. As with the utility's gas meters, the regulator is upstream of the discharge of the meter and is covered by federal regulations, not by the *National Fuel Gas Code*.

**5.8.2** Line gas pressure regulators shall be listed in accordance with ANSI Z21.80/CSA 3.7, *Line Pressure Regulators*.

### 5.8.3 Location.

The gas pressure regulator shall be accessible for servicing.

Opening a door or removing a cover plate that may be attached with screws or other operable fasteners provides access to regulators. Note that *accessible* is defined in 3.3.1 as follows:

> **3.3.1 Accessible.** Having access to but which first requires the removal of a panel, door, or similar covering of the item described.

Prior to the 1999 edition, regulators were required to be "readily accessible," which is defined in 3.3.1.1 as accessible without the use of tools. The requirement was changed because access to regulators is not required in emergencies.

### 5.8.4 Regulator Protection.

Pressure regulators shall be protected against physical damage.

### 5.8.5 Venting.

**5.8.5.1 Line Gas Pressure Regulators.** Line gas pressure regulators shall comply with the following:

(1) An independent vent to the exterior of the building, sized in accordance with the regulator manufacturer's instructions, shall be provided where the location of a regulator is such that a ruptured diaphragm will cause a hazard.
    (a) Where more than one regulator is at a location, each regulator shall have a separate vent to the outdoors or, if approved by the authority having jurisdiction, the vent lines shall be permitted to be manifolded in accordance with accepted engineering practices to minimize back pressure in the event of diaphragm failure. *(See 5.9.7 for information on properly locating the vent.)*
    (b) Materials for vent piping shall be in accordance with Section 5.6.

Paragraph 5.8.5.1 provides the requirements for vents from line gas regulators installed in buildings. The basic rule is that a piped vent to the outdoors is required where rupture of a diaphragm can cause a hazard. Although "hazard" is not defined, a hazard occurs if the release of gas can reach the flammable range and ignite.

Manifolding (or combining more than one vent in one pipe) of regulator vents is allowed by the code, but requires that manifolding of vents be approved to ensure that the vent piping is of adequate size. Approval is defined in 3.2.1 as "acceptable to the authority having jurisdiction." The manifolding of such vents presents the following two potential problems:

◀ **FAQ**
Does the code permit manifolding of regulator vents?

1. A diaphragm on one of the connected regulators could fail. If the manifolded vent is not large enough, the flow from the failed diaphragm can build a back pressure in the manifold. The back pressure is applied to the top of the diaphragm of the other regulators, which has the same effect as tightening the regulator spring; that is, it creates a higher discharge pressure. A higher discharge pressure can increase the pressure of the piping system, which can create a hazardous condition.
2. The manifolded vent termination could become plugged and a diaphragm could subsequently fail. This failure could result in high pressure gas in the vent line that will drive all of the remaining regulators wide open and completely negate their installation.

Any approval of vent line manifolding should address these potential problems. Even single regulator vents must be sized per the manufacturer's installation instructions. Very long runs may restrict the free air movement to the top of the diaphragm, causing fluctuating gas pressure delivery to the burner. Many manufacturers specify a larger diameter vent on long vent runs.

Note that vent piping used must be constructed using materials specified in Section 5.6. This requirement limits vent piping to metallic piping above ground. A new provision in 5.6.4.2 allows PVC vents conforming to UL 651, *Schedule 40 and Schedule 80 Rigid PVC Conduit and Fittings*, to be used, but not indoors.

*Exception: A regulator and vent limiting means combination listed as complying with ANSI Z21.80/CSA 3.7, Line Pressure Regulators, shall be permitted to be used without a vent to the outdoors.*

***EXHIBIT 5.14*** *Gas Line Regulator with Vent-Limiting Means. (Courtesy of Maxitrol)*

ANSI Z21.80/CSA 3.7, *Line Pressure Regulators*, provides requirements for line gas regulators and optional vent limiters. Where vent limiters are used, they limit vented gas in the unlikely event of a diaphragm rupture so that a flammable concentration is not reached. Line gas regulators, like all pressure regulators, incorporate a flexible diaphragm that moves up and down to regulate pressure. The space above the diaphragm must be free to move, and the space above the diaphragm has a vent to the atmosphere to avoid compressing the air above the diaphragm, which would interfere with the operation of the regulator. No gas flows out of the vent during normal operation. Gas can only flow out of the vent if the diaphragm fails. The vent limiter is an orifice that restricts this flow of gas out of the regulator.

The requirement that the installation be made in a ventilated location is not intended to require mechanical ventilation, but rather to prohibit installation in spaces where flammable gas–air mixtures can accumulate. Line gas pressure regulators installed in places where gas appliances are installed under this code, for instance, will meet the requirement of this paragraph. The regulator and vent limiter are listed as a combination, and the vent limiter cannot be removed and used on a different regulator.

Exhibit 5.14 illustrates an example of a line gas pressure regulator with vent-limiting means addressed in the exception to 5.8.5.1(1). Note the vent hole in the exhibit. Common vent-limiting means can be installation-sensitive. The manufacturer's instructions should be consulted to ensure proper installation and functioning. For example, vent-limiting means intended for upright horizontal installation cannot function properly in any other position.

(2) The vent shall be designed to prevent the entry of water, insects, or other foreign materials that could cause blockage.

NEW (3) The regulator vent shall terminate at least 3 ft (0.9 m) from a source of ignition.

(4) At locations where regulators might be submerged during floods, a special antiflood-type breather vent fitting shall be installed, or the vent line shall be extended above the height of the expected flood waters.

(5) A regulator shall not be vented to the appliance flue or exhaust system.

The requirement in 5.8.5.1(5) recognizes that the flue or exhaust system cannot safely handle the amount of gas that is likely to be discharged through a failed diaphragm and that sources of ignition may be present in such locations.

**5.8.5.2 Appliance Pressure Regulators.** For venting of gas appliance pressure regulators, see 9.1.19.

### 5.8.6 Bypass Piping.

Valved and regulated bypasses shall be permitted to be placed around gas line pressure regulators where continuity of service is imperative.

If a bypass is necessary to ensure continued system operation in the event of failure of a line gas pressure regulator, 5.8.6 ensures that the bypass includes a means of regulating the bypass flow when the regulator is not operable. Bypasses are not normally installed in residential gas piping systems, as brief shutdowns to replace a malfunctioning regulator are acceptable. They are sometimes installed in industrial systems where continuity of service is imperative.

### 5.8.7 Identification.

Line pressure regulators at multiple regulator installations shall be marked by a metal tag or other permanent means designating the building or the part of the building being supplied.

See the commentary following 5.7.5 for information on permanent means of marking regulator installations, addressed in 5.8.7.

## 5.9 Overpressure Protection Devices

Section 5.9 spells out requirements for overpressure protection of gas piping systems, even though some of the components that can provide such protection are not within the scope of this code. For instance, U.S. Department of Transportation regulations cover natural gas service regulators, and NFPA 58 covers LP-Gas second-stage regulators. Therefore, consultation with the gas supplier may be necessary to ensure compliance with Section 5.9.

### 5.9.1 General.

Overpressure protection devices shall be provided to prevent the pressure in the piping system from exceeding that value that would cause unsafe operation of any connected and properly adjusted appliances. *(See 5.9.5.)*

The reference in 5.9.1 to 5.9.5 highlights the requirement that a pressure relief device, if used, be set to prevent damage to equipment. (See Exhibit 5.15.) Most new residential and commercial gas utilization equipment rated for a maximum operating inlet pressure of 14 in. w.c. (3.5 kPa) will withstand an inlet pressure of 2 psi to 2.5 psi (14 kPa to 17 kPa) without failing, although pressure control and leakage rates can be compromised. In addition, these controls will function normally (no permanent effects) once the overpressure is removed.

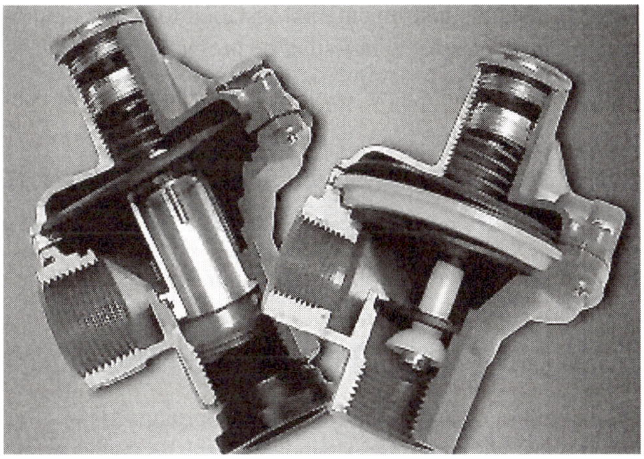

***EXHIBIT 5.15*** *Cut-Away of Pressure Relief Device. (Courtesy of Bryan Donkin, RMG – Canada Limited)*

Accordingly, limiting any possible overpressure to 2 psi to 2.5 psi (14 kPa to 17 kPa) in the case of these controls generally is accepted as adequate. When the maximum overpressure is needed for a specific piece of equipment, the equipment manufacturer's data should be obtained.

**5.9.1.1** The requirements of this section shall be met and a piping system deemed to have overpressure protection where a service or line pressure regulator plus one other device are installed such that the following occur:

(1) Each device limits the pressure to a value that does not exceed the maximum working pressure of the downstream system.
(2) The individual failure of either device does not result in overpressure of the downstream system.

Note that all the requirements in 5.9.1.1 must be met. These requirements ensure that a single-point failure will not result in a dangerous system overpressure. Several common means exist for providing the protection specified.

Systems operating at inches-of-water-column pressures use regulators [usually 7 in. w.c. (1.7 kPa) for natural gas, 11 in. w.c. (2.7 kPa) for LP-Gas] equipped with either an integral or separate relief valve that will open upon an overpressure situation and flow enough gas to keep the outlet pressure of the failed regulator below 2 psi (14 kPa). If either the regulator or the relief valve fails, the connected equipment will not be damaged. Both must fail simultaneously for a dangerous overpressure to occur.

Piping systems operating at 2 psi (14 kPa) use line gas pressure regulators to reduce pressure to appliances, and a similar relief valve is used at the service regulator. This relief valve has a capacity sufficient to maintain the outlet pressure of the failed service regulator below the maximum inlet pressure of the line gas pressure regulator [generally, 10 psi (69 kPa) for common models]. Therefore, if the service regulator fails, the line gas pressure regulator is still able to handle the increased pressure. Additionally, if the line gas pressure regulator fails completely, the equipment will be able to handle the resulting 2 psi (14 kPa) pressure without failure.

Similar systems in which the gas pressure is further elevated [for example, to 5 psi (34 kPa)] must then have an additional element of overpressure protection, because a failure of the line gas pressure regulator would subject the downstream equipment to a pressure of 5 psi (34 kPa). ANSI Z21.80/CSA 3.7, *Line Pressure Regulators*, requires that line gas pressure regulators installed in such systems have an overpressure shutoff device that automatically shuts off the gas if a high downstream pressure is detected.

**5.9.1.2** The pressure regulating, limiting, and relieving devices shall be properly maintained, inspection procedures shall be devised or suitable instrumentation installed to detect failures or malfunctions of such devices, and replacements or repairs shall be promptly made.

**5.9.1.3** A pressure relieving or limiting device shall not be required where the following conditions exist:

(1) The gas does not contain materials that could seriously interfere with the operation of the service or line pressure regulator.
(2) The operating pressure of the gas source is 60 psi (414 kPa) or less.
(3) The service or line pressure regulator has all of the following design features or characteristics:
   (a) Pipe connections to the service or line regulator do not exceed 2 in. nominal diameter.
   (b) The regulator is self-contained with no external static or control piping.
   (c) The regulator has a single port valve with an orifice diameter no greater than that recommended by the manufacturer for the maximum gas pressure at the regulator inlet.
   (d) The valve seat is made of resilient material designed to withstand abrasion of the gas, impurities in the gas, and cutting by the valve and to resist permanent deformation where it is pressed against the valve port.
   (e) The regulator is capable, under normal operating conditions, of regulating the downstream pressure within the necessary limits of accuracy and of limiting the discharge pressure under no-flow conditions to not more than 150 percent of the discharge pressure maintained under flow conditions.

**FAQ ▶** When would overpressure protection devices not be required on gas piping systems?

The requirements of 5.9.1.3 recognize that some systems are not likely to fail completely. Therefore, the code allows systems meeting all of the restrictions to operate without overpressure protection devices. Special attention to maintaining unobstructed (from insects, ice, etc.) regulator vents should be given on systems of this type because regulator vent blockage can cause an overpressure situation.

### 5.9.2 Devices.

**5.9.2.1** Pressure relieving or pressure limiting devices shall be one of the following:

(1) Spring-loaded relief device
(2) Pilot-loaded back pressure regulator used as a relief valve designed so that failure of the pilot system or external control piping causes the regulator relief valve to open
(3) A monitoring regulator installed in series with the service or line pressure regulator
(4) A series regulator installed upstream from the service or line regulator and set to continuously limit the pressure on the inlet of the service or line regulator to the maximum working pressure of the downstream piping system
(5) An automatic shutoff device installed in series with the service or line pressure regulator and set to shut off when the pressure on the downstream piping system reaches the maximum working pressure or some other predetermined pressure less than the maximum working pressure. This device shall be designed so that it will remain closed until manually reset.

The manual reset requirement in 5.9.2.1(5) is necessary because a system could shut down with open manual valves (e.g., on a range) and then come back on automatically some time later, creating a dangerous condition.

(6) A liquid seal relief device that can be set to open accurately and consistently at the desired pressure

**5.9.2.2** The devices in 5.9.2.1 shall be installed either as an integral part of the service or line pressure regulator or as separate units. Where separate pressure relieving or pressure limiting devices are installed, they shall comply with 5.9.3 through 5.9.8.

### 5.9.3 Construction and Installation.

All pressure relieving or pressure limiting devices shall meet the following requirements:

(1) Be constructed of materials so that the operation of the device is not impaired by corrosion of external parts by the atmosphere or of internal parts by the gas.
(2) Be designed and installed so they can be operated to determine whether the valve is free. The devices shall also be designed and installed so they can be tested to determine the pressure at which they operate and be examined for leakage when in the closed position.

### 5.9.4 External Control Piping.

External control piping shall be protected from falling objects, excavations, or other causes of damage and shall be designed and installed so that damage to any control piping shall not render both the regulator and the overpressure protective device inoperative.

Falling objects, which are referred to in 5.9.4, can include ice and snow shedding from roofs.

### 5.9.5 Setting.

Each pressure limiting or pressure relieving device shall be set so that the pressure shall not exceed a safe level beyond the maximum allowable working pressure for the piping and appliances connected.

See the commentary following 5.9.1.1(2) for a discussion of overpressure parameters.

### 5.9.6 Unauthorized Operation.

Precautions shall be taken to prevent unauthorized operation of any shutoff valve that makes a pressure relieving valve or pressure limiting device inoperative. The following are acceptable methods for complying with this provision:

(1) Lock the valve in the open position. Instruct authorized personnel in the importance of leaving the shutoff valve open and of being present while the shutoff valve is closed so that it can be locked in the open position before leaving the premises.
(2) Install duplicate relief valves, each having adequate capacity to protect the system, and arrange the isolating valves or three-way valve so that only one safety device can be rendered inoperative at a time.

### 5.9.7 Vents.

The manufacturer's data should be consulted to properly size all vent stacks to comply with the requirements of 5.9.7. A stack diameter larger than the device outlet diameter may be required. The vent termination should be located so that the discharge does not re-enter the building through a door, window, or other building opening. The blocking effect of screening the vent opening should be taken into consideration when sizing the terminal outlet.

**5.9.7.1** The discharge stacks, vents, or outlet parts of all pressure relieving and pressure limiting devices shall be located so that gas is safely discharged to the outdoors. Discharge stacks or vents shall be designed to prevent the entry of water, insects, or other foreign material that could cause blockage.

**5.9.7.2** The discharge stack or vent line shall be at least the same size as the outlet of the pressure relieving device.

### 5.9.8 Size of Fittings, Pipe, and Openings.

The fittings, pipe, and openings located between the system to be protected and the pressure relieving device shall be sized to prevent hammering of the valve and to prevent impairment of relief capacity.

Consideration must be given to the piping between the pressure-relieving device and the gas system it is protecting. Excessive use of fittings or long lengths of piping could hamper the operation of the valve.

## 5.10 Back Pressure Protection

### 5.10.1 Where to Install.

Industrial applications may involve equipment that uses a mixture of gases in its processes. If the system uses compressed air, back pressure protection is required because of the possibility of injecting air into the gas piping system. For example, natural gas and oxygen may be used in handheld brazing torches. The torch assembly at the point of use could easily overpressure one side of the gas supply, creating a hazard. Appropriate check valves (5.10.2) installed prior to torch assembly, as required by 5.10.1, alleviate this concern.

Venturi-type standby fuel systems require that the served gas piping system be provided with backflow protection, because a failure of the Venturi gas valves could cause an overpressure of the gas piping system.

**FAQ ►** Why would zero governor-equipped mixing equipment not need back pressure protection?

Gas systems connected to zero governor-equipped mixing equipment require no further back pressure prevention devices, as permitted by 5.10.1, because the standby gas will not flow until the blower is operating to "pull" the gas through the zero governor. This setup is very resistant to failure.

A zero governor is a regulator that requires a positive action to allow flow. A simple example of a zero governor is a diving regulator. The diver must inhale to initiate the flow of air from the SCBA cylinder. No air will flow from the cylinder when the diver is not inhaling.

## REFERENCES CITED IN COMMENTARY

The following publications are available from the National Fire Protection Association, 1 Batterymarch Park, Quincy, MA 02169-7471, 1-800-344-3555, www.nfpa.org.

NFPA 58, *Liquefied Petroleum Gas Code*, 2008 edition.

*NFPA Pocket Guide to Fuel Gas Storage and Use*, 2004.

The following publications are available from CSA-America, 8501 East Pleasant Valley Road, Cleveland, OH 44131, 216-524-4990, www.csa-america.org.

ANSI LC 1/CSA 6.26, *Fuel Gas Piping Systems Using Corrugated Stainless Steel Tubing*, 2005.

ANSI LC-4, *Press-Connect Copper and Copper Alloy Fittings for Use in Fuel Gas Distribution Systems*, 2007.

ANSI Z21.80/CSA 3.7, *Line Pressure Regulators*, 2005.

The following publication is available from Underwriters Laboratories, Inc., 333 Pfingsten Road, Northbrook, IL 60062-2096, www.ul.com.

UL 651, *Schedule 40 and Schedule 80 Rigid PVC Conduit and Fittings*, 2005.

The following publication is available from the U.S. Government Printing Office, Washington, DC 20401, www.gpo.gov.

Title 49, Code of Federal Regulations, Part 192.

# Pipe Sizing

CHAPTER 6

Chapter 6 contains pipe and tubing sizing tables. The tables have been coordinated to have the following common features that make them easier to use.

- All tables have a Title Block with the type of gas, inlet pressure, pressure drop, and gas specific gravity.
- The length of pipe included in the table is 2000 ft (610 m) for smooth wall pipe and 300 ft (90 m) or 500 ft (150 m) for CSST and all polyethylene at under 2 psi (14 kPa).
- All table entries use 3 significant digits.
- Very small table entries of less than 10 cfh (0.3 $m^3$/hr) natural gas or 10,000 Btu/hr use N/A (not applicable).

Tables for natural gas at pressures of 5 psi (34 kPa) and higher, included in the code prior to the 2002 edition, were deleted from the code because the committee believed that gas piping systems operating above 5 psi (34 kPa) are designed using engineering methods rather than the tables. The tables for natural gas at pressures of 10, 20, and 50 psi (69, 140, and 340 kPa) have been retained in this handbook for use if needed. (See Commentary Tables 6.1, 6.2, and 6.3 at the end of this chapter.)

Section 6.1 specifies the pipe sizing methods that can be used to size piping systems. These include the longest length method (6.1.1), the branch length method (6.1.2), and the hybrid pressure method for systems using multiple gas pressures (6.1.3).

Section 6.2 provides a reference to the tables for sizing gas piping systems using natural gas. Section 6.3 provides a reference to the tables for sizing gas piping systems using propane. Section 6.4 provides sizing equations for low-pressure (6.4.1) and high-pressure systems (6.4.2).

## 6.1 Pipe Sizing Methods

Where the pipe size is to be determined using any of the methods in 6.1.1 through 6.1.3, the diameter of each pipe segment shall be obtained from the pipe sizing tables in Section 6.2 or from the sizing equations in Section 6.4. *(See calculation examples in Annex C.)* For SI units, 1 $ft^3$ = 0.028 $m^3$, 1 ft = 0.305 m, 1 in. w.c. = 0.249 kPa, 1 psi = 6.894 kPa, 1000 Btu/hr = 0.293 kW.

### 6.1.1* Longest Length Method.

The pipe size of each section of gas piping shall be determined using the longest length of piping from the point of delivery to the most remote outlet and the load of the section.

**A.6.1.1** The Longest Length Method is the traditional method used to determine the equivalent piping length $L$ that is then used along with the pipe sizing tables to determine the appropriate pipe diameter size.

The longest length method, which is addressed in 6.1.1, is not new to the code. This method is the only sizing method using tables that were included in editions previous to the 2002 edition of the code. This sizing method is the simplest and is effective for all piping systems, but it can result in using larger pipe sizes than needed.

### 6.1.2* Branch Length Method.

Pipe shall be sized as follows:

(1) Pipe size of each section of the longest pipe run from the point of delivery to the most remote outlet shall be determined using the longest run of piping and the load of the section.
(2) The pipe size of each section of branch piping not previously sized shall be determined using the length of piping from the point of delivery to the most remote outlet in each branch and the load of the section.

**A.6.1.2** The Branch Length Method is an alternate sizing method that could permit slightly smaller pipe diameters in some segments of a piping system when compared with the Longest Length Method.

The branch length method can reduce the branch pipe sizes on systems with long runs between branches. The method is a little more complicated than the longest length method because the size of the main line and the branch lines must be determined using different lengths. In the longest length method, all sizes are determined using the same, longest, length.

### 6.1.3 Hybrid Pressure.

The pipe size for each section of higher pressure gas piping shall be determined using the longest length of piping from the point of delivery to the most remote line pressure regulator. The pipe size from the line pressure regulator to each outlet shall be determined using the length of piping from the regulator to the most remote outlet served by the regulator.

**FAQ ▶** Why does the hybrid pressure method prove advantageous on systems with very long runs of piping?

The diameter of a fuel gas piping system is proportional to the pressure of the fuel gas. By distributing fuel gas at a higher pressure, smaller pipe or tubing can be used, with corresponding cost savings, or existing piping may be able to be reused if appliances are added to an existing system. For example, at less than 2 psi (14 kPa), a 100 ft (30 m) length run of 1 in. (25 mm) Schedule 40 steel pipe has a capacity of 195 $ft^3/hr$ (5.5 $m^3/hr$) [per Table 6.2(b)], or 195,000 Btu/hr (57 kW). At 2 psi (14 kPa), the same pipe has a capacity of 1710 $ft^3/hr$ (48 $m^3/hr$) [per Table 6.2(c)] or about 1,700,000 Btu/hr (290 kW), which is roughly a 9 times increase in capacity. Of course, a regulator will be needed to reduce the pressure to that required by the appliance.

The illustration in Exhibit 6.1 on p. 99 shows a marked reduction in pipe size between the longest length method and the hybrid pressure method.

## 6.2 Tables for Sizing Gas Piping Systems Using Natural Gas

Table 6.2(a) through Table 6.2(v) shall be used to size gas piping in conjunction with one of the methods described in 6.1.1 through 6.1.3.

Examples of the use of the pipe sizing tables in Section 6.2 are found in Annex C of the code. Calculation worksheets that can be copied and used are included in Supplement 6 of this handbook.

Commentary Tables 6.1, 6.2, and 6.3 on pp. 98–99 can be used to size gas piping systems operating at 10 psi (69 kPa), 20 psi (140 kPa), or 50 psi (340 kPa). These tables were last included in the 1999 edition, but were deleted because most piping systems under the code are at 5 psi (34 kPa) or less. The tables from the 1999 edition have been included here in the event that they are needed. The values in the tables are unchanged from those in previous editions.

***TABLE 6.2(a)*** *Schedule 40 Metallic Pipe*

| | |
|---|---|
| *Gas:* | *Natural* |
| *Inlet Pressure:* | *Less than 2 psi* |
| *Pressure Drop:* | *0.3 in. w.c.* |
| *Specific Gravity:* | *0.60* |

| | *Pipe Size (in.)* | | | | | | | | | | | | | |
|---|---|---|---|---|---|---|---|---|---|---|---|---|---|---|
| ***Nominal:*** | *½* | *¾* | *1* | *1¼* | *1½* | *2* | *2½* | *3* | *4* | *5* | *6* | *8* | *10* | *12* |
| ***Actual ID:*** | *0.622* | *0.824* | *1.049* | *1.380* | *1.610* | *2.067* | *2.469* | *3.068* | *4.026* | *5.047* | *6.065* | *7.981* | *10.020* | *11.938* |
| ***Length (ft)*** | ***Capacity in Cubic Feet of Gas per Hour*** | | | | | | | | | | | | | |
| 10 | 131 | 273 | 514 | 1,060 | 1,580 | 3,050 | 4,860 | 8,580 | 17,500 | 31,700 | 51,300 | 105,000 | 191,000 | 303,000 |
| 20 | 90 | 188 | 353 | 726 | 1,090 | 2,090 | 3,340 | 5,900 | 12,000 | 21,800 | 35,300 | 72,400 | 132,000 | 208,000 |
| 30 | 72 | 151 | 284 | 583 | 873 | 1,680 | 2,680 | 4,740 | 9,660 | 17,500 | 28,300 | 58,200 | 106,000 | 167,000 |
| 40 | 62 | 129 | 243 | 499 | 747 | 1,440 | 2,290 | 4,050 | 8,270 | 15,000 | 24,200 | 49,800 | 90,400 | 143,000 |
| 50 | 55 | 114 | 215 | 442 | 662 | 1,280 | 2,030 | 3,590 | 7,330 | 13,300 | 21,500 | 44,100 | 80,100 | 127,000 |
| 60 | 50 | 104 | 195 | 400 | 600 | 1,160 | 1,840 | 3,260 | 6,640 | 12,000 | 19,500 | 40,000 | 72,600 | 115,000 |
| 70 | 46 | 95 | 179 | 368 | 552 | 1,060 | 1,690 | 3,000 | 6,110 | 11,100 | 17,900 | 36,800 | 66,800 | 106,000 |
| 80 | 42 | 89 | 167 | 343 | 514 | 989 | 1,580 | 2,790 | 5,680 | 10,300 | 16,700 | 34,200 | 62,100 | 98,400 |
| 90 | 40 | 83 | 157 | 322 | 482 | 928 | 1,480 | 2,610 | 5,330 | 9,650 | 15,600 | 32,100 | 58,300 | 92,300 |
| 100 | 38 | 79 | 148 | 304 | 455 | 877 | 1,400 | 2,470 | 5,040 | 9,110 | 14,800 | 30,300 | 55,100 | 87,200 |
| 125 | 33 | 70 | 131 | 269 | 403 | 777 | 1,240 | 2,190 | 4,460 | 8,080 | 13,100 | 26,900 | 48,800 | 77,300 |
| 150 | 30 | 63 | 119 | 244 | 366 | 704 | 1,120 | 1,980 | 4,050 | 7,320 | 11,900 | 24,300 | 44,200 | 70,000 |
| 175 | 28 | 58 | 109 | 224 | 336 | 648 | 1,030 | 1,820 | 3,720 | 6,730 | 10,900 | 22,400 | 40,700 | 64,400 |
| 200 | 26 | 54 | 102 | 209 | 313 | 602 | 960 | 1,700 | 3,460 | 6,260 | 10,100 | 20,800 | 37,900 | 59,900 |
| 250 | 23 | 48 | 90 | 185 | 277 | 534 | 851 | 1,500 | 3,070 | 5,550 | 8,990 | 18,500 | 33,500 | 53,100 |
| 300 | 21 | 43 | 82 | 168 | 251 | 484 | 771 | 1,360 | 2,780 | 5,030 | 8,150 | 16,700 | 30,400 | 48,100 |
| 350 | 19 | 40 | 75 | 154 | 231 | 445 | 709 | 1,250 | 2,560 | 4,630 | 7,490 | 15,400 | 28,000 | 44,300 |
| 400 | 18 | 37 | 70 | 143 | 215 | 414 | 660 | 1,170 | 2,380 | 4,310 | 6,970 | 14,300 | 26,000 | 41,200 |
| 450 | 17 | 35 | 66 | 135 | 202 | 389 | 619 | 1,090 | 2,230 | 4,040 | 6,540 | 13,400 | 24,400 | 38,600 |
| 500 | 16 | 33 | 62 | 127 | 191 | 367 | 585 | 1,030 | 2,110 | 3,820 | 6,180 | 12,700 | 23,100 | 36,500 |
| 550 | 15 | 31 | 59 | 121 | 181 | 349 | 556 | 982 | 2,000 | 3,620 | 5,870 | 12,100 | 21,900 | 34,700 |
| 600 | 14 | 30 | 56 | 115 | 173 | 333 | 530 | 937 | 1,910 | 3,460 | 5,600 | 11,500 | 20,900 | 33,100 |
| 650 | 14 | 29 | 54 | 110 | 165 | 318 | 508 | 897 | 1,830 | 3,310 | 5,360 | 11,000 | 20,000 | 31,700 |
| 700 | 13 | 27 | 52 | 106 | 159 | 306 | 488 | 862 | 1,760 | 3,180 | 5,150 | 10,600 | 19,200 | 30,400 |
| 750 | 13 | 26 | 50 | 102 | 153 | 295 | 470 | 830 | 1,690 | 3,060 | 4,960 | 10,200 | 18,500 | 29,300 |
| 800 | 12 | 26 | 48 | 99 | 148 | 285 | 454 | 802 | 1,640 | 2,960 | 4,790 | 9,840 | 17,900 | 28,300 |
| 850 | 12 | 25 | 46 | 95 | 143 | 275 | 439 | 776 | 1,580 | 2,860 | 4,640 | 9,530 | 17,300 | 27,400 |
| 900 | 11 | 24 | 45 | 93 | 139 | 267 | 426 | 752 | 1,530 | 2,780 | 4,500 | 9,240 | 16,800 | 26,600 |
| 950 | 11 | 23 | 44 | 90 | 135 | 259 | 413 | 731 | 1,490 | 2,700 | 4,370 | 8,970 | 16,300 | 25,800 |
| 1,000 | 11 | 23 | 43 | 87 | 131 | 252 | 402 | 711 | 1,450 | 2,620 | 4,250 | 8,720 | 15,800 | 25,100 |
| 1,100 | 10 | 21 | 40 | 83 | 124 | 240 | 382 | 675 | 1,380 | 2,490 | 4,030 | 8,290 | 15,100 | 23,800 |
| 1,200 | NA | 20 | 39 | 79 | 119 | 229 | 364 | 644 | 1,310 | 2,380 | 3,850 | 7,910 | 14,400 | 22,700 |
| 1,300 | NA | 20 | 37 | 76 | 114 | 219 | 349 | 617 | 1,260 | 2,280 | 3,680 | 7,570 | 13,700 | 21,800 |
| 1,400 | NA | 19 | 35 | 73 | 109 | 210 | 335 | 592 | 1,210 | 2,190 | 3,540 | 7,270 | 13,200 | 20,900 |
| 1,500 | NA | 18 | 34 | 70 | 105 | 203 | 323 | 571 | 1,160 | 2,110 | 3,410 | 7,010 | 12,700 | 20,100 |
| 1,600 | NA | 18 | 33 | 68 | 102 | 196 | 312 | 551 | 1,120 | 2,030 | 3,290 | 6,770 | 12,300 | 19,500 |
| 1,700 | NA | 17 | 32 | 66 | 98 | 189 | 302 | 533 | 1,090 | 1,970 | 3,190 | 6,550 | 11,900 | 18,800 |
| 1,800 | NA | 16 | 31 | 64 | 95 | 184 | 293 | 517 | 1,050 | 1,910 | 3,090 | 6,350 | 11,500 | 18,300 |
| 1,900 | NA | 16 | 30 | 62 | 93 | 178 | 284 | 502 | 1,020 | 1,850 | 3,000 | 6,170 | 11,200 | 17,700 |
| 2,000 | NA | 16 | 29 | 60 | 90 | 173 | 276 | 488 | 1,000 | 1,800 | 2,920 | 6,000 | 10,900 | 17,200 |

NA: A flow of less than 10 cfh.

Note: All table entries are rounded to 3 significant digits.

*TABLE 6.2(b) Schedule 40 Metallic Pipe*

| | |
|---|---|
| *Gas:* | *Natural* |
| *Inlet Pressure:* | *Less than 2 psi* |
| *Pressure Drop:* | *0.5 in. w.c.* |
| *Specific Gravity:* | *0.60* |

| | *Pipe Size (in.)* | | | | | | | | | | | | | |
|---|---|---|---|---|---|---|---|---|---|---|---|---|---|---|
| *Nominal:* | ½ | ¾ | 1 | 1¼ | 1½ | 2 | 2½ | 3 | 4 | 5 | 6 | 8 | 10 | 12 |
| *Actual ID:* | 0.622 | 0.824 | 1.049 | 1.380 | 1.610 | 2.067 | 2.469 | 3.068 | 4.026 | 5.047 | 6.065 | 7.981 | 10.020 | 11.938 |
| *Length (ft)* | *Capacity in Cubic Feet of Gas per Hour* | | | | | | | | | | | | | |
| 10 | 172 | 360 | 678 | 1,390 | 2,090 | 4,020 | 6,400 | 11,300 | 23,100 | 41,800 | 67,600 | 139,000 | 252,000 | 399,000 |
| 20 | 118 | 247 | 466 | 957 | 1,430 | 2,760 | 4,400 | 7,780 | 15,900 | 28,700 | 46,500 | 95,500 | 173,000 | 275,000 |
| 30 | 95 | 199 | 374 | 768 | 1,150 | 2,220 | 3,530 | 6,250 | 12,700 | 23,000 | 37,300 | 76,700 | 139,000 | 220,000 |
| 40 | 81 | 170 | 320 | 657 | 985 | 1,900 | 3,020 | 5,350 | 10,900 | 19,700 | 31,900 | 65,600 | 119,000 | 189,000 |
| 50 | 72 | 151 | 284 | 583 | 873 | 1,680 | 2,680 | 4,740 | 9,660 | 17,500 | 28,300 | 58,200 | 106,000 | 167,000 |
| 60 | 65 | 137 | 257 | 528 | 791 | 1,520 | 2,430 | 4,290 | 8,760 | 15,800 | 25,600 | 52,700 | 95,700 | 152,000 |
| 70 | 60 | 126 | 237 | 486 | 728 | 1,400 | 2,230 | 3,950 | 8,050 | 14,600 | 23,600 | 48,500 | 88,100 | 139,000 |
| 80 | 56 | 117 | 220 | 452 | 677 | 1,300 | 2,080 | 3,670 | 7,490 | 13,600 | 22,000 | 45,100 | 81,900 | 130,000 |
| 90 | 52 | 110 | 207 | 424 | 635 | 1,220 | 1,950 | 3,450 | 7,030 | 12,700 | 20,600 | 42,300 | 76,900 | 122,000 |
| 100 | 50 | 104 | 195 | 400 | 600 | 1,160 | 1,840 | 3,260 | 6,640 | 12,000 | 19,500 | 40,000 | 72,600 | 115,000 |
| 125 | 44 | 92 | 173 | 355 | 532 | 1,020 | 1,630 | 2,890 | 5,890 | 10,600 | 17,200 | 35,400 | 64,300 | 102,000 |
| 150 | 40 | 83 | 157 | 322 | 482 | 928 | 1,480 | 2,610 | 5,330 | 9,650 | 15,600 | 32,100 | 58,300 | 92,300 |
| 175 | 37 | 77 | 144 | 296 | 443 | 854 | 1,360 | 2,410 | 4,910 | 8,880 | 14,400 | 29,500 | 53,600 | 84,900 |
| 200 | 34 | 71 | 134 | 275 | 412 | 794 | 1,270 | 2,240 | 4,560 | 8,260 | 13,400 | 27,500 | 49,900 | 79,000 |
| 250 | 30 | 63 | 119 | 244 | 366 | 704 | 1,120 | 1,980 | 4,050 | 7,320 | 11,900 | 24,300 | 44,200 | 70,000 |
| 300 | 27 | 57 | 108 | 221 | 331 | 638 | 1,020 | 1,800 | 3,670 | 6,630 | 10,700 | 22,100 | 40,100 | 63,400 |
| 350 | 25 | 53 | 99 | 203 | 305 | 587 | 935 | 1,650 | 3,370 | 6,100 | 9,880 | 20,300 | 36,900 | 58,400 |
| 400 | 23 | 49 | 92 | 189 | 283 | 546 | 870 | 1,540 | 3,140 | 5,680 | 9,190 | 18,900 | 34,300 | 54,300 |
| 450 | 22 | 46 | 86 | 177 | 266 | 512 | 816 | 1,440 | 2,940 | 5,330 | 8,620 | 17,700 | 32,200 | 50,900 |
| 500 | 21 | 43 | 82 | 168 | 251 | 484 | 771 | 1,360 | 2,780 | 5,030 | 8,150 | 16,700 | 30,400 | 48,100 |
| 550 | 20 | 41 | 78 | 159 | 239 | 459 | 732 | 1,290 | 2,640 | 4,780 | 7,740 | 15,900 | 28,900 | 45,700 |
| 600 | 19 | 39 | 74 | 152 | 228 | 438 | 699 | 1,240 | 2,520 | 4,560 | 7,380 | 15,200 | 27,500 | 43,600 |
| 650 | 18 | 38 | 71 | 145 | 218 | 420 | 669 | 1,180 | 2,410 | 4,360 | 7,070 | 14,500 | 26,400 | 41,800 |
| 700 | 17 | 36 | 68 | 140 | 209 | 403 | 643 | 1,140 | 2,320 | 4,190 | 6,790 | 14,000 | 25,300 | 40,100 |
| 750 | 17 | 35 | 66 | 135 | 202 | 389 | 619 | 1,090 | 2,230 | 4,040 | 6,540 | 13,400 | 24,400 | 38,600 |
| 800 | 16 | 34 | 63 | 130 | 195 | 375 | 598 | 1,060 | 2,160 | 3,900 | 6,320 | 13,000 | 23,600 | 37,300 |
| 850 | 16 | 33 | 61 | 126 | 189 | 363 | 579 | 1,020 | 2,090 | 3,780 | 6,110 | 12,600 | 22,800 | 36,100 |
| 900 | 15 | 32 | 59 | 122 | 183 | 352 | 561 | 992 | 2,020 | 3,660 | 5,930 | 12,200 | 22,100 | 35,000 |
| 950 | 15 | 31 | 58 | 118 | 178 | 342 | 545 | 963 | 1,960 | 3,550 | 5,760 | 11,800 | 21,500 | 34,000 |
| 1,000 | 14 | 30 | 56 | 115 | 173 | 333 | 530 | 937 | 1,910 | 3,460 | 5,600 | 11,500 | 20,900 | 33,100 |
| 1,100 | 14 | 28 | 53 | 109 | 164 | 316 | 503 | 890 | 1,810 | 3,280 | 5,320 | 10,900 | 19,800 | 31,400 |
| 1,200 | 13 | 27 | 51 | 104 | 156 | 301 | 480 | 849 | 1,730 | 3,130 | 5,070 | 10,400 | 18,900 | 30,000 |
| 1,300 | 12 | 26 | 49 | 100 | 150 | 289 | 460 | 813 | 1,660 | 3,000 | 4,860 | 9,980 | 18,100 | 28,700 |
| 1,400 | 12 | 25 | 47 | 96 | 144 | 277 | 442 | 781 | 1,590 | 2,880 | 4,670 | 9,590 | 17,400 | 27,600 |
| 1,500 | 11 | 24 | 45 | 93 | 139 | 267 | 426 | 752 | 1,530 | 2,780 | 4,500 | 9,240 | 16,800 | 26,600 |
| 1,600 | 11 | 23 | 44 | 89 | 134 | 258 | 411 | 727 | 1,480 | 2,680 | 4,340 | 8,920 | 16,200 | 25,600 |
| 1,700 | 11 | 22 | 42 | 86 | 130 | 250 | 398 | 703 | 1,430 | 2,590 | 4,200 | 8,630 | 15,700 | 24,800 |
| 1,800 | 10 | 22 | 41 | 84 | 126 | 242 | 386 | 682 | 1,390 | 2,520 | 4,070 | 8,370 | 15,200 | 24,100 |
| 1,900 | 10 | 21 | 40 | 81 | 122 | 235 | 375 | 662 | 1,350 | 2,440 | 3,960 | 8,130 | 14,800 | 23,400 |
| 2,000 | NA | 20 | 39 | 79 | 119 | 229 | 364 | 644 | 1,310 | 2,380 | 3,850 | 7,910 | 14,400 | 22,700 |

NA: A flow of less than 10 cfh.
Note: All table entries are rounded to 3 significant digits.

***TABLE 6.2(c)*** *Schedule 40 Metallic Pipe*

| | | | | | | | | |
|---|---|---|---|---|---|---|---|---|
| | | | | | | | **Gas:** | **Natural** |
| | | | | | | | **Inlet Pressure:** | **2.0 psi** |
| | | | | | | | **Pressure Drop:** | **1.0 psi** |
| | | | | | | | **Specific Gravity:** | **0.60** |

| | Pipe Size (in.) | | | | | | | | |
|---|---|---|---|---|---|---|---|---|---|
| **Nominal:** | ½ | ¾ | 1 | 1¼ | 1½ | 2 | 2½ | 3 | 4 |
| **Actual ID:** | 0.622 | 0.824 | 1.049 | 1.380 | 1.610 | 2.067 | 2.469 | 3.068 | 4.026 |
| **Length (ft)** | Capacity in Cubic Feet of Gas per Hour | | | | | | | | |
| 10 | 1,510 | 3,040 | 5,560 | 11,400 | 17,100 | 32,900 | 52,500 | 92,800 | 189,000 |
| 20 | 1,070 | 2,150 | 3,930 | 8,070 | 12,100 | 23,300 | 37,100 | 65,600 | 134,000 |
| 30 | 869 | 1,760 | 3,210 | 6,590 | 9,880 | 19,000 | 30,300 | 53,600 | 109,000 |
| 40 | 753 | 1,520 | 2,780 | 5,710 | 8,550 | 16,500 | 26,300 | 46,400 | 94,700 |
| 50 | 673 | 1,360 | 2,490 | 5,110 | 7,650 | 14,700 | 23,500 | 41,500 | 84,700 |
| 60 | 615 | 1,240 | 2,270 | 4,660 | 6,980 | 13,500 | 21,400 | 37,900 | 77,300 |
| 70 | 569 | 1,150 | 2,100 | 4,320 | 6,470 | 12,500 | 19,900 | 35,100 | 71,600 |
| 80 | 532 | 1,080 | 1,970 | 4,040 | 6,050 | 11,700 | 18,600 | 32,800 | 67,000 |
| 90 | 502 | 1,010 | 1,850 | 3,810 | 5,700 | 11,000 | 17,500 | 30,900 | 63,100 |
| 100 | 462 | 934 | 1,710 | 3,510 | 5,260 | 10,100 | 16,100 | 28,500 | 58,200 |
| 125 | 414 | 836 | 1,530 | 3,140 | 4,700 | 9,060 | 14,400 | 25,500 | 52,100 |
| 150 | 372 | 751 | 1,370 | 2,820 | 4,220 | 8,130 | 13,000 | 22,900 | 46,700 |
| 175 | 344 | 695 | 1,270 | 2,601 | 3,910 | 7,530 | 12,000 | 21,200 | 43,300 |
| 200 | 318 | 642 | 1,170 | 2,410 | 3,610 | 6,960 | 11,100 | 19,600 | 40,000 |
| 250 | 279 | 583 | 1,040 | 2,140 | 3,210 | 6,180 | 9,850 | 17,400 | 35,500 |
| 300 | 253 | 528 | 945 | 1,940 | 2,910 | 5,600 | 8,920 | 15,800 | 32,200 |
| 350 | 232 | 486 | 869 | 1,790 | 2,670 | 5,150 | 8,210 | 14,500 | 29,600 |
| 400 | 216 | 452 | 809 | 1,660 | 2,490 | 4,790 | 7,640 | 13,500 | 27,500 |
| 450 | 203 | 424 | 759 | 1,560 | 2,330 | 4,500 | 7,170 | 12,700 | 25,800 |
| 500 | 192 | 401 | 717 | 1,470 | 2,210 | 4,250 | 6,770 | 12,000 | 24,400 |
| 550 | 182 | 381 | 681 | 1,400 | 2,090 | 4,030 | 6,430 | 11,400 | 23,200 |
| 600 | 174 | 363 | 650 | 1,330 | 2,000 | 3,850 | 6,130 | 10,800 | 22,100 |
| 650 | 166 | 348 | 622 | 1,280 | 1,910 | 3,680 | 5,870 | 10,400 | 21,200 |
| 700 | 160 | 334 | 598 | 1,230 | 1,840 | 3,540 | 5,640 | 9,970 | 20,300 |
| 750 | 154 | 322 | 576 | 1,180 | 1,770 | 3,410 | 5,440 | 9,610 | 19,600 |
| 800 | 149 | 311 | 556 | 1,140 | 1,710 | 3,290 | 5,250 | 9,280 | 18,900 |
| 850 | 144 | 301 | 538 | 1,100 | 1,650 | 3,190 | 5,080 | 8,980 | 18,300 |
| 900 | 139 | 292 | 522 | 1,070 | 1,600 | 3,090 | 4,930 | 8,710 | 17,800 |
| 950 | 135 | 283 | 507 | 1,040 | 1,560 | 3,000 | 4,780 | 8,460 | 17,200 |
| 1,000 | 132 | 275 | 493 | 1,010 | 1,520 | 2,920 | 4,650 | 8,220 | 16,800 |
| 1,100 | 125 | 262 | 468 | 960 | 1,440 | 2,770 | 4,420 | 7,810 | 15,900 |
| 1,200 | 119 | 250 | 446 | 917 | 1,370 | 2,640 | 4,220 | 7,450 | 15,200 |
| 1,300 | 114 | 239 | 427 | 878 | 1,320 | 2,530 | 4,040 | 7,140 | 14,600 |
| 1,400 | 110 | 230 | 411 | 843 | 1,260 | 2,430 | 3,880 | 6,860 | 14,000 |
| 1,500 | 106 | 221 | 396 | 812 | 1,220 | 2,340 | 3,740 | 6,600 | 13,500 |
| 1,600 | 102 | 214 | 382 | 784 | 1,180 | 2,260 | 3,610 | 6,380 | 13,000 |
| 1,700 | 99 | 207 | 370 | 759 | 1,140 | 2,190 | 3,490 | 6,170 | 12,600 |
| 1,800 | 96 | 200 | 358 | 736 | 1,100 | 2,120 | 3,390 | 5,980 | 12,200 |
| 1,900 | 93 | 195 | 348 | 715 | 1,070 | 2,060 | 3,290 | 5,810 | 11,900 |
| 2,000 | 91 | 189 | 339 | 695 | 1,040 | 2,010 | 3,200 | 5,650 | 11,500 |

Note: All table entries are rounded to 3 significant digits.

*TABLE 6.2(d) Schedule 40 Metallic Pipe*

| | |
|---|---|
| *Gas:* | *Natural* |
| *Inlet Pressure:* | *3.0 psi* |
| *Pressure Drop:* | *2.0 psi* |
| *Specific Gravity:* | *0.60* |

| | *Pipe Size (in.)* | | | | | | | | |
|---|---|---|---|---|---|---|---|---|---|
| *Nominal:* | ½ | ¾ | 1 | 1¼ | 1½ | 2 | 2½ | 3 | 4 |
| *Actual ID:* | 0.622 | 0.824 | 1.049 | 1.380 | 1.610 | 2.067 | 2.469 | 3.068 | 4.026 |
| *Length (ft)* | *Capacity in Cubic Feet of Gas per Hour* | | | | | | | | |
| 10 | 2,350 | 4,920 | 9,270 | 19,000 | 28,500 | 54,900 | 87,500 | 155,000 | 316,000 |
| 20 | 1,620 | 3,380 | 6,370 | 13,100 | 19,600 | 37,700 | 60,100 | 106,000 | 217,000 |
| 30 | 1,300 | 2,720 | 5,110 | 10,500 | 15,700 | 30,300 | 48,300 | 85,400 | 174,000 |
| 40 | 1,110 | 2,320 | 4,380 | 8,990 | 13,500 | 25,900 | 41,300 | 73,100 | 149,000 |
| 50 | 985 | 2,060 | 3,880 | 7,970 | 11,900 | 23,000 | 36,600 | 64,800 | 132,000 |
| 60 | 892 | 1,870 | 3,520 | 7,220 | 10,800 | 20,800 | 33,200 | 58,700 | 120,000 |
| 70 | 821 | 1,720 | 3,230 | 6,640 | 9,950 | 19,200 | 30,500 | 54,000 | 110,000 |
| 80 | 764 | 1,600 | 3,010 | 6,180 | 9,260 | 17,800 | 28,400 | 50,200 | 102,000 |
| 90 | 717 | 1,500 | 2,820 | 5,800 | 8,680 | 16,700 | 26,700 | 47,100 | 96,100 |
| 100 | 677 | 1,420 | 2,670 | 5,470 | 8,200 | 15,800 | 25,200 | 44,500 | 90,800 |
| 125 | 600 | 1,250 | 2,360 | 4,850 | 7,270 | 14,000 | 22,300 | 39,500 | 80,500 |
| 150 | 544 | 1,140 | 2,140 | 4,400 | 6,590 | 12,700 | 20,200 | 35,700 | 72,900 |
| 175 | 500 | 1,050 | 1,970 | 4,040 | 6,060 | 11,700 | 18,600 | 32,900 | 67,100 |
| 200 | 465 | 973 | 1,830 | 3,760 | 5,640 | 10,900 | 17,300 | 30,600 | 62,400 |
| 250 | 412 | 862 | 1,620 | 3,330 | 5,000 | 9,620 | 15,300 | 27,100 | 55,300 |
| 300 | 374 | 781 | 1,470 | 3,020 | 4,530 | 8,720 | 13,900 | 24,600 | 50,100 |
| 350 | 344 | 719 | 1,350 | 2,780 | 4,170 | 8,020 | 12,800 | 22,600 | 46,100 |
| 400 | 320 | 669 | 1,260 | 2,590 | 3,870 | 7,460 | 11,900 | 21,000 | 42,900 |
| 450 | 300 | 627 | 1,180 | 2,430 | 3,640 | 7,000 | 11,200 | 19,700 | 40,200 |
| 500 | 283 | 593 | 1,120 | 2,290 | 3,430 | 6,610 | 10,500 | 18,600 | 38,000 |
| 550 | 269 | 563 | 1,060 | 2,180 | 3,260 | 6,280 | 10,000 | 17,700 | 36,100 |
| 600 | 257 | 537 | 1,010 | 2,080 | 3,110 | 5,990 | 9,550 | 16,900 | 34,400 |
| 650 | 246 | 514 | 969 | 1,990 | 2,980 | 5,740 | 9,150 | 16,200 | 33,000 |
| 700 | 236 | 494 | 931 | 1,910 | 2,860 | 5,510 | 8,790 | 15,500 | 31,700 |
| 750 | 228 | 476 | 897 | 1,840 | 2,760 | 5,310 | 8,470 | 15,000 | 30,500 |
| 800 | 220 | 460 | 866 | 1,780 | 2,660 | 5,130 | 8,180 | 14,500 | 29,500 |
| 850 | 213 | 445 | 838 | 1,720 | 2,580 | 4,960 | 7,910 | 14,000 | 28,500 |
| 900 | 206 | 431 | 812 | 1,670 | 2,500 | 4,810 | 7,670 | 13,600 | 27,700 |
| 950 | 200 | 419 | 789 | 1,620 | 2,430 | 4,670 | 7,450 | 13,200 | 26,900 |
| 1,000 | 195 | 407 | 767 | 1,580 | 2,360 | 4,550 | 7,240 | 12,800 | 26,100 |
| 1,100 | 185 | 387 | 729 | 1,500 | 2,240 | 4,320 | 6,890 | 12,200 | 24,800 |
| 1,200 | 177 | 369 | 695 | 1,430 | 2,140 | 4,120 | 6,570 | 11,600 | 23,700 |
| 1,300 | 169 | 353 | 666 | 1,370 | 2,050 | 3,940 | 6,290 | 11,100 | 22,700 |
| 1,400 | 162 | 340 | 640 | 1,310 | 1,970 | 3,790 | 6,040 | 10,700 | 21,800 |
| 1,500 | 156 | 327 | 616 | 1,270 | 1,900 | 3,650 | 5,820 | 10,300 | 21,000 |
| 1,600 | 151 | 316 | 595 | 1,220 | 1,830 | 3,530 | 5,620 | 10,000 | 20,300 |
| 1,700 | 146 | 306 | 576 | 1,180 | 1,770 | 3,410 | 5,440 | 9,610 | 19,600 |
| 1,800 | 142 | 296 | 558 | 1,150 | 1,720 | 3,310 | 5,270 | 9,320 | 19,000 |
| 1,900 | 138 | 288 | 542 | 1,110 | 1,670 | 3,210 | 5,120 | 9,050 | 18,400 |
| 2,000 | 134 | 280 | 527 | 1,080 | 1,620 | 3,120 | 4,980 | 8,800 | 18,000 |

Note: All table entries are rounded to 3 significant digits.

*TABLE 6.2(e) Schedule 40 Metallic Pipe*

| | | | | | | | Gas: | Natural | |
|---|---|---|---|---|---|---|---|---|---|
| | | | | | | | Inlet Pressure: | 5.0 psi | |
| | | | | | | | Pressure Drop: | 3.5 psi | |
| | | | | | | | Specific Gravity: | 0.60 | |
| | Pipe Size (in.) | | | | | | | | |
| Nominal: | 1/2 | 3/4 | 1 | 1 1/4 | 1 1/2 | 2 | 2 1/2 | 3 | 4 |
| Actual ID: | 0.622 | 0.824 | 1.049 | 1.380 | 1.610 | 2.067 | 2.469 | 3.068 | 4.026 |
| Length (ft) | Capacity in Cubic Feet of Gas per Hour | | | | | | | | |
| 10 | 3,190 | 6,430 | 11,800 | 24,200 | 36,200 | 69,700 | 111,000 | 196,000 | 401,000 |
| 20 | 2,250 | 4,550 | 8,320 | 17,100 | 25,600 | 49,300 | 78,600 | 139,000 | 283,000 |
| 30 | 1,840 | 3,720 | 6,790 | 14,000 | 20,900 | 40,300 | 64,200 | 113,000 | 231,000 |
| 40 | 1,590 | 3,220 | 5,880 | 12,100 | 18,100 | 34,900 | 55,600 | 98,200 | 200,000 |
| 50 | 1,430 | 2,880 | 5,260 | 10,800 | 16,200 | 31,200 | 49,700 | 87,900 | 179,000 |
| 60 | 1,300 | 2,630 | 4,800 | 9,860 | 14,800 | 28,500 | 45,400 | 80,200 | 164,000 |
| 70 | 1,200 | 2,430 | 4,450 | 9,130 | 13,700 | 26,400 | 42,000 | 74,300 | 151,000 |
| 80 | 1,150 | 2,330 | 4,260 | 8,540 | 12,800 | 24,700 | 39,300 | 69,500 | 142,000 |
| 90 | 1,060 | 2,150 | 3,920 | 8,050 | 12,100 | 23,200 | 37,000 | 65,500 | 134,000 |
| 100 | 979 | 1,980 | 3,620 | 7,430 | 11,100 | 21,400 | 34,200 | 60,400 | 123,000 |
| 125 | 876 | 1,770 | 3,240 | 6,640 | 9,950 | 19,200 | 30,600 | 54,000 | 110,000 |
| 150 | 786 | 1,590 | 2,910 | 5,960 | 8,940 | 17,200 | 27,400 | 48,500 | 98,900 |
| 175 | 728 | 1,470 | 2,690 | 5,520 | 8,270 | 15,900 | 25,400 | 44,900 | 91,600 |
| 200 | 673 | 1,360 | 2,490 | 5,100 | 7,650 | 14,700 | 23,500 | 41,500 | 84,700 |
| 250 | 558 | 1,170 | 2,200 | 4,510 | 6,760 | 13,000 | 20,800 | 36,700 | 74,900 |
| 300 | 506 | 1,060 | 1,990 | 4,090 | 6,130 | 11,800 | 18,800 | 33,300 | 67,800 |
| 350 | 465 | 973 | 1,830 | 3,760 | 5,640 | 10,900 | 17,300 | 30,600 | 62,400 |
| 400 | 433 | 905 | 1,710 | 3,500 | 5,250 | 10,100 | 16,100 | 28,500 | 58,100 |
| 450 | 406 | 849 | 1,600 | 3,290 | 4,920 | 9,480 | 15,100 | 26,700 | 54,500 |
| 500 | 384 | 802 | 1,510 | 3,100 | 4,650 | 8,950 | 14,300 | 25,200 | 51,500 |
| 550 | 364 | 762 | 1,440 | 2,950 | 4,420 | 8,500 | 13,600 | 24,000 | 48,900 |
| 600 | 348 | 727 | 1,370 | 2,810 | 4,210 | 8,110 | 12,900 | 22,900 | 46,600 |
| 650 | 333 | 696 | 1,310 | 2,690 | 4,030 | 7,770 | 12,400 | 21,900 | 44,600 |
| 700 | 320 | 669 | 1,260 | 2,590 | 3,880 | 7,460 | 11,900 | 21,000 | 42,900 |
| 750 | 308 | 644 | 1,210 | 2,490 | 3,730 | 7,190 | 11,500 | 20,300 | 41,300 |
| 800 | 298 | 622 | 1,170 | 2,410 | 3,610 | 6,940 | 11,100 | 19,600 | 39,900 |
| 850 | 288 | 602 | 1,130 | 2,330 | 3,490 | 6,720 | 10,700 | 18,900 | 38,600 |
| 900 | 279 | 584 | 1,100 | 2,260 | 3,380 | 6,520 | 10,400 | 18,400 | 37,400 |
| 950 | 271 | 567 | 1,070 | 2,190 | 3,290 | 6,330 | 10,100 | 17,800 | 36,400 |
| 1,000 | 264 | 551 | 1,040 | 2,130 | 3,200 | 6,150 | 9,810 | 17,300 | 35,400 |
| 1,100 | 250 | 524 | 987 | 2,030 | 3,030 | 5,840 | 9,320 | 16,500 | 33,600 |
| 1,200 | 239 | 500 | 941 | 1,930 | 2,900 | 5,580 | 8,890 | 15,700 | 32,000 |
| 1,300 | 229 | 478 | 901 | 1,850 | 2,770 | 5,340 | 8,510 | 15,000 | 30,700 |
| 1,400 | 220 | 460 | 866 | 1,780 | 2,660 | 5,130 | 8,180 | 14,500 | 29,500 |
| 1,500 | 212 | 443 | 834 | 1,710 | 2,570 | 4,940 | 7,880 | 13,900 | 28,400 |
| 1,600 | 205 | 428 | 806 | 1,650 | 2,480 | 4,770 | 7,610 | 13,400 | 27,400 |
| 1,700 | 198 | 414 | 780 | 1,600 | 2,400 | 4,620 | 7,360 | 13,000 | 26,500 |
| 1,800 | 192 | 401 | 756 | 1,550 | 2,330 | 4,480 | 7,140 | 12,600 | 25,700 |
| 1,900 | 186 | 390 | 734 | 1,510 | 2,260 | 4,350 | 6,930 | 12,300 | 25,000 |
| 2,000 | 181 | 379 | 714 | 1,470 | 2,200 | 4,230 | 6,740 | 11,900 | 24,300 |

Note: All table entries are rounded to 3 significant digits.

*TABLE 6.2(f) Semirigid Copper Tubing*

| | | | | | | | | |
|---|---|---|---|---|---|---|---|---|
| | | | | | | | **Gas:** | **Natural** |
| | | | | | | | **Inlet Pressure:** | **Less than 2 psi** |
| | | | | | | | **Pressure Drop:** | **0.3 in. w.c.** |
| | | | | | | | **Specific Gravity:** | **0.60** |

| | Tube Size (in.) | | | | | | | | |
|---|---|---|---|---|---|---|---|---|---|
| **Nominal: K & L:** | 1/4 | 3/8 | 1/2 | 5/8 | 3/4 | 1 | 1 1/4 | 1 1/2 | 2 |
| **Nominal: ACR:** | 3/8 | 1/2 | 5/8 | 3/4 | 7/8 | 1 1/8 | 1 3/8 | — | — |
| **Outside:** | 0.375 | 0.500 | 0.625 | 0.750 | 0.875 | 1.125 | 1.375 | 1.625 | 2.125 |
| **Inside:*** | 0.305 | 0.402 | 0.527 | 0.652 | 0.745 | 0.995 | 1.245 | 1.481 | 1.959 |
| **Length (ft)** | **Capacity in Cubic Feet of Gas per Hour** | | | | | | | | |
| 10 | 20 | 42 | 85 | 148 | 210 | 448 | 806 | 1,270 | 2,650 |
| 20 | 14 | 29 | 58 | 102 | 144 | 308 | 554 | 873 | 1,820 |
| 30 | 11 | 23 | 47 | 82 | 116 | 247 | 445 | 701 | 1,460 |
| 40 | 10 | 20 | 40 | 70 | 99 | 211 | 381 | 600 | 1,250 |
| 50 | NA | 17 | 35 | 62 | 88 | 187 | 337 | 532 | 1,110 |
| 60 | NA | 16 | 32 | 56 | 79 | 170 | 306 | 482 | 1,000 |
| 70 | NA | 14 | 29 | 52 | 73 | 156 | 281 | 443 | 924 |
| 80 | NA | 13 | 27 | 48 | 68 | 145 | 262 | 413 | 859 |
| 90 | NA | 13 | 26 | 45 | 64 | 136 | 245 | 387 | 806 |
| 100 | NA | 12 | 24 | 43 | 60 | 129 | 232 | 366 | 761 |
| 125 | NA | 11 | 22 | 38 | 53 | 114 | 206 | 324 | 675 |
| 150 | NA | 10 | 20 | 34 | 48 | 103 | 186 | 294 | 612 |
| 175 | NA | NA | 18 | 31 | 45 | 95 | 171 | 270 | 563 |
| 200 | NA | NA | 17 | 29 | 41 | 89 | 159 | 251 | 523 |
| 250 | NA | NA | 15 | 26 | 37 | 78 | 141 | 223 | 464 |
| 300 | NA | NA | 13 | 23 | 33 | 71 | 128 | 202 | 420 |
| 350 | NA | NA | 12 | 22 | 31 | 65 | 118 | 186 | 387 |
| 400 | NA | NA | 11 | 20 | 28 | 61 | 110 | 173 | 360 |
| 450 | NA | NA | 11 | 19 | 27 | 57 | 103 | 162 | 338 |
| 500 | NA | NA | 10 | 18 | 25 | 54 | 97 | 153 | 319 |
| 550 | NA | NA | NA | 17 | 24 | 51 | 92 | 145 | 303 |
| 600 | NA | NA | NA | 16 | 23 | 49 | 88 | 139 | 289 |
| 650 | NA | NA | NA | 15 | 22 | 47 | 84 | 133 | 277 |
| 700 | NA | NA | NA | 15 | 21 | 45 | 81 | 128 | 266 |
| 750 | NA | NA | NA | 14 | 20 | 43 | 78 | 123 | 256 |
| 800 | NA | NA | NA | 14 | 20 | 42 | 75 | 119 | 247 |
| 850 | NA | NA | NA | 13 | 19 | 40 | 73 | 115 | 239 |
| 900 | NA | NA | NA | 13 | 18 | 39 | 71 | 111 | 232 |
| 950 | NA | NA | NA | 13 | 18 | 38 | 69 | 108 | 225 |
| 1,000 | NA | NA | NA | 12 | 17 | 37 | 67 | 105 | 219 |
| 1,100 | NA | NA | NA | 12 | 16 | 35 | 63 | 100 | 208 |
| 1,200 | NA | NA | NA | 11 | 16 | 34 | 60 | 95 | 199 |
| 1,300 | NA | NA | NA | 11 | 15 | 32 | 58 | 91 | 190 |
| 1,400 | NA | NA | NA | 10 | 14 | 31 | 56 | 88 | 183 |
| 1,500 | NA | NA | NA | NA | 14 | 30 | 54 | 84 | 176 |
| 1,600 | NA | NA | NA | NA | 13 | 29 | 52 | 82 | 170 |
| 1,700 | NA | NA | NA | NA | 13 | 28 | 50 | 79 | 164 |
| 1,800 | NA | NA | NA | NA | 13 | 27 | 49 | 77 | 159 |
| 1,900 | NA | NA | NA | NA | 12 | 26 | 47 | 74 | 155 |
| 2,000 | NA | NA | NA | NA | 12 | 25 | 46 | 72 | 151 |

NA: A flow of less than 10 cfh.

Note: All table entries are rounded to 3 significant digits.

*Table capacities are based on Type K copper tubing inside diameter (shown), which has the smallest inside diameter of the copper tubing products.

*TABLE 6.2(g) Semirigid Copper Tubing*

| | | | | | | *Gas:* | *Natural* | | |
|---|---|---|---|---|---|---|---|---|---|
| | | | | | | *Inlet Pressure:* | *Less than 2 psi* | | |
| | | | | | | *Pressure Drop:* | *0.5 in. w.c.* | | |
| | | | | | | *Specific Gravity:* | *0.60* | | |
| | *Tube Size (in.)* | | | | | | | | |
| *Nominal: K & L:* | ¼ | ⅜ | ½ | ⅝ | ¾ | 1 | 1¼ | 1½ | 2 |
| *Nominal: ACR:* | ⅜ | ½ | ⅝ | ¾ | ⅞ | 1⅛ | 1⅜ | — | — |
| *Outside:* | 0.375 | 0.500 | 0.625 | 0.750 | 0.875 | 1.125 | 1.375 | 1.625 | 2.125 |
| *Inside:** | 0.305 | 0.402 | 0.527 | 0.652 | 0.745 | 0.995 | 1.245 | 1.481 | 1.959 |
| *Length (ft)* | *Capacity in Cubic Feet of Gas per Hour* | | | | | | | | |
| 10 | 27 | 55 | 111 | 195 | 276 | 590 | 1,060 | 1,680 | 3,490 |
| 20 | 18 | 38 | 77 | 134 | 190 | 406 | 730 | 1,150 | 2,400 |
| 30 | 15 | 30 | 61 | 107 | 152 | 326 | 586 | 925 | 1,930 |
| 40 | 13 | 26 | 53 | 92 | 131 | 279 | 502 | 791 | 1,650 |
| 50 | 11 | 23 | 47 | 82 | 116 | 247 | 445 | 701 | 1,460 |
| 60 | 10 | 21 | 42 | 74 | 105 | 224 | 403 | 635 | 1,320 |
| 70 | NA | 19 | 39 | 68 | 96 | 206 | 371 | 585 | 1,220 |
| 80 | NA | 18 | 36 | 63 | 90 | 192 | 345 | 544 | 1,130 |
| 90 | NA | 17 | 34 | 59 | 84 | 180 | 324 | 510 | 1,060 |
| 100 | NA | 16 | 32 | 56 | 79 | 170 | 306 | 482 | 1,000 |
| 125 | NA | 14 | 28 | 50 | 70 | 151 | 271 | 427 | 890 |
| 150 | NA | 13 | 26 | 45 | 64 | 136 | 245 | 387 | 806 |
| 175 | NA | 12 | 24 | 41 | 59 | 125 | 226 | 356 | 742 |
| 200 | NA | 11 | 22 | 39 | 55 | 117 | 210 | 331 | 690 |
| 250 | NA | NA | 20 | 34 | 48 | 103 | 186 | 294 | 612 |
| 300 | NA | NA | 18 | 31 | 44 | 94 | 169 | 266 | 554 |
| 350 | NA | NA | 16 | 28 | 40 | 86 | 155 | 245 | 510 |
| 400 | NA | NA | 15 | 26 | 38 | 80 | 144 | 228 | 474 |
| 450 | NA | NA | 14 | 25 | 35 | 75 | 135 | 214 | 445 |
| 500 | NA | NA | 13 | 23 | 33 | 71 | 128 | 202 | 420 |
| 550 | NA | NA | 13 | 22 | 32 | 68 | 122 | 192 | 399 |
| 600 | NA | NA | 12 | 21 | 30 | 64 | 116 | 183 | 381 |
| 650 | NA | NA | 12 | 20 | 29 | 62 | 111 | 175 | 365 |
| 700 | NA | NA | 11 | 20 | 28 | 59 | 107 | 168 | 350 |
| 750 | NA | NA | 11 | 19 | 27 | 57 | 103 | 162 | 338 |
| 800 | NA | NA | 10 | 18 | 26 | 55 | 99 | 156 | 326 |
| 850 | NA | NA | 10 | 18 | 25 | 53 | 96 | 151 | 315 |
| 900 | NA | NA | NA | 17 | 24 | 52 | 93 | 147 | 306 |
| 950 | NA | NA | NA | 17 | 24 | 50 | 90 | 143 | 297 |
| 1,000 | NA | NA | NA | 16 | 23 | 49 | 88 | 139 | 289 |
| 1,100 | NA | NA | NA | 15 | 22 | 46 | 84 | 132 | 274 |
| 1,200 | NA | NA | NA | 15 | 21 | 44 | 80 | 126 | 262 |
| 1,300 | NA | NA | NA | 14 | 20 | 42 | 76 | 120 | 251 |
| 1,400 | NA | NA | NA | 13 | 19 | 41 | 73 | 116 | 241 |
| 1,500 | NA | NA | NA | 13 | 18 | 39 | 71 | 111 | 232 |
| 1,600 | NA | NA | NA | 13 | 18 | 38 | 68 | 108 | 224 |
| 1,700 | NA | NA | NA | 12 | 17 | 37 | 66 | 104 | 217 |
| 1,800 | NA | NA | NA | 12 | 17 | 36 | 64 | 101 | 210 |
| 1,900 | NA | NA | NA | 11 | 16 | 35 | 62 | 98 | 204 |
| 2,000 | NA | NA | NA | 11 | 16 | 34 | 60 | 95 | 199 |

NA: A flow of less than 10 cfh.

Note: All table entries are rounded to 3 significant digits.

*Table capacities are based on Type K copper tubing inside diameter (shown), which has the smallest inside diameter of the copper tubing products.

*TABLE 6.2(h)* *Semirigid Copper Tubing*

| | |
|---|---|
| ***Gas:*** | ***Natural*** |
| ***Inlet Pressure:*** | ***Less than 2 psi*** |
| ***Pressure Drop:*** | ***1.0 in. w.c.*** |
| ***Specific Gravity:*** | ***0.60*** |

***INTENDED USE: Tube Sizing Between House Line Regulator and the Appliance.***

| | | ***Tube Size (in.)*** | | | | | | | | |
|---|---|---|---|---|---|---|---|---|---|---|
| ***Nominal:*** | ***K & L:*** | 1/4 | 3/8 | 1/2 | 5/8 | 3/4 | 1 | 1 1/4 | 1 1/2 | 2 |
| | ***ACR:*** | 3/8 | 1/2 | 5/8 | 3/4 | 7/8 | 1 1/8 | 1 3/8 | — | — |
| | ***Outside:*** | 0.375 | 0.500 | 0.625 | 0.750 | 0.875 | 1.125 | 1.375 | 1.625 | 2.125 |
| | ***Inside:**** | 0.305 | 0.402 | 0.527 | 0.652 | 0.745 | 0.995 | 1.245 | 1.481 | 1.959 |
| ***Length (ft)*** | | ***Capacity in Cubic Feet of Gas per Hour*** | | | | | | | | |
| 10 | | 39 | 80 | 162 | 283 | 402 | 859 | 1,550 | 2,440 | 5,080 |
| 20 | | 27 | 55 | 111 | 195 | 276 | 590 | 1,060 | 1,680 | 3,490 |
| 30 | | 21 | 44 | 89 | 156 | 222 | 474 | 853 | 1,350 | 2,800 |
| 40 | | 18 | 38 | 77 | 134 | 190 | 406 | 730 | 1,150 | 2,400 |
| 50 | | 16 | 33 | 68 | 119 | 168 | 359 | 647 | 1,020 | 2,130 |
| 60 | | 15 | 30 | 61 | 107 | 152 | 326 | 586 | 925 | 1,930 |
| 70 | | 13 | 28 | 57 | 99 | 140 | 300 | 539 | 851 | 1,770 |
| 80 | | 13 | 26 | 53 | 92 | 131 | 279 | 502 | 791 | 1,650 |
| 90 | | 12 | 24 | 49 | 86 | 122 | 262 | 471 | 742 | 1,550 |
| 100 | | 11 | 23 | 47 | 82 | 116 | 247 | 445 | 701 | 1,460 |
| 125 | | NA | 20 | 41 | 72 | 103 | 219 | 394 | 622 | 1,290 |
| 150 | | NA | 18 | 37 | 65 | 93 | 198 | 357 | 563 | 1,170 |
| 175 | | NA | 17 | 34 | 60 | 85 | 183 | 329 | 518 | 1,080 |
| 200 | | NA | 16 | 32 | 56 | 79 | 170 | 306 | 482 | 1,000 |
| 250 | | NA | 14 | 28 | 50 | 70 | 151 | 271 | 427 | 890 |
| 300 | | NA | 13 | 26 | 45 | 64 | 136 | 245 | 387 | 806 |
| 350 | | NA | 12 | 24 | 41 | 59 | 125 | 226 | 356 | 742 |
| 400 | | NA | 11 | 22 | 39 | 55 | 117 | 210 | 331 | 690 |
| 450 | | NA | 10 | 21 | 36 | 51 | 110 | 197 | 311 | 647 |
| 500 | | NA | NA | 20 | 34 | 48 | 103 | 186 | 294 | 612 |
| 550 | | NA | NA | 19 | 32 | 46 | 98 | 177 | 279 | 581 |
| 600 | | NA | NA | 18 | 31 | 44 | 94 | 169 | 266 | 554 |
| 650 | | NA | NA | 17 | 30 | 42 | 90 | 162 | 255 | 531 |
| 700 | | NA | NA | 16 | 28 | 40 | 86 | 155 | 245 | 510 |
| 750 | | NA | NA | 16 | 27 | 39 | 83 | 150 | 236 | 491 |
| 800 | | NA | NA | 15 | 26 | 38 | 80 | 144 | 228 | 474 |
| 850 | | NA | NA | 15 | 26 | 36 | 78 | 140 | 220 | 459 |
| 900 | | NA | NA | 14 | 25 | 35 | 75 | 135 | 214 | 445 |
| 950 | | NA | NA | 14 | 24 | 34 | 73 | 132 | 207 | 432 |
| 1,000 | | NA | NA | 13 | 23 | 33 | 71 | 128 | 202 | 420 |
| 1,100 | | NA | NA | 13 | 22 | 32 | 68 | 122 | 192 | 399 |
| 1,200 | | NA | NA | 12 | 21 | 30 | 64 | 116 | 183 | 381 |
| 1,300 | | NA | NA | 12 | 20 | 29 | 62 | 111 | 175 | 365 |
| 1,400 | | NA | NA | 11 | 20 | 28 | 59 | 107 | 168 | 350 |
| 1,500 | | NA | NA | 11 | 19 | 27 | 57 | 103 | 162 | 338 |
| 1,600 | | NA | NA | 10 | 18 | 26 | 55 | 99 | 156 | 326 |
| 1,700 | | NA | NA | 10 | 18 | 25 | 53 | 96 | 151 | 315 |
| 1,800 | | NA | NA | NA | 17 | 24 | 52 | 93 | 147 | 306 |
| 1,900 | | NA | NA | NA | 17 | 24 | 50 | 90 | 143 | 297 |
| 2,000 | | NA | NA | NA | 16 | 23 | 49 | 88 | 139 | 289 |

NA: A flow of less than 10 cfh.

Note: All table entries are rounded to 3 significant digits.

*Table capacities are based on Type K copper tubing inside diameter (shown), which has the smallest inside diameter of the copper tubing products.

*TABLE 6.2(i) Semirigid Copper Tubing*

| | Gas: | Natural |
|---|---|---|
| | Inlet Pressure: | Less than 2.0 psi |
| | Pressure Drop: | 17.0 in. w.c. |
| | Specific Gravity: | 0.60 |

| | | Tube Size (in.) | | | | | | | | |
|---|---|---|---|---|---|---|---|---|---|---|
| Nominal: | K & L: | ¼ | ⅜ | ½ | ⅝ | ¾ | 1 | 1¼ | 1½ | 2 |
| | ACR: | ⅜ | ½ | ⅝ | ¾ | ⅞ | 1⅛ | 1⅜ | — | — |
| | Outside: | 0.375 | 0.500 | 0.625 | 0.750 | 0.875 | 1.125 | 1.375 | 1.625 | 2.125 |
| | Inside:* | 0.305 | 0.402 | 0.527 | 0.652 | 0.745 | 0.995 | 1.245 | 1.481 | 1.959 |
| Length (ft) | | Capacity in Cubic Feet of Gas per Hour | | | | | | | | |
| 10 | | 190 | 391 | 796 | 1,390 | 1,970 | 4,220 | 7,590 | 12,000 | 24,900 |
| 20 | | 130 | 269 | 547 | 956 | 1,360 | 2,900 | 5,220 | 8,230 | 17,100 |
| 30 | | 105 | 216 | 439 | 768 | 1,090 | 2,330 | 4,190 | 6,610 | 13,800 |
| 40 | | 90 | 185 | 376 | 657 | 932 | 1,990 | 3,590 | 5,650 | 11,800 |
| 50 | | 79 | 164 | 333 | 582 | 826 | 1,770 | 3,180 | 5,010 | 10,400 |
| 60 | | 72 | 148 | 302 | 528 | 749 | 1,600 | 2,880 | 4,540 | 9,460 |
| 70 | | 66 | 137 | 278 | 486 | 689 | 1,470 | 2,650 | 4,180 | 8,700 |
| 80 | | 62 | 127 | 258 | 452 | 641 | 1,370 | 2,460 | 3,890 | 8,090 |
| 90 | | 58 | 119 | 243 | 424 | 601 | 1,280 | 2,310 | 3,650 | 7,590 |
| 100 | | 55 | 113 | 229 | 400 | 568 | 1,210 | 2,180 | 3,440 | 7,170 |
| 125 | | 48 | 100 | 203 | 355 | 503 | 1,080 | 1,940 | 3,050 | 6,360 |
| 150 | | 44 | 90 | 184 | 321 | 456 | 974 | 1,750 | 2,770 | 5,760 |
| 175 | | 40 | 83 | 169 | 296 | 420 | 896 | 1,610 | 2,540 | 5,300 |
| 200 | | 38 | 77 | 157 | 275 | 390 | 834 | 1,500 | 2,370 | 4,930 |
| 250 | | 33 | 69 | 140 | 244 | 346 | 739 | 1,330 | 2,100 | 4,370 |
| 300 | | 30 | 62 | 126 | 221 | 313 | 670 | 1,210 | 1,900 | 3,960 |
| 350 | | 28 | 57 | 116 | 203 | 288 | 616 | 1,110 | 1,750 | 3,640 |
| 400 | | 26 | 53 | 108 | 189 | 268 | 573 | 1,030 | 1,630 | 3,390 |
| 450 | | 24 | 50 | 102 | 177 | 252 | 538 | 968 | 1,530 | 3,180 |
| 500 | | 23 | 47 | 96 | 168 | 238 | 508 | 914 | 1,440 | 3,000 |
| 550 | | 22 | 45 | 91 | 159 | 226 | 482 | 868 | 1,370 | 2,850 |
| 600 | | 21 | 43 | 87 | 152 | 215 | 460 | 829 | 1,310 | 2,720 |
| 650 | | 20 | 41 | 83 | 145 | 206 | 441 | 793 | 1,250 | 2,610 |
| 700 | | 19 | 39 | 80 | 140 | 198 | 423 | 762 | 1,200 | 2,500 |
| 750 | | 18 | 38 | 77 | 135 | 191 | 408 | 734 | 1,160 | 2,410 |
| 800 | | 18 | 37 | 74 | 130 | 184 | 394 | 709 | 1,120 | 2,330 |
| 850 | | 17 | 35 | 72 | 126 | 178 | 381 | 686 | 1,080 | 2,250 |
| 900 | | 17 | 34 | 70 | 122 | 173 | 370 | 665 | 1,050 | 2,180 |
| 950 | | 16 | 33 | 68 | 118 | 168 | 359 | 646 | 1,020 | 2,120 |
| 1,000 | | 16 | 32 | 66 | 115 | 163 | 349 | 628 | 991 | 2,060 |
| 1,100 | | 15 | 31 | 63 | 109 | 155 | 332 | 597 | 941 | 1,960 |
| 1,200 | | 14 | 29 | 60 | 104 | 148 | 316 | 569 | 898 | 1,870 |
| 1,300 | | 14 | 28 | 57 | 100 | 142 | 303 | 545 | 860 | 1,790 |
| 1,400 | | 13 | 27 | 55 | 96 | 136 | 291 | 524 | 826 | 1,720 |
| 1,500 | | 13 | 26 | 53 | 93 | 131 | 280 | 505 | 796 | 1,660 |
| 1,600 | | 12 | 25 | 51 | 89 | 127 | 271 | 487 | 768 | 1,600 |
| 1,700 | | 12 | 24 | 49 | 86 | 123 | 262 | 472 | 744 | 1,550 |
| 1,800 | | 11 | 24 | 48 | 84 | 119 | 254 | 457 | 721 | 1,500 |
| 1,900 | | 11 | 23 | 47 | 81 | 115 | 247 | 444 | 700 | 1,460 |
| 2,000 | | 11 | 22 | 45 | 79 | 112 | 240 | 432 | 681 | 1,420 |

Note: All table entries are rounded to 3 significant digits.

*Table capacities are based on Type K copper tubing inside diameter (shown), which has the smallest inside diameter of the copper tubing products.

*TABLE 6.2(j) Semirigid Copper Tubing*

| | | | | | | | Gas: | Natural | |
|---|---|---|---|---|---|---|---|---|---|
| | | | | | | | Inlet Pressure: | 2.0 psi | |
| | | | | | | | Pressure Drop: | 1.0 psi | |
| | | | | | | | Specific Gravity: | 0.60 | |
| | Tube Size (in.) | | | | | | | | |
| Nominal: K & L: | ¼ | ⅜ | ½ | ⅝ | ¾ | 1 | 1¼ | 1½ | 2 |
| ACR: | ⅜ | ½ | ⅝ | ¾ | ⅞ | 1⅛ | 1⅜ | — | — |
| Outside: | 0.375 | 0.500 | 0.625 | 0.750 | 0.875 | 1.125 | 1.375 | 1.625 | 2.125 |
| Inside:* | 0.305 | 0.402 | 0.527 | 0.652 | 0.745 | 0.995 | 1.245 | 1.481 | 1.959 |
| Length (ft) | Capacity in Cubic Feet of Gas per Hour | | | | | | | | |
| 10 | 245 | 506 | 1,030 | 1,800 | 2,550 | 5,450 | 9,820 | 15,500 | 32,200 |
| 20 | 169 | 348 | 708 | 1,240 | 1,760 | 3,750 | 6,750 | 10,600 | 22,200 |
| 30 | 135 | 279 | 568 | 993 | 1,410 | 3,010 | 5,420 | 8,550 | 17,800 |
| 40 | 116 | 239 | 486 | 850 | 1,210 | 2,580 | 4,640 | 7,310 | 15,200 |
| 50 | 103 | 212 | 431 | 754 | 1,070 | 2,280 | 4,110 | 6,480 | 13,500 |
| 60 | 93 | 192 | 391 | 683 | 969 | 2,070 | 3,730 | 5,870 | 12,200 |
| 70 | 86 | 177 | 359 | 628 | 891 | 1,900 | 3,430 | 5,400 | 11,300 |
| 80 | 80 | 164 | 334 | 584 | 829 | 1,770 | 3,190 | 5,030 | 10,500 |
| 90 | 75 | 154 | 314 | 548 | 778 | 1,660 | 2,990 | 4,720 | 9,820 |
| 100 | 71 | 146 | 296 | 518 | 735 | 1,570 | 2,830 | 4,450 | 9,280 |
| 125 | 63 | 129 | 263 | 459 | 651 | 1,390 | 2,500 | 3,950 | 8,220 |
| 150 | 57 | 117 | 238 | 416 | 590 | 1,260 | 2,270 | 3,580 | 7,450 |
| 175 | 52 | 108 | 219 | 383 | 543 | 1,160 | 2,090 | 3,290 | 6,850 |
| 200 | 49 | 100 | 204 | 356 | 505 | 1,080 | 1,940 | 3,060 | 6,380 |
| 250 | 43 | 89 | 181 | 315 | 448 | 956 | 1,720 | 2,710 | 5,650 |
| 300 | 39 | 80 | 164 | 286 | 406 | 866 | 1,560 | 2,460 | 5,120 |
| 350 | 36 | 74 | 150 | 263 | 373 | 797 | 1,430 | 2,260 | 4,710 |
| 400 | 33 | 69 | 140 | 245 | 347 | 741 | 1,330 | 2,100 | 4,380 |
| 450 | 31 | 65 | 131 | 230 | 326 | 696 | 1,250 | 1,970 | 4,110 |
| 500 | 30 | 61 | 124 | 217 | 308 | 657 | 1,180 | 1,870 | 3,880 |
| 550 | 28 | 58 | 118 | 206 | 292 | 624 | 1,120 | 1,770 | 3,690 |
| 600 | 27 | 55 | 112 | 196 | 279 | 595 | 1,070 | 1,690 | 3,520 |
| 650 | 26 | 53 | 108 | 188 | 267 | 570 | 1,030 | 1,620 | 3,370 |
| 700 | 25 | 51 | 103 | 181 | 256 | 548 | 986 | 1,550 | 3,240 |
| 750 | 24 | 49 | 100 | 174 | 247 | 528 | 950 | 1,500 | 3,120 |
| 800 | 23 | 47 | 96 | 168 | 239 | 510 | 917 | 1,450 | 3,010 |
| 850 | 22 | 46 | 93 | 163 | 231 | 493 | 888 | 1,400 | 2,920 |
| 900 | 22 | 44 | 90 | 158 | 224 | 478 | 861 | 1,360 | 2,830 |
| 950 | 21 | 43 | 88 | 153 | 217 | 464 | 836 | 1,320 | 2,740 |
| 1,000 | 20 | 42 | 85 | 149 | 211 | 452 | 813 | 1,280 | 2,670 |
| 1,100 | 19 | 40 | 81 | 142 | 201 | 429 | 772 | 1,220 | 2,540 |
| 1,200 | 18 | 38 | 77 | 135 | 192 | 409 | 737 | 1,160 | 2,420 |
| 1,300 | 18 | 36 | 74 | 129 | 183 | 392 | 705 | 1,110 | 2,320 |
| 1,400 | 17 | 35 | 71 | 124 | 176 | 376 | 678 | 1,070 | 2,230 |
| 1,500 | 16 | 34 | 68 | 120 | 170 | 363 | 653 | 1,030 | 2,140 |
| 1,600 | 16 | 33 | 66 | 116 | 164 | 350 | 630 | 994 | 2,070 |
| 1,700 | 15 | 31 | 64 | 112 | 159 | 339 | 610 | 962 | 2,000 |
| 1,800 | 15 | 30 | 62 | 108 | 154 | 329 | 592 | 933 | 1,940 |
| 1,900 | 14 | 30 | 60 | 105 | 149 | 319 | 575 | 906 | 1,890 |
| 2,000 | 14 | 29 | 59 | 102 | 145 | 310 | 559 | 881 | 1,830 |

Note: All table entries are rounded to 3 significant digits.

*Table capacities are based on Type K copper tubing inside diameter (shown), which has the smallest inside diameter of the copper tubing products.

***TABLE 6.2(k)*** *Semirigid Copper Tubing*

| | |
|---|---|
| *Gas:* | *Natural* |
| *Inlet Pressure:* | *2.0 psi* |
| *Pressure Drop:* | *1.5 psi* |
| *Specific Gravity:* | *0.60* |

***INTENDED USE: Pipe Sizing Between Point of Delivery and the House Line Regulator. Total Load Supplied by a Single House Line Regulator Not Exceeding 150 Cubic Feet per Hour.**** 

| | Tube Size (in.) | | | | | | | | |
|---|---|---|---|---|---|---|---|---|---|
| **Nominal: K & L:** | ¼ | ⅜ | ½ | ⅝ | ¾ | 1 | 1¼ | 1½ | 2 |
| **ACR:** | ⅜ | ½ | ⅝ | ¾ | ⅞ | 1⅛ | 1⅜ | — | — |
| **Outside:** | 0.375 | 0.500 | 0.625 | 0.750 | 0.875 | 1.125 | 1.375 | 1.625 | 2.125 |
| **Inside:**[†] | 0.305 | 0.402 | 0.527 | 0.652 | 0.745 | 0.995 | 1.245 | 1.481 | 1.959 |
| **Length (ft)** | **Capacity in Cubic Feet of Gas per Hour** | | | | | | | | |
| 10 | 303 | 625 | 1,270 | 2,220 | 3,150 | 6,740 | 12,100 | 19,100 | 39,800 |
| 20 | 208 | 430 | 874 | 1,530 | 2,170 | 4,630 | 8,330 | 13,100 | 27,400 |
| 30 | 167 | 345 | 702 | 1,230 | 1,740 | 3,720 | 6,690 | 10,600 | 22,000 |
| 40 | 143 | 295 | 601 | 1,050 | 1,490 | 3,180 | 5,730 | 9,030 | 18,800 |
| 50 | 127 | 262 | 532 | 931 | 1,320 | 2,820 | 5,080 | 8,000 | 16,700 |
| 60 | 115 | 237 | 482 | 843 | 1,200 | 2,560 | 4,600 | 7,250 | 15,100 |
| 70 | 106 | 218 | 444 | 776 | 1,100 | 2,350 | 4,230 | 6,670 | 13,900 |
| 80 | 98 | 203 | 413 | 722 | 1,020 | 2,190 | 3,940 | 6,210 | 12,900 |
| 90 | 92 | 190 | 387 | 677 | 961 | 2,050 | 3,690 | 5,820 | 12,100 |
| 100 | 87 | 180 | 366 | 640 | 907 | 1,940 | 3,490 | 5,500 | 11,500 |
| 125 | 77 | 159 | 324 | 567 | 804 | 1,720 | 3,090 | 4,880 | 10,200 |
| 150 | 70 | 144 | 294 | 514 | 729 | 1,560 | 2,800 | 4,420 | 9,200 |
| 175 | 64 | 133 | 270 | 472 | 670 | 1,430 | 2,580 | 4,060 | 8,460 |
| 200 | 60 | 124 | 252 | 440 | 624 | 1,330 | 2,400 | 3,780 | 7,870 |
| 250 | 53 | 110 | 223 | 390 | 553 | 1,180 | 2,130 | 3,350 | 6,980 |
| 300 | 48 | 99 | 202 | 353 | 501 | 1,070 | 1,930 | 3,040 | 6,320 |
| 350 | 44 | 91 | 186 | 325 | 461 | 984 | 1,770 | 2,790 | 5,820 |
| 400 | 41 | 85 | 173 | 302 | 429 | 916 | 1,650 | 2,600 | 5,410 |
| 450 | 39 | 80 | 162 | 283 | 402 | 859 | 1,550 | 2,440 | 5,080 |
| 500 | 36 | 75 | 153 | 268 | 380 | 811 | 1,460 | 2,300 | 4,800 |
| 550 | 35 | 72 | 146 | 254 | 361 | 771 | 1,390 | 2,190 | 4,560 |
| 600 | 33 | 68 | 139 | 243 | 344 | 735 | 1,320 | 2,090 | 4,350 |
| 650 | 32 | 65 | 133 | 232 | 330 | 704 | 1,270 | 2,000 | 4,160 |
| 700 | 30 | 63 | 128 | 223 | 317 | 676 | 1,220 | 1,920 | 4,000 |
| 750 | 29 | 60 | 123 | 215 | 305 | 652 | 1,170 | 1,850 | 3,850 |
| 800 | 28 | 58 | 119 | 208 | 295 | 629 | 1,130 | 1,790 | 3,720 |
| 850 | 27 | 57 | 115 | 201 | 285 | 609 | 1,100 | 1,730 | 3,600 |
| 900 | 27 | 55 | 111 | 195 | 276 | 590 | 1,060 | 1,680 | 3,490 |
| 950 | 26 | 53 | 108 | 189 | 268 | 573 | 1,030 | 1,630 | 3,390 |
| 1,000 | 25 | 52 | 105 | 184 | 261 | 558 | 1,000 | 1,580 | 3,300 |
| 1,100 | 24 | 49 | 100 | 175 | 248 | 530 | 954 | 1,500 | 3,130 |
| 1,200 | 23 | 47 | 95 | 167 | 237 | 505 | 910 | 1,430 | 2,990 |
| 1,300 | 22 | 45 | 91 | 160 | 227 | 484 | 871 | 1,370 | 2,860 |
| 1,400 | 21 | 43 | 88 | 153 | 218 | 465 | 837 | 1,320 | 2,750 |
| 1,500 | 20 | 42 | 85 | 148 | 210 | 448 | 806 | 1,270 | 2,650 |
| 1,600 | 19 | 40 | 82 | 143 | 202 | 432 | 779 | 1,230 | 2,560 |
| 1,700 | 19 | 39 | 79 | 138 | 196 | 419 | 753 | 1,190 | 2,470 |
| 1,800 | 18 | 38 | 77 | 134 | 190 | 406 | 731 | 1,150 | 2,400 |
| 1,900 | 18 | 37 | 74 | 130 | 184 | 394 | 709 | 1,120 | 2,330 |
| 2,000 | 17 | 36 | 72 | 126 | 179 | 383 | 690 | 1,090 | 2,270 |

Note: All table entries are rounded to 3 significant digits.

*When this table is used to size the tubing upstream of a line pressure regulator, the pipe or tubing downstream of the line pressure regulator shall be sized using a pressure drop no greater than 1 in. w.c.

†Table capacities are based on Type K copper tubing inside diameter (shown), which has the smallest inside diameter of the copper tubing products.

***TABLE 6.2(l)*** *Semirigid Copper Tubing*

| | |
|---|---|
| ***Gas:*** | ***Natural*** |
| ***Inlet Pressure:*** | ***5.0 psi*** |
| ***Pressure Drop:*** | ***3.5 psi*** |
| ***Specific Gravity:*** | ***0.60*** |

| | ***Tube Size (in.)*** | | | | | | | | |
|---|---|---|---|---|---|---|---|---|---|
| ***Nominal: K & L:*** | ¼ | ⅜ | ½ | ⅝ | ¾ | *1* | *1¼* | *1½* | *2* |
| ***ACR:*** | ⅜ | ½ | ⅝ | ¾ | ⅞ | *1⅛* | *1⅜* | — | — |
| ***Outside:*** | ***0.375*** | ***0.500*** | ***0.625*** | ***0.750*** | ***0.875*** | ***1.125*** | ***1.375*** | ***1.625*** | ***2.125*** |
| ***Inside:**** | ***0.305*** | ***0.402*** | ***0.527*** | ***0.652*** | ***0.745*** | ***0.995*** | ***1.245*** | ***1.481*** | ***1.959*** |
| ***Length (ft)*** | ***Capacity in Cubic Feet of Gas per Hour*** | | | | | | | | |
| 10 | 511 | 1,050 | 2,140 | 3,750 | 5,320 | 11,400 | 20,400 | 32,200 | 67,100 |
| 20 | 351 | 724 | 1,470 | 2,580 | 3,650 | 7,800 | 14,000 | 22,200 | 46,100 |
| 30 | 282 | 582 | 1,180 | 2,070 | 2,930 | 6,270 | 11,300 | 17,800 | 37,000 |
| 40 | 241 | 498 | 1,010 | 1,770 | 2,510 | 5,360 | 9,660 | 15,200 | 31,700 |
| 50 | 214 | 441 | 898 | 1,570 | 2,230 | 4,750 | 8,560 | 13,500 | 28,100 |
| 60 | 194 | 400 | 813 | 1,420 | 2,020 | 4,310 | 7,750 | 12,200 | 25,500 |
| 70 | 178 | 368 | 748 | 1,310 | 1,860 | 3,960 | 7,130 | 11,200 | 23,400 |
| 80 | 166 | 342 | 696 | 1,220 | 1,730 | 3,690 | 6,640 | 10,500 | 21,800 |
| 90 | 156 | 321 | 653 | 1,140 | 1,620 | 3,460 | 6,230 | 9,820 | 20,400 |
| 100 | 147 | 303 | 617 | 1,080 | 1,530 | 3,270 | 5,880 | 9,270 | 19,300 |
| 125 | 130 | 269 | 547 | 955 | 1,360 | 2,900 | 5,210 | 8,220 | 17,100 |
| 150 | 118 | 243 | 495 | 866 | 1,230 | 2,620 | 4,720 | 7,450 | 15,500 |
| 175 | 109 | 224 | 456 | 796 | 1,130 | 2,410 | 4,350 | 6,850 | 14,300 |
| 200 | 101 | 208 | 424 | 741 | 1,050 | 2,250 | 4,040 | 6,370 | 13,300 |
| 250 | 90 | 185 | 376 | 657 | 932 | 1,990 | 3,580 | 5,650 | 11,800 |
| 300 | 81 | 167 | 340 | 595 | 844 | 1,800 | 3,250 | 5,120 | 10,700 |
| 350 | 75 | 154 | 313 | 547 | 777 | 1,660 | 2,990 | 4,710 | 9,810 |
| 400 | 69 | 143 | 291 | 509 | 722 | 1,540 | 2,780 | 4,380 | 9,120 |
| 450 | 65 | 134 | 273 | 478 | 678 | 1,450 | 2,610 | 4,110 | 8,560 |
| 500 | 62 | 127 | 258 | 451 | 640 | 1,370 | 2,460 | 3,880 | 8,090 |
| 550 | 58 | 121 | 245 | 429 | 608 | 1,300 | 2,340 | 3,690 | 7,680 |
| 600 | 56 | 115 | 234 | 409 | 580 | 1,240 | 2,230 | 3,520 | 7,330 |
| 650 | 53 | 110 | 224 | 392 | 556 | 1,190 | 2,140 | 3,370 | 7,020 |
| 700 | 51 | 106 | 215 | 376 | 534 | 1,140 | 2,050 | 3,240 | 6,740 |
| 750 | 49 | 102 | 207 | 362 | 514 | 1,100 | 1,980 | 3,120 | 6,490 |
| 800 | 48 | 98 | 200 | 350 | 497 | 1,060 | 1,910 | 3,010 | 6,270 |
| 850 | 46 | 95 | 194 | 339 | 481 | 1,030 | 1,850 | 2,910 | 6,070 |
| 900 | 45 | 92 | 188 | 328 | 466 | 1,000 | 1,790 | 2,820 | 5,880 |
| 950 | 43 | 90 | 182 | 319 | 452 | 967 | 1,740 | 2,740 | 5,710 |
| 1,000 | 42 | 87 | 177 | 310 | 440 | 940 | 1,690 | 2,670 | 5,560 |
| 1,100 | 40 | 83 | 169 | 295 | 418 | 893 | 1,610 | 2,530 | 5,280 |
| 1,200 | 38 | 79 | 161 | 281 | 399 | 852 | 1,530 | 2,420 | 5,040 |
| 1,300 | 37 | 76 | 154 | 269 | 382 | 816 | 1,470 | 2,320 | 4,820 |
| 1,400 | 35 | 73 | 148 | 259 | 367 | 784 | 1,410 | 2,220 | 4,630 |
| 1,500 | 34 | 70 | 143 | 249 | 353 | 755 | 1,360 | 2,140 | 4,460 |
| 1,600 | 33 | 68 | 138 | 241 | 341 | 729 | 1,310 | 2,070 | 4,310 |
| 1,700 | 32 | 65 | 133 | 233 | 330 | 705 | 1,270 | 2,000 | 4,170 |
| 1,800 | 31 | 63 | 129 | 226 | 320 | 684 | 1,230 | 1,940 | 4,040 |
| 1,900 | 30 | 62 | 125 | 219 | 311 | 664 | 1,200 | 1,890 | 3,930 |
| 2,000 | 29 | 60 | 122 | 213 | 302 | 646 | 1,160 | 1,830 | 3,820 |

Note: All table entries are rounded to 3 significant digits.

*Table capacities are based on Type K copper tubing inside diameter (shown), which has the smallest inside diameter of the copper tubing products.

NEW

*TABLE 6.2(m) Corrugated Stainless Steel Tubing (CSST)*

| | |
|---|---|
| *Gas:* | *Natural* |
| *Inlet Pressure:* | *Less than 2 psi* |
| *Pressure Drop:* | *0.5 in. w.c.* |
| *Specific Gravity:* | *0.60* |

| | *Tube Size (EHD)* | | | | | | | | | | | | | |
|---|---|---|---|---|---|---|---|---|---|---|---|---|---|---|
| ***Flow Designation:*** | ***13*** | ***15*** | ***18*** | ***19*** | ***23*** | ***25*** | ***30*** | ***31*** | ***37*** | ***39*** | ***46*** | ***48*** | ***60*** | ***62*** |
| ***Length (ft)*** | ***Capacity in Cubic Feet of Gas per Hour*** | | | | | | | | | | | | | |
| 5 | 46 | 63 | 115 | 134 | 225 | 270 | 471 | 546 | 895 | 1,037 | 1,790 | 2,070 | 3,660 | 4,140 |
| 10 | 32 | 44 | 82 | 95 | 161 | 192 | 330 | 383 | 639 | 746 | 1,260 | 1,470 | 2,600 | 2,930 |
| 15 | 25 | 35 | 66 | 77 | 132 | 157 | 267 | 310 | 524 | 615 | 1,030 | 1,200 | 2,140 | 2,400 |
| 20 | 22 | 31 | 58 | 67 | 116 | 137 | 231 | 269 | 456 | 536 | 888 | 1,050 | 1,850 | 2,080 |
| 25 | 19 | 27 | 52 | 60 | 104 | 122 | 206 | 240 | 409 | 482 | 793 | 936 | 1,660 | 1,860 |
| 30 | 18 | 25 | 47 | 55 | 96 | 112 | 188 | 218 | 374 | 442 | 723 | 856 | 1,520 | 1,700 |
| 40 | 15 | 21 | 41 | 47 | 83 | 97 | 162 | 188 | 325 | 386 | 625 | 742 | 1,320 | 1,470 |
| 50 | 13 | 19 | 37 | 42 | 75 | 87 | 144 | 168 | 292 | 347 | 559 | 665 | 1,180 | 1,320 |
| 60 | 12 | 17 | 34 | 38 | 68 | 80 | 131 | 153 | 267 | 318 | 509 | 608 | 1,080 | 1,200 |
| 70 | 11 | 16 | 31 | 36 | 63 | 74 | 121 | 141 | 248 | 295 | 471 | 563 | 1,000 | 1,110 |
| 80 | 10 | 15 | 29 | 33 | 60 | 69 | 113 | 132 | 232 | 277 | 440 | 527 | 940 | 1,040 |
| 90 | 10 | 14 | 28 | 32 | 57 | 65 | 107 | 125 | 219 | 262 | 415 | 498 | 887 | 983 |
| 100 | 9 | 13 | 26 | 30 | 54 | 62 | 101 | 118 | 208 | 249 | 393 | 472 | 843 | 933 |
| 150 | 7 | 10 | 20 | 23 | 42 | 48 | 78 | 91 | 171 | 205 | 320 | 387 | 691 | 762 |
| 200 | 6 | 9 | 18 | 21 | 38 | 44 | 71 | 82 | 148 | 179 | 277 | 336 | 600 | 661 |
| 250 | 5 | 8 | 16 | 19 | 34 | 39 | 63 | 74 | 133 | 161 | 247 | 301 | 538 | 591 |
| 300 | 5 | 7 | 15 | 17 | 32 | 36 | 57 | 67 | 95 | 148 | 226 | 275 | 492 | 540 |

EHD: Equivalent hydraulic diameter. A measure of the relative hydraulic efficiency between different tubing sizes. The greater the value of EHD, the greater the gas capacity of the tubing.

Notes:

(1) Table includes losses for four 90 degree bends and two end fittings. Tubing runs with larger numbers of bends and/or fittings shall be increased by an equivalent length of tubing to the following equation: $L = 1.3n$, where $L$ is additional length (ft) of tubing and $n$ is the number of additional fittings and/or bends.

(2) All table entries are rounded to 3 significant digits.

***TABLE 6.2(n)*** *Corrugated Stainless Steel Tubing (CSST)*

| | | | | | | | | | | | | |
|---|---|---|---|---|---|---|---|---|---|---|---|---|
| | | | | | | | | | | ***Gas:*** | ***Natural*** | |
| | | | | | | | | | | ***Inlet Pressure:*** | ***Less than 2 psi*** | |
| | | | | | | | | | | ***Pressure Drop:*** | ***3.0 in. w.c.*** | |
| | | | | | | | | | | ***Specific Gravity:*** | ***0.60*** | |

| | ***Tube Size (EHD)*** | | | | | | | | | | | | |
|---|---|---|---|---|---|---|---|---|---|---|---|---|---|
| ***Flow Designation:*** | ***13*** | ***15*** | ***18*** | ***19*** | ***23*** | ***25*** | ***30*** | ***31*** | ***37*** | ***46*** | ***48*** | ***60*** | ***62*** |
| ***Length (ft)*** | ***Capacity in Cubic Feet of Gas per Hour*** | | | | | | | | | | | | |
| 5 | 120 | 160 | 277 | 327 | 529 | 649 | 1,180 | 1,370 | 2,140 | 4,430 | 5,010 | 8,800 | 10,100 |
| 10 | 83 | 112 | 197 | 231 | 380 | 462 | 828 | 958 | 1,530 | 3,200 | 3,560 | 6,270 | 7,160 |
| 15 | 67 | 90 | 161 | 189 | 313 | 379 | 673 | 778 | 1,250 | 2,540 | 2,910 | 5,140 | 5,850 |
| 20 | 57 | 78 | 140 | 164 | 273 | 329 | 580 | 672 | 1,090 | 2,200 | 2,530 | 4,460 | 5,070 |
| 25 | 51 | 69 | 125 | 147 | 245 | 295 | 518 | 599 | 978 | 1,960 | 2,270 | 4,000 | 4,540 |
| 30 | 46 | 63 | 115 | 134 | 225 | 270 | 471 | 546 | 895 | 1,790 | 2,070 | 3,660 | 4,140 |
| 40 | 39 | 54 | 100 | 116 | 196 | 234 | 407 | 471 | 778 | 1,550 | 1,800 | 3,180 | 3,590 |
| 50 | 35 | 48 | 89 | 104 | 176 | 210 | 363 | 421 | 698 | 1,380 | 1,610 | 2,850 | 3,210 |
| 60 | 32 | 44 | 82 | 95 | 161 | 192 | 330 | 383 | 639 | 1,260 | 1,470 | 2,600 | 2,930 |
| 70 | 29 | 41 | 76 | 88 | 150 | 178 | 306 | 355 | 593 | 1,170 | 1,360 | 2,420 | 2,720 |
| 80 | 27 | 38 | 71 | 82 | 141 | 167 | 285 | 331 | 555 | 1,090 | 1,280 | 2,260 | 2,540 |
| 90 | 26 | 36 | 67 | 77 | 133 | 157 | 268 | 311 | 524 | 1,030 | 1,200 | 2,140 | 2,400 |
| 100 | 24 | 34 | 63 | 73 | 126 | 149 | 254 | 295 | 498 | 974 | 1,140 | 2,030 | 2,280 |
| 150 | 19 | 27 | 52 | 60 | 104 | 122 | 206 | 240 | 409 | 793 | 936 | 1,660 | 1,860 |
| 200 | 17 | 23 | 45 | 52 | 91 | 106 | 178 | 207 | 355 | 686 | 812 | 1,440 | 1,610 |
| 250 | 15 | 21 | 40 | 46 | 82 | 95 | 159 | 184 | 319 | 613 | 728 | 1,290 | 1,440 |
| 300 | 13 | 19 | 37 | 42 | 75 | 87 | 144 | 168 | 234 | 559 | 665 | 1,180 | 1,320 |

EHD: Equivalent hydraulic diameter. A measure of the relative hydraulic efficiency between different tubing sizes. The greater the value of EHD, the greater the gas capacity of the tubing.

Notes:

(1) Table includes losses for four 90 degree bends and two end fittings. Tubing runs with larger numbers of bends and/or fittings shall be increased by an equivalent length of tubing to the following equation: $L = 1.3n$, where $L$ is additional length (ft) of tubing and $n$ is the number of additional fittings and/or bends.

(2) All table entries are rounded to 3 significant digits.

*TABLE 6.2(o) Corrugated Stainless Steel Tubing (CSST)*

| | | | | | | | | | | | | | |
|---|---|---|---|---|---|---|---|---|---|---|---|---|---|
| | | | | | | | | | | *Gas:* | | *Natural* | |
| | | | | | | | | | | *Inlet Pressure:* | | *Less than 2 psi* | |
| | | | | | | | | | | *Pressure Drop:* | | *6.0 in. w.c.* | |
| | | | | | | | | | | *Specific Gravity:* | | *0.60* | |
| | *Tube Size (EHD)* | | | | | | | | | | | | |
| *Flow Designation:* | *13* | *15* | *18* | *19* | *23* | *25* | *30* | *31* | *37* | *46* | *48* | *60* | *62* |
| *Length (ft)* | *Capacity in Cubic Feet of Gas per Hour* | | | | | | | | | | | | |
| 5 | 173 | 229 | 389 | 461 | 737 | 911 | 1,690 | 1,950 | 3,000 | 6,280 | 7,050 | 12,400 | 14,260 |
| 10 | 120 | 160 | 277 | 327 | 529 | 649 | 1,180 | 1,370 | 2,140 | 4,430 | 5,010 | 8,800 | 10,100 |
| 15 | 96 | 130 | 227 | 267 | 436 | 532 | 960 | 1,110 | 1,760 | 3,610 | 4,100 | 7,210 | 8,260 |
| 20 | 83 | 112 | 197 | 231 | 380 | 462 | 828 | 958 | 1,530 | 3,120 | 3,560 | 6,270 | 7,160 |
| 25 | 74 | 99 | 176 | 207 | 342 | 414 | 739 | 855 | 1,370 | 2,790 | 3,190 | 5,620 | 6,400 |
| 30 | 67 | 90 | 161 | 189 | 313 | 379 | 673 | 778 | 1,250 | 2,540 | 2,910 | 5,140 | 5,850 |
| 40 | 57 | 78 | 140 | 164 | 273 | 329 | 580 | 672 | 1,090 | 2,200 | 2,530 | 4,460 | 5,070 |
| 50 | 51 | 69 | 125 | 147 | 245 | 295 | 518 | 599 | 978 | 1,960 | 2,270 | 4,000 | 4,540 |
| 60 | 46 | 63 | 115 | 134 | 225 | 270 | 471 | 546 | 895 | 1,790 | 2,070 | 3,660 | 4,140 |
| 70 | 42 | 58 | 106 | 124 | 209 | 250 | 435 | 505 | 830 | 1,660 | 1,920 | 3,390 | 3,840 |
| 80 | 39 | 54 | 100 | 116 | 196 | 234 | 407 | 471 | 778 | 1,550 | 1,800 | 3,180 | 3,590 |
| 90 | 37 | 51 | 94 | 109 | 185 | 221 | 383 | 444 | 735 | 1,460 | 1,700 | 3,000 | 3,390 |
| 100 | 35 | 48 | 89 | 104 | 176 | 210 | 363 | 421 | 698 | 1,380 | 1,610 | 2,850 | 3,210 |
| 150 | 28 | 39 | 73 | 85 | 145 | 172 | 294 | 342 | 573 | 1,130 | 1,320 | 2,340 | 2,630 |
| 200 | 24 | 34 | 63 | 73 | 126 | 149 | 254 | 295 | 498 | 974 | 1,140 | 2,030 | 2,280 |
| 250 | 21 | 30 | 57 | 66 | 114 | 134 | 226 | 263 | 447 | 870 | 1,020 | 1,820 | 2,040 |
| 300 | 19 | 27 | 52 | 60 | 104 | 122 | 206 | 240 | 409 | 793 | 936 | 1,660 | 1,860 |

EHD: Equivalent hydraulic diameter. A measure of the relative hydraulic efficiency between different tubing sizes. The greater the value of EHD, the greater the gas capacity of the tubing.

Notes:

(1) Table includes losses for four 90 degree bends and two end fittings. Tubing runs with larger numbers of bends and/or fittings shall be increased by an equivalent length of tubing to the following equation: $L = 1.3n$, where $L$ is additional length (ft) of tubing and $n$ is the number of additional fittings and/or bends.

(2) All table entries are rounded to 3 significant digits.

NEW *TABLE 6.2(p) Corrugated Stainless Steel Tubing (CSST)*

| | |
|---|---|
| *Gas:* | *Natural* |
| *Inlet Pressure:* | *2.0 psi* |
| *Pressure Drop:* | *1.0 psi* |
| *Specific Gravity:* | *0.60* |

| | *Tube Size (EHD)* | | | | | | | | | | | | | |
|---|---|---|---|---|---|---|---|---|---|---|---|---|---|---|
| *Flow Designation:* | *13* | *15* | *18* | *19* | *23* | *25* | *30* | *31* | *37* | *39* | *46* | *48* | *60* | *62* |
| *Length (ft)* | *Capacity in Cubic Feet of Gas per Hour* | | | | | | | | | | | | | |
| 10 | 270 | 353 | 587 | 700 | 1,100 | 1,370 | 2,590 | 2,990 | 4,510 | 5,037 | 9,600 | 10,700 | 18,600 | 21,600 |
| 25 | 166 | 220 | 374 | 444 | 709 | 876 | 1,620 | 1,870 | 2,890 | 3,258 | 6,040 | 6,780 | 11,900 | 13,700 |
| 30 | 151 | 200 | 342 | 405 | 650 | 801 | 1,480 | 1,700 | 2,640 | 2,987 | 5,510 | 6,200 | 10,900 | 12,500 |
| 40 | 129 | 172 | 297 | 351 | 567 | 696 | 1,270 | 1,470 | 2,300 | 2,605 | 4,760 | 5,380 | 9,440 | 10,900 |
| 50 | 115 | 154 | 266 | 314 | 510 | 624 | 1,140 | 1,310 | 2,060 | 2,343 | 4,260 | 4,820 | 8,470 | 9,720 |
| 75 | 93 | 124 | 218 | 257 | 420 | 512 | 922 | 1,070 | 1,690 | 1,932 | 3,470 | 3,950 | 6,940 | 7,940 |
| 80 | 89 | 120 | 211 | 249 | 407 | 496 | 892 | 1,030 | 1,640 | 1,874 | 3,360 | 3,820 | 6,730 | 7,690 |
| 100 | 79 | 107 | 189 | 222 | 366 | 445 | 795 | 920 | 1,470 | 1,685 | 3,000 | 3,420 | 6,030 | 6,880 |
| 150 | 64 | 87 | 155 | 182 | 302 | 364 | 646 | 748 | 1,210 | 1,389 | 2,440 | 2,800 | 4,940 | 5,620 |
| 200 | 55 | 75 | 135 | 157 | 263 | 317 | 557 | 645 | 1,050 | 1,212 | 2,110 | 2,430 | 4,290 | 4,870 |
| 250 | 49 | 67 | 121 | 141 | 236 | 284 | 497 | 576 | 941 | 1,090 | 1,890 | 2,180 | 3,850 | 4,360 |
| 300 | 44 | 61 | 110 | 129 | 217 | 260 | 453 | 525 | 862 | 999 | 1,720 | 1,990 | 3,520 | 3,980 |
| 400 | 38 | 52 | 96 | 111 | 189 | 225 | 390 | 453 | 749 | 871 | 1,490 | 1,730 | 3,060 | 3,450 |
| 500 | 34 | 46 | 86 | 100 | 170 | 202 | 348 | 404 | 552 | 783 | 1,330 | 1,550 | 2,740 | 3,090 |

EHD: Equivalent hydraulic diameter. A measure of the relative hydraulic efficiency between different tubing sizes. The greater the value of EHD, the greater the gas capacity of the tubing.

Notes:

(1) Table does not include effect of pressure drop across the line regulator. Where regulator loss exceeds $^3/_4$ psi, do not use this table. Consult with regulator manufacturer for pressure drops and capacity factors. Pressure drops across a regulator may vary with flow rate.

(2) CAUTION: Capacities shown in table may exceed maximum capacity for a selected regulator. Consult with regulator or tubing manufacturer for guidance.

(3) Table includes losses for four 90 degree bends and two end fittings. Tubing runs with larger number of bends and/or fittings shall be increased by an equivalent length of tubing according to the following equation: $L = 1.3n$, where $L$ is additional length (ft) of tubing and $n$ is the number of additional fittings and/or bends.

(4) All table entries are rounded to 3 significant digits.

## *TABLE 6.2(q) Corrugated Stainless Steel Tubing (CSST)*

| | |
|---|---|
| *Gas:* | *Natural* |
| *Inlet Pressure:* | *5.0 psi* |
| *Pressure Drop:* | *3.5 psi* |
| *Specific Gravity:* | *0.60* |

| | *Tube Size (EHD)* | | | | | | | | | | | | | |
|---|---|---|---|---|---|---|---|---|---|---|---|---|---|---|
| ***Flow Designation:*** | ***13*** | ***15*** | ***18*** | ***19*** | ***23*** | ***25*** | ***30*** | ***31*** | ***37*** | ***39*** | ***46*** | ***48*** | ***60*** | ***62*** |
| ***Length (ft)*** | ***Capacity in Cubic Feet of Gas per Hour*** | | | | | | | | | | | | | |
| 10 | 523 | 674 | 1,080 | 1,300 | 2,000 | 2,530 | 4,920 | 5,660 | 8,300 | 9,140 | 18,100 | 19,800 | 34,400 | 40,400 |
| 25 | 322 | 420 | 691 | 827 | 1,290 | 1,620 | 3,080 | 3,540 | 5,310 | 5,911 | 11,400 | 12,600 | 22,000 | 25,600 |
| 30 | 292 | 382 | 632 | 755 | 1,180 | 1,480 | 2,800 | 3,230 | 4,860 | 5,420 | 10,400 | 11,500 | 20,100 | 23,400 |
| 40 | 251 | 329 | 549 | 654 | 1,030 | 1,280 | 2,420 | 2,790 | 4,230 | 4,727 | 8,970 | 10,000 | 17,400 | 20,200 |
| 50 | 223 | 293 | 492 | 586 | 926 | 1,150 | 2,160 | 2,490 | 3,790 | 4,251 | 8,020 | 8,930 | 15,600 | 18,100 |
| 75 | 180 | 238 | 403 | 479 | 763 | 944 | 1,750 | 2,020 | 3,110 | 3,506 | 6,530 | 7,320 | 12,800 | 14,800 |
| 80 | 174 | 230 | 391 | 463 | 740 | 915 | 1,690 | 1,960 | 3,020 | 3,400 | 6,320 | 7,090 | 12,400 | 14,300 |
| 100 | 154 | 205 | 350 | 415 | 665 | 820 | 1,510 | 1,740 | 2,710 | 3,057 | 5,650 | 6,350 | 11,100 | 12,800 |
| 150 | 124 | 166 | 287 | 339 | 548 | 672 | 1,230 | 1,420 | 2,220 | 2,521 | 4,600 | 5,200 | 9,130 | 10,500 |
| 200 | 107 | 143 | 249 | 294 | 478 | 584 | 1,060 | 1,220 | 1,930 | 2,199 | 3,980 | 4,510 | 7,930 | 9,090 |
| 250 | 95 | 128 | 223 | 263 | 430 | 524 | 945 | 1,090 | 1,730 | 1,977 | 3,550 | 4,040 | 7,110 | 8,140 |
| 300 | 86 | 116 | 204 | 240 | 394 | 479 | 860 | 995 | 1,590 | 1,813 | 3,240 | 3,690 | 6,500 | 7,430 |
| 400 | 74 | 100 | 177 | 208 | 343 | 416 | 742 | 858 | 1,380 | 1,581 | 2,800 | 3,210 | 5,650 | 6,440 |
| 500 | 66 | 89 | 159 | 186 | 309 | 373 | 662 | 766 | 1,040 | 1,422 | 2,500 | 2,870 | 5,060 | 5,760 |

EHD: Equivalent hydraulic diameter. A measure of the relative hydraulic efficiency between different tubing sizes. The greater the value of EHD, the greater the gas capacity of the tubing.

Notes:

(1) Table does not include effect of pressure drop across line regulator. Where regulator loss exceeds 1 psi, do not use this table. Consult with regulator manufacturer for pressure drops and capacity factors. Pressure drop across regulator may vary with the flow rate.

(2) CAUTION: Capacities shown in table may exceed maximum capacity of selected regulator. Consult with tubing manufacturer for guidance.

(3) Table includes losses for four 90 degree bends and two end fittings. Tubing runs with larger numbers of bends and/or fittings shall be increased by an equivalent length of tubing to the following equation: $L = 1.3n$, where $L$ is additional length (ft) of tubing and $n$ is the number of additional fittings and/or bends.

(4) All table entries are rounded to 3 significant digits.

*TABLE 6.2(r) Polyethylene Plastic Pipe*

| | | | | | | |
|---|---|---|---|---|---|---|
| | | | | *Gas:* | *Natural* | |
| | | | | *Inlet Pressure:* | *Less than 2 psi* | |
| | | | | *Pressure Drop:* | *0.3 in. w.c.* | |
| | | | | *Specific Gravity:* | *0.60* | |
| | *Pipe Size (in.)* | | | | | |
| **Nominal OD:** | ½ | ¾ | 1 | 1¼ | 1½ | 2 |
| **Designation:** | SDR 9.33 | SDR 11.0 | SDR 11.00 | SDR 10.00 | SDR 11.00 | SDR 11.00 |
| **Actual ID:** | 0.660 | 0.860 | 1.077 | 1.328 | 1.554 | 1.943 |
| **Length (ft)** | *Capacity in Cubic Feet of Gas per Hour* | | | | | |
| 10 | 153 | 305 | 551 | 955 | 1,440 | 2,590 |
| 20 | 105 | 210 | 379 | 656 | 991 | 1,780 |
| 30 | 84 | 169 | 304 | 527 | 796 | 1,430 |
| 40 | 72 | 144 | 260 | 451 | 681 | 1,220 |
| 50 | 64 | 128 | 231 | 400 | 604 | 1,080 |
| 60 | 58 | 116 | 209 | 362 | 547 | 983 |
| 70 | 53 | 107 | 192 | 333 | 503 | 904 |
| 80 | 50 | 99 | 179 | 310 | 468 | 841 |
| 90 | 46 | 93 | 168 | 291 | 439 | 789 |
| 100 | 44 | 88 | 159 | 275 | 415 | 745 |
| 125 | 39 | 78 | 141 | 243 | 368 | 661 |
| 150 | 35 | 71 | 127 | 221 | 333 | 598 |
| 175 | 32 | 65 | 117 | 203 | 306 | 551 |
| 200 | 30 | 60 | 109 | 189 | 285 | 512 |
| 250 | 27 | 54 | 97 | 167 | 253 | 454 |
| 300 | 24 | 48 | 88 | 152 | 229 | 411 |
| 350 | 22 | 45 | 81 | 139 | 211 | 378 |
| 400 | 21 | 42 | 75 | 130 | 196 | 352 |
| 450 | 19 | 39 | 70 | 122 | 184 | 330 |
| 500 | 18 | 37 | 66 | 115 | 174 | 312 |

Note: All table entries are rounded to 3 significant digits.

*TABLE 6.2(s) Polyethylene Plastic Pipe*

| | | | | | | |
|---|---|---|---|---|---|---|
| | | | | *Gas:* | *Natural* | |
| | | | | *Inlet Pressure:* | *Less than 2 psi* | |
| | | | | *Pressure Drop:* | *0.5 in. w.c.* | |
| | | | | *Specific Gravity:* | *0.60* | |
| | *Pipe Size (in.)* | | | | | |
| *Nominal OD:* | ½ | ¾ | 1 | 1¼ | 1½ | 2 |
| *Designation:* | *SDR 9.33* | *SDR 11.0* | *SDR 11.00* | *SDR 10.00* | *SDR 11.00* | *SDR 11.00* |
| *Actual ID:* | *0.660* | *0.860* | *1.077* | *1.328* | *1.554* | *1.943* |
| *Length (ft)* | *Capacity in Cubic Feet of Gas per Hour* | | | | | |
| 10 | 201 | 403 | 726 | 1,260 | 1,900 | 3,410 |
| 20 | 138 | 277 | 499 | 865 | 1,310 | 2,350 |
| 30 | 111 | 222 | 401 | 695 | 1,050 | 1,880 |
| 40 | 95 | 190 | 343 | 594 | 898 | 1,610 |
| 50 | 84 | 169 | 304 | 527 | 796 | 1,430 |
| 60 | 76 | 153 | 276 | 477 | 721 | 1,300 |
| 70 | 70 | 140 | 254 | 439 | 663 | 1,190 |
| 80 | 65 | 131 | 236 | 409 | 617 | 1,110 |
| 90 | 61 | 123 | 221 | 383 | 579 | 1,040 |
| 100 | 58 | 116 | 209 | 362 | 547 | 983 |
| 125 | 51 | 103 | 185 | 321 | 485 | 871 |
| 150 | 46 | 93 | 168 | 291 | 439 | 789 |
| 175 | 43 | 86 | 154 | 268 | 404 | 726 |
| 200 | 40 | 80 | 144 | 249 | 376 | 675 |
| 250 | 35 | 71 | 127 | 221 | 333 | 598 |
| 300 | 32 | 64 | 115 | 200 | 302 | 542 |
| 350 | 29 | 59 | 106 | 184 | 278 | 499 |
| 400 | 27 | 55 | 99 | 171 | 258 | 464 |
| 450 | 26 | 51 | 93 | 160 | 242 | 435 |
| 500 | 24 | 48 | 88 | 152 | 229 | 411 |

Note: All table entries are rounded to 3 significant digits.

*TABLE 6.2(t) Polyethylene Plastic Pipe*

| | | | | Gas: | Natural | |
|---|---|---|---|---|---|---|
| | | | | Inlet Pressure: | 2.0 psi | |
| | | | | Pressure Drop: | 1.0 psi | |
| | | | | Specific Gravity: | 0.60 | |
| | Pipe Size (in.) | | | | | |
| Nominal OD: | ½ | ¾ | 1 | 1¼ | 1½ | 2 |
| Designation: | SDR 9.33 | SDR 11.0 | SDR 11.00 | SDR 10.00 | SDR 11.00 | SDR 11.00 |
| Actual ID: | 0.660 | 0.860 | 1.077 | 1.328 | 1.554 | 1.943 |
| Length (ft) | Capacity in Cubic Feet of Gas per Hour | | | | | |
| 10 | 1,860 | 3,720 | 6,710 | 11,600 | 17,600 | 31,600 |
| 20 | 1,280 | 2,560 | 4,610 | 7,990 | 12,100 | 21,700 |
| 30 | 1,030 | 2,050 | 3,710 | 6,420 | 9,690 | 17,400 |
| 40 | 878 | 1,760 | 3,170 | 5,490 | 8,300 | 14,900 |
| 50 | 778 | 1,560 | 2,810 | 4,870 | 7,350 | 13,200 |
| 60 | 705 | 1,410 | 2,550 | 4,410 | 6,660 | 12,000 |
| 70 | 649 | 1,300 | 2,340 | 4,060 | 6,130 | 11,000 |
| 80 | 603 | 1,210 | 2,180 | 3,780 | 5,700 | 10,200 |
| 90 | 566 | 1,130 | 2,050 | 3,540 | 5,350 | 9,610 |
| 100 | 535 | 1,070 | 1,930 | 3,350 | 5,050 | 9,080 |
| 125 | 474 | 949 | 1,710 | 2,970 | 4,480 | 8,050 |
| 150 | 429 | 860 | 1,550 | 2,690 | 4,060 | 7,290 |
| 175 | 395 | 791 | 1,430 | 2,470 | 3,730 | 6,710 |
| 200 | 368 | 736 | 1,330 | 2,300 | 3,470 | 6,240 |
| 250 | 326 | 652 | 1,180 | 2,040 | 3,080 | 5,530 |
| 300 | 295 | 591 | 1,070 | 1,850 | 2,790 | 5,010 |
| 350 | 272 | 544 | 981 | 1,700 | 2,570 | 4,610 |
| 400 | 253 | 506 | 913 | 1,580 | 2,390 | 4,290 |
| 450 | 237 | 475 | 856 | 1,480 | 2,240 | 4,020 |
| 500 | 224 | 448 | 809 | 1,400 | 2,120 | 3,800 |
| 550 | 213 | 426 | 768 | 1,330 | 2,010 | 3,610 |
| 600 | 203 | 406 | 733 | 1,270 | 1,920 | 3,440 |
| 650 | 194 | 389 | 702 | 1,220 | 1,840 | 3,300 |
| 700 | 187 | 374 | 674 | 1,170 | 1,760 | 3,170 |
| 750 | 180 | 360 | 649 | 1,130 | 1,700 | 3,050 |
| 800 | 174 | 348 | 627 | 1,090 | 1,640 | 2,950 |
| 850 | 168 | 336 | 607 | 1,050 | 1,590 | 2,850 |
| 900 | 163 | 326 | 588 | 1,020 | 1,540 | 2,770 |
| 950 | 158 | 317 | 572 | 990 | 1,500 | 2,690 |
| 1,000 | 154 | 308 | 556 | 963 | 1,450 | 2,610 |
| 1,100 | 146 | 293 | 528 | 915 | 1,380 | 2,480 |
| 1,200 | 139 | 279 | 504 | 873 | 1,320 | 2,370 |
| 1,300 | 134 | 267 | 482 | 836 | 1,260 | 2,270 |
| 1,400 | 128 | 257 | 463 | 803 | 1,210 | 2,180 |
| 1,500 | 124 | 247 | 446 | 773 | 1,170 | 2,100 |
| 1,600 | 119 | 239 | 431 | 747 | 1,130 | 2,030 |
| 1,700 | 115 | 231 | 417 | 723 | 1,090 | 1,960 |
| 1,800 | 112 | 224 | 404 | 701 | 1,060 | 1,900 |
| 1,900 | 109 | 218 | 393 | 680 | 1,030 | 1,850 |
| 2,000 | 106 | 212 | 382 | 662 | 1,000 | 1,800 |

Note: All table entries are rounded to 3 significant digits.

***TABLE 6.2(u)*** *Polyethylene Plastic Tubing*

| | | |
|---|---|---|
| | *Gas:* | *Natural* |
| | *Inlet Pressure:* | *Less than 2.0 psi* |
| | *Pressure Drop:* | *0.3 in. w.c.* |
| | *Specific Gravity:* | *0.60* |
| | *Plastic Tubing Size (CTS) (in.)* | |
| *Nominal OD:* | ½ | ¾ |
| *Designation:* | *SDR 7.00* | *SDR 11.00* |
| *Actual ID:* | *0.445* | *0.927* |
| *Length (ft)* | *Capacity in Cubic Feet of Gas per Hour* | |
| 10 | 54 | 372 |
| 20 | 37 | 256 |
| 30 | 30 | 205 |
| 40 | 26 | 176 |
| 50 | 23 | 156 |
| 60 | 21 | 141 |
| 70 | 19 | 130 |
| 80 | 18 | 121 |
| 90 | 17 | 113 |
| 100 | 16 | 107 |
| 125 | 14 | 95 |
| 150 | 13 | 86 |
| 175 | 12 | 79 |
| 200 | 11 | 74 |
| 225 | 10 | 69 |
| 250 | NA | 65 |
| 275 | NA | 62 |
| 300 | NA | 59 |
| 350 | NA | 54 |
| 400 | NA | 51 |
| 450 | NA | 47 |
| 500 | NA | 45 |

CTS: Copper tube size.
NA: A flow of less than 10 cfh.
Note: All table entries are rounded to 3 significant digits.

***TABLE 6.2(v)*** *Polyethylene Plastic Tubing*

| | | |
|---|---|---|
| | *Gas:* | *Natural* |
| | *Inlet Pressure:* | *Less than 2.0 psi* |
| | *Pressure Drop:* | *0.5 in. w.c.* |
| | *Specific Gravity:* | *0.60* |
| | *Plastic Tubing Size (CTS) (in.)* | |
| *Nominal OD:* | ½ | ¾ |
| *Designation:* | *SDR 7.00* | *SDR 11.00* |
| *Actual ID:* | *0.445* | *0.927* |
| *Length (ft)* | *Capacity in Cubic Feet of Gas per Hour* | |
| 10 | 72 | 490 |
| 20 | 49 | 337 |
| 30 | 39 | 271 |
| 40 | 34 | 232 |
| 50 | 30 | 205 |
| 60 | 27 | 186 |
| 70 | 25 | 171 |
| 80 | 23 | 159 |
| 90 | 22 | 149 |
| 100 | 21 | 141 |
| 125 | 18 | 125 |
| 150 | 17 | 113 |
| 175 | 15 | 104 |
| 200 | 14 | 97 |
| 225 | 13 | 91 |
| 250 | 12 | 86 |
| 275 | 11 | 82 |
| 300 | 11 | 78 |
| 350 | 10 | 72 |
| 400 | NA | 67 |
| 450 | NA | 63 |
| 500 | NA | 59 |

CTS: Copper tube size.
NA: A flow of less than 10 cfh.
Note: All table entries are rounded to 3 significant digits.

***COMMENTARY TABLE 6.1*** *Schedule 40 Metallic Pipe (10 psi)*

| | |
|---|---|
| ***Gas:*** | ***Natural*** |
| ***Inlet Pressure:*** | ***10.0 psi*** |
| ***Pressure Drop:*** | ***10%*** |
| ***Specific Gravity:*** | ***0.60*** |

| | ***Pipe Size (in.)*** | | | | | | | | | | | | |
|---|---|---|---|---|---|---|---|---|---|---|---|---|---|
| ***Nominal:*** | ***1*** | ***1 1/4*** | ***1 1/2*** | ***2*** | ***2 1/2*** | ***3*** | ***3 1/2*** | ***4*** | ***5*** | ***6*** | ***8*** | ***10*** | ***12*** |
| ***Actual ID:*** | ***1.049*** | ***1.380*** | ***1.610*** | ***2.067*** | ***2.469*** | ***3.068*** | ***3.548*** | ***4.026*** | ***5.047*** | ***6.065*** | ***7.981*** | ***10.020*** | ***11.938*** |
| ***Length (ft)*** | ***Capacity in Cubic Feet of Gas per Hour*** | | | | | | | | | | | | |
| 50 | 3,259 | 6,690 | 10,024 | 19,305 | 30,769 | 54,395 | 79,642 | 110,948 | 200,720 | 325,013 | 667,777 | 1,212,861 | 1,920,112 |
| 100 | 2,240 | 4,598 | 6,889 | 13,268 | 21,148 | 37,385 | 54,737 | 76,254 | 137,954 | 223,379 | 458,959 | 833,593 | 1,319,682 |
| 150 | 1,798 | 3,692 | 5,532 | 10,655 | 16,982 | 30,022 | 43,956 | 61,235 | 110,782 | 179,382 | 368,561 | 669,404 | 1,059,751 |
| 200 | 1,539 | 3,160 | 4,735 | 9,119 | 14,535 | 25,695 | 37,621 | 52,409 | 94,815 | 153,527 | 315,440 | 572,924 | 907,010 |
| 250 | 1,364 | 2,801 | 4,197 | 8,082 | 12,882 | 22,773 | 33,343 | 46,449 | 84,033 | 136,068 | 279,569 | 507,772 | 803,866 |
| 300 | 1,236 | 2,538 | 3,802 | 7,323 | 11,672 | 20,634 | 30,211 | 42,086 | 76,140 | 123,288 | 253,310 | 460,078 | 728,361 |
| 400 | 1,058 | 2,172 | 3,254 | 6,268 | 9,990 | 17,660 | 25,857 | 36,020 | 65,166 | 105,518 | 216,800 | 393,767 | 623,383 |
| 500 | 938 | 1,925 | 2,884 | 5,555 | 8,854 | 15,652 | 22,916 | 31,924 | 57,755 | 93,519 | 192,146 | 348,988 | 552,493 |
| 1,000 | 644 | 1,323 | 1,982 | 3,818 | 6,085 | 10,757 | 15,750 | 21,941 | 39,695 | 64,275 | 132,061 | 239,858 | 379,725 |
| 1,500 | 517 | 1,062 | 1,592 | 3,066 | 4,886 | 8,638 | 12,648 | 17,620 | 31,876 | 51,615 | 106,050 | 192,614 | 304,933 |
| 2,000 | 443 | 909 | 1,362 | 2,624 | 4,182 | 7,393 | 10,825 | 15,080 | 27,282 | 44,176 | 90,765 | 164,853 | 260,983 |

***COMMENTARY TABLE 6.2*** *Schedule 40 Metallic Pipe (20 psi)*

| | |
|---|---|
| ***Gas:*** | ***Natural*** |
| ***Inlet Pressure:*** | ***20.0 psi*** |
| ***Pressure Drop:*** | ***10%*** |
| ***Specific Gravity:*** | ***0.60*** |

| | ***Pipe Size (in.)*** | | | | | | | | | | | | |
|---|---|---|---|---|---|---|---|---|---|---|---|---|---|
| ***Nominal:*** | ***1*** | ***1 1/4*** | ***1 1/2*** | ***2*** | ***2 1/2*** | ***3*** | ***3 1/2*** | ***4*** | ***5*** | ***6*** | ***8*** | ***10*** | ***12*** |
| ***Actual ID:*** | ***1.049*** | ***1.380*** | ***1.610*** | ***2.067*** | ***2.469*** | ***3.068*** | ***3.548*** | ***4.026*** | ***5.047*** | ***6.065*** | ***7.981*** | ***10.020*** | ***11.938*** |
| ***Length (ft)*** | ***Capacity of Pipe in Cubic Feet of Gas per Hour*** | | | | | | | | | | | | |
| 50 | 5,674 | 11,649 | 17,454 | 33,615 | 53,577 | 94,714 | 138,676 | 193,187 | 349,503 | 565,926 | 1,162,762 | 2,111,887 | 3,343,383 |
| 100 | 3,900 | 8,006 | 11,996 | 23,103 | 36,823 | 65,097 | 95,311 | 132,777 | 240,211 | 388,958 | 799,160 | 1,451,488 | 2,297,888 |
| 150 | 3,132 | 6,429 | 9,633 | 18,553 | 29,570 | 52,275 | 76,538 | 106,624 | 192,898 | 312,347 | 641,754 | 1,165,596 | 1,845,285 |
| 200 | 2,680 | 5,503 | 8,245 | 15,879 | 25,308 | 44,741 | 65,507 | 91,257 | 165,096 | 267,329 | 549,258 | 997,600 | 1,579,326 |
| 250 | 2,375 | 4,877 | 7,307 | 14,073 | 22,430 | 39,653 | 58,058 | 80,879 | 146,322 | 236,928 | 486,797 | 884,154 | 1,399,727 |
| 300 | 2,152 | 4,419 | 6,621 | 12,751 | 20,323 | 35,928 | 52,604 | 73,282 | 132,578 | 214,674 | 441,074 | 801,108 | 1,268,254 |
| 400 | 1,842 | 3,782 | 5,667 | 10,913 | 17,394 | 30,750 | 45,023 | 62,720 | 113,470 | 183,733 | 377,502 | 685,645 | 1,085,462 |
| 500 | 1,633 | 3,352 | 5,022 | 9,672 | 15,416 | 27,253 | 39,903 | 55,538 | 100,566 | 162,840 | 334,573 | 607,674 | 962,025 |
| 1,000 | 1,122 | 2,304 | 3,452 | 6,648 | 10,595 | 18,731 | 27,425 | 38,205 | 69,118 | 111,919 | 229,950 | 417,651 | 661,194 |
| 1,500 | 901 | 1,850 | 2,772 | 5,338 | 8,509 | 15,042 | 22,023 | 30,680 | 55,505 | 89,875 | 184,658 | 335,388 | 530,962 |
| 2,000 | 771 | 1,583 | 2,372 | 4,569 | 7,282 | 12,874 | 18,849 | 26,258 | 47,505 | 76,921 | 158,043 | 287,049 | 454,435 |

***COMMENTARY TABLE 6.3** Schedule 40 Metallic Pipe (50 psi)*

| | | | | | | | | | | | | |
|---|---|---|---|---|---|---|---|---|---|---|---|---|
| | | | | | | | | | | **Gas:** | **Natural** | |
| | | | | | | | | | | **Inlet Pressure:** | **50.0 psi** | |
| | | | | | | | | | | **Pressure Drop:** | **10%** | |
| | | | | | | | | | | **Specific Gravity:** | **0.60** | |

| | **Pipe Size (in.)** | | | | | | | | | | | | |
|---|---|---|---|---|---|---|---|---|---|---|---|---|---|
| **Nominal:** | 1 | 1¼ | 1½ | 2 | 2½ | 3 | 3½ | 4 | 5 | 6 | 8 | 10 | 12 |
| **Actual ID:** | 1.049 | 1.380 | 1.610 | 2.067 | 2.469 | 3.068 | 3.548 | 4.026 | 5.047 | 6.065 | 7.981 | 10.020 | 11.938 |
| **Length (ft)** | **Capacity of Pipe in Cubic Feet of Gas per Hour** | | | | | | | | | | | | |
| 50 | 12,993 | 26,676 | 39,970 | 76,977 | 122,690 | 216,893 | 317,564 | 442,393 | 800,352 | 1,295,955 | 2,662,693 | 4,836,161 | 7,656,252 |
| 100 | 8,930 | 18,335 | 27,471 | 52,906 | 84,324 | 149,070 | 218,260 | 304,054 | 550,077 | 890,703 | 1,830,054 | 3,323,866 | 5,262,099 |
| 150 | 7,171 | 14,723 | 22,060 | 42,485 | 67,715 | 119,708 | 175,271 | 244,166 | 441,732 | 715,266 | 1,469,598 | 2,669,182 | 4,225,651 |
| 200 | 6,138 | 12,601 | 18,881 | 36,362 | 57,955 | 102,455 | 150,009 | 208,975 | 378,065 | 612,175 | 1,257,785 | 2,284,474 | 3,616,611 |
| 250 | 5,440 | 11,168 | 16,733 | 32,227 | 51,365 | 90,804 | 132,950 | 185,211 | 335,072 | 542,559 | 1,114,752 | 2,024,687 | 3,205,335 |
| 300 | 4,929 | 10,119 | 15,162 | 29,200 | 46,540 | 82,275 | 120,463 | 167,814 | 303,600 | 491,598 | 1,010,046 | 1,834,514 | 2,904,266 |
| 400 | 4,218 | 8,661 | 12,976 | 24,991 | 39,832 | 70,417 | 103,100 | 143,627 | 259,842 | 420,744 | 864,469 | 1,570,106 | 2,485,676 |
| 500 | 3,739 | 7,676 | 11,501 | 22,149 | 35,303 | 62,409 | 91,376 | 127,294 | 230,293 | 372,898 | 766,163 | 1,391,556 | 2,203,009 |
| 1,000 | 2,570 | 5,276 | 7,904 | 15,223 | 24,263 | 42,893 | 62,802 | 87,489 | 158,279 | 256,291 | 526,579 | 956,409 | 1,514,115 |
| 1,500 | 2,063 | 4,236 | 6,348 | 12,225 | 19,484 | 34,445 | 50,432 | 70,256 | 127,104 | 205,810 | 422,862 | 768,030 | 1,215,888 |
| 2,000 | 1,766 | 3,626 | 5,433 | 10,463 | 16,676 | 29,480 | 43,164 | 60,130 | 108,784 | 176,147 | 361,915 | 657,334 | 1,040,643 |

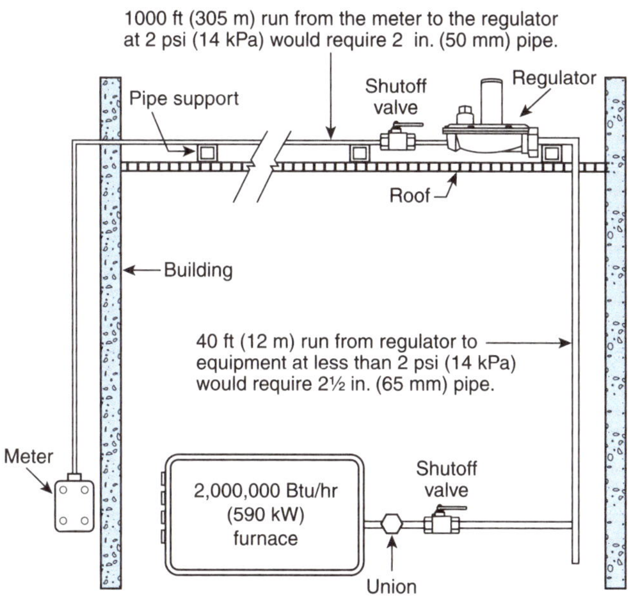


***EXHIBIT 6.1** Advantages of the Hybrid Pressure Method.*

## 6.3 Tables for Sizing Gas Piping Systems Using Propane

Table 6.3(a) through Table 6.3(m) shall be used to size gas piping in conjunction with one of the methods described in 6.1.1 through 6.1.3.

Examples of the use of the pipe sizing tables in Section 6.3 are found in Annex C of the code. Calculation worksheets that can be copied and used are included in Supplement 6 of this handbook.

*TABLE 6.3(a) Schedule 40 Metallic Pipe*

| | |
|---|---|
| Gas: | Undiluted Propane |
| Inlet Pressure: | 10.0 psi |
| Pressure Drop: | 1.0 psi |
| Specific Gravity: | 1.50 |

NEW

**INTENDED USE: Pipe Sizing Between First-Stage (High-Pressure) Regulator and Second-Stage (Low-Pressure) Regulator.**

| | Pipe Size (in.) | | | | | | | | |
|---|---|---|---|---|---|---|---|---|---|
| Nominal Inside: | ½ | ¾ | 1 | 1¼ | 1½ | 2 | 2 ½ | 3 | 4 |
| Actual: | 0.622 | 0.824 | 1.049 | 1.380 | 1.610 | 2.067 | 2.469 | 3.068 | 4.026 |
| Length (ft) | Capacity in Thousands of Btu per Hour | | | | | | | | |
| 10 | 3,320 | 6,950 | 13,100 | 26,900 | 40,300 | 77,600 | 124,000 | 219,000 | 446,000 |
| 20 | 2,280 | 4,780 | 9,000 | 18,500 | 27,700 | 53,300 | 85,000 | 150,000 | 306,000 |
| 30 | 1,830 | 3,840 | 7,220 | 14,800 | 22,200 | 42,800 | 68,200 | 121,000 | 246,000 |
| 40 | 1,570 | 3,280 | 6,180 | 12,700 | 19,000 | 36,600 | 58,400 | 103,000 | 211,000 |
| 50 | 1,390 | 2,910 | 5,480 | 11,300 | 16,900 | 32,500 | 51,700 | 91,500 | 187,000 |
| 60 | 1,260 | 2,640 | 4,970 | 10,200 | 15,300 | 29,400 | 46,900 | 82,900 | 169,000 |
| 70 | 1,160 | 2,430 | 4,570 | 9,380 | 14,100 | 27,100 | 43,100 | 76,300 | 156,000 |
| 80 | 1,080 | 2,260 | 4,250 | 8,730 | 13,100 | 25,200 | 40,100 | 70,900 | 145,000 |
| 90 | 1,010 | 2,120 | 3,990 | 8,190 | 12,300 | 23,600 | 37,700 | 66,600 | 136,000 |
| 100 | 956 | 2,000 | 3,770 | 7,730 | 11,600 | 22,300 | 35,600 | 62,900 | 128,000 |
| 125 | 848 | 1,770 | 3,340 | 6,850 | 10,300 | 19,800 | 31,500 | 55,700 | 114,000 |
| 150 | 768 | 1,610 | 3,020 | 6,210 | 9,300 | 17,900 | 28,600 | 50,500 | 103,000 |
| 175 | 706 | 1,480 | 2,780 | 5,710 | 8,560 | 16,500 | 26,300 | 46,500 | 94,700 |
| 200 | 657 | 1,370 | 2,590 | 5,320 | 7,960 | 15,300 | 24,400 | 43,200 | 88,100 |
| 250 | 582 | 1,220 | 2,290 | 4,710 | 7,060 | 13,600 | 21,700 | 38,300 | 78,100 |
| 300 | 528 | 1,100 | 2,080 | 4,270 | 6,400 | 12,300 | 19,600 | 34,700 | 70,800 |
| 350 | 486 | 1,020 | 1,910 | 3,930 | 5,880 | 11,300 | 18,100 | 31,900 | 65,100 |
| 400 | 452 | 945 | 1,780 | 3,650 | 5,470 | 10,500 | 16,800 | 29,700 | 60,600 |
| 450 | 424 | 886 | 1,670 | 3,430 | 5,140 | 9,890 | 15,800 | 27,900 | 56,800 |
| 500 | 400 | 837 | 1,580 | 3,240 | 4,850 | 9,340 | 14,900 | 26,300 | 53,700 |
| 550 | 380 | 795 | 1,500 | 3,070 | 4,610 | 8,870 | 14,100 | 25,000 | 51,000 |
| 600 | 363 | 759 | 1,430 | 2,930 | 4,400 | 8,460 | 13,500 | 23,900 | 48,600 |
| 650 | 347 | 726 | 1,370 | 2,810 | 4,210 | 8,110 | 12,900 | 22,800 | 46,600 |
| 700 | 334 | 698 | 1,310 | 2,700 | 4,040 | 7,790 | 12,400 | 21,900 | 44,800 |
| 750 | 321 | 672 | 1,270 | 2,600 | 3,900 | 7,500 | 12,000 | 21,100 | 43,100 |
| 800 | 310 | 649 | 1,220 | 2,510 | 3,760 | 7,240 | 11,500 | 20,400 | 41,600 |
| 850 | 300 | 628 | 1,180 | 2,430 | 3,640 | 7,010 | 11,200 | 19,800 | 40,300 |
| 900 | 291 | 609 | 1,150 | 2,360 | 3,530 | 6,800 | 10,800 | 19,200 | 39,100 |
| 950 | 283 | 592 | 1,110 | 2,290 | 3,430 | 6,600 | 10,500 | 18,600 | 37,900 |
| 1,000 | 275 | 575 | 1,080 | 2,230 | 3,330 | 6,420 | 10,200 | 18,100 | 36,900 |
| 1,100 | 261 | 546 | 1,030 | 2,110 | 3,170 | 6,100 | 9,720 | 17,200 | 35,000 |
| 1,200 | 249 | 521 | 982 | 2,020 | 3,020 | 5,820 | 9,270 | 16,400 | 33,400 |
| 1,300 | 239 | 499 | 940 | 1,930 | 2,890 | 5,570 | 8,880 | 15,700 | 32,000 |
| 1,400 | 229 | 480 | 903 | 1,850 | 2,780 | 5,350 | 8,530 | 15,100 | 30,800 |
| 1,500 | 221 | 462 | 870 | 1,790 | 2,680 | 5,160 | 8,220 | 14,500 | 29,600 |
| 1,600 | 213 | 446 | 840 | 1,730 | 2,590 | 4,980 | 7,940 | 14,000 | 28,600 |
| 1,700 | 206 | 432 | 813 | 1,670 | 2,500 | 4,820 | 7,680 | 13,600 | 27,700 |
| 1,800 | 200 | 419 | 789 | 1,620 | 2,430 | 4,670 | 7,450 | 13,200 | 26,900 |
| 1,900 | 194 | 407 | 766 | 1,570 | 2,360 | 4,540 | 7,230 | 12,800 | 26,100 |
| 2,000 | 189 | 395 | 745 | 1,530 | 2,290 | 4,410 | 7,030 | 12,400 | 25,400 |

Note: All table entries are rounded to 3 significant digits.

*TABLE 6.3(b) Schedule 40 Metallic Pipe*

| | |
|---|---|
| *Gas:* | *Undiluted Propane* |
| *Inlet Pressure:* | *10.0 psi* |
| *Pressure Drop:* | *3.0 psi* |
| *Specific Gravity:* | *1.50* |

*INTENDED USE: Pipe Sizing Between First-Stage (High-Pressure) Regulator and Second-Stage (Low-Pressure) Regulator.*

| | *Pipe Size (in.)* | | | | | | | | |
|---|---|---|---|---|---|---|---|---|---|
| *Nominal Inside:* | ½ | ¾ | 1 | 1¼ | 1½ | 2 | 2 ½ | 3 | 4 |
| *Actual:* | 0.622 | 0.824 | 1.049 | 1.380 | 1.610 | 2.067 | 2.469 | 3.068 | 4.026 |
| *Length (ft)* | *Capacity in Thousands of Btu per Hour* | | | | | | | | |
| 10 | 5,890 | 12,300 | 23,200 | 47,600 | 71,300 | 137,000 | 219,000 | 387,000 | 789,000 |
| 20 | 4,050 | 8,460 | 15,900 | 32,700 | 49,000 | 94,400 | 150,000 | 266,000 | 543,000 |
| 30 | 3,250 | 6,790 | 12,800 | 26,300 | 39,400 | 75,800 | 121,000 | 214,000 | 436,000 |
| 40 | 2,780 | 5,810 | 11,000 | 22,500 | 33,700 | 64,900 | 103,000 | 183,000 | 373,000 |
| 50 | 2,460 | 5,150 | 9,710 | 19,900 | 29,900 | 57,500 | 91,600 | 162,000 | 330,000 |
| 60 | 2,230 | 4,670 | 8,790 | 18,100 | 27,100 | 52,100 | 83,000 | 147,000 | 299,000 |
| 70 | 2,050 | 4,300 | 8,090 | 16,600 | 24,900 | 47,900 | 76,400 | 135,000 | 275,000 |
| 80 | 1,910 | 4,000 | 7,530 | 15,500 | 23,200 | 44,600 | 71,100 | 126,000 | 256,000 |
| 90 | 1,790 | 3,750 | 7,060 | 14,500 | 21,700 | 41,800 | 66,700 | 118,000 | 240,000 |
| 100 | 1,690 | 3,540 | 6,670 | 13,700 | 20,500 | 39,500 | 63,000 | 111,000 | 227,000 |
| 125 | 1,500 | 3,140 | 5,910 | 12,100 | 18,200 | 35,000 | 55,800 | 98,700 | 201,000 |
| 150 | 1,360 | 2,840 | 5,360 | 11,000 | 16,500 | 31,700 | 50,600 | 89,400 | 182,000 |
| 175 | 1,250 | 2,620 | 4,930 | 10,100 | 15,200 | 29,200 | 46,500 | 82,300 | 167,800 |
| 200 | 1,160 | 2,430 | 4,580 | 9,410 | 14,100 | 27,200 | 43,300 | 76,500 | 156,100 |
| 250 | 1,030 | 2,160 | 4,060 | 8,340 | 12,500 | 24,100 | 38,400 | 67,800 | 138,400 |
| 300 | 935 | 1,950 | 3,680 | 7,560 | 11,300 | 21,800 | 34,800 | 61,500 | 125,400 |
| 350 | 860 | 1,800 | 3,390 | 6,950 | 10,400 | 20,100 | 32,000 | 56,500 | 115,300 |
| 400 | 800 | 1,670 | 3,150 | 6,470 | 9,690 | 18,700 | 29,800 | 52,600 | 107,300 |
| 450 | 751 | 1,570 | 2,960 | 6,070 | 9,090 | 17,500 | 27,900 | 49,400 | 100,700 |
| 500 | 709 | 1,480 | 2,790 | 5,730 | 8,590 | 16,500 | 26,400 | 46,600 | 95,100 |
| 550 | 673 | 1,410 | 2,650 | 5,450 | 8,160 | 15,700 | 25,000 | 44,300 | 90,300 |
| 600 | 642 | 1,340 | 2,530 | 5,200 | 7,780 | 15,000 | 23,900 | 42,200 | 86,200 |
| 650 | 615 | 1,290 | 2,420 | 4,980 | 7,450 | 14,400 | 22,900 | 40,500 | 82,500 |
| 700 | 591 | 1,240 | 2,330 | 4,780 | 7,160 | 13,800 | 22,000 | 38,900 | 79,300 |
| 750 | 569 | 1,190 | 2,240 | 4,600 | 6,900 | 13,300 | 21,200 | 37,400 | 76,400 |
| 800 | 550 | 1,150 | 2,170 | 4,450 | 6,660 | 12,800 | 20,500 | 36,200 | 73,700 |
| 850 | 532 | 1,110 | 2,100 | 4,300 | 6,450 | 12,400 | 19,800 | 35,000 | 71,400 |
| 900 | 516 | 1,080 | 2,030 | 4,170 | 6,250 | 12,000 | 19,200 | 33,900 | 69,200 |
| 950 | 501 | 1,050 | 1,970 | 4,050 | 6,070 | 11,700 | 18,600 | 32,900 | 67,200 |
| 1,000 | 487 | 1,020 | 1,920 | 3,940 | 5,900 | 11,400 | 18,100 | 32,000 | 65,400 |
| 1,100 | 463 | 968 | 1,820 | 3,740 | 5,610 | 10,800 | 17,200 | 30,400 | 62,100 |
| 1,200 | 442 | 923 | 1,740 | 3,570 | 5,350 | 10,300 | 16,400 | 29,000 | 59,200 |
| 1,300 | 423 | 884 | 1,670 | 3,420 | 5,120 | 9,870 | 15,700 | 27,800 | 56,700 |
| 1,400 | 406 | 849 | 1,600 | 3,280 | 4,920 | 9,480 | 15,100 | 26,700 | 54,500 |
| 1,500 | 391 | 818 | 1,540 | 3,160 | 4,740 | 9,130 | 14,600 | 25,700 | 52,500 |
| 1,600 | 378 | 790 | 1,490 | 3,060 | 4,580 | 8,820 | 14,100 | 24,800 | 50,700 |
| 1,700 | 366 | 765 | 1,440 | 2,960 | 4,430 | 8,530 | 13,600 | 24,000 | 49,000 |
| 1,800 | 355 | 741 | 1,400 | 2,870 | 4,300 | 8,270 | 13,200 | 23,300 | 47,600 |
| 1,900 | 344 | 720 | 1,360 | 2,780 | 4,170 | 8,040 | 12,800 | 22,600 | 46,200 |
| 2,000 | 335 | 700 | 1,320 | 2,710 | 4,060 | 7,820 | 12,500 | 22,000 | 44,900 |

Note: All table entries are rounded to 3 significant digits.

*TABLE 6.3(c) Schedule 40 Metallic Pipe*

| | | | | | | | | | |
|---|---|---|---|---|---|---|---|---|---|
| | | | | | | *Gas:* | | *Undiluted Propane* | |
| | | | | | | *Inlet Pressure:* | | *2.0 psi* | |
| | | | | | | *Pressure Drop:* | | *1.0 psi* | |
| | | | | | | *Specific Gravity:* | | *1.50* | |
| *INTENDED USE: Pipe Sizing Between 2 psig Service and Line Pressure Regulator.* | | | | | | | | | |
| | *Pipe Size (in.)* | | | | | | | | |
| *Nominal:* | ½ | ¾ | 1 | 1¼ | 1½ | 2 | 2 ½ | 3 | 4 |
| *Actual ID:* | 0.622 | 0.824 | 1.049 | 1.380 | 1.610 | 2.067 | 2.469 | 3.068 | 4.026 |
| *Length (ft)* | *Capacity in Thousands of Btu per Hour* | | | | | | | | |
| 10 | 2,680 | 5,590 | 10,500 | 21,600 | 32,400 | 62,400 | 99,500 | 176,000 | 359,000 |
| 20 | 1,840 | 3,850 | 7,240 | 14,900 | 22,300 | 42,900 | 68,400 | 121,000 | 247,000 |
| 30 | 1,480 | 3,090 | 5,820 | 11,900 | 17,900 | 34,500 | 54,900 | 97,100 | 198,000 |
| 40 | 1,260 | 2,640 | 4,980 | 10,200 | 15,300 | 29,500 | 47,000 | 83,100 | 170,000 |
| 50 | 1,120 | 2,340 | 4,410 | 9,060 | 13,600 | 26,100 | 41,700 | 73,700 | 150,000 |
| 60 | 1,010 | 2,120 | 4,000 | 8,210 | 12,300 | 23,700 | 37,700 | 66,700 | 136,000 |
| 70 | 934 | 1,950 | 3,680 | 7,550 | 11,300 | 21,800 | 34,700 | 61,400 | 125,000 |
| 80 | 869 | 1,820 | 3,420 | 7,020 | 10,500 | 20,300 | 32,300 | 57,100 | 116,000 |
| 90 | 815 | 1,700 | 3,210 | 6,590 | 9,880 | 19,000 | 30,300 | 53,600 | 109,000 |
| 100 | 770 | 1,610 | 3,030 | 6,230 | 9,330 | 18,000 | 28,600 | 50,600 | 103,000 |
| 125 | 682 | 1,430 | 2,690 | 5,520 | 8,270 | 15,900 | 25,400 | 44,900 | 91,500 |
| 150 | 618 | 1,290 | 2,440 | 5,000 | 7,490 | 14,400 | 23,000 | 40,700 | 82,900 |
| 175 | 569 | 1,190 | 2,240 | 4,600 | 6,890 | 13,300 | 21,200 | 37,400 | 76,300 |
| 200 | 529 | 1,110 | 2,080 | 4,280 | 6,410 | 12,300 | 19,700 | 34,800 | 71,000 |
| 250 | 469 | 981 | 1,850 | 3,790 | 5,680 | 10,900 | 17,400 | 30,800 | 62,900 |
| 300 | 425 | 889 | 1,670 | 3,440 | 5,150 | 9,920 | 15,800 | 27,900 | 57,000 |
| 350 | 391 | 817 | 1,540 | 3,160 | 4,740 | 9,120 | 14,500 | 25,700 | 52,400 |
| 400 | 364 | 760 | 1,430 | 2,940 | 4,410 | 8,490 | 13,500 | 23,900 | 48,800 |
| 450 | 341 | 714 | 1,340 | 2,760 | 4,130 | 7,960 | 12,700 | 22,400 | 45,800 |
| 500 | 322 | 674 | 1,270 | 2,610 | 3,910 | 7,520 | 12,000 | 21,200 | 43,200 |
| 550 | 306 | 640 | 1,210 | 2,480 | 3,710 | 7,140 | 11,400 | 20,100 | 41,100 |
| 600 | 292 | 611 | 1,150 | 2,360 | 3,540 | 6,820 | 10,900 | 19,200 | 39,200 |
| 650 | 280 | 585 | 1,100 | 2,260 | 3,390 | 6,530 | 10,400 | 18,400 | 37,500 |
| 700 | 269 | 562 | 1,060 | 2,170 | 3,260 | 6,270 | 9,990 | 17,700 | 36,000 |
| 750 | 259 | 541 | 1,020 | 2,090 | 3,140 | 6,040 | 9,630 | 17,000 | 34,700 |
| 800 | 250 | 523 | 985 | 2,020 | 3,030 | 5,830 | 9,300 | 16,400 | 33,500 |
| 850 | 242 | 506 | 953 | 1,960 | 2,930 | 5,640 | 9,000 | 15,900 | 32,400 |
| 900 | 235 | 490 | 924 | 1,900 | 2,840 | 5,470 | 8,720 | 15,400 | 31,500 |
| 950 | 228 | 476 | 897 | 1,840 | 2,760 | 5,310 | 8,470 | 15,000 | 30,500 |
| 1,000 | 222 | 463 | 873 | 1,790 | 2,680 | 5,170 | 8,240 | 14,600 | 29,700 |
| 1,100 | 210 | 440 | 829 | 1,700 | 2,550 | 4,910 | 7,830 | 13,800 | 28,200 |
| 1,200 | 201 | 420 | 791 | 1,620 | 2,430 | 4,680 | 7,470 | 13,200 | 26,900 |
| 1,300 | 192 | 402 | 757 | 1,550 | 2,330 | 4,490 | 7,150 | 12,600 | 25,800 |
| 1,400 | 185 | 386 | 727 | 1,490 | 2,240 | 4,310 | 6,870 | 12,100 | 24,800 |
| 1,500 | 178 | 372 | 701 | 1,440 | 2,160 | 4,150 | 6,620 | 11,700 | 23,900 |
| 1,600 | 172 | 359 | 677 | 1,390 | 2,080 | 4,010 | 6,390 | 11,300 | 23,000 |
| 1,700 | 166 | 348 | 655 | 1,340 | 2,010 | 3,880 | 6,180 | 10,900 | 22,300 |
| 1,800 | 161 | 337 | 635 | 1,300 | 1,950 | 3,760 | 6,000 | 10,600 | 21,600 |
| 1,900 | 157 | 327 | 617 | 1,270 | 1,900 | 3,650 | 5,820 | 10,300 | 21,000 |
| 2,000 | 152 | 318 | 600 | 1,230 | 1,840 | 3,550 | 5,660 | 10,000 | 20,400 |

Note: All table entries are rounded to 3 significant digits.

NEW

*TABLE 6.3(d) Schedule 40 Metallic Pipe*

| | | | | | | | | | |
|---|---|---|---|---|---|---|---|---|---|
| | | | | | | *Gas:* | *Undiluted Propane* | | |
| | | | | | | *Inlet Pressure:* | *11.0 in. w.c.* | | |
| | | | | | | *Pressure Drop:* | *0.5 in. w.c.* | | |
| | | | | | | *Specific Gravity:* | *1.50* | | |
| *INTENDED USE: Pipe Sizing Between Single- or Second-Stage (Low-Pressure) Regulator and Appliance.* | | | | | | | | | |
| | *Pipe Size (in.)* | | | | | | | | |
| *Nominal Inside:* | ½ | ¾ | 1 | 1¼ | 1½ | 2 | 2 ½ | 3 | 4 |
| *Actual:* | 0.622 | 0.824 | 1.049 | 1.380 | 1.610 | 2.067 | 2.469 | 3.068 | 4.026 |
| *Length (ft)* | *Capacity in Thousands of Btu per Hour* | | | | | | | | |
| 10 | 291 | 608 | 1,150 | 2,350 | 3,520 | 6,790 | 10,800 | 19,100 | 39,000 |
| 20 | 200 | 418 | 787 | 1,620 | 2,420 | 4,660 | 7,430 | 13,100 | 26,800 |
| 30 | 160 | 336 | 632 | 1,300 | 1,940 | 3,750 | 5,970 | 10,600 | 21,500 |
| 40 | 137 | 287 | 541 | 1,110 | 1,660 | 3,210 | 5,110 | 9,030 | 18,400 |
| 50 | 122 | 255 | 480 | 985 | 1,480 | 2,840 | 4,530 | 8,000 | 16,300 |
| 60 | 110 | 231 | 434 | 892 | 1,340 | 2,570 | 4,100 | 7,250 | 14,800 |
| 80 | 101 | 212 | 400 | 821 | 1,230 | 2,370 | 3,770 | 6,670 | 13,600 |
| 100 | 94 | 197 | 372 | 763 | 1,140 | 2,200 | 3,510 | 6,210 | 12,700 |
| 125 | 89 | 185 | 349 | 716 | 1,070 | 2,070 | 3,290 | 5,820 | 11,900 |
| 150 | 84 | 175 | 330 | 677 | 1,010 | 1,950 | 3,110 | 5,500 | 11,200 |
| 175 | 74 | 155 | 292 | 600 | 899 | 1,730 | 2,760 | 4,880 | 9,950 |
| 200 | 67 | 140 | 265 | 543 | 814 | 1,570 | 2,500 | 4,420 | 9,010 |
| 250 | 62 | 129 | 243 | 500 | 749 | 1,440 | 2,300 | 4,060 | 8,290 |
| 300 | 58 | 120 | 227 | 465 | 697 | 1,340 | 2,140 | 3,780 | 7,710 |
| 350 | 51 | 107 | 201 | 412 | 618 | 1,190 | 1,900 | 3,350 | 6,840 |
| 400 | 46 | 97 | 182 | 373 | 560 | 1,080 | 1,720 | 3,040 | 6,190 |
| 450 | 42 | 89 | 167 | 344 | 515 | 991 | 1,580 | 2,790 | 5,700 |
| 500 | 40 | 83 | 156 | 320 | 479 | 922 | 1,470 | 2,600 | 5,300 |
| 550 | 37 | 78 | 146 | 300 | 449 | 865 | 1,380 | 2,440 | 4,970 |
| 600 | 35 | 73 | 138 | 283 | 424 | 817 | 1,300 | 2,300 | 4,700 |
| 650 | 33 | 70 | 131 | 269 | 403 | 776 | 1,240 | 2,190 | 4,460 |
| 700 | 32 | 66 | 125 | 257 | 385 | 741 | 1,180 | 2,090 | 4,260 |
| 750 | 30 | 64 | 120 | 246 | 368 | 709 | 1,130 | 2,000 | 4,080 |
| 800 | 29 | 61 | 115 | 236 | 354 | 681 | 1,090 | 1,920 | 3,920 |
| 850 | 28 | 59 | 111 | 227 | 341 | 656 | 1,050 | 1,850 | 3,770 |
| 900 | 27 | 57 | 107 | 220 | 329 | 634 | 1,010 | 1,790 | 3,640 |
| 950 | 26 | 55 | 104 | 213 | 319 | 613 | 978 | 1,730 | 3,530 |
| 1,000 | 25 | 53 | 100 | 206 | 309 | 595 | 948 | 1,680 | 3,420 |
| 1,100 | 25 | 52 | 97 | 200 | 300 | 578 | 921 | 1,630 | 3,320 |
| 1,200 | 24 | 50 | 95 | 195 | 292 | 562 | 895 | 1,580 | 3,230 |
| 1,300 | 23 | 48 | 90 | 185 | 277 | 534 | 850 | 1,500 | 3,070 |
| 1,400 | 22 | 46 | 86 | 176 | 264 | 509 | 811 | 1,430 | 2,930 |
| 1,500 | 21 | 44 | 82 | 169 | 253 | 487 | 777 | 1,370 | 2,800 |
| 1,600 | 20 | 42 | 79 | 162 | 243 | 468 | 746 | 1,320 | 2,690 |
| 1,700 | 19 | 40 | 76 | 156 | 234 | 451 | 719 | 1,270 | 2,590 |
| 1,800 | 19 | 39 | 74 | 151 | 226 | 436 | 694 | 1,230 | 2,500 |
| 1,900 | 18 | 38 | 71 | 146 | 219 | 422 | 672 | 1,190 | 2,420 |
| 2,000 | 18 | 37 | 69 | 142 | 212 | 409 | 652 | 1,150 | 2,350 |

Note: All table entries are rounded to 3 significant digits.

*TABLE 6.3(e) Semirigid Copper Tubing*

| | |
|---|---|
| *Gas:* | *Undiluted Propane* |
| *Inlet Pressure:* | *10.0 psi* |
| *Pressure Drop:* | *1.0 psi* |
| *Specific Gravity:* | *1.50* |

**INTENDED USE: Tube Sizing Between First-Stage (High-Pressure) Regulator and Second-Stage (Low-Pressure) Regulator.**

| | Tube Size (in.) | | | | | | | | |
|---|---|---|---|---|---|---|---|---|---|
| Nominal: K & L: | ¼ | ⅜ | ½ | ⅝ | ¾ | 1 | 1¼ | 1½ | 2 |
| Nominal: ACR: | ⅜ | ½ | ⅝ | ¾ | ⅞ | 1⅛ | 1⅜ | — | — |
| Outside: | 0.375 | 0.500 | 0.625 | 0.750 | 0.875 | 1.125 | 1.375 | 1.625 | 2.125 |
| Inside:* | 0.305 | 0.402 | 0.527 | 0.652 | 0.745 | 0.995 | 1.245 | 1.481 | 1.959 |
| Length (ft) | Capacity in Thousands of Btu per Hour | | | | | | | | |
| 10 | 513 | 1,060 | 2,150 | 3,760 | 5,330 | 11,400 | 20,500 | 32,300 | 67,400 |
| 20 | 352 | 727 | 1,480 | 2,580 | 3,670 | 7,830 | 14,100 | 22,200 | 46,300 |
| 30 | 283 | 584 | 1,190 | 2,080 | 2,940 | 6,290 | 11,300 | 17,900 | 37,200 |
| 40 | 242 | 500 | 1,020 | 1,780 | 2,520 | 5,380 | 9,690 | 15,300 | 31,800 |
| 50 | 215 | 443 | 901 | 1,570 | 2,230 | 4,770 | 8,590 | 13,500 | 28,200 |
| 60 | 194 | 401 | 816 | 1,430 | 2,020 | 4,320 | 7,780 | 12,300 | 25,600 |
| 70 | 179 | 369 | 751 | 1,310 | 1,860 | 3,980 | 7,160 | 11,300 | 23,500 |
| 80 | 166 | 343 | 699 | 1,220 | 1,730 | 3,700 | 6,660 | 10,500 | 21,900 |
| 90 | 156 | 322 | 655 | 1,150 | 1,630 | 3,470 | 6,250 | 9,850 | 20,500 |
| 100 | 147 | 304 | 619 | 1,080 | 1,540 | 3,280 | 5,900 | 9,310 | 19,400 |
| 125 | 131 | 270 | 549 | 959 | 1,360 | 2,910 | 5,230 | 8,250 | 17,200 |
| 150 | 118 | 244 | 497 | 869 | 1,230 | 2,630 | 4,740 | 7,470 | 15,600 |
| 175 | 109 | 225 | 457 | 799 | 1,130 | 2,420 | 4,360 | 6,880 | 14,300 |
| 200 | 101 | 209 | 426 | 744 | 1,060 | 2,250 | 4,060 | 6,400 | 13,300 |
| 250 | 90 | 185 | 377 | 659 | 935 | 2,000 | 3,600 | 5,670 | 11,800 |
| 300 | 81 | 168 | 342 | 597 | 847 | 1,810 | 3,260 | 5,140 | 10,700 |
| 350 | 75 | 155 | 314 | 549 | 779 | 1,660 | 3,000 | 4,730 | 9,840 |
| 400 | 70 | 144 | 292 | 511 | 725 | 1,550 | 2,790 | 4,400 | 9,160 |
| 450 | 65 | 135 | 274 | 480 | 680 | 1,450 | 2,620 | 4,130 | 8,590 |
| 500 | 62 | 127 | 259 | 453 | 643 | 1,370 | 2,470 | 3,900 | 8,120 |
| 550 | 59 | 121 | 246 | 430 | 610 | 1,300 | 2,350 | 3,700 | 7,710 |
| 600 | 56 | 115 | 235 | 410 | 582 | 1,240 | 2,240 | 3,530 | 7,350 |
| 650 | 54 | 111 | 225 | 393 | 558 | 1,190 | 2,140 | 3,380 | 7,040 |
| 700 | 51 | 106 | 216 | 378 | 536 | 1,140 | 2,060 | 3,250 | 6,770 |
| 750 | 50 | 102 | 208 | 364 | 516 | 1,100 | 1,980 | 3,130 | 6,520 |
| 800 | 48 | 99 | 201 | 351 | 498 | 1,060 | 1,920 | 3,020 | 6,290 |
| 850 | 46 | 96 | 195 | 340 | 482 | 1,030 | 1,850 | 2,920 | 6,090 |
| 900 | 45 | 93 | 189 | 330 | 468 | 1,000 | 1,800 | 2,840 | 5,910 |
| 950 | 44 | 90 | 183 | 320 | 454 | 970 | 1,750 | 2,750 | 5,730 |
| 1,000 | 42 | 88 | 178 | 311 | 442 | 944 | 1,700 | 2,680 | 5,580 |
| 1,100 | 40 | 83 | 169 | 296 | 420 | 896 | 1,610 | 2,540 | 5,300 |
| 1,200 | 38 | 79 | 161 | 282 | 400 | 855 | 1,540 | 2,430 | 5,050 |
| 1,300 | 37 | 76 | 155 | 270 | 383 | 819 | 1,470 | 2,320 | 4,840 |
| 1,400 | 35 | 73 | 148 | 260 | 368 | 787 | 1,420 | 2,230 | 4,650 |
| 1,500 | 34 | 70 | 143 | 250 | 355 | 758 | 1,360 | 2,150 | 4,480 |
| 1,600 | 33 | 68 | 138 | 241 | 343 | 732 | 1,320 | 2,080 | 4,330 |
| 1,700 | 32 | 66 | 134 | 234 | 331 | 708 | 1,270 | 2,010 | 4,190 |
| 1,800 | 31 | 64 | 130 | 227 | 321 | 687 | 1,240 | 1,950 | 4,060 |
| 1,900 | 30 | 62 | 126 | 220 | 312 | 667 | 1,200 | 1,890 | 3,940 |
| 2,000 | 29 | 60 | 122 | 214 | 304 | 648 | 1,170 | 1,840 | 3,830 |

Note: All table entries are rounded to 3 significant digits.

*Table capacities are based on Type K copper tubing inside diameter (shown), which has the smallest inside diameter of the copper tubing products.

*TABLE 6.3(f) Semirigid Copper Tubing*

NEW

| | | | | | | | | | |
|---|---|---|---|---|---|---|---|---|---|
| | | | | | Gas: | | Undiluted Propane | | |
| | | | | | Inlet Pressure: | | 11.0 in. w.c. | | |
| | | | | | Pressure Drop: | | 0.5 in. w.c. | | |
| | | | | | Specific Gravity: | | 1.50 | | |
| INTENDED USE: Tube Sizing Between Single- or Second-Stage (Low-Pressure) Regulator and Appliance. | | | | | | | | | |
| | Tube Size (in.) | | | | | | | | |
| Nominal: K & L: | ¼ | ⅜ | ½ | ⅝ | ¾ | 1 | 1¼ | 1½ | 2 |
| Nominal: ACR: | ⅜ | ½ | ⅝ | ¾ | ⅞ | 1⅛ | 1⅜ | — | — |
| Outside: | 0.375 | 0.500 | 0.625 | 0.750 | 0.875 | 1.125 | 1.375 | 1.625 | 2.125 |
| Inside:* | 0.305 | 0.402 | 0.527 | 0.652 | 0.745 | 0.995 | 1.245 | 1.481 | 1.959 |
| Length (ft) | Capacity in Thousands of Btu per Hour | | | | | | | | |
| 10 | 45 | 93 | 188 | 329 | 467 | 997 | 1,800 | 2,830 | 5,890 |
| 20 | 31 | 64 | 129 | 226 | 321 | 685 | 1,230 | 1,950 | 4,050 |
| 30 | 25 | 51 | 104 | 182 | 258 | 550 | 991 | 1,560 | 3,250 |
| 40 | 21 | 44 | 89 | 155 | 220 | 471 | 848 | 1,340 | 2,780 |
| 50 | 19 | 39 | 79 | 138 | 195 | 417 | 752 | 1,180 | 2,470 |
| 60 | 17 | 35 | 71 | 125 | 177 | 378 | 681 | 1,070 | 2,240 |
| 70 | 16 | 32 | 66 | 115 | 163 | 348 | 626 | 988 | 2,060 |
| 80 | 15 | 30 | 61 | 107 | 152 | 324 | 583 | 919 | 1,910 |
| 90 | 14 | 28 | 57 | 100 | 142 | 304 | 547 | 862 | 1,800 |
| 100 | 13 | 27 | 54 | 95 | 134 | 287 | 517 | 814 | 1,700 |
| 125 | 11 | 24 | 48 | 84 | 119 | 254 | 458 | 722 | 1,500 |
| 150 | 10 | 21 | 44 | 76 | 108 | 230 | 415 | 654 | 1,360 |
| 175 | NA | 20 | 40 | 70 | 99 | 212 | 382 | 602 | 1,250 |
| 200 | NA | 18 | 37 | 65 | 92 | 197 | 355 | 560 | 1,170 |
| 250 | NA | 16 | 33 | 58 | 82 | 175 | 315 | 496 | 1,030 |
| 300 | NA | 15 | 30 | 52 | 74 | 158 | 285 | 449 | 936 |
| 350 | NA | 14 | 28 | 48 | 68 | 146 | 262 | 414 | 861 |
| 400 | NA | 13 | 26 | 45 | 63 | 136 | 244 | 385 | 801 |
| 450 | NA | 12 | 24 | 42 | 60 | 127 | 229 | 361 | 752 |
| 500 | NA | 11 | 23 | 40 | 56 | 120 | 216 | 341 | 710 |
| 550 | NA | 11 | 22 | 38 | 53 | 114 | 205 | 324 | 674 |
| 600 | NA | 10 | 21 | 36 | 51 | 109 | 196 | 309 | 643 |
| 650 | NA | NA | 20 | 34 | 49 | 104 | 188 | 296 | 616 |
| 700 | NA | NA | 19 | 33 | 47 | 100 | 180 | 284 | 592 |
| 750 | NA | NA | 18 | 32 | 45 | 96 | 174 | 274 | 570 |
| 800 | NA | NA | 18 | 31 | 44 | 93 | 168 | 264 | 551 |
| 850 | NA | NA | 17 | 30 | 42 | 90 | 162 | 256 | 533 |
| 900 | NA | NA | 17 | 29 | 41 | 87 | 157 | 248 | 517 |
| 950 | NA | NA | 16 | 28 | 40 | 85 | 153 | 241 | 502 |
| 1,000 | NA | NA | 16 | 27 | 39 | 83 | 149 | 234 | 488 |
| 1,100 | NA | NA | 15 | 26 | 37 | 78 | 141 | 223 | 464 |
| 1,200 | NA | NA | 14 | 25 | 35 | 75 | 135 | 212 | 442 |
| 1,300 | NA | NA | 14 | 24 | 34 | 72 | 129 | 203 | 423 |
| 1,400 | NA | NA | 13 | 23 | 32 | 69 | 124 | 195 | 407 |
| 1,500 | NA | NA | 13 | 22 | 31 | 66 | 119 | 188 | 392 |
| 1,600 | NA | NA | 12 | 21 | 30 | 64 | 115 | 182 | 378 |
| 1,700 | NA | NA | 12 | 20 | 29 | 62 | 112 | 176 | 366 |
| 1,800 | NA | NA | 11 | 20 | 28 | 60 | 108 | 170 | 355 |
| 1,900 | NA | NA | 11 | 19 | 27 | 58 | 105 | 166 | 345 |
| 2,000 | NA | NA | 11 | 19 | 27 | 57 | 102 | 161 | 335 |

NA: A flow of less than 10,000 Btu/hr.

Note: All table entries are rounded to 3 significant digits.

*Table capacities are based on Type K copper tubing inside diameter (shown), which has the smallest inside diameter of the copper tubing products.

*TABLE 6.3(g) Semirigid Copper Tubing*

| | |
|---|---|
| *Gas:* | *Undiluted Propane* |
| *Inlet Pressure:* | *2.0 psi* |
| *Pressure Drop:* | *1.0 psi* |
| *Specific Gravity:* | *1.50* |

*INTENDED USE: Tube Sizing Between 2 psig Service and Line Pressure Regulator.*

| | Tube Size (in.) | | | | | | | | |
|---|---|---|---|---|---|---|---|---|---|
| Nominal: K & L: | 1/4 | 3/8 | 1/2 | 5/8 | 3/4 | 1 | 1 1/4 | 1 1/2 | 2 |
| Nominal: ACR: | 3/8 | 1/2 | 5/8 | 3/4 | 7/8 | 1 1/8 | 1 3/8 | — | — |
| Outside: | 0.375 | 0.500 | 0.625 | 0.750 | 0.875 | 1.125 | 1.375 | 1.625 | 2.125 |
| Inside:* | 0.305 | 0.402 | 0.527 | 0.652 | 0.745 | 0.995 | 1.245 | 1.481 | 1.959 |
| Length (ft) | Capacity in Thousands of Btu per Hour | | | | | | | | |
| 10 | 413 | 852 | 1,730 | 3,030 | 4,300 | 9,170 | 16,500 | 26,000 | 54,200 |
| 20 | 284 | 585 | 1,190 | 2,080 | 2,950 | 6,310 | 11,400 | 17,900 | 37,300 |
| 30 | 228 | 470 | 956 | 1,670 | 2,370 | 5,060 | 9,120 | 14,400 | 29,900 |
| 40 | 195 | 402 | 818 | 1,430 | 2,030 | 4,330 | 7,800 | 12,300 | 25,600 |
| 50 | 173 | 356 | 725 | 1,270 | 1,800 | 3,840 | 6,920 | 10,900 | 22,700 |
| 60 | 157 | 323 | 657 | 1,150 | 1,630 | 3,480 | 6,270 | 9,880 | 20,600 |
| 70 | 144 | 297 | 605 | 1,060 | 1,500 | 3,200 | 5,760 | 9,090 | 18,900 |
| 80 | 134 | 276 | 562 | 983 | 1,390 | 2,980 | 5,360 | 8,450 | 17,600 |
| 90 | 126 | 259 | 528 | 922 | 1,310 | 2,790 | 5,030 | 7,930 | 16,500 |
| 100 | 119 | 245 | 498 | 871 | 1,240 | 2,640 | 4,750 | 7,490 | 15,600 |
| 125 | 105 | 217 | 442 | 772 | 1,100 | 2,340 | 4,210 | 6,640 | 13,800 |
| 150 | 95 | 197 | 400 | 700 | 992 | 2,120 | 3,820 | 6,020 | 12,500 |
| 175 | 88 | 181 | 368 | 644 | 913 | 1,950 | 3,510 | 5,540 | 11,500 |
| 200 | 82 | 168 | 343 | 599 | 849 | 1,810 | 3,270 | 5,150 | 10,700 |
| 250 | 72 | 149 | 304 | 531 | 753 | 1,610 | 2,900 | 4,560 | 9,510 |
| 300 | 66 | 135 | 275 | 481 | 682 | 1,460 | 2,620 | 4,140 | 8,610 |
| 350 | 60 | 124 | 253 | 442 | 628 | 1,340 | 2,410 | 3,800 | 7,920 |
| 400 | 56 | 116 | 235 | 411 | 584 | 1,250 | 2,250 | 3,540 | 7,370 |
| 450 | 53 | 109 | 221 | 386 | 548 | 1,170 | 2,110 | 3,320 | 6,920 |
| 500 | 50 | 103 | 209 | 365 | 517 | 1,110 | 1,990 | 3,140 | 6,530 |
| 550 | 47 | 97 | 198 | 346 | 491 | 1,050 | 1,890 | 2,980 | 6,210 |
| 600 | 45 | 93 | 189 | 330 | 469 | 1,000 | 1,800 | 2,840 | 5,920 |
| 650 | 43 | 89 | 181 | 316 | 449 | 959 | 1,730 | 2,720 | 5,670 |
| 700 | 41 | 86 | 174 | 304 | 431 | 921 | 1,660 | 2,620 | 5,450 |
| 750 | 40 | 82 | 168 | 293 | 415 | 888 | 1,600 | 2,520 | 5,250 |
| 800 | 39 | 80 | 162 | 283 | 401 | 857 | 1,540 | 2,430 | 5,070 |
| 850 | 37 | 77 | 157 | 274 | 388 | 829 | 1,490 | 2,350 | 4,900 |
| 900 | 36 | 75 | 152 | 265 | 376 | 804 | 1,450 | 2,280 | 4,750 |
| 950 | 35 | 72 | 147 | 258 | 366 | 781 | 1,410 | 2,220 | 4,620 |
| 1,000 | 34 | 71 | 143 | 251 | 356 | 760 | 1,370 | 2,160 | 4,490 |
| 1,100 | 32 | 67 | 136 | 238 | 338 | 721 | 1,300 | 2,050 | 4,270 |
| 1,200 | 31 | 64 | 130 | 227 | 322 | 688 | 1,240 | 1,950 | 4,070 |
| 1,300 | 30 | 61 | 124 | 217 | 309 | 659 | 1,190 | 1,870 | 3,900 |
| 1,400 | 28 | 59 | 120 | 209 | 296 | 633 | 1,140 | 1,800 | 3,740 |
| 1,500 | 27 | 57 | 115 | 201 | 286 | 610 | 1,100 | 1,730 | 3,610 |
| 1,600 | 26 | 55 | 111 | 194 | 276 | 589 | 1,060 | 1,670 | 3,480 |
| 1,700 | 26 | 53 | 108 | 188 | 267 | 570 | 1,030 | 1,620 | 3,370 |
| 1,800 | 25 | 51 | 104 | 182 | 259 | 553 | 1,000 | 1,570 | 3,270 |
| 1,900 | 24 | 50 | 101 | 177 | 251 | 537 | 966 | 1,520 | 3,170 |
| 2,000 | 23 | 48 | 99 | 172 | 244 | 522 | 940 | 1,480 | 3,090 |

Note: All table entries are rounded to 3 significant digits.

*Table capacities are based on Type K copper tubing inside diameter (shown), which has the smallest inside diameter of the copper tubing products.

*TABLE 6.3(h) Corrugated Stainless Steel Tubing (CSST)*

| | |
|---|---|
| *Gas:* | *Undiluted Propane* |
| *Inlet Pressure:* | *11.0 in. w.c.* |
| *Pressure Drop:* | *0.5 in. w.c.* |
| *Specific Gravity:* | *1.50* |

***INTENDED USE: CSST Sizing Between Single- or Second-Stage (Low-Pressure) Regulator and Appliance Shutoff Valve.***

| | *Tube Size (EHD)* | | | | | | | | | | | | | |
|---|---|---|---|---|---|---|---|---|---|---|---|---|---|---|
| ***Flow Designation:*** | ***13*** | ***15*** | ***18*** | ***19*** | ***23*** | ***25*** | ***30*** | ***31*** | ***37*** | ***39*** | ***46*** | ***48*** | ***60*** | ***62*** |
| ***Length (ft)*** | ***Capacity in Thousands of Btu per Hour*** | | | | | | | | | | | | | |
| 5 | 72 | 99 | 181 | 211 | 355 | 426 | 744 | 863 | 1,420 | 1,638 | 2,830 | 3,270 | 5,780 | 6,550 |
| 10 | 50 | 69 | 129 | 150 | 254 | 303 | 521 | 605 | 971 | 1,179 | 1,990 | 2,320 | 4,110 | 4,640 |
| 15 | 39 | 55 | 104 | 121 | 208 | 248 | 422 | 490 | 775 | 972 | 1,620 | 1,900 | 3,370 | 3,790 |
| 20 | 34 | 49 | 91 | 106 | 183 | 216 | 365 | 425 | 661 | 847 | 1,400 | 1,650 | 2,930 | 3,290 |
| 25 | 30 | 42 | 82 | 94 | 164 | 192 | 325 | 379 | 583 | 762 | 1,250 | 1,480 | 2,630 | 2,940 |
| 30 | 28 | 39 | 74 | 87 | 151 | 177 | 297 | 344 | 528 | 698 | 1,140 | 1,350 | 2,400 | 2,680 |
| 40 | 23 | 33 | 64 | 74 | 131 | 153 | 256 | 297 | 449 | 610 | 988 | 1,170 | 2,090 | 2,330 |
| 50 | 20 | 30 | 58 | 66 | 118 | 137 | 227 | 265 | 397 | 548 | 884 | 1,050 | 1,870 | 2,080 |
| 60 | 19 | 26 | 53 | 60 | 107 | 126 | 207 | 241 | 359 | 502 | 805 | 961 | 1,710 | 1,900 |
| 70 | 17 | 25 | 49 | 57 | 99 | 117 | 191 | 222 | 330 | 466 | 745 | 890 | 1,590 | 1,760 |
| 80 | 15 | 23 | 45 | 52 | 94 | 109 | 178 | 208 | 307 | 438 | 696 | 833 | 1,490 | 1,650 |
| 90 | 15 | 22 | 44 | 50 | 90 | 102 | 169 | 197 | 286 | 414 | 656 | 787 | 1,400 | 1,550 |
| 100 | 14 | 20 | 41 | 47 | 85 | 98 | 159 | 186 | 270 | 393 | 621 | 746 | 1,330 | 1,480 |
| 150 | 11 | 15 | 31 | 36 | 66 | 75 | 123 | 143 | 217 | 324 | 506 | 611 | 1,090 | 1,210 |
| 200 | 9 | 14 | 28 | 33 | 60 | 69 | 112 | 129 | 183 | 283 | 438 | 531 | 948 | 1,050 |
| 250 | 8 | 12 | 25 | 30 | 53 | 61 | 99 | 117 | 163 | 254 | 390 | 476 | 850 | 934 |
| 300 | 8 | 11 | 23 | 26 | 50 | 57 | 90 | 107 | 147 | 234 | 357 | 434 | 777 | 854 |

EHD: Equivalent hydraulic diameter. A measure of the relative hydraulic efficiency between different tubing sizes. The greater the value of EHD, the greater the gas capacity of the tubing.

Notes:

(1) Table includes losses for four 90 degree bends and two end fittings. Tubing runs with larger numbers of bends and/or fittings shall be increased by an equivalent length of tubing to the following equation: $L = 1.3n$, where $L$ is additional length (ft) of tubing and $n$ is the number of additional fittings and/or bends.

(2) All table entries are rounded to 3 significant digits.

*TABLE 6.3(i) Corrugated Stainless Steel Tubing (CSST)*

| Gas: | Undiluted Propane |
|---|---|
| Inlet Pressure: | 2.0 psi |
| Pressure Drop: | 1.0 psi |
| Specific Gravity: | 1.50 |

**INTENDED USE: CSST Sizing Between 2 psig Service and Line Pressure Regulator.**

| | Tube Size (EHD) | | | | | | | | | | | | | |
|---|---|---|---|---|---|---|---|---|---|---|---|---|---|---|
| **Flow Designation:** | **13** | **15** | **18** | **19** | **23** | **25** | **30** | **31** | **37** | **39** | **46** | **48** | **60** | **62** |
| **Length (ft)** | **Capacity in Thousands of Btu per Hour** | | | | | | | | | | | | | |
| 10 | 426 | 558 | 927 | 1,110 | 1,740 | 2,170 | 4,100 | 4,720 | 7,130 | 7,958 | 15,200 | 16,800 | 29,400 | 34,200 |
| 25 | 262 | 347 | 591 | 701 | 1,120 | 1,380 | 2,560 | 2,950 | 4,560 | 5,147 | 9,550 | 10,700 | 18,800 | 21,700 |
| 30 | 238 | 316 | 540 | 640 | 1,030 | 1,270 | 2,330 | 2,690 | 4,180 | 4,719 | 8,710 | 9,790 | 17,200 | 19,800 |
| 40 | 203 | 271 | 469 | 554 | 896 | 1,100 | 2,010 | 2,320 | 3,630 | 4,116 | 7,530 | 8,500 | 14,900 | 17,200 |
| 50 | 181 | 243 | 420 | 496 | 806 | 986 | 1,790 | 2,070 | 3,260 | 3,702 | 6,730 | 7,610 | 13,400 | 15,400 |
| 75 | 147 | 196 | 344 | 406 | 663 | 809 | 1,460 | 1,690 | 2,680 | 3,053 | 5,480 | 6,230 | 11,000 | 12,600 |
| 80 | 140 | 189 | 333 | 393 | 643 | 768 | 1,410 | 1,630 | 2,590 | 2,961 | 5,300 | 6,040 | 10,600 | 12,200 |
| 100 | 124 | 169 | 298 | 350 | 578 | 703 | 1,260 | 1,450 | 2,330 | 2,662 | 4,740 | 5,410 | 9,530 | 10,900 |
| 150 | 101 | 137 | 245 | 287 | 477 | 575 | 1,020 | 1,180 | 1,910 | 2,195 | 3,860 | 4,430 | 7,810 | 8,890 |
| 200 | 86 | 118 | 213 | 248 | 415 | 501 | 880 | 1,020 | 1,660 | 1,915 | 3,340 | 3,840 | 6,780 | 7,710 |
| 250 | 77 | 105 | 191 | 222 | 373 | 448 | 785 | 910 | 1,490 | 1,722 | 2,980 | 3,440 | 6,080 | 6,900 |
| 300 | 69 | 96 | 173 | 203 | 343 | 411 | 716 | 829 | 1,360 | 1,578 | 2,720 | 3,150 | 5,560 | 6,300 |
| 400 | 60 | 82 | 151 | 175 | 298 | 355 | 616 | 716 | 1,160 | 1,376 | 2,350 | 2,730 | 4,830 | 5,460 |
| 500 | 53 | 72 | 135 | 158 | 268 | 319 | 550 | 638 | 1,030 | 1,237 | 2,100 | 2,450 | 4,330 | 4,880 |

EHD: Equivalent hydraulic diameter. A measure of the relative hydraulic efficiency between different tubing sizes. The greater the value of EHD, the greater the gas capacity of the tubing.

Notes:

(1) Table does not include effect of pressure drop across the line regulator. Where regulator loss exceeds $^1/_2$ psi (based on 13 in. w.c. outlet pressure), do not use this table. Consult with regulator manufacturer for pressure drops and capacity factors. Pressure drops across a regulator may vary with flow rate.

(2) CAUTION: Capacities shown in table may exceed maximum capacity for a selected regulator. Consult with regulator or tubing manufacturer for guidance.

(3) Table includes losses for four 90 degree bends and two end fittings. Tubing runs with larger number of bends and/or fittings shall be increased by an equivalent length of tubing according to the following equation: $L = 1.3n$, where $L$ is additional length (ft) of tubing and $n$ is the number of additional fittings and/or bends.

(4) All table entries are rounded to 3 significant digits.

*TABLE 6.3(j) Corrugated Stainless Steel Tubing (CSST)*

| | |
|---|---|
| *Gas:* | *Undiluted Propane* |
| *Inlet Pressure:* | *5.0 psi* |
| *Pressure Drop:* | *3.5 psi* |
| *Specific Gravity:* | *1.50* |

| | *Tube Size (EHD)* | | | | | | | | | | | | | |
|---|---|---|---|---|---|---|---|---|---|---|---|---|---|---|
| ***Flow Designation:*** | ***13*** | ***15*** | ***18*** | ***19*** | ***23*** | ***25*** | ***30*** | ***31*** | ***37*** | ***39*** | ***46*** | ***48*** | ***60*** | ***62*** |
| ***Length (ft)*** | ***Capacity in Thousands of Btu per Hour*** | | | | | | | | | | | | | |
| 10 | 826 | 1,070 | 1,710 | 2,060 | 3,150 | 4,000 | 7,830 | 8,950 | 13,100 | 14,441 | 28,600 | 31,200 | 54,400 | 63,800 |
| 25 | 509 | 664 | 1,090 | 1,310 | 2,040 | 2,550 | 4,860 | 5,600 | 8,400 | 9,339 | 18,000 | 19,900 | 34,700 | 40,400 |
| 30 | 461 | 603 | 999 | 1,190 | 1,870 | 2,340 | 4,430 | 5,100 | 7,680 | 8,564 | 16,400 | 18,200 | 31,700 | 36,900 |
| 40 | 396 | 520 | 867 | 1,030 | 1,630 | 2,030 | 3,820 | 4,400 | 6,680 | 7,469 | 14,200 | 15,800 | 27,600 | 32,000 |
| 50 | 352 | 463 | 777 | 926 | 1,460 | 1,820 | 3,410 | 3,930 | 5,990 | 6,717 | 12,700 | 14,100 | 24,700 | 28,600 |
| 75 | 284 | 376 | 637 | 757 | 1,210 | 1,490 | 2,770 | 3,190 | 4,920 | 5,539 | 10,300 | 11,600 | 20,300 | 23,400 |
| 80 | 275 | 363 | 618 | 731 | 1,170 | 1,450 | 2,680 | 3,090 | 4,770 | 5,372 | 9,990 | 11,200 | 19,600 | 22,700 |
| 100 | 243 | 324 | 553 | 656 | 1,050 | 1,300 | 2,390 | 2,760 | 4,280 | 4,830 | 8,930 | 10,000 | 17,600 | 20,300 |
| 150 | 196 | 262 | 453 | 535 | 866 | 1,060 | 1,940 | 2,240 | 3,510 | 3,983 | 7,270 | 8,210 | 14,400 | 16,600 |
| 200 | 169 | 226 | 393 | 464 | 755 | 923 | 1,680 | 1,930 | 3,050 | 3,474 | 6,290 | 7,130 | 12,500 | 14,400 |
| 250 | 150 | 202 | 352 | 415 | 679 | 828 | 1,490 | 1,730 | 2,740 | 3,124 | 5,620 | 6,390 | 11,200 | 12,900 |
| 300 | 136 | 183 | 322 | 379 | 622 | 757 | 1,360 | 1,570 | 2,510 | 2,865 | 5,120 | 5,840 | 10,300 | 11,700 |
| 400 | 117 | 158 | 279 | 328 | 542 | 657 | 1,170 | 1,360 | 2,180 | 2,498 | 4,430 | 5,070 | 8,920 | 10,200 |
| 500 | 104 | 140 | 251 | 294 | 488 | 589 | 1,050 | 1,210 | 1,950 | 2,247 | 3,960 | 4,540 | 8,000 | 9,110 |

EHD: Equivalent hydraulic diameter. A measure of the relative hydraulic efficiency between different tubing sizes. The greater the value of EHD, the greater the gas capacity of the tubing.

Notes:

(1) Table does not include effect of pressure drop across the line regulator. Where regulator loss exceeds $^1/_2$ psi (based on 13 in. w.c. outlet pressure), do not use this table. Consult with regulator manufacturer for pressure drops and capacity factors. Pressure drops across a regulator may vary with flow rate.

(2) CAUTION: Capacities shown in table may exceed maximum capacity for a selected regulator. Consult with regulator or tubing manufacturer for guidance.

(3) Table includes losses for four 90 degree bends and two end fittings. Tubing runs with larger number of bends and/or fittings shall be increased by an equivalent length of tubing according to the following equation: $L = 1.3n$, where $L$ is additional length (ft) of tubing and $n$ is the number of additional fittings and/or bends.

(4) All table entries are rounded to 3 significant digits.

***TABLE 6.3(k)*** *Polyethylene Plastic Pipe*

| | |
|---|---|
| ***Gas:*** | ***Undiluted Propane*** |
| ***Inlet Pressure:*** | ***11.0 in. w.c.*** |
| ***Pressure Drop:*** | ***0.5 in. w.c.*** |
| ***Specific Gravity:*** | ***1.50*** |

***INTENDED USE: PE Pipe Sizing Between Integral Second-Stage Regulator at Tank or Second-Stage (Low-Pressure) Regulator and Building.***

| | ***Pipe Size (in.)*** | | | | | |
|---|---|---|---|---|---|---|
| ***Nominal OD:*** | ½ | ¾ | ***1*** | ***1¼*** | ***1½*** | ***2*** |
| ***Designation:*** | ***SDR 9.33*** | ***SDR 11.0*** | ***SDR 11.00*** | ***SDR 10.00*** | ***SDR 11.00*** | ***SDR 11.00*** |
| ***Actual ID:*** | ***0.660*** | ***0.860*** | ***1.077*** | ***1.328*** | ***1.554*** | ***1.943*** |
| ***Length (ft)*** | ***Capacity in Thousands of Btu per Hour*** | | | | | |
| 10 | 340 | 680 | 1,230 | 2,130 | 3,210 | 5,770 |
| 20 | 233 | 468 | 844 | 1,460 | 2,210 | 3,970 |
| 30 | 187 | 375 | 677 | 1,170 | 1,770 | 3,180 |
| 40 | 160 | 321 | 580 | 1,000 | 1,520 | 2,730 |
| 50 | 142 | 285 | 514 | 890 | 1,340 | 2,420 |
| 60 | 129 | 258 | 466 | 807 | 1,220 | 2,190 |
| 70 | 119 | 237 | 428 | 742 | 1,120 | 2,010 |
| 80 | 110 | 221 | 398 | 690 | 1,040 | 1,870 |
| 90 | 103 | 207 | 374 | 648 | 978 | 1,760 |
| 100 | 98 | 196 | 353 | 612 | 924 | 1,660 |
| 125 | 87 | 173 | 313 | 542 | 819 | 1,470 |
| 150 | 78 | 157 | 284 | 491 | 742 | 1,330 |
| 175 | 72 | 145 | 261 | 452 | 683 | 1,230 |
| 200 | 67 | 135 | 243 | 420 | 635 | 1,140 |
| 250 | 60 | 119 | 215 | 373 | 563 | 1,010 |
| 300 | 54 | 108 | 195 | 338 | 510 | 916 |
| 350 | 50 | 99 | 179 | 311 | 469 | 843 |
| 400 | 46 | 92 | 167 | 289 | 436 | 784 |
| 450 | 43 | 87 | 157 | 271 | 409 | 736 |
| 500 | 41 | 82 | 148 | 256 | 387 | 695 |

Note: All table entries are rounded to 3 significant digits.

NEW

*TABLE 6.3(l)* *Polyethylene Plastic Pipe*

| | |
|---|---|
| *Gas:* | *Undiluted Propane* |
| *Inlet Pressure:* | *2.0 psi* |
| *Pressure Drop:* | *1.0 psi* |
| *Specific Gravity:* | *1.50* |

***INTENDED USE: PE Pipe Sizing Between 2 psi Service Regulator and Line Pressure Regulator.***

| | *Pipe Size (in.)* | | | | | |
|---|---|---|---|---|---|---|
| ***Nominal OD:*** | ½ | ¾ | 1 | 1¼ | 1½ | 2 |
| ***Designation:*** | ***SDR 9.33*** | ***SDR 11.0*** | ***SDR 11.00*** | ***SDR 10.00*** | ***SDR 11.00*** | ***SDR 11.00*** |
| ***Actual ID:*** | ***0.660*** | ***0.860*** | ***1.077*** | ***1.328*** | ***1.554*** | ***1.943*** |
| ***Length (ft)*** | ***Capacity in Thousands of Btu per Hour*** | | | | | |
| 10 | 3,130 | 6,260 | 11,300 | 19,600 | 29,500 | 53,100 |
| 20 | 2,150 | 4,300 | 7,760 | 13,400 | 20,300 | 36,500 |
| 30 | 1,730 | 3,450 | 6,230 | 10,800 | 16,300 | 29,300 |
| 40 | 1,480 | 2,960 | 5,330 | 9,240 | 14,000 | 25,100 |
| 50 | 1,310 | 2,620 | 4,730 | 8,190 | 12,400 | 22,200 |
| 60 | 1,190 | 2,370 | 4,280 | 7,420 | 11,200 | 20,100 |
| 70 | 1,090 | 2,180 | 3,940 | 6,830 | 10,300 | 18,500 |
| 80 | 1,010 | 2,030 | 3,670 | 6,350 | 9,590 | 17,200 |
| 90 | 952 | 1,910 | 3,440 | 5,960 | 9,000 | 16,200 |
| 100 | 899 | 1,800 | 3,250 | 5,630 | 8,500 | 15,300 |
| 125 | 797 | 1,600 | 2,880 | 4,990 | 7,530 | 13,500 |
| 150 | 722 | 1,450 | 2,610 | 4,520 | 6,830 | 12,300 |
| 175 | 664 | 1,330 | 2,400 | 4,160 | 6,280 | 11,300 |
| 200 | 618 | 1,240 | 2,230 | 3,870 | 5,840 | 10,500 |
| 250 | 548 | 1,100 | 1,980 | 3,430 | 5,180 | 9,300 |
| 300 | 496 | 994 | 1,790 | 3,110 | 4,690 | 8,430 |
| 350 | 457 | 914 | 1,650 | 2,860 | 4,320 | 7,760 |
| 400 | 425 | 851 | 1,530 | 2,660 | 4,020 | 7,220 |
| 450 | 399 | 798 | 1,440 | 2,500 | 3,770 | 6,770 |
| 500 | 377 | 754 | 1,360 | 2,360 | 3,560 | 6,390 |
| 550 | 358 | 716 | 1,290 | 2,240 | 3,380 | 6,070 |
| 600 | 341 | 683 | 1,230 | 2,140 | 3,220 | 5,790 |
| 650 | 327 | 654 | 1,180 | 2,040 | 3,090 | 5,550 |
| 700 | 314 | 628 | 1,130 | 1,960 | 2,970 | 5,330 |
| 750 | 302 | 605 | 1,090 | 1,890 | 2,860 | 5,140 |
| 800 | 292 | 585 | 1,050 | 1,830 | 2,760 | 4,960 |
| 850 | 283 | 566 | 1,020 | 1,770 | 2,670 | 4,800 |
| 900 | 274 | 549 | 990 | 1,710 | 2,590 | 4,650 |
| 950 | 266 | 533 | 961 | 1,670 | 2,520 | 4,520 |
| 1,000 | 259 | 518 | 935 | 1,620 | 2,450 | 4,400 |
| 1,100 | 246 | 492 | 888 | 1,540 | 2,320 | 4,170 |
| 1,200 | 234 | 470 | 847 | 1,470 | 2,220 | 3,980 |
| 1,300 | 225 | 450 | 811 | 1,410 | 2,120 | 3,810 |
| 1,400 | 216 | 432 | 779 | 1,350 | 2,040 | 3,660 |
| 1,500 | 208 | 416 | 751 | 1,300 | 1,960 | 3,530 |
| 1,600 | 201 | 402 | 725 | 1,260 | 1,900 | 3,410 |
| 1,700 | 194 | 389 | 702 | 1,220 | 1,840 | 3,300 |
| 1,800 | 188 | 377 | 680 | 1,180 | 1,780 | 3,200 |
| 1,900 | 183 | 366 | 661 | 1,140 | 1,730 | 3,110 |
| 2,000 | 178 | 356 | 643 | 1,110 | 1,680 | 3,020 |

Note: All table entries are rounded to 3 significant digits.

***TABLE 6.3(m)*** *Polyethylene Plastic Tubing*

| | | |
|---|---|---|
| | *Gas:* | *Undiluted Propane* |
| | *Inlet Pressure:* | *11.0 in. w.c.* |
| | *Pressure Drop:* | *0.5 in. w.c* |
| | *Specific Gravity:* | *1.50* |
| ***INTENDED USE: PE Tube Sizing Between Integral Second-Stage Regulator at Tank or Second-Stage (Low-Pressure) Regulator and Building.*** | | |
| | ***Plastic Tubing Size (CTS) (in.)*** | |
| ***Nominal OD:*** | ***1/2*** | ***1*** |
| ***Designation:*** | ***SDR 7.00*** | ***SDR 11.00*** |
| ***Actual ID:*** | ***0.445*** | ***0.927*** |
| ***Length (ft)*** | ***Capacity in Thousands of Btu per Hour*** | |
| 10 | 121 | 828 |
| 20 | 83 | 569 |
| 30 | 67 | 457 |
| 40 | 57 | 391 |
| 50 | 51 | 347 |
| 60 | 46 | 314 |
| 70 | 42 | 289 |
| 80 | 39 | 269 |
| 90 | 37 | 252 |
| 100 | 35 | 238 |
| 125 | 31 | 211 |
| 150 | 28 | 191 |
| 175 | 26 | 176 |
| 200 | 24 | 164 |
| 225 | 22 | 154 |
| 250 | 21 | 145 |
| 275 | 20 | 138 |
| 300 | 19 | 132 |
| 350 | 18 | 121 |
| 400 | 16 | 113 |
| 450 | 15 | 106 |
| 500 | 15 | 100 |

CTS: Copper tube size.

Note: All table entries are rounded to 3 significant digits.

## 6.4 Sizing Equations

The inside diameter of smooth wall pipe or tubing shall be determined by the sizing equations in 6.4.1 and 6.4.2 using the equivalent pipe length determined by the methods in 6.1.1 through 6.1.3.

See Section C.5 in Annex C of the code for an example of the use of sizing equations.

### 6.4.1* Low-Pressure Gas Formula.

Less than 1.5 psi (10.3 kPa):

$$D = \frac{Q^{0.381}}{19.17\left(\dfrac{\Delta H}{Cr \times L}\right)^{0.206}}$$

**A.6.4.1** The Low-Pressure Formula is the standard flow formula located in Annex C but rearranged to solve for the pipe diameter.

### 6.4.2* High-Pressure Gas Formula.

1.5 psi (10.3 kPa) and above:

$$D = \frac{Q^{0.381}}{18.93\left[\dfrac{(P_1^2 - P_2^2) \cdot Y}{Cr \times L}\right]^{0.206}}$$

where:

$D$ = inside diameter of pipe (in.)

$Q$ = input rate appliance(s) (cubic feet per hour at 60°F and 30 in. mercury column)

$P_1$ = upstream pressure [psia ($P_1$ + 14.7)]

$P_2$ = downstream pressure [psia ($P_2$ + 14.7)]

$L$ = equivalent length of pipe (ft)

$\Delta H$ = pressure drop [in. w.c. (27.7 in. $H_2O$ = 1 psi)]

See Table 6.4.2 for values of *Cr* and *Y*.

***TABLE 6.4.2** Cr and Y for Natural Gas and Undiluted Propane at Standard Conditions*

| | Formula Factors | |
|---|---|---|
| **Gas** | ***Cr*** | ***Y*** |
| Natural gas | 0.6094 | 0.9992 |
| Undiluted propane | 1.2462 | 0.9910 |

**A.6.4.2** The High-Pressure Formula is the standard flow formula located in Annex C but rearranged to solve for the pipe diameter.

# Gas Piping Installation

CHAPTER 7

Chapter 7 includes requirements for the installation of piping systems. Chapter 7 complements Chapter 5, which specifies piping materials. Testing and purging of gas piping systems are covered in Chapter 8. Chapter 7 covers the following:

- Piping installed underground, including protection against mechanical damage, corrosion, freezing, piping through foundation walls and beneath buildings, and the use of plastic pipe (See Section 7.1.)
- Piping and pipe supports installed aboveground indoors (and not concealed), including prohibited locations for gas piping, and outdoors (See Section 7.2.)
- Concealed piping in buildings, including restrictions on piping and valves in concealed spaces, and piping in floors (See Section 7.3.)
- Piping in vertical chases at pressures exceeding 5 psi (34 kPa), including location of pressure regulators (See Section 7.4.)
- Gas pipe turns, including equipment and methods to make bends in piping (See Section 7.5.)
- Drips, which are needed when the gas supplied contains liquids, and sediment traps, which prevent debris from clogging equipment and controls (See Section 7.6.)
- Outlets, including detailed requirements for the location and capping of outlets (See Section 7.7.)
- Branch pipe connection, which allows for future expansion of gas piping systems (See Section 7.8.)
- Manual gas shutoff valves, which are required at gas pressure regulators and at buildings served by gas, for each individual unit supplied by gas (Also included are requirements for the exterior location and marking of emergency shutoff valves.) (See Section 7.9.)
- Prohibited devices that could restrict the flow of gas (See Section 7.10.)
- Systems containing gas–air mixtures outside the flammable range for which requirements are provided to prevent the mixture from entering the flammable range (See Section 7.11.)
- Systems containing flammable gas–air mixtures, including minimum requirements for piping and equipment (See Section 7.12.)
- Electrical bonding and grounding, electrical circuits, and electrical connections, which are consistent with *NFPA 70®*, *National Electrical Code®* (*NEC®*) (See Section 7.13, Section 7.14, and Section 7.15.)

## 7.1 Piping Underground

### 7.1.1 Clearances.

Underground gas piping shall be installed with sufficient clearance from any other underground structure to avoid contact therewith, to allow maintenance, and to protect against damage from proximity to other structures. In addition, underground plastic piping shall be installed with sufficient clearance or shall be insulated from any source of heat so as to prevent the heat from impairing the serviceability of the pipe.

The requirement in 7.1.1 recognizes the possible hazard from the pipe's being in contact with, or in close proximity to, other underground structures. Such contact can be a hazard and can cause harm to the pipe during backfilling or later, due to settlement of the structure(s).

A potential hazard to underground gas piping is the possibility of a lightning strike causing failure of underground gas piping. This hazard is not addressed in the code, because of the very small number of known incidents. In areas where lightning occurs more frequently, installers may consider spacing gas pipe (both metal and plastic) at a distance from other underground pipes, cables, and conduits serving the building where multiple services are installed in a common trench. In the United States, the National Oceanographic and Atmospheric Administration is a source of information on the frequency of lightning strikes.

### 7.1.2 Protection Against Damage.

Means shall be provided to prevent excessive stressing of the piping where vehicular traffic is heavy or soil conditions are unstable and settling of piping or foundation walls could occur. Piping shall be buried or covered in a manner so as to protect the piping from physical damage. Piping shall be protected from physical damage where it passes through flower beds, shrub beds, and other such cultivated areas where such damage is reasonably expected.

**7.1.2.1 Cover Requirements.** Underground piping systems shall be installed with a minimum of 12 in. (300 mm) of cover.

**(A)** The minimum cover shall be increased to 18 in. (460 mm) if external damage to the pipe or tubing from external forces is likely to result.

**(B)** Where a minimum of 12 in. (300 mm) of cover cannot be provided, the pipe shall be installed in conduit or bridged (shielded).

The intent of all of the coverage for underground piping is to provide protection for the installed piping from corrosion or physical damage. Note that changes in other paragraphs (3.3.76.1, 7.1.6.1, 9.3.8.8, 12.4.3.6, and 12.9.2) have changed the word "grade" to "finished ground level." In applying this requirement the finished grade level, after any landscaping, should meet the burial requirement in this paragraph.

**FAQ ►** Why would gas piping installed under an asphalt driveway need special consideration?

Protection from heavy vehicle loads must be ensured. In general, deeper burial [at a depth of 18 in. (460 mm) or more] or casing of the pipe carrying the gas is the means used to comply with the requirement in 7.1.2.1.

Deeper burial applies to cultivated areas, with depth of cover requirements depending on the anticipated risk, which can vary from hand garden tools to tillage equipment pulled by large tractors. Gas companies with distribution pipes in the area are a good resource with which to consult when questions about burial depth of piping are raised.

Damage to pipes is not limited to plastic pipe. Copper tubing is severed easily with a shovel, and digging tools can damage the coating on steel pipe, leading to corrosion. Digging with powered digging equipment, such as a backhoe, can damage steel pipe.

There are two ways to protect the pipe from physical damage. One is to protect the pipe from damage caused by digging, so the protection from digging tools must be of sufficient strength to protect the pipe. Such protection can be provided by installing the gas pipe inside a larger pipe, which acts as a shield, or by locating a protective material such as a steel plate with corrosion protection above the pipe. When a larger pipe is used to encase and protect a gas line, steps must be taken to prevent the accumulation of water. The presence of water in the shielding pipe could accelerate corrosion or crush the gas line if freezing occurs. In most cases, the pipe is buried at a sufficient depth so that protective devices are not needed.

The second means is to provide excavators with an indicator that a pipe is buried below. Many gas utilities use a "terra tape" (a yellow plastic tape printed with a warning that gas pipe is buried below) placed a few inches above the pipe to warn future diggers that a gas pipe is buried below. (See Exhibit 7.1.) The tape is an indicator and not a protective device.

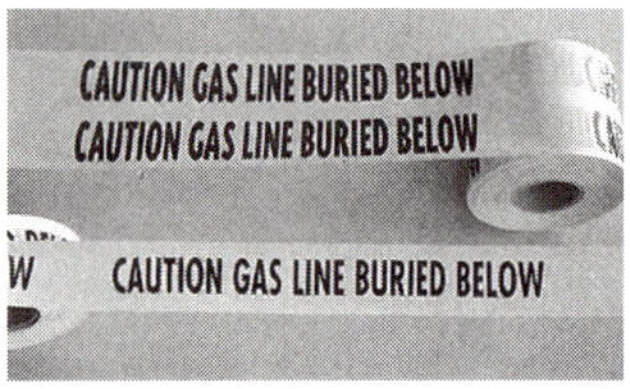


*EXHIBIT 7.1 Terra Tape.*

In the United States, the possibility of excavation work damaging underground piping can be minimized by calling any of the regional or statewide "One Call," also known as "Dig Safe," centers before excavation work is begun. The call center notifies utility companies of excavation work near their underground installations and notifies them to mark the locations prior to excavation. All underground utilities, including gas, electricity, telephone, and cable TV companies, are members of Dig Safe.

The Dig Safe program covers gas lines installed by gas transmission companies and by local gas utilities to buildings. It does not include gas lines run from a building to another building, such as a garage, or to an outdoor appliance, such as a gas grill or pool heater. To avoid accidents, excavators should be made aware when underground gas lines exist that were not installed by gas utility companies.

More information on "Dig Safe" programs can be obtained from www.call811.com, a web site maintained by the Common Ground Alliance. The web site lists local call numbers, provides information on when to call, and explains how the program works. The web site also offers free promotional outreach material, such as the advertisement shown in Exhibit 7.2. A national "811" number is being implemented; however the time frame has not been finalized at the time of publication.

***EXHIBIT 7.2*** *Advertisement for the Call 811 Program. (Courtesy of Common Ground Alliance)*

If protective devices are needed, they must be of sufficient strength to protect the pipe and must be made of a material that will not corrode over time. Plastic pipe or PVC is often used as a protective material. Split along its length, plastic pipe or PVC provides the necessary strength and does not corrode underground.

**7.1.2.2 Trenches.** The trench shall be graded so that the pipe has a firm, substantially continuous bearing on the bottom of the trench.

**7.1.2.3 Backfilling.** Where flooding of the trench is done to consolidate the backfill, care shall be exercised to see that the pipe is not floated from its firm bearing on the trench bottom.

### 7.1.3* Protection Against Corrosion.

Gas piping in contact with earth or other material that could corrode the piping shall be protected against corrosion in an approved manner. When dissimilar metals are joined underground, an insulating coupling or fitting shall be used. Piping shall not be laid in contact with cinders. Uncoated threaded or socket welded joints shall not be used in piping in contact with soil or where internal or external crevice corrosion is known to occur.

**A.7.1.3** For information on corrosion protection of underground pipe, see NACE RP 0169, *Control of External Corrosion on Underground or Submerged Metallic Piping Systems.* Information on installation, maintenance, and corrosion protection might be available from the gas supplier.

Corrosion protection methods for metal pipe must always be approved by the authority having jurisdiction prior to installation. (See the definition of the term *approved* in 3.2.1.) Selection of a corrosion protection system should be based either on experience or on recommendation of experts in the field, which includes manufacturers of corrosion protection systems.

With the availability of polyethylene and other corrosion-resistant piping systems, the need to provide corrosion protection for steel as required by 7.1.3 often makes steel pipe the last choice for underground installation. Steel pipe that is installed underground or in contact with water deteriorates through galvanic corrosion, which is an electrochemical process.

This electrochemical process can be prevented by stopping the flow of electricity between the pipe and the soil. Stopping electricity flow is accomplished by coating the pipe with an electrically insulating material. Factory-applied coatings include polyethylene over a mastic primer and fusion-bonded epoxy. Field-applied coatings usually consist of either hot- or cold-applied tapes over a complementary primer.

The second step in properly protecting a steel pipe underground is to maintain an electrical charge on the pipe, relative to the surrounding soil, that is adequate to ensure that the corrosion-causing transfer of electrons will not occur. Maintaining this electrical charge is called cathodic protection and is accomplished by an impressed voltage supplied by either a passive sacrificial anode system or an active rectifier system. Sizing and spacing of either the anodes or the rectifiers are important. Manufacturers of these components are a good source of guidance.

**FAQ ►**
Why is cathodic protection highly recommended in underground steel pipe installations?

Partial steel pipe corrosion solutions, such as coating the pipe but not using cathodic protection, usually fare worse than unprotected pipe. Uncoated pipe will corrode over its entire surface area, whereas corrosion of coated pipe lacking cathodic protection will be concentrated at any pinhole or other imperfection (called a "holiday") in the coating. Because obtaining and maintaining a perfect coating application is virtually impossible, protection of underground steel pipe should include cathodic protection. Any cathodically protected pipe should be electrically isolated from upstream and downstream components to prevent the loss of the required electrical charge. This isolation is accomplished with the use of dielectric unions or flange-insulating kits.

In reality, it is not practical to protect short runs of buried steel pipe using cathodic protection, which is the reason that alternate materials are used.

The requirement for protection against corrosion also recognizes that threaded and welded steel pipe joints are especially prone to the effects of corrosion and should not be used where corrosion is anticipated. Similarly, the use of cinders as backfill is prohibited because they accelerate corrosion.

### 7.1.4* Protection Against Freezing.

Where the formation of hydrates or ice is known to occur, piping shall be protected against freezing.

**A.7.1.4** The gas supplier can be consulted for recommendations.

Ice problems are all but nonexistent with pipeline natural gas and propane, but they can still exist in gas from alternative sources such as local wells, landfills, or waste treatment plants. The user is to consult the gas supplier about the possibility of ice, which can block lines and prevent gas flow.

### 7.1.5 Piping Through Foundation Wall.

NEW

Underground piping installed through the outer foundation or basement wall of a building shall be encased in a protective sleeve or protected by an approved device or method. The space between the gas piping and the building or sleeve shall be sealed to prevent entry of gas and water.

Subsection 7.1.5 is included in the code because gas leaks in outdoor underground piping can migrate along the buried piping and into the basement instead of dispersing in the atmosphere above grade. This condition can be extremely hazardous and can result in the building exploding. The requirement to seal the casing against water entry is intended to block this leakage path. See Exhibit 7.3 for an illustration of a below-grade foundation wall penetration.

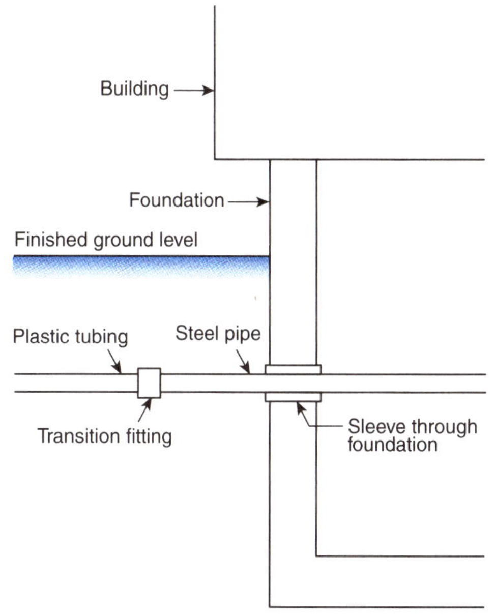


***EXHIBIT 7.3*** *Piping Through a Foundation Wall. (Courtesy of J. J. Drechsler)*

Caulking should not be used as a foundation wall seal where the gap is larger than that recommended by the caulking manufacturer. Pipe can move relative to the foundation, and the caulking can fail if it is sealing a large gap.

The use of an inert wrapping material provides an additional method of protection. The material must be approved and be suitable for the application. The material must also be able to seal around the pipe and the wall to prevent the entry of gas or water.

### 7.1.6 Piping Underground Beneath Buildings.

Where the installation of gas piping underground beneath buildings is unavoidable, the piping shall be encased in an approved conduit designed to withstand the superimposed loads and installed in accordance with 7.1.6.1 or 7.1.6.2.

Subsection 7.1.6 permits the installation of the following:

- Pipe under a building that originates and terminates inside a building, such as a pipe that extends gas piping in a building under a concrete floor to another part of the building
- Pipe that originates outside a building and terminates inside a building, such as a gas service that enters under a slab or basement

Gas leaking from a pipe installed underneath a building can migrate into the building. The code requires that any such pipes be cased and that the annular space between the pipe and the case be vented to the outdoors. (See Exhibit 7.4.) The authority having jurisdiction may require that the sleeve be tested in the same manner as the gas pipe it contains. Although the resulting installation is perfectly safe, the installation is difficult and the repairs are more so. For these reasons, practicality demands that most pipes run through buildings rather than under them.

*EXHIBIT 7.4 Vented Sleeve.*

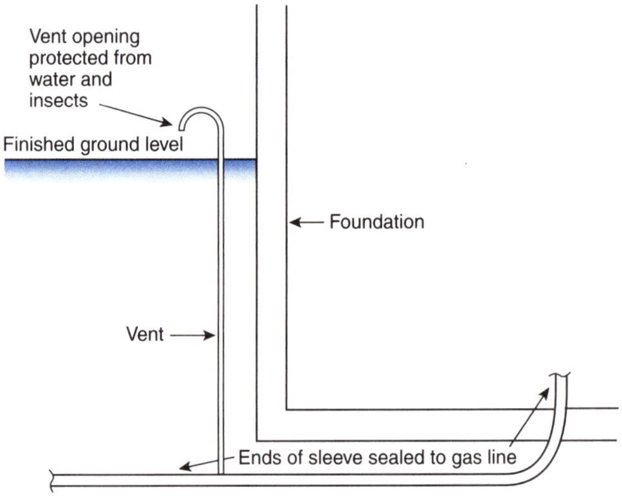


NEW **7.1.6.1 Conduit with One End Terminating Outdoors.** The conduit shall extend into an accessible portion of the building and, at the point where the conduit terminates in the building, the space between the conduit and the gas piping shall be sealed to prevent the possible entrance of any gas leakage. Where the end sealing is of a type that retains the full pressure of the pipe, the conduit shall be designed for the same pressure as the pipe. The conduit shall extend at least 4 in. (100 mm) outside the building, be vented outdoors above finished ground level, and be installed so as to prevent the entrance of water and insects.

The need to install gas piping under buildings has been recognized in NFPA 54 since 1959. The requirement in 7.1.6.1 allows gas piping to be installed under buildings to enter through the slab and requires that the pipe under the building be encased in a conduit. The requirement for a conduit recognizes that gas leakage under a building can create a hazardous condition and provides a path to vent such leakage to the atmosphere.

**7.1.6.2 Conduit with Both Ends Terminating Indoors.** Where the conduit originates and terminates within the same building, the conduit shall originate and terminate in an accessible portion of the building and shall not be sealed.

In the 2006 edition, a new option was added that recognizes the need to supply fuel gas to appliances located in the center of a first floor room, without piping dropping from the ceiling where piping the gas directly from outside the building is not possible or is impractical. The most common use is for island ranges in kitchens. This option allows shorter pipe runs in many cases and eliminates the need for a vent to the outdoors.

### 7.1.7 Plastic Pipe.

Plastic pipe is to be used outdoors, underground only, and must be installed 12 in. (300 mm) or 18 in. (460 mm) below finished grade. Plastic pipe may not be used inside buildings for fuel gas. Note that in new construction, landscaping may change the finished grade. The prohibition of the use of any plastic pipe recognizes that it will fail under fire conditions, and the released gas will accelerate any building fire.

Risers and wall head adapters (also called service head adapters) are also covered by 7.1.7. These fittings are used to install polyethylene pipe and tubing. Polyethylene has been used successfully for more than 30 years in gas distribution service upstream of the gas meter.

Plastic pipe and tubing is more susceptible to inadvertent damage during installation than most metallic pipe. For this reason, special attention should be given to proper compaction below the pipe, to the elimination of shear points on connections during backfilling, and to the materials in the backfill, making certain that angular and large materials are not used near the pipe.

Note that while the code uses the word *plastic* in this section, only polyethylene or polyamide materials are allowed. Polyvinyl chloride (PVC) cannot be used for gas piping.

**7.1.7.1 Connection of Plastic Piping.** Plastic pipe shall be installed outdoors, underground only.

The use of anode-type risers and transition fittings with anodes (Exhibits 7.5 and 7.6) has been allowed for several editions of the code. In both of these applications, the plastic-to-steel transition is made below grade and meets the requirement in 7.1.7.1.

*Exception No. 1: Plastic pipe shall be permitted to terminate aboveground where an anodeless riser is used.*

Anodeless risers are constructed with the plastic pipe running inside steel all the way to the top end, where the pressure seal and the transition from plastic to steel are located. These

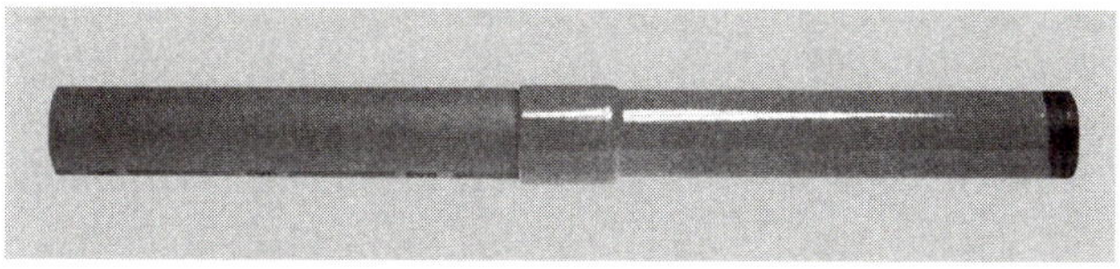

**EXHIBIT 7.5** *Transition Fitting. (Courtesy of Ray Murray, Inc.)*

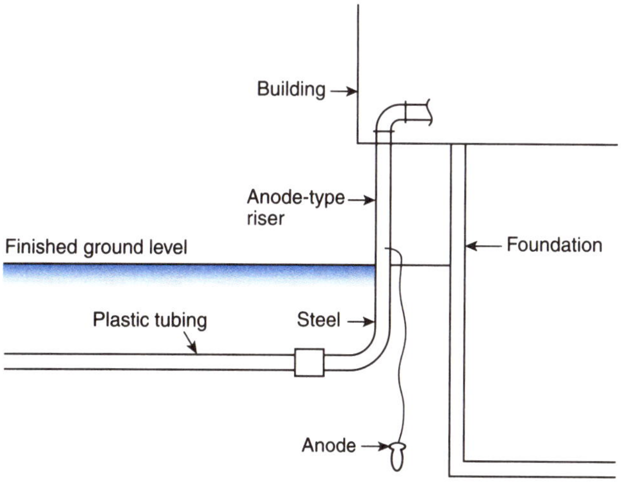


**EXHIBIT 7.6** *Anode-Type Riser Installation.*

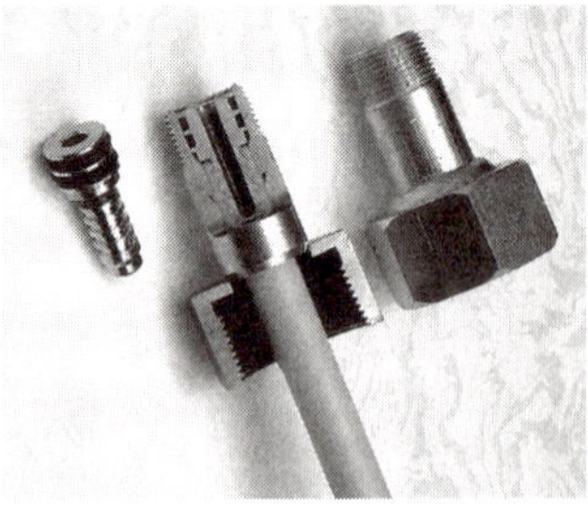

*EXHIBIT 7.7 Service/Wall Head Adapters (Cutaway). (Courtesy of R. W. Lyall & Company, Inc.)*

risers are accepted by the code because their design addresses the concerns regarding plastic pipe terminating aboveground.

*Exception No. 2: Plastic pipe shall be permitted to terminate with a wall head adapter aboveground in buildings, including basements, where the plastic pipe is inserted in a piping material permitted for use in buildings.*

Exception No. 2 permits plastic gas piping to be terminated aboveground and outside of buildings. This exception permits the use of fittings, known as wall head adapters, to terminate buried plastic pipe inside of buildings in basement areas. Wall head adapters, which are shown in Exhibit 7.7, often are used by gas utilities for insertion renewals of steel service lines. Within the scope of this code, wall head adapters can be used for renewals of steel pipes from meters at the property line or in renewals of existing underground steel pipes between buildings, as well as for making a clean-sleeved, belowgrade foundation wall penetration.

See Exhibit 7.8 for a typical steel line renewal using a wall head adapter. The line is renewed by inserting plastic pipe through the old steel pipe all the way from the outer end into the basement (i.e., insertion renewal). The inserted plastic in the basement is terminated by installing an appropriately sized wall head adapter to the end of the plastic pipe or tube in accordance with the manufacturer's installation instructions. The wall head adapter/plastic pipe assembly is then pushed back toward and threaded onto the old steel service pipe that extends through the basement wall. Next, the house piping is fitted to the male threads of the wall head adapter. The plastic pipe and house piping are pressure tested. At this point the plastic is hooked up at the other end and purged, completing the renewal installation.

*EXHIBIT 7.8 Basement Termination of Plastic Pipe.*

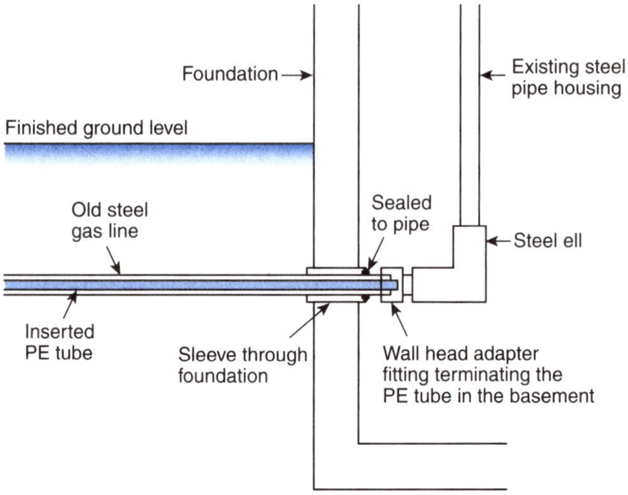


**7.1.7.2 Connections Between Metallic and Plastic Piping.** Connections made between metallic and plastic piping shall be made with fittings conforming to one of the following:

(1) ASTM D 2513, *Standard Specification for Thermoplastic Gas Pressure Pipe, Tubing, and Fittings*, Category I transition fittings
(2) ASTM F 1973, *Standard Specification for Factory Assembled Anodeless Risers and Transition Fittings in Polyethylene (PE) and Polyamide 11 (PA11) Fuel Gas Distribution Systems*
(3) ASTM F 2509, *Standard Specification for Field-Assembled Anodeless Riser Kits for Use on Outside Diameter Controlled Polyethylene Gas Distribution Pipe and Tubing*

**7.1.7.3 Tracer Wire.** An electrically continuous corrosion-resistant tracer wire (minimum AWG 14) or tape shall be buried with the plastic pipe to facilitate locating. One end shall be brought aboveground at a building wall or riser.

Because pipe locators cannot locate plastic pipe or tube, the requirement in 7.1.7.3 for tracer wire or tape is included. Tracer wire is electrical wire, AWG 14 or larger. The tracer wire should be continuous, or when joined together, the splice must be protected from corrosion. The wire must be able to conduct current all the way to the end. The ends of the tracer wire must terminate aboveground at the riser locations. Where tracer wire is used, special equipment sends a current through the wire, creating a magnetic field that can then be located using a special detector.

◀ **FAQ**
Why must tracer wire be installed where underground plastic gas pipe is installed?

Tracer tape is plastic tape with a continuous foil backing that can be detected by a metal detector. Where metalized tape is used, a metal detector is usually used to locate the tape.

## 7.2 Installation of Piping

Section 7.2 includes requirements for the installation of pipe that is not concealed in buildings. Installation of concealed pipe is covered in Section 7.3, and installation of buried pipe is covered in Section 7.1. Included in Section 7.2 are requirements for protection of the pipe; a caution not to damage building structural supports, pipe support, and prohibited locations; and requirements for piping on roofs.

**7.2.1** Piping installed aboveground shall be securely supported and located where it is protected from physical damage *(also see 7.1.4)*. Where passing through an exterior wall, the piping shall also be protected against corrosion by coating or wrapping with an inert material approved for such applications. Where piping is encased in a protective pipe sleeve, the annular space between the gas piping and the sleeve shall be sealed at the wall to prevent the entry of water, insects, or rodents.

Piping that passes through an outside wall must be protected from corrosion, because this area can collect water. In addition, piping in contact with building materials can corrode due to chemical reactions with those materials or due to galvanic corrosion from contact with metallic siding. The inert wrapping material must be approved for these applications. Refer to commentary following 7.1.5, where the use of caulking is discussed.

The requirement in 7.2.1 that the pipe be sealed at the wall is often overlooked. Usually the pipe is protected only when the pipe passes through a concrete wall. However, when the pipe passes through a noncorrosive wall, such as wood, it must be sealed to prevent the entry of insects and water.

### 7.2.2 Building Structure.

**7.2.2.1** The installation of gas piping shall not cause structural stresses within building components to exceed allowable design limits.

The support of gas piping must be properly designed with a view to the structural integrity of the building elements. The bottom chord of wooden trusses may not have any load-carrying capacity, for example. Heavy steel pipe may need support spanning several support members. Guidance on support should come from design professionals.

**7.2.2.2** Approval shall be obtained before any beams or joists are cut or notched.

The intent of 7.2.2.2 is to ensure that the installation of gas piping does not cause damage to the building. Note that the term *approved* is defined in 3.2.1 as follows:

> **3.2.1 Approved.** Acceptable to the authority having jurisdiction.

Before granting approval to cut or notch beams or joists, the authority having jurisdiction should be satisfied that the installer has consulted with the building architect, structural engineer, building official, or builder.

### 7.2.3 Other than Dry Gas.

Drips, sloping, protection from freezing, and branch pipe connections, as provided for in 7.1.4, 7.6.1, and Section 7.8, shall be provided when other than dry gas is distributed and climatic conditions make such provisions necessary.

Dry gas is gas that does not contain components that will condense, resulting in liquid in the gas piping. The liquids can be water, various flammable liquids, and oils. Natural gas from wells has various components, including flammable liquids and other flammable gases that can condense (e.g., butane). These are normally removed before natural gas enters the pipeline distribution system. Natural gas is normally "dry" and drips are not normally required. Abnormal pipeline operations and certain natural gas supplies can result in liquids forming in pipelines. These liquids can enter a building gas piping system and obstruct gas flow. If this occurs, the natural gas utility company may require a drip to separate any liquid from the building gas piping system. To the editor's knowledge, the occurrence of liquids in natural gas building piping systems is highly unusual.

It would be more unusual for propane piping systems in buildings to contain liquids and require drips. Propane is normally withdrawn from the storage tank as a vapor, which will not contain liquids. Some propane systems incorporate vaporizers, which boil the propane to supply a sufficient amount of vapor. Improper vaporizer operation can result in liquid contaminants, if present in the propane tank, entering the building propane gas piping system. Experience is usually the best guide in this area, but the gas supplier may be able to provide additional information.

### 7.2.4 Gas Piping to Be Sloped.

Piping for other than dry gas conditions shall be sloped not less than $^{1}/_{4}$ in. in 15 ft (7 mm in 4.6 m) to prevent traps.

Piping must be sloped in order to prevent low spots or traps when transporting gases that contain significant water vapor, as shown in Exhibit 7.9. Pipeline quality natural gas is considered to be dry, but pipeline compressor oils and other liquids have been known to enter building piping systems. The installer determines the best way to install the pipe to prevent traps. In general, the horizontal piping should be sloped so that any accumulation of liquids will be trapped by the drips or sediment traps.

*EXHIBIT 7.9 Sloping of Gas Piping for Other than Dry Gas. (Courtesy of J. J. Drechsler)*

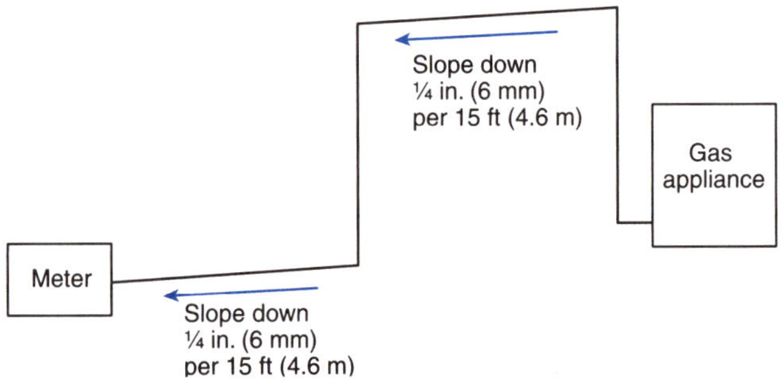


### 7.2.5* Prohibited Locations.

Gas piping inside any building shall not be installed in or through a clothes chute, chimney or gas vent, dumbwaiter, elevator shaft, or air duct, other than combustion air ducts.

**A.7.2.5** The intent is that gas piping, shutoff valves required by this code, and regulators be allowed to be installed in accessible portions of plenums, accessible ducts used to supply combustion and ventilation air in accordance with Section 5.3, and accessible spaces between a fixed ceiling and dropped ceiling.

The intent of 7.2.5 is to prevent the migration of gas leakage throughout the building. Piping can be installed in combustion and ventilation ducts because they connect directly to the area containing the appliance. If there is any leakage, it will not be distributed throughout the building.

Prior to the 2002 edition, the code prohibited the installation of valves in the space between a fixed ceiling and a dropped ceiling. The requirement was deleted because it created many problems for installers, and it did not appear to affect safety. Although there were reports of many installations with valves installed in the space formerly prohibited, there were no reports of accidents or fires.

### 7.2.6 Hangers, Supports, and Anchors.

**7.2.6.1** Piping shall be supported with metal pipe hooks, metal pipe straps, metal bands, metal brackets, metal hangers, or building structural components, suitable for the size of piping, of adequate strength and quality, and located at intervals so as to prevent or damp out excessive vibration. Piping shall be anchored to prevent undue strains on connected appliances and equipment and shall not be supported by other piping. Pipe hangers and supports shall conform to the requirements of ANSI/MSS SP-58, *Pipe Hangers and Supports — Materials, Design and Manufacture.*

**7.2.6.2** Spacings of supports in gas piping installations shall not be greater than shown in Table 7.2.6.2. Spacing of supports of CSST shall be in accordance with the CSST manufacturer's instructions.

***TABLE 7.2.6.2*** *Support of Piping*

| ***Steel Pipe, Nominal Size of Pipe (in.)*** | ***Spacing of Supports (ft)*** | ***Nominal Size of Tubing Smooth Wall (in. O.D.)*** | ***Spacing of Supports (ft)*** |
|---|---|---|---|
| $\frac{1}{2}$ | 6 | $\frac{1}{2}$ | 4 |
| $\frac{3}{4}$ or 1 | 8 | $\frac{5}{8}$ or $\frac{3}{4}$ | 6 |
| $1\frac{1}{4}$ or larger (horizontal) | 10 | $\frac{7}{8}$ or 1 (horizontal) | 8 |
| $1\frac{1}{4}$ or larger (vertical) | Every floor level | 1 or larger (vertical) | Every floor level |

For SI units, 1 ft = 0.305 m.

Table 7.2.6.2 lists the maximum distances allowable for sections of piping between supports. It includes steel pipe and smooth-wall tubing. The support of CSST is not covered in this table, and the manufacturer's installation instructions should be used where supports are needed for CSST. Threaded pipe should also be supported at each change of direction to help support the fittings used.

The distances specified in 7.2.6.2 should always be followed. The pipe and tubing at vertical drops to appliances should be supported by hangers in anticipation of the gas piping at an appliance being disconnected for servicing. When installing vertical gas piping, it is important that the piping be supported and secured to a wall.

◀ **FAQ**
Do the support requirements of Table 7.2.6.2 apply to vertical gas piping?

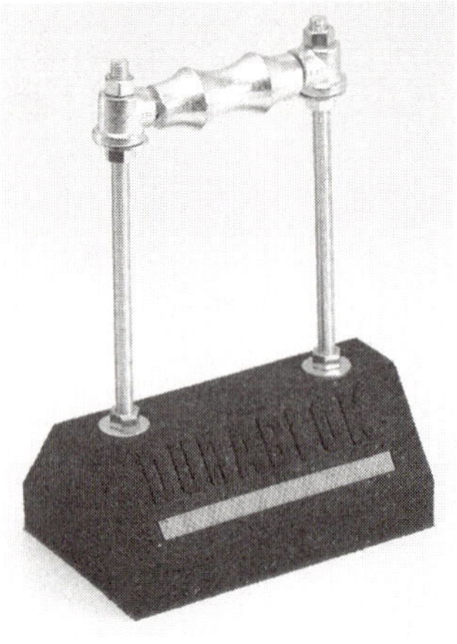

***EXHIBIT 7.10*** *Roof Base with Roller-Type Support. (Courtesy of Cooper B-Line)*

**7.2.6.3** Supports, hangers, and anchors shall be installed so as not to interfere with the free expansion and contraction of the piping between anchors. All parts of the supporting system shall be designed and installed so they are not disengaged by movement of the supported piping.

Long, straight piping runs should be supported in a manner that allows for expansion and contraction. Pipe firmly affixed to a support and subjected to temperature fluctuations such as those that occur outside the building could dislodge the support, causing an unsafe condition. These installations may feature roller-type supports to accommodate expansion. See Exhibit 7.10 for an example of a support for a roof.

**7.2.6.4 Piping on Roof Tops.** Gas piping installed on the roof surfaces shall be elevated above the roof surface and shall be supported in accordance with Table 7.2.6.2.

The support interval required for rooftop piping installations is the same as the support intervals required for all indoor piping. The committee addressed support for these installations in the 2006 edition, because rooftop installations are increasingly used for "big box" retailing, commercial buildings, and other applications. Previous editions of the code were silent on the subject. Note that the requirement does not specify the height to which the pipe needs to be elevated above the roof; such height requirements vary across the country, depending on local weather and construction practices.

### 7.2.7 Removal of Pipe.

Where piping containing gas is to be removed, the line shall be first disconnected from all sources of gas and then thoroughly purged with air, water, or inert gas before any cutting or welding is done. *(See Section 8.3.)*

The potential for open fittings should be considered when removing gas piping. If a valve exists that will stop flow to pipe that is being removed, it should be verified that it is stopping the flow completely, and it should be plugged or capped after the pipe is removed. If piping is removed while the gas is shut off, the modified piping system must be inspected in accordance with Chapter 8, Inspection, Testing, and Purging.

Although 7.2.7 permits the use of air for purging, an inert gas or water may provide an increase in safety for larger-volume systems, because a flammable gas–air mixture is never reached. Cutting is not limited to flame cutting because mechanical methods, such as cutting with a manual or power saw, can produce sparks that can ignite a gas–air mixture. Additional safety information can be found in NFPA 51B, *Standard for Fire Prevention During Welding, Cutting, and Other Hot Work.*

### 7.2.8 CSST.

CSST piping systems shall be installed in accordance with this code and the manufacturer's installation instructions.

This requirement clarifies the coverage of CSST. Prior to the 2006 edition of the code, CSST was required to be tested and was listed in Chapter 6 (the piping materials chapter), but no installation guidance specific to CSST was provided in Chapter 7. In the 2006 edition, 5.6.4.3 was revised to require that CSST be listed, and this paragraph was added to cover the installation in Chapter 7. Note that CSST is treated as a listed system and is the only listed piping system. Other piping systems, such as steel pipe and copper tubing, are not listed.

## 7.3 Concealed Piping in Buildings

Section 7.3 covers the installation of piping in concealed spaces in buildings. Piping that is concealed by fixed construction is of greater concern than exposed piping because of the potential for damage, such as by nail puncture (especially to tubing) during alterations or when pictures are hung on a wall where any damage will not be seen.

### 7.3.1 General.

Gas piping in concealed locations shall be installed in accordance with this section.

### 7.3.2 Connections.

Where gas piping is to be concealed, unions, tubing fittings, right and left couplings, bushings, swing joints, and compression couplings made by combinations of fittings shall not be used. Connections shall be of the following type:

(1) Pipe fittings such as elbows, tees, and couplings
(2) Joining tubing by brazing *(see 5.6.8.2)*
(3) Fittings listed for use in concealed spaces that have been demonstrated to sustain, without leakage, any forces due to temperature expansion or contraction, vibration, or fatigue based on their geographic location, application, or operation
(4) Where necessary to insert fittings in gas pipe that has been installed in a concealed location, the pipe shall be reconnected by welding, flanges, or the use of a ground joint union with the nut center-punched to prevent loosening by vibration

The prohibited list of fittings in 7.3.2 consists of those that have more potential to leak than threaded steel fittings or are unnecessarily complicated for concealed installation. Brazed tubing joints are as strong as the tubing itself and therefore are acceptable.

◀ **FAQ**
Why are the fittings listed in 7.3.2 prohibited in concealed installations?

Paragraph 7.3.2(3) deals with the connections that must be concealed, such as those necessary for installation of listed gas convenience outlets. These fittings are a part of a listed system designed to meet the concerns that prohibit other tubing fittings in concealed locations. For example, fittings and connections for CSST systems would be allowed when installed per the manufacturer's instructions.

### 7.3.3 Piping in Partitions.

Concealed gas piping shall not be located in solid partitions.

Solid partitions are building spaces between walls that are completely solid. They can be solid wood, masonry blocks with concrete filling the spaces, or partitions with rigid solid insulation.

Gas piping (pipe and tubing) installed in solid partitions is prohibited because it can be damaged by nails or saws penetrating the partition in the event of construction or nails being hammered into the partition because the pipe or tubing is rigidly held in place.

◀ **FAQ**
Why is gas piping prohibited in solid partitions?

### 7.3.4 Tubing in Partitions.

NEW

This provision shall not apply to tubing that pierces walls, floors, or partitions. Tubing installed vertically and horizontally inside hollow walls or partitions without protection along its entire concealed length shall meet the following requirements:

Subsection 7.3.4 applies to tubing (and not pipe) that lies within a wall or partition that is not solid.

(1) A steel striker barrier not less than 0.0508 in. (1.3 mm) thick, or equivalent, is installed between the tubing and the finished wall and extends at least 4 in. (100 mm) beyond concealed penetrations of plates, firestops, wall studs, and so on.

The protection specified in 7.3.4(1) differs from that required for CSST by ANSI/LC-1, *Standard for Interior Fuel Gas Piping Systems Using Corrugated Stainless Steel Tubing*, which requires protection based on the distance of the penetration to the edge of the stud, joist, or plate, and the size of the CSST. These requirements are summarized in Commentary Table 7.1.

***COMMENTARY TABLE 7.1** Length of Striker Plate at CSST Penetrations*

| *Distance Between Finished Wall Surface and the CSST (in.)* | *CSST Size (in.)* | |
|---|---|---|
| | *≤1* | *>1* |
| | *Length of Striker Plate Beyond Penetration* | |
| <2 | 5 | Along entire concealed length where 2 × 4 studs are used** |
| ≥2 to ≤3 | 0* | |
| >3 | Not required | |

*Strike plate with no extension required.

**Where studs or joist 2 × 6 or larger are used, use the ≤1 in. column.

ANSI/LC-1 also requires that the width of the striker plate be at least 1.5 times the outside diameter of the tubing. When installing CSST, 7.2.8 requires that CSST be installed in accordance with this code and the manufacturer's installation requirements of the listing standard (LC-1 in this case). If the manufacturer's instructions differ from the requirements of this code for installing tubing, the more stringent one should be followed. Steel thickness of 0.058 in. (1.3 mm) is approximately 16 gauge.

Why is it prohibited to secure multiple runs of tubing together in a hollow wall?

FAQ ▶

(2) The tubing is installed in single runs and is not rigidly secured.

Strapping multiple pieces of tubing would make the assembly more rigid and less likely to be moved by a nail penetrating the hollow wall. If a nail penetrates the tubing, leakage will occur.

FAQ ▶

Why are plates required for tubing penetrating wall studs or plates?

Plates are required near penetrations. At this point, nails and screws are more apt to be able to pin down and penetrate a tube because it is restrained by the penetration. The required overlap of the plates protects against drywall installers missing the stud or plate with their screw or nail. This hazard is also the basis for the requirement that tubing away from penetrations must not be secured. Unsecured tubing can move. Thus, nails and screws will have difficulty penetrating the tubing because the tubing can be moved to the side by a nail or screw.

### 7.3.5 Piping in Floors.

**7.3.5.1 Industrial Occupancies.** In industrial occupancies, gas piping in solid floors such as concrete shall be laid in channels in the floor and covered to permit access to the piping with a minimum of damage to the building. Where piping in floor channels could be exposed to excessive moisture or corrosive substances, the piping shall be protected in an approved manner.

**7.3.5.2 Other Occupancies.** In other than industrial occupancies and where approved by the authority having jurisdiction, gas piping embedded in concrete floor slabs constructed with Portland cement shall be surrounded with a minimum of $1^1/_2$ in. (38 mm) of concrete and shall not be in physical contact with other metallic structures such as reinforcing rods or electrically neutral conductors. All piping, fittings, and risers shall be protected against corrosion in accordance with 5.6.6. Piping shall not be embedded in concrete slabs containing quickset additives or cinder aggregate.

Concrete containing cinder aggregate or quickset additives, usually calcium chloride, is not permitted because such products can attack the metals used to make gas piping.

It is important to remember that different requirements apply to piping in floors in industrial occupancies and in other occupancies. Access to the piping in industrial occupancies is necessary because materials that can corrode piping are often used in industrial occupancies.

## 7.4 Piping in Vertical Chases

Where gas piping exceeding 5 psi (34 kPa) is located within vertical chases in accordance with 5.5.1(2), the requirements of 7.4.1 through 7.4.3 shall apply.

### 7.4.1 Pressure Reduction.

Where pressure reduction is required in branch connections for compliance with 5.5.1, such reduction shall take place either inside the chase or immediately adjacent to the outside wall of the chase. Regulator venting and downstream overpressure protection shall comply with 5.8.5 and Section 5.9. The regulator shall be accessible for service and repair and vented in accordance with one of the following:

(1) Where the fuel gas is lighter than air, regulators equipped with a vent limiting means shall be permitted to be vented into the chase. Regulators not equipped with a vent limiting means shall be permitted to be vented either directly to the outdoors or to a point within the top 1 ft (0.3 m) of the chase.
(2) Where the fuel gas is heavier than air, the regulator vent shall be vented only directly to the outdoors.

### 7.4.2 Chase Construction.

Chase construction shall comply with local building codes with respect to fire resistance and protection of horizontal and vertical openings.

### 7.4.3* Ventilation.

A chase shall be ventilated to the outdoors and only at the top. The opening(s) shall have a minimum free area (in square inches) equal to the product of one-half of the maximum pressure in the piping (in psi) times the largest nominal diameter of that piping (in inches), or the cross-sectional area of the chase, whichever is smaller. Where more than one fuel gas piping system is present, the free area for each system shall be calculated and the largest area used.

**A.7.4.3** Only vertical chases are recognized by the coverage. It is believed that welded joints for a horizontal gas line would be preferable to a horizontal chase.

Subsection 7.4.3 contains specific provisions that are necessary to make safe installations in vertical chases. It is important to note that, because a chase is ventilated at the top, any leaked gas inside of the chase goes outdoors, not into the building.

## 7.5 Gas Pipe Turns

Changes in direction of gas pipe shall be made by the use of fittings, factory bends, or field bends.

### 7.5.1 Metallic Pipe.

Metallic pipe bends shall comply with the following:

(1) Bends shall be made only with bending tools and procedures intended for that purpose.
(2) All bends shall be smooth and free from buckling, cracks, or other evidence of mechanical damage.
(3) The longitudinal weld of the pipe shall be near the neutral axis of the bend.

Paragraph 7.5.1(3) allows a method of bending welding pipe without stressing the longitudinal weld. The neutral axis of the bend is the area where the metal is not stretched or compressed during the bending operation. The axis lies at the sides of the pipe at the approximate midpoint between the inside and the outside of the bend. (See Exhibit 7.11.)

*EXHIBIT 7.11 Examples of Bending Welding Pipe.*

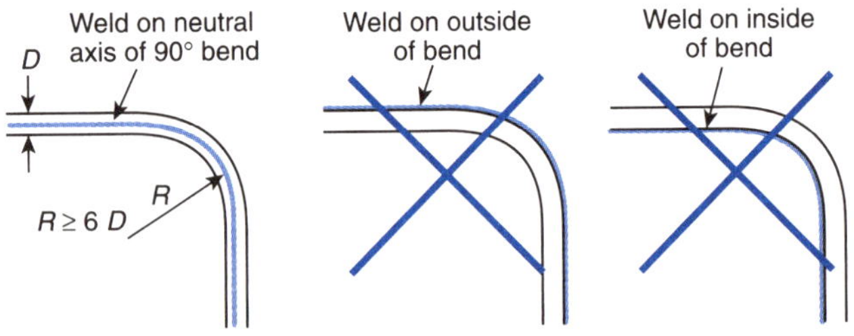


(4) Pipe shall not be bent through an arc of more than 90 degrees.
(5) The inside radius of a bend shall be not less than 6 times the outside diameter of the pipe.

### 7.5.2 Plastic Pipe.

Plastic pipe bends shall comply with the following:

(1) The pipe shall not be damaged, and the internal diameter of the pipe shall not be effectively reduced.
(2) Joints shall not be located in pipe bends.
(3) The radius of the inner curve of such bends shall not be less than 25 times the inside diameter of the pipe.
(4) Where the piping manufacturer specifies the use of special bending tools or procedures, such tools or procedures shall be used.

**FAQ ▶** What is the smallest radius bend for plastic pipe allowed by the code?

Paragraph 7.5.2(3) requires the radius to be at least 25 times the inside diameter of plastic pipe. If the pipe is 2 in. (50 mm) nominal size, the inside diameter is 1.943 in. (48.4 mm), per Table 6.2(t). The bend minimum bend radius is 25 × 1.943 in. (25 × 48.4 mm), or 48.6 in. (1.23 m), which is about 4 ft.

### 7.5.3 Elbows.

Factory-made welding elbows or transverse segments cut therefrom shall have an arc length measured along the crotch of at least 1 in. (25 mm) for pipe sizes 2 in. and larger.

## 7.6 Drips and Sediment Traps

A drip and a sediment trap are identical assemblies of piping but serve different purposes. Drips are used where the gas is "wet" or contains liquids. Sediment traps are installed to remove debris in the piping system. The major source of debris results from the installation of the piping itself.

### 7.6.1 Provide Drips Where Necessary.

For other than dry gas conditions, a drip shall be provided at any point in the line of pipe where condensate could collect. Where required by the authority having jurisdiction or the serving gas supplier, a drip shall also be provided at the outlet of the meter. This drip shall be installed so as to constitute a trap wherein an accumulation of condensate shuts off the flow of gas before it runs back into the meter.

The phrase "for other than dry gas conditions" in 7.6.1 clarifies that drips are needed when "other than dry gas conditions exist" and are not necessary when only dry gas is expected to be present. Note that sediment traps are still required in 9.6.7.

Drips were originally required in manufactured gas systems because water was fogged into the low pressure manufactured gas distribution system to keep the packing used in the bell joint pipe fittings wet. When the packing dried, it leaked. In the 1950s in the United States, manufactured gas was replaced by natural gas, and distribution systems were upgraded to handle the higher pressure in most natural gas distribution systems. As these changes were made, fogging was no longer needed. Today, it is believed that fogging is no longer done in fuel gas distribution systems in the United States.

Drips collect condensate from gases that contain condensable products (usually water) or oil added in the gas transmission process. Drips are required by the code if required by the gas supplier or the authority having jurisdiction. Usually, drips are located near the outlet of the meter or the service entrance and at other locations where condensate could collect. The drip is installed at the bottom of a downward-flowing line by placing a tee at the bottom of the line, installing a nipple and cap in the run of the tee, and continuing the pipe run out the side of the tee. Drips must be located so that they can be emptied. Drip locations also must be protected from freezing.

A drip can be required if the natural gas supplier expects condensate in the gas received from the transmission company. Drips are not required when the fuel gas is LP-Gas. Exhibit 7.12 illustrates a typical drip trap design.

Sediment traps are similar to drip traps, but they are intended for removal of debris that results from construction or corrosion.

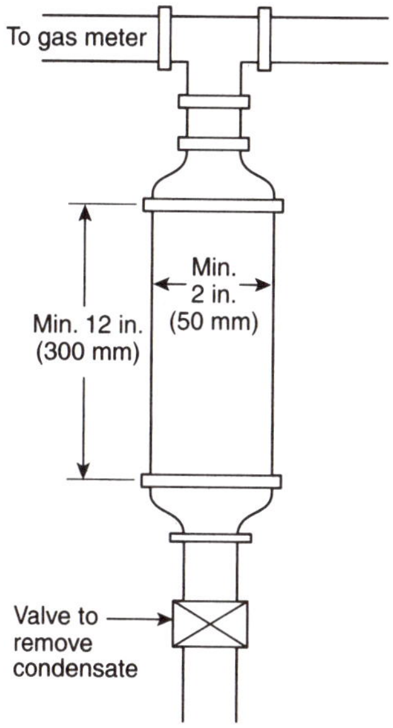


*EXHIBIT 7.12 Drip Trap. (Courtesy of J. J. Drechsler)*

### 7.6.2 Location of Drips.

All drips shall be installed only in such locations that they are readily accessible to permit cleaning or emptying. A drip shall not be located where the condensate is likely to freeze.

### 7.6.3 Sediment Traps.

See 9.6.7.

Sediment traps are covered in 7.6.3 with a reference to 9.6.7 (so that the requirements are not duplicated). Sediment traps, sometimes confused with drip traps, are installed to collect solid foreign particles to prevent such material from entering close-fitting parts or small passageways (e.g., valves and orifices). Although some fuel gases can contain foreign solids, this situation is not likely to be a problem in either utility gas or LP-Gas because of the methods and equipment used in handling these products. Dirt and pipe material cuttings that are placed unavoidably in the system itself (usually during construction) and are present in limited amounts are the target of sediment traps. Thus, sediment traps seldom need to be opened for service or cleaning. See Figure 9.6.7 for a depiction of a sediment trap.

Many appliance manufacturers are incorporating sediment traps in their appliances, and a number of the standards for gas appliances require the installation of a sediment trap.

Subsection 9.6.7 requires the installation of a sediment trap at the time of installation of most appliances if the appliance is not already equipped with one. It also states that "illuminating appliances, ranges, clothes dryers, decorative appliances for installation in vented fireplaces, gas fireplaces, and outdoor grills shall not be required to be so equipped" with a sediment trap.

◀ **FAQ**
**Why are sediment traps not required on appliances that are not automatically operated?**

These appliances are manually ignited, either with a match or by an electrical igniter. If the appliance does not light because sediment restricts the flow of fuel gas, the user will be aware of that in a relatively short time and will take corrective action. For example, if a load of wet clothing is not dried, the user will call for repair. A central heating furnace, boiler, or water heater is more critical, and a sediment trap is therefore required to prevent a shutdown, which could result in freezing of building piping. Sediment traps are required for automatically (thermostatically) operated appliances, such as central heating furnaces and boilers, water heaters, and space heaters.

## 7.7 Outlets

### 7.7.1 Location and Installation.

**7.7.1.1** The outlet fittings or piping shall be securely fastened in place.

**7.7.1.2** Outlets shall not be located behind doors.

**7.7.1.3** Outlets shall be located far enough from floors, walls, patios, slabs, and ceilings to permit the use of wrenches without straining, bending, or damaging the piping.

**7.7.1.4** The unthreaded portion of gas piping outlets shall extend not less than 1 in. (25 mm) through finished ceilings or indoor or outdoor walls.

**FAQ ▶** Why is a specified minimum length of unthreaded pipe projected past the penetrations of walls, ceilings, and floors?

Minimum projections of pipe are specified, in the case of walls and ceilings, to allow a wrench to grip the pipe. In the case of floors, the extra projection will tend to protect the threads from flooding and mechanical damage.

**7.7.1.5** The unthreaded portion of gas piping outlets shall extend not less than 2 in. (50 mm) above the surface of floors or outdoor patios or slabs.

**7.7.1.6** The provisions of 7.7.1.4 and 7.7.1.5 shall not apply to listed quick-disconnect devices of the flush-mounted type or listed gas convenience outlets. Such devices shall be installed in accordance with the manufacturers' installation instructions.

Because gas convenience outlets are listed as an installed assembly, the location and installation concerns are addressed by their listing, and the provisions of 7.7.1.6 need not apply.

### 7.7.2 Cap All Outlets.

**7.7.2.1** Each outlet, including a valve, shall be closed gastight with a threaded plug or cap immediately after installation and shall be left closed until the appliance or equipment is connected thereto. When an appliance or equipment is disconnected from an outlet and the outlet is not to be used again immediately, it shall be capped or plugged gastight.

The importance of this requirement cannot be stressed enough. Fires can be caused by gas leakage from open pipes in buildings. The incidents are usually of one of the following three types:

- Occurs during construction where an open end of a gas piping system is left unplugged prior to turning the gas on. Often there is an intent to plug the opening the next day, but it does not happen. If the gas system is not pressure tested, the open end may not be found.
- Occurs following the removal of an appliance, gas connector, and shutoff valve during a change of renters, often after the gas has been shut off.
- Occurs where an appliance has been removed, and the gas valve is turned off, but not sealed with a plug or cap. The valve, usually a $^1/_4$-turn gas cock, is accidentally opened by contact with moving furniture or other accidental contact with the valve handle.

All these accidents can be prevented by capping or plugging open lines if the appliance is not immediately connected, or after disconnection.

While data on the frequency of these types of fires cannot be reliably determined, anecdotes suggest that collectively these are a small yet significant part of fuel gas fires, due to the severity of these types of fires.

*Exception No. 1: Laboratory appliances installed in accordance with 9.6.2(1) shall be permitted.*

In Exception No. 1, laboratory equipment, such as a Bunsen burner, which is connected and disconnected from hose-end valves as a normal, everyday operation, is exempted from the requirement for a plug or cap. Due to its construction, it is not possible to plug a Bunsen burner, nor is it practical to plug a lab gas turret. The editor of this handbook is not aware of lab fires caused by this Exception, which has been in the code for many decades.

*Exception No. 2: The use of a listed quick-disconnect device with integral shutoff or listed gas convenience outlet shall be permitted.*

Listed quick-disconnect devices and gas convenience outlets can go uncapped or unplugged, because they incorporate valves that are integral with the device that will automatically shut off the gas either prior to or during the disconnect.

**7.7.2.2** Appliance shutoff valves installed in fireplaces shall be removed and the piping capped gastight where the fireplace is used for solid fuel burning.

A plug or cap is required for all gas openings in fireplaces to ensure that gas will not accidentally flow from an outlet that is not connected for use. Closing a valve is not enough to satisfy this requirement because the valve can be opened inadvertently or accidentally. No temporary or makeshift closure is permitted because anything except a proper plug or cap could leak. Because of the significant consequences of gas flow from an open outlet, there are no exceptions to this requirement, and it applies to gas piping installers who install outlets before an appliance that can be connected is present.

When a fireplace is converted from burning gas to burning solid fuel, the removal of the appliance shutoff valve from the fireplace is required. Gas shutoff valves will leak if they are subjected to the heat of a wood fire. Whenever possible, all gas piping in a fireplace to be used to burn solid fuel should also be removed as a preventive measure. Leaving the gas piping in the fireplace would subject the piping to extreme temperatures, which over time could cause the piping connections to leak.

## 7.8 Branch Pipe Connection

When a branch outlet is placed on a main supply line before it is known what size pipe will be connected to it, the outlet shall be of the same size as the line that supplies it.

Section 7.8 needs clarification because this requirement has been misinterpreted in the past to mean that the branch must be the same size as the largest main supply line. For example, it is sometimes claimed that if the main supply line is a 2 in. (50 mm) line and is reduced to a 1 in. (25 mm) line downstream, a branch from the 1 in. (25 mm) line must be 2 in. (50 mm). This type of size reduction increase is not the intent of this requirement.

◄ **FAQ**
Is a branch connection placed in a gas line before the size of the branch is known required to be the size of the largest main gas supply line?

In this case, a 1 in. (25 mm) branch is sufficient, because 1 in. (25 mm) is the size of the supply line where the branch is made. The reader must remember that when any modification is made to a fuel distribution system, the system must be recalculated to be certain that it can provide the necessary fuel without an excessive pressure drop. If the system cannot provide fuel without an excessive pressure drop, then larger or additional pipe must be installed.

## 7.9 Manual Gas Shutoff Valves

*(Also see 9.6.4.)*

### 7.9.1 Valves at Regulators.

An accessible gas shutoff valve shall be provided upstream of each gas pressure regulator. Where two gas pressure regulators are installed in series in a single gas line, a manual valve shall not be required at the second regulator.

The purpose of the accessible gas shutoff valve is to be able to remove the regulator without shutting off other branches of the piping system that may exist, which would affect appliances not served by the regulator being removed. If there are only two regulators in series, such as in an LP-Gas piping system serving a home, the valve at the LP-Gas tank meets the intent of this requirement even though it is installed outside the scope of NFPA 54.

### 7.9.2 Valves Controlling Multiple Systems.

**7.9.2.1 Accessibility of Gas Valves.** Main gas shutoff valves controlling several gas piping systems shall be readily accessible for operation and installed so as to be protected from physical damage. They shall be marked with a metal tag or other permanent means attached by the installing agency so that the gas piping systems supplied through them can be readily identified.

The main gas shutoff valve serving multiple gas systems is usually located at the gas meter and must be protected from damage. It must also be readily accessible, allowing emergency responders to shut off the flow of gas in an emergency. The term *readily accessible* is defined in 3.3.1.1 as follows:

> **3.3.1.1 Readily Accessible.** Having direct access without the need of removing or moving any panel, door, or similar covering of the item described.

Note that the code is silent as to the need for a tool to close the valve. The valves on natural gas meters are usually not provided with a hand wheel or other operator. Fire fighters in areas served by piped gas systems have a tool that fits the valve used on gas meter assemblies.

Main gas shutoff valves must be marked so that they can be easily identified. Main gas shutoff valves at propane tanks are usually not marked, because the propane tank itself is sufficient. When propane tanks are installed underground, the local emergency responders should be contacted to determine whether identification is required. Meters used with piped gas systems are self-identifying, and marking is usually not required.

**7.9.2.2 Shutoff Valves for Multiple House Lines.** In multiple-tenant buildings supplied through a master meter, through one service regulator where a meter is not provided, or where meters or service regulators are not readily accessible from the appliance or equipment location, an individual shutoff valve for each apartment or tenant line shall be provided at a convenient point of general accessibility. In a common system serving a number of individual buildings, shutoff valves shall be installed at each building.

The reason for requiring a shutoff valve for each line is to allow for the system to be separated for repair or maintenance. In a case where only part of the system has to be shut down for repair, the remaining system can continue to operate. Shutoff valves must be plainly marked so that in the future, service persons can determine which lines service which units or buildings.

The requirement of a shutoff valve for each building provides emergency responders with the ability to shut off the flow of gas to a specific building involved in fire.

**7.9.2.3 Emergency Shutoff Valves.** An exterior shutoff valve to permit turning off the gas supply to each building in an emergency shall be provided. The emergency shutoff valves shall be plainly marked as such and their locations posted as required by the authority having jurisdiction.

In natural gas installations, the emergency shutoff valve is part of the utility's service and meter installation. The natural gas supplier provides a shutoff valve for the supply to their gas meter, under U.S. Department of Transportation regulations (49 CFR 192). The emergency shutoff valve required in 7.9.2.3 can be the same valve that is required in 7.9.2.2 for multiple house lines.

◀ **FAQ**
Can the shutoff valve in front of the meter serve as the emergency shutoff valve?

If the natural gas supplier installs a shutoff valve and gas meter (or shutoff valve if no meter is provided), the shutoff valve can also serve as the emergency shutoff valve. If the fuel line is run from this location to another building underground, the line must be brought up outside and another shutoff valve installed. This second shutoff will be the emergency shutoff valve for the building it serves. In any piping system, there must be an accessible exterior shutoff valve.

The installations just described are presumed to be obvious enough to emergency responders to meet the requirement of being "plainly marked." The requirement of posting the location is not treated uniformly by different gas suppliers and local fire departments. Where an installation is not obvious (e.g., an underground propane tank or a meter/regulator concealed by vegetation), a sign or other marking may be needed or required by the authority having jurisdiction. The intent of this provision is to enable emergency responders to shut off the gas supply to a building in the event of a fire. The local fire department is normally the authority having jurisdiction for this requirement.

In LP-Gas systems, the emergency shutoff valve is normally the propane tank shutoff valve.

NEW

**7.9.2.4 Shutoff Valve for Laboratories.** Each laboratory space containing two or more gas outlets installed on tables, benches, or in hoods in educational, research, commercial, and industrial occupancies shall have a single shutoff valve through which all such gas outlets are supplied. The shutoff valve shall be accessible, located within the laboratory or adjacent to the laboratory's egress door, and identified.

The requirement in 7.9.2.4 is new in the 2009 edition. It recognizes that the master lab shutoff has become a commonly installed feature in most laboratories. In many laboratories, especially in educational occupancies, the gas turrets are not used every day, and the ability to disable them is a desirable safety feature.

## 7.10 Prohibited Devices

No device shall be placed inside the gas piping or fittings that reduces the cross-sectional area or otherwise obstructs the free flow of gas, except where proper allowance in the piping system design has been made for such a device and where approved by the authority having jurisdiction.

The intent of Section 7.10 is to prohibit restrictive devices in gas piping that could reduce the flow of gas. This is important because restrictions, if not taken into account when the gas piping systems are designed or sized, can reduce the flow of gas to the appliances, which may result in insufficient gas for proper appliance operation.

The committee also recognizes that some devices that can restrict the flow of gas are a necessary part of most gas piping systems. Examples of these devices include the following:

- An orifice plate (part of an orifice meter set assembly), which can be used to measure gas flow to a large gas user such as a boiler

- A gas meter
- A gas valve with other than a full, round, inside cross-section, which includes most valves used in gas piping systems

Devices should be sized adequately to allow for the pressure drop it produces. In residential gas piping systems that are sized using the tables in Chapter 13, sufficient capacity is provided to allow for one or more gas cock valves or other similar restrictive devices.

## 7.11 Systems Containing Gas–Air Mixtures Outside the Flammable Range

Where gas–air mixing machines are employed to produce mixtures above or below the flammable range, they shall be provided with stops to prevent adjustment of the mixture to within or approaching the flammable range.

## 7.12 Systems Containing Flammable Gas–Air Mixtures

### 7.12.1 Required Components.

A central premix system with a flammable mixture in the blower or compressor shall consist of the following components:

(1) Gas-mixing machine in the form of an automatic gas–air proportioning device combined with a downstream blower or compressor
(2) Flammable mixture piping, minimum Schedule 40
(3) Automatic firecheck(s)
(4) Safety blowout(s) or backfire preventers for systems utilizing flammable mixture lines above $2^1/_2$ in. nominal pipe size or the equivalent

### 7.12.2 Optional Components.

The following components shall also be permitted to be utilized in any type of central premix system:

(1) Flowmeter(s)
(2) Flame arrester(s)

### 7.12.3 Additional Requirements.

Gas-mixing machines shall have nonsparking blowers and shall be constructed so that a flashback does not rupture machine casings.

### 7.12.4* Special Requirements for Mixing Blowers.

A mixing blower system shall be limited to applications with minimum practical lengths of mixture piping, limited to a maximum mixture pressure of 10 in. w.c. (2.5 kPa) and limited to gases containing no more than 10 percent hydrogen. The blower shall be equipped with a gas control valve at its air entrance arranged so that gas is admitted to the airstream, entering the blower in proper proportions for correct combustion by the type of burners employed, the said gas control valve being of either the zero governor or mechanical ratio valve type that controls the gas and air adjustment simultaneously. No valves or other obstructions shall be installed between the blower discharge and the burner or burners.

**A.7.12.4** The mixing blower is acknowledged as a special case because of its inability to tolerate control valves or comparable restrictions between mixing blower(s) and burner(s). With these limitations, mixing blower installations are not required to utilize safety blowouts, backfire preventers, explosion heads, flame arresters, or automatic firechecks that introduce pressure losses.

## 7.12.5 Installation of Gas-Mixing Machines.

**7.12.5.1* Location.** The gas-mixing machine shall be located in a large well-ventilated area or in a small detached building or cutoff room provided with room construction and explosion vents in accordance with sound engineering principles. Such rooms or belowgrade installations shall have adequate positive ventilation.

**A.7.12.5.1** For information on venting of deflagrations, see NFPA 68, *Standard on Explosion Protection by Deflagration Venting*.

**7.12.5.2 Electrical Requirements.** Where gas-mixing machines are installed in well-ventilated areas, the type of electrical equipment shall be in accordance with *NFPA 70, National Electrical Code*, for general service conditions unless other hazards in the area prevail. Where gas-mixing machines are installed in small detached buildings or cutoff rooms, the electrical equipment and wiring shall be installed in accordance with *NFPA 70* for hazardous locations (Articles 500 and 501, Class I, Division 2).

**7.12.5.3 Air Intakes.** Air intakes for gas-mixing machines using compressors or blowers shall be taken from outdoors whenever practical.

**7.12.5.4* Controls.** Controls for gas-mixing machines shall include interlocks and a safety shutoff valve of the manual reset type in the gas supply connection to each machine arranged to automatically shut off the gas supply in the event of high or low gas pressure. Except for open burner installations only, the controls shall be interlocked so that the blower or compressor stops operating following a gas supply failure. Where a system employs pressurized air, means shall be provided to shut off the gas supply in the event of air failure.

**A.7.12.5.4** Additional interlocks might be necessary for safe operation of appliances supplied by the gas-mixing machine.

**7.12.5.5 Installation in Parallel.** Centrifugal gas-mixing machines in parallel shall be reviewed by the user and equipment manufacturer before installation, and means or plans for minimizing these effects of downstream pulsation and equipment overload shall be prepared and utilized as needed.

Gas–air mixers are used to premix gas and air where required by a few appliances. They are found only in industrial installations. Because the gas is flammable, 7.12.5.5 requires extensive precautions.

If a gas–air mixer is installed in a well-ventilated area of a building, no electrical area classification requirements are needed, but if the area is not well ventilated, the possibility of the accumulation of a flammable gas–air mixture exists. The electrical equipment in the area must be designed with this in mind. The reader is also advised to review NFPA 68, *Standard on Explosion Protection by Deflagration Venting*, which provides requirements for sizing of deflagration vents and other design information. Note that as of the 2007 edition of NFPA 68, this document is now a standard. Formerly, it was a recommended practice.

## 7.12.6 Use of Automatic Firechecks, Safety Blowouts, or Backfire Preventers.

Automatic firechecks and safety blowouts or backfire preventers shall be provided in piping systems distributing flammable air–gas mixtures from gas-mixing machines to protect the piping and the machines in the event of flashback, in accordance with the following:

(1)* Approved automatic firechecks shall be installed upstream as close as practicable to the burner inlets following the firecheck manufacturers' instructions.

**A.7.12.6(1)** Two basic methods are generally used. One calls for a separate firecheck at each burner, the other a firecheck at each group of burners. The second method is generally more practical if a system consists of many closely spaced burners.

An approved automatic firecheck should be installed as near as practical upstream from a flame arrester used for local protection where test burners or lighting torches are employed.

See Exhibit 7.13 for examples of a firecheck and a blowout.

***EXHIBIT 7.13*** *Firecheck and Blowout.*

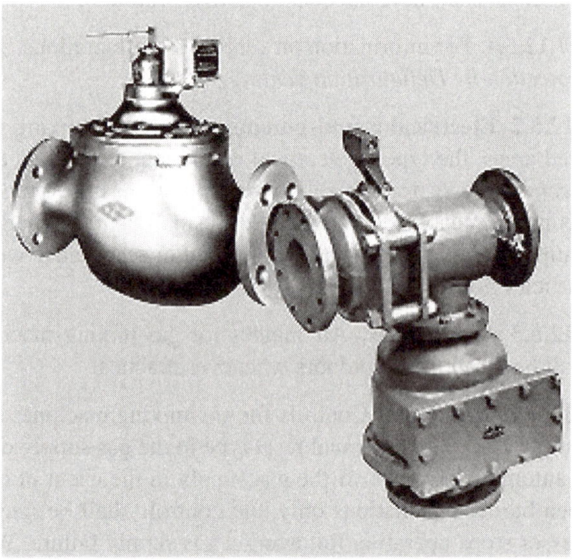

(2) A separate manually operated gas valve shall be provided at each automatic firecheck for shutting off the flow of gas–air mixture through the firecheck after a flashback has occurred. The valve shall be located upstream as close as practical to the inlet of the automatic firecheck. Caution: these valves shall not be reopened after a flashback has occurred until the firecheck has cooled sufficiently to prevent re-ignition of the flammable mixture and has been reset properly.

(3) A safety blowout or backfiring preventer shall be provided in the mixture line near the outlet of each gas-mixing machine where the size of the piping is larger than $2^1/_2$ in. NPS, or equivalent, to protect the mixing equipment in the event of an explosion passing through an automatic firecheck. The manufacturers' instructions shall be followed when installing these devices, particularly after a disc has burst. The discharge from the safety blowout or backfire preventer shall be located or shielded so that particles from the ruptured disc cannot be directed toward personnel. Wherever there are interconnected installations of gas-mixing machines with safety blowouts or backfire preventers, provision shall be made to keep the mixture from other machines from reaching any ruptured disc opening. Check valves shall not be used for this purpose.

(4) Large-capacity premix systems provided with explosion heads (rupture disc) to relieve excessive pressure in pipelines shall be located at and vented to a safe outdoor location. Provisions shall be provided for automatically shutting off the supply of the gas–air mixture in the event of rupture.

Any installation of a large-capacity premix system, which is addressed in 7.12.6(4), may require the advice and assistance of an experienced person or organization familiar with the design of such systems due to their complexity and the potential for fire.

## 7.13 Electrical Bonding and Grounding

### 7.13.1 Pipe and Tubing Other than CSST.

NEW

Each aboveground portion of a gas piping system other than CSST that is likely to become energized shall be electrically continuous and bonded to an effective ground-fault current path. Gas piping, other than CSST, shall be considered to be bonded when it is connected to appliances that are connected to the appliance grounding conductor of the circuit supplying that appliance.

The requirements for electrical bonding of gas piping systems were revised in the 2009 edition. The key phrase in 7.13.1 is *likely to become energized*. There is more than one way that gas piping can become electrically energized. However, the primary focus of the code is a failure of appliance wiring or an electrical component within the equipment that energizes the appliance and, potentially, any gas piping connected to it. This focus is based on life safety concerns. If this type of failure occurs, the equipment grounding conductor provided to the appliance (typical in a three-wire circuit) will provide a low-impedance pathway back to the service enclosure and the circuit breaker. The sizing of this bonding connection is, therefore, based on the size of the branch circuit that is used to energize the equipment. If there are gas appliances with no electrically powered components, then the gas piping system is not likely to become energized in this manner, and no bonding is required. However, it is also possible for the gas piping system to become energized by other means such as unintended contact with higher voltage lines and lightning strikes. The bonding requirements for these situations must be more robust, and this is typically beyond the capacity of the equipment grounding conductor and is not addressed in the *NEC*.

### 7.13.2 CSST.

NEW

CSST gas piping systems shall be bonded to the electrical service grounding electrode system at the point where the gas service enters the building. The bonding jumper shall not be smaller than 6 AWG copper wire or equivalent.

Two other changes in the 2009 edition of the code were designed to specifically recognize that metallic gas piping systems are sometimes connected to a lightning protection system. Specific recognition of lightning protection systems was added in the 2009 edition for two reasons. First, because of the new awareness of lightning damage as indicated by the number of CSST incidents, and second because some code enforcers considered the bonding connection to a lightning protection system to be prohibited by the requirements of 7.13.3. The intent of 7.13.3 is that underground gas piping not be intentionally used as a grounding electrode to protect any appliances or other electrical devices connected to the piping system. However, bonding of the gas piping system is permitted, but it must be attached downstream of the gas meter. Refer to NFPA 780 for information on the design and installation of lightning protection systems.

Gas piping does not require a separate bonding connection unless one of the following situations occurs:

◀ **FAQ**
When does gas piping need to have a separate bonding connection?

- There are gas appliances with electrical connections that are connected to ungrounded wiring systems (two-pronged plugs).
- There are sources of high voltage electricity outside the piping system that could energize the gas piping system. This situation is highly unusual or unlikely.
- The gas piping material is CSST (see 7.13.2).

### 7.13.3 Prohibited Use.

NEW

Gas piping shall not be used as a grounding conductor or electrode. This does not preclude the bonding of metallic piping to a grounding system.

**FAQ ▶**
**Why are the bonding requirements for CSST different from those for other fuel gas piping materials?**

The electrical bonding requirements for CSST and other metal piping systems have been separated in the 2009 edition. While the requirements in 7.13.1 for pipe and tubing (other than CSST) are identical to those in the previous edition of the code, CSST systems now require additional bonding to address its vulnerability to damage during electrical storms. Subsection 7.13.2 is an electrical requirement and is intended to provide additional clarification between the bonding requirements stipulated in the *National Fuel Gas Code* and the *NEC*.

The change follows the committee being made aware of several instances where CSST was damaged due to lightning strikes. The incidents appear to be caused by both direct and indirect strikes, where lightning strikes the ground nearby or a part of a building. The resulting electrical forces can energize the gas piping and, depending on the level of bonding, can lead to differences in voltage buildup between the CSST and other electrically conductive systems in the building (such as copper water pipe, coax cable, and electric wiring). When energized in this fashion, any differences in electrical potential can lead to arcing between the metallic pathways to ground that are in close proximity to each other. This arcing can result in the creation of a perforation through the wall of the CSST, which is thinner when compared to other gas piping materials. Gas leaking from this hole can be ignited, and in the presence of combustible material, it can result in a building fire.

The new requirement calls for a minimum 6 AWG copper wire (or equivalent when other than copper wire is used), which will provide a lower impedance and more efficient pathway to earth than available through the equipment grounding conductor. The equipment grounding conductor size is based on the branch circuit size, and is usually a 12 or 14 copper wire. The new provisions also require that the bonding connection be made at the gas service entrance to the building rather than at the appliance. This location will permit much of the energy from an indirect lightning strike to be diverted off the gas piping. The use of a 6 AWG size is a common practice (for residential applications) within the *NEC* for the bonding of many other metallic systems.

By increasing the size of the bonding conductor, the following benefits are obtained:

1. The voltage impressed on the gas pipe will be reduced.
2. The potential difference between parallel metallic pathways to ground, such as electric wire and copper water piping (that are also similarly bonded), can be minimized.

These two factors are significant contributors toward minimizing the occurrence of arcing and are expected to reduce the number of CSST fires.

NEW

### 7.13.4* Lightning Protection Systems.

Where a lightning protection system is installed, the bonding of the gas piping shall be in accordance with NFPA 780, *Standard for the Installation of Lightning Protection Systems.*

NEW

**A.7.13.4** NFPA 780, Section 4.14, requires that all grounding media including underground metallic piping systems be interconnected to provide a common ground potential. These underground piping systems are not permitted to be substituted for grounding electrodes but must be bonded to the lightning protection grounding system. Where galvanic corrosion is of concern, the bond may be made via a spark gap or gas discharge tube.

The specific reference to NFPA 780, *Standard for the Installation of Lightning Protection Systems,* was added to recognize that in buildings with lightning protection systems, all metal systems must be bonded to the lightning protection system. It has been reported that some enforcement officials have not allowed such connection and reference the *NEC*. Lightning protection systems are not covered by the *NEC*, and the bonding is needed and is not detrimental to the gas piping system.

## 7.14 Electrical Circuits

Electrical circuits shall not utilize gas piping or components as conductors.

*Exception: Low-voltage (50 V or less) control circuits, ignition circuits, and electronic flame detection device circuits shall be permitted to make use of piping or components as a part of an electric circuit.*

Section 7.14 continues the prohibition on electrical circuits that could develop a significant spark as well as a severe shock hazard. This section recognizes that if a gas pipe conducts high-voltage electricity and that gas pipe is disconnected, the person disconnecting the pipe can receive a severe shock. Electrical systems operating at 50 volts or less are exempt because they do not present a shock hazard.

## 7.15 Electrical Connections

**7.15.1** All electrical connections between wiring and electrically operated control devices in a piping system shall conform to the requirements of *NFPA 70, National Electrical Code. (See Section 7.13.)*

**7.15.2** Any essential safety control depending on electric current as the operating medium shall be of a type that shuts off (fail safe) the flow of gas in the event of current failure.

Electrical requirements for fuel gas systems are presented in Section 7.15 as they apply to piping systems and in Section 9.7 as they apply to appliances. The reference to the *NEC* is made to guide those who are not familiar with electrical requirements.

The requirement for essential safety controls covers items such as control valves installed in a gas line that are operated in an emergency. Such valves are sometimes installed in commercial and industrial buildings to stop the flow of gas when a fire alarm system activates. These valves are not prohibited, but they must stop the flow of gas in the event of failure of electric current. If such valves are installed in gas piping systems serving manually operated appliances such as ranges, provisions should be made so that the flow of gas cannot be restarted with gas valves in the open position.

## REFERENCES CITED IN COMMENTARY


The following publications are available from the National Fire Protection Association, 1 Batterymarch Park, Quincy, MA 02169-7471, 1-800-344-3555, www.nfpa.org.

NFPA 51B, *Standard for Fire Prevention During Welding, Cutting, and Other Hot Work*, 2009 edition.

NFPA 68, *Standard on Explosion Protection by Deflagration Venting*, 2007 edition.

*NFPA 70®*, *National Electrical Code®*, 2008 edition.

NFPA 780, *Standard for the Installation of Lightning Protection Systems*, 2008 edition.

The following publication is available from CSA America, Inc., 8501 East Pleasant Valley Road, Cleveland, OH 44131, 216-524-4990, www.csa-america.org.

ANSI/AGA LC 1, *Standard for Interior Fuel Gas Piping Systems Using Corrugated Stainless Steel Tubing*, 2007.

The following publication is available from the U.S. Government Printing Office, Washington, DC 20401, www.gpo.gov.

Title 49, Code of Federal Regulations, Part 192, "Transportation of Natural and Other Gas by Pipeline: Minimum Federal Safety Standards."

# Inspection, Testing, and Purging

CHAPTER 8

Chapter 8 of the code covers the final step in the process after installation of the gas piping system: testing and placing the pipe in service. Chapter 8 contains requirements for the following:

- Pressure testing and inspecting new and modified gas piping systems. Chapter 8 covers the piping system only and does not include appliances (covered in Chapter 9) and gas connectors (covered in Section 9.6). (See Section 8.1.)
- Piping system, appliance, and equipment leakage testing. Included are requirements for the leak test that is conducted before turning on the gas in a new or modified piping system, when leakage is suspected, and when the gas meter is replaced. The test is also conducted when an out-of-gas condition occurs. (See Section 8.2.)
- Purging or removing air from gas piping systems that are new or have been modified. (See Section 8.3.)

## 8.1 Pressure Testing and Inspection

Section 8.1 is the essence of the safety message in the code for piping. Before being put into operation, piping installations must be inspected and tested to determine that the total project complies with the code. The piping installation includes all fixed piping from the point of delivery to and including the equipment shutoff valves. Appliances and appliance connectors are not part of the fixed piping system and are not included in this test. The pressure test normally is conducted once, prior to initial operation of the piping system, and again only when the piping system is modified.

Section 8.1 is not intended to be used when a piping system is checked for leaks after initial testing. Section 8.2 covers these leak tests.

### 8.1.1* General.

**A.8.1.1** Because it is sometimes necessary to divide a piping system into test sections and install test heads, connecting piping, and other necessary appurtenances for testing, it is not required that the tie-in sections of pipe be pressure-tested. Tie-in connections, however, should be tested with a noncorrosive leak detection fluid after gas has been introduced and the pressure has been increased sufficiently to give some indications whether leaks exist.

The test procedure used should be capable of disclosing all leaks in the section being tested and should be selected after giving due consideration to the volumetric content of the section and to its location.

Under no circumstances should a valve in a line be used as a bulkhead between gas in one section of the piping system and test medium in an adjacent section, unless two valves are installed in series with a valved "telltale" located between these valves. A valve should not be subjected to the test pressure unless it can be determined that the valve, including the valve closing mechanism, is designed to safely withstand the test pressure.

**8.1.1.1** Prior to acceptance and initial operation, all piping installations shall be inspected and pressure tested to determine that the materials, design, fabrication, and installation practices comply with the requirements of this code.

Paragraph 8.1.1.1 applies to new piping systems and modifications to existing piping systems. Refer to Section 8.2 for existing piping systems.

**8.1.1.2** Inspection shall consist of visual examination, during or after manufacture, fabrication, assembly, or pressure tests as appropriate. Supplementary types of nondestructive inspection techniques, such as magnetic-particle, radiographic, and ultrasonic, shall not be required unless specifically listed herein or in the engineering design.

The supplemental tests listed in 8.1.1.2 are not normally required for flammable gas piping at low pressures. If the piping is in a public building or a building containing unusual hazards, there may be additional criteria for verification that all systems are installed properly, and one of these supplemental tests, or other tests, may be required.

**8.1.1.3** Where repairs or additions are made following the pressure test, the affected piping shall be tested. Minor repairs and additions are not required to be pressure tested, provided that the work is inspected and connections are tested with a noncorrosive leak-detecting fluid or other leak-detecting methods approved by the authority having jurisdiction.

An example of added piping that would fall under 8.1.1.3 is the additional pipe and fittings used after the required test of the system to hook up to the meter, from the building stub out. These joints should be checked with an appropriate leak-detecting fluid or electronic leak detector when the fuel gas is introduced into the system. Soap solution should not be used, as it may be corrosive.

**8.1.1.4** Where new branches are installed to new appliance(s), only the newly installed branch(es) shall be required to be pressure tested. Connections between the new piping and the existing piping shall be tested with a noncorrosive leak-detecting fluid or approved leak-detecting methods.

The code's intent is that all new piping should be pressure tested. However, the code also recognizes that testing minor areas or testing the minimum amount of material between two substantial areas that have been tested already is not needed, provided that the work is reasonably and carefully inspected and connections are tested with a leak-detecting solution or other leak-detecting methods. The requirement to use a noncorrosive, leak-detecting fluid recognizes that some users were employing solutions that contained ammonia and other corrosive compounds.

**8.1.1.5** A piping system shall be tested as a complete unit or in sections. Under no circumstances shall a valve in a line be used as a bulkhead between gas in one section of the piping system and test medium in an adjacent section, unless two valves are installed in series with a valved "telltale" located between these valves. A valve shall not be subjected to the test pressure unless it can be determined that the valve, including the valve closing mechanism, is designed to safely withstand the pressure.

**FAQ ▶** Can a gas piping system be separated into sections? If yes, how should the sections be separated?

A piping system might have to be tested in sections for any of several reasons. This code section allows piping systems to be tested in sections, with a valve separating the sections when adjacent sections are not being tested simultaneously. If two sections are tested simultaneously, two valves and a "valved telltale" are required. The first reason for this requirement is that the valve might not close completely, so testing could indicate a leak when there is none. Second, the valve could be opened inadvertently, allowing the test medium to contaminate the gas, which could lead to operational difficulties with the appliance or could result in a flammable gas–air mixture in the piping system. Overpressurization could also result. Using two valves with a valved "telltale" that is left open to atmosphere during the test is an acceptable method of separating adjacent sections being tested simultaneously. If one valve were used, leakage from one section could be perceived to come from the other section. Caution should be taken to ensure that if gas is leaked from the "telltale," it does not become a hazard. See Exhibit 8.1, which shows an example of how two sections of piping could be connected. The arrangement with one valve is not permitted.

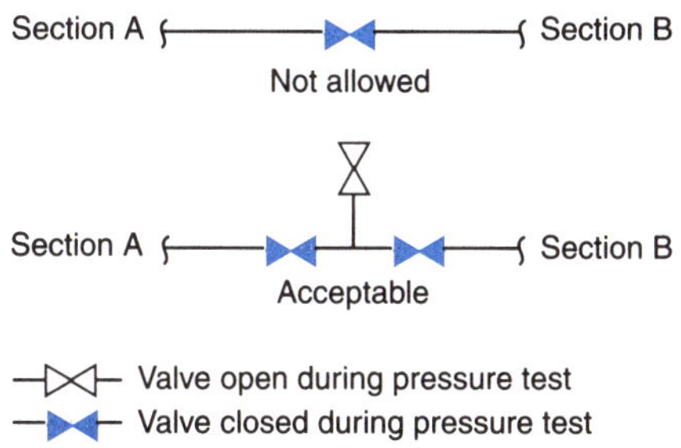


*EXHIBIT 8.1 Example of Valving of a Piping System with More than One Section.*

**8.1.1.6** Regulator and valve assemblies fabricated independently of the piping system in which they are to be installed shall be permitted to be tested with inert gas or air at the time of fabrication.

Paragraph 8.1.1.6 permits the use of regulator and valve assemblies that are assembled and tested elsewhere. By implication, any tested subassembly should be acceptable without field testing. While not common in residential construction, chemical plants often use prefabricated pipe assemblies due to the large amount of piping to be installed. A leak check must be performed after the gas is turned on, in accordance with 8.2.3.

### 8.1.2 Test Medium.

The test medium shall be air, nitrogen, carbon dioxide, or an inert gas.

OXYGEN SHALL NEVER BE USED.

The prohibition on the use of oxygen as a test medium is based on the fact that a mixture of fuel gas and pure oxygen, if ignited, can cause the pipe to explode. That oxygen-enriched fuel gas can ignite is not a theoretical possibility but a real danger. For example, in a building housing jewelry manufacturers, a gas pipe exploded when oxygen was allowed to enter fuel gas piping.

The term *inert gas* is intended to include gases that will not burn when mixed with air. Common inert gases include argon, nitrogen, carbon dioxide, and helium.

### 8.1.3 Test Preparation.

**8.1.3.1** Pipe joints, including welds, shall be left exposed for examination during the test.

*Exception: Covered or concealed pipe end joints that have been previously tested in accordance with this code.*

**8.1.3.2** Expansion joints shall be provided with temporary restraints, if required, for the additional thrust load under test.

The requirement of 8.1.3.2 recognizes that expansion joints can fail during testing if not properly restrained for a pressure beyond their normal operating range. Manufacturers' information can be helpful in planning a system to secure expansion joints for a pressure test.

**8.1.3.3** Appliances and equipment that are not to be included in the test shall be either disconnected from the piping or isolated by blanks, blind flanges, or caps. Flanged joints at which blinds are inserted to blank off other equipment during the test shall not be required to be tested.

**8.1.3.4** Where the piping system is connected to appliances or equipment designed for operating pressures of less than the test pressure, such appliances or equipment shall be isolated from the piping system by disconnecting them and capping the outlet(s).

Paragraph 8.1.3.4 recognizes that the 3 psi (20 kPa) minimum test pressure (8.1.4.2) can damage appliances that operate at about $^{1}/_{2}$ psi [14 in. w.c. (3 kPa)]. In cases where the piping to the appliance has been isolated according to 8.1.3.3, the piping must be reconnected after the test. The piping between the shutoff or cap and the appliance must be checked to be sure that there are no leaks. The leak detection methods described in Annex D are widely used. Leak solution or electronic leak detectors can also be used. See Section 8.2 for additional details.

**8.1.3.5** Where the piping system is connected to appliances or equipment designed for operating pressures equal to or greater than the test pressure, such appliances or equipment shall be isolated from the piping system by closing the individual appliance or equipment shutoff valve(s).

**8.1.3.6** All testing of piping systems shall be done with due regard for the safety of employees and the public during the test. Bulkheads, anchorage, and bracing suitably designed to resist test pressures shall be installed if necessary. Prior to testing, the interior of the pipe shall be cleared of all foreign material.

Safety must be a primary consideration when a gas test is conducted. Residential systems tested at 3 psig (21 kPag) require cleaning and do not normally require any additional safety precautions. In systems that are significantly larger or are operated at higher pressures, 8.1.3.6 outlines some provisions that may be prudent, especially if the building is occupied during the test. Paragraph 8.1.3.6 also addresses the clearing of foreign material from pipe systems, which can be done with compressed air, nitrogen, or other nonflammable gas prior to connecting any appliances.

### 8.1.4 Test Pressure.

To conduct a pressure test as required by 8.1.4, a manometer or pressure gauge is installed temporarily in the piping system being tested. Exhibit 8.2 shows the location of a gauge on a system to be served by natural gas. The pressure gauge has been installed where the gas meter will go. In this exhibit, the natural gas utilities service line to the building is at the left, with the service regulator. The horizontal bar at the top is a "meter bar" and does not carry gas. The pressure gauge can be seen below the right side of the meter bar.

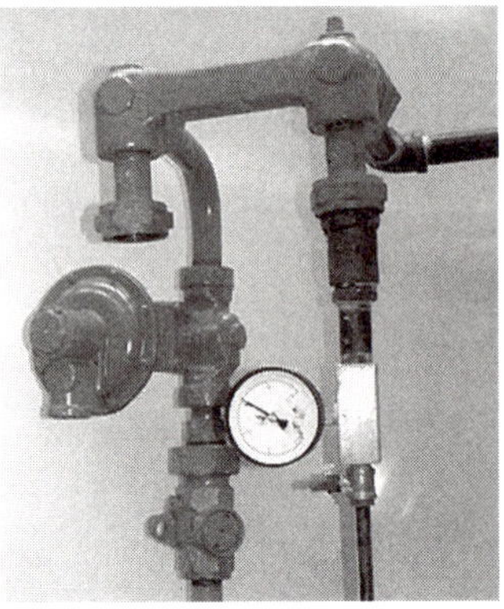

*EXHIBIT 8.2 Pressure Gauge Installation. (Courtesy of NSTAR)*

- The test is being conducted at 3 psi (20 kPa), because the piping system will operate at 7 in. w.c. (1.4 kPa). Paragraph 8.1.4.2 requires the test pressure to be 1.5 times the operating pressure, with a minimum of 3 psig (21 kPag).
- The gauge has a maximum reading of 5 psi (34 kPa), and each gauge mark is 0.1 psi (0.02 kPa). Paragraph 8.1.4.1 limits the maximum gauge reading to 5 times the test pressure, or 15 psig (100 kPag) in this case where the test pressure is 3 psig (21 kPag).
- When the system passes the test, the pressure gauge is removed, and a meter is installed.

**8.1.4.1** Test pressure shall be measured with a manometer or with a pressure measuring device designed and calibrated to read, record, or indicate a pressure loss due to leakage during the pressure test period. The source of pressure shall be isolated before the pressure tests are made. Mechanical gauges used to measure test pressures shall have a range such that the highest end of the scale is not greater than 5 times the test pressure.

Paragraph 8.1.4.1 requires that mechanical gauges used to measure test pressures include a scale not to exceed 5 times the test pressure. This limit provides accuracy approximately equal to a water manometer, the traditional pressure-measuring instrument used for low-pressure gas piping systems.

◀ **FAQ**
How do I determine what pressure range the scale on my test gauge should have?

For example, when testing a 2 psi (14 kPa) system, the test pressure will be 3 psi (20 kPa) [1.5 times 2 psi (14 kPa)] and a pressure test on a 7 in. or an 11 in. w.c. (1.4 or 2.7 kPa) system will also be 3 psi (20 kPa) (minimum test pressure). The gauge can have a maximum reading of 15 psi (100 kPa) [3 psi (34 kPa) times 5]. Thus, a gauge with a maximum reading of 10 psi (69 kPa) can be used, but one with a 20 psi (140 kPa) maximum cannot.

**8.1.4.2** The test pressure to be used shall be no less than 1½ times the proposed maximum working pressure, but not less than 3 psi (20 kPa), irrespective of design pressure. Where the test pressure exceeds 125 psi (862 kPa), the test pressure shall not exceed a value that produces a hoop stress in the piping greater than 50 percent of the specified minimum yield strength of the pipe.

**8.1.4.3*** Test duration shall be not less than ½ hour for each 500 $ft^3$ (14 $m^3$) of pipe volume or fraction thereof. When testing a system having a volume less than 10 $ft^3$ (0.28 $m^3$) or a system in a single-family dwelling, the test duration shall be a minimum of 10 minutes. The duration of the test shall not be required to exceed 24 hours.

The duration of the pressure test of a piping system installed in a residential occupancy is 10 minutes. Few residential gas piping systems require a test of longer than 10 minutes because most residential gas piping systems have a volume of less than 500 $ft^3$ (14 $m^3$), which is equivalent to approximately 16 mi (26 km) of ½ in. pipe or about 4 mi (6 km) of 2 in. pipe. The 10-minute test limit for residential gas piping systems clarifies the test duration requirement for those who test piping systems and for enforcement officials. The longer time requirement was never intended for the smaller, residential gas piping systems.

**A.8.1.4.3** During pressure tests conducted over long periods of time, such as overnight, the effects of temperature on pressure should be considered. Temperature drops can cause a drop in pressure great enough to be indicated by the test gauge. These temperature drops may cause test evaluators to think that a leak exists in the piping system when in fact the pressure drop was caused by a decrease in the ambient temperature. See Example 5 in C.8.5.

### 8.1.5 Detection of Leaks and Defects.

**8.1.5.1** The piping system shall withstand the test pressure specified without showing any evidence of leakage or other defects. Any reduction of test pressures as indicated by pressure gauges shall be deemed to indicate the presence of a leak unless such reduction can be readily attributed to some other cause.

Other possible causes of a drop in the pressure after a test include sharp drops in temperature. The likelihood of this phenomenon causing a pressure change is small, except on very large systems or systems with long test times. A pressure drop almost always means a leak. When trying to locate leaks, it must be remembered that the test apparatus itself can be the culprit.

◀ **FAQ**
Will a pressure test cause a piping system to fail if there is a rise in the test pressure?

A temperature rise can result in an increase in the pressure. If a pressure test is started, and the reading is delayed, a change in pressure may be observed if the temperature changes significantly. Normally, an inspector would view an increase in pressure as an indication that the system is gastight. Care should be taken that the pressure does not "peg" the upper limit of the gauge. The excess pressure could damage the gauge, and an inspector may view the test as faulty because the pressure is out of the measurable range of the gauge during the required test duration. Refer to Annex C, C.8.5 of the code, for a sample calculation of the pressure change due to a change in temperature. It shows that a piping system tested at 20 psi (140 kPa) and 70°F (21°C) will drop to 18 psig (124 kPag) at 40°F (4°C).

**8.1.5.2** The leakage shall be located by means of an approved gas detector, a noncorrosive leak detection fluid, or other approved leak detection methods. **Matches, candles, open flames, or other methods that provide a source of ignition shall not be used.**

**8.1.5.3** Where leakage or other defects are located, the affected portion of the piping system shall be repaired or replaced and retested. *(See 8.1.1.3.)*

NEW

## 8.2 Piping System Leak Check

Note that the title of Section 8.2 has been changed from "Piping System, Appliance, and Equipment Leakage Check." This title change was made in conjunction with a change in the definition of the term *leak check*, which recognizes that the leak check is done on a piping system, not appliances. The definition now reads as follows:

> **3.3.64 Leak Check.** An operation performed on a gas piping system to verify that the system does not leak.

While the leak check can be done with appliances connected, appliances can be isolated and the test redone if the leak check fails. Some appliances are allowed to have very small "leaks" by the standards to which they are manufactured and tested. Such small leaks will not accumulate flammable amounts of fuel gas.

Section 8.2 describes a method for checking for leakage throughout the system of piping and appliances. It is equally applicable to new and existing gas systems. For systems of new piping, it is conducted after the pressure test described in Section 8.1. Normally, the test for leakage is conducted when one of the following occurs:

- A system of new or modified gas piping is placed into service for the first time.
- A gas leakage is suspected.
- A gas meter is replaced.
- An appliance or appliance connector is replaced.
- An out-of-gas condition occurs.

The pressure test and leak check are often confused with each other. A pressure test is required for new piping installations and additions to piping installations, while a leak check is required whenever the gas system is initially placed into service or when the gas is turned back on after being turned off.

The test for leakage differs from the pressure test in Section 8.1 in that it requires no special preparations.

The medium used for a leak check is fuel gas at its normal supply pressure. The gas is applied to the total system (i.e., piping, equipment, and equipment connections and valves). A manometer or pressure gauge that measures inches of water column (in. w.c.) can be used for the test. Exhibit 8.3 illustrates a manometer at atmospheric pressure and one at 7 in. w.c.

See Annex D, Suggested Method of Checking for Leakage, for a recommended leak check procedure. The procedure in Annex D is only one possible method. Other test methods can be used. It is recommended that a written procedure for the method be developed and that steps are taken to ensure that all employees follow the method so that companies test every system identically.

Note that the methodology of Annex D does not include the use of leak detect solution or electronic leak detectors. The reason for this exclusion is that leaks in an existing or newly installed gas system may not be detectable using either leak detect solution or electronic leak detectors. One example of this could be with buried or hidden piping.

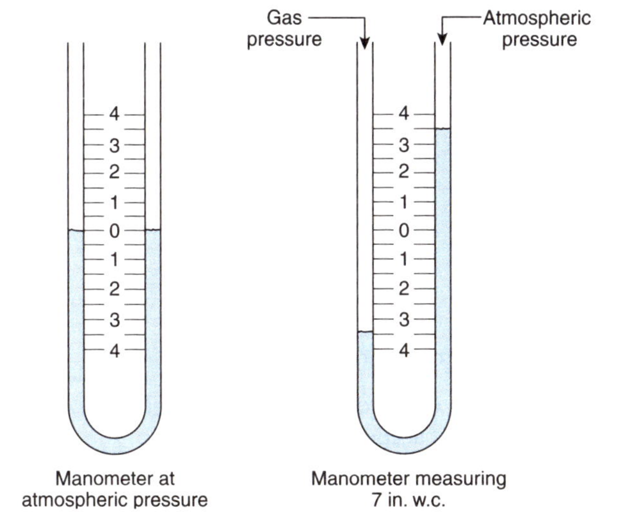


***EXHIBIT 8.3*** *Manometers.*

## 8.2.1 Test Gases.

Leak checks using fuel gas shall be permitted in piping systems that have been pressure tested in accordance with Section 8.1.

The requirement in 8.2.1 clarifies that fuel gas may be used to test for leaks in piping systems after the system has been pressure tested in accordance with Section 8.1.

## 8.2.2 Turning Gas On.

During the process of turning gas on into a system of new gas piping, the entire system shall be inspected to determine that there are no open fittings or ends and that all valves at unused outlets are closed and plugged or capped.

Note that the title of 8.2.2 was changed in the 2009 edition of the code from "Before Turning Gas On," and the first sentence was changed from "Before gas is introduced into a system. . . ." These changes were made to clarify that the visual inspection for open ends and valves that are not plugged or capped must be done at the time the gas is turned on. The wording of the text in the previous edition could be interpreted to allow the inspection for open ends to be done days or weeks before gas is first introduced.

The requirement in 8.2.2 applies only to new systems. Open fittings or ends can occur both in new construction and in existing systems. In new systems, it is possible that the pipe installer not having the proper size plug or cap may leave a system outlet open, assuming the appliance will be installed before the system is first pressurized. In existing systems, occupants may remove appliances after the gas has been shut off, and the valve may also be removed. There have been incidents in which a gas range, dryer, or other appliance has been removed, along with the gas connector and the shutoff valve, after the gas service has been discontinued. If the new occupant is not aware that a gas appliance previously existed at the location, which is more probable if remodeling has been done, an uncapped end would remain open after gas is turned on for the new occupant.

## 8.2.3* Leak Check.

Immediately after the gas is turned on into a new system or into a system that has been initially restored after an interruption of service, the piping system shall be checked for leakage. Where leakage is indicated, the gas supply shall be shut off until the necessary repairs have been made.

Note that the requirement in 8.2.3 applies to both new and existing systems and requires that a leak check be done following an interruption of service, as well as for new systems. The leak check is not intended to determine that the system is free of all leaks, only of leaks that will not be dissipated before reaching a flammable level by normal air exchange in a building.

The requirement in 8.2.3 states that if leakage is indicated by the test, repairs must be made. The leak check is not a test that is intended to find extremely small leaks. Thus, the leak check should not detect the very small amount of gas leakage, such as the amount allowed by the standard for gas valves incorporated in gas appliances, which is in the range of a few bubbles per hour, and is not a safety issue.

**A.8.2.3** See Annex D for a suggested method.

### 8.2.4 Placing Appliances and Equipment in Operation.

Appliances and equipment shall not be placed in operation until after the piping system has been tested in accordance with 8.2.3 and purged in accordance with 8.3.2.

Although gas piping systems are purged before being operated, caution should always be used when lighting equipment. A chance exists that not all of the air in the system is gone, and delayed ignition can result. Reasonable precautions include wearing safety glasses and keeping face, hands, and arms away from combustion chambers.

## 8.3* Purging

**A.8.3** The process of purging a gas pipeline of fuel gas and replacing the fuel gas with air or charging a gas pipeline that is full of air with fuel gas requires that a significant amount of combustible mixture not be developed within the pipeline or released within a confined space.

### 8.3.1 Removal from Service.

When gas piping is to be opened for an addition, a modification, or service, the section to be worked on shall be turned off from the gas supply at the nearest convenient point and the line pressure vented to the outdoors or to ventilated areas of sufficient size to prevent accumulation of flammable mixtures. The remaining gas in this section of pipe shall be displaced with an inert gas as required by Table 8.3.1.

***TABLE 8.3.1*** *Length of Piping Requiring Purging with Inert Gas for Servicing or Modification*

| *Nominal Pipe Size (in.)* | *Length of Piping Requiring Purging (ft)* |
|---|---|
| $2^1/_2$ | > 50 |
| 3 | > 30 |
| 4 | > 15 |
| 6 | > 10 |
| 8 or larger | Any length |

For SI units, 1 ft = 0.305 m.

**FAQ ▶** When is purging a gas system required?

Purging is required when the system is being added to, modified, or serviced, and it exceeds the diameters and lengths given in Table 8.3.1. For example, a gas system with 2 in. (50 mm) or smaller piping, which includes the majority of single-residence systems, does not require purging. Purging requirements start at systems containing $2^1/_2$ in. (65 mm) and larger pipes.

### 8.3.2 Placing in Operation.

When piping full of air is placed in operation, the air in the piping shall be displaced with fuel gas, except where such piping is required by Table 8.3.2 to be purged with an inert gas prior to introduction of fuel gas. The air can be safely displaced with fuel gas, provided that a moderately rapid and continuous flow of fuel gas is introduced at one end of the line and air is vented out at the other end. The fuel gas flow shall be continued without interruption until the vented gas is free of air. The point of discharge shall not be left unattended during purging. After purging, the vent shall then be closed. Where required by Table 8.3.2, the air in the piping shall first be displaced with an inert gas, and the inert gas shall then be displaced with fuel gas.

***TABLE 8.3.2*** *Length of Piping Requiring Purging with Inert Gas Before Being Placed in Operation*

| *Nominal Pipe Size (in.)* | *Length of Piping Requiring Purging (ft)* |
|---|---|
| 3 | > 30 |
| 4 | > 15 |
| 6 | > 10 |
| 8 or larger | Any length |

For SI units, 1 ft = 0.305 m.

### 8.3.3 Discharge of Purged Gases.

The open end of piping systems being purged shall not discharge into confined spaces or areas where there are sources of ignition unless precautions are taken to perform this operation in a safe manner by ventilation of the space, control of purging rate, and elimination of all hazardous conditions.

### 8.3.4 Placing Appliances and Equipment in Operation.

After the piping system has been placed in operation, all appliances and equipment shall be purged and then placed in operation, as necessary.

The goal of Section 8.3 is to provide instructions to minimize the possibility of developing a combustible mixture either in the piping or while gas is exiting from the piping during purging. Accordingly, if a system is to be shut down for service, removing the gas by filling the piping with a nonflammable atmosphere is necessary.

Purged gas is usually vented outdoors, and then the system is filled with inert gas. If the pipe volume is small, inert gas does not need to be used, which simplifies the gas purging process. The inert gas step can be eliminated for small systems because of the small volume of potentially flammable mix that could be produced.

Placing the system back in service presents the same possibility of a flammable mix being formed within the pipe system. First, larger systems must be filled with inert gas, then with fuel gas. Again, smaller systems can be purged directly with fuel gas, which simplifies the procedure. Regardless of the method, the purge outlet must be attended during the procedure. Although discharge of purged gases is allowed indoors with appropriate restrictions, outdoor discharge eliminates any associated hazard and is the preferred method when practical.

# Appliance, Equipment, and Accessory Installation

## CHAPTER 9

Chapter 9 of the code, along with Chapter 10, covers requirements for appliance installation and applies to all gas appliances. Chapter 10, Installation of Specific Appliances, provides additional requirements for specific types of appliances that are used primarily in residential and commercial installations. Chapter 9 includes references to Chapter 12, Venting of Appliances, to remind the installer that appliances must be properly vented.

Section 9.1 provides the following general requirements covering many appliances and installations:

- Approval of appliances and accessories (See 9.1.1.)
- Adding or converting of appliances to existing systems (See 9.1.2.)
- Type of gas being used (See 9.1.3.)
- Safety shutoff devices for LP-Gas appliances used indoors (See 9.1.4.)
- Use of air or oxygen under pressure (See 9.1.5.)
- Protection of appliances from fumes or gases, other than products of combustion (See 9.1.6.)
- Process air requirements (See 9.1.7.)
- Building structural members (See 9.1.8.)
- Flammable vapors (See 9.1.9.)
- Installation in residential garages, commercial garages, and aircraft hangars (See 9.1.10 through 9.1.12.)
- Physical protection of appliances (See 9.1.13.)
- Venting of flue gases (See 9.1.14.)
- Extra devices or attachments (See 9.1.15.)
- Adequacy of pipe sizing and prevention of strain on piping (See 9.1.16 and 9.1.17.)
- Installation and venting of appliance pressure regulators, and bleed lines for diaphragm valves (See 9.1.18 through 9.1.20.)
- Combination of appliances and equipment (See 9.1.21.)
- Installation instructions (See 9.1.22.)
- Protection of outdoor appliances (See 9.1.23.)

Section 9.2, Accessibility and Clearance, provides requirements for installers to allow for access to appliances for maintenance and clearance to combustible construction.

Section 9.3, Air for Combustion and Ventilation, provides the following requirements for sufficient air for gas to burn properly, dilution of products of combustion, and ventilation of appliance rooms and spaces:

- General requirements (See 9.3.1.)
- Requirements for indoor combustion air (See 9.3.2.)
- Requirements for outdoor combustion air (See 9.3.3.)
- Combination of indoor and outdoor combustion air (See 9.3.4.)
- Engineered installations (See 9.3.5.)
- Mechanical combustion air supply (See 9.3.6.)
- Louvers, grilles, and screens (See 9.3.7.)
- Combustion air ducts (See 9.3.8.)

Section 9.4, Appliances on Roofs, provides requirements for roof installations, and Section 9.5, Appliances in Attics, provides requirements for appliances in attics.

Section 9.6, Appliance and Equipment Connections to Building Piping, provides the following requirements for connection of appliances to the piping system:

- Materials used to connect appliances to a building piping system. (See 9.6.1.)
- Materials to connect movable appliances to a building piping system. (See 9.6.3.)
- Appliance shutoff valves (See 9.6.4.)
- Quick-disconnect devices and gas convenience outlets (See 9.6.5 and 9.6.6.)
- Sediment traps (See 9.6.7.)
- Installation of piping so as not to interfere with appliances (See 9.6.8.)

Section 9.7, Electrical, includes requirements for electrical connections, ignition and control devices, circuits, and continuous power for electrically controlled appliances.

Section 9.8, Room Temperature Thermostats, provides requirements for location of thermostats for thermostatically controlled space conditioning appliances.

## 9.1 General

### 9.1.1* Appliances, Equipment, and Accessories to Be Approved.

Appliances, equipment, and accessories shall be approved.

**A.9.1.1** The American Gas Association, American National Standards Institute, Inc., and the National Fire Protection Association do not approve, inspect, or certify any installations, procedures, appliances, equipment, or materials; nor do they approve or evaluate testing laboratories. In determining acceptability of installations, procedures, appliances, equipment, or materials, the authority having jurisdiction can base acceptance on compliance with AGA, ANSI, CSA, or NFPA, or other appropriate standards. In the absence of such standards, said authority can require evidence of proper installation, procedure, or use. The authority having jurisdiction can also refer to the listings or labeling practices *(see 3.2.3, Labeled, and 3.2.4, Listed)* of an organization concerned with product evaluations and is thus in a position to determine compliance with appropriate standards for the current production of listed items. Additional information regarding the coordination of appliance design, construction, and maintenance can be found in Annex B.

The authority having jurisdiction (AHJ), which may be the local building or fire official, insurer, property owner, or commanding officer (in the military), approves the materials, equipment, appliances, and method of construction as the result of investigations and tests or by reason of accepted principles or tests by national authorities, technical organizations, and scientific organizations.

**FAQ ►**
**How can the authority having jurisdiction determine whether a specific appliance or piece of equipment is approved?**

The authority having jurisdiction often has lists of acceptable organizations that provide listing services. These services provide details on the terms of the listing and labeling, identifying the equipment and appliances that meet the standards appropriate to its use. A label carrying the identifying mark of the listing agency affixed to the appliance, or inclusion in the installation manual provides the AHJ with a fast way of approving equipment and appliances. See Exhibit 10.8 in Chapter 10 for an example of an appliance label with an agency logo. See also the definitions of labeled (3.2.3) and listed (3.2.4) in Chapter 3.

Subsection 9.1.4 provides criteria for approving unlisted equipment and appliances and provides three criteria for acceptance. Additionally, Annex B, Coordination of Appliance and Equipment Design, Construction, and Maintenance, is provided to assist an installer in developing criteria for acceptance of nonconventional gas systems that have been properly designed and installed. Annex B is especially useful when applied to industrial systems.

**9.1.1.1** Approved shall mean "acceptable to the authority having jurisdiction."

**9.1.1.2** Listed appliances, equipment, and accessories shall be installed in accordance with Chapter 8 and the manufacturers' installation instructions.

**9.1.1.3** Acceptance of unlisted appliances, equipment, and accessories shall be on the basis of a sound engineering evaluation.

**9.1.1.4** The unlisted appliance, equipment, or accessory shall be safe and suitable for the proposed service and shall be recommended for the service by the manufacturer.

### 9.1.2 Added or Converted Appliances.

When additional or replacement appliances or equipment is installed or an appliance is converted to gas from another fuel, the location in which the appliances or equipment is to be operated shall be checked to verify the following:

(1) Air for combustion and ventilation is provided where required, in accordance with the provisions of Section 9.3. Where existing facilities are not adequate, they shall be upgraded to meet Section 9.3 specifications.
(2) The installation components and appliances meet the clearances to combustible material provisions of 9.2.2. It shall be determined that the installation and operation of the additional or replacement appliances do not render the remaining appliances unsafe for continued operation.
(3) The venting system is constructed and sized in accordance with the provisions of Chapter 12. Where the existing venting system is not adequate, it shall be upgraded to comply with Chapter 12.

The requirements in 9.1.2 are meant to ensure that installation procedures are thoroughly reviewed when an appliance is installed, replaced, or converted to gas from another fuel. The references specify the need to verify that air for combustion and ventilation is adequate, that the clearance to combustible material is sufficient, and that the venting system meets existing requirements. If any of the requirements are not met, the portion not meeting code requirements must be upgraded to present code standards.

Paragraph 9.1.2(3) emphasizes that installations are not "grandfathered" when the appliance is replaced or converted. In many instances, the location has undergone renovation since the original installation. In the case of new appliances, the design and operation of the new appliance may be different from the existing appliance. New, more energy-efficient appliances have different installation requirements, and new appliances can have different requirements for clearance to combustible materials.

The requirement of 9.1.2(3) is especially important because the venting requirements have changed significantly since the 1988 edition of the code. These changes resulted from the introduction of higher-efficiency appliances. In most cases, the venting appliances that replace appliances installed under the 1988 or previous editions of the code must meet the new vent sizing tables because many of the newer appliances are more efficient and produce lower vent gas temperatures. In some cases, this may require lining of chimneys or replacing chimneys with other venting systems.

### 9.1.3 Type of Gas(es).

It shall be determined whether the appliance has been designed for use with the gas to which it will be connected. No attempt shall be made to convert the appliance from the gas specified on the rating plate for use with a different gas without consulting the installation instruction, the serving gas supplier, or the appliance manufacturer for complete instructions.

**FAQ ►** The existing furnace uses propane; natural gas is now available. Can the appliance be converted to natural gas?

Subsection 9.1.3 highlights an important safety concern: that the proper gas is supplied to appliances. Appliance rating plates specify the gas for which a particular appliance is specifically designed. (See Exhibit 9.5, Typical Appliance Label, later in this chapter.) Some appliances are not designed to be converted, whereas others include parts and installation instructions for conversion from one gas to another that specify a specific conversion kit part number. If there are no such instructions, the manufacturer or serving gas supplier should be contacted to determine whether the unit can be converted safely and how to make the conversion. If the burners are satisfactory for use with both gases, changing only the pilot orifice(s) and regulator should be necessary. [Natural gas appliances usually operate at 3.5 in. w.c. to 5 in. w.c. (0.9 kPa to 1.2 kPa) manifold pressure. Most LP-Gas appliances operate at 10 in. w.c. (2.5 kPa) to 11 in. w.c. (2.7 kPa) manifold pressure.] Refer to Annex F, Flow of Gas Through Fixed Orifices, for additional information on orifice sizing.

### 9.1.4 Safety Shutoff Devices for Unlisted LP-Gas Appliances Used Indoors.

Unlisted appliances for use with undiluted LP-Gases and installed indoors, except attended laboratory equipment, shall be equipped with safety shutoff devices of the complete shutoff type.

Subsection 9.1.4 requires a complete (100 percent) shutoff-type safety device for unlisted LP-Gas appliances. A safety shutoff device of the complete shutoff type is one that will stop the flow of gas to both the main burner(s) and the pilot burner in the event of pilot failure. A flame sensor that shuts off gas flow to the burner if flame is extinguished can be used.

### 9.1.5 Use of Air or Oxygen Under Pressure.

Where air or oxygen under pressure is used in connection with the gas supply, effective means such as a back-pressure regulator and relief valve shall be provided to prevent air or oxygen from passing back into the gas piping. Where oxygen is used, installation shall be in accordance with NFPA 51, *Standard for the Design and Installation of Oxygen–Fuel Gas Systems for Welding, Cutting, and Allied Processes.*

A backup of air or oxygen into a gas pipe can result in a flammable mixture in the pipe. This mixture can be extremely dangerous and must be avoided to prevent an explosion within the gas pipe. Subsection 9.1.5 requires that, when oxygen is used, the installation must be in accordance with NFPA 51, *Standard for the Design and Installation of Oxygen–Fuel Gas Systems for Welding, Cutting, and Allied Processes.*

In many applications, air or oxygen is used at a greater pressure than the fuel gas. If the air or oxygen enters the gas supply, the air or oxygen could travel all the way back to the gas supply source. If air or oxygen travels back to the source in a natural gas distribution system, it could result in a very dangerous situation in which the gas supply to other customers contains air or oxygen.

Installations where the use of air or oxygen combined with natural gas or propane is likely to be found include jewelry stores that do repairs, muffler shops, and other metalworking facilities.

### 9.1.6* Protection of Appliances from Fumes or Gases Other than Products of Combustion.

**A.9.1.6** Halogenated hydrocarbons are particularly injurious and corrosive after contact with flames or hot surfaces.

**9.1.6.1** Where corrosive or flammable process fumes or gases are present, means for their safe disposal shall be provided. Such fumes or gases include carbon monoxide, hydrogen sulfide, ammonia, chlorine, and halogenated hydrocarbons.

**9.1.6.2** Non-direct-vent appliances installed in beauty shops, barber shops, or other facilities where chemicals that generate corrosive or flammable products such as aerosol sprays are routinely used shall be located in a mechanical room separate or partitioned off from other areas with provisions for combustion and dilution air from outdoors. Direct vent appliances shall be in accordance with the appliance manufacturer's installation instructions.

Corrosive or flammable process fumes or gases, addressed in 9.1.6.1, can be disposed of safely only by a separate exhaust system and an adequate supply of fresh makeup air. Providing a system of isolation for the gas-fired appliance may be necessary to ensure that uncontaminated air is available for its combustion process.

For many years, halogenated hydrocarbons were used extensively in aerosol spray cans containing various household and commercial products, such as hair spray. These products are still in use, although their widespread use has been curtailed for environmental reasons. However, chlorine and fluorine compounds are still present in many products (e.g., water softener salt, laundry bleaches, swimming pool chemicals, detergents, adhesives, paints, varnishes, paint strippers, waxes, plastics, and many refrigerants). When burned, these compounds form acids that corrode the heat exchanger and vent system. The only way to avoid the subsequent problem of excessive corrosion of the heat exchanger surface and vent system is to isolate the heating or water-heating appliance so that it uses only outside air for combustion, ventilation, and draft hood dilution.

◀ **FAQ**
**Why might a gas water heater and gas dryer installed in a beauty salon need special consideration?**

Direct-vent appliances are not required to meet these requirements, because these appliances are designed to obtain their air for combustion from the outside atmosphere and discharge flue gases outside as well, provided that the combustion air source is not located in an area where these chemicals are used (e.g., swimming pools).

### 9.1.7 Process Air.

In addition to air needed for combustion in commercial or industrial processes, process air shall be provided as required for cooling of appliances, equipment, or material; for controlling dew point, heating, drying, oxidation, dilution, safety exhaust, odor control, and air for compressors; and for comfort and proper working conditions for personnel.

The requirement in 9.1.7 applies only to commercial and industrial installations. It directs the installer to consider additional air requirements of appliances — for example, the additional air required by clothes dryers. When installed in quantities, such as in a commercial laundry, it is critical to consider the air used outside of the combustion process. Manufacturers provide the process air requirements in the installation instructions for their appliances.

### 9.1.8 Building Structural Members.

**9.1.8.1** Structural members of a building shall not pass through appliances having an operating temperature in excess of 500°F (260°C).

**9.1.8.2** Structural members passing through appliances having an operating temperature of 500°F (260°C) or less shall be of noncombustible material. Building columns, girders, beams, or trusses shall not be installed within appliances, unless insulation and ventilation are provided to avoid all deterioration in strength and linear expansion of the building structure in either a vertical or a horizontal direction.

**9.1.8.3** Appliances and equipment shall be furnished either with load distributing bases or with a sufficient number of supports to prevent damage to either the building structure or appliance and equipment.

**9.1.8.4** At the locations selected for installation of appliances and equipment, the dynamic and static load carrying capacities of the building structure shall be checked to determine whether they are adequate to carry the additional loads. The appliances and equipment shall be supported and shall be connected to the piping so as not to exert undue stress on the connections.

Paragraphs 9.1.8.1 through 9.1.8.4 cover important installation concerns that may not be obvious to the gas appliance installer. Local building codes should be reviewed for the installation requirements of certain appliances, especially in multifamily residential occupancies. Some codes specify appliance location, provide fire protection requirements for mechanical rooms, and may include specific requirements for rooftop installations.

**FAQ ▶**
**Should the building structure be considered when adding a new appliance?**

When adding a large appliance to an existing building, such as a new rooftop heating and cooling unit, the installer should check building plans and obtain verification that the building can support both the weight and forces developed during operation or that structural support has been added.

### 9.1.9 Flammable Vapors.

Appliances shall not be installed in areas where the open use, handling, or dispensing of flammable liquids occurs, unless the design, operation, or installation reduces the potential of ignition of the flammable vapors. Appliances installed in compliance with 9.1.10 through 9.1.12 shall be considered to comply with the intent of this provision.

Areas of buildings in which flammable liquids are openly used, handled, or dispensed as discussed in 9.1.9 include any area where non-water-based paints and varnish are used. Appliance options include the use of direct vent appliances or an indirect steam or hot water coil in a makeup air unit with a remote boiler to ensure that the boiler is not in the proximity of the flammable vapors that are generated by the painting process.

Liquids that readily generate flammable vapors include gasoline, mineral spirits, acetone, denatured alcohol, oil-based paint and lacquer thinners, and camp stove fuel. For information on safe, recommended storage enclosures for these materials, see NFPA 30, *Flammable and Combustible Liquids Code*.

Installers must verify that the room in which an appliance is being installed is not an area where flammable vapors will be present routinely. Although the future uses (or misuses) of a room cannot be predicted, gasoline, paint thinners, and other household solvents must be verified as not being stored and used in the room. A small spill of gasoline or other flammable liquid will create an invisible vapor cloud near the floor, because the vapors generated usually are heavier than air. This cloud can be drawn into an appliance through its combustion air intake, ignite, and flash back to the liquid pool.

Residential gas water heaters designed to resist the ignition of flammable vapors are available in all types and sizes in the United States. (See the commentary following 9.1.10.1.) These new water heaters, listed to ANSI Z21.10.1/CSA 4.1, *Gas Water Heaters — Volume I — Storage Water Heaters with Input Ratings of 75,000 Btu per Hour or Less*, are required to meet a flammable vapor ignition test as part of their listing. The installation requirement under 9.1.9 is different from the installation requirement for residential garages under 9.1.10. Whereas 9.1.10.1 mandates that all appliances installed in residential garages be installed so that their burners and burner ignition devices are at least 18 in. (460 mm) above the floor, 9.1.9 requires that gas appliances not be installed in areas where flammable vapors are likely to be present (except for residential garages) "unless the design, operation, or installation reduces the potential of ignition of the flammable vapors." Residential water heaters listed in the revised ANSI Z21.10.1 standard, which contains the flammable vapor ignition test, should not be installed in these locations. The standard's flammable resistance requirements are meant to provide protection from accidental spills in residential locations. The current flammable-resistant water heater design renders the water heater unusable after activation and would require that the entire water heater be replaced. Commercial locations where routine handling of flammable liquids occurs due to business activities are an entirely different hazard class not addressed by the ANSI water heater standard.

The design used to enable the new water heaters to pass the flammable vapor ignition resistance (FVIR) test is based on the principle first used in the mine safety lamp. In a mine safety lamp, a fine mesh screen encloses an oil lamp flame. If flammable gas enters the screened area, the gas ignites but the flame does not propagate. See Exhibit 9.1 for an example of the construction of an FVIR water heater.

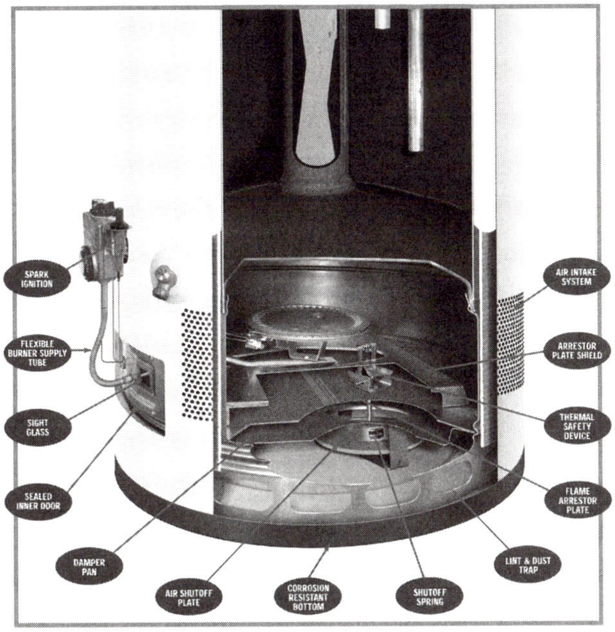


*EXHIBIT 9.1 Cutaway of an FVIR Water Heater. [Courtesy of Rheem (EMCO Corp.)]*

### 9.1.10 Installation in Residential Garages.

**9.1.10.1** Appliances in residential garages and in adjacent spaces that open to the garage and are not part of the living space of a dwelling unit shall be installed so that all burners and burner ignition devices are located not less than 18 in. (460 mm) above the floor unless listed as flammable vapor ignition resistant.

Paragraph 9.1.10.1 addresses the possibility of flammable liquids being spilled in residential garages where gas appliances are installed. Household solvents, oil-based paint and thinners, and gasoline for use as fuel for automobiles, lawn mowers, and so forth, are often stored in residential garages. Spills can occur when these products are dispensed from their storage containers or during automobile maintenance. These flammable liquids vaporize readily, and their vapors remain near the floor. Available evidence indicates that elevating burners and ignition devices 18 in. (460 mm) or more above the floor locates them high enough so the vapors will not be ignited. Exhibit 9.2 illustrates the safe installation of a nonflammable vapor ignition–resistant appliance in a residential garage.

**◀ FAQ**
Can a water heater that is listed as an FVIR heater be installed on a garage floor?

Much concern has been voiced regarding the ignition of flammable vapors by gas appliances — specifically, residential water heaters that are installed on a floor. The standard for water heaters, ANSI Z21.10.1/CSA 4.1, *Gas Water Heaters — Volume I — Storage Water Heaters with Input Ratings of 75,000 Btu per Hour or Less*, was revised in May 1999. Changes to the standard require that all storage-type residential water heaters pass a flammable vapor ignition resistance test in order to be listed. The manufacture of instantaneous-type water heaters is covered under a different standard. Corresponding changes have not been made to require resistance to the ignition of flammable vapors in these instantaneous-type heaters because they have not been associated with the ignition of flammable vapors. All new listed storage-type water heaters of 75,000 Btu per hour or less, with the exception of recreational vehicles, produced for the U.S. market are resistant to the ignition of flammable vapors.

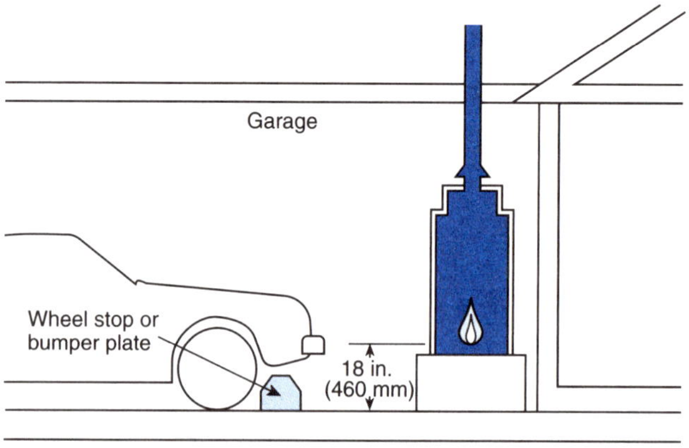


*EXHIBIT 9.2 Water Heater Installation in a Residential Garage. (Courtesy of American Gas Association)*

**9.1.10.2** Such appliances shall be located or protected so they are not subject to physical damage by a moving vehicle.

As noted in 9.1.10.2, appliances in the path of an automobile or other vehicle must be protected from accidental contact. Wheel stops, such as those shown in Exhibit 9.2, bollards, or walls must be used to prevent accidental displacement of fuel-fired appliances.

**9.1.10.3** Where appliances are installed in a separate, enclosed space having access only from outside of the garage, such appliances shall be permitted to be installed at floor level, providing the required combustion air is taken from the exterior of the garage.

Paragraph 9.1.10.3 allows an appliance to be installed in a room that is constructed in a garage but does not have a door that opens into the garage. This absence of a door to the garage prevents any flammable vapors that may be present in the garage from interacting with the combustion air for the appliance. If the room has an opening into the garage, appliances cannot be installed on the garage floor.

## 9.1.11 Installation in Commercial Garages.

**9.1.11.1 Parking Structures.** Appliances installed in enclosed, basement, and underground parking structures shall be installed in accordance with NFPA 88A, *Standard for Parking Structures*.

The installation of appliances in parking garages that are located in any enclosed structure, basement, or underground, both with and without fuel-dispensing facilities, must comply with the requirements of NFPA 88A, *Standard for Parking Structures*, as required by 9.1.11.1.

NFPA 88A requires that all gas-fueled heating appliances in enclosed parking garages be approved and be located at least 18 in. (500 mm) above the floor or be protected by a partition that is at least 18 in. (500 mm) high. (See 5.2.2 and 5.2.3 of NFPA 88A.)

NFPA 88A also requires continuous, mechanical ventilation in enclosed structures, basements, and underground parking garages that store flammable liquids. NFPA 88A covers the following fuels: CNG, LNG, LP-Gas; all Class I flammable liquids, including gasoline; and Class II flammable liquids, including diesel fuel, fuel oil, and kerosene.

NFPA 88A does not require the installation of heaters at least 8 ft (2.4 m) above the floor to provide clearance for passenger vehicles in parking garages. NFPA 30A, *Code for Motor Fuel Dispensing Facilities and Repair Garages*, requires the 8 ft (2.4 m) installation height for repair garages. The appliance manufacturer's instructions should be consulted for the specific requirements for the installation of overhead radiant heaters to avoid overheating vehicles that could be parked underneath them.

**9.1.11.2 Repair Garages.** Appliances installed in repair garages shall be installed in a detached building or room, separated from repair areas by walls or partitions, floors, or floor ceiling assemblies that are constructed so as to prohibit the transmission of vapors and having a fire resistance rating of not less than 1 hour, and that have no openings in the wall separating the repair area within 8 ft (2.4 m) of the floor. Wall penetrations shall be firestopped. Air for combustion purposes shall be obtained from the outdoors. The heating room shall not be used for the storage of combustible materials.

*Exception No. 1: Overhead heaters where installed not less than 8 ft (2.4 m) above the floor shall be permitted.*

*Exception No. 2: Heating appliances for vehicle repair areas where there is no dispensing or transferring of Class I or Class II flammable or combustible liquids or LP-Gas shall be installed in accordance with NFPA 30A, Code for Motor Fuel Dispensing Facilities and Repair Garages.*

Paragraph 9.1.11.2 requires that appliances installed in repair garages be located either in a detached building or in a room separated from the repair garage, with air for combustion and ventilation obtained from outside the building.

Unit heaters can be installed in a commercial garage. However, unit heaters are required to be installed at least 8 ft (2.4 m) above the floor. (See Exception No. 1 to 9.1.11.2.) Unit heaters in repair garages may be installed where there is no dispensing or transferring of flammable liquids or liquid petroleum gas.

◀ **FAQ**
Can a unit heater be installed in a commercial garage?

Paragraph 9.1.11.2 is in agreement with NFPA 30A, *Code for Motor Fuel Dispensing Facilities and Repair Garages*, which has the following requirements:

> **7.5* Heating, Ventilating, and Air-Conditioning.**
>
> **7.5.1*** Forced air heating, air-conditioning, and ventilating systems serving a fuel dispensing area inside a building or a repair garage shall not be interconnected with any such systems serving other occupancies in the building. Such systems shall be installed in accordance with NFPA 90A, *Standard for the Installation of Air-Conditioning and Ventilating Systems.*
>
> **7.5.2** Return air openings in areas of repair garages used for the repair or servicing of vehicles or in a fuel dispensing area shall be not less than 455 mm (18 in.) above floor level measured to the bottom of the openings.
>
> **7.5.3** Combined ventilation and heating systems shall not recirculate air from areas that are below grade level.
>
> **7.5.4** Exhaust duct openings shall be located so that they effectively remove vapor accumulations at floor level from all parts of the floor area.
>
> **7.6 Heat-Producing Appliances.**
>
> **7.6.1** Heat-producing appliances shall be installed in accordance with the requirements of Section 7.6. They shall be permitted to be installed in the conventional manner except as provided in Section 7.6.
>
> **7.6.2** Heat-producing appliances shall be of an approved type. Solid fuel stoves, improvised furnaces, salamanders, or space heaters shall not be permitted in areas of repair garages used for repairing or servicing of vehicles or in a fuel dispensing area.
>
> *Exception No. 1: Unit heaters, when installed in accordance with this chapter, need not meet this requirement.*
>
> *Exception No. 2: Heat-producing equipment for any lubrication room or service room where there is no dispensing or transferring of Class I or Class II liquids or LP-Gas, when installed in accordance with this chapter, need not meet this requirement.*
>
> **7.6.3** Heat-producing appliances shall be permitted to be installed in a special room that is separated from areas that are classified as Division 1 or Division 2, in accordance with Chap-

ter 8, by walls that are constructed to prevent the transmission of vapors, that have a fire resistance rating of at least 1 hour, and that have no openings in the walls that lead to a classified area within 2.4 m (8 ft) of the floor. Specific small openings through the wall, such as for piping and electrical conduit, shall be permitted, provided the gaps and voids are filled with a fire-resistant material to resist transmission of vapors. All air for combustion purposes shall be taken from outside the building. This room shall not be used for storage of combustible materials, except for fuel storage as permitted by the standards referenced in 7.6.9.

**7.6.4** Heat-producing appliances using gas or oil fuel shall be permitted to be installed in a lubrication or service room where there is no dispensing or transferring of Class I liquids, including the open draining of automotive gasoline tanks, provided the bottom of the combustion chamber is at least 455 mm (18 in.) above the floor and the appliances are protected from physical damage.

**7.6.5** Heat-producing appliances using gas or oil fuel listed for use in garages shall be permitted to be installed in lubrication rooms, service rooms, or fuel dispensing areas where Class I liquids are dispensed or transferred, provided the equipment is installed at least 2.4 m (8 ft) above the floor.

**7.6.6*** Where major repairs are conducted on CNG-fueled vehicles or LNG-fueled vehicles, open flame heaters or heating equipment with exposed surfaces having a temperature in excess of 399°C (750°F) shall not be permitted in areas subject to ignitible concentrations of gas.

**7.6.7** Electrical heat-producing appliances shall meet the requirements of Chapter 8.

**7.6.8** Fuels used shall be of the type and quality specified by the manufacturer of the heating appliance. Crankcase drainings shall not be used in oil-fired appliances, unless the appliances are specifically approved for such use.

**7.6.9** Heat-producing appliances shall be installed to meet the requirements of NFPA 31, *Standard for the Installation of Oil-Burning Equipment*; NFPA 54, *National Fuel Gas Code*; NFPA 82, *Standard on Incinerators and Waste and Linen Handling Systems and Equipment*; NFPA 90A, *Standard for the Installation of Air-Conditioning and Ventilating Systems*; and NFPA 211, *Standard for Chimneys, Fireplaces, Vents, and Solid Fuel–Burning Appliances*, as applicable, except as hereinafter specifically provided.

* Additional explanation is available in Annex A of NFPA 30B.

### 9.1.12 Installation in Aircraft Hangars.

Heaters in aircraft hangars shall be installed in accordance with NFPA 409, *Standard on Aircraft Hangars*.

### 9.1.13 Appliance Physical Protection.

Where it is necessary to locate appliances close to a passageway traveled by vehicles or machinery, guardrails or bumper plates shall be installed to protect the equipment from damage.

See Exhibit 9.2 for an illustration of a bumper plate.

### 9.1.14 Venting of Flue Gases.

Appliances shall be vented in accordance with the provisions of Chapter 12.

### 9.1.15 Extra Device or Attachment.

No device or attachment shall be installed on any appliance that could in any way impair the combustion of gas.

Subsection 9.1.15 addresses the concern that a gas flame could be interfered with. The appliance manufacturer's instructions provide the information needed to determine whether there are any accessories approved for use on a particular appliance.

### 9.1.16 Adequate Capacity of Piping.

When additional appliances are being connected to a gas piping system, the existing piping shall be checked to determine whether it has adequate capacity. *(See Section 5.4.)* Where inadequate, the existing system shall be enlarged as necessary, or separate gas piping of adequate capacity shall be run from the point of delivery to the appliance.

### 9.1.17 Avoiding Strain on Gas Piping.

Appliances shall be supported and connected to the piping so as not to exert undue strain on the connections.

The intent of 9.1.17 is that the weight of an appliance should not be supported by the gas piping.

### 9.1.18 Gas Appliance Pressure Regulators.

Where the gas supply pressure is higher than that at which the appliance is designed to operate or varies beyond the design pressure limits of the appliance, a gas appliance pressure regulator shall be installed.

Most gas appliances are designed for an inlet pressure of 7 in. w.c (1.7 kPa) or 11 in. w.c. (2.7 kPa), the normal building pressure for natural gas and propane, respectively. Where a building piping system operates at higher pressure, a pressure regulator called an appliance pressure regulator is needed. Many residential, commercial, and industrial gas piping systems deliver gas at a pressure of 2 psi (13.8 kPa) or higher to the proximity of the appliance, requiring the appliance installer to connect a "pounds to inches" regulator ahead of the gas appliance.

Building propane systems generally are provided with two regulators, one at the tank (first stage) and one outside the building (second stage). The second-stage propane regulator usually delivers gas downstream from the regulator at a nominal 11 in. w.c. (2.7 kPa) pressure. If the second stage regulator supplies gas to the proximity of the appliance of greater than $^1/_2$ psig (3.4 kPa), a third appliance or line regulator must be used to reduce the inlet pressure to the nominal 11 in w.c. (2.7 kPa). If a propane appliance is not equipped with a regulator, the manufacturer's instructions should be consulted to ensure that the appliance is suitable for use at the nominal 11 in. w.c. (2.7 kPa) of pressure from the LP-Gas system tank.

### 9.1.19 Venting of Gas Appliance Pressure Regulators.

Venting of gas appliance pressure regulators shall comply with the following requirements:

(1) Appliance pressure regulators requiring access to the atmosphere for successful operation shall be equipped with vent piping leading outdoors or, if the regulator vent is an integral part of the appliance, into the combustion chamber adjacent to a continuous pilot, unless constructed or equipped with a vent limiting means to limit the escape of gas from the vent opening in the event of diaphragm failure.

Pressure regulators incorporate a diaphragm that moves up and down and compresses a spring as pressure increases. As the diaphragm moves, it adjusts the position of the valve that controls the flow of gas into the regulator. The space above the diaphragm must be able to move up and down freely, and the space above the diaphragm is vented to the atmosphere to allow such free movement. Normally only air is vented from the space above the regulator, but if the diaphragm fails natural gas or propane will be discharged.

Any gas that is relieved by a gas appliance pressure regulator must be vented safely to the outdoors to prevent the accumulation of a potentially dangerous concentration of gas indoors. At a minimum, the size of the vent line should not be smaller than the vent connection on the regulator.

Regulator vent lines must be made using piping materials permitted by Section 5.6, which allows both metal piping and a specific plastic pipe for regulator vents outdoors in 5.6.4.2:

**5.6.4.2* Regulator Vent Piping.** Plastic pipe, tubing, and fittings used to connect regulator vents to remote vent terminations shall be PVC conforming to ANSI/UL 651, Schedule 40 and 80 Rigid PVC Conduit and Fittings. PVC vent piping shall not be installed indoors.

Therefore, a specific PVC conduit may be used, but only outdoors. When the regulator vent is an integral part of the gas-fired appliance, it can be vented adjacent to the continuous burning pilot in the combustion chamber.

Some gas appliance pressure regulators are equipped with a vent-limiting device that restricts the flow of gas to less than 2.5 $ft^3/hr$ (0.07 $m^3/hr$) for natural gas and 1 $ft^3/hr$ (0.028 $m^3/hr$) for LP-Gas or propane gas. This leakage rate is considered not likely to create a dangerous concentration of gas. Note that the vent limiter must be supplied with the regulator and cannot be changed by the installer.

The code provides neither restrictions nor guidance to those who manifold the vents of multiple regulators together. Information on sizing of discharge piping from pressure regulators can be found in American Petroleum Institute Recommended Practice RP 520, *Sizing, Selection, and Installation of Pressure-Relieving Devices in Refineries, Part II, Installation.*

Prohibiting termination of bleed lines inside a building is imperative because if the rubber diaphragm fails, unburned gas will be released into the building, creating a potentially hazardous condition.

(2) Vent limiting means shall be employed on listed appliance pressure regulators only.

**FAQ ▶**
**Can a vent limiter be removed from a regulator and installed on another regulator?**

Paragraph 9.1.19(2) requires that listed regulators and vent limiters be installed only on the regulator for which they are specified. Field installation of vent limiters must comply with the manufacturers' instructions, which will deal with this subject in greater depth. A vent limiter usually cannot be removed and reinstalled on a different regulator. If questions arise, the regulator manufacturer's installation instructions should be consulted.

Many vent limiters are designed to operate in a specific position. If this is the case, a marking will be placed on the regulator or vent limiter to indicate the position.

(3) In the case of vents leading outdoors, means shall be employed to prevent water from entering this piping and also to prevent blockage of vents by insects and foreign matter.

As noted in 9.1.19(3), regulator vents must not be blocked. Insects, insect nests, ice, or freezing rain can block outdoor vents from regulators. If a regulator vent becomes blocked, the downstream pressure (pressure in the building piping) will be increased, which may result in gas leaking into the building. Pointing the end of the vent line down, flaring it approximately 40 degrees, and screening the opening usually prevents this blockage.

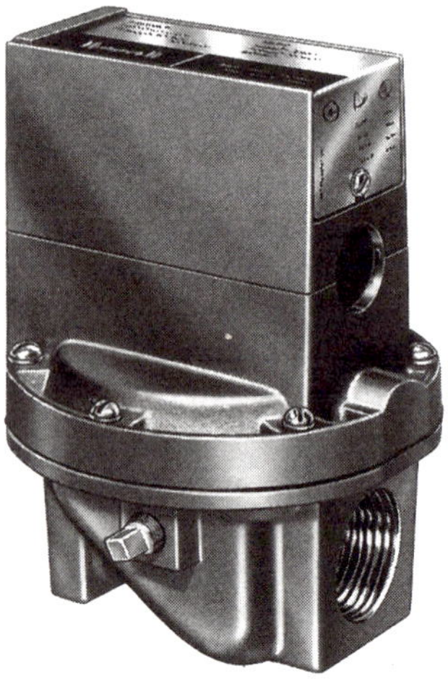

***EXHIBIT 9.3*** *Diaphragm Valve. (Courtesy of Honeywell, Inc.)*

(4) Under no circumstances shall a regulator be vented to the appliance flue or exhaust system.

(5) In the case of vents entering the combustion chamber, the vent shall be located so the escaping gas is readily ignited by the pilot and the heat liberated thereby does not adversely affect the normal operation of the safety shutoff system. The terminus of the vent shall be securely held in a fixed position relative to the pilot. For manufactured gas, the need for a flame arrester in the vent piping shall be determined.

(6) A vent line(s) from an appliance pressure regulator and a bleed line(s) from a diaphragm-type valve shall not be connected to a common manifold terminating in a combustion chamber. Vent lines shall not terminate in positive-pressure-type combustion chambers.

Manifolding of both a bleed line from a diaphragm-type valve, as shown in Exhibit 9.3, and a vent line from a gas appliance regulator that terminates in a combustion chamber represents a potential hazard. The danger is that the gas vented by the diaphragm valve could be drawn into the regulator through the regulator vent line until a flammable air–gas mixture is reached. The continuous pilot could ignite this mixture and produce a small explosion that could extinguish the pilot and rupture the pressure regulator diaphragm. This rupture would then allow unburned gas to flow continuously through the vent line into the combustion chamber, causing an overrate condition for the main burner and pilot and subjecting the gas controls to excessive gas pressures. Exhibit 9.4 illustrates the operation of a diaphragm valve.

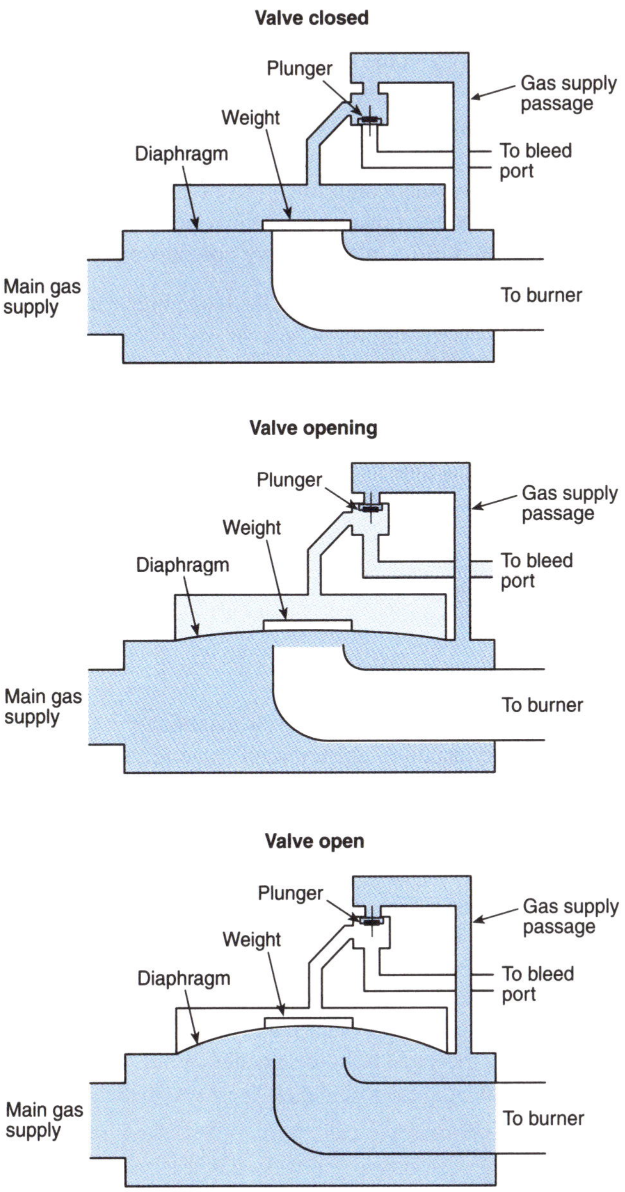


*EXHIBIT 9.4 Operation of a Diaphragm Valve. (Courtesy of Honeywell, Inc.)*

The prohibition of vent lines terminating in positive pressure–type combustion chambers recognizes that more appliances are being designed for positive pressure combustion chambers. This type of design can result in back pressure on the diaphragm of regulators and diaphragm valves and can cause the pressure regulator to deliver a higher pressure, which can lead to appliance failure.

### 9.1.20 Bleed Lines for Diaphragm-Type Valves.

Bleed lines shall comply with the following requirements:

(1) Diaphragm-type valves shall be equipped to convey bleed gas to the outdoors or into the combustion chamber adjacent to a continuous pilot.
(2) In the case of bleed lines leading outdoors, means shall be employed to prevent water from entering this piping and also to prevent blockage of vents by insects and foreign matter.
(3) Bleed lines shall not terminate in the appliance flue or exhaust system.
(4) In the case of bleed lines entering the combustion chamber, the bleed line shall be located so the bleed gas is readily ignited by the pilot and the heat liberated thereby does not adversely affect the normal operation of the safety shutoff system. The terminus of the bleed line shall be securely held in a fixed position relative to the pilot. For manufactured gas, the need for a flame arrester in the bleed line piping shall be determined.
(5) A bleed line(s) from a diaphragm-type valve and a vent line(s) from an appliance pressure regulator shall not be connected to a common manifold terminating in a combustion chamber. Bleed lines shall not terminate in positive-pressure-type combustion chambers.

For the same reasons given in the commentary following 9.1.19(6), the bleed line from the diaphragm valve must not be interconnected by the manifold to the vent line from a gas pressure regulator. (See Exhibit 9.3 and Exhibit 9.4.)

### 9.1.21 Combination of Appliances and Equipment.

Any combination of appliances, equipment, attachments, or devices used together in any manner shall comply with the standards that apply to the individual appliance and equipment.

Subsection 9.1.21 is intended to prevent the use of any combination of appliances that could create a dangerous situation. A list of standards is included in Chapter 2 and Annex M of the code. Also, the specific appliance manufacturer should be consulted for further information.

### 9.1.22 Installation Instructions.

The installing agency shall conform to the appliance and equipment manufacturers' recommendations in completing an installation. The installing agency shall leave the manufacturers' installation, operating, and maintenance instructions in a location on the premises where they are readily available for reference and guidance of the authority having jurisdiction, service personnel, and the owner or operator.

Leaving the manufacturer's installation, service, and maintenance instructions and the owner's/user's manual on the job site is important so that the authority having jurisdiction, service personnel, and the owner or user can use these materials, as required by 9.1.22.

**FAQ ▶** Why are the appliance manufacturer's installation instructions required to be on the job site?

The installation instructions contain specific installation and service requirements for the appliance that may be needed by service and repair personnel and that may not be readily available. This information cannot be found in this code or in local codes. While many appliance manufacturers post appliance instructions on their web pages, most service technicians do not have access to the Internet during service calls.

### 9.1.23 Protection of Outdoor Appliances.

Appliances not listed for outdoor installation but installed outdoors shall be provided with protection to the degree that the environment requires. Appliances listed for outdoor installation shall be permitted to be installed without protection in accordance with the manufacturer's installation instructions. *(See 9.2.1.)*

Appliances not listed for outdoor installation can be enclosed satisfactorily if the appliance installation meets the requirements of this code for air, for combustion and ventilation, and for clearance to combustible surfaces or if the installation meets the requirements included in the manufacturer's instructions. The clearance requirements for outdoor appliance installations in relation to walls, ceilings, and combustibles should also be considered.

## 9.2 Accessibility and Clearance

### 9.2.1 Accessibility for Service.

All appliances shall be located with respect to building construction and other equipment so as to permit access to the appliance. Sufficient clearance shall be maintained to permit cleaning of heating surfaces; the replacement of filters, blowers, motors, burners, controls, and vent connections; the lubrication of moving parts where necessary; the adjustment and cleaning of burners and pilots; and the proper functioning of explosion vents, if provided. For attic installation, the passageway and servicing area adjacent to the appliance shall be floored.

The requirements in 9.2.1 are intended to highlight the importance of keeping service clearances of appliances in mind when an appliance is installed. Manufacturer's instructions and/or local building codes should be consulted for width and length of passageways to appliances installed in attics, on roofs, and in crawl spaces. Note that Section 9.5 covers appliances in attics.

Generally, passageways must be floored and limited to not less than 24 in. (0.6 m) wide or to the width of the appliance for removal and 20 ft (6 m) in length. In addition, there should be at least 30 in. (0.76 m) of clearance in front of an appliance to provide adequate service space and access to the electrical controls and other operating components.

Appliances approved for closet installation can have reduced clearance if service is accomplished through a door or removable panel and the reduced clearance is not less than that specified on the rating plate or in Chapter 10 of the code, Installation of Specific Appliances. (See 9.4.3 for access to appliances on roofs and Section 9.5 for installation in attics.)

### 9.2.2 Clearance to Combustible Materials.

Appliances and their vent connectors shall be installed with clearances from combustible material so their operation does not create a hazard to persons or property. Minimum clearances between combustible walls and the back and sides of various conventional types of appliances and their vent connectors are specified in Chapters 10 and 12. *(Reference can also be made to* NFPA 211, *Standard for Chimneys, Fireplaces, Vents, and Solid Fuel–Burning Appliances.)*

Listed appliances have a label attached to the appliance that describes required clearances. Exhibit 9.5 shows an example of an appliance label. The lower right part of the label displays the clearance requirements. The manufacturer's instructions will also have this information. Chapter 10 of the code provides clearances for many types of unlisted appliances.

◄ **FAQ**
Where can information on required clearances for a specific appliance be found?

### 9.2.3 Installation on Carpeting.

Appliances shall not be installed on carpeting, unless the appliances are listed for such installation.

As stated in 9.2.3, an appliance is prohibited from being installed on carpeting unless the appliance is listed for such use. If it is necessary to install an appliance on a carpet, refer to the commentary following 10.3.3, which provides details for a method of protecting combustible flooring.

*EXHIBIT 9.5 Typical Appliance Label. (Courtesy of Utica Boilers — An ECR International Brand)*

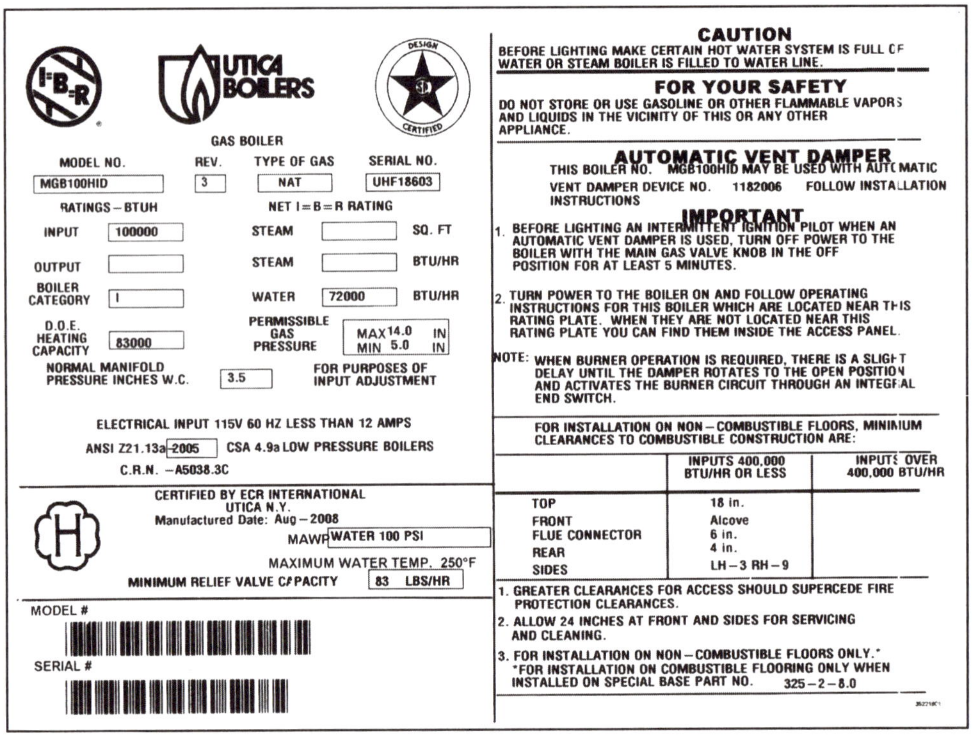
UTICA BOILERS
GAS BOILER

MODEL NO. MGB100HID | REV. 3 | TYPE OF GAS NAT | SERIAL NO. UHF18603

RATINGS – BTUH | NET I=B=R RATING
INPUT 100000 | STEAM SQ. FT
OUTPUT | STEAM BTU/HR
BOILER CATEGORY I | WATER 72000 BTU/HR
D.O.E. HEATING CAPACITY 83000 | PERMISSIBLE GAS PRESSURE MAX 14.0 IN MIN 5.0 IN
NORMAL MANIFOLD PRESSURE INCHES W.C. 3.5 | FOR PURPOSES OF INPUT ADJUSTMENT

ELECTRICAL INPUT 115V 60 HZ LESS THAN 12 AMPS
ANSI Z21.13a-2005 CSA 4.9a LOW PRESSURE BOILERS
C.R.N. –A5038.3C

CERTIFIED BY ECR INTERNATIONAL
UTICA N.Y.
Manufactured Date: Aug – 2008
MAWP WATER 100 PSI
MAXIMUM WATER TEMP. 250°F
MINIMUM RELIEF VALVE CAPACITY 83 LBS/HR

MODEL #
SERIAL #

CAUTION
BEFORE LIGHTING MAKE CERTAIN HOT WATER SYSTEM IS FULL OF WATER OR STEAM BOILER IS FILLED TO WATER LINE.

FOR YOUR SAFETY
DO NOT STORE OR USE GASOLINE OR OTHER FLAMMABLE VAPORS AND LIQUIDS IN THE VICINITY OF THIS OR ANY OTHER APPLIANCE.

AUTOMATIC VENT DAMPER
THIS BOILER NO. MGB100HID MAY BE USED WITH AUTOMATIC VENT DAMPER DEVICE NO. 1182006 FOLLOW INSTALLATION INSTRUCTIONS

IMPORTANT
1. BEFORE LIGHTING AN INTERMITTENT IGNITION PILOT WHEN AN AUTOMATIC VENT DAMPER IS USED, TURN OFF POWER TO THE BOILER WITH THE MAIN GAS VALVE KNOB IN THE OFF POSITION FOR AT LEAST 5 MINUTES.
2. TURN POWER TO THE BOILER ON AND FOLLOW OPERATING INSTRUCTIONS FOR THIS BOILER WHICH ARE LOCATED NEAR THIS RATING PLATE. WHEN THEY ARE NOT LOCATED NEAR THIS RATING PLATE YOU CAN FIND THEM INSIDE THE ACCESS PANEL.
NOTE: WHEN BURNER OPERATION IS REQUIRED, THERE IS A SLIGHT DELAY UNTIL THE DAMPER ROTATES TO THE OPEN POSITION AND ACTIVATES THE BURNER CIRCUIT THROUGH AN INTEGRAL END SWITCH.

FOR INSTALLATION ON NON – COMBUSTIBLE FLOORS, MINIMUM CLEARANCES TO COMBUSTIBLE CONSTRUCTION ARE:

| | INPUTS 400,000 BTU/HR OR LESS | INPUTS OVER 400,000 BTU/HR |
|---|---|---|
| TOP | 18 in. | |
| FRONT | Alcove | |
| FLUE CONNECTOR | 6 in. | |
| REAR | 4 in. | |
| SIDES | LH – 3 RH – 9 | |

1. GREATER CLEARANCES FOR ACCESS SHOULD SUPERCEDE FIRE PROTECTION CLEARANCES.
2. ALLOW 24 INCHES AT FRONT AND SIDES FOR SERVICING AND CLEANING.
3. FOR INSTALLATION ON NON – COMBUSTIBLE FLOORS ONLY.*
*FOR INSTALLATION ON COMBUSTIBLE FLOORING ONLY WHEN INSTALLED ON SPECIAL BASE PART NO. 325 – 2 – 8.0

## 9.3* Air for Combustion and Ventilation

The importance of combustion, dilution, and ventilation air, as shown in Exhibit 9.6, cannot be overemphasized. Typical gas-fired natural draft (draft hood) furnaces require approximately 21 $ft^3$ (0.6 $m^3$) of air (i.e., combustion, vent dilution, and ventilation) for every cubic foot of gas burned. Although modern fan-assisted combustion system furnaces do not need dilution air, they still require approximately 15 $ft^3$ (0.4 $m^3$) of air for each cubic foot of gas burned. This amount of air can range between 1500 $ft^3$/hr and 2100 $ft^3$/hr (43 $m^3$/hr and 60 $m^3$/hr) for each 100,000 Btu/hr (29.3 kW) of gas input. This air is needed to support the combustion and venting process of the appliance and to provide ventilation cooling for the casing and internal controls. This section of the code covers methods to provide adequate air for complete combustion of fuel gas, ventilation of spaces appliances are installed in, and for dilution of vent gases. See Exhibit 9.7.

**A.9.3** Operation of exhaust fans, ventilation systems, clothes dryers, or fireplaces can create conditions requiring special attention to avoid unsatisfactory operation of installed appliances.

The main objective of Section 9.3 is to offer methods of providing sufficient combustion air for complete combustion of fuel gas. Refer to Exhibit 9.8 for an illustration of the use of these methods to supply combustion air.

Section 9.3 provides the following five methods to supply combustion air:

1. 100 percent indoor air
2. 100 percent outdoor air
3. Combination of indoor and outdoor air
4. Engineered systems
5. Mechanical air supply

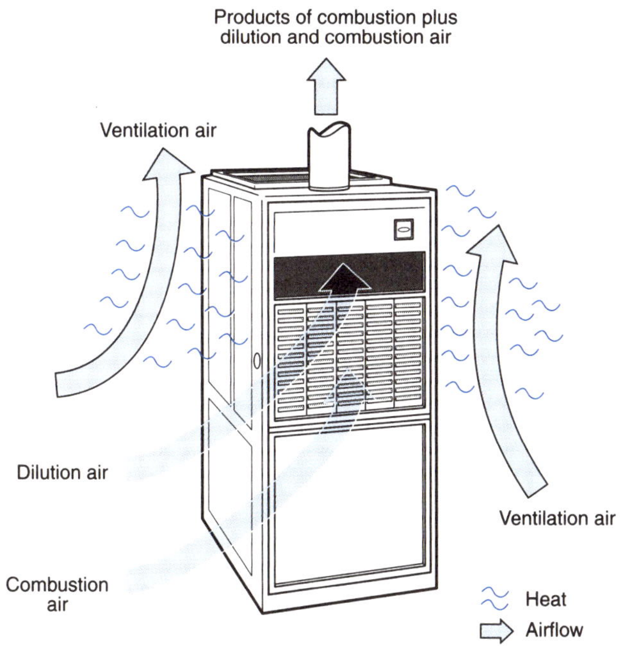


*EXHIBIT 9.6 Combustion, Dilution, and Ventilation Air. (Courtesy of Carrier Corporation)*

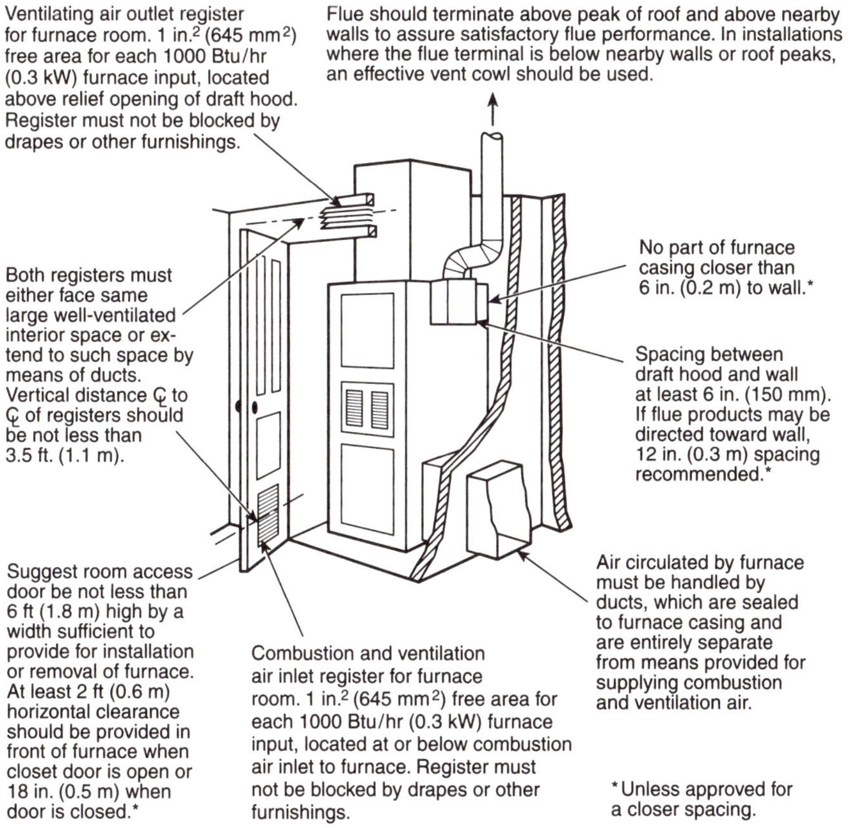


*EXHIBIT 9.7 Openings Necessary to Supply Air for Combustion When an Appliance Is Installed in a Confined Space. [From NFPA 54 and ANSI Z21.30 (1950 edition)]*

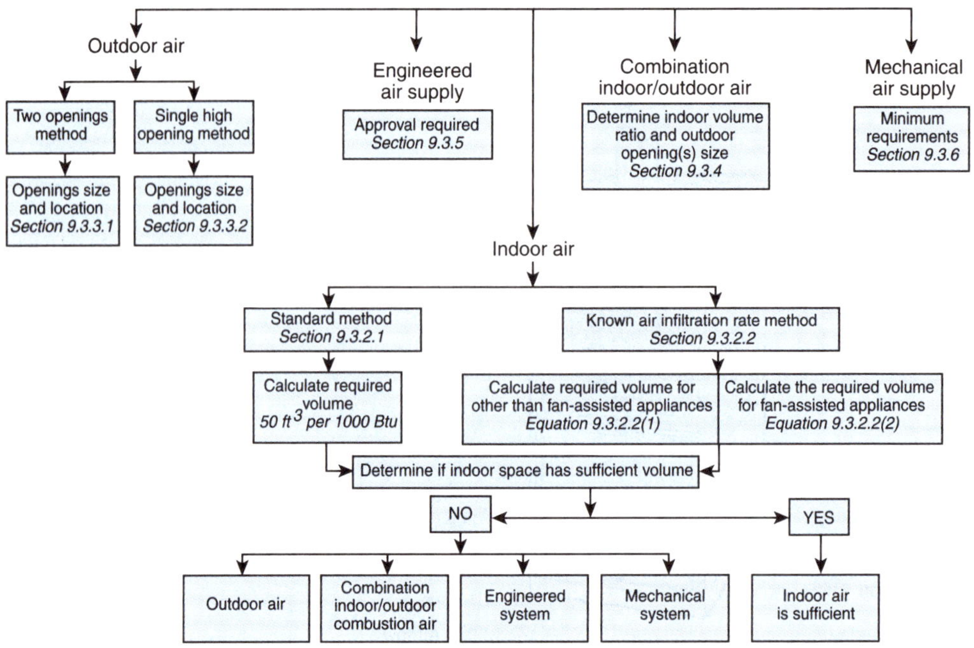


***EXHIBIT 9.8*** *Combustion Air Options. (Courtesy of American Gas Association)*

The code contains two ways to calculate the required indoor air volume to meet the combustion air needs of appliances: the standard method and the KAIR method.

**Standard Method.** The standard method, which has been in the code since the 1950s, can be used for buildings that have a 0.40 air change per hour (ACH) or higher air infiltration rate.

**KAIR Method.** The KAIR method, which was added in the 2002 edition to address buildings with low air infiltration (to conserve energy), can be used when the structure's air infiltration rate is known or estimated, or when the structure has less than 0.40 ACH.

An important feature of KAIR is that it recognizes the difference in combustion air volume requirements between the two appliance types — fan-assisted appliances and other than fan-assisted appliances, which is "all other appliances." The amount of required indoor air for each appliance type is different due to the different amounts of dilution air that each uses. If an installation includes both types of appliances, the required air volume is calculated separately for the fan-assisted appliance and for the other than fan-assisted appliance.

**Purpose of the KAIR Method.** The KAIR method solves a problem that has been addressed in steps since the 1984 edition of the code. Prior to the 1984 edition, the code requirements for indoor combustion air did not provide guidance for "unusually tight construction." In the 1984 edition, a requirement was added, stating that if indoor combustion air was used and if the construction was unusually tight, outside combustion air had to be provided. This requirement addressed valid concerns that buildings were being built to minimize air infiltration due to the energy crisis of the 1970s. Although this addition provided no guidance on what unusually tight construction was, the addition provided a positive step toward addressing the meaning of the term.

**FAQ ▶** Why was the term *unusually tight construction* removed from the code in the 2002 edition?

In the 1988 edition, a definition of unusually tight construction was added, which included the following three items:

1. Walls and ceilings exposed to the outside atmosphere have a continuous water vapor retarder with a rating of 1 perm (a unit of permeability) or less with openings gasketed or sealed.

2. Weatherstripping has been added on openable windows and doors.
3. Caulking or sealants are applied to areas such as joints around window and door frames, between sole plates and floors, between wall-ceiling joints, between wall panels, at penetrations for plumbing, electrical, and gas lines, and at other openings.

This definition provided the needed guidance for a time. By the mid-1990s, however, the type of construction described as unusually tight was, in fact, normal construction required by building and energy codes. A further revision to the definition was added in the 1999 edition. This revision added a fourth item to the definition of *unusually tight construction* — that is, that the building have an average air infiltration rate of less than 0.35 air changes per hour.

The fourth item provided additional guidance and changed the definition to not automatically include all new construction, but to base the need for outdoor combustion air on the actual air infiltration rate.

The 2002 edition of the code went further by eliminating the definitions of unusually tight construction, confined space, and unconfined space and providing two methods to determine the amount of indoor volume needed for buildings with varying air infiltration rates. The code was now clearer as to what conditions were needed to mandate the use of additional outside combustion air, which many jurisdictions had not applied uniformly. Some jurisdictions considered all new structures built to a model or state energy code to be inherently unusually tight. No research has been found that can conclusively link the construction details associated with the unusually tight construction definition to a level of reduced air infiltration that would be too low to support combustion appliances.

The terms *confined space* and *unconfined space* were eliminated because they did not denote one single value [50 $ft^3$ per 1000 Btu (4.8 $m^3/kW$)] but rather varying volumes, depending on the method used and the amount of air infiltration. A new term, *required volume*, was used to better describe the code's intent.

The combustion air tables in Annex A of the 2006 edition — A.9.3.2.1, A.9.3.2.1(a), and A.9.3.2.1(b) — provide users with the calculated values without the need to perform calculations for the required air volumes under the standard and KAIR methods. Annex I provides three installation examples to help the code user apply the combustion air coverage. Annex J contains an example of combination of indoor and outdoor combustion and ventilation opening design.

**History of Combustion Air Requirements.** As noted in the history of this code, the first gas safety standard (covering appliances), ASA K-2, was published in 1927. The requirements for air for combustion were very brief: "No appliance designed to burn gas at a rate greater than an ordinary lighting burner shall be installed in a room which is not adequately ventilated."

The increased use of gas for heating and the popularity of "closet furnaces" following World War II resulted in AGA Research Bulletin 53, "The Effects of Confined Space Installation on Central Gas Space Heating Equipment Performance," which was published in 1947.

The study resulted in a requirement, which was included in the 1950 edition of ASA Z21.30 and NFPA 54, for two air openings into the confined space, each sized at 1 $in.^2$ per 1000 Btu/hr (650 $mm^2/kW$) and communicating with an area having "adequate infiltration." (See Exhibit 9.7.)

If the confined space was within a building of unusually tight construction, then air for combustion and ventilation was to be obtained from outdoors (e.g., crawl or attic space), and the size of the openings could be reduced by 50 percent. [At that time, the code defined confined space as a building area containing less than 50 $ft^3$ for each 1000 $ft^3$ (4.8 $m^3/kW$) of room volume.]

In subsequent years, additional work by AGA Laboratories, published in AGA Research Bulletin 67, "Combustion and Ventilation Air Supply to Gas Equipment in Small Rooms," and work by the University of Illinois Engineer Experiment Station, published in AGA Bulletin 427, "Outdoor Air Supply and Ventilation of a Furnace Closet Used with a Warm Air Heating System," reaffirmed the need for two openings in a confined space. These publications provided more detailed information for the standards. In 1969, the most significant change in this section of the code was to require that the openings be located within 12 in. (300 mm) of the top and 12 in. (300 mm) of the bottom of the enclosure. Consequently, five detailed drawings were added in 1959 for clarification. (One drawing was dropped in the 1980 edition of the code.)

An additional drawing was added in the 1999 edition, and all five drawings were relocated to Appendix A in the 1999 edition. The present requirement is that the minimum dimension of the air openings be not less than 3 in. (80 mm).

Significant changes in the design of gas appliances have occurred since the initial development of combustion air requirements. Increases in efficiency and the use of fan-assisted combustion systems prompted the Gas Research Institute to initiate a study of combustion air requirements. Report Number GRI 93/0316, "Analysis of Combustion Air Openings to the Outdoors: Preliminary Results," was issued in 1994. This report found that a single opening to the outdoors also could be used if certain qualifications were met [i.e., that the opening be within 12 in. (300 mm) of the ceiling and minimum specified clearances be provided around the appliance]. (See 9.3.3.2.)

The use of two openings, high and low, to the outdoors still presented practical problems, especially in an era when building envelopes were being tightened. In the 1996 edition, an alternative allowing one opening [located within 12 in. (300 mm) of the ceiling] was added based on the GRI report just cited.

## 9.3.1 General.

The requirements of 9.3.1 are outlined in Exhibit 9.8. The exhibit contains an overview of the methods that can be used to provide air for combustion for appliances.

**9.3.1.1** Air for combustion, ventilation, and dilution of flue gases for appliances installed in buildings shall be obtained by application of one of the methods covered in 9.3.2 through 9.3.6. Where the requirements of 9.3.2 are not met, outdoor air shall be introduced in accordance with methods covered in 9.3.3 through 9.3.6.

*Exception No. 1: This provision shall not apply to direct vent appliances.*

*Exception No. 2: Type 1 clothes dryers that are provided with make-up air in accordance with 10.4.3.*

Type 1 clothes dryer exhaust fans typically range from 180 cfm to 200 cfm (5.1 $m^3$/min to 5.6 $m^3$/min). Larger-capacity models may exhaust up to 250 cfm (7.1 $m^3$/min). The air exhaust rates include air for removal of moisture, as well as the air needed for combustion of the fuel gas. For example, models with an exhaust rate of 180 cfm (5.1 $m^3$/min) require 10,800 $ft^3$ (306 $m^3$) per hour of makeup air. A gas clothes dryer's energy input, regardless of exhaust rate, ranges up to 35,000 Btu/hr (3.2 kW/hr), which at a rate of 10 to 1 for complete combustion is 350 $ft^3$/hr (9.9 $m^3$/hr). Therefore, a dryer's combustion air is only 3 percent of the total air required for proper dryer operation.

**FAQ ▶** If there is sufficient makeup air for the clothes dryer per the manufacturer's installation instructions, is additional combustion air required?

Subsection 10.4.3.1 requires that makeup air be provided and that it be in accordance with the manufacturer's instructions. Because this amount of air would include sufficient combustion air, the combustion air requirements of 9.3.1.1 do not apply.

**9.3.1.2** Appliances of other than natural draft design and other than Category I vented appliances shall be provided with combustion, ventilation, and dilution air in accordance with the appliance manufacturer's instructions.

"Appliances of other than natural draft design" refers to appliances with power vents, either integral with the appliance or added; clothes dryers; high-efficiency furnaces (95 percent or higher); direct vent appliances; and so on. The manufacturer's installation instructions include air openings or other means to provide adequate air for safe appliance operation. Direct vent appliances do not draw air for combustion and dilution from the building when they are installed in accordance with the manufacturer's installation instructions.

**9.3.1.3** Appliances shall be located so as not to interfere with proper circulation of combustion, ventilation, and dilution air.

**9.3.1.4** Where used, a draft hood or a barometric draft regulator shall be installed in the same room or enclosure as the appliance served so as to prevent any difference in pressure between the hood or regulator and the combustion air supply.

A draft hood and barometric draft regulator (barometric damper) are defined as follows:

> **3.3.33 Draft Hood.** A nonadjustable device built into an appliance, or made a part of the vent connector from an appliance, that is designed to (1) provide for the ready escape of the flue gases from the appliance in the event of no draft, backdraft, or stoppage beyond the draft hood, (2) prevent a backdraft from entering the appliance, and (3) neutralize the effect of stack action of the chimney or gas vent upon the operation of the appliance.
>
> **3.3.88.2.1 Barometric Draft Regulator.** A balanced damper device attached to a chimney, vent connector, breeching, or flue gas manifold to control chimney draft.

Draft hoods and barometric draft regulators allow air to enter the vent close to an appliance. Where they are used, these devices allow air (called dilution air) to enter the flue and mix with the products of combustion. As covered in 9.3.1.4, these devices must be in the same room or enclosure with the appliance to operate properly. If a device is located in a different room or floor or outside an appliance enclosure, the device could be affected by a pressure different from the pressure in the room or space, due to a furnace fan or exhaust fan, and the draft hood or barometric draft regulator could provide a path for the products of combustion to spill into the room in which the appliance is installed.

**9.3.1.5** Where exhaust fans, clothes dryers, and kitchen ventilation systems interfere with the operation of appliances, makeup air shall be provided. NEW

The requirement in 9.3.1.5 was revised in the 2009 edition to clarify the requirement. Formerly it required that makeup air be considered where exhaust fans, kitchen ventilation systems, clothes dryers, and fireplaces were installed. The revised text requires makeup air only where the operation of appliances is interfered with.

The operation of exhaust fans, kitchen ventilators, clothes dryers, fireplaces, and other systems that exhaust air from a building can deprive a gas appliance of combustion air. In extreme cases, an exhaust fan can pull air in through the vent of natural draft appliances. This situation results in the combustion products spilling from the draft hood or flue collar into the room in which the appliance is located. Coupled with a lack of combustion and ventilation air, incomplete combustion can result in flame rollout, and high levels of carbon monoxide can be formed.

Section 9.3 contains provisions intended to provide adequate combustion and ventilation air for gas appliances based on an air infiltration rate of 0.5 ACH. Because of interest in energy conservation, consumers are attempting to reduce heat losses by adding weatherstripping, caulking, and insulation to their residences. Although these measures are effective in reducing energy consumption, they also can reduce the air infiltration rate of the building, possibly to levels below those used to determine proper combustion air. If this happens, additional air must be provided for proper operation of appliances. Air will enter the building through any opening available, and the vent of a gas appliance may provide the easiest available opening. An example of this is shown in Exhibit 9.9.

## 9.3.2 Indoor Combustion Air.

The required volume of indoor air shall be determined in accordance with the method in 9.3.2.1 or 9.3.2.2 except that where the air infiltration rate is known to be less than 0.40 *ACH*

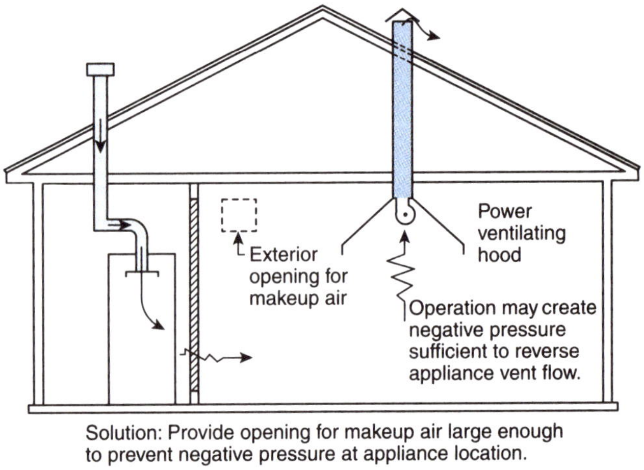


***EXHIBIT 9.9*** *Mechanical Exhausting. (Courtesy of Carrier Corporation)*

(air change per hour), the method in 9.3.2.2 shall be used. The total required volume shall be the sum of the required volume calculated for all appliances located within the space. Rooms communicating directly with the space in which the appliances are installed through openings not furnished with doors, and through combustion air openings sized and located in accordance with 9.3.2.3, are considered a part of the required volume.

Indoor combustion air is determined using one of the two methods — the standard method or the KAIR method. The standard method, which is covered in 9.3.2.1, has been in the code since the 1950s and remains unchanged. It is limited to buildings with an air infiltration rate of 0.4 ACH or more. The KAIR method, which is covered in 9.3.2.2 and was added to the code in 2002, can be used to calculate the required indoor volume for all structures, including those built with very low air infiltration rates.

The code does not require the determination of a structure's air infiltration. The standard method can be used for many installations. However, if the structure's ACH is known to be less than 0.40 ACH, then the standard method for obtaining combustion air cannot be used, and the KAIR method must be used.

Supplement 5 of this handbook contains a procedure and an example on how to calculate air infiltration based on an American Society of Heating, Refrigerating, and Air-Conditioning Engineers (ASHRAE) method. This method is provided to assist those who need to know the air exchange of a building but do not wish to measure the air exchange rate, or those who cannot measure the air exchange rate because the building has not yet been built. This calculation can be performed based on a building's blueprints or on an existing building. The ACH for an existing residential building may also be measured with a pressurization test, commonly called a blower-door test. A value for ACH may also be available that is based either on a nationally recognized ventilation standard or on a value established by an adopting jurisdiction based on local experience and applied to structures located in that jurisdiction. The ASHRAE Standard 62.1, *Ventilation for Acceptable Indoor Air Quality,* recommends that structures have a minimum of 0.35 ACH with at least 15 cfm per occupant. Exhibits 9.10 and 9.11 show that most building spaces meet the recommended ACH.

**No measures were taken to make my building unusually tight. Would the standard method be the appropriate method to determine combustion air?**

**9.3.2.1* Standard Method.** The minimum required volume shall be 50 $ft^3$/1000 Btu/hr (4.8 $m^3$/kW).

**FAQ ►** The standard method can be used for a vast majority of residential and light commercial buildings with an ACH rate lower than 0.4. If the air exchange rate is not known, and the building does not have special provisions to make it airtight beyond those in the applicable building code, then the standard method can most probably be used. To verify the acceptability of the standard method, check the appliances for proper operation after installation.

ASHRAE provides data on air infiltration in buildings in the *ASHRAE Handbook — Fundamentals*. Exhibit 9.10 and Exhibit 9.11, reprinted from the *ASHRAE Handbook — Fundamentals*, show average seasonal infiltration rates in two national studies. In Exhibit 9.10, samples are biased toward new, energy-efficient houses and have a median value of 0.50 ACH. Exhibit 9.11 represents older, low-income housing, and samples have a median value of 0.90 ACH.

The standard method is sometimes referred to as the 1/20 rule, because 1000 Btu/hr ÷ 50 $ft^3$ = 20 Btu/hr per $ft^3$ (207 $W/m^2$) or, for each cubic foot of room volume, 20 Btu/hr can be installed.

Exhibit 9.12 shows a room in which a furnace and water heater are installed. The calculation worksheet shown in Worksheet 9.1 shows how to determine whether the space shown in Exhibit 9.12 is confined or unconfined.

The volume of the room connecting by an archway could be used in the calculation. The room connecting with a door could not. Only permanent openings that cannot be blocked, such as by a door, qualify an adjoining space to be included in the calculation for combustion air. See the commentary following 9.3.2.3 for more information on the required openings for combining room volumes.

**◀ FAQ**

**The gas appliances are in a storage area that connects to two other rooms. One room connects by an archway, the other room connects by a door that stands open all the time. Can the volume of these rooms be included in the calculation for combustion air?**

**A.9.3.2.1** See Table A.9.3.2.1.

**9.3.2.2* Known Air Infiltration Rate Method.** Where the air infiltration rate of a structure is known, the minimum required volume shall be determined as follows:

(1) For appliances other than fan assisted, calculate using the following equation:

$$\text{Required Volume}_{other} \geq \frac{21\ \text{ft}^3}{ACH}\left(\frac{I_{other}}{1000\ \text{Btu/hr}}\right)$$

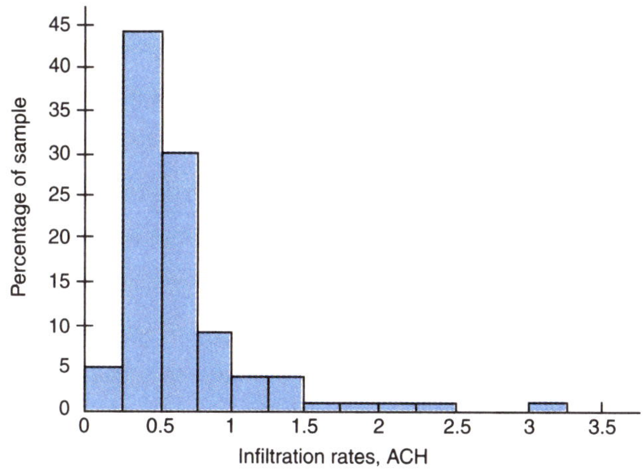


*EXHIBIT 9.10 Histogram of Infiltration Values: New Construction. (Source: Copyright 2005, © American Society of Heating, Refrigerating and Air-Conditioning Engineers, Inc. (www.ashrae.org). Reprinted by permission from 2005 ASHRAE Handbook — Fundamentals, Fig. 10. This text may not be copied or distributed in either paper or digital form without ASHRAE's permission.)*

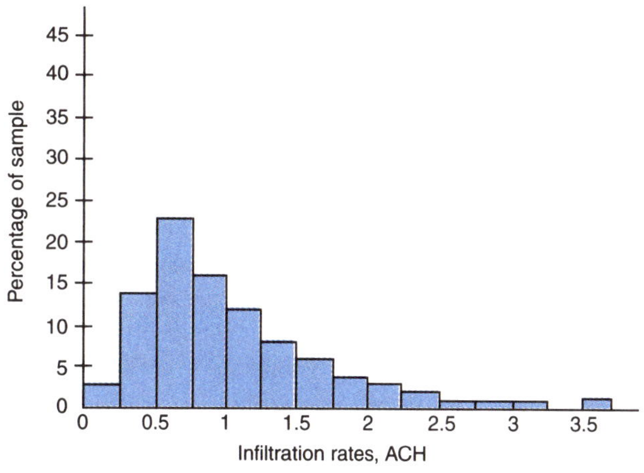


*EXHIBIT 9.11 Histogram of Infiltration Values: Low-Income Housing. (Source: Copyright 2005, © American Society of Heating, Refrigerating and Air-Conditioning Engineers, Inc. (www.ashrae.org). Reprinted by permission from 2005 ASHRAE Handbook — Fundamentals, Fig. 11. This text may not be copied or distributed in either paper or digital form without ASHRAE's permission.)*

*EXHIBIT 9.12 Room with Furnace and Water Heater.*

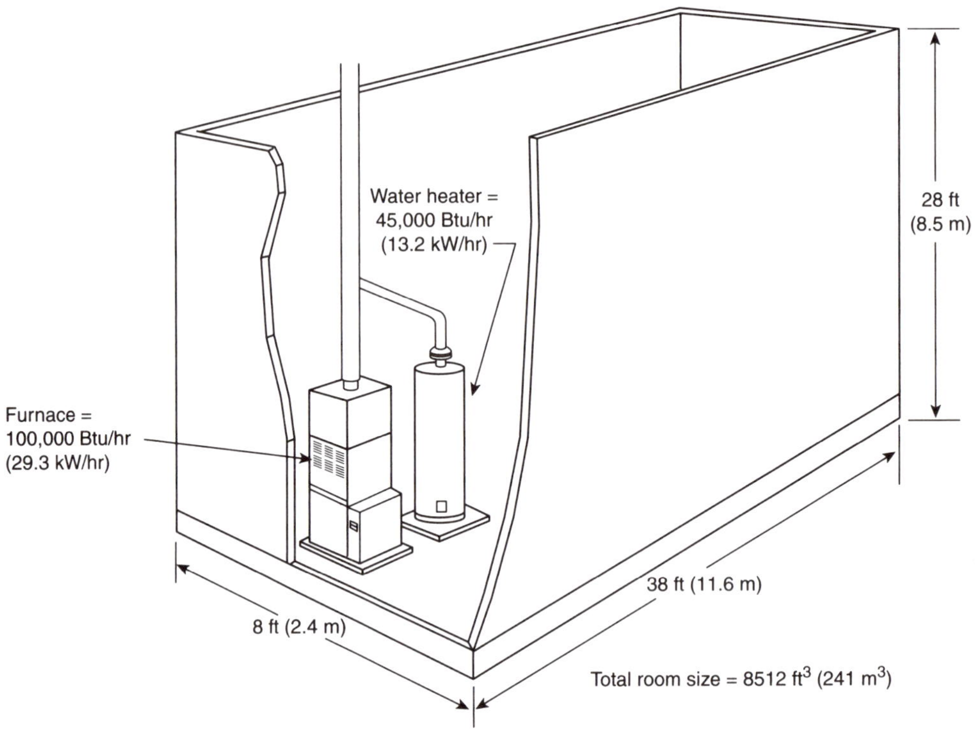


(2) For fan-assisted appliances, calculate using the following equation:

$$\text{Required Volume}_{fan} \geq \frac{15\ \text{ft}^3}{ACH}\left(\frac{I_{fan}}{1000\ \text{Btu/hr}}\right)$$

where:

$I_{other}$ = all appliances other than fan-assisted input (Btu/hr)

$I_{fan}$ = fan-assisted appliance input (Btu/hr)

$ACH$ = air change per hour (percent of volume of space exchanged per hour, expressed as a decimal)

The KAIR method requires a calculation of the required air volume based on two appliance types, other than fan assisted [see 9.3.2.2(1)] and fan assisted [see 9.3.2.2(2)]. The formula for each appliance type is based on the total combustion air needs of each type of appliance, which differ due to the amount of dilution air required. If installations include both types of appliance, then a separate calculation is done for each type of appliance. These calculations are combined to determine the total required air volume. These calculations determine the amount of air for combustion and ventilation required by the appliance for complete combustion. Combustion air includes three components:

1. The amount of air for complete combustion, 10 $ft^3$ per $ft^3$ (10 $m^3$ per $m^3$) gas burned
2. Excess air to help ensure complete combustion, typically 5 $ft^3$ per $ft^3$ (5 $m^3$ per $m^3$) of gas
3. Dilution air for proper venting of draft hood equipped appliances, typically 6 $ft^3$ per $ft^3$ (6 $m^3$ per $m^3$)

The historical assumptions for excess and dilution air are not for fan-assisted appliances, which were introduced in 1990. Fan-assisted appliances have neither a draft hood nor a flue collar, and therefore, dilution air is limited to leakage from vent fittings. Research by Battelle, published in *Combustion-Air Issues Related to Residential Gas Appliances in Confined Spaces and Unusually Tight Construction,* has verified this assumption.

## CALCULATION WORKSHEET: COMBUSTION AIR, STANDARD METHOD

**Step 1:**

- Calculate the room volume.

Room volume: Room length: 38 ft
Room width: 28 ft
Room height: 8 ft

Room volume: = Length × width × height
= 8512 ft³

**Step 2:**

- Calculate the total input of all appliances in the room.
- Enter the input rating of all appliances in Table 1. (Per 9.3.1.1, Exception 2, dryers are not included.)
- Total the column.
- Divide the total by 1000 (of Btu/hr).

**Table 1 Appliances Table**

| Appliance | Input rating (Btu/hr) |
|---|---|
| Furnace | 100,000 |
| Water heater | 45,000 |
| Space heater | |
| Range | |
| Other | |
| Total | 145,000 |
| Total/1000 | 145 |

**Step 3:**

- Calculate the required volume. Divide room volume (Step 1) by total/1000 (Step 2).
- If less than 50, additional air is needed.
- If greater than or equal to 50, no additional air is needed.

Required volume: $= \frac{8512\ ft^3}{145}$

= 58.7

Additional air needed? (Check one) Yes ❑ No ☑

## ALTERNATE CALCULATION METHOD

**Step 1:**

- Calculate the room volume.

Room volume = 8512 ft³ (from Step 1 above)

**Step 2:**

- Calculate the maximum appliance input.

Maximum appliance input: = Room volume × 20
= 8512 × 20
= 170,240 Btu/hr

**Step 3:**

- Determine if additional air is needed.
- If less than max., no additional air is needed.
- If greater than or equal to max., additional air is needed.

Total appliance input: = 145,000 Btu/hr

Additional air needed? (Check one) Yes ❑ No ☑

**Job:** 25 Main Street **Prepared by:** TL **Date:** 1/1/09



***WORKSHEET 9.1*** *Sample Calculation for Determining Combustion Air Using the Standard Method.*

*TABLE A.9.3.2.1 Standard Method: Required Volume, All Appliances*

| Appliance Input (Btu/hr) | Required Volume ($ft^3$) | Appliance Input (Btu/hr) | Required Volume ($ft^3$) |
|---|---|---|---|
| 5,000 | 250 | 130,000 | 6,500 |
| 10,000 | 500 | 135,000 | 6,750 |
| 15,000 | 750 | 140,000 | 7,000 |
| 20,000 | 1,000 | 145,000 | 7,250 |
| 25,000 | 1,250 | 150,000 | 7,500 |
| 30,000 | 1,500 | 160,000 | 8,000 |
| 35,000 | 1,750 | 170,000 | 8,500 |
| 40,000 | 2,000 | 180,000 | 9,000 |
| 45,000 | 2,250 | 190,000 | 9,500 |
| 50,000 | 2,500 | 200,000 | 10,000 |
| 55,000 | 2,750 | 210,000 | 10,500 |
| 60,000 | 3,000 | 220,000 | 11,000 |
| 65,000 | 3,250 | 230,000 | 11,500 |
| 70,000 | 3,500 | 240,000 | 12,000 |
| 75,000 | 3,750 | 250,000 | 12,500 |
| 80,000 | 4,000 | 260,000 | 13,000 |
| 85,000 | 4,250 | 270,000 | 13,500 |
| 90,000 | 4,500 | 280,000 | 14,000 |
| 95,000 | 4,750 | 290,000 | 14,500 |
| 100,000 | 5,000 | 300,000 | 15,000 |
| 105,000 | 5,250 | | |
| 110,000 | 5,500 | | |
| 115,000 | 5,750 | | |
| 120,000 | 6,000 | | |
| 125,000 | 6,250 | | |

For SI units, 1 $ft^3$ = 0.028 $m^3$, 1000 Btu/hr = 0.293 kW. NEW

Worksheet 9.2 shows an example of using the KAIR method to calculate the required air volume of a room with other than a fan-assisted appliance [9.3.2.2(1)] and with a fan-assisted appliance [9.3.2.2(2)].

(3) For purposes of this calculation, an infiltration rate greater than 0.60 *ACH* shall not be used in the equations in 9.3.2.2(1) and 9.3.2.2(2). (using the appliances shown in Exhibit 9.12)

An infiltration rate greater than 0.60 ACH, required by 9.3.2.2(3), is a practical upper limit. Because the KAIR method can be applied to structures with an air infiltration rate greater than 0.40 ACH, a limitation was added as a safety precaution to prevent appliances from being installed in rooms that are too small.

The KAIR equations result in a required volume (room size) that is inversely proportional to the structure's air infiltration rate. Therefore, as the amount of air infiltration increases, the required volume decreases. If the KAIR method were allowed to be applied to structures with high air exchange rates without a limit, the calculated required volume could result in a room size approaching that of a closet. This scenario could be in conflict with 10.3.2.2, which requires central heating furnaces and low-pressure boilers to be listed for closet installation when installed in rooms that are NOT large, such as alcoves and closets, in comparison with the size of the appliance.

## CALCULATION WORKSHEET: COMBUSTION AIR, KAIR METHOD

**Step 1:**

- Enter the input ratings of all other than fan-assisted appliances in Table 1.
- Total the column.

**Table 1 Ratings for Non-Fan-Assisted Appliances**

| Appliance | Input rating (Btu/hr) |
|---|---|
| Furnace | |
| Water heater | 45,000 |
| Space heater | |
| Refrigerator | |
| Other | |
| Other | |
| Total($I_{other}$) | 45,000 |

**Step 2:**

- Enter the air infiltration rate of the space. This can be determined by measurement or estimation. See Supplement 5 for an estimation method.
- If the air infiltration rate exceeds 0.6, use 0.6.

Air infiltration rate: 0.5 ACH (max. 0.6)

**Step 3:**

- Calculate the required volume for non-fan-assisted appliances using the equation.
- Enter the required volume here and in Step 6.

Required volume:

$$\frac{21\text{ ft}^3}{\text{ACH}}\left(\frac{I_{other}}{100{,}000\text{ Btu/hr}}\right) = \frac{21\text{ ft}^3}{0.5}\left(\frac{45{,}000\text{ Btu/hr}}{100{,}000\text{ Btu/hr}}\right)$$

$$42\ (45) = 1890\text{ ft}^3$$

**Step 4:**

- Enter the input ratings of all fan-assisted appliances in Table 2.
- Total the column.

**Table 2 Ratings for Fan-Assisted Appliances**

| Appliance | Input rating (Btu/hr) |
|---|---|
| Furnace | 100,000 |
| Furnacc | |
| Space heater | |
| Other | |
| Other | |
| Other | |
| Total ($I_{fan}$) | 100,000 |

**Step 5:**

- Calculate the required volume for fan-assisted appliances using the equation.
- Enter the required volume here and below.

Required volume:

$$\frac{15\text{ ft}^3}{\text{ACH}}\left(\frac{I_{fan}}{100{,}000\text{ Btu/hr}}\right) = \frac{15\text{ ft}^3}{0.5}\left(\frac{1{,}000\text{ Btu/hr}}{100{,}000\text{ Btu/hr}}\right)$$

$$15\ (100) = 3000\text{ ft}^3$$

**Step 6:**

- Add the two required volumes calculated to find the total required volume for all appliances.

Required volume:
(Non-fan assisted) = 1890 ft$^3$
Required volume
(Fan assisted) = 3000 ft$^3$
Total required volume = 4890 ft$^3$

**Step 7:**

- Enter the available room volume.

Available volume = 30 x 20 x 8
Room volume = 4800 ft$^3$

**Step 8:**

- If the total required volume is greater than the available room volume, air openings are required. Check Yes and determine the opening size and location. If the total required volume is less than the available room volume, no air openings are required. Check No.

Air openings required? (Check one)

Yes ☑ No ☐

**Job:** 25 Main Street **Prepared by:** TL **Date:** 1/1/09



***WORKSHEET 9.2*** *Sample Calculation for Determining Combustion Air Using the KAIR Method.*

**A.9.3.2.2** See Table A.9.3.2.2(a) and Table A.9.3.2.2(b).

**9.3.2.3 Indoor Opening Size and Location.** Openings used to connect indoor spaces shall be sized and located in accordance with the following:

(1)* *Combining spaces on the same story.* Each opening shall have a minimum free area of 1 in.$^2$/1000 Btu/hr (2200 mm$^2$/kW) of the total input rating of all appliances in the space but not less than 100 in.$^2$ (0.06 m$^2$). One opening shall commence within 12 in. (300 mm) of the top, and one opening shall commence within 12 in. (300 mm) of the bottom, of the enclosure *[see Figure A.9.3.2.3(1)]*. The minimum dimension of air openings shall not be less than 3 in. (80 mm).

**A.9.3.2.3(1)** See Figure A.9.3.2.3(1).

(2) *Combining spaces in different stories.* The volumes of spaces in different stories shall be considered as communicating spaces where such spaces are connected by one or more openings in doors or floors having a total minimum free area of 2 in.$^2$/1000 Btu/hr (4400 mm$^2$/kW) of total input rating of all appliances.

***TABLE A.9.3.2.2(a)*** *Known Air Infiltration Rate Method: Minimum Space Volume for Appliances Other than Fan-Assisted for Specified Infiltration Rates (ACH)*

| *Appliance Input* | *Space Volume (ft$^3$)* | | | *Appliance Input* | *Space Volume (ft$^3$)* | | |
|---|---|---|---|---|---|---|---|
| *(Btu/hr)* | *0.25 ACH* | *0.30 ACH* | *0.35 ACH* | *(Btu/hr)* | *0.25 ACH* | *0.30 ACH* | *0.35 ACH* |
| 5,000 | 420 | 350 | 300 | 130,000 | 10,920 | 9,100 | 7,800 |
| 10,000 | 840 | 700 | 600 | 135,000 | 11,340 | 9,450 | 8,100 |
| 15,000 | 1,260 | 1,050 | 900 | 140,000 | 11,760 | 9,800 | 8,400 |
| 20,000 | 1,680 | 1,400 | 1,200 | 145,000 | 12,180 | 10,150 | 8,700 |
| 25,000 | 2,100 | 1,750 | 1,500 | 150,000 | 12,600 | 10,500 | 9,000 |
| 30,000 | 2,520 | 2,100 | 1,800 | 160,000 | 13,440 | 11,200 | 9,600 |
| 35,000 | 2,940 | 2,450 | 2,100 | 170,000 | 14,280 | 11,900 | 10,200 |
| 40,000 | 3,360 | 2,800 | 2,400 | 180,000 | 15,120 | 12,600 | 10,800 |
| 45,000 | 3,780 | 3,150 | 2,700 | 190,000 | 15,960 | 13,300 | 11,400 |
| 50,000 | 4,200 | 3,500 | 3,000 | 200,000 | 16,800 | 14,000 | 12,000 |
| 55,000 | 4,620 | 3,850 | 3,300 | 210,000 | 17,640 | 14,700 | 12,600 |
| 60,000 | 5,040 | 4,200 | 3,600 | 220,000 | 18,480 | 15,400 | 13,200 |
| 65,000 | 5,460 | 4,550 | 3,900 | 230,000 | 19,320 | 16,100 | 13,800 |
| 70,000 | 5,880 | 4,900 | 4,200 | 240,000 | 20,160 | 16,800 | 14,400 |
| 75,000 | 6,300 | 5,250 | 4,500 | 250,000 | 21,000 | 17,500 | 15,000 |
| 80,000 | 6,720 | 5,600 | 4,800 | 260,000 | 21,840 | 18,200 | 15,600 |
| 85,000 | 7,140 | 5,950 | 5,100 | 270,000 | 22,680 | 18,900 | 16,200 |
| 90,000 | 7,560 | 6,300 | 5,400 | 280,000 | 23,520 | 19,600 | 16,800 |
| 95,000 | 7,980 | 6,650 | 5,700 | 290,000 | 24,360 | 20,300 | 17,400 |
| 100,000 | 8,400 | 7,000 | 6,000 | 300,000 | 25,200 | 21,000 | 18,000 |
| 105,000 | 8,820 | 7,350 | 6,300 | | | | |
| 110,000 | 9,240 | 7,700 | 6,600 | | | | |
| 115,000 | 9,660 | 8,050 | 6,900 | | | | |
| 120,000 | 10,080 | 8,400 | 7,200 | | | | |
| 125,000 | 10,500 | 8,750 | 7,500 | | | | |

For SI units, 1 ft$^3$ = 0.028 m$^3$, 1000 Btu/hr = 0.293 kW.
ACH: Air change per hour.

*TABLE A.9.3.2.2(b) Known Air Infiltration Rate Method: Minimum Space Volume for Fan-Assisted Appliance, for Specified Infiltration Rates (ACH)*

| Appliance Input (Btu/hr) | Space Volume ($ft^3$) | | | Appliance Input (Btu/hr) | Space Volume ($ft^3$) | | |
|---|---|---|---|---|---|---|---|
| | 0.25 ACH | 0.30 ACH | 0.35 ACH | | 0.25 ACH | 0.30 ACH | 0.35 ACH |
| 5,000 | 300 | 250 | 214 | 130,000 | 7,800 | 6,500 | 5,571 |
| 10,000 | 600 | 500 | 429 | 135,000 | 8,100 | 6,750 | 5,786 |
| 15,000 | 900 | 750 | 643 | 140,000 | 8,400 | 7,000 | 6,000 |
| 20,000 | 1,200 | 1,000 | 857 | 145,000 | 8,700 | 7,250 | 6,214 |
| 25,000 | 1,500 | 1,250 | 1,071 | 150,000 | 9,000 | 7,500 | 6,429 |
| 30,000 | 1,800 | 1,500 | 1,286 | 160,000 | 9,600 | 8,000 | 6,857 |
| 35,000 | 2,100 | 1,750 | 1,500 | 170,000 | 10,200 | 8,500 | 7,286 |
| 40,000 | 2,400 | 2,000 | 1,714 | 180,000 | 10,800 | 9,000 | 7,714 |
| 45,000 | 2,700 | 2,250 | 1,929 | 190,000 | 11,400 | 9,500 | 8,143 |
| 50,000 | 3,000 | 2,500 | 2,143 | 200,000 | 12,000 | 10,000 | 8,571 |
| 55,000 | 3,300 | 2,750 | 2,357 | 210,000 | 12,600 | 10,500 | 9,000 |
| 60,000 | 3,600 | 3,000 | 2,571 | 220,000 | 13,200 | 11,000 | 9,429 |
| 65,000 | 3,900 | 3,250 | 2,786 | 230,000 | 13,800 | 11,500 | 9,857 |
| 70,000 | 4,200 | 3,500 | 3,000 | 240,000 | 14,400 | 12,000 | 10,286 |
| 75,000 | 4,500 | 3,750 | 3,214 | 250,000 | 15,000 | 12,500 | 10,714 |
| 80,000 | 4,800 | 4,000 | 3,429 | 260,000 | 15,600 | 13,000 | 11,143 |
| 85,000 | 5,100 | 4,250 | 3,643 | 270,000 | 16,200 | 13,500 | 11,571 |
| 90,000 | 5,400 | 4,500 | 3,857 | 280,000 | 16,800 | 14,000 | 12,000 |
| 95,000 | 5,700 | 4,750 | 4,071 | 290,000 | 17,400 | 14,500 | 12,429 |
| 100,000 | 6,000 | 5,000 | 4,286 | 300,000 | 18,000 | 15,000 | 12,857 |
| 105,000 | 6,300 | 5,250 | 4,500 | | | | |
| 110,000 | 6,600 | 5,500 | 4,714 | | | | |
| 115,000 | 6,900 | 5,750 | 4,929 | | | | |
| 120,000 | 7,200 | 6,000 | 5,143 | | | | |
| 125,000 | 7,500 | 6,250 | 5,357 | | | | |

For SI units, 1 $ft^3$ = 0.028 $m^3$, 1000 Btu/hr = 0.293 kW.
ACH: Air change per hour.

When appliances are installed in a room that has less than the required volume, the required volume can be obtained by "combining" the room with an adjacent room. This method is usually preferred to using outdoor air, which can bring large amounts of unheated air into a room and cool it. To combine rooms, openings are cut to the sizes specified in 9.3.2.3(1).

Paragraph 9.3.2.3(2) specifically addresses connecting rooms located one above the other. Compliance with 9.3.2.3(2) requires one or more openings, with an area equal to the two openings required in 9.3.2.3(1). For example, combining spaces on different stories might involve an appliance in a basement. The basement may not be of the required volume, and the installer may elect to use space on the first floor.

Separation of the high and the low opening, as used when combining spaces on the same story, does not apply to the vertical connection of spaces. The vertical opening provision combines the total area, as required for both the high and low opening on the same story, into one opening for vertical connections.

What is meant by "minimum free area" when referring to openings in doors, floors, or ceilings?
◀ **FAQ**

The openings in buildings are often finished with grilles or louvers, as shown in Exhibit 9.13. Consideration must be given to the blocking effect of the vanes incorporated in the louver or grille. This blocking effect could be as much as 80 percent on grilles designed for

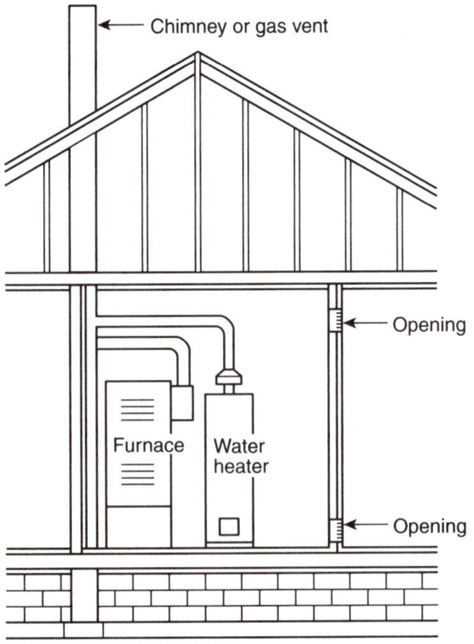


***Figure A.9.3.2.3(1)*** *All Combustion Air from Adjacent Indoor Spaces Through Indoor Combustion Air Openings.*

*EXHIBIT 9.13 Louver-Free Area. (Courtesy of Carrier Corporation)*

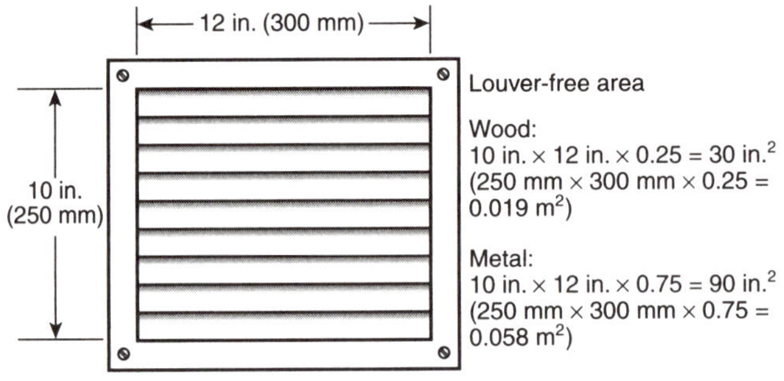


privacy and low sound transmission. The openings covered by such a louver would have to be correspondingly bigger (5 times the area) to provide the minimum free area required by these code sections. See the commentary on 9.3.7 for more information on grilles and louvers.

### 9.3.3 Outdoor Combustion Air.

Outdoor combustion air shall be provided through opening(s) to the outdoors in accordance with the methods in 9.3.3.1 or 9.3.3.2. The minimum dimension of air openings shall not be less than 3 in. (80 mm).

The minimum dimension of 3 in. (80 mm) required by 9.3.3 is intended to prevent the use of a long, narrow opening. Leaves and lint can more easily block narrow openings than wider openings.

**9.3.3.1 Two Permanent Openings Method.** Two permanent openings, one commencing within 12 in. (300 mm) of the top and one commencing within 12 in. (300 mm) of the bottom, of the enclosure shall be provided. The openings shall communicate directly, or by ducts, with the outdoors or spaces that freely communicate with the outdoors, as follows:

(1)* Where directly communicating with the outdoors or where communicating to the outdoors through vertical ducts, each opening shall have a minimum free area of 1 in.$^2$/4000 Btu/hr (550 mm$^2$/kW) of total input rating of all appliances in the enclosure. *[See Figure A.9.3.3.1(1)(a) and Figure A.9.3.3.1(1)(b).]*

**A.9.3.3.1(1)** See Figure A.9.3.3.1(1)(a) and Figure A.9.3.3.1(1)(b).

(2)* Where communicating with the outdoors through horizontal ducts, each opening shall have a minimum free area of 1 in.$^2$/2000 Btu/hr (1100 mm$^2$/kW) of total input rating of all appliances in the enclosure. *[See Figure A.9.3.3.1(2).]*

Where all combustion and ventilation air is taken from outdoors, either directly or by ducts, one of two methods is permitted. The first method requires two openings or two ducts, either vertical or horizontal, commencing within 12 in. (300 mm) of the ceiling and floor. The second method allows one opening and is covered in 9.3.3.2.

Ducts must be the same area as the openings to which they connect. Vertical ducts or openings are sized at a minimum free area of 1 in.$^2$ per 4000 Btu/hr (550 mm$^2$/kW) of the total appliance input installed in the space. Paragraph 9.3.3.1(2) requires horizontal ducts to have a minimum free area of 1 in.$^2$ per 2000 Btu/hr (1100 mm$^2$/kW) of the total appliance input installed in the space. These openings or ducts can connect to spaces (e.g., crawl or attic) that freely communicate with the outdoors.

Vertical ducts allow a convection effect to occur, where warmer air gravitates out the upper duct opening while the lower duct introduces cooler air. This convection effect provides a higher ACH in vertical ducts than in horizontal ducts, where this effect is less pronounced.

Worksheet 9.3 shows how to determine the size of air openings.

◀ **FAQ**
**Why are horizontal ducts used for combustion air twice as large as vertical ducts used for the same purpose?**

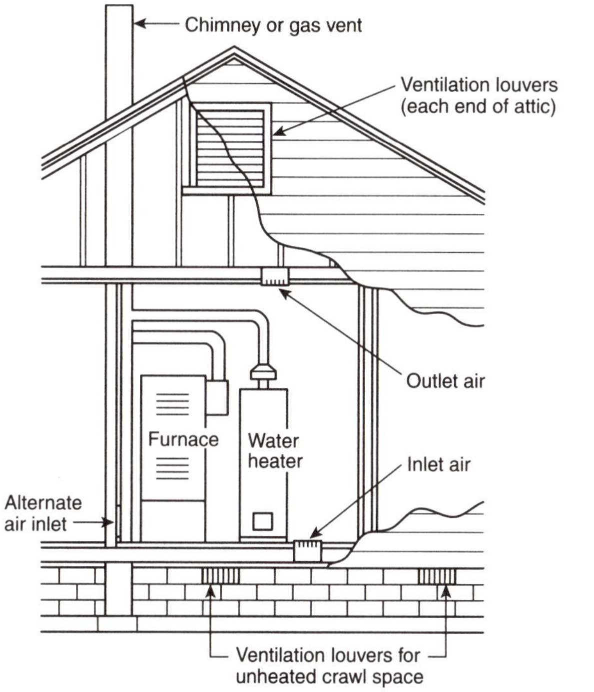


*Figure A.9.3.3.1(1)(a) All Combustion Air from Outdoors — Inlet Air from Ventilated Crawl Space and Outlet Air to Ventilated Attic.*

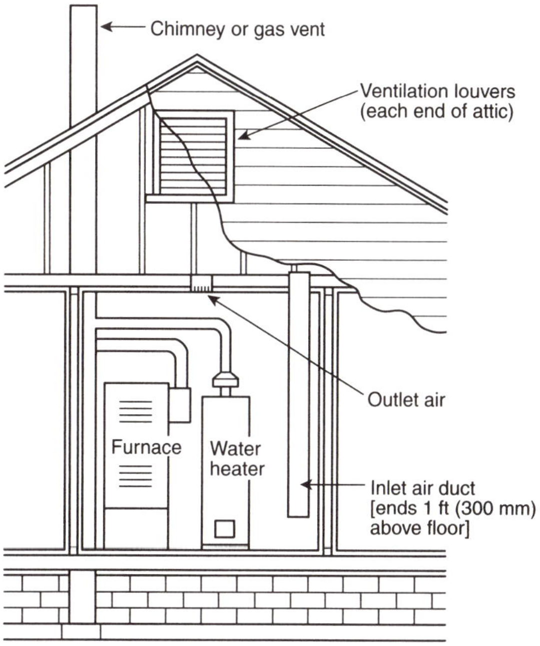


*Figure A.9.3.3.1(1)(b) All Combustion Air from Outdoors Through Ventilated Attic.*

## CALCULATION WORKSHEET: SIZING COMBUSTION AIR OPENINGS

**Step 1:**

- Enter the input ratings of all appliances in Table 1.
- Total the column.
- Proceed to either Part A, Part B, or Part C.

**Table 1 Ratings for Appliances**

| Appliance | Input rating (Btu/hr) |
|---|---|
| Furnace | 100,000 |
| Water heater | 45,000 |
| Space heater | |
| | |
| | |
| Total | 145,000 |

### A. All air from outdoors via two permanent openings (or vertical ducts).

**Step 2:**

- Where all air is to be taken from the outdoors, divide the total input of all gas appliances in the space by 4000.

Total input: 145,000 /4000 = 36.25 in.²/opening

**Step 3:**

- Select a duct with the area needed.
- Use Table 2 to calculate for square or round ducts. For other sizes or rectangular shapes, the duct size can be calculated.

Square duct: 6 in. × 6 in. = 36 in.²

Round duct: 7 in. round = 38.5 in.²

**Table 2 Sizes for Square and Round Ducts**

| Area of square ducts | | Area of round ducts | |
|---|---|---|---|
| Side (in.) | Area (in.) | Diameter (in.) | Area (in.²) |
| 3 | 9 | 3 | 7.1 |
| 4 | 16 | 4 | 12.6 |
| 5 | 25 | 5 | 19.6 |
| 6 | 36 | 6 | 28.3 |
| 7 | 49 | 7 | 38.5 |
| 8 | 64 | 8 | 50.2 |
| 10 | 100 | 10 | 78.5 |
| 12 | 144 | 12 | 113.0 |

### B. All air from the outdoors via two horizontal ducts.

**Step 2:**

- Where all air is to be taken from the outdoors, divide the total input of all gas appliances in the space by 2000.

Total input: 145,000 /2000 = 73 in.²/opening

**Step 3:**

- Select a duct with the area needed.
- Use Table 2 to calculate for square or round ducts. For other sizes or rectangular shapes, the duct size can be calculated.

Square duct: 10 in. × 10 in. = 100 in.²

Round duct: 10 in. round = 78.5 in.²

### C. All air from the outdoors via one opening.

**Step 2:**

- Where all air is to be taken from the outdoors using one opening, divide the total input of all appliances in the space by 3000.

Total input: 145,000 /3000 = 48 in.²

Provide one 7 in. square duct or one 8 in. round duct.

**Step 3:**

- Check minimum clearances: side and back, 1 in.; front, 6 in.

Side and back: 1 in.
Front: 6 in.

Clearances OK? (Check one) Yes ☑ No ☐

**Step 4:**

- Check that total vent connector area is less than or equal to the opening area (Step 2).

Furnace connector diameter = 5 in.
Divide by 2 = 2.5 in.
Squared = 6.25 × 3.14 = 19.63 in.²
Water heater connector diameter = 4 in.
Divide by 2 = 2 in.
Squared = 4.00 × 3.14 = 12.56 in.²
Total vent connector area = 32.19 in.²

**Job:** 25 Main Street **Prepared by:** TL **Date:** 1/1/09



*WORKSHEET 9.3 Sample Calculation for Determining Size of Combustion Air Openings.*

**A.9.3.3.1(2)** See Figure A.9.3.3.1(2).

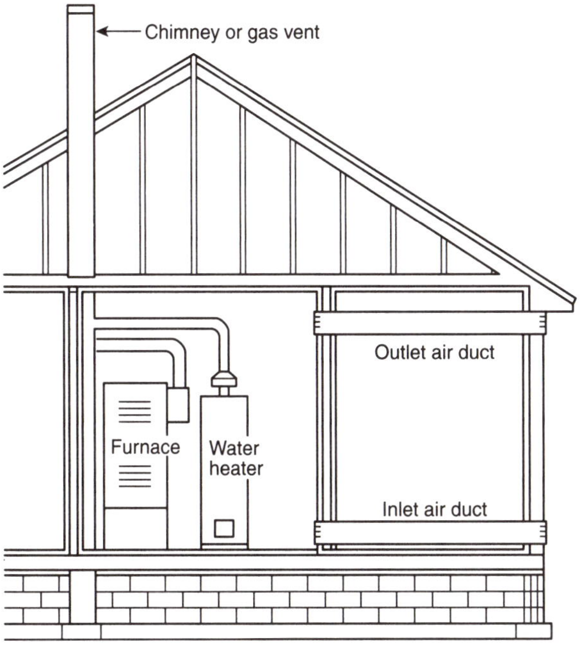


*Figure A.9.3.3.1(2) All Combustion Air from Outdoors Through Horizontal Ducts.*

**9.3.3.2* One Permanent Opening Method.** One permanent opening, commencing within 12 in. (300 mm) of the top of the enclosure, shall be provided. The appliance shall have clearances of at least 1 in. (25 mm) from the sides and back and 6 in. (150 mm) from the front of the appliance. The opening shall directly communicate with the outdoors or shall communicate through a vertical or horizontal duct to the outdoors or spaces that freely communicate with the outdoors *(see Figure A.9.3.3.2)* and shall have a minimum free area of the following:

(1) 1 in.$^2$/3000 Btu/hr (700 mm$^2$/kW) of the total input rating of all appliances located in the enclosure
(2) Not less than the sum of the areas of all vent connectors in the space

Gas Research Institute Report Number GRI 93/0316, "Analysis of Combustion Air Openings to the Outdoors: Preliminary Results," identifies the second method for sizing outdoor air openings, which allows one high opening to be used rather than one high and one low opening. This method enables the combustion and ventilation air openings to be smaller, thereby reducing freezing water in pipes and limiting low temperatures in appliance rooms.

One permanent opening or a vertical or horizontal duct can be used, provided that it directly communicates to the outdoors or connects through spaces (e.g., crawl space or attic) that freely communicate with the outdoors. The opening or duct must be within 12 in. (300 mm) of the ceiling of the enclosure. Also, GRI 93/0316 requires that the gas appliance(s) have a minimum clearance of 1 in. (25 mm) on each side and back and 6 in. (150 mm) in the front. The opening or duct is sized at a minimum free area of 1 in.$^2$ per 3000 Btu/hr (730 mm$^2$/kW) of the aggregate input rating of all appliances installed in the space, and the free area must not be less than the sum of the areas of all the vent connectors installed in the space, as required by 9.3.3.2(2).

Worksheet 9.4 shows how to size a single air opening.

## CALCULATION WORKSHEET: REDUCED BUILDING AREA OPENINGS

**Step 1:**

- Calculate the room volume.

Room volume: Room length: 15 ft
Room width: 10 ft
Room height: 8 ft

Room volume: = Length × width × height
= 1200 $ft^3$

**Step 2:**

- Calculate the total input of all appliances in the room.
- Enter the input rating of all appliances in Table 1.
- Total the column.
- Divide the total (room volume) by 1000 (of Btu/hr).

**Table 1 Appliances Table**

| Appliance | Input rating (Btu/hr) |
|---|---|
| Furnace | 100,000 |
| Water heater | 45,000 |
| Space heater | |
| Range | |
| Other | |
| Total | 145,000 |
| Total/1000 | 145 |

**Step 3:**

- Calculate the required volume.
- Enter the available room volume (Step 1).

Required volume:
= 50 $ft^3$ per 1000 Btu/hr (total/1000)
= 50 × 145
= 7250

Available room volume:
= 1200 $ft^3$

**Step 4:**

- Calculate the percentage of required volume that is available in the room.

Percentage of volume available: $= \frac{\text{Available volume}}{\text{Required volume}} \times 100$
$= \frac{1{,}200}{7{,}250} \times 100$
= 17 %

**Step 5:**

- Calculate the percentage of volume required.

Percentage of volume available: = 100% – 17 %
= 83 %

**Step 6:**

- Calculate the reduced opening area.

Reduced opening = Required area × percentage reduction
= 36.25 $in.^2$ × 83 %
= 30.09 $in.^2$ or
or two 5 1/2 $in.^2$ by 5 1/2 $in.^2$ openings.

**CAUTION: Do not apply to one opening.**

**Job:** 25 Main Street **Prepared by:** TL **Date:** 1/1/09



*WORKSHEET 9.4 Sample Calculation for Determining Size of Reduced Building Area Openings.*

**A.9.3.3.2** See Figure A.9.3.3.2.

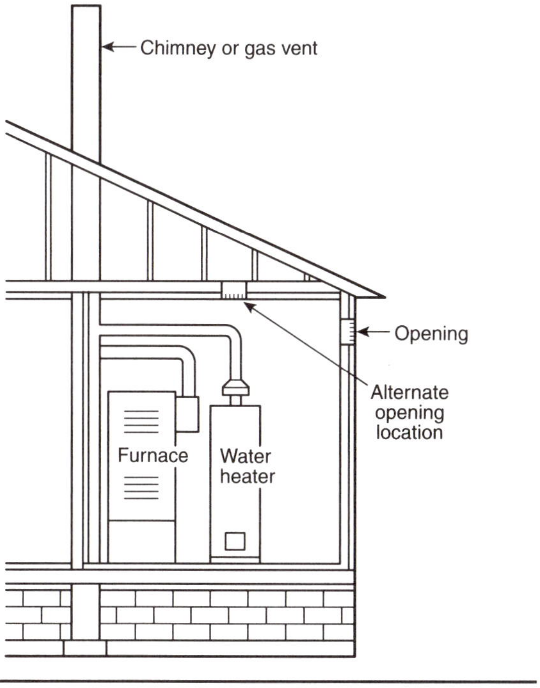


*Figure A.9.3.3.2 All Combustion Air from Outdoors Through Single Combustion Air Opening.*

### 9.3.4 Combination Indoor and Outdoor Combustion Air.

The use of a combination of indoor and outdoor combustion air shall be in accordance with the following *(see example calculation in Annex J)*:

(1) *Indoor Openings.* Where used, openings connecting the interior spaces shall comply with 9.3.2.3.
(2) *Outdoor Opening(s) Location.* Outdoor opening(s) shall be located in accordance with 9.3.3.
(3) *Outdoor Opening(s) Size.* The outdoor opening(s) size shall be calculated in accordance with the following:
    (a) The ratio of the interior spaces shall be the available volume of all communicating spaces divided by the required volume.
    (b) The outdoor size reduction factor shall be 1 minus the ratio of interior spaces.
    (c) The minimum size of outdoor opening(s) shall be the full size of outdoor opening(s) calculated in accordance with 9.3.3, multiplied by the reduction factor. The minimum dimension of air openings shall not be less than 3 in. (80 mm).

Using a combination of indoor and outdoor air to meet the air requirement for combustion in 9.3.4 has advantages in many installations. If the required volume cannot be obtained inside the building, it can be used to minimize the size of openings to the outdoors. Smaller openings mean less unheated air in the winter and less non-air-conditioned air leakage in the summer.

The requirements in 9.3.4 for sizing the outdoor air openings are similar to the requirements in 9.3.3 for outside combustion air only, but they allow proportionally smaller openings. The two high/low or single high openings to the outdoors can be reduced in proportion to the fraction with which the communicating volume complies, with the volume determined using the standard method or KAIR method. Note that the size calculation method in 9.3.4(3) to size outdoor openings cannot be used to reduce the size of openings to other rooms or spaces within the building.

For example, if the communicating volume amounted to 25 $ft^3$ per 1000 Btu/hr (2.4 $m^3$ per kW), which is 50 percent of what is needed, then the openings to the outdoors could be reduced 50 percent. This situation is shown schematically in Exhibit 9.14.

*EXHIBIT 9.14 Free Area Reduction.*

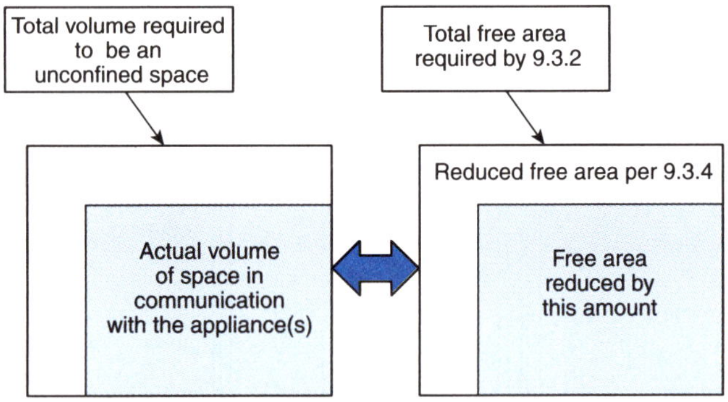


The provision is easy to apply if the following principles are understood:

1. The volume of the room that directly holds the appliances is included in the total communicating volume. All adjoining rooms that comply with 9.3.3 also may be considered to be part of the communicating volume. Thus, it may be necessary to provide two high/low openings between rooms that meet the requirements of 9.3.3.1.
2. The sizes of the high/low openings are based on the normal sizes required by 9.3.2.3(1).
3. The calculation should begin by determining the required volume, the actual communicating volume, and the normal free area of the openings to the outdoors as required by 9.3.3.

An example of how to calculate the size of the opening to the exterior is found in Worksheet 9.3. An example of the use of a combination indoor and outdoor air for combustion is shown in Annex J.

## 9.3.5 Engineered Installations.

Engineered combustion air installations shall provide an adequate supply of combustion, ventilation, and dilution air and shall be approved by the authority having jurisdiction.

Subsection 9.3.5 recognizes that some commercial and industrial gas appliances use fans to power the vent, making engineering calculations necessary for installations where the requirements of 9.3.3 are not relevant. It is imperative that only engineers thoroughly familiar with all aspects of the subject perform the calculations. Appliance manufacturers often provide assistance.

The free area required for makeup air called for in 9.3.3.1 assumes that the driving force moving air into the room where appliances are installed is the room pressure reduction caused by the operation of the vent. In other words, the removal of the heated products of combustion by the vent reduces the amount of air in the room in which the appliance is installed. The result is a small reduction in room pressure, forcing air into the room through all available openings. Large commercial and industrial appliances frequently incorporate forced air fans, which provide greater driving force than "natural" vents. This greater force moves more air through a given size opening, permitting a smaller opening than 9.3.3 allows.

With an emphasis on energy conservation, many builders and homeowners are sealing their homes to minimize air infiltration. This tight construction can lead to many problems, such as reduced combustion air and increased indoor air pollution. The use of heat recovery ventilators (HRVs) is one method to minimize heat loss due to infiltration and to maintain sufficient air for combustion and ventilation. See Exhibit 9.15 for an illustration of an HRV. The HRV draws fresh, outside air into the dwelling and exhausts stale, indoor air to the outdoors. These two air streams pass through a heat exchanger, warming or cooling the incoming outdoor air using the energy from the outgoing, conditioned indoor air, thereby minimizing energy loss.

HRVs can make up for a lack of combustion and ventilation air. For each cubic foot of gas burned, a natural draft (draft hood) furnace requires about 21 $ft^3$/hr or 0.35 $ft^3$/min (0.6 $m^3$/hr or 0.17 L/sec), and a fan-assisted combustion system furnace needs about 15 $ft^3$/hr or 0.25 $ft^3$/min (0.4 $m^3$/hr or 0.11 L/sec) of air. Therefore, a 100,000 Btu/hr (29.3 kW/hr) natural draft (draft hood) furnace would require 35 $ft^3$/min (1 $m^3$/hr) to meet its combustion and ventilation air requirements.

If an HRV is used to supply combustion and ventilation air, the HRV must be interlocked with the gas appliance(s) to ensure that it is operating prior to the main burner operation. If the HRV fails to operate, or if its airflow becomes restricted during gas appliance operation, the main burners must not be allowed to operate. Exhibit 9.16 shows a sail switch (in a duct with moving air) in a closed position, confirming combustion airflow and allowing the control valve of the interlocked appliance to operate. Should the air movement stop, the sail would drop and open the electrical circuit, causing the control valve to stop gas from flowing to the appliance.

◄ **FAQ**

**When an HRV is used to provide the combustion air (in lieu of openings) for appliances, is anything additional required for its installation?**

### 9.3.6 Mechanical Combustion Air Supply.

Where all combustion air is provided by a mechanical air supply system, the combustion air shall be supplied from outdoors at the minimum rate of 0.35 $ft^3$/min/1000 Btu/hr (0.034 $m^3$/min/kW) for all appliances located within the space.

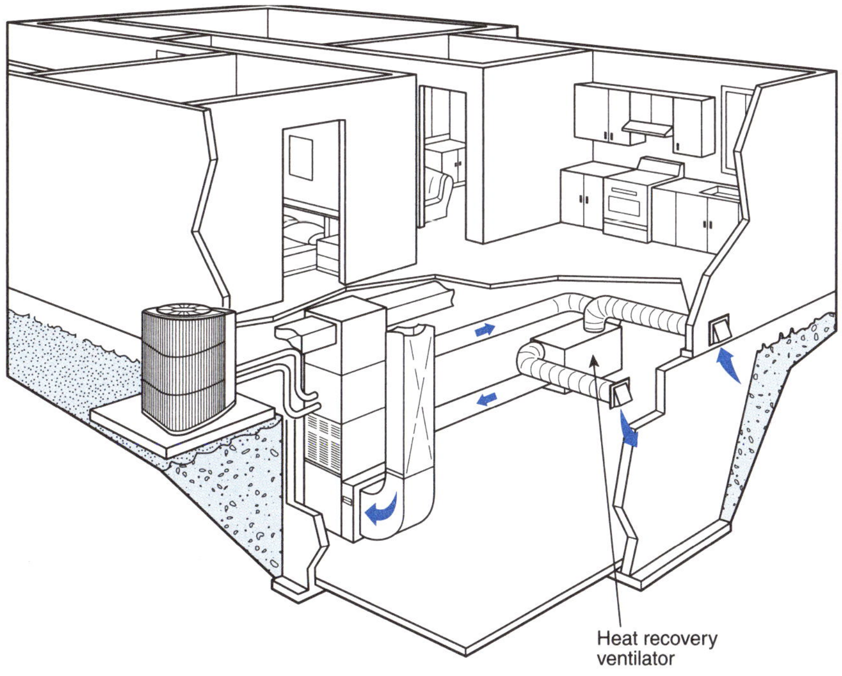


*EXHIBIT 9.15* *Heat Recovery Ventilator with Forced Air System. (Courtesy of Carrier Corporation)*

*EXHIBIT 9.16 Sail Switch Operating in Closed Position.*

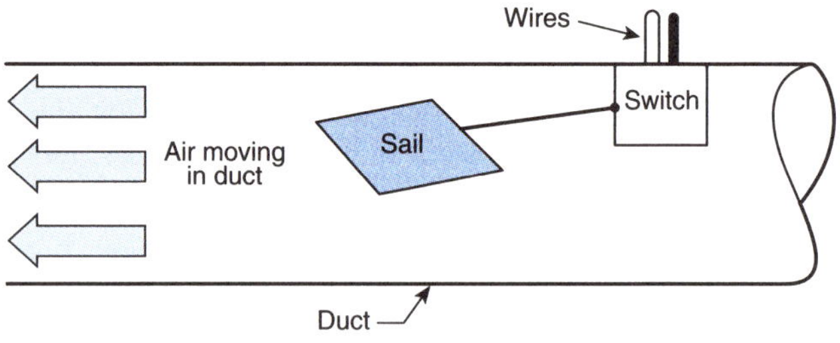


Fans are an acceptable method of providing combustion air. When used, the mechanical system referred to in 9.3.6 must supply at least 0.35 cfm per 1000 Btu/hr (0.034 $m^3$/min per kW) input and be interlocked with the main burner to prevent its operation in the event of air supply failure. (See Exhibit 9.16 for one method to interlock.) The minimum amount of air required, 0.35 cfm/1000 Btu/hr (0.034 $m^3$/min per kW), equals 21 cfh (0.6 $m^3$/hr) (0.35 cfm ÷ 60 min/hr = 21 cfh), which is the constant in the KAIR equation shown in 9.3.2.2(1).

The air supplied by fans must be available to the gas appliance. Additional air must be provided to equal air withdrawal by exhaust fans in the building. If additional air is not provided, the room in which the gas appliance is located may have insufficient air available from the supply fan. As the gas appliance operates, it will cause negative pressure in the space, and air will enter through any opening available, which could be the vent itself. Air entering through any available opening must be prevented by supplying sufficient air for the gas appliance and all building exhaust fans that can operate when the gas appliance is operating.

**9.3.6.1** Where exhaust fans are installed, additional air shall be provided to replace the exhausted air.

**9.3.6.2** Each of the appliances served shall be interlocked to the mechanical air supply system to prevent main burner operation where the mechanical air supply system is not in operation.

**9.3.6.3** Where combustion air is provided by the building's mechanical ventilation system, the system shall provide the specified combustion air rate in addition to the required ventilation air.

## 9.3.7 Louvers, Grilles, and Screens.

**9.3.7.1 Louvers and Grilles.** The required size of openings for combustion, ventilation, and dilution air shall be based on the net free area of each opening. Where the free area through a design of louver, grille, or screen is known, it shall be used in calculating the size opening required to provide the free area specified. Where the louver and grille design and free area are not known, it shall be assumed that wood louvers have 25 percent free area, and metal louvers and grilles have 75 percent free area. Nonmotorized louvers and grilles shall be fixed in the open position.

**9.3.7.2 Minimum Screen Mesh Size.** Screens shall not be smaller than $^1/_4$ in. mesh.

The minimum screen size requirement in 9.3.7.2 prevents fine mesh screens from being installed. Fine mesh screens could catch insects and debris and become blocked.

**9.3.7.3 Motorized Louvers.** Motorized louvers shall be interlocked with the appliance so they are proven in the full open position prior to main burner ignition and during main burner operation. Means shall be provided to prevent the main burner from igniting should the louver fail to open during burner startup and to shut down the main burner if the louvers close during burner operation.

Motorized louvers prevent the entrance of rain, snow, and unconditioned air. Motorized louvers are used where outside air is not wanted in a building when the appliance is not operating. Reasons for not wanting outside air include cold outside temperature, which can freeze pipes, or to retain heat when the appliance is not operating in the winter in cold climates. All outside openings should be screened to be no smaller than $^1/_4$ in. (6.4 mm). Smaller screen size could collect airborne material that could block the screen.

When calculating the free area of louvers, the restricting effect of the louver must be taken into account. For example, louvers made of wood can have a free area of only 25 percent and need an opening size that is four times larger for the required free area. Exhibit 9.13 shows how to calculate the free area.

Louvers can be fixed in the open position, or they must be interlocked with the appliance to ensure that they are open during main burner operation, as required by 9.3.7.3. Louvers also must prevent main burner ignition if they fail to open. If automatic means are used to open and close the louvers, the louvers must be proven to be open during burner ignition and operation. If the louvers fail to open or close during operation, the main burners must not operate.

Louvers must be in the open position to ensure adequate combustion and ventilation air. Without adequate air, there is a strong possibility of incomplete combustion, which can produce toxic gases such as carbon monoxide. Carbon monoxide is an odorless, tasteless, invisible gas that can be poisonous in high concentrations or over an extended period of time.

### 9.3.8 Combustion Air Ducts.

Combustion air ducts shall comply with 9.3.8.1 through 9.3.8.8.

The requirements in 9.3.8 for combustion air ducts provide minimum specifications for combustion air duct materials and installation. Subsection 9.3.8 specifies that the combustion air duct must have adequate resistance to corrosion and be dedicated to providing combustion air for one space. The high and low opening ducts to a space must be separate over their entire run. The requirement that the combustion air duct must not slope down and away from the appliance is intended to prevent a "trap" from being formed by the duct, which could impede the duct's flow.

**9.3.8.1** Ducts shall be constructed of galvanized steel or a material having equivalent corrosion resistance, strength, and rigidity.

*Exception: Within dwellings units, unobstructed stud and joist spaces shall not be prohibited from conveying combustion air, provided that not more than one fireblock is removed.*

**9.3.8.2** Ducts shall terminate in an unobstructed space, allowing free movement of combustion air to the appliances.

**9.3.8.3** Ducts shall serve a single space.

**9.3.8.4** Ducts shall not serve both upper and lower combustion air openings where both such openings are used. The separation between ducts serving upper and lower combustion air openings shall be maintained to the source of combustion air.

**9.3.8.5** Ducts shall not be screened where terminating in an attic space.

**9.3.8.6** Horizontal upper combustion air ducts shall not slope downward toward the source of combustion air.

**9.3.8.7** The remaining space surrounding a chimney liner, gas vent, special gas vent, or plastic piping installed within a masonry, metal, or factory built chimney shall not be used to supply combustion air.

As noted in 9.3.8.7, the space surrounding a chimney liner or gas vent must not be used as a source of combustion air unless it is used for a direct vent appliance that is specifically listed for such use. If a chimney is no longer used for venting but has a liner or B vent installed in it, the surrounding space could be used to supply combustion air to a direct vent appliance.

FAQ ▶ Why is using the air space in a chimney, around a gas vent, prohibited as a source of combustion air?

Using this surrounding space is prohibited, unless for a direct vent appliance specifically listed for such use, because of the possibility of leakage from the liner or vent being drawn into the building. The use of this space is of special concern because the chimney liner or gas vent cannot be inspected. Direct vent appliances are less of a problem, and they are allowed to use the space surrounding a chimney liner or gas vent installed in an abandoned chimney if listed for such installation.

*Exception: Direct vent appliances designed for installation in a solid fuel burning fireplace where installed in accordance with the manufacturer's installation instructions.*

NEW **9.3.8.8** Combustion air intake openings located on the exterior of the building shall have the lowest side of the combustion air intake openings located at least 12 in. (300 mm) vertically from the adjoining finished ground level.

The requirement in 9.3.8.8 was revised in the 2009 edition to change the measurement of elevation of the outdoor termination of combustion air openings. The elevation height of 12 in. (300 mm) remains unchanged, but the distance now must be measured from the finished ground level. This change recognizes that in new construction projects, landscaping usually takes place after all construction is done. The installer must anticipate what the finished grade will be, which requires some judgment or research on the part of the installer.

There is a possibility of snow covering the inlet. In areas of high snowfall, the installer should consider a higher minimum elevation above grade, based on experience with local snowfalls.

## 9.4 Appliances on Roofs

Section 9.4 provides requirements for gas appliances installed on roofs. Such installations are common in commercial and industrial installations as a way to make more interior building space available. Section 9.4 covers the following topics:

- Special construction features for outdoor appliances
- Location on the roof
- Roof access
- Lighting and electrical outlets

### 9.4.1 General.

**9.4.1.1** Appliances on roofs shall be designed or enclosed so as to withstand climatic conditions in the area in which they are installed. Where enclosures are provided, each enclosure shall permit easy entry and movement, shall be of reasonable height, and shall have at least a 30 in. (760 mm) clearance between the entire service access panel(s) of the appliance and the wall of the enclosure.

**9.4.1.2** Roofs on which appliances are to be installed shall be capable of supporting the additional load or shall be reinforced to support the additional load.

**9.4.1.3** All access locks, screws, and bolts shall be of corrosion-resistant material.

Many gas-fired appliances for rooftop installations are listed for outdoor installation with their own integral vent system. An enclosure, if provided for a conventional furnace, must follow the guidelines given in 9.4.1. Any permanent structure on the roof must be designed in accordance with local building codes.

### 9.4.2 Installation of Appliances on Roofs.

**9.4.2.1** Appliances shall be installed in accordance with the manufacturers' installation instructions.

**9.4.2.2** Appliances shall be installed on a well-drained surface of the roof. At least 6 ft (1.8 m) of clearance shall be available between any part of the appliance and the edge of a roof or similar hazard, or rigidly fixed rails, guards, parapets, or other building structures at least 42 in. (1.1 m) in height shall be provided on the exposed side.

**9.4.2.3** All appliances requiring an external source of electrical power for its operation shall be provided with the following:

(1) A readily accessible electrical disconnecting means within sight of the appliance that completely de-energizes the appliance
(2) A 120 V ac grounding-type receptacle outlet on the roof adjacent to the appliance on the supply side of the disconnect switch

**9.4.2.4** Where water stands on the roof at the appliance or in the passageways to the appliance, or where the roof is of a design having a water seal, a suitable platform, walkway, or both shall be provided above the water line. Such platform(s) or walkway(s) shall be located adjacent to the appliance and control panels so that the appliance can be safely serviced where water stands on the roof.

Appliances installed on sloping roofs with a pitch greater than 4 in. in 12 in. should be mounted on a platform, such as the one shown in Exhibit 9.17. A service platform of at least 30 in. (760 mm) in depth should be provided on the side(s) of a unit that has access panels for service and maintenance. Local building codes may have more restrictive requirements for clearances to the edge of the roof and for the location and placement of guardrails.

***EXHIBIT 9.17*** *Example of an Appliance Roof Installation. (Courtesy of Carrier Corporation)*

Where appliances are installed on sloping roofs, the installer must provide a means of access to the appliance in all anticipated weather conditions for the area. Refer to the local building code for any requirements for walkways and railings on sloping roofs. To walk on a sloped roof when it is icy or snow covered is unsafe.

### 9.4.3 Access to Appliances on Roofs.

Some types of buildings have specific requirements for access to roofs. For example, in most cases, single-story dwellings may use a portable ladder to provide access to the roof. For example, *NFPA 5000®*, *Building Construction and Safety Code®*, has the following requirement:

**4.6 Roof Access.**

**4.6.1** Structures shall provide means of access to the roof in accordance with Section 4.6 in cases where the roof is flat, or the roof has a pitch less than 3 in 12, and the structure either is four or more stories in height or the roof is 40 ft (12 m) or more above grade for more than 50 percent of the building perimeter.

**4.6.2** Where required by 4.6.1, at least one means of access to the roof shall be provided. Additional access shall be provided at the rate of one access for each 100,000 $ft^2$ (9300 $m^2$) of roof area.

**4.6.3** Where roof access is required, at least one means of access shall be provided by stairs complying with 11.2.2.3. Additional roof access using ladders in accordance with 11.2.9 shall be permitted.

**4.6.4*** Where roof access is required, such access shall be provided with a door that is readily operable from both sides by fire department personnel.

**9.4.3.1** Appliances located on roofs or other elevated locations shall be accessible.

**9.4.3.2** Buildings of more than 15 ft (4.6 m) in height shall have an inside means of access to the roof, unless other means acceptable to the authority having jurisdiction are used.

The building height of 15 ft (4.6 m) in 9.4.3.2 is a somewhat arbitrary figure, but this requirement means that that all two-story buildings must have a "ready" means of access to the roof. Some jurisdictions permit an exterior ladder but, in the interest of security, require either that the lower portion be removable and be kept on the premises or that a locking door be placed over the lower 8 ft (2.4 m) of the ladder. Because building codes vary on roof access requirements, a building official should be consulted if questions arise, to ensure that the installation meets both the *National Fuel Gas Code* requirements and local building code requirements.

**9.4.3.3** The inside means of access shall be a permanent or foldaway inside stairway or ladder, terminating in an enclosure, scuttle, or trapdoor. Such scuttles or trapdoors shall be at least 22 in. × 24 in. (560 mm × 610 mm) in size, shall open easily and safely under all conditions, especially snow, and shall be constructed so as to permit access from the roof side unless deliberately locked on the inside. At least 6 ft (1.8 m) of clearance shall be available between the access opening and the edge of the roof or similar hazard, or rigidly fixed rails or guards a minimum of 42 in. (1.1 m) in height shall be provided on the exposed side. Where parapets or other building structures are utilized in lieu of guards or rails, they shall be a minimum of 42 in. (1.1 m) in height.

If access from the roof scuttle is to a sloping roof or to a roof that is designed to hold water, then a walkway with a level 30 in. (760 mm) working space may be required to provide ready access to the appliance. If the unit is closer than 6 ft (1.8 m) to the edge of the roof, guards, such as railings or parapets, must be provided to ensure the safety of service personnel.

The reader should note that the 22 in. × 24 in. (560 mm × 610 mm) opening in 9.4.3.3 is a minimum requirement and may not be large enough for replacing an appliance or appliance parts. The installer should keep this in mind when sizing any roof access opening, and the local building and mechanical codes should be checked because either may require a larger opening.

**9.4.3.4** Permanent lighting shall be provided at the roof access. The switch for such lighting shall be located inside the building near the access means leading to the roof.

### 9.4.4 Additional Provisions.

Also see 9.1.23, 9.2.1, and 12.4.3.

In 9.4.4, the committee is recommending that readers also review the following areas of the code:

- **Protection of Outdoor Appliances** (see 9.1.23), which requires either protection of the appliance from the elements, or listing for outdoor use and installation following the manufacturer's installation instructions
- **Accessibility for Service** (see 9.2.1), which requires adequate clearance for maintenance and parts replacement
- **Mechanical Draft Systems** (see 12.4.3), which requires in 12.4.3.6 that the exit terminals of mechanical draft systems be at least 7 ft (2.1 m) above finished ground level

## 9.5 Appliances in Attics

Section 9.5 is intended to provide guidance to installers of appliances in attics.

### 9.5.1 Attic Access.

An attic in which an appliance is installed shall be accessible through an opening and passageway at least as large as the largest component of the appliance and not less than 22 in. × 30 in. (560 mm × 760 mm).

**9.5.1.1** Where the height of the passageway is less than 6 ft (1.8 m), the distance from the passageway access to the appliance shall not exceed 20 ft (6.1 m) measured along the centerline of the passageway.

Paragraph 9.5.1.1 specifies that the access opening be within 20 ft (6.1 m) of the appliance to prevent excessive travel in a restricted area.

**9.5.1.2** The passageway shall be unobstructed and shall have solid flooring not less than 24 in. (610 mm) wide from the entrance opening to the appliance.

Subsection 9.5.1 provides for an access opening of at least 22 in. × 30 in. (560 mm × 760 mm) or as large as the largest component of the appliance. This provision allows for an opening between trusses set 24 in. (610 mm) on center.

See Exhibit 9.18 for an illustration of an attic installation with adequate access to the appliance.

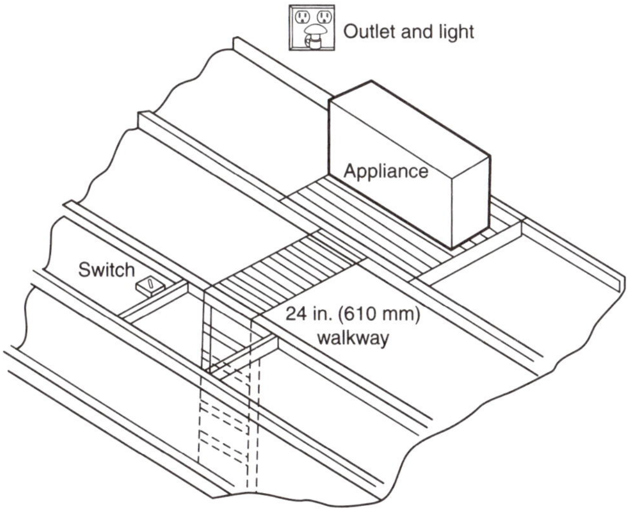


***EXHIBIT 9.18** Attic Access and Installation.*

### 9.5.2 Work Platform.

A level working platform not less than 30 in. × 30 in. (760 mm × 760 mm) shall be provided in front of the service side of the appliance.

While the installation of a work platform may seem to be a matter of common sense, one is not always provided. Service and repairs can be difficult and costly without a work platform.

### 9.5.3 Lighting and Convenience Outlet.

A permanent 120 V receptacle outlet and a lighting fixture shall be installed near the appliance. The switch controlling the lighting fixture shall be located at the entrance to the passageway.

A convenience electrical outlet and a light switch at the access opening are required at an appliance. This requirement allows service personnel to use power tools if they are needed and to be able to see the appliance clearly.

## 9.6 Appliance and Equipment Connections to Building Piping

### 9.6.1 Connecting Appliances and Equipment.

Appliances and equipment shall be connected to the building piping in compliance with 9.6.4 through 9.6.6 by one of the following:

There are a number of ways to connect appliances to a gas piping system in a building, and these are listed in 9.6.1, along with any limitations or restrictions. Some appliances are connected directly to the piping system (i.e., boilers and furnaces), but the majority of appliances use a connector, which provides the ability to move the appliance without disconnecting it from the piping system.

(1) Rigid metallic pipe and fittings.
(2) Semirigid metallic tubing and metallic fittings. Aluminum alloy tubing shall not be used in exterior locations.

Semirigid metallic tubing can be used to connect appliances to building piping systems. Traditionally, copper tubing was the only material available. With the introduction of corrugated stainless steel tubing (CSST), a second material is now available. CSST was added in the 2002 edition of the code to the list of materials that can be used to connect appliances to fixed piping systems. [See 9.6.1(5).]

(3) A listed connector in compliance with ANSI Z21.24/CSA 6.10, *Standard for Connectors for Gas Appliances*. The connector shall be used in accordance with the manufacturer's installation instructions and shall be in the same room as the appliance. Only one connector shall be used per appliance.

Gas connectors are listed in accordance with ANSI Z21.24, *Standard for Connectors for Gas Appliances*, or ANSI Z21.69, *Standard for Connectors for Movable Gas Appliances* (required in 9.6.1.1). Gas connectors are the most common means of connecting residential appliances such as ranges and gas dryers. Other codes require listed connectors for water heaters in many seismically active areas.

(4) A listed connector in compliance with ANSI Z21.75/CSA 6.27, *Connectors for Outdoor Gas Appliances and Manufactured Homes*. Only one connector shall be used per appliance.
(5) CSST where installed in accordance with the manufacturer's installation instructions.

Can CSST be used as a connector?

**FAQ ►** NFPA 54 does not prohibit the use of CSST as a connector. Because CSST is a listed product, its use must follow the terms of the listing, which are placed in the manufacturer's installation instructions. If there are any questions on the use of CSST as a connector, consult the manufacturer's instructions.

(6) Listed nonmetallic gas hose connectors in accordance with 9.6.2.
(7) In 9.6.1(2), 9.6.1(3), 9.6.1(4), 9.6.1(5), and 9.6.1(6), the connector or tubing shall be installed so as to be protected against physical and thermal damage. Aluminum alloy tubing and connectors shall be coated to protect against external corrosion where they are in contact with masonry, plaster, or insulation or are subject to repeated wettings by such liquids as water (except rain water), detergents, or sewage.
(8) Materials addressed in 9.6.1(2), 9.6.1(3), 9.6.1(4), 9.6.1(5), and 9.6.1(6) shall not be installed through an opening in an appliance housing, cabinet, or casing, unless the tubing or connector is protected against damage.

The requirement in 9.6.1(8) addresses a number of fires where thin wall piping materials and connectors passing through appliance housings were reported to have failed, resulting in gas leakage. While there is no clear evidence of the cause of the failures, the code requires connection of these materials to be outside the "envelope" of the appliance if not protected. Installing a grommet to prevent chafing of the connector is a method of protecting thin wall tubing and connectors against damage. Installers of furnaces and boilers should take note of this requirement.

**9.6.1.1 Commercial Cooking Appliances.** Commercial cooking appliances that are moved for cleaning and sanitation purposes shall be connected in accordance with the connector manufacturer's installation instructions using a listed appliance connector complying with ANSI Z21.69/CSA 6.22, *Connectors for Movable Gas Appliances*. The commercial cooking appliance connector installation shall be configured in accordance with the manufacturer's installation instructions. NEW

The requirement in 9.6.1.1 was revised in the 2009 edition with the addition of the last sentence. The change recognizes that the manufacturers of appliance connectors for commercial appliances recommend specific orientations of the pipe terminations to minimize strain on connectors.

Connectors for commercial cooking appliances differ from those for residential appliances and can be identified as being listed to ANSI Z21.69. These connectors are intended for appliances that are moved frequently for cleaning, such as fryers in fast food restaurants. These connectors typically come with a cable that restrains the rolling appliance and prevents overextension and possible damage to the connector. Residential connectors are not designed for such service and can present a hazard if improperly used for commercial cooking appliances.

**9.6.1.2** Movement of appliances with casters shall be limited by a restraining device installed in accordance with the connector and appliance manufacturer's installation instructions. NEW

The new requirement in 9.6.1.2 was added in the 2009 edition to recognize that appliances with casters are moved, and that this movement must be limited to prevent stretching the connector, and possibly shortening its life.

**9.6.1.3* Suspended Low-Intensity Infrared Tube Heaters.** Suspended low-intensity infrared tube heaters shall be connected to the building piping system with a connector listed for the application in accordance with ANSI Z21.24/CGA 6.10, *Standard for Connectors for Gas Appliances*. NEW

**(A)** The connector shall be installed in accordance with the tube heater installation instructions and shall be in the same room as the appliance. NEW

**(B)** Only one connector shall be used per appliance. NEW

The new requirement in 9.6.1.3 recognizes several reported failures in piping systems and piping connections for suspended low-intensity infrared tube heaters. In these incidents, the appliances were connected using rigid pipe or semirigid metallic tubing. These failures occurred due to vibrations created by the heaters, causing fatigue in the piping. The new provision in 9.6.1.3 requires the use of a specific connector to minimize this problem.

NEW **A.9.6.1.3** The expansion and contraction of the heater and the vibration from the blower motor may lead to work hardening of the rigid pipe or semirigid metallic tubing, which may ultimately lead to fractures and leakage. Connectors for this type of heater should have adequate flexibility, temperature rating, and vibration resistance to accommodate the characteristics of the heater. Such flexible connectors for suspended heaters should meet the following criteria:

(1) Be determined to be appropriate for the application.
(2) Be specified by the heater manufacturer.
(3) Be installed in accordance with the manufacturer's installation instructions.

### 9.6.2 Use of Nonmetallic Gas Hose Connectors.

Listed gas hose connectors shall be used in accordance with the manufacturer's installation instructions and as follows:

(1) *Indoor.* Indoor gas hose connectors shall be used only to connect laboratory, shop, and ironing appliances requiring mobility during operation.
  (a) An appliance shutoff valve shall be installed where the connector is attached to the building piping.
  (b) The connector shall be of minimum length and shall not exceed 6 ft (1.8 m).
  (c) The connector shall not be concealed and shall not extend from one room to another or pass through wall partitions, ceilings, or floors.
(2) *Outdoor.* Outdoor gas hose connectors are permitted to connect portable outdoor appliances.
  (a) An appliance shutoff valve, a listed quick-disconnect device, or a listed gas convenience outlet shall be installed where the connector is attached to the supply piping and in such a manner so as to prevent the accumulation of water or foreign matter.
  (b) This connection shall be made only in the outdoor area where the appliance is to be used.

Indoor gas hose connectors may be used only to connect appliances requiring mobility, such as laboratory, shop, and ironing appliances. The connector must be listed and connected to a listed manual shutoff valve immediately before the gas hose. It is not necessary to plug or cap the outlet of the shutoff valve when the appliance is disconnected, but the connection should be plugged to prevent accidental gas leakage.

### 9.6.3 Connection of Portable and Mobile Industrial Appliances.

**9.6.3.1** Where portable industrial appliances or appliances requiring mobility or subject to vibration are connected to the building gas piping system by the use of a flexible hose, the hose shall be suitable and safe for the conditions under which it can be used.

**9.6.3.2** Where industrial appliances requiring mobility are connected to the rigid piping by the use of swivel joints or couplings, the swivel joints or couplings shall be suitable for the service required and only the minimum number required shall be installed.

**9.6.3.3** Where industrial appliances subject to vibration are connected to the building piping system by the use of all metal flexible connectors, the connectors shall be suitable for the service required.

**9.6.3.4** Where flexible connections are used, they shall be of the minimum practical length and shall not extend from one room to another or pass through any walls, partitions, ceilings, or floors. Flexible connections shall not be used in any concealed location. They shall be protected against physical or thermal damage and shall be provided with gas shutoff valves in readily accessible locations in rigid piping upstream from the flexible connections.

**9.6.4 Appliance Shutoff Valves and Connections.**

Each appliance connected to a piping system shall have an accessible, approved manual shutoff valve with a nondisplaceable valve member, or a listed gas convenience outlet. Appliance shutoff valves and convenience outlets shall serve a single appliance only and shall be installed in accordance with 9.6.4.1.

The requirement in 9.6.4 for a manual shutoff valve is intended to isolate an appliance for servicing, removal, or replacement without shutting off the gas supply to other appliances. This valve is not an emergency shutoff valve. The shutoff valve located outside the building provides for that need. (See 7.9.2.3.) The manual shutoff may be used to isolate certain appliances from the building piping system during pressure testing that is required in 8.1.3.6. Because the appliance shutoff valve is not an emergency shutoff valve, it does not have to be readily accessible. Exhibit 9.19 illustrates a common shutoff valve location.

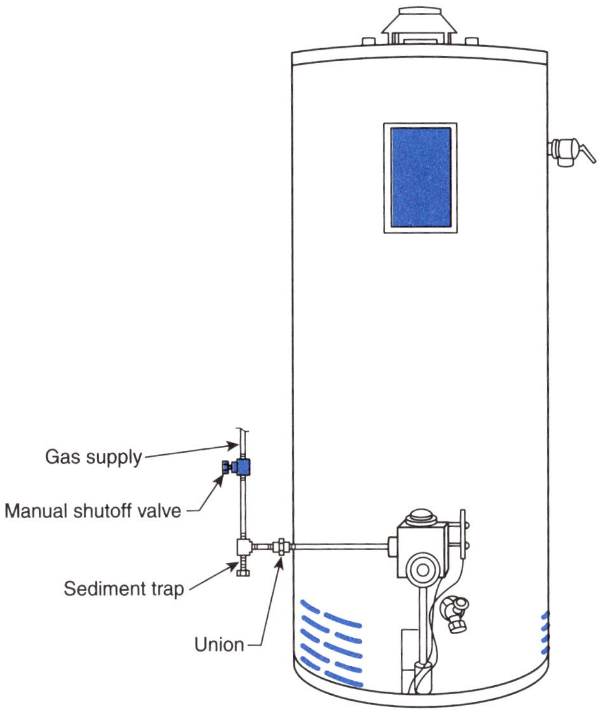


*EXHIBIT 9.19 Shutoff Valve Location.*

The appliance shutoff valve must be an approved manual shutoff valve and, if a connector is used, it must be installed before the connector and within 6 ft (1.8 m) of the appliance. Shutoff valves that are 1 in. National Pipe Thread and smaller must be listed. (See Section 5.12.)

The code permits shutoff valves that serve decorative gas appliances to be installed in fireplaces if listed for such use.

A listed decorative gas appliance has a door beneath the hearth that has the control valve and related piping. Can the appliance shutoff valve be located in this space?

◄ FAQ

**9.6.4.1** The shutoff valve shall be located within 6 ft (1.8 m) of the appliance it serves except as permitted in 9.6.4.2 or 9.6.4.3.

**(A)** Where a connector is used, the valve shall be installed upstream of the connector. A union or flanged connection shall be provided downstream from the valve to permit removal of appliance controls.

**(B)** Shutoff valves serving decorative appliances shall be permitted to be installed in fireplaces if listed for such use.

The requirements in 9.6.4.1(A) and (B) provide installers with another option for the location of appliance shutoff valves. Prior to the 2006 edition of the code, appliance shutoff valves had to be within 6 ft (1.8 m) of the appliances they served. The requirements in 9.6.4.1(A) and 9.6.4.1(B) recognize the difficulty in locating appliance shutoff valves (typically serving gas logs or space heaters) within 6 ft (1.8 m) of appliances installed in vented fireplaces. The 6 ft (1.8 m) limit is eliminated, and there is no minimum distance from the appliance shutoff valve to an appliance installed in a vented fireplace. While this requirement is a significant change from the requirement prior to the 2006 edition of the code, it should be noted that the location of the valve is limited to the run of piping, and service personnel can trace the pipe to locate the valve. Also note that piping beyond 6 ft (1.8 m) from the appliance must meet the requirements for piping materials in the code, and that multiple connectors cannot be used.

**9.6.4.2** Shutoff valves serving appliances installed in vented fireplaces and ventless firebox enclosures shall not be required to be located within 6 ft (1.8 m) of the appliance where such valves are readily accessible and permanently identified. The piping from the shutoff valve to within 6 ft (1.8 m) of the appliance shall be designed, sized, and installed in accordance with Chapters 5, 6, and 7.

**9.6.4.3** Where installed at a manifold, the appliance shutoff valve shall be located within 50 ft (15 m) of the appliance served and shall be readily accessible and permanently identified. The piping from the manifold to within 6 ft (1.8 m) of the appliance shall be designed, sized, and installed in accordance with Chapters 5, 6, and 7.

The requirement in 9.6.4.3 provides an option that allows all appliance shutoff valves in a building to be located at a manifold no more than 50 ft (15 m) from the most remote appliance. This system has been successfully used in Canada for many years. As the appliance shutoff valve is not an emergency valve, safety is not compromised.

## 9.6.5 Quick-Disconnect Devices.

A listed quick-disconnect device often is used as one of the end fittings of a listed gas appliance connector or listed gas hose. The device functions as a manual gas shutoff in that, when it is manually disconnected, the device is designed to shut off the flow of gas before the disconnection is made. The more common designs use a spring-loaded poppet valve, which is internal to the device. When installed indoors, a manual shutoff valve is required upstream to provide the additional safety of a positive and visual shutoff for the gas. The manual shutoff valve must be an approved gas valve.

**9.6.5.1** Quick-disconnect devices used to connect appliances to the building piping shall be listed to ANSI Z21.41/CSA 6.9, *Quick-Disconnect Devices for Use with Gas Fuel Appliances.*

The requirement for a listed quick-disconnect device is a conservative one, because it is anticipated that these devices will be used regularly to connect and disconnect gas appliances. The committee is concerned that quick-disconnect devices be thoroughly investigated for their intended use and have a long life. Quick-disconnect devices are used to connect portable gas cooking appliances, such as gas grills, and are used for some industrial appliances.

**9.6.5.2** Where installed indoors, an approved manual shutoff valve with a nondisplaceable valve member shall be installed upstream of the quick-disconnect device.

### 9.6.6 Gas Convenience Outlets.

Gas convenience outlets shall be listed in accordance with ANSI Z21.90, *Gas Convenience Outlets and Optional Enclosures*, and installed in accordance with the manufacturers' installation instructions.

Subsection 9.6.6 permits the use of gas convenience outlets that are listed in accordance with ANSI Z21.90, *Gas Convenience Outlets and Optional Enclosures*. These devices anticipate portable, "plug-in" gas appliances, which have yet to be made commercially available. A gas convenience outlet is shown in Exhibit 9.20. The requirement was revised in the 2009 edition with reference to ANSI Z21.90 added. This provides a specific requirement for gas convenience outlets.

### 9.6.7 Sediment Trap.

Where a sediment trap is not incorporated as a part of the appliance, a sediment trap shall be installed downstream of the appliance shutoff valve as close to the inlet of the appliance as practical at the time of appliance installation. The sediment trap shall be either a tee fitting with a capped nipple in the bottom outlet as illustrated in Figure 9.6.7 or other device recognized as an effective sediment trap. Illuminating appliances, ranges, clothes dryers, decorative appliances for installation in vented fireplaces, gas fireplaces, and outdoor grills shall not be required to be so equipped.

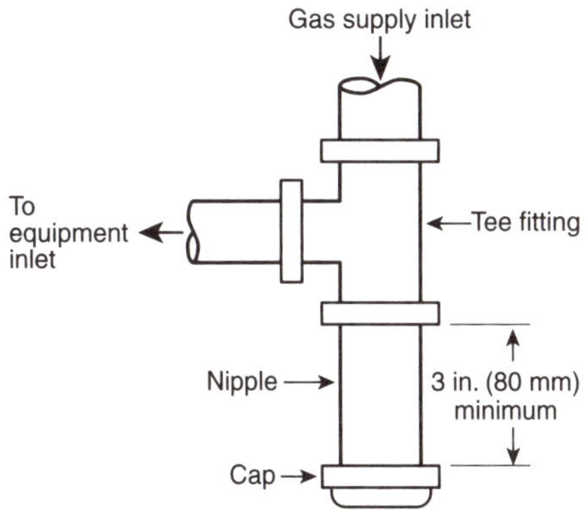


***Figure 9.6.7*** *Method of Installing a Tee Fitting Sediment Trap.*

***EXHIBIT 9.20*** *Gas Convenience Outlet and Fitting. (Courtesy of NAHB Research Center)*

Sediment traps are required where automatically operated appliances are installed, if they are not incorporated in the appliance. Refer to commentary following 7.6.3.

### 9.6.8 Installation of Piping.

Piping shall be installed in a manner not to interfere with inspection, maintenance, or servicing of the appliances.

## 9.7 Electrical

### 9.7.1 Electrical Connections.

Electrical connections between appliances and the building wiring, including the grounding of the appliances, shall conform to *NFPA 70, National Electrical Code*.

Article 422 in *NFPA 70*®, *National Electrical Code*® (*NEC*®), covers the requirement for appliances, and *NEC* Article 440 covers the installation of appliances that also contain hermetic refrigeration motor compressors, which are found in many combination "gas-electric" rooftop units that provide heating and cooling. *NEC* Article 250, Parts V and VI, covers bonding and grounding requirements for appliances.

### 9.7.2 Electrical Ignition and Control Devices.

Electrical ignition, burner control, and electrical vent damper devices shall not permit unsafe operation of the appliance in the event of electrical power interruption or when the power is restored.

**FAQ ▶** What is meant by the term *unsafe operation* in 9.7.2?

While the term *unsafe operation* is not defined in this code, the flow of unignited gas when power is restored after a power outage is an unsafe situation and is not acceptable. Any appliance equipped with a flame safety control, such as a furnace or water heater, will not allow flow of gas to a main burner following a shutoff of the gas supply without ignition, a safe condition. Appliances without flame safety devices, such as gas range burners, can create an unsafe condition if the flow of gas is stopped where an electrically operated valve is installed and electrical power is curtailed. This situation can occur if remotely operated gas valves are installed as part of a fire protection system. If a building has an alarm system that shuts off the flow of gas when fire is detected, and the flow of gas is restored, a gas range that had been left with the burners on will flow gas into the building when the fire alarm system is reset unless steps are taken to turn the burners off.

### 9.7.3 Electrical Circuit.

The electrical circuit employed for operating the automatic main gas control valve, automatic pilot, room temperature thermostat, limit control, or other electrical devices used with the appliances shall be in accordance with the wiring diagrams certified or approved by the original appliance manufacturer.

Appliance wiring diagrams are included with appliances, usually inside one of the access panels and also typically shown in the installation instructions. Most manufacturers also will show the necessary wiring connections for the installation of optional accessories, such as external vent dampers, humidifiers, air cleaners, and other auxiliary controls. Accessory installation instructions should be left with the appliance for future reference.

### 9.7.4 Continuous Power.

All appliances using electrical controls shall have the controls connected into a permanently live electrical circuit — that is, one that is not controlled by a light switch. Central heating appliances shall be provided with a separate electrical circuit.

The *National Electrical Code* requires a separate electrical circuit and overcurrent protection for central heating appliances, as well as a means for disconnecting the power source at or within sight of the appliance. This separate circuit also can include circulating water pumps, valves, humidifiers, electronic air cleaners, and other accessories normally associated with an appliance. Certain appliances can be connected with a flexible cord that has a grounding-type attachment plug. The grounding-type attachment plug also serves as a disconnecting means for future service and maintenance.

In 9.7.4, note the phrase "a permanently live electrical circuit — that is, one that is not controlled by a light switch." This phrase means that no switches can be included in the wiring from the fuse or circuit breaker to the appliance (the branch circuit wiring). No wall switches are allowed. If the appliance is equipped with a switch, or the manufacturer's installation instructions require the installation of a switch on the appliance, switches are allowed, as they are not part of the branch circuit.

In areas where oil is a common source of heating fuel, wall switches with red cover plates marked "Oil Burner Emergency Switch" are installed in buildings heated by fuel oil, as required by NFPA 31, *Standard for the Installation of Oil-Burning Equipment*, to stop the flow of oil to the burner. When converting an oil furnace or boiler to gas, the emergency switch must be removed. Emergency switches in the branch circuit wiring must not be installed when installing a new gas furnace or boiler. The user should note that this does not prevent the installation of a service switch on the appliance if required by the appliance manufacturer's installation instructions. This prohibition is clearly stated in 9.7.4, which requires a "permanently live electrical circuit." An emergency switch, if installed, would make the electrical circuit not permanently live.

◄ **FAQ**
**Can the electrical circuit, including the emergency shutoff switch, from an oil-fired furnace be reused for a new gas installation?**

The emergency shutoff switch provides safety for oil-fired heating equipment, because failure of the boiler or furnace to shut off can prevent the oil pump from shutting off, thereby filling the equipment and surrounding area with oil. An oil pool fire is possible. With gas-fueled appliances, the emergency shutoff valve required in 7.9.2.3 provides equivalent safety.

## 9.8 Room Temperature Thermostats

### 9.8.1 Locations.

Room temperature thermostats shall be installed in accordance with the manufacturer's instructions.

Room thermostats generally are located on inside walls to avoid drafts from the heating or cooling system supply registers and should avoid pipes in walls, lights, TVs, or other heat sources that can affect their operation. Usually, thermostats are installed 4 ft (1.2 m) to 5 ft (1.5 m) above the floor and can include supplemental switches to control the functions of the system. The appliance manufacturer's instructions should be consulted for additional details.

Each appliance should have its own thermostat unless the appliance design specifically states that one thermostat can serve more than one appliance. If a thermostat serves more than one appliance, this dual use can cause the appliances it serves to operate erratically. Each heating appliance should have its own thermostat so that it can properly heat the area it is designed to heat. If the thermostat were connected to additional units, it would cause the different areas to become underheated or overheated, because the thermostat could not accurately record the temperature for both areas. A typical digital clock thermostat is shown in Exhibit 9.21.

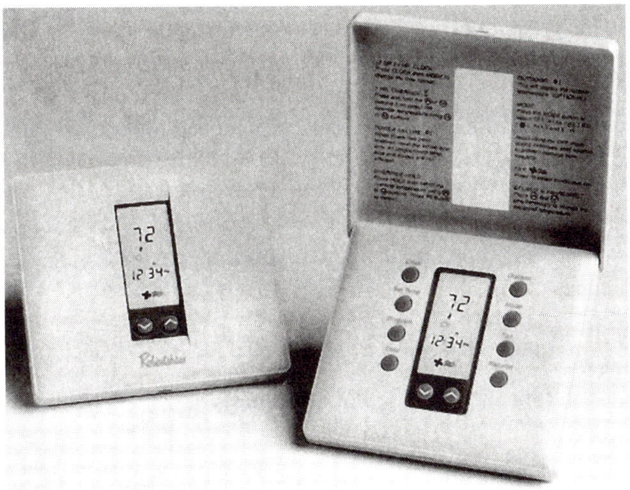

***EXHIBIT 9.21*** *Digital Thermostat. (Courtesy of Invensys Climate Controls)*

### 9.8.2 Drafts.

Any hole in the plaster or panel through which the wires pass from the thermostat to the appliance being controlled shall be sealed so as to prevent drafts from affecting the thermostat.

The intent of the requirement in 9.8.2 is to increase the system's overall efficiency by reducing unnecessary appliance operating time.

## REFERENCES CITED IN COMMENTARY

The following publications are available from the National Fire Protection Association, 1 Batterymarch Park, Quincy, MA 02169-7471, 1-800-344-3555, www.nfpa.org.

NFPA 30, *Flammable and Combustible Liquids Code,* 2008 edition.
NFPA 30A, *Code for Motor Fuel Dispensing Facilities and Repair Garages,* 2008 edition.
NFPA 31, *Standard for the Installation of Oil-Burning Equipment,* 2006 edition.
NFPA 51, *Standard for the Design and Installation of Oxygen–Fuel Gas Systems for Welding, Cutting, and Allied Processes,* 2007 edition.
*NFPA 70®, National Electrical Code®*, 2008 edition.
NFPA 88A, *Standard for Parking Structures,* 2007 edition.
*NFPA 5000®, Building Construction and Safety Code®*, 2009 edition.

The following publication is available from the American Petroleum Institute, 1220 L Street NW, Washington, DC 20005, 202-682-8000, www.api.org/Standards/.

API RP 520, *Sizing, Selection, and Installation of Pressure-Relieving Devices in Refineries, Part II, Installation*, 2003.

GRI publications are now available from the Gas Technology Institute, 1700 South Mount Prospect Road, Des Plaines, IL 60018-1804, 847-298-8191, www.gastechnology.org.

Report Number GRI 93/0316, "Analysis of Combustion Air Openings to the Outdoors: Preliminary Results," 1994.
AGA Research Bulletin 53, "The Effects of Confined Space Installation on Central Gas Space Heating Equipment Performance," 1947.
AGA Research Bulletin 67, "Combustion and Ventilation Air Supply to Gas Equipment in Small Rooms," 1954.
AGA Research Bulletin 427, "Outdoor Air Supply and Ventilation of a Furnace Closet Used with a Warm Air Heating System," 1954.

The following publications are available from the Canadian Standards Association, 8501 East Pleasant Valley Road, Cleveland, OH 44131-5575, 216-524-4990, www.csa-america.org.

ANSI Z21.10.1/CSA 4.1, *Gas Water Heaters — Volume I — Storage Water Heaters with Input Ratings of 75,000 Btu per Hour or Less*, 2006.
ANSI Z21.24, *Standard for Connectors for Gas Appliances,* 2006.
ANSI Z21.30, *Installation of Gas Piping and Gas Appliances in Buildings,* 1950.
ANSI Z21.69, *Standard for Connectors for Movable Gas Appliances,* 2006.
ANSI Z21.90/CSA 6.24, *Gas Convenience Outlets and Optional Enclosures,* 2006.

The following publications are available from the American Society of Heating, Refrigeration and Air Conditioning Engineers, Inc., 1791 Tullie Circle, NE, Atlanta, GA 30329-2305, 404-636-8400, www.ashrae.org.

*ASHRAE Handbook — Fundamentals*, 2005 edition.
ASHRAE 62.1, *Ventilation for Acceptable Indoor Air Quality,* 2007.

The following publication is available from Battelle, 5 King Avenue, Columbus, OH 43201-2693.

*Combustion-Air Issues Related to Residential Gas Appliances in Confined Spaces and Unusually Tight Construction,* January 2001.

# Installation of Specific Appliances

CHAPTER 10

Chapter 10 of the code covers requirements for installing specific gas-fired appliances and applies to appliances used in nonindustrial applications. See Chapter 9, Appliance, Equipment, and Accessory Installation, for general requirements applying to the installation of residential, commercial, and industrial gas appliances.

Section 10.1, General, provides the overall requirement that listed appliances be installed in accordance with the terms of their listing. This requirement recognizes that the installation instructions for listed equipment provide clearances based on testing of the specific appliance. In many cases, these clearances are less restrictive than those required for unlisted equipment of the same type. Section 10.1 also covers restrictions on the installation of gas-fired appliances in bedrooms and bathrooms.

Sections 10.2 through 10.3 provide requirements for the following specific appliances:

- Gas-fired air-conditioning equipment (See Section 10.2.)
- Central heating furnaces and boilers (See Section 10.3.)
- Clothes dryers (See Section 10.4.)
- Conversion burners (See Section 10.5.)
- Decorative appliances for installation in vented fireplaces and vented gas fireplaces (See Sections 10.6 and 10.7.)
- Industrial air heaters (See Sections 10.8 and 10.9.)
- Duct and floor furnaces (See Sections 10.10 and 10.11.)
- Commercial food service equipment and household cooking appliances (See Sections 10.12 through 10.15, 10.19, and 10.20.)
- Illuminating appliances (See Section 10.16.)
- Commercial-industrial incinerators (See Section 10.17.)
- Infrared heaters (See Section 10.18.)
- Pool heaters (See Section 10.21.)
- Refrigerators (See Section 10.22.)
- Room (space) heaters (See Section 10.23.)
- Stationary gas engines (See Section 10.24.)
- Gas-fired toilets (See Section 10.25.)
- Unit heaters (See Section 10.26.)
- Wall furnaces (See Section 10.27.)
- Water heaters (See Section 10.28.)
- Compressed natural gas (CNG) vehicular fuel systems (See Section 10.29.)
- Appliances for installation in manufactured housing (See Section 10.30.)
- Fuel cell power plants (See Section 10.31.)
- Outdoor open flame decorative appliances (See Section 10.32.)

## 10.1 General

### 10.1.1 Application.

This chapter is applicable primarily to nonindustrial-type appliances and installations and, unless specifically indicated, does not apply to industrial appliances and installations. Listed appliances shall be installed in accordance with the manufacturers' installation instructions or, as elsewhere specified in this chapter, as applicable to the appliance. Unlisted appliances shall be installed as specified in this chapter as applicable to the appliances. For additional information concerning particular gas appliances and accessories, including industrial types, reference can be made to the standards listed in Chapter 2 and Annex M.

Chapter 10 of this code, for the most part, is descended directly from the pre-1974 ASA Z21.30/NFPA 54, *Standard for the Installation of Gas Appliances and Piping*, which covered residential and some commercial installations. These appliances, such as ranges, furnaces, and dryers, are normally mass produced and their installations are very similar. Industrial equipment and larger commercial equipment vary widely and can be unique, so generalized installation requirements are not practical. The references to Chapter 2 and Annex M in 10.1.1 can assist in determining the applicability of industrial-type gas equipment and accessories.

Chapter 10 contains requirements that apply to a range of appliances. The reader's attention is directed particularly to the following statement in 10.1.1, which applies to all of Chapter 10:

> Listed appliances shall be installed in accordance with the manufacturers' installation instructions or, as elsewhere specified in this chapter, as applicable to the appliance.

**FAQ ▶**
**Could the manufacturer's installation instructions supersede the code requirements in Chapter 10?**

Listing of the appliance allows instruction specific to the appliance tested to be included in the manufacturer's installation instructions and may supersede the specific requirements in Chapter 10. This is allowed because listed appliances are tested by an independent listing agency, and each installation requirement is verified for safety.

### 10.1.2* Installation in a Bedroom or Bathroom.

Appliances shall not be installed so their combustion, ventilation, and dilution air are obtained only from a bedroom or bathroom unless the bedroom or bathroom has the required volume in accordance with 9.3.2.

**A.10.1.2** Also see prohibited installations in 10.6.1, 10.7.1, 10.8.2, 10.9.2, and 10.23.1.

The requirement in 10.1.2 was added to the 1984 edition after considerable discussion on whether gas-fired appliances should be allowed to be installed in bedrooms and bathrooms. Originally, the code prohibited most gas appliances in bedrooms or bathrooms, and in closets or small spaces having access only through a bedroom or bathroom, for the following reasons:

1. Such rooms are usually kept closed, are small in size, and may not provide adequate air for combustion, draft hood dilution, and ventilation.
2. This prohibition was consistent with other codes.
3. There is an increased possibility of moisture and lint accumulation within the appliance in such rooms.
4. The potential for contact with hot surfaces and for ignition of fabric is increased in small rooms.
5. Proper ventilation for the products of combustion can be lacking in small rooms.
6. Hair spray and other aerosol products that can accelerate corrosion of the equipment are likely to be present in bedrooms and bathrooms. (Note: Hair spray is no longer a problem since the propellant has been changed.)

The requirement was revised in the 1988 edition of the code to provide a more positive statement on small spaces. The revised requirement permitted the installation of direct vent appliances (which do not draw air from the room) in bedrooms and bathrooms, and to provide for access to furnace closets (which can draw outside air for combustion and ventilation) located in or having access through bedrooms.

Bedrooms and bathrooms almost always have doors. Therefore, appliances that take air for combustion and ventilation from the room in which they are installed must not be installed in these rooms, with the exception of certain low-input heating appliances covered in 10.23.1. Paragraph A.10.1.2 draws attention to other paragraphs having specific prohibitions on the installation of appliances in bedrooms and bathrooms under certain conditions. Prohibited installations for the appliances indicated are covered in the following subsections:

- Decorative appliances for installation in vented fireplaces (see 10.6.1)
- Vented gas fireplaces (see 10.7.1)
- Non-recirculating direct gas-fired industrial air heaters (see 10.8.2)
- Recirculating direct gas-fired industrial air heaters (see 10.9.2)
- Room heaters (see 10.23.1)

Most of these prohibitions exempt rooms or spaces in which air for combustion and ventilation meets the volume requirements of 9.3.2.

In 1994, the code was revised to permit the installation of listed, wall-mounted room heaters with limited input ratings and oxygen depletion safety shutoff systems in bedrooms and bathrooms that meet the requirements of Section 10.23, and where these heaters are acceptable to the authority having jurisdiction. Small space heaters are the normal means of heating homes in warmer parts of the United States, especially in the Southeast. The 1994 revision permitted older, open flame heaters to be replaced with newer, safer appliances by allowing them to be installed where permitted by the authority having jurisdiction. Since then, many states have reversed their position of prohibiting such appliances.

### 10.1.3 Room Size in Comparison Calculation.

Where the room size in comparison with the size of the appliance is to be calculated, the total volume of the appliance is determined from exterior dimensions and is to include fan compartments and burner vestibules, where used. Where the actual ceiling height of a room is greater than 8 ft (2.4 m), the volume of the room is figured on the basis of a ceiling height of 8 ft (2.4 m).

Subsection 10.1.3 contains an important requirement concerning the life of appliances. Appliances installed in small spaces can heat the space with the waste heat generated by the appliance. High temperature can lead to reduced life of electric motors due to lubrication failure and can shorten the life of electrical and electronic components. The important concept of room size in comparison with appliance size is discussed further in the commentary following 10.2.3(1).

## 10.2 Air-Conditioning Appliances (Gas-Fired Air Conditioners and Heat Pumps)

In accordance with Section 5.4, the gas piping serving either the heating or cooling equipment covered in 10.2.1 must be sized on the basis of the larger gas demand of the two appliances. The diversity provision in 5.4.2.2 permits the pipe size to be based on the larger, rather than the combined, gas demand once it has been determined that the heating and cooling appliances cannot be operated simultaneously.

◄ **FAQ**
**The heating and cooling appliance uses gas for both functions with different Btu/hr inputs. Should the Btu/hr input ratings of both functions be combined to size the gas supply?**

### 10.2.1 Independent Gas Piping.

Gas piping serving heating appliances shall be permitted to also serve cooling appliances where heating and cooling appliances cannot be operated simultaneously. *(See Section 5.4.)*

### 10.2.2 Connection of Gas Engine–Powered Air Conditioners.

To protect against the effects of normal vibration in service, gas engines shall not be rigidly connected to the gas supply piping.

Because an engine is prone to vibration, 10.2.2 requires a flexible piping connection to the gas supply. It is important that the engine be secured to a firm foundation so that it cannot move. An appliance shutoff valve must be located on the fixed pipe upstream of the connector. During installation, it is important that the connector be oriented so that there is no stress on the connector and so that any vibration does not adversely affect the connector.

### 10.2.3 Clearances for Indoor Installation.

The installation of air-conditioning appliances shall comply with the following requirements:

(1) Listed air-conditioning appliances installed in a room large in comparison with size of appliance shall be installed with clearances in accordance with the manufacturer's instructions. *[See Table 10.2.3(a) and 3.3.90, Room Large in Comparison with Size of Equipment.]*

In the 2002 edition, the term *plenum* was changed to *furnace plenum* [used in Table 10.2.3(a) and elsewhere in the code]. This change recognized that the term *plenum* is used in other codes (e.g., in building codes to mean certain above-ceiling spaces and in other NPFA codes with different meanings). The term *furnace plenum* is more specific. This use of the term is consistent with its use in NFPA 31, *Standard for the Installation of Oil-Burning Equipment*; NFPA 211, *Standard for Chimneys, Fireplaces, Vents, and Solid Fuel-Burning Appliances*; and NFPA 501, *Standard on Manufactured Housing*. The consistency minimizes confusion among designers, installers, and enforcement officials who use multiple codes.

Note that the installation provisions in 10.2.3(1) for listed gas air-conditioning appliances installed in rooms large in comparison with the size of the appliance require that they be installed in accordance with the manufacturer's instructions. The term *room large in comparison with size of appliance* is defined in 3.3.90 as "rooms having a volume equal to at least

***TABLE 10.2.3(a)*** *Clearances to Combustible Material for Unlisted Furnaces, Boilers, and Air Conditioners Installed in Rooms that Are Large in Comparison with the Size of Appliances*

| | Minimum Clearance (in.) | | | | | |
|---|---|---|---|---|---|---|
| ***Appliance*** | ***Above and Sides of Furnace Plenum*** | ***Top of Boiler*** | ***Jacket Sides and Rear*** | ***Front*** | ***Draft Hood and Barometric Draft Regulator*** | ***Single-Wall Vent Connector*** |
| I Automatically fired, forced air or gravity system, equipped with temperature limit control that cannot be set higher than 250°F (121°C) | 6 | — | 6 | 18 | 6 | 18 |
| II Automatically fired heating boilers — steam boilers operating at not over 15 psi (103 kPa) and hot water boilers operating at 250°F (121°C) or less | 6 | 6 | 6 | 18 | 18 | 18 |
| III Central heating boilers and furnaces, other than in I or II | 18 | 18 | 18 | 18 | 18 | 18 |
| IV Air-conditioning appliances | 18 | 18 | 18 | 18 | 18 | 18 |

Note: See 10.2.3 for additional requirements for air-conditioning appliances and 10.3.1 for additional requirements for central heating boilers and furnaces.

12 times the total volume of a furnace or air-conditioning appliance and at least 16 times the total volume of a boiler." Refer to the commentary following 3.3.90 for more information on the term.

(2) Air-conditioning appliances installed in rooms that are NOT large (such as alcoves and closets) in comparison with the size of the appliance shall be listed for such installations and installed in accordance with the manufacturer's instructions. Listed clearances shall not be reduced by the protection methods described in Table 10.2.3(b), regardless of whether the enclosure is of combustible or noncombustible material.

Equipment installed in alcoves and closets must be listed specifically for such installations. Also note that the last sentence of 10.2.3(2) prohibits clearance reduction for appliances installed in rooms *not* large in comparison with the size of the equipment. This prohibition recognizes that, in these small rooms, closets, and alcoves, the minimum clearances are needed for air circulation to keep appliances from overheating.

◀ **FAQ**
Can unlisted equipment be installed in rooms not large in comparison to the equipment?

(3) Unlisted air-conditioning appliances shall be installed with clearances from combustible material of not less than 18 in. (460 mm) above the appliance and at the sides, front, and rear and 9 in. (230 mm) from the draft hood.

(4) Air-conditioning appliances (listed and unlisted) installed in rooms that are large in comparison with the size of the appliance shall be permitted to be installed with reduced clearances to combustible material, provided the combustible material or appliance is protected as described in Table 10.2.3(b) *[see 10.2.3(2)].*

The requirement in 10.2.3(4) specifically permits the clearance reduction systems described in Table 10.2.3(b) to be used with all (listed and unlisted) gas-fired air-conditioning appliances, thus reinforcing usage of the clearance reduction table [see Table 10.2.3(b)] for any appliance.

(5) Where the furnace plenum is adjacent to plaster on metal lath or noncombustible material attached to combustible material, the clearance shall be measured to the surface of the plaster or other noncombustible finish where the clearance specified is 2 in. (50 mm) or less.

When used alone, plaster or other insulating material that is applied to combustible material, such as wooden studs, is not considered to be adequate protection for combustible construction. For this reason, the clearance must be measured to the surface of the plaster or other finish if the specified clearance is 2 in. (50 mm) or less.

◀ **FAQ**
The walls and ceiling of the appliance room are plaster, a noncombustible. Can I use the distance to the wood underneath the plaster to calculate the required clearance?

(6) Listed air-conditioning appliances shall have the clearance from supply ducts within 3 ft (0.9 m) of the furnace plenum be not less than that specified from the furnace plenum. No clearance is necessary beyond this distance.

### 10.2.4 Assembly and Installation.

Air-conditioning appliances shall be installed in accordance with the manufacturer's instructions. Unless the appliance is listed for installation on a combustible surface such as a floor or roof, or unless the surface is protected in an approved manner, it shall be installed on a surface of noncombustible construction with noncombustible material and surface finish and with no combustible material against the underside thereof.

Prior to the 1988 edition, the *National Fuel Gas Code* contained a footnote referring the reader to Appendix E of the 1976 edition of the *National Building Code* for additional information on methods of installation on combustible floors. The 1976 edition of the *National Building Code* was the last edition of that code written by the American Insurance Association. After 1976, the project was taken over by Building Officials and Code Administrators International and was discontinued. Appendix E did not appear in the *National Building Code*

*TABLE 10.2.3(b) Reduction of Clearances with Specified Forms of Protection*

| *Type of protection applied to and covering all surfaces of combustible material within the distance specified as the required clearance with no protection [See Figure 10.3.2.2(a) through Figure 10.3.2.2(c).]Allowable Clearances with Specified Protection (in.)* | *Where the required clearance with no protection from appliance, vent connector, or single-wall metal pipe is:* | | | | | | | | | |
|---|---|---|---|---|---|---|---|---|---|---|
| | *36 in.* | | *18 in.* | | *12 in.* | | *9 in.* | | *6 in.* | |
| | *Allowable Clearances with Specified Protection (in.)* | | | | | | | | | |
| | *Use Col. 1 for clearances above appliance or horizontal connector. Use Col. 2 for clearances from appliance, vertical connector, and single-wall metal pipe.* | | | | | | | | | |
| | *Above (Col. 1)* | *Sides and Rear (Col. 2)* | *Above (Col. 1)* | *Sides and Rear (Col. 2)* | *Above (Col. 1)* | *Sides and Rear (Col. 2)* | *Above (Col. 1)* | *Sides and Rear (Col. 2)* | *Above (Col. 1)* | *Sides and Rear (Col. 2)* |
| (1) $3\frac{1}{2}$ in. thick masonry wall without ventilated air space | — | 24 | — | 12 | — | 9 | — | 6 | — | 5 |
| (2) $\frac{1}{2}$ in. insulation board over 1 in. glass fiber or mineral wool batts | 24 | 18 | 12 | 9 | 9 | 6 | 6 | 5 | 4 | 3 |
| (3) 0.024 in. (nominal 24 gauge) sheet metal over 1 in. glass fiber or mineral wool batts reinforced with wire on rear face with ventilated air space | 18 | 12 | 9 | 6 | 6 | 4 | 5 | 3 | 3 | 3 |
| (4) $3\frac{1}{2}$ in. thick masonry wall with ventilated air space | — | 12 | — | 6 | — | 6 | — | 6 | — | 6 |
| (5) 0.024 in. (nominal 24 gauge) sheet metal with ventilated air space | 18 | 12 | 9 | 6 | 6 | 4 | 5 | 3 | 3 | 2 |
| (6) $\frac{1}{2}$ in. thick insulation board with ventilated air space | 18 | 12 | 9 | 6 | 6 | 4 | 5 | 3 | 3 | 3 |
| (7) 0.024 in. (nominal 24 gauge) sheet metal with ventilated air space over 0.024 in. (nominal 24 gauge) sheet metal with ventilated air space | 18 | 12 | 9 | 6 | 6 | 4 | 5 | 3 | 3 | 3 |
| (8) 1 in. glass fiber or mineral wool batts sandwiched between two sheets 0.024 in. (nominal 24 gauge) sheet metal with ventilated air space | 18 | 12 | 9 | 6 | 6 | 4 | 5 | 3 | 3 | 3 |

For SI units, 1 in. = 25.4 mm.

Notes:

(1) Reduction of clearances from combustible materials shall not interfere with combustion air, draft hood clearance and relief, and accessibility of servicing.

(2) All clearances shall be measured from the outer surface of the combustible material to the nearest point on the surface of the appliance, disregarding any intervening protection applied to the combustible material.

(3) Spacers and ties shall be of noncombustible material. No spacer or tie shall be used directly opposite the appliance or connector.

(4) Where all clearance reduction systems use a ventilated air space, adequate provision for air circulation shall be provided as described. *[See Figure 10.3.2.2(b) and Figure 10.3.2.2(c).]*

(5) At least 1 in. (25 mm) shall be between clearance reduction systems and combustible walls and ceilings for reduction systems using a ventilated air space.

(6) Where a wall protector is mounted on a single flat wall away from corners, it shall have a minimum 1 in. (25 mm) air gap. To provide adequate air circulation, the bottom and top edges, or only the side and top edges, or all edges shall be left open.

(7) Mineral wool batts (blanket or board) shall have a minimum density of 8 $lb/ft^3$ (128 $kg/m^3$) and a minimum melting point of 1500°F (816°C).

(8) Insulation material used as part of a clearance reduction system shall have a thermal conductivity of 1.0 Btu $in./ft^2/hr$-°F (0.144 W/m-K) or less.

(9) At least 1 in. (25 mm) shall be between the appliance and the protector. In no case shall the clearance between the appliance and the combustible surface be reduced below that allowed in Table 10.2.3(b).

(10) All clearances and thicknesses are minimum; larger clearances and thicknesses are acceptable.

(11) Listed single-wall connectors shall be installed in accordance with the manufacturers' installation instructions.

after 1976. Part of Appendix E from the 1976 edition is reprinted here to supply the user of this handbook with further information. By referring to the requirements for room heaters, the *National Building Code* covered the installation of residential air-conditioning equipment (addressed in 10.2.4), clothes dryers, and water heaters. The placement requirements were stated as follows:

> **(a)** Room heaters, except as permitted by the provisions of Sections (b) through (e), shall be placed on the ground; or on approved limited-combustible assemblies having a fire resistance rating of not less than 2 hours, with floors constructed of noncombustible material; or on concrete slabs or masonry arches that do not have combustible materials attached to the underside. Any floor covering on floor-ceiling assemblies, slabs, or arches shall extend not less than 6 in. (15 cm) beyond the appliances on all sides, and, where solid fuel is used, they shall extend not less than 18 in. (46 cm) at the front or side where ashes are removed.
> **(b)** Room heaters, that are tested and listed by a nationally recognized testing laboratory for installation on floors constructed of combustible materials, are permitted to be placed on floors, other than as required by the provisions of Section (a), provided they are installed in accordance with the requirements of the listing and conditions of approval.
> **(c)** Room heaters, which are set on legs or simulated legs that provide not less than 4 in. (10 cm) open space under the base of the appliance, are permitted to be placed on floors other than as required by the provisions of Section (a), provided the floor under the appliance is protected with sheet metal of not less than No. 24 gage, or by other noncombustible material that will reflect heat and is durable. Where solid fuel is used, the protection shall extend not less than 18 in. (46 cm) beyond the appliance at the front or side where ashes are removed. With radiating type gas burning room heaters which make use of metal, asbestos or ceramic material to direct radiation to the front of the device, the floor protection shall extend out at the front not less than 36 in. (91 cm) when the heater has not been tested and listed by a nationally recognized testing laboratory for installation on a combustible floor.
> **(d)** Room heaters which are set on legs that provide not less than 18 in. (46 cm) open space under the base of the appliance, or which have no burners within 18 in. (46 cm) of the floor, are permitted to be placed on floors other than as required by the provisions of Section (a) without special floor protection, provided there is at least one sheet metal baffle not less than No. 24 gage between the burners and the floor.
> **(e)** Room heaters, other than those described in the provisions of Sections (b), (c), and (d), are permitted to be placed on floors other than as required by the provisions of Section (a), provided the floor under the appliance is protected with hollow masonry not less than 4 in. (10 cm) thick covered with sheet metal not less than No. 24 gage. The masonry shall be laid with ends unsealed and joints matched in such a way as to provide a free circulation of air from side to side through the masonry. Where solid fuel is used, the floor, for 18 in. (46 cm) beyond the front of the appliance or side where ashes are removed, shall be protected with sheet metal not less than No. 24 gage or with other noncombustible materials that will not allow hot ashes to ignite the floor or floor covering.

Note that Appendix E of the *National Building Code* specified a sheet metal thickness of 24 gauge. The *National Fuel Gas Code* specifies a thickness of 0.024 in. (0.6 mm) and notes in parenthesis that this is nominal 24 gauge. According to the 1973 fifth edition of the *Chemical Engineers' Handbook,* 24 U.S. Standard Gauge (for sheet or plate metal and wrought iron) measures 0.0250 in. (0.64 mm) thick. Different gauges exist (e.g., wire, sheet, hoop), and any sheet metal used should be specified to have a minimum thickness of 0.024 in. (0.6 mm).

### 10.2.5 Furnace Plenums and Air Ducts.

A furnace plenum supplied as a part of the air-conditioning appliance shall be installed in accordance with the manufacturer's instructions. Where a furnace plenum is not supplied with the appliance, any fabrication and installation instructions provided by the manufacturer shall be followed. The method of connecting supply and return ducts shall facilitate proper circulation of air. Where the air conditioner is installed within a room not large in comparison with the size of the appliance, the air circulated by the appliance shall be handled by ducts that are sealed to the casing of the appliance and that separate the circulating air from the combustion and ventilation air.

In addition to the standards referenced in 10.3.7.1 — NFPA 90A, *Standard for the Installation of Air-Conditioning and Ventilating Systems,* and NFPA 90B, *Standard for the Installation of Warm Air Heating and Air-Conditioning Systems* — installers and inspectors are referred to UL 181, *Standard for Factory-Made Air Ducts and Air Connectors,* and UL 263, *Standard for Fire Tests of Building Construction and Materials*. These are the primary standards for factory-made air ducts and divide factory-made nonmetallic ducts into three classes, depending on their surface-burning characteristics and flame spread characteristics. The referenced standards specify where these ducts can be used.

An additional standard for the construction of metal ducts is the Sheet Metal and Air Conditioning Contractors National Association (SMACNA) publication, *HVAC Duct Construction Standards — Metal and Flexible,* for low-pressure and high-pressure duct construction. Low-pressure metal ducts generally are those types that handle air distribution systems with internal pressures of up to 2 in. w.c. (0.5 kPa), either positive or negative. This standard also shows the construction methods for high-pressure ducts of up to 10 in. w.c. (2.5 kPa) and includes methods for joining the sections and hanging the ducts to ensure compliance with most building codes. In addition, the standard contains methods for sealing duct sections to ensure that they are essentially airtight from the appliance to the discharge grille or diffuser.

### 10.2.6* Refrigeration Coils.

See 10.3.7 and 10.3.8.

**A.10.2.6** Reference can be made to NFPA 90A, *Standard for the Installation of Air-Conditioning and Ventilating Systems*, or to NFPA 90B, *Standard for the Installation of Warm Air Heating and Air-Conditioning Systems*.

### 10.2.7 Switches in Electrical Supply Line.

Means for interrupting the electrical supply to the air-conditioning appliance and to its associated cooling tower (if supplied and installed in a location remote from the air conditioner) shall be provided within sight of and not over 50 ft (15 m) from the air conditioner and cooling tower.

**FAQ ►** Does the condenser require a separate disconnect on any furnace that has air-conditioning with a separate condensing unit outdoors?

In correlation with Section 440.14 in the 2008 edition of *NFPA 70®*, *National Electrical Code®* (*NEC®*), 10.2.7 requires that the disconnecting means for the electrical circuit supplying an air-conditioning unit and its associated cooling tower (or condenser) be located within sight from and not more than 50 ft (15 m) from the air-conditioning equipment or cooling tower that the circuit supplies. This location requirement corresponds with the term *in sight from* as defined in the *NEC.*

The purpose of this disconnecting means is to afford service personnel with the ability to open the supply circuit from a location that is in close proximity to the equipment being serviced or maintained. Quite often, the electrical equipment (such as a panelboard, switchboard, or motor control center) where the supply circuit originates is not located in the same area as the air-conditioning equipment, thus necessitating the additional disconnecting means within sight of the air-conditioning equipment. In those instances when the air-conditioning equipment is located within sight of the electrical equipment that supplies that unit, the circuit breaker or fusible switch provides the required disconnecting means. This disconnecting means requirement applies to all air-conditioning equipment, whether installed indoors or outdoors.

In some cases, manufacturers will specify an HACR circuit breaker as an acceptable form of branch circuit overcurrent protection. "HACR" stands for heating, air-conditioning, and refrigeration, and this marking on a circuit breaker indicates that the circuit breaker meets the requirements of standards governing overcurrent protection of multimotors and combination loads inherent in air-conditioning equipment. If a local disconnect is required to meet the "within sight" rule, that disconnect can be used as the means to provide this specified type of overcurrent protection.

## 10.3 Central Heating Boilers and Furnaces

Section 10.3 contains installation requirements for listed gas furnaces, boilers, and air conditioners, and this section is consistent with Table 10.2.3(a) in regard to clearance to combustibles. The table covers clearances for unlisted central heating boilers, furnaces, and air conditioners only. Clearances for listed central boilers, furnaces, and air conditioners are provided in the manufacturer's installation instructions.

Chapter 3, Definitions, includes definitions of the different types of central heating boilers and furnaces. The reader should note that this code specifically covers boilers operating at low pressures and at temperatures not exceeding 250°F (121°C). The reader should also note that there are several different types of furnaces, including upflow furnaces, horizontal furnaces, downflow furnaces, forced-air furnaces with cooling units, gravity furnaces, direct vent furnaces, and enclosed central furnaces. See Exhibit 10.1 through Exhibit 10.7 for some examples of the types of units that are covered by this code.

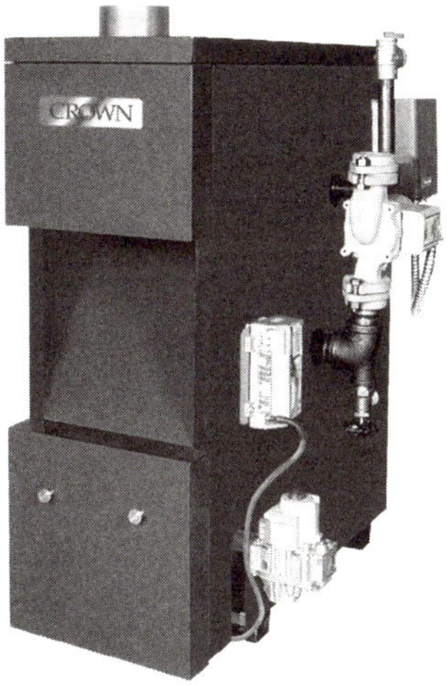


***EXHIBIT 10.1*** *Hot Water Boiler. (Courtesy of Crown Boiler Company)*

***EXHIBIT 10.2*** *Outdoor Installation of Central Furnace with Cooling Unit. (Courtesy of Modine Manufacturing Company)*

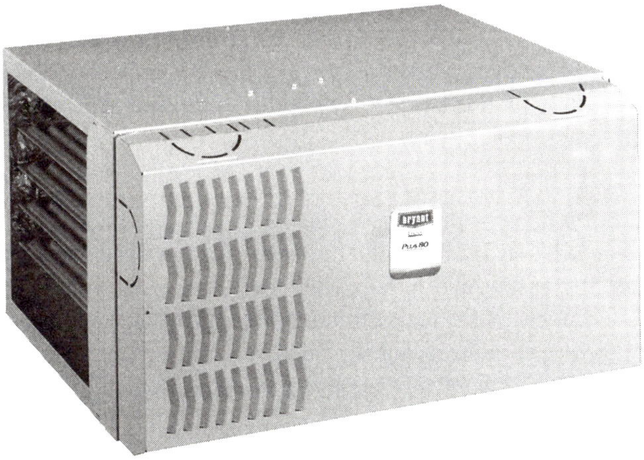

***EXHIBIT 10.3*** *Horizontal Airflow Furnace. (Courtesy of Carrier Corporation)*

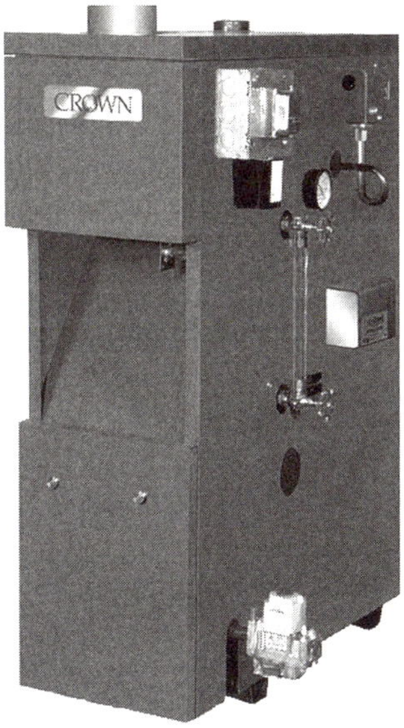


***EXHIBIT 10.4*** *Steam Boiler. (Courtesy of Crown Boiler Company)*

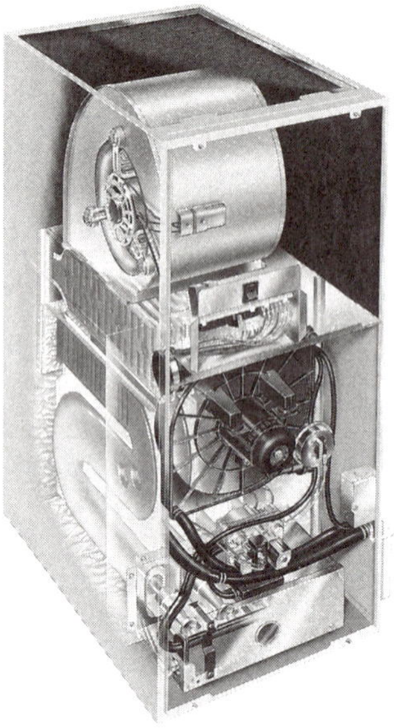

*EXHIBIT 10.5* *Downflow Furnace. (Courtesy of Carrier Corporation)*

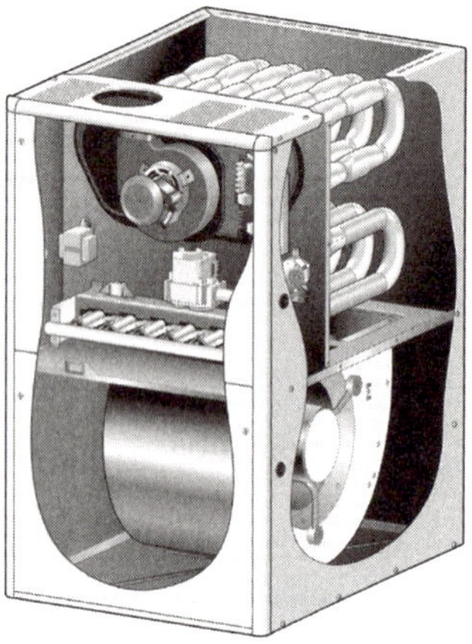

*EXHIBIT 10.6* *Upflow Furnace. (Courtesy of NORDYNE, Inc.)*

*EXHIBIT 10.7* *Upflow Furnace Installed. (Courtesy of Empire Comfort Systems)*

## 10.3.1 Location.

Central heating furnace and low-pressure boiler installations in bedrooms or bathrooms shall comply with one of the following:

NEW

(1) Central heating furnaces and low-pressure boilers shall be installed in a closet equipped with a weather-stripped door with no openings, and with a self-closing device. All combustion air shall be obtained from the outdoors in accordance with 9.3.3.
(2) Central heating furnaces and low-pressure boilers shall be of the direct vent type.

Note that Item (1) has been revised to clarify that a solid door is not required. The previous text, "weather-stripped solid door," did not convey the committee's intent that the door have no louvers or other openings and that a standard interior closet door was acceptable.

**FAQ ►**
The only location available for the furnace is a closet off a bedroom. Is this permitted?

A furnace is permitted in a closet off a bedroom, with the restrictions noted in 10.3.1. These requirements clarify the installation of boilers and furnaces in bedrooms and bathrooms and are consistent with an option for the installation of water heaters in closets opening to bedrooms and bathrooms. Another option is a direct vent boiler or furnace, which may safely be installed in or near a bedroom or a bathroom.

## 10.3.2 Clearance.

Providing adequate clearances and ventilation space around appliances is an important issue that is covered in 10.3.2. Central heating boilers and furnaces and their vent connectors must be installed so that continued or intermittent operation will not cause a fire where installed. By its nature, heating equipment gets warm — even hot sometimes — during operation. If adequate clearances and ventilation around the appliance are not provided, a potential for fire exists. In addition, elevated temperatures that are not high enough to start a fire can exceed the recommended operating temperature of electrical and mechanical boiler and furnace components, leading to reduced service life.

**10.3.2.1** Listed central heating furnaces and low-pressure boilers installed in a room large in comparison with the size of the appliance shall be installed with clearances in accordance with the manufacturer's instructions. *(See 3.3.90, Room Large in Comparison with Size of Equipment.)*

Note that the term *room large in comparison with size of appliance* is defined in 3.3.90. See the commentary following 10.2.3 for the definition of this term.

**10.3.2.2** Central heating furnaces and low-pressure boilers installed in rooms that are NOT large (such as alcoves and closets) in comparison with the size of the appliance shall be listed for such installations. Listed clearances shall not be reduced by the protection methods described in Table 10.2.3(b) and illustrated in Figure 10.3.2.2(a) through Figure 10.3.2.2(c), regardless of whether the enclosure is of combustible or noncombustible material.

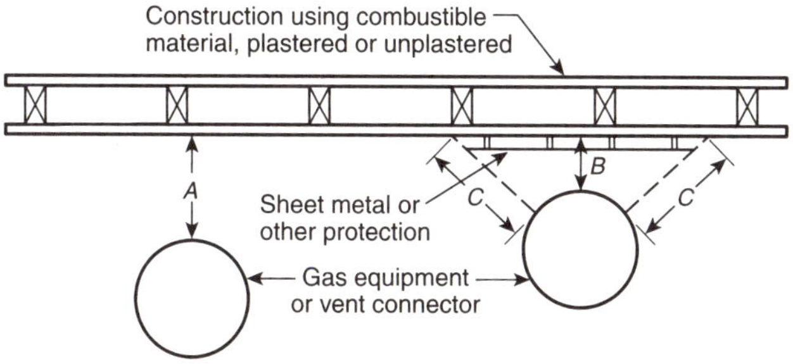


Notes:

(1) *A* equals the clearance with no protection specified in Tables 10.2.3(a) and 12.5.1 and in the sections applying to various types of equipment.

(2) *B* equals the reduced clearance permitted in accordance with Table 10.2.3(b).

(3) The protection applied to the construction using combustible material shall extend far enough in each direction to make *C* equal to *A*.

***FIGURE 10.3.2.2(a)*** *Extent of Protection Necessary to Reduce Clearances from Gas Appliance or Vent Connectors.*

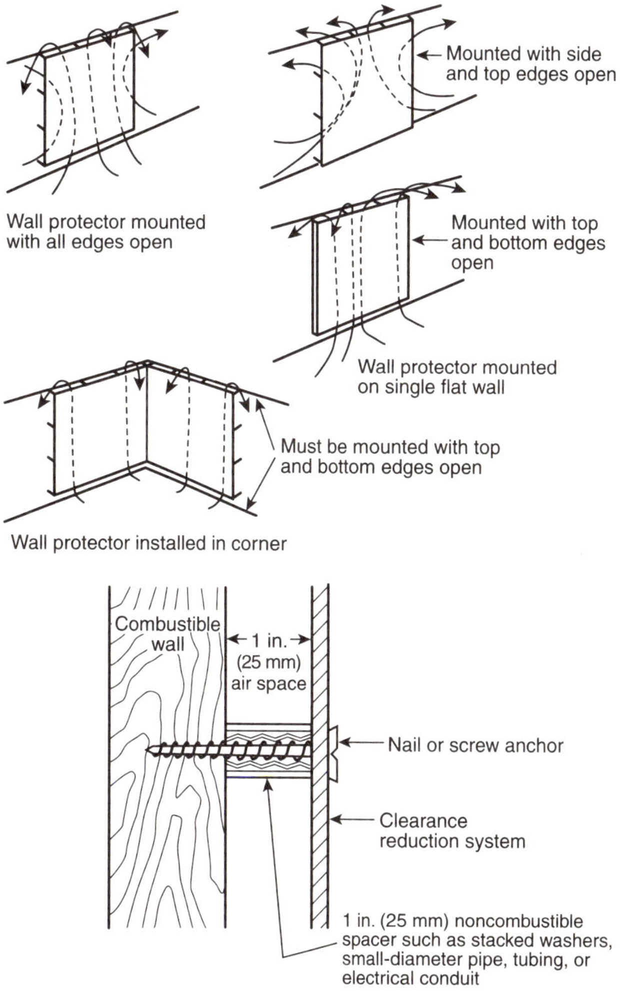


Notes:

(1) Masonry walls can be attached to combustible walls using wall ties.

(2) Spacers should not be used directly behind appliance or connector.

***FIGURE 10.3.2.2(b)*** *Wall Protector Clearance Reduction System.*

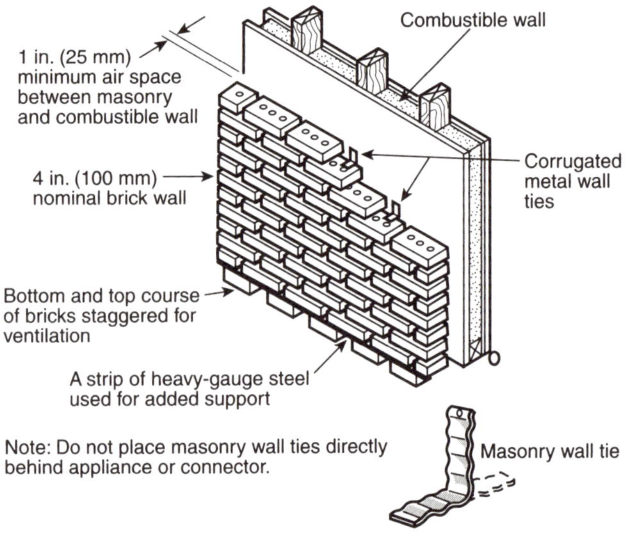


*FIGURE 10.3.2.2(c) Masonry Clearance Reduction System.*

**FAQ ▶**
Can Table 10.2.3(b) be used to reduce the clearance for a listed alcove furnace in a room not large in comparison to the appliance?

Central heating furnaces and boilers that are to be installed in alcoves and closets must be specifically listed for such installation. Note that the listed clearance must not be reduced by the application of methods given in Table 10.2.3(b). A typical rating nameplate for an appliance that is approved for closet installation at the specified clearances is shown in Exhibit 10.8.

Note that the last sentence of 10.3.2.2 prohibits clearance reduction for appliances installed in rooms *not* large in comparison with the size of the appliance. Such small rooms, closets, and alcoves require the minimum clearances for air circulation, in order to keep appliances from overheating.

**10.3.2.3** Unlisted central heating furnaces and low-pressure boilers installed in rooms that are large in comparison with the size of the appliance shall be installed with clearances not less than those specified in Table 10.2.3(a).

***EXHIBIT 10.8*** *Typical Nameplate. (Courtesy of Carrier Corporation)*

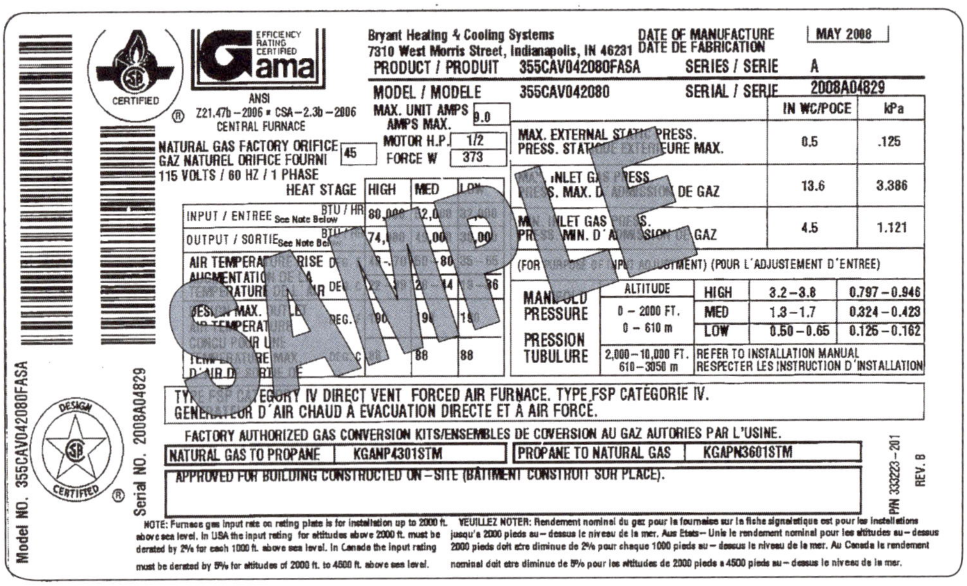

**10.3.2.4** Central heating furnaces and low-pressure boilers (listed and unlisted) installed in rooms that are large in comparison with the size of the appliance shall be permitted to be installed with reduced clearances to combustible material provided the combustible material or appliance is protected as described in Table 10.2.3(b) *(see 10.3.2).*

If the installation is in a room large in comparison with the size of the appliance, 10.3.2.4 permits the use of the clearance reduction systems of Table 10.2.3(b). By allowing clearance reduction for central heating furnaces and low-pressure boilers in rooms large in comparison with the size of the appliance, and prohibiting clearance reduction in rooms not large in comparison with the size of the appliance, 10.3.2.4 recognizes that in smaller rooms the room air will be heated by use of the clearance reduction system, which uses air to remove heat from nearby combustible materials.

**10.3.2.5** Front clearance shall be sufficient for servicing the burner and the furnace or boiler.

Clearance space must be provided for both routine and major service work. Note that 9.5.1 covers access to appliances installed in attics, and applies in addition to the requirements of 10.3.2.5 where appliances are installed in attics. The largest serviceable component must be able to be removed into the area provided for servicing the appliance.

**10.3.2.6** Where the furnace plenum is adjacent to plaster on metal lath or noncombustible material attached to combustible material, the clearance shall be measured to the surface of the plaster or other noncombustible finish where the clearance specified is 2 in. (50 mm) or less.

Paragraph 10.3.2.6 covers clearance to the furnace plenum, not to other parts of the furnace. Sheetrock, lath, or plaster applied to combustible materials is not adequate protection to prevent ignition of the combustible construction. If clearance reduction is required, one of the systems in Table 10.2.3(b) should be used.

**10.3.2.7** The clearances to these appliances shall not interfere with combustion air, draft hood clearance and relief, and accessibility for servicing. *(See 9.2.1, Section 9.3, and 12.13.7.)*

**10.3.2.8** Listed central heating furnaces shall have the clearance from supply ducts within 3 ft (0.9 m) of the furnace plenum be not less than that specified from the furnace plenum. No clearance is necessary beyond this distance.

**10.3.2.9** Unlisted central heating furnaces with temperature limit controls that cannot be set higher than 250°F (121°C) shall have the clearance from supply ducts within 6 ft (1.8 m) of the furnace plenum be not less than 6 in. (150 mm). No clearance is necessary beyond this distance.

**10.3.2.10** Central heating furnaces other than those listed in 10.3.2.8 or 10.3.2.9 shall have clearances from the supply ducts of not less than 18 in. (460 mm) from the furnace plenum for the first 3 ft (0.9 m), then 6 in. (150 mm) for the next 3 ft (0.9 m), and 1 in. (25 mm) beyond 6 ft (1.8 m).

### 10.3.3 Assembly and Installation.

A central heating boiler or furnace shall be installed in accordance with the manufacturer's instructions in one of the following manners:

(1) On a floor of noncombustible construction with noncombustible flooring and surface finish and with no combustible material against the underside thereof
(2) On fire-resistive slabs or arches having no combustible material against the underside thereof

*Exception No. 1: Appliances listed for installation on a combustible floor.*

*Exception No. 2: Installation on a floor protected in an approved manner.*

Many furnaces and boilers, particularly those intended for residential use, are listed for installation on combustible flooring. Subsection 10.3.3 allows furnaces and boilers listed for installation on noncombustible flooring to be listed for optional installation on combustible flooring if a listed accessory base is added. An example of such a device is shown for a boiler in Exhibit 10.9. The base must be used and installed as shown when the furnace is installed on combustible flooring.

***EXHIBIT 10.9*** *Installation of a Central Heating Boiler on a Combustible Floor. (Courtesy of Laars Heating Systems Co.)*

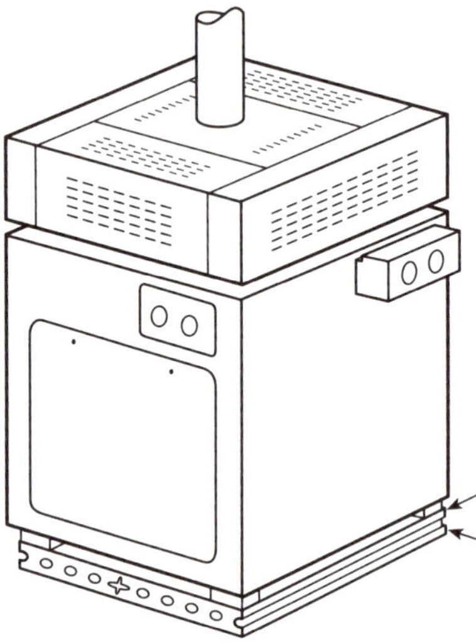


As stated earlier, prior to the 1988 edition, the *National Fuel Gas Code* contained a footnote referring the reader to Appendix E of the 1976 edition of the *National Building Code* for information on installation of central heating boilers and furnaces not listed for installation on combustible floors. The 1976 edition of the *National Building Code* was the last edition of that code written by the American Insurance Association. After 1976, the project was taken over by the Building Officials and Code Administrators International. Appendix E did not appear in the *National Building Code* after 1976. Part of Appendix E from the 1976 edition is reprinted here to supply the user of this handbook with further information. The *National Building Code* covered the installation of residential type boilers and central furnaces as follows:

**(a)** Residential type boilers and central furnaces, except as permitted by the provisions of Section (b) through (g), shall be placed on concrete bases supported on compacted soil, crushed rock, or gravel; on approved limited-combustible floor-ceiling assemblies with noncombustible floors and having a fire resistance rating of not less than 2 hours; or on concrete slabs or masonry arches, that do not have any combustible materials attached to the underside. Any floor covering on floor-ceiling assemblies, slabs, or arches shall be of noncombustible materials. Concrete bases, concrete slabs, masonry arches, and floor-ceiling assemblies and their supports shall be designed and constructed to support the appliances and shall extend not less than 12 in. (30 cm) beyond the appliance on all sides, and, where solid fuel is used, the base or floor shall extend not less than 18 in. (46 cm) at the front or side where ashes are removed.

**(b)** Residential type boilers and central furnaces, that are tested and listed by a nationally recognized testing laboratory for installation on floors constructed of combustible materials, are permitted to be placed on floors other than as required by the provisions of Section (a), provided they are installed in accordance with the requirements of the listing and conditions of approval.

**(c)** Central furnaces are permitted to be placed on floors other than as required by the provisions of Section (a), provided they are designed so that the fan chamber occupies the entire

area beneath the firing chamber and forms a well ventilated air space, between the firing chamber and the floor, not less than 18 in. (46 cm) in height, with at least one sheet metal baffle not less than No. 24 gage between the firing chamber and the floor.
**(d)** Residential type boilers of the water-base type are permitted to be placed on floors, other than as required by the provisions of Section (a), provided the water chamber extends under the firing chamber or under the ash pit and the firebox. Where solid fuel is used, the floor shall be covered as required by the provisions of Section (h).
**(e)** Residential type boilers and central furnaces, which are set on legs that provide an open space of not less than 4 in. (10 cm) under the base of the appliance, are permitted to be placed on the floors, other than as required by the provisions of Section (a), provided the appliance is designed so the flames or hot gases do not come in contact with its base, and further provided the floor under the appliance is protected with asbestos millboard not less than $^1/_4$ in. (6.4 mm) thick, covered with sheet metal not less than No. 24 gage. The floor protection shall extend not less than 6 in. (15 cm) beyond the appliance on all sides. Where solid fuel is used, the floor shall be covered as required by the provisions of Section (h).
**(f)** Residential type boilers and central furnaces are permitted to be placed on floors, other than as required by the provisions of Section (a), provided the appliance is designed so the flames or hot gases do not come in contact with its base, and further provided the floor under the appliance is protected with hollow masonry units not less than 4 in. (10 cm) thick, covered with sheet metal not less than No. 24 gage. The masonry shall be laid with ends unsealed and joints matched to provide a free circulation of air through the core spaces of the masonry. Where solid fuel is used, the floor shall be covered as required by the provisions of Section (h).
**(g)** Residential type boilers and central furnaces, which are designed so the flames or hot gases are in contact with the base, are permitted to be placed on floors, other than as required by the provisions of Section (a), provided the floor under the appliance is protected by two courses of 4 in. (10 cm) hollow masonry units, that are laid with ends unsealed and joints matched to provide a free circulation of air through the core spaces of the masonry. The masonry shall be covered with a steel plate not less than $^3/_{16}$ in. (4.8 mm) thick. Where solid fuel is used, the floor shall be covered as required by the provisions of Section (h).
**(h)** Where appliances which burn solid fuel are placed on floors other than as required by the provisions of Section (a), the floor, for not less than 18 in. (46 cm) beyond the front of the appliance or side where ashes are removed, shall be covered with asbestos millboard not less than $^1/_4$ in. (6.4 mm) thick, covered with sheet metal not less than No. 24 gage, or with materials providing equivalent protection to the floor and durability from hot ashes.
**(i)** Downflow type central furnaces shall not be placed on floors other than specified in Section (a) unless the appliance rests upon hollow masonry units not less than 4 in. (10 cm) thick. Masonry units shall be laid with the ends unsealed and joints matched to provide a free circulation of air through the core spaces of the masonry. Downflow type furnaces, which are tested and listed by a nationally recognized testing laboratory for installation on a floor constructed of combustible material, are permitted to be mounted in accordance with the requirements of the listing and conditions of approval. Downflow type furnaces shall be installed so that there are no open passages in the floor through which flame or hot gases from a fire originating in the space below the floor can travel to the room above.

Refer to the commentary following 10.2.4 for further discussion of sheet metal thickness.

## 10.3.4 Temperature or Pressure Limiting Devices.

Steam and hot water boilers, respectively, shall be provided with approved automatic limiting devices for shutting down the burner(s) to prevent boiler steam pressure or boiler water temperature from exceeding the maximum allowable working pressure or temperature. Safety limit controls shall not be used as operating controls.

Subsection 10.3.4 covers both automatic temperature limiting devices and automatic pressure limiting devices for installation on steam and hot water boilers. If the manufacturer provides these devices, they must be installed in accordance with the boiler manufacturer's instructions. If provided by the installer, they must be installed in accordance with the control manufacturer's instructions. Neither the safety pressure control nor the temperature limiting control can be used as an "operating control" for the appliance. These safety controls intentionally go beyond the high limit of temperature and pressure control devices normally operating the appliance to provide two completely separate devices. Refer to the manufacturer's installation instructions for the necessary control wiring to the room thermostat or operating control specified in the piping system. A supplemental, low-temperature control can be used if the appliance manufacturer permits one.

### 10.3.5* Low-Water Cutoff.

All water boilers and steam boilers shall be provided with an automatic means to shut off the fuel supply to the burner(s) if the boiler water level drops below the lowest safe water line. In lieu of the low-water cutoff, water tube or coil-type boilers which require forced circulation to prevent overheating and failure shall have an approved flow sensing device arranged to shut down the boiler when the flow rate is inadequate to protect the boiler against overheating.

This requirement was revised in the 2009 edition to require all boilers to have a safety feature to shut off the boiler if the water level is below the minimum safe level. This device should be supplied as part of the boiler. Depending on the design of the boiler either a low-water cutoff switch or a water flow sensor can be used to comply with this requirement. This safety feature has been required on oil fired boilers covered by NFPA 31 for many years. Previous editions of NFPA 54 had required a low water safety device only if the boiler was installed "below the level of radiation," or below any of the water or steam radiators fed by the boiler.

**A.10.3.5** Paragraph 10.3.5 differs from that published in ANSI Z223.1. This paragraph was revised based on an amendment approved in the NFPA process. A similar change has been proposed as an amendment to ANSI Z223.1.

*EXHIBIT 10.10 Temperature and Pressure Relief Valve with Rating Plate.*

### 10.3.6* Steam Safety and Pressure Relief Valves.

Steam and hot water boilers shall be equipped, respectively, with listed or approved steam safety or pressure relief valves of appropriate discharge capacity and conforming with ASME requirements. A shutoff valve shall not be placed between the relief valve and the boiler or on discharge pipes between such valves and the atmosphere.

The Btu/hr rating of the valve should be equal to or greater than the Btu/hr rating of the boiler. An undersized valve will compromise safety.

**A.10.3.6** For details of requirements on low-pressure heating boiler safety devices, refer to ASME *Boiler and Pressure Vessel Code,* Section IV, "Rules for Construction of Heating Boilers."

**10.3.6.1** Relief valves shall be piped to discharge near the floor.

**10.3.6.2** The entire discharged piping shall be at least the same size as the relief valve discharge piping.

**10.3.6.3** Discharge piping shall not contain threaded end connection at its termination point.

The steam or hot water boiler manufacturer generally provides, or specifies, pressure relief valves. These pressure relief valves must not be altered. The valves should be installed at the location specified in the manufacturer's installation instructions.

Exhibit 10.10 shows a temperature and pressure relief valve with a blowup of the maximum ratings of the valve. In this case, the maximum Btu/hr rating of the heating appliance should not be more than 100,000 Btu/hr (29 kW). The prohibition of valves in piping to or from these safety devices prevents the devices from being defeated by being capped. For example, if a pressure relief valve is observed to be leaking, maintenance workers who are not familiar with boilers might cap the leaking valve to stop the leak, with disastrous results. In addition, following the termination and size requirements for added safety is important.

The three specific requirements in 10.3.6.1, 10.3.6.2, and 10.3.6.3 have been added to the minimum requirements for pressure safety to ensure the following:

1. In the unlikely event of discharge from a pressure relief valve, the discharge does not create an additional hazard to people in the area.
2. The safe operation of pressure relief devices is not compromised by downstream flow restrictions.
3. A valve or cap is not installed unintentionally.

### 10.3.7 Furnace Plenums and Air Ducts.

Refer to the commentary following 10.2.5 for a list of standards covering duct construction.

**10.3.7.1** Furnace plenums and air ducts shall be installed in accordance with NFPA 90A, *Standard for the Installation of Air-Conditioning and Ventilating Systems*, or NFPA 90B, *Standard for the Installation of Warm Air Heating and Air-Conditioning Systems*.

**10.3.7.2** A furnace plenum supplied as a part of a furnace shall be installed in accordance with the manufacturer's instructions.

**10.3.7.3*** Where a furnace plenum is not supplied with the furnace, any fabrication and installation instructions provided by the manufacturer shall be followed. The method of connecting supply and return ducts shall facilitate proper circulation of air.

**A.10.3.7.3** Reference can be made to NFPA 90A, *Standard for the Installation of Air-Conditioning and Ventilating Systems*, or to NFPA 90B, *Standard for the Installation of Warm Air Heating and Air-Conditioning Systems*.

**10.3.7.4** Where a furnace is installed so supply ducts carry air circulated by the furnace to areas outside the space containing the furnace, the return air shall also be handled by a duct(s) sealed to the furnace casing and terminating outside the space containing the furnace.

Installing the furnace as required in 10.3.7.4 separates the furnace's circulating air system physically from its combustion system. The reason for this physical separation is that, if the circulating air is carried in ducts out of, but not back into, the space in which a forced-air furnace is installed, the air will be replaced due to the partial vacuum created. The most likely source of this air is a building opening, such as a window (which is likely to be closed if the furnace is operating) or the chimney or vent. The flow in the chimney or vent can be reversed, even if the space is large or the doors to the space are louvered. In a small utility room, this reversal of the vent flow is probable. When the flow is reversed, combustion products will enter the circulating air system through the open return air duct connection on the furnace and will circulate throughout the dwelling. For this reason, ANSI Z21.47, *Gas-Fired Central Furnaces,* requires an interlock to prevent furnace operation if the blower (filter) access panel is not in place. (See Exhibit 10.11.)

### 10.3.8 Refrigeration Coils.

The installation of refrigeration coils shall comply with the following requirements:

(1) A refrigeration coil shall not be installed in conjunction with a forced air furnace where circulation of cooled air is provided by the furnace blower, unless the blower has sufficient capacity to overcome the external static resistance imposed by the duct system and cooling coil and the air throughput necessary for heating or cooling, whichever is greater.

**FAQ ▶**
**Can a refrigeration coil be installed in a duct downstream of a forced air furnace?**

A refrigeration coil can be installed with a forced air furnace only if the furnace blower can provide sufficient airflow.

The general requirement assumes that the furnace blower can deliver sufficient airflow at least 0.5 in. w.c. (120 Pa) external static pressure resistance to overcome the resistance imposed by the supply and return air duct systems plus the addition of a refrigeration coil. The requirement in 10.3.8(1) provides enough blower capacity for a nominal 0.15 in. w.c. (37 Pa) external static pressure for the supply duct system, 0.30 in. w.c. (75 Pa) for the cooling coil, and 0.05 in. w.c. (12.5 Pa) for the return air duct system and filters. Failure to provide sufficient airflow to overcome the external static pressure can lead to problems that include the following:

- Condensate freezing on the refrigeration coil during air-conditioning mode, which can block airflow through the system
- Overheating of the furnace heating element, which can cause premature failure of the furnace

Consult the furnace manufacturer's specifications to determine available airflow rates and external static pressures and recommendations for refrigeration coil applications. Consult the air conditioner manufacturer's specifications to determine the required airflow rate and external static pressure resistance of the coil.

(2) Furnaces shall not be located upstream from cooling units, unless the cooling unit is designed or equipped so as not to develop excessive temperature or pressure.

Direct-expansion cooling coils with a specified condensing unit are listed as part of the air-conditioning unit assembly. The requirements for listing generally include the protective devices on the system necessary to prevent the development of excessive temperatures and pressures in the cooling coil, as stated in 10.3.8(2), when the coil is installed downstream from the furnace. In direct expansion systems, the vapor in the cooling coil expands into the

**EXHIBIT 10.11** *Typical Installation of a Horizontal Furnace Showing Plenum and Ducts. (Courtesy of Carrier Corporation)*

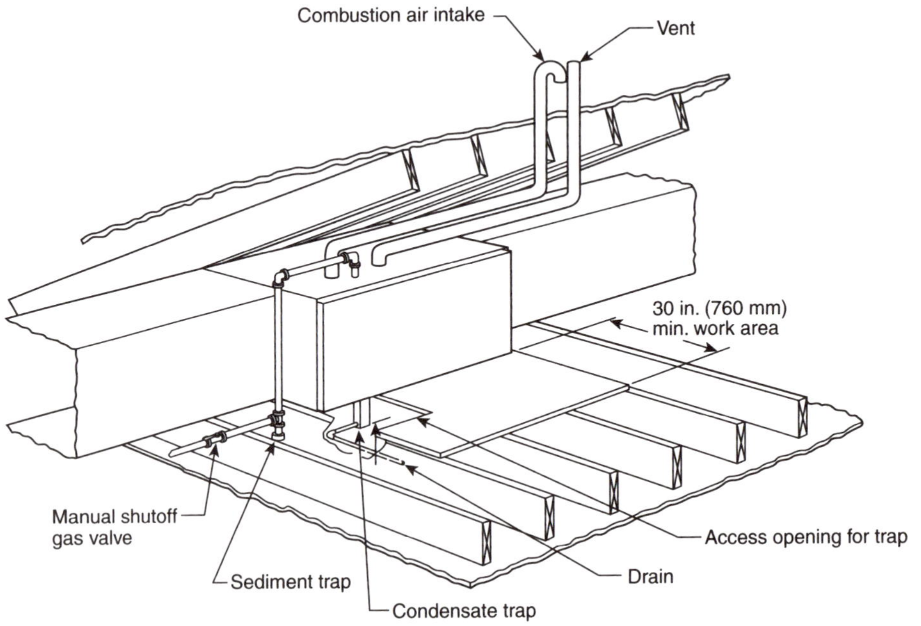


Note: Local codes may require a drain pan under the furnace and condensate trap when a condensing furnace is installed above finished ceilings.

interconnected piping and other portions of the air-conditioning unit. Provisions for expansion of the water in the piping system and remote chiller will protect the chilled water coil.

(3) Refrigeration coils shall be installed in parallel with or on the downstream side of central furnaces to avoid condensation in the heating element, unless the furnace has been specifically listed for downstream installation. With a parallel flow arrangement, the dampers or other means used to control flow of air shall be sufficiently tight to prevent any circulation of cooled air through the furnace.

Subsection 10.3.8(3) requires that only a furnace specifically constructed and designed for installation downstream from a cooling coil be so installed, and that the heat exchanger and burners be of corrosion-resistant material. If the furnace is listed for downstream installation, the tests conducted on the furnace will show that there is no danger of condensate dripping on the burners and pilots during the cooling cycle or that the heat exchanger will corrode. When a parallel flow system is designed, there should be face and bypass dampers in front of both the cooling coil and the heating unit, to ensure that there is no recirculation of chilled air back over the heat exchanger.

Exhibit 10.12 illustrates the permissible location of refrigeration coils in forced air furnaces. Exhibit 10.12(a) shows the refrigeration coil upstream to the furnace fan, which is not permitted by the code. Exhibit 10.12(b) shows the refrigeration coil downstream from the furnace fan, which is permitted by the code. Note that the installation in Exhibit 10.12(b) would not require a listed furnace and would meet the code requirements if blower capacity is sufficient.

(4) Means shall be provided for disposal of condensate and to prevent dripping of condensate on the heating element.

A typical coil for installation in a plenum above an upflow furnace, complying with the condensate drain requirement of 10.3.8(4), is shown in Exhibit 10.12(b). The "A-type" coil is constructed so that the condensate forming on the coil will travel down the edge of the coil fins to the drain pan connection. Note also that this particular coil has a secondary drain

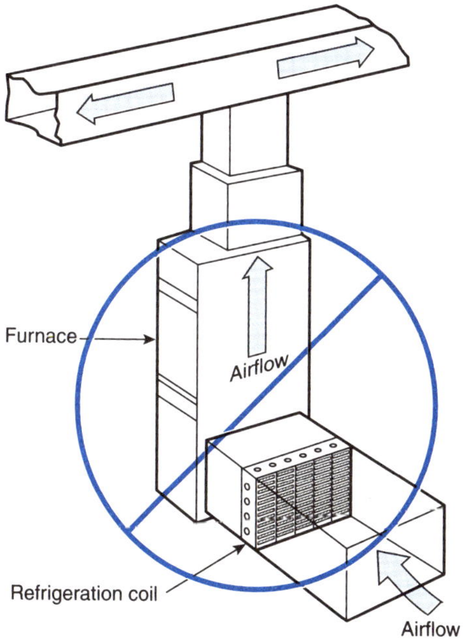


(a) Refrigeration coil upstream to furnace fan

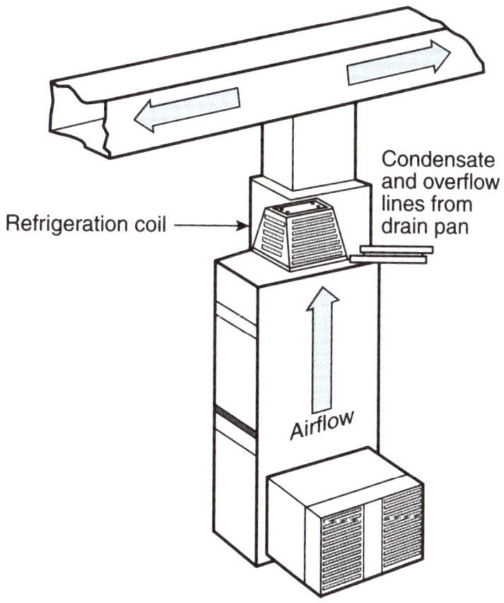


(b) Refrigeration coil downstream of furnace fan

*EXHIBIT 10.12 Refrigeration Coils in Forced Air Furnaces: (a) Location of Coils Not Permitted and (b) Location of Coils Permitted.*

connection to provide for disposal of condensate in the event that the primary drain connection becomes plugged. Condensate drains should follow local plumbing or mechanical codes.

### 10.3.9 Cooling Units Used with Heating Boilers.

**10.3.9.1** Boilers, where used in conjunction with refrigeration systems, shall be installed so that the chilled medium is piped in parallel with the heating boiler with appropriate valves to prevent the chilled medium from entering the heating boiler.

**10.3.9.2** Where hot water heating boilers are connected to heating coils located in air-handling units where they can be exposed to refrigerated air circulation, such boiler piping systems shall be equipped with flow control valves or other automatic means to prevent gravity circulation of the boiler water during the cooling cycle.

## 10.4 Clothes Dryers

Clothes dryers covered by this code are classified into the following two types:

1. Type 1: Primarily for residential use
2. Type 2: Designed for use by businesses frequented by the public

Both types may be coin-operated. (See 3.3.19.) Exhibit 10.13 illustrates Type 1 clothes dryer installation requirements.

*EXHIBIT 10.13 Type 1 Clothes Dryer Installation Requirements.*

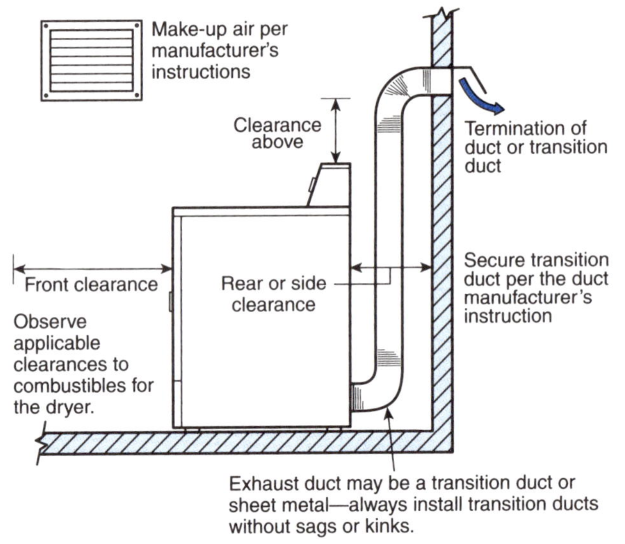


### 10.4.1 Clearance.

The installation of clothes dryers shall comply with the following requirements:

(1) Listed Type 1 clothes dryers shall be installed with a minimum clearance of 6 in. (150 mm) from adjacent combustible material. Clothes dryers listed for installation at reduced clearances shall be installed in accordance with the manufacturer's installation instructions. Type 1 clothes dryers installed in closets shall be specifically listed for such installation.

The manufacturer's instructions or rating nameplate on the unit specifies the required clearance from combustible materials. Exhibit 10.14 illustrates the typical installation of a Type 1

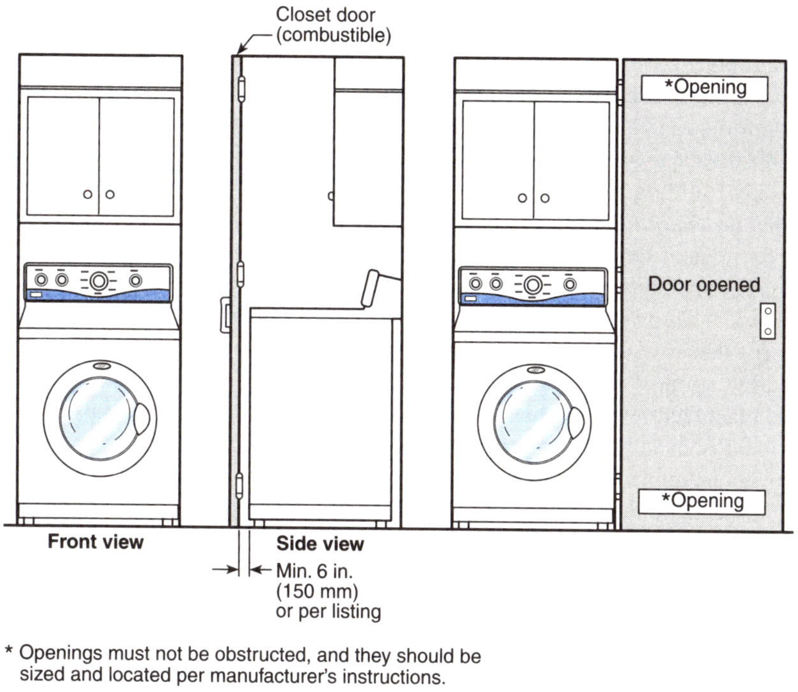


***EXHIBIT 10.14*** *Installation of a Type 1 Clothes Dryer.*

clothes dryer in an alcove or closet. Clearance must be provided for adequate air supply and ease of installation and service. All dryers must be exhausted to the outdoors. As already stated, for listed dryers, the minimum installation clearances defined by the appliance manufacturer must be followed. Unlisted dryer clearance must be at least 18 in. (460 mm) from any combustible material. Chapter 3, Definitions, should be consulted for the difference between a Type 1 and a Type 2 dryer.

(2) Listed Type 2 clothes dryers shall be installed with clearances of not less than shown on the marking plate and in the manufacturer's instructions. Type 2 clothes dryers designed and marked "For use only in noncombustible locations" shall not be installed elsewhere.
(3) Unlisted clothes dryers shall be installed with clearances to combustible material of not less than 18 in. (460 mm). Combustible floors under unlisted clothes dryers shall be protected in an approved manner.

The commentary following 10.2.4 gives additional information on gas appliance installation on combustible floors.

### 10.4.2 Exhausting to the Outdoors.

Type 1 and Type 2 clothes dryers shall be exhausted to the outdoors.

◀ **FAQ**
**Requirements are provided for the exhausting of clothes dryers. Is this different from venting?**

Exhausting of clothes dryers sometimes is confused with venting of appliances. Exhausting is the removal from a building of both the products of combustion and the water that has been removed from the clothing. Clothes dryer exhausts include both the products of combustion and moist air. Venting of an appliance, such as a furnace or water heater, is the removal of the products of combustion from a building. Because of the significant difference between the exhaust gas from a clothes dryer and the vent gas from a furnace, water heater, or other gas appliance, the removal methods are given different names and have different requirements.

The manufacturer's instructions must be followed for exhaust duct installation and location of the termination. Again, all clothes dryers must be exhausted to the outdoors. The requirement for exhausting Type 2 dryers to the outdoors has been in the *National Fuel Gas*

*Code* for many editions. The requirement to exhaust Type 1 dryers to the outdoors was added in the 1999 edition of the code.

The requirement in 10.4.2 for exhausting of all clothes dryers to the outdoors is a practical one. Indoor exhausting of clothes dryers in most climates adds humidity to the point of being uncomfortable in the building. In cold climates, the humidity that accumulates in a building with indoor exhaust of a clothes dryer would make for a very uncomfortable, and possibly moldy, building. Lint buildup would also occur.

**FAQ ▶** The Type 1 clothes dryer exhaust duct is 30 ft (9 m) long with two elbows. Is this allowed?

Note that the *National Fuel Gas Code* does not limit the length of dryer ducts, other than a general requirement to install listed dryers in accordance with manufacturer's installation instructions. The standard for gas dryer construction and testing, ANSI Z21.5.1/CSA 7.1, *Gas Clothes Dryers — Volume I, Type I Clothes Dryers*, requires that gas dryers be tested with a 15 ft (4.6 m) exhaust with two 90-degree elbows, unless the manufacturer specifies another exhaust configuration. If tested with a different configuration, this information will be indicated in the manufacturer's installation instructions. Some mechanical codes have limits on dryer duct lengths, usually about 15 ft (4.6 m) with two 90-degree elbows.

In renovation and new construction, plan approval and construction can occur before the dryer is purchased, and the exhaust is usually installed before the dryer is available. Also, the clothes dryer is not usually considered a permanent part of the building, and owners or renters often take their clothes dryers with them when they move.

Fires associated with gas dryers do occur. Note from Commentary Table 10.1 that dryers rank quite low on the list of gas-fueled appliances in terms of reported home fires. Electric clothes dryers also cause fires, about 11,000 fires each year, according to the NFPA report, *Home Heating Fire Patterns and Trends*. Contrast the estimated 2001 number of reported home structure fires involving gas dryers (2,200) with the corresponding number for electric dryers (8,500) and for dryers involving other or unclassified fuel or power (200). The ratio of number of fires (8,500 versus 2,200) for electric versus gas dryers is 4.0, slightly higher than the 3.6 ratio in household usage of the two types of dryers.

***COMMENTARY TABLE 10.1** Reported Home Structure Fires (2001) Involving Gas-Fueled Appliances*

| *Appliance* | *Number of Fires* |
|---|---|
| Range or stove | 13,400 |
| Oven or rotisserie | 4,100 |
| Water heater | 3,700 |
| Furnace | 2,800 |
| Dryer | 2,200 |
| Fixed space heater | 1,900 |
| Grill | 600 |

Note: The data from Commentary Table 10.1 are presented to provide an overview of the number of fires involving gas-fueled appliances. These statistics do not indicate or imply levels of risk relative to usage.

Sources of Data: John R. Hall, Jr., *Home Cooking Fire Patterns and Trends*, NFPA, Fire Analysis and Research Division, Quincy, MA, January 2005 and John R. Hall, Jr., *Home Heating Fire Patterns and Trends*, NFPA, Fire Analysis and Research Division, Quincy, MA, November 2004.

The risk of death from fires involving ranges and stoves is higher for gas than for electric, but the risk of other types of losses (incidents, injuries, and property damage) is higher for electric than for gas. The risk of death from fires involving ovens and rotisseries was not separately analyzed, but the comments on risk of death from fires for ranges and stoves would probably apply. The risk of death from fires involving water heaters is much higher for gas than for electric. The risk of death from fires involving furnaces is lower for gas than for oil, and comparable for gas and electric. The risk of death from fires involving dryers was not separately analyzed. The risk of fire involving fixed space heaters for gas is higher than for electric or wood.

### 10.4.3 Provisions for Make-Up Air.

**10.4.3.1** Make-up air shall be provided for Type 1 clothes dryers in accordance with the manufacturers' installation instructions.

The quantity of make-up air required for a Type 1 clothes dryer exceeds that of other gas appliances. The manufacturer's installation instructions provide opening requirements to allow such air make-up. Note that Section 9.3, Air for Combustion and Ventilation, was revised in the 2002 edition to exclude clothes dryers with make-up air provided in accordance with this section from the air for combustion calculation. This exclusion recognized that a large amount of make-up air essential for proper performance includes the small amount required for combustion.

Five to ten times the air required for an appliance of similar heat input should be allowed. If several gas appliances are installed in a space with insufficient air for combustion and ventilation (see Section 9.3, Air for Combustion and Ventilation), or in a limited space (e.g., a utility room or service porch), adequate combustion and ventilation air must be provided. Type 1 dryers that can be installed in closets will be identified as such by the manufacturer's installation instructions. In replacing a dryer in a small space, one listed for installation in a closet must be selected.

**10.4.3.2** Provision for make-up air shall be provided for Type 2 clothes dryers, with a minimum free area *(see 9.3.7)* of 1 in.$^2$/1000 Btu/hr (2200 mm$^2$/kW) total input rating of the dryer(s) installed.

Type 2 dryers are required to have at least 1 in.$^2$ (650 mm$^2$) of make-up air opening per 1000 Btu/hr (2200 mm$^2$/kW) input rating. In large rooms, installing a separate ventilation inlet duct can fulfill this requirement. A ventilation system normally should be provided in a commercial public laundry and may be required by a mechanical code if one is adopted and used. The manufacturer's installation instructions should be consulted. The manufacturer's instructions may specify larger openings for ventilation air, and in order to meet the requirements of both codes, the larger opening must be used.

### 10.4.4 Exhaust Ducts for Type 1 Clothes Dryers.

**10.4.4.1** A clothes dryer exhaust duct shall not be connected into any vent connector, gas vent, chimney, crawl space, attic, or other similar concealed space.

**10.4.4.2** Ducts for exhausting clothes dryers shall not be assembled with screws or other fastening means that extend into the duct and that would catch lint and reduce the efficiency of the exhaust system.

The manufacturer's instructions specify the type of material that can be used for a duct assembly. Many manufacturers recommend that only single-wall metal pipe or flexible metal ducts be used, not flexible plastic or other types of ducts.

Sheet metal screws or other fastenings that can extend into the duct and catch lint are prohibited, and either external clamps or tape (such as duct tape) must be used to make all duct connections and fittings.

◀ **FAQ**
Can sheet metal screws be used to connect dryer exhaust duct?

**10.4.4.3** Exhaust ducts shall be constructed of rigid metallic material. Transition ducts used to connect the dryer to the exhaust duct shall be listed for that application or installed in accordance with the clothes dryer manufacturer's installation instructions.

*EXHIBIT 10.15 Transition Duct. (Courtesy of Deflecto Corporation)*

The requirement in 10.4.4.3 clearly specifies that Type 1 dryer ducts, as shown in Exhibit 10.15, must be constructed of rigid metal vent material. Prior to the 1992 edition of the code, Type 1 duct material was not specified and the subject was left to the dryer manufacturer's instructions. Plastic and foil-type materials, which are more susceptible to damage and reduced airflow, were commonly used. The requirement recognizes that flexibility is needed in the duct installation to allow moving the dryer to access the duct connection, and the requirement allows only listed transition duct or transition duct recommended by the dryer manufacturer.

### 10.4.5 Exhaust Ducts for Type 2 Clothes Dryers.

**10.4.5.1** Exhaust ducts for Type 2 clothes dryers shall comply with 10.4.4.

**10.4.5.2** Exhaust ducts for Type 2 clothes dryers shall be constructed of sheet metal or other noncombustible material. Such ducts shall be equivalent in strength and corrosion resistance to ducts made of galvanized sheet steel not less than 0.0195 in. (0.5 mm) thick.

**10.4.5.3** Type 2 clothes dryers shall be equipped or installed with lint-controlling means.

**10.4.5.4** Exhaust ducts for unlisted Type 2 clothes dryers shall be installed with a minimum clearance of 6 in. (150 mm) from adjacent combustible material. Where exhaust ducts for Type 2 clothes dryers are installed with reduced clearances, the adjacent combustible material shall be protected in accordance with Table 10.2.3(b).

**10.4.5.5** Where ducts pass through walls, floors, or partitions, the space around the duct shall be sealed with noncombustible material.

**10.4.5.6** Multiple installations of Type 2 clothes dryers shall be made in a manner to prevent adverse operation due to back pressures that might be created in the exhaust systems.

The requirements in 10.4.5 for exhaust ducts for Type 2 dryers are straightforward. Corrosion-resistant sheet metal is required, and a minimum thickness is specified. Corrosion resistance recognizes the high moisture content of dryer exhaust gases, which can condense and corrode steel exhausts. Lint control means are required, recognizing the higher lint production of Type 2 clothes dryers compared to Type 1. Penetrations through building walls and floors must be sealed with noncombustible material, recognizing the higher temperature at which Type 2 dryer exhausts can operate compared with Type 1 dryer exhausts. Installations where multiple dryers exhaust into a single exhaust duct are required to have a means to prevent the operation of several dryers' overcoming the exhaust of a single dryer, and forcing the exhaust air back into the room. A backdraft damper installed on the individual dryer's exhaust is one way to comply with this requirement.

### 10.4.6 Multiple-Family or Public Use.

All clothes dryers installed for multiple-family or public use shall be equipped with approved safety shutoff devices and shall be installed as specified for a Type 2 clothes dryer under 10.4.5.

The requirements for the installation of Type 2 dryer ducts include the requirements for Type 1 ducts with several provisions added. Both types of ducts must be metal, but Type 2 ducts may have a minimum thickness of 0.0195 in. (0.5 mm). Type 2 duct systems must have a means to control lint, usually a lint trap, and minimum clearances to Type 2 ducts are specified.

## 10.5 Conversion Burners

Installation of conversion burners shall conform to ANSI Z21.8, *Installation of Domestic Gas Conversion Burners.*

A conversion burner is used to replace a burner assembly designed for another fuel, such as heating oil. This installation allows a boiler or furnace to be converted to gas without replacing the entire appliance. Exhibit 10.16 illustrates a typical gas conversion burner.

*EXHIBIT 10.16 Gas Conversion Burner. (Courtesy of Wayne Combustion Systems)*

## 10.6 Decorative Appliances for Installation in Vented Fireplaces

Decorative appliances (sometimes called "gas logs") for installation in vented fireplaces, which are covered in Section 10.6 and shown in Exhibit 10.17, are self-contained, freestanding, fuel-gas-burning appliances designed for installation only in vented fireplaces. These decorative appliances rely on fireplace venting to vent their products of combustion.

*EXHIBIT 10.17 Decorative Gas Appliance Installed in a Wood-Burning Fireplace. (Courtesy of DESA International)*

The primary function of a decorative appliance lies in the aesthetic effect of its flames. The four different types of decorative appliances are described as follows:

1. Coal basket: An open-flame appliance with a metal basket filled with simulated coals.
2. Fireplace insert: An open-flame, radiant-type appliance mounted in a decorative metal panel that covers the fireplace opening. This appliance has a provision for venting to the fireplace chimney.
3. Gas log: An open-flame appliance consisting of a metal frame or base supporting simulated logs.
4. Radiant appliance: An open-flame appliance designed primarily to convert the energy in fuel gas to radiant heat by means of refractory or similarly radiating materials. A radiant heater typically has no external jacket.

### 10.6.1* Prohibited Installations.

Decorative appliances for installation in vented fireplaces shall not be installed in bathrooms or bedrooms unless the appliance is listed and the bedroom or bathroom has the required volume in accordance with 9.3.2.

Subsection 10.1.2 permits the installation of gas appliances in bedrooms or bathrooms only if the required room volume specified in 9.3.2 is met. This criterion was changed in 10.6.1 and elsewhere in Chapter 10 to recognize the change in approach to providing sufficient air for combustion that was made in the 2002 edition. The commentary following 9.3.2 gives a complete description of the changes.

Because many homes now have bedrooms and bathrooms large enough to meet the requirement of 9.3.2, the code now permits these installations if the decorative appliance is listed.

**A.10.6.1** For information on decorative appliances for installation in vented fireplaces, see ANSI Z21.60/CGA 2.26, *Decorative Gas Appliances for Installation in Solid-Fuel Burning Fireplaces.*

Paragraphs A.10.6.1 and A.10.7.1 assist code users in identifying decorative appliances for installation in vented fireplaces and vented decorative appliances. The appliance nameplate lists the standard to which the appliance is listed, so the identification can be made easily.

### 10.6.2 Installation.

A decorative appliance for installation in a vented fireplace shall be installed only in a vented fireplace having a working chimney flue and constructed of noncombustible materials. These appliances shall not be thermostatically controlled.

Decorative appliances are not permitted to have thermostatic controls because they are not to be used as a source of heat. If a room needs a thermostatically controlled device, a room heater can be used. The installation of room heaters is covered in Section 10.23.

**10.6.2.1** A listed decorative appliance for installation in a vented fireplace shall be installed in accordance with the manufacturer's installation instructions.

**10.6.2.2** A decorative appliance for installation in a vented fireplace, where installed in a manufactured home, shall be listed for installation in manufactured homes.

**10.6.2.3** An unlisted decorative appliance for installation in a vented fireplace shall be installed in a fireplace having a permanent free opening, based on appliance input rating and chimney height equal to or greater than that specified in Table 10.6.2.3.

***TABLE 10.6.2.3*** *Free Opening Area of Chimney Damper for Venting Flue Gases from Unlisted Decorative Appliances for Installation in Vented Fireplaces*

| *Chimney Height (ft)* | *Minimum Permanent Free Opening (in.$^2$)** | | | | | | |
|---|---|---|---|---|---|---|---|
| | *8* | *13* | *20* | *29* | *39* | *51* | *64* |
| | *Appliance Input Rating (Btu/hr)* | | | | | | |
| 6 | 7,800 | 14,000 | 23,200 | 34,000 | 46,400 | 62,400 | 80,000 |
| 8 | 8,400 | 15,200 | 25,200 | 37,000 | 50,400 | 68,000 | 86,000 |
| 10 | 9,000 | 16,800 | 27,600 | 40,400 | 55,800 | 74,400 | 96,400 |
| 15 | 9,800 | 18,200 | 30,200 | 44,600 | 62,400 | 84,000 | 108,800 |
| 20 | 10,600 | 20,200 | 32,600 | 50,400 | 68,400 | 94,000 | 122,200 |
| 30 | 11,200 | 21,600 | 36,600 | 55,200 | 76,800 | 105,800 | 138,600 |

For SI units, 1 ft = 0.305 m, 1 in.$^2$ = 645 mm$^2$, 1000 Btu/hr = 0.293 kW.

* The first six minimum permanent free openings (8 in.$^2$ to 51 in.$^2$) correspond approximately to the cross-sectional areas of chimneys having diameters of 3 in. through 8 in., respectively. The 64 in.$^2$ opening corresponds to the cross-sectional area of standard 8 in. × 8 in. chimney tile.

**FAQ ►** Can a gas log be installed in an existing wood fireplace?

A gas log can be installed in an existing wood fireplace, but the damper on the solid fuel fireplace must be altered or removed because a gas log cannot be allowed to operate in a fireplace with a closed damper. There is no interlock between the existing damper and the new decorative appliance, so the damper must be either permanently blocked open, removed, or cut to allow sufficient venting area. Table 10.6.2.3 gives the necessary free opening for a chimney damper serving an unlisted decorative appliance installed in a vented wood-burning fireplace.

### 10.6.3 Fireplace Screens.

A fireplace screen shall be installed with a decorative appliance for installation in a vented fireplace.

A screen is required to be installed to provide protection for people and pets from inadvertent contact with the appliance. Tempered glass enclosures designed for fireplaces may be used if they provide adequate combustion air for the appliance.

## 10.7 Gas Fireplaces, Vented

A vented gas fireplace, addressed in Section 10.7, is a vented appliance whose only function lies in the aesthetic effect of its flames (see Exhibit 10.18). These appliances were formerly called vented decorative appliances. The name change follows an identical name change in the standard used for listing these appliances, ANSI Z21.50/CSA 2.2, *Vented Gas Fireplaces*.

***EXHIBIT 10.18*** *Vented Gas Fireplace. (Courtesy of Empire Comfort Systems)*

Simple explanations of these terms follow:

◀ **FAQ**
What is the difference between *decorative appliances for installation in vented fireplaces* and *vented gas fireplaces*?

- Decorative appliances are intended for installation in vented fireplaces. Gas logs that are installed in existing wood-burning fireplaces are an example.
- Vented gas fireplaces are complete, enclosed units that are vented to the outdoors via a vent or by insertion in a wood-burning fireplace.

The manufacturer's instructions must be followed for the installation of listed appliances, as well as requirements in 10.7.2; in Chapter 12, Venting of Appliances; and in Table 10.6.2.3 for the essential venting requirements.

### 10.7.1* Prohibited Installations.

Vented gas fireplaces shall not be installed in bathrooms or bedrooms unless the appliance is listed and the bedroom or bathroom has the required volume in accordance with 9.3.2.

*Exception: Direct vent gas fireplaces.*

Subsection 10.7.1 allows the installation of vented gas fireplaces in bedrooms and bathrooms when the room volume requirements of 9.3.2 are met. Decorative appliances installed in vented gas fireplaces must also meet the requirements of 9.3.2. See the commentary following 10.6.1.

Direct vent gas fireplaces can be installed whether or not the combustion air requirements of 9.3.2 are met, because direct vent appliances do not use air from the room for combustion.

**A.10.7.1** For information on vented gas fireplaces, see ANSI Z21.50/CGA 2.22, *Vented Gas Fireplaces.*

The commentary following A.10.6.1 discusses the identification of decorative appliances, which are listed to ANSI Z21.60/CSA 2.26, *Decorative Gas Appliances for Installation in Solid-Fuel Burning Fireplaces.*

### 10.7.2 Installation.

The installation of vented gas fireplaces shall comply with the following requirements:

(1) Listed vented gas fireplaces shall be installed in accordance with the manufacturer's installation instructions and where installed in or attached to combustible material shall be specifically listed for such installation.
(2) Unlisted vented gas fireplaces shall not be installed in or attached to combustible material and shall also comply with the following:
   (a) They shall have a clearance at the sides and rear of not less than 18 in. (460 mm).
   (b) Combustible floors under unlisted vented gas fireplaces shall be protected in an approved manner.
   (c) Unlisted appliances of other than the direct vent type shall be equipped with a draft hood and shall be properly vented in accordance with Chapter 12.
   (d) Appliances that use metal, asbestos, or ceramic material to direct radiation to the front of the appliance shall have a clearance of 36 in. (910 mm) in front and, if constructed with a double back of metal or ceramic, shall be installed with a minimum clearance of 18 in. (460 mm) at the sides and 12 in. (300 mm) at the rear.

Refer to the commentary following 10.2.4 and 10.3.2 for information on the installation of gas utilization appliances on combustible floors, addressed in 10.7.2(2).

(3) Panels, grilles, and access doors that are required to be removed for normal servicing operations shall not be attached to the building.
(4) Direct vent gas fireplaces shall be installed with the vent air intake terminal in the outdoors and in accordance with the manufacturers' instructions.

### 10.7.3 Combustion and Circulating Air.

Combustion and circulating air shall be provided in accordance with Section 9.3.

Section 9.3, Air for Combustion and Ventilation, was revised in the 2002 edition to delete the terms *confined space* and *unconfined space*. These terms were replaced with *room volume calculations*. Refer to Section 9.3 for a description of the requirements.

## 10.8 Non-Recirculating Direct Gas-Fired Industrial Air Heaters

Non-recirculating direct gas-fired industrial air heaters, which obtain all air for combustion from the outdoors, are covered in Section 10.8. Refer to Section 10.9, Recirculating Direct Gas-Fired Industrial Air Heaters, for installation requirements for gas-fired industrial air heaters, which recirculate building air. Note that these two sections provide the only installation instructions for appliances that are installed in industrial occupancies. These non-recirculating industrial heaters are included here because of their widespread use and confusion as to their installation.

### 10.8.1 Application.

Direct gas-fired industrial air heaters of the non-recirculating type shall be listed in accordance with ANSI Z83.4/CSA 3.7, *Non-Recirculating Direct Gas-Fired Industrial Air Heaters*.

Note that non-recirculating direct gas-fired industrial air heaters must be listed to ANSI Z83.4/CSA 3.7, *Non-Recirculating Direct Gas-Fired Industrial Air Heaters.*

### 10.8.2 Prohibited Installations.

The installation of a direct-fired make-up air heater must be in accordance with its listing and with the general requirements outlined in Section 10.8. These heaters are intended for commercial and industrial applications only and are not intended to heat rooms used for sleeping, as stated in 10.8.2.1.

Note that 10.8.3.2 specifically limits the installation of non-recirculating direct gas-fired industrial air heaters to commercial and industrial occupancies.

**10.8.2.1** Non-recirculating direct gas-fired industrial air heaters shall not serve any area containing sleeping quarters.

**10.8.2.2** Non-recirculating direct gas-fired industrial air heaters shall not recirculate room air.

### 10.8.3 Installation.

**10.8.3.1** Non-recirculating direct gas-fired industrial air heaters shall be installed in accordance with the manufacturer's instructions.

**10.8.3.2** Non-recirculating direct gas-fired industrial air heaters shall be installed only in industrial or commercial occupancies.

**10.8.3.3** Non-recirculating direct gas-fired industrial air heaters shall be permitted to provide fresh air ventilation.

Building ventilation requirements normally are specified in a mechanical code. The requirement in 10.8.3.3 is intended to specifically allow the non-recirculating direct gas-fired industrial air heater to be the source of fresh ventilation air required by mechanical code ventilation requirements.

**10.8.3.4** Non-recirculating direct gas-fired industrial air heaters shall be provided with access for removal of burners; for replacement of motors, controls, filters, and other working parts; and for adjustment and lubrication of parts requiring maintenance.

### 10.8.4 Clearance from Combustible Materials.

Non-recirculating direct gas-fired industrial air heaters shall be installed with a clearance from combustible materials of not less than that shown on the rating plate and the manufacturer's instructions.

### 10.8.5 Air Supply.

All air to the non-recirculating direct gas-fired industrial air heater shall be ducted directly from outdoors. Where outdoor air dampers or closing louvers are used, they shall be verified to be in the open position prior to main burner operation.

### 10.8.6 Atmospheric Vents or Gas Reliefs or Bleeds.

Non-recirculating direct gas-fired industrial air heaters with valve train components equipped with atmospheric vents, gas reliefs, or bleeds shall have their vent lines, gas reliefs, or bleeds lead to a safe point outdoors. Means shall be employed on these lines to prevent water from entering and to prevent blockage from insects and foreign matter. An atmospheric vent line shall not be required to be provided on a valve train component equipped with a listed vent limiter.

### 10.8.7 Relief Openings.

The design of the installation shall include adequate provisions to permit the non-recirculating direct gas-fired industrial air heater to operate at its rated airflow without overpressurizing the space served by the heater by taking into account the structure's designed infiltration rate, properly designed relief openings, or an interlocked powered exhaust system, or a combination of these methods.

**10.8.7.1** The structure's designed infiltration rate and the size of relief opening(s) shall be determined by approved engineering methods.

**10.8.7.2** Louver or counterbalanced gravity damper relief openings shall be permitted. Where motorized dampers or closable louvers are used, they shall be proved to be in their open position prior to main burner operation.

**FAQ ▶** Can a non-recirculating direct gas-fired heater be installed in a factory with existing relief vents?

With direct-fired heaters blowing heated air into a building, positive building pressure can be a problem for both the heater and the building. High building pressure can restrict airflow through the heater, resulting in reduced airflow and improper combustion. High building pressure can also cause other building air-handling systems to operate improperly. Subsection 10.8.7 addresses these concerns by requiring an opening to the outdoors when the heater is operating. Local authorities may require an air balance report to verify that the heater is operating at its rated airflow.

### 10.8.8 Purging.

Inlet ducting, when used, shall be purged with at least four air changes prior to an ignition attempt.

## 10.9 Recirculating Direct Gas-Fired Industrial Air Heaters

Section 10.9 covers recirculating gas-fired industrial air heaters, which are intended for use in industrial and commercial occupancies, such as factories and warehouses. They must not be installed in areas containing sleeping quarters. Subsection 10.9.3 also prohibits their uses in assembly occupancies.

### 10.9.1 Application.

Direct gas-fired industrial air heaters of the recirculating type shall be listed in accordance with ANSI Z83.18, *Recirculating Direct Gas-Fired Industrial Air Heaters*.

Note that 10.9.1 requires that recirculating direct gas-fired industrial air heaters be listed to ANSI Z83.18, *Recirculating Direct Gas-Fired Industrial Air Heaters*.

### 10.9.2 Prohibited Installations.

**10.9.2.1** Recirculating direct gas-fired industrial air heaters shall not serve any area containing sleeping quarters.

**10.9.2.2*** Recirculating direct gas-fired industrial air heaters shall not recirculate room air in buildings that contain flammable solids, liquids, or gases; explosive materials; or substances that can become toxic when exposed to flame or heat.

**A.10.9.2.2** Recirculation of room air can be hazardous in the presence of flammable solids, liquids, gases, explosive materials (e.g., grain dust, coal dust, gun powder), and substances (e.g., refrigerants, aerosols) that can become toxic when exposed to flame or heat.

Paragraph 10.9.2.2 is a practical requirement that prohibits the recirculation of room air in buildings that contain flammable liquids, solids, and gases, as well as substances that can become toxic when exposed to flame. The term *flammable* as applied to liquids in 10.9.2.2 is a broad application and is not limited to the term *flammable liquid* as defined by NFPA 30, *Flammable and Combustible Liquids Code*. NFPA 30 defines *flammable liquid* as a liquid with a flash point below 100°F (37.8°C). The term *flammable liquid* in 10.9.2.2 also includes *Class II combustible liquids,* as defined by NFPA 30, which have a flash point of 100°F to below 140°F (37.8°C to below 60°C). Class II combustible liquids can be heated to above 100°F (37.8°C) by localized room heating processes. Heating above 100°F (37.8°C) results in the generation of flammable vapors. Also, the temperature points selected by NFPA 30 are valid at sea level. If the building is above sea level, flammable vapors will develop at lower temperatures.

### 10.9.3 Installation.

Installation of direct gas-fired industrial air heaters shall comply with the following requirements:

(1) Recirculating direct gas-fired industrial air heaters shall be installed in accordance with the manufacturer's instructions.
(2) Recirculating direct gas-fired industrial air heaters shall be installed only in industrial or commercial occupancies.

Note that the installation prohibitions for recirculating direct gas-fired industrial air heaters are similar to those for non-recirculating direct gas-fired heaters. That is, they must be installed only in industrial and commercial occupancies. Where recirculating direct gas-fired industrial air heaters are installed, all requirements in building and mechanical codes for minimum fresh air and ventilation must be complied with.

### 10.9.4 Clearance from Combustible Materials.

Recirculating direct gas-fired industrial air heaters shall be installed with a clearance from combustible materials of not less than that shown on the rating plate and the manufacturer's instructions.

### 10.9.5 Air Supply.

Ventilation air to the recirculating direct gas-fired industrial air heater shall be ducted directly from outdoors. Air to the recirculating direct gas-fired industrial air heater in excess of the minimum ventilation air specified on the heater's rating plate shall be taken from the building, ducted directly from outdoors, or a combination of both. Where outdoor air dampers or closing louvers are used, they shall be verified to be in the open position prior to main burner operation.

### 10.9.6 Atmospheric Vents, Gas Reliefs, or Bleeds.

Recirculating direct gas-fired industrial air heaters with valve train components equipped with atmospheric vents, gas reliefs, or bleeds shall have their vent lines, gas reliefs, or bleeds lead to a safe point outdoors. Means shall be employed on these lines to prevent water from entering and to prevent blockage from insects and foreign matter. An atmospheric vent line shall not be required to be provided on a valve train component equipped with a listed vent limiter.

### 10.9.7 Relief Openings.

The design of the installation shall include adequate provisions to permit the recirculating direct gas-fired industrial air heater to operate at its rated airflow without overpressurizing the space served by the heater, by taking into account the structure's designed infiltration rate, properly designed relief openings, an interlocked powered exhaust system, or a combination of these methods.

**10.9.7.1** The structure's designed infiltration rate and the size of relief opening(s) shall be determined by approved engineering methods.

**10.9.7.2** Louver or counterbalanced gravity damper relief openings shall be permitted. Where motorized dampers or closable louvers are used, they shall be proved to be in their open position prior to main burner operation.

An exhaust system or properly designed relief openings must be provided to exhaust the air that is brought into an area. Paragraph 10.9.7.2 addresses the buildup of static pressure within the building (area), which could reduce the amount of airflow through the heater and affect its combustion process.

### 10.9.8 Purging.

Inlet ducting, when used, shall be purged with at least four air changes prior to an ignition attempt.

## 10.10 Duct Furnaces

Section 10.10 covers duct furnaces, which are designed to fit into ducts and look like a section of duct. Listed duct furnaces are available for indoor installation only, for outdoor installation only, or for both indoor and outdoor installation. The installation of outdoor furnaces must follow the manufacturer's installation instructions for weather protection, which are not provided in the code.

Certain duct furnaces also can be constructed of corrosion-resistant material and will be listed for installation downstream from a refrigeration coil. Subsection 10.10.6.4 allows duct furnaces to be installed downstream from evaporative coolers or air washers if the heat exchanger is made of corrosion-resistant material.

### 10.10.1 Clearances.

The installation of duct furnaces shall comply with the following clearance requirements:

(1) Listed duct furnaces shall be installed with clearances of at least 6 in. (150 mm) between adjacent walls, ceilings, and floors of combustible material and the furnace draft hood and shall comply with the following:
   (a) Furnaces listed for installation at lesser clearances shall be installed in accordance with the manufacturer's installation instructions.
   (b) In no case shall the clearance be such as to interfere with combustion air and accessibility. *(See 9.2.1 and Section 9.3.)*

(2) Unlisted duct furnaces shall be installed with clearances to combustible material in accordance with the clearances specified for unlisted furnaces and boilers in Table 10.2.3(a). Combustible floors under unlisted duct furnaces shall be protected in an approved manner.

Refer to the commentary following 10.2.4 and 10.3.2 for information on the installation of gas appliances on combustible floors, addressed in 10.10.1(2).

### 10.10.2 Installation of Duct Furnaces.

NEW

Duct furnaces shall be installed in accordance with the manufacturers' instructions.

This requirement was revised editorially in the 2009 edition for clarity.

### 10.10.3 Access Panels.

The ducts connected to duct furnaces shall have removable access panels on both the upstream and downstream sides of the furnace.

### 10.10.4 Location of Draft Hood and Controls.

The controls, combustion air inlet, and draft hoods for duct furnaces shall be located outside the ducts. The draft hood shall be located in the same enclosure from which combustion air is taken.

### 10.10.5 Circulating Air.

Where a duct furnace is installed so that supply ducts carry air circulated by the furnace to areas outside the space containing the furnace, the return air shall also be handled by a duct(s) sealed to the furnace casing and terminating outside the space containing the furnace. The duct furnace shall be installed on the positive-pressure side of the circulating air blower.

The last sentence of 10.10.5 clearly states that all duct furnaces must be installed on the positive-pressure side of the circulation air blower. This is important to ensure separation of the products of combustion and the circulating air. If the duct furnace were installed upstream of the circulating fan, the negative pressure created by the fan could draw some of the products of combustion into the circulating air stream.

### 10.10.6 Duct Furnaces Used with Refrigeration Systems.

**10.10.6.1** A duct furnace shall not be installed in conjunction with a refrigeration coil where circulation of cooled air is provided by the blower.

*Exception: Where the blower has sufficient capacity to overcome the external static resistance imposed by the duct system, furnace, and the cooling coil and the air throughput necessary for heating or cooling, whichever is greater.*

**10.10.6.2** Duct furnaces used in conjunction with cooling appliances shall be installed in parallel with or on the upstream side of cooling coils to avoid condensation within heating elements. With a parallel flow arrangement, the dampers or other means used to control the flow of air shall be sufficiently tight to prevent any circulation of cooled air through the unit.

*Exception: Where the duct furnace has been specifically listed for downstream installation.*

Paragraph 10.10.6.2 addresses a duct furnace installed in a parallel arrangement with a cooling coil. Such a furnace must be installed with tight-fitting face dampers in front of both the duct furnace and the cooling coil to divert the airflow as necessary. Recirculated cold air over the heat exchanger will cause corrosion and premature failure.

**10.10.6.3** Where duct furnaces are to be located upstream from cooling units, the cooling unit shall be so designed or equipped as to not develop excessive temperatures or pressures.

**10.10.6.4** Where a duct furnace is installed downstream of an evaporative cooler or air washer, the heat exchanger shall be constructed of corrosion-resistant materials. Stainless steel, ceramic-coated steel, and an aluminum-coated steel in which the bond between the steel and the aluminum is an iron–aluminum alloy are considered to be corrosion resistant. Air washers operating with chilled water that deliver air below the dew point of the ambient air at the duct furnace are considered as refrigeration systems.

### 10.10.7 Installation in Commercial Garages and Aircraft Hangars.

Duct furnaces installed in garages for more than three motor vehicles or in aircraft hangars shall be of a listed type and shall be installed in accordance with 9.1.11 and 9.1.12.

## 10.11 Floor Furnaces

Section 10.11 covers floor furnaces, which are space heaters that are usually used in residential occupancies and installed below a floor that provides heated air through a grate in the floor. The use of floor furnaces has declined because central heat has become the most common form of heat in residential buildings. These furnaces are still manufactured but are not used as widely as they were in the past.

A floor furnace is a completely self-contained unit, which is suspended from the floor of the space being heated. The unit takes its air for combustion from areas other than the heated space. The typical installation of a floor furnace is illustrated in Exhibit 10.19.

*EXHIBIT 10.19 Installation of a Floor Furnace. (Courtesy of the American Gas Association)*

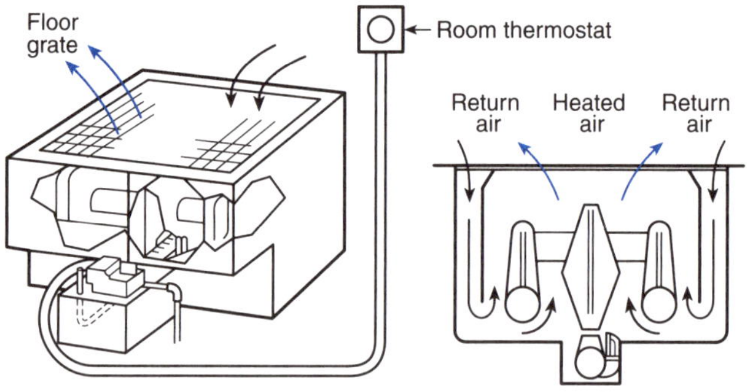


A floor furnace can be equipped with a fan to provide the primary means for circulation of air, or it can be a gravity-type furnace that depends on the circulation of air by convection. A gravity furnace that is equipped with a booster-type fan is considered also to be a gravity flow furnace when the fan does not restrict free circulation of air and when the fan is not in operation.

### 10.11.1 Installation.

The installation of floor furnaces shall comply with the following requirements:

(1) Listed floor furnaces shall be installed in accordance with the manufacturers' installation instructions.
(2) Unlisted floor furnaces shall not be installed in combustible floors.
(3) Thermostats controlling floor furnaces shall not be located in a room or space that can be separated from the room or space in which the register of the floor furnace is located.

### 10.11.2 Temperature Limit Controls.

**10.11.2.1** Listed automatically operated floor furnaces shall be equipped with temperature limit controls.

**10.11.2.2** Unlisted automatically operated floor furnaces shall be equipped with a temperature limit control arranged to shut off the flow of gas to the burner in the event the temperature at the warm air outlet register exceeds 350°F (177°C) above room temperature.

Historically, floor furnaces did not have controls to limit the temperature at the warm air register, as required by 10.11.2.2. The standard for floor furnaces, ANSI Z21.86, *Gas-Fired Gravity Space Heating Equipment*, now requires additional controls to limit register temperatures, minimizing the burn hazard potential for occupants. Therefore, the requirement is included in the code to be sure it is not overlooked where unlisted floor furnaces are installed.

### 10.11.3 Combustion and Circulating Air.

Combustion and circulating air shall be provided in accordance with Section 9.3.

Since the burners and draft hood are generally below the floor, adequate ventilation of the underfloor space to provide air for combustion, draft hood dilution, and ventilation is covered in 10.11.3. Normal construction methods in buildings with crawl spaces usually will provide more than adequate ventilation and combustion air, but it is not uncommon to find such spaces blocked in an effort to maintain warmer floors.

### 10.11.4 Placement.

The following provisions apply to furnaces that serve one story:

(1) *Floors.* Floor furnaces shall not be installed in the floor of any doorway, stairway landing, aisle, or passageway of any enclosure, public or private, or in an exitway from any such room or space.
(2) *Walls and Corners.* The register of a floor furnace with a horizontal warm air outlet shall not be placed closer than 6 in. (150 mm) from the nearest wall. A distance of at least 18 in. (460 mm) from two adjoining sides of the floor furnace register to walls shall be provided to eliminate the necessity of occupants walking over the warm air discharge. The remaining sides shall be a minimum of 6 in. (150 mm) from a wall. Wall register models shall not be placed closer than 6 in. (150 mm) to a corner.
(3) *Draperies.* The furnace shall be placed so that a door, drapery, or similar object cannot be nearer than 12 in. (300 mm) to any portion of the register of the furnace.

### 10.11.5 Bracing.

The space provided for the furnace shall be framed with doubled joists and with headers not lighter than the joists.

Subsection 10.11.5 highlights an important concern to the floor furnace installer. Structural beams and floor joists must not be cut. Double joists and headers must be installed in the opening where floor joists are cut to install the furnace.

◀ **FAQ**
**It is proposed to add a floor furnace to an existing room as part of remodeling. Are there any special structural requirements?**

### 10.11.6 Support.

Means shall be provided to support the furnace when the floor register is removed.

### 10.11.7 Clearance.

The lowest portion of the floor furnace shall have at least a 6 in. (150 mm) clearance from the general ground level. A reduced clearance to a minimum of 2 in. (50 mm) shall be permitted, provided the lower 6 in. (150 mm) portion of the floor furnace is sealed by the manufacturer to prevent entrance of water. Where these clearances are not present, the ground below and to the sides shall be excavated to form a "basin-like" pit under the furnace so that the required clearance is provided beneath the lowest portion of the furnace. A 12 in. (300 mm) clearance shall be provided on all sides except the control side, which shall have an 18 in. (460 mm) clearance.

### 10.11.8 Access.

The space in which any floor furnace is installed shall be accessible by an opening in the foundation not less than 24 in. × 18 in. (610 mm × 460 mm) or by a trapdoor not less than 24 in. × 24 in. (610 mm × 610 mm) in any cross section thereof, and a passageway not less than 24 in. × 18 in. (610 mm × 460 mm) in any cross section thereof.

### 10.11.9 Seepage Pan.

Where the excavation exceeds 12 in. (300 mm) in depth or water seepage is likely to collect, a watertight copper pan, concrete pit, or other suitable material shall be used, unless adequate drainage is provided or the appliance is sealed by the manufacturer to meet this condition. A copper pan shall be made of not less than 16 oz/ft$^2$ (4.9 kg/m$^2$) sheet copper. The pan shall be anchored in place so as to prevent floating, and the walls shall extend at least 4 in. (100 mm) above the ground level with at least a 6 in. (150 mm) clearance on all sides, except on the control side, which shall have at least an 18 in. (460 mm) clearance.

### 10.11.10 Wind Protection.

Floor furnaces shall be protected, where necessary, against severe wind conditions.

### 10.11.11 Upper Floor Installations.

Listed floor furnaces shall be permitted to be installed in an upper floor, provided the furnace assembly projects below into a utility room, closet, garage, or similar nonhabitable space. In such installations, the floor furnace shall be enclosed completely (entirely separated from the nonhabitable space) with means for air intake to meet the provisions of Section 9.3, with access for servicing, minimum furnace clearances of 6 in. (150 mm) to all sides and bottom, and with the enclosure constructed of Portland cement plaster or metal lath or other noncombustible material.

### 10.11.12 First Floor Installation.

Listed floor furnaces installed in the first or ground floors of buildings shall not be required to be enclosed unless the basements of these buildings have been converted to apartments or sleeping quarters, in which case the floor furnace shall be enclosed as specified for upper floor installations and shall project into a nonhabitable space.

For examples of floor-mounted food service equipment, see Exhibit 10.20 through Exhibit 10.23.

## 10.12 Food Service Appliance, Floor-Mounted

### 10.12.1 Clearance for Listed Appliances.

Listed floor-mounted food service appliances, such as ranges for hotels and restaurants, deep fat fryers, unit broilers, kettles, steam cookers, steam generators, and baking and roasting ovens, shall be installed at least 6 in. (150 mm) from combustible material except that at least a 2 in. (50 mm) clearance shall be maintained between a draft hood and combustible material. Floor-mounted food service appliances listed for installation at lesser clearances shall be installed in accordance with the manufacturer's installation instructions. Appliances designed and marked "For use only in noncombustible locations" shall not be installed elsewhere.

**EXHIBIT 10.20** *Gas Convection Oven. (Courtesy of Vulcan & Wolf)*

**EXHIBIT 10.21** *Commercial Gas Range. (Courtesy of Vulcan & Wolf)*

**EXHIBIT 10.23** *Dry Hot Food Table.*

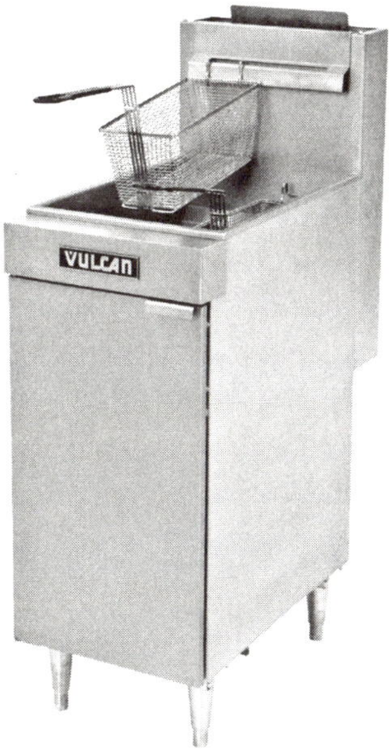


**EXHIBIT 10.22** *Deep Fat Fryer. (Courtesy of Vulcan & Wolf)*

### 10.12.2 Clearance for Unlisted Appliances.

Unlisted floor-mounted food service appliances shall be installed to provide a clearance to combustible material of not less than 18 in. (460 mm) from the sides and rear of the appliance and from the vent connector and not less than 48 in. (1.2 m) above cooking tops and at the front of the appliance. Clearances for unlisted appliances installed in partially enclosed areas such as alcoves shall not be reduced. Reduced clearances for unlisted appliances installed in rooms that are not partially enclosed shall be in accordance with Table 10.2.3(b).

### 10.12.3 Mounting on Combustible Floor.

**10.12.3.1** Listed floor-mounted food service appliances that are listed specifically for installation on floors constructed of combustible material shall be permitted to be mounted on combustible floors in accordance with the manufacturer's installation instructions.

**10.12.3.2** Floor-mounted food service appliances that are not listed for mounting on a combustible floor shall be mounted in accordance with 10.12.4 or be mounted in accordance with one of the following:

(1) Where the appliance is set on legs that provide not less than 18 in. (460 mm) open space under the base of the appliance or where it has no burners and no portion of any oven or broiler within 18 in. (460 mm) of the floor, it shall be permitted to be mounted on a combustible floor without special floor protection, provided at least one sheet metal baffle is between the burner and the floor.
(2) Where the appliance is set on legs that provide not less than 8 in. (200 mm) open space under the base of the appliance, it shall be permitted to be mounted on combustible floors, provided the floor under the appliance is protected with not less than $^3/_8$ in. (9.5 mm) insulating millboard covered with sheet metal not less than 0.0195 in. (0.5 mm) thick. The preceding specified floor protection shall extend not less than 6 in. (150 mm) beyond the appliance on all sides.
(3) Where the appliance is set on legs that provide not less than 4 in. (100 mm) under the base of the appliance, it shall be permitted to be mounted on combustible floors, provided the floor under the appliance is protected with hollow masonry not less than 4 in. (100 mm) in thickness covered with sheet metal not less than 0.0195 in. (0.5 mm) thick. Such masonry courses shall be laid with ends unsealed and joints matched in such a way as to provide for free circulation of air through the masonry.
(4) Where the appliance does not have legs at least 4 in. (100 mm) high, it shall be permitted to be mounted on combustible floors, provided the floor under the appliance is protected by two courses of 4 in. (100 mm) hollow clay tile, or equivalent, with courses laid at right angles and with ends unsealed and joints matched in such a way as to provide for free circulation of air through such masonry courses, and covered with steel plate not less than $^3/_{16}$ in. (4.8 mm) in thickness.

### 10.12.4 Installation on Noncombustible Floor.

**10.12.4.1** Listed floor-installed food service appliances that are designed and marked "For use only in noncombustible locations" shall be installed on floors of noncombustible construction with noncombustible flooring and surface finish and with no combustible material against the underside thereof, or on noncombustible slabs or arches having no combustible material against the underside thereof.

**10.12.4.2** Such construction shall in all cases extend not less than 12 in. (300 mm) beyond the appliance on all sides.

### 10.12.5 Combustible Material Adjacent to Cooking Top.

Any portion of combustible material adjacent to a cooking top section of a food service range, even though listed for close-to-wall installation, that is not shielded from the wall by a high shelf, warming closet, and so on, shall be protected as specified in 10.12.2 for a distance of at least 2 ft (0.6 m) above the surface of the cooking top.

### 10.12.6 Use with Casters.

NEW

Floor-mounted appliances with casters shall be listed for such construction and shall be installed in accordance with the manufacturer's installation instructions for limiting the movement of the appliance to prevent strain on the connection.

The requirement was revised in the 2009 edition to recognize that appliances with casters, which are usually provided to allow the intended movement of the appliance for sanitation. The new text recognizes that the appliance manufacturer includes specific instructions on connecting the appliance to the building piping system and may require a cable or other restraint to protect the connector from overextension.

### 10.12.7 Level Installation.

Floor-mounted food service appliances shall be installed level on a firm foundation.

### 10.12.8* Ventilation.

Means shall be provided to properly ventilate the space in which a food service appliance is installed to permit proper combustion of the gas.

**A.10.12.8** Where exhaust fans are used for ventilation, precautions might be necessary to avoid interference with the operation of the appliance.

In commercial kitchens, compliance with 10.12.8 is normally achieved with vent fans connected to hoods. Vent fans remove the products of combustion, water vapor, grease, and odors generated during cooking. Provision for make-up air, such as wall louvers, must be provided to replace the air removed by these exhaust fans. Kitchen hoods are normally checked for proper air balance to provide optimal removal of the by-products of cooking. A properly balanced hood provides more than sufficient combustion air to an appliance operating in its vicinity.

## 10.13 Food Service Appliances, Counter Appliances

**◀ FAQ**
A hot plate is installed in a kitchen pass-through with a wood-trimmed arch that is 40 in. (1 m) at the top of the arch, above the hot plate. Does this installation meet code?

Subsection 10.13.1 requires 48 in. (1.2 m) above the surface of a hot plate to any combustibles. Therefore, this installation would not meet code requirements.

### 10.13.1 Vertical Clearance.

A vertical distance of not less than 48 in. (1.2 m) shall be provided between the top of all food service hot plates and griddles and combustible material.

### 10.13.2 Clearance for Listed Appliances.

Listed food service counter appliances such as hot plates and griddles, food and dish warmers, and coffee brewers and urns, where installed on combustible surfaces, shall be set on their own bases or legs and shall be installed with a minimum horizontal clearance of 6 in. (150 mm) from combustible material, except that at least a 2 in. (50 mm) clearance shall be maintained between a draft hood and combustible material. Food service counter appliances listed for installation at lesser clearances shall be installed in accordance with the manufacturer's installation instructions.

### 10.13.3 Clearance for Unlisted Appliances.

Unlisted food service hot plates and griddles shall be installed with a horizontal clearance from combustible material of not less than 18 in. (460 mm). Unlisted gas food service counter appliances, including coffee brewers and urns, waffle bakers, and hot water immersion sterilizers, shall be installed with a horizontal clearance from combustible material of not less than 12 in. (300 mm). Reduced clearances for gas food service counter appliances shall be in accordance with Table 10.2.3(b). Unlisted food and dish warmers shall be installed with a horizontal clearance from combustible material of not less than 6 in. (150 mm).

### 10.13.4 Mounting of Unlisted Appliances.

Unlisted food service counter appliances shall not be set on combustible material unless they have legs that provide not less than 4 in. (100 mm) of open space below the burners and the combustible surface is protected with insulating millboard at least $^1/_4$ in. (6 mm) thick covered with sheet metal not less than 0.0122 in. (0.3 mm) thick, or with equivalent protection.

## 10.14 Hot Plates and Laundry Stoves

**10.14.1** Listed domestic hot plates and laundry stoves installed on combustible surfaces shall be set on their own legs or bases. They shall be installed with minimum horizontal clearances of 6 in. (150 mm) from combustible material.

Refer to the commentary following 10.2.4 and 10.3.2 for information on the installation of gas appliances, including hot plates and laundry stoves, on combustible floors.

**10.14.2** Unlisted domestic hot plates and laundry stoves shall be installed with horizontal clearances to combustible material of not less than 12 in. (300 mm). Combustible surfaces under unlisted domestic hot plates and laundry stoves shall be protected in an approved manner.

**10.14.3** The vertical distance between tops of all domestic hot plates and laundry stoves and combustible material shall be at least 30 in. (760 mm).

## 10.15 Household Cooking Appliances

Household cooking appliances are used for domestic food preparation and provide surface cooking, oven cooking, broiling, or a combination of these functions. Broilers can be either enclosed or open, with the radiant heat source above the cooking surface in an enclosed broiler or below the cooking surface in an open broiler. Domestic cooking appliances, such as the range shown in Exhibit 10.24, can be either built-in or floor-mounted.

Refer to the commentary following 10.2.4 and 10.3.2 for information on the installation of gas utilization appliances, including household cooking appliances, on combustible floors.

### 10.15.1 Floor-Mounted Units.

**10.15.1.1 Clearance from Combustible Material.** The clearances specified as follows shall not interfere with combustion air, accessibility for operation, and servicing:

(1) Listed floor-mounted household cooking appliances, where installed on combustible floors, shall be set on their own bases or legs and shall be installed in accordance with the manufacturer's installation instructions.
(2) Listed household cooking appliances with listed gas room heater sections shall be installed so that the warm air discharge side shall have a minimum clearance of 18 in. (460 mm) from adjacent combustible material. A minimum clearance of 36 in. (910 mm) shall be provided between the top of the heater section and the bottom of cabinets.

*EXHIBIT 10.24 Household Cooking Range. (Courtesy of Maytag)*

(3) Listed household cooking appliances that include a solid or liquid fuel-burning section shall be spaced from combustible material and otherwise installed in accordance with the manufacturer's installation instructions for the supplementary fuel section of the appliance.

(4) Unlisted floor-mounted household cooking appliances shall be installed with at least a 6 in. (150 mm) clearance at the back and sides to combustible material. Combustible floors under unlisted appliances shall be protected in an approved manner.

Modern kitchens may use kitchen appliances, such as ranges, that were once confined to commercial kitchens. Care should be given when using one of these, as the design requirements are less stringent than for residential appliances. The required clearances of commercial cooking appliances are normally greater than for residential appliances, some of which may require considerably more room to the back and sides. In addition, surface temperatures may be higher than for residential appliances, which can be a factor if children are expected to be in the home.

**10.15.1.2 Vertical Clearance Above Cooking Top.** Household cooking appliances shall have a vertical clearance above the cooking top of not less than 30 in. (760 mm) to combustible material or metal cabinets. A minimum clearance of 24 in. (610 mm) is permitted when one of the following is installed:

(1) The underside of the combustible material or metal cabinet above the cooking top is protected with not less than $^1/_4$ in. (6 mm) insulating millboard covered with sheet metal not less than 0.0122 in. (0.3 mm) thick.

(2) A metal ventilating hood of sheet metal not less than 0.0122 in. (0.3 mm) thick is installed above the cooking top with a clearance of not less than $^1/_4$ in. (6 mm) between the hood and the underside of the combustible material or metal cabinet, and the hood is at least as wide as the appliance and is centered over the appliance.

(3) A listed cooking appliance or microwave oven is installed over a listed cooking appliance and conforms to the terms of the upper appliance's manufacturer's installation instructions.

**FAQ ▶**
After installing the microwave/fan unit above the stove, there is less than 24 in. (0.6 m) between the fan and the stove. Is this allowed?

Paragraph 10.15.1.2(3) addresses the conflict between the code and manufacturer's installation requirements for separate, listed microwave ovens installed above household cooking appliances. Listed cooking appliances and microwave ovens placed above listed cooking appliances are permitted, but they must be installed according to the terms of the upper appliance listing and the manufacturer's instructions. The instructions must be followed, because heat from the lower cooking appliance can adversely affect the operation of the upper appliance.

**10.15.1.3 Level Installation.** Cooking appliances shall be installed so that the cooking top or oven racks are level.

### 10.15.2 Built-In Units.

See the commentary following 10.15.1.2(3).

**10.15.2.1 Installation.** Listed built-in household cooking appliances shall be installed in accordance with the manufacturer's installation instructions. The installation shall not interfere with combustion air, accessibility for operation, and servicing. Unlisted built-in household cooking appliances shall not be installed in, or adjacent to, combustible material.

**10.15.2.2 Vertical Clearance.** Built-in top (or surface) cooking appliances shall have a vertical clearance above the cooking top of not less than 30 in. (760 mm) to combustible material or metal cabinets. A minimum clearance of 24 in. (610 mm) shall be permitted when one of the following is installed:

(1) The underside of the combustible material or metal cabinet above the cooking top is protected with not less than $^1/_4$ in. (6 mm) insulating millboard covered with sheet metal not less than 0.0122 in. (0.3 mm) thick.
(2) A metal ventilating hood of sheet metal not less than 0.0122 in. (0.3 mm) thick is installed above the cooking top with a clearance of not less than $^1/_4$ in. (6 mm) between the hood and the underside of the combustible material or metal cabinet, and the hood is at least as wide as the appliance and is centered over the appliance.
(3) A listed cooking appliance or microwave oven is installed over a listed cooking appliance and conforms to the terms of the upper appliance manufacturer's installation instructions.

**10.15.2.3 Horizontal Clearance.** The minimum horizontal distance from the center of the burner head(s) of a listed top (or surface) cooking appliance to vertical combustible walls extending above the top panel shall be not less than that distance specified by the permanent marking on the appliance.

**10.15.2.4 Level Installation.** Built-in household cooking appliances shall be installed so that the cooking top, broiler pan, or oven racks are level.

## 10.16 Illuminating Appliances

Section 10.16 covers gas lights, which are rare in buildings today, but they are used for lighting effect and in remote buildings with no electric supply. Gas lights use a very small amount of gas compared to other appliances; therefore, obtaining the required room volume is rarely difficult. Table 5.4.2.1, Approximate Gas Input for Typical Appliances, includes gas lights with an entry of 2500 Btu/hr, the lowest input rate in the table.

The major safety concern around gas lights is clearance to combustibles, which is reflected in the code requirements. Exhibit 10.25 provides examples of gas lights.

### 10.16.1 Clearances for Listed Appliances.

Listed illuminating appliances shall be installed in accordance with the manufacturer's installation instructions.

### 10.16.2 Clearances for Unlisted Appliances.

**10.16.2.1 Enclosed Type.** Clearance shall comply with the following:

(1) Unlisted enclosed illuminating appliances installed outdoors shall be installed with clearances in any direction from combustible material of not less than 12 in. (300 mm).
(2) Unlisted enclosed illuminating appliances installed indoors shall be installed with clearances in any direction from combustible material of not less than 18 in. (460 mm).

**10.16.2.2 Open-Flame Type.** Clearance shall comply with the following:

(1) Unlisted open-flame illuminating appliances installed outdoors shall have clearances from combustible material not less than that specified in Table 10.16.2.2. The distance from ground level to the base of the burner shall be a minimum of 7 ft (2.1 m) where installed within 2 ft (0.6 m) of walkways. Lesser clearances shall be permitted to be used where acceptable to the authority having jurisdiction.
(2) Unlisted open-flame illuminating appliances installed outdoors shall be equipped with a limiting orifice or other limiting devices that maintain a flame height consistent with the clearance from combustible material, as given in Table 10.16.2.2.
(3) Appliances designed for flame heights in excess of 30 in. (760 mm) shall be permitted to be installed if acceptable to the authority having jurisdiction. Such appliances shall be equipped with a safety shutoff device or automatic ignition.
(4) Unlisted open-flame illuminating appliances installed indoors shall have clearances from combustible material acceptable to the authority having jurisdiction.

*TABLE 10.16.2.2 Clearances for Unlisted Outdoor Open-Flame Illuminating Appliances*

| *Flame Height Above Burner Head (in.)* | *Minimum Clearance from Combustible Material (ft)** | |
|---|---|---|
| | *Horizontal* | *Vertical* |
| 12 | 2 | 6 |
| 18 | 3 | 8 |
| 24 | 3 | 10 |
| 30 | 4 | 12 |

For SI units, 1 in. = 25.4 mm, 1 ft = 0.305 m.
*Measured from the nearest portion of the burner head.

### 10.16.3 Mounting on Buildings.

Illuminating appliances designed for wall or ceiling mounting shall be securely attached to substantial structures in such a manner that they are not dependent on the gas piping for support.

### 10.16.4 Mounting on Posts.

Illuminating appliances designed for post mounting shall be securely and rigidly attached to a post. Posts shall be rigidly mounted. The strength and rigidity of posts greater than 3 ft (0.9 m) in height shall be at least equivalent to that of a $2^1/_2$ in. (64 mm) diameter post constructed

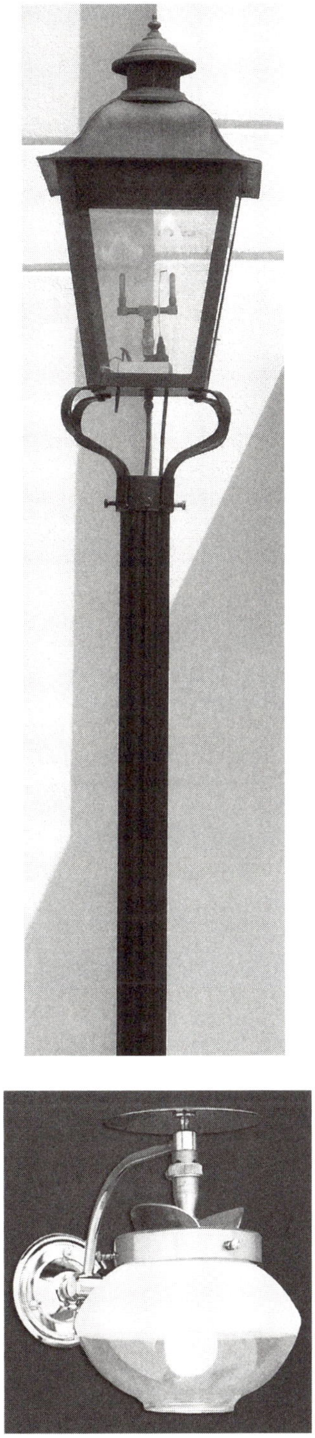

**EXHIBIT 10.25** Gas Lights: (Top) Outdoor Gas Light and (Bottom) Indoor Gas Light. (Outdoor Gas Light Courtesy of Charleston Lighting & Mfg.; Indoor Gas Light Courtesy of Galaxy Gas)

of 0.064 in. (1.6 mm) thick steel or a 1 in. Schedule 40 steel pipe. Posts 3 ft (0.9 m) or less in height shall not be smaller than a $^3/_4$ in. Schedule 40 steel pipe. Drain openings shall be provided near the base of posts where water collecting inside the posts is possible.

### 10.16.5 Appliance Pressure Regulators.

Where an appliance pressure regulator is not supplied with an illuminating appliance and the service line is not equipped with a service pressure regulator, an appliance pressure regulator shall be installed in the line serving one or more illuminating appliances.

## 10.17 Incinerators, Commercial-Industrial

Commercial-industrial-type incinerators shall be constructed and installed in accordance with NFPA 82, *Standard on Incinerators and Waste and Linen Handling Systems and Equipment*.

Section 10.17 covers requirements for commercial and industrial incinerators. In the 2002 edition of the code, former Section 10.18, Domestic Incinerators, was deleted because gas-fueled domestic incinerators were no longer being produced, and the standard for their construction was dropped in 1981. Coverage was continued in the code for many years to allow for reinstalled units. Environmental regulations have essentially banned gas-fueled domestic incinerators.

## 10.18 Infrared Heaters

Section 10.18 covers infrared heaters, which are heaters that direct infrared energy into a specific area. Such heaters can be either vented or unvented. Model building and mechanical codes often have minimum mounting height requirements for these heaters. Exhibit 10.26 shows an infrared heater designed for residential use, and Exhibit 10.27 shows an infrared heater designed for commercial or industrial use.

***EXHIBIT 10.26*** *Infrared Space Heater. (Courtesy of Detroit Radiant Products Company)*

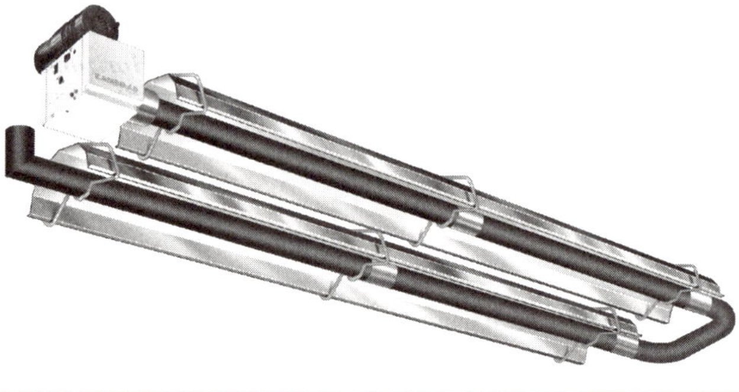

***EXHIBIT 10.27*** *Low Intensity, Tubular Infrared Heater. (Courtesy of AmbiRad Limited)*

### 10.18.1 Support.

Suspended-type infrared heaters shall be fixed in position independent of gas and electric supply lines. Hangers and brackets shall be of noncombustible material. Heaters subject to vibration shall be provided with vibration-isolating hangers.

The requirements in 10.18.1 are for materials to support hanging infrared heaters and include steel pipe, steel channels, or fabricated hangers of at least 16 gauge material. Heater manufacturer's installation instructions provide specific recommendations. Additional bracing to protect against seismic forces is also recommended in geographic areas that are seismically active.

### 10.18.2 Clearance.

The installation of infrared heaters shall meet the following clearance requirements:

(1) Listed heaters shall be installed with clearances from combustible material in accordance the manufacturer's installation instructions.
(2) Unlisted heaters shall be installed in accordance with clearances from combustible material acceptable to the authority having jurisdiction.
(3) In locations used for the storage of combustible materials, signs shall be posted to specify the maximum permissible stacking height to maintain required clearances from the heater to the combustibles.

### 10.18.3 Combustion and Ventilation Air.

**10.18.3.1** Where unvented infrared heaters are used, natural or mechanical means shall be provided to supply and exhaust at least 4 $ft^3$/min/1000 Btu/hr (0.38 $m^3$/min/kW) input of installed heaters.

**10.18.3.2** Exhaust openings for removing flue products shall be above the level of the heaters.

Note that the requirement of 10.18.3 covering air for combustion and removal of the products of combustion from the room in which an infrared heater is installed allows natural or mechanical means of ventilation. Where mechanical ventilation is used, the air requirements are clear and the capacity of mechanical ventilation can easily be determined. Where natural ventilation is used, the paragraph is silent on how to provide 4 cfm/1000 Btu/hr of combustion and ventilation air. The intent of the code is that compliance with Section 9.3, Air for Combustion and Ventilation, will provide the required air.

### 10.18.4 Installation in Commercial Garages and Aircraft Hangars.

Overhead heaters installed in garages for more than three motor vehicles or in aircraft hangars shall be of a listed type and shall be installed in accordance with 9.1.11 and 9.1.12.

## 10.19 Open-Top Broiler Units

Open-top broiler units are similar to outdoor gas grills, except that they are designed for countertop installation indoors in conjunction with a ventilating hood.

### 10.19.1 Listed Units.

Listed open-top broiler units shall be installed in accordance with the manufacturer's installation instructions.

### 10.19.2 Unlisted Units.

Unlisted open-top broiler units shall be installed in accordance with the manufacturers' instructions but shall not be installed in combustible material.

### 10.19.3 Protection Above Domestic Units.

Domestic open-top broiler units shall be provided with a metal ventilating hood not less than 0.0122 in. (0.3 mm) thick with a clearance of not less than $^1/_4$ in. (6 mm) between the hood and the underside of combustible material or metal cabinets. A clearance of at least 24 in. (610 mm) shall be maintained between the cooking top and the combustible material or metal cabinet, and the hood shall be at least as wide as the open-top broiler unit and centered over the unit. Listed domestic open-top broiler units incorporating an integral exhaust system and listed for use without a ventilating hood need not be provided with a ventilating hood if installed in accordance with 10.15.1.2(1).

**FAQ ▶** The open-top broiler being installed has vent openings alongside the grill/griddle that pull the cooking odors downward. Is a hood still required?

Most listed domestic open-top broilers are required to be installed with at least a 24 in. (610 mm) clearance between the cooking top and the ventilating hood. Open-top broiler units that incorporate an integral exhaust system, which usually includes a fan and filter located below the broiler level, are designed to pull the flue gases and cooking vapors downward and discharge them to the outside. These units do not require a hood, but they must maintain a clearance of at least 24 in. (610 mm) above the cooking top.

### 10.19.4 Commercial Units.

Commercial open-top broiler units shall be provided with ventilation in accordance with NFPA 96, *Standard for Ventilation Control and Fire Protection of Commercial Cooking Operations*.

As noted in 10.19.4, NFPA 96, *Standard for Ventilation Control and Fire Protection of Commercial Cooking Operations,* contains detailed requirements for the design and installation of hoods, duct systems, grease-removal devices, auxiliary equipment, and fire-extinguishing equipment, for commercial open-top broiler units.

## 10.20 Outdoor Cooking Appliances

The definition of *outdoor cooking appliance* was revised in the 2006 edition to read:

> **3.3.6.10 Outdoor Cooking Appliance.** A gas-fired cooking appliance for outdoor use only that is provided with a means of support by the manufacturer and is connected to a fixed gas piping system.

Previously the definition was limited to post-mounted appliances. Outdoor cooking appliances are very popular, and types other than post-mounted are being used. Note that the definition is limited to appliances connected to a fixed gas piping system, which can supply any fuel gas appropriate for the appliance. A typical outdoor gas grill is shown in Exhibit 10.28. Outdoor cooking appliances fueled by propane cylinders are covered under NFPA 58, *Liquefied Petroleum Gas Code*.

***EXHIBIT 10.28*** *Outdoor Gas Grill. [Courtesy of Modern Home Products (MHP)]*

### 10.20.1 Listed Units.

Listed outdoor cooking appliances shall be installed in accordance with the manufacturer's installation instructions.

### 10.20.2 Unlisted Units.

Unlisted outdoor cooking appliances shall be installed outdoors with clearances to combustible material of not less than 36 in. (910 mm) at the sides and back and not less than 48 in. (1220 mm) at the front. In no case shall the appliance be located under overhead combustible construction.

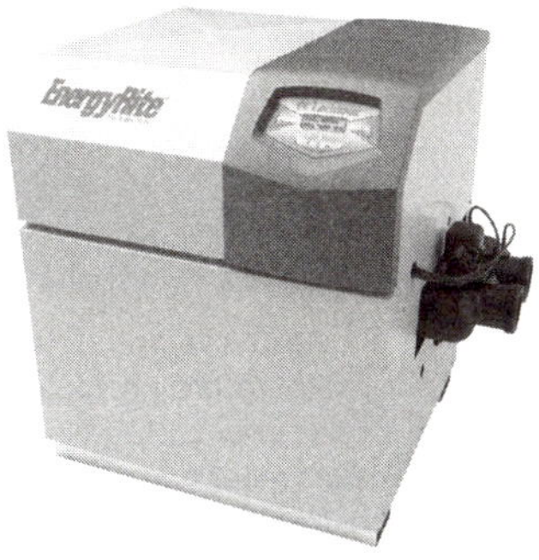


***EXHIBIT 10.29*** *Pool Heater. (Courtesy of Lochinvar Corp.)*

## 10.21 Pool Heaters

A pool heater, such as the one shown in Exhibit 10.29, is an appliance designed for heating non-potable water stored at atmospheric pressure, such as water in swimming pools, therapeutic pools, and similar applications. The unit can be either a coil-type heater, where the heat exchanger consists primarily of water tubes whose inside diameters are less than $1\frac{1}{4}$ in. (3.2 cm), or an indirect-type heater, which uses water in a primary heat exchanger to transmit heat from the gas combustion process to the pool water by means of a secondary heat exchanger.

Pool heaters are listed for indoor installation only, for outdoor installations only, or for either indoor or outdoor installation. Pool heaters installed indoors must be vented and have air for combustion and ventilation that is provided in accordance with Chapter 9, Appliance, Equipment, and Accessory Installation. Pool heaters installed outdoors must have the vent termination and any protective cover installed as required by the manufacturer's instructions.

### 10.21.1 Location.

A pool heater shall be located or protected so as to minimize accidental contact of hot surfaces by persons.

### 10.21.2 Clearance.

The installation of pool heaters shall meet the following requirements:

(1) In no case shall the clearances be such as to interfere with combustion air, draft hood or vent terminal clearance and relief, and accessibility for servicing.

Outdoor installations of pool heaters must meet model code restrictions on vent termination. This can be especially critical on pool heaters near property lines, where minimum distances are often required by these codes.

(2) A listed pool heater shall be installed in accordance with the manufacturer's installation instructions.
(3) An unlisted pool heater shall be installed with a minimum clearance of 12 in. (300 mm) on all sides and the rear. A combustible floor under an unlisted pool heater shall be protected in an approved manner.

The commentary following 10.2.4 and 10.3.2 discusses the installation of pool heaters and similar appliances on a combustible floor.

### 10.21.3 Temperature or Pressure Limiting Devices.

**10.21.3.1** An unlisted pool heater shall be provided with overtemperature protection or overtemperature and overpressure protection by means of an approved device(s).

**10.21.3.2** Where a pool heater is provided with overtemperature protection only and is installed with any device in the discharge line of the heater that can restrict the flow of water from the heater to the pool (such as a check valve, shutoff valve, therapeutic pool valving, or flow nozzles), a pressure relief valve shall be installed either in the heater or between the heater and the restrictive device.

**FAQ ►**
**The pool heater has an integral overtemperature protection shutoff but only a plug where the pressure relief valve should go. Is a separate pressure relief valve required?**

A separate pressure relief valve is required on a pool heater only if there are valves that could isolate the heater from the pool, limiting the ability of the heated water inside the heater to expand. Open systems with no valves do not require relief valves on the heater.

### 10.21.4 Bypass Valves.

Where an integral bypass system is not provided as a part of the pool heater, a bypass line and valve shall be installed between the inlet and outlet piping for use in adjusting the flow of water through the heater.

### 10.21.5 Venting.

A pool heater listed for outdoor installation shall be installed with the venting means supplied by the manufacturer and in accordance with the manufacturer's instructions. *(See 12.3.5, 12.3.6, 12.4.3, and Section 12.9.)*

## 10.22 Refrigerators

Gas refrigerators were used widely in rural areas prior to the expansion of electricity to these locations. They are still used in remote areas that are not served by electricity, in areas where the supply of electricity is interrupted frequently, by people whose religious beliefs do not permit the use of electricity (e.g., the Amish), and in recreational vehicles. The use of these appliances in recreational vehicles is not covered in this code but is covered in ANSI A119.2/NFPA 1192, *Standard on Recreational Vehicles*. Section 10.22 covers gas refrigerators in installations covered by the code.

### 10.22.1 Clearance.

Refrigerators shall be provided with clearances for ventilation at the top and back in accordance with the manufacturers' instructions. Where such instructions are not available, at least 2 in. (50 mm) shall be provided between the back of the refrigerator and the wall and at least 12 in. (300 mm) above the top.

### 10.22.2 Venting or Ventilating Kits Approved for Use with a Refrigerator.

Where an accessory kit is used for conveying air for burner combustion or unit cooling to the refrigerator from areas outside the room in which it is located, or for conveying combustion products diluted with air containing waste heat from the refrigerator to areas outside the room in which it is located, the kit shall be installed in accordance with the refrigerator manufacturer's instructions.

A gas refrigerator generally has a low enough Btu input that it can be installed conveniently in any desired space. Table 5.4.2.1, Approximate Gas Input for Typical Appliances, lists gas refrigerators at 3000 Btu/hr, the second lowest input. Because of this low input, no special provisions are required for air for combustion and ventilation. The clearances required are similar to those required for electric refrigerators.

If the manufacturer provides an accessory kit for venting and ventilation, 10.22.2 requires that it be installed in accordance with the manufacturer's instructions.

## 10.23 Room Heaters

By design, room heaters are either unvented, as shown in Exhibit 10.30, or vented, as shown in Exhibit 10.31. These room heaters can be either the circulating type or the radiant type. Some circulating-type room heaters are equipped with fans to distribute the heat more uniformly in the rooms being heated.

***EXHIBIT 10.30*** *Unvented Room Heater. (Courtesy of DESA™)*

***EXHIBIT 10.31*** *Vented Room Heater. (Courtesy of Louisville Tin & Stove Co.)*

### 10.23.1* Prohibited Installations.

Unvented room heaters shall not be installed in bathrooms or bedrooms.

Subsection 10.23.1 clearly prohibits the installation of unvented room heaters in bedrooms and bathrooms (with two specific exceptions) for the following reasons:

1. Bedrooms and bathrooms are often small spaces that provide inadequate air for combustion and ventilation and thus present a carbon monoxide or an asphyxiation hazard.
2. No matter where an unvented room heater is located, in these small rooms the potential for accidental ignition of flammable materials (e.g., towels, curtains, and clothing) always exists.

See also the commentary following 10.1.2.

*Exception No. 1: Where approved by the authority having jurisdiction, one listed wall-mounted, unvented room heater equipped with an oxygen depletion safety shutoff system shall be permitted to be installed in a bathroom, provided that the input rating does not exceed 6000 Btu/hr (1760 W/hr) and combustion and ventilation air is provided as specified in 10.1.2.*

*Exception No. 2: Where approved by the authority having jurisdiction, one listed wall-mounted unvented room heater equipped with an oxygen depletion safety shutoff system shall be permitted to be installed in a bedroom, provided that the input rating does not exceed 10,000 Btu/hr (2930 W/hr) and combustion and ventilation air is provided as specified in 10.1.2.*

The two exceptions to 10.23.1 permit, with approval of the authority having jurisdiction, the installation of listed, wall-mounted, unvented room heaters — of limited capacity and with an oxygen depletion safety shutoff system — in bedrooms and bathrooms, provided that they comply with 10.1.2.

If a heater of greater capacity than permitted in the exceptions is required, or if the approval of the authority having jurisdiction is not received, heaters of the direct vent type can be used. Direct vent heaters take all of the air for combustion from outside the building and discharge the products of combustion back outside the building.

Also note that 10.1.2 specifies that appliances must be installed so that combustion and ventilation air is not obtained from a bedroom or a bathroom unless the bedroom or bathroom volume complies with 9.3.2.

ANSI Z21.11.2, *Gas Fired Room Heaters — Volume II — Unvented Room Heaters,* requires an oxygen depletion sensor. The sensor is a very simple device that is highly effective. It consists of a temperature sensor located near the base of a pilot flame. If the level of oxygen in the room is depleted (or reduced), the flame will "lift" — or move up from the gas mixture leaving the pilot — to the point where the temperature sensor is not in the flame. This lifting will cool the sensor and stop the flow of gas to the burner, which will shut down the appliance. This sensor is an integral part of the appliance and is usually not field adjustable.

**A.10.23.1** It is recommended that space heating appliances installed in all bedrooms or rooms generally kept closed be of the direct vent type. *(See Section 10.27.)*

NEW **10.23.2** Unvented room heaters shall be listed in accordance with ANSI Z21.11.2, *Gas-Fired Room Heaters-Volume II, Unvented Room Heaters*, and shall be installed in accordance with the manufacturer's installation instructions.

This new requirement provides a specific standard to which the room heater installed in bedrooms and bathrooms must be listed.

### 10.23.3 Installations in Institutions.

Room heaters shall not be installed in the following occupancies:

(1) Residential board and care
(2) Health care

### 10.23.4 Clearance.

A room heater shall be placed so as not to cause a hazard to walls, floors, curtains, furniture, doors when open, and so on, and to the free movements of persons within the room. Heaters designed and marked "For use in noncombustible fireplace only" shall not be installed elsewhere. Listed room heaters shall be installed in accordance with the manufacturer's installation instructions. In no case shall the clearances be such as to interfere with combustion air and accessibility. Unlisted room heaters shall be installed with clearances from combustible material not less than the following:

(1) *Circulating Type.* Room heaters having an outer jacket surrounding the combustion chamber, arranged with openings at top and bottom so that air circulates between the inner and outer jacket, and without openings in the outer jacket to permit direct radiation, shall have clearance at sides and rear of not less than 12 in. (300 mm).
(2) *Radiating Type.* Room heaters other than those of the circulating type described in 10.23.4(1) shall have clearance at sides and rear of not less than 18 in. (460 mm), except that heaters that make use of metal, asbestos, or ceramic material to direct radiation to the front of the heater shall have a clearance of 36 in. (910 mm) in front and, if con-

structed with a double back of metal or ceramic, shall be permitted to be installed with a clearance of 18 in. (460 mm) at sides and 12 in. (300 mm) at rear. Combustible floors under unlisted room heaters shall be protected in an approved manner.

### 10.23.5 Wall-Type Room Heaters.

Wall-type room heaters shall not be installed in or attached to walls of combustible material unless listed for such installation.

## 10.24 Stationary Gas Engines

The installation of gas engines shall conform to NFPA 37, *Standard for the Installation and Use of Stationary Combustion Engines and Gas Turbines*.

Section 10.24 specifies the use of NFPA 37, *Standard for the Installation and Use of Stationary Combustion Engines and Gas Turbines,* which applies to the installation and operation of stationary combustion engines and gas turbines of up to 7500 horsepower per unit. Engines that are used to drive fire pumps and those that are used in essential electrical systems in health care facilities are included in NFPA 37, but additional requirements are found in the standards covering specific installations, including NFPA 20, *Standard for the Installation of Stationary Pumps for Fire Protection,* and NFPA 99, *Standard for Health Care Facilities*.

The installation of fuel gas piping supplying engines is included in NFPA 37, but specific requirements are not included. NFPA 37 references the *National Fuel Gas Code* for piping of up to 125 psi (862 kPa) and ASME B31.3, *Process Piping,* for piping above 125 psi (862 kPa). Additional requirements for installation, engine protective devices, valves, and exhaust piping and chimneys (including clearance to combustibles) are included in NFPA 37.

**10.24.1** Stationary gas engines shall not be rigidly connected to the gas supply piping.

This new requirement recognizes that stationary gas engines vibrate and that rigid piping must not be used to prevent pipe failure. The use of a flexible connector is required.

## 10.25 Gas-Fired Toilets

Section 10.25 covers gas-fired toilets, which are often used at construction sites and in areas that are remote from water supplies (e.g., hunting and skiing cabins and office trailers). Gas-fired toilets also are used in areas without sewer service, in areas where soils will not accept septic systems, and in environmentally sensitive areas. Exhibit 10.32 illustrates a typical gas-fired toilet installation.

### 10.25.1 Clearance.

A listed gas-fired toilet shall be installed in accordance with the manufacturer's installation instructions, provided that the clearance is in any case sufficient to afford ready accessibility for use, cleanout, and necessary servicing.

### 10.25.2 Installation on Combustible Floors.

Listed gas-fired toilets installed on combustible floors shall be listed for such installation.

The title was changed in the 2009 edition to more clearly indicate the requirement in the paragraph.

*EXHIBIT 10.32 Installation of a Gas-Fired Toilet. (Courtesy of Storburn)*

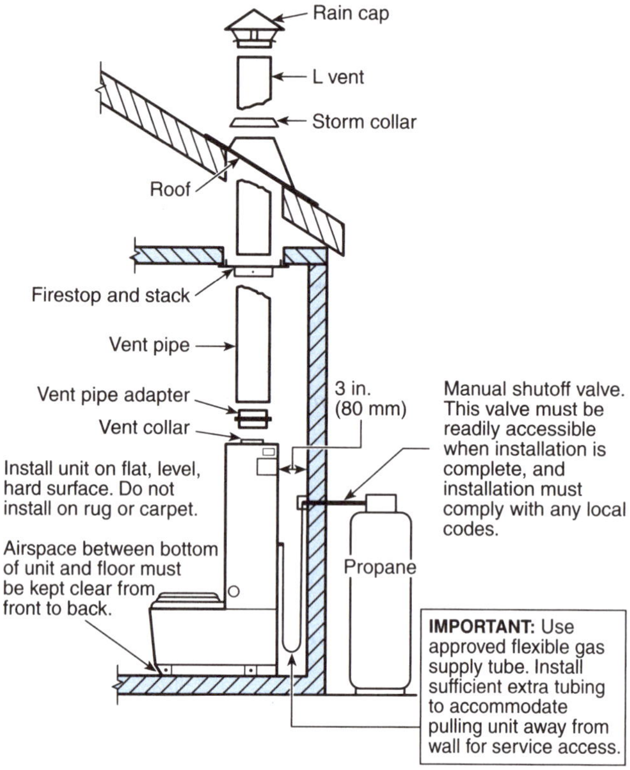


### 10.25.3 Installation.

Vents or vent connectors that are capable of being contacted during casual use of the room in which the toilet is installed shall be protected or shielded to prevent such contact.

**FAQ ▶** A gas-fired toilet has a vent collar that connects it to the Type B vent that continues through a roof. The vent starts at approximately 48 in. (1.2 m) off the floor. Is anything more required for the venting system?

Subsection 10.25.3 requires protection for any part of the venting system that may be contacted during use. Because 48 in. (1.2 m) is well within the zone that may see such contact, protection must be installed for the Type B vent to prevent contact with the vent up to a height that people can be expected to come in contact with.

## 10.26 Unit Heaters

Section 10.26 addresses unit heaters. Exhibit 10.33 illustrates a self-contained, automatically controlled, vented, fuel-gas-burning appliance that has an integral means for the circulation of air.

A high-static-pressure unit heater can deliver heated air at a pressure of 0.2 in. w.c. (50 Pa) or greater static pressure. The unit heater may be equipped with provisions for attaching an outlet air duct, and, if designed for indoor installation in a place remote from the space to be heated, it also will be equipped with provisions for attaching an inlet air duct. A low-static-pressure unit heater usually circulates air by means of an integral propeller fan, and it can be equipped with louvers or face extensions on the discharge side, in accordance with the manufacturer's specifications. A low-static-pressure unit heater is intended for installation in the space to be heated and must not be used with circulating air ducts.

*EXHIBIT 10.33 Unit Heater. (Courtesy of Reznor/Thomas & Betts)*

### 10.26.1 Support.

Suspended-type unit heaters shall be safely and adequately supported, with due consideration given to their weight and vibration characteristics. Hangers and brackets shall be of noncombustible material.

Manufacturers generally specify suspension methods for unit heaters, depending on the weight of the heater. Subsection 10.26.1 recognizes this need and is written in performance language. Normally, steel pipe and the use of pipe couplings or bushings will provide satisfactory support and will enable the unit to be removed for service. Additional bracing to protect against seismic forces is also recommended in geographic areas that are seismically active.

### 10.26.2 Clearance.

The requirements for installation clearance are covered in 10.26.2. Model mechanical codes often establish minimum installation height and vent termination requirements for suspended unit heaters. Unit heaters often vent out the back, so a unit near an outside wall facing into the space may not have sufficient space to correctly vent the heater if there are parapet walls above the roof line. (See 12.7.2, Gas Vent Termination.)

**10.26.2.1 Suspended-Type Unit Heaters.** Suspended-type unit heaters shall meet the following requirements:

(1) A listed unit heater shall be installed with clearances from combustible material of not less than 18 in. (460 mm) at the sides, 12 in. (300 mm) at the bottom, and 6 in. (150 mm) above the top where the unit heater has an internal draft hood, or 1 in. (25 mm) above the top of the sloping side of a vertical draft hood. A unit heater listed for reduced clearances shall be installed in accordance with the manufacturer's installation instructions.
(2) Unlisted unit heaters shall be installed with clearances to combustible material of not less than 18 in. (460 mm).
(3) Clearances for servicing shall be in accordance with the manufacturers' recommendations contained in the installation instructions.

**10.26.2.2 Floor-Mounted-Type Unit Heaters.** Floor-mounted-type unit heaters shall meet the following requirements:

(1) A listed unit heater shall be installed with clearances from combustible material at the back and one side only of not less than 6 in. (150 mm). Where the flue gases are vented horizontally, the 6 in. (150 mm) clearance shall be measured from the draft hood or vent instead of the rear wall of the unit heater. A unit heater listed for reduced clearances shall be installed in accordance with the manufacturer's installation instructions.
(2) Floor-mounted-type unit heaters installed on combustible floors shall be listed for such installation.
(3) Combustible floors under unlisted floor-mounted unit heaters shall be protected in an approved manner.
(4) Clearances for servicing shall be in accordance with the manufacturers' recommendations contained in the installation instructions.

### 10.26.3 Combustion and Circulating Air.

Combustion and circulating air shall be provided in accordance with Section 9.3.

Normally, unit heaters are installed in large commercial or industrial areas for spot heating. Thus, adequate air for combustion and ventilation is usually available. Unit heaters, and all other vented gas appliances, require sufficient air for complete combustion of gas. As noted in 10.26.3, refer to Section 9.3 for the requirements for the required volume to provide the needed air for combustion and ventilation.

### 10.26.4 Ductwork.

A unit heater shall not be attached to a warm air duct system unless listed and marked for such installation.

### 10.26.5 Installation in Commercial Garages and Aircraft Hangars.

Unit heaters installed in garages for more than three motor vehicles or in aircraft hangars shall be of a listed type and shall be installed in accordance with 9.1.11 and 9.1.12.

## 10.27 Wall Furnaces

Wall furnaces, which are covered in Section 10.27, are self-contained vented appliances, complete with grilles or their equivalent and designed for incorporation in or permanent attachment to the structure of a building. Wall furnaces furnish radiant heat or heated air, circulated by gravity or by a fan, directly into the space to be heated through openings in the casing. Exhibit 10.34 shows an example of a typical direct vent wall furnace. Exhibit 10.35 illustrates air circulation in a direct vent wall furnace.

Some wall furnaces are listed with accessory duct extensions, or boots, which do not extend more than 10 in. (250 mm) beyond the appliance casing, for extension through walls of nominal thickness. The manufacturer must supply these boots as an integral part of the appliance.

A direct vent wall furnace is a system composed of an appliance, combustion air and flue gas connections between the appliance and outdoor atmosphere, and a vent cap supplied by the manufacturer. It is constructed so that all air for combustion is obtained from outdoors and all flue gases are discharged to the outdoors. Wall furnaces do not include floor furnaces, unit heaters, or central furnaces. A vented wall furnace installation is illustrated in Exhibit 10.36.

*EXHIBIT 10.34 Direct Vent Wall Furnace. (Courtesy of Empire Comfort Systems)*

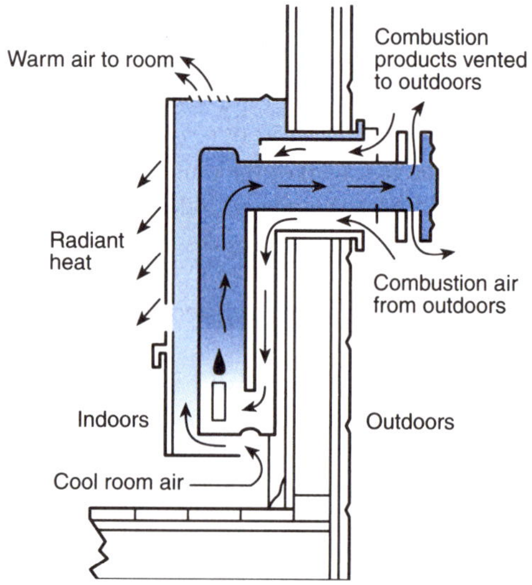


*EXHIBIT 10.35 Air Circulation in a Direct Vent Wall Furnace.*

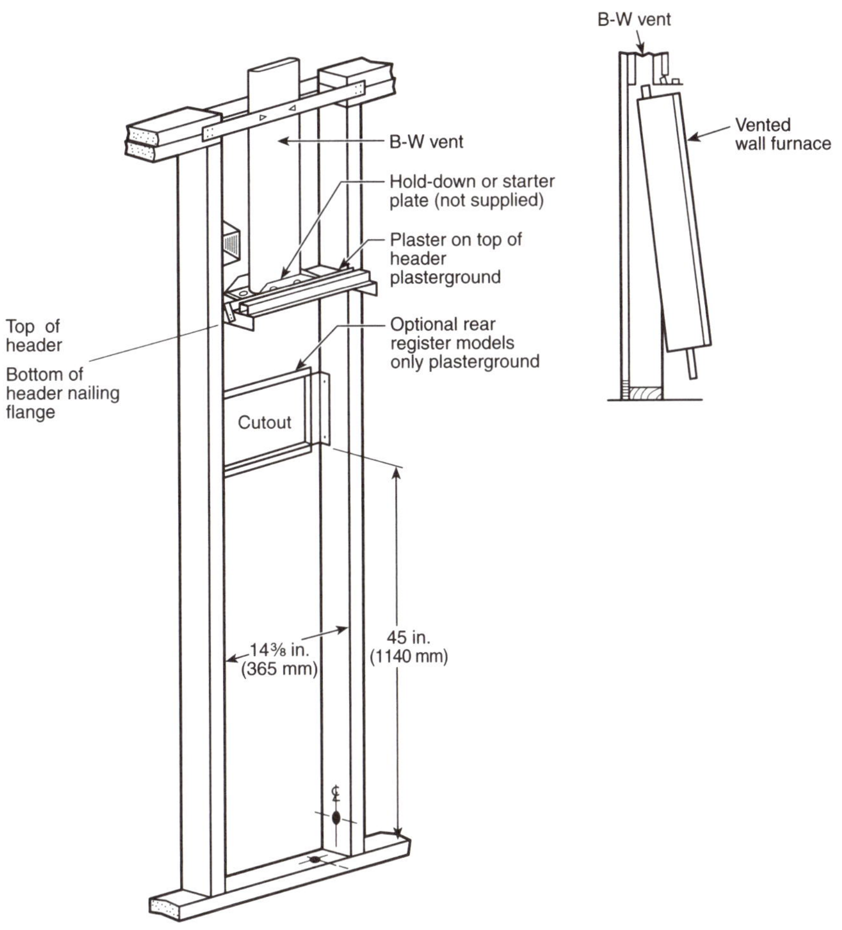


***EXHIBIT 10.36*** *Typical Installation of a Vented Wall Furnace.*

## 10.27.1 Installation.

**10.27.1.1** Listed wall furnaces shall be installed in accordance with the manufacturer's installation instructions. Wall furnaces installed in or attached to combustible material shall be listed for such installation.

**10.27.1.2** Unlisted wall furnaces shall not be installed in or attached to combustible material.

**10.27.1.3** Vented wall furnaces connected to a Type B-W gas vent system listed only for a single story shall be installed only in single-story buildings or the top story of multistory buildings. Vented wall furnaces connected to a Type B-W gas vent system listed for installation in multistory buildings shall be permitted to be installed in single-story or multistory buildings. Type B-W gas vents shall be attached directly to a solid header plate that serves as a firestop at that point and that shall be permitted to be an integral part of the vented wall furnace. The stud space in which the vented wall furnace is installed shall be ventilated at the first ceiling level by installation of the ceiling plate spacers furnished with the gas vent. Firestop spacers shall be installed at each subsequent ceiling or floor level penetrated by the vent. *(See Figure 10.27.1.3 for Type B-W gas vent installation.)*

**The wall furnace is designed for an in-the-wall installation. A Type B gas vent does not fit the wall inner dimensions. What should be installed?**

**◀ FAQ**

Type B-W gas vent systems are designed for use with wall furnaces and are installed in the wall cavity. Special header plates and spacers are provided as an integral part of the venting

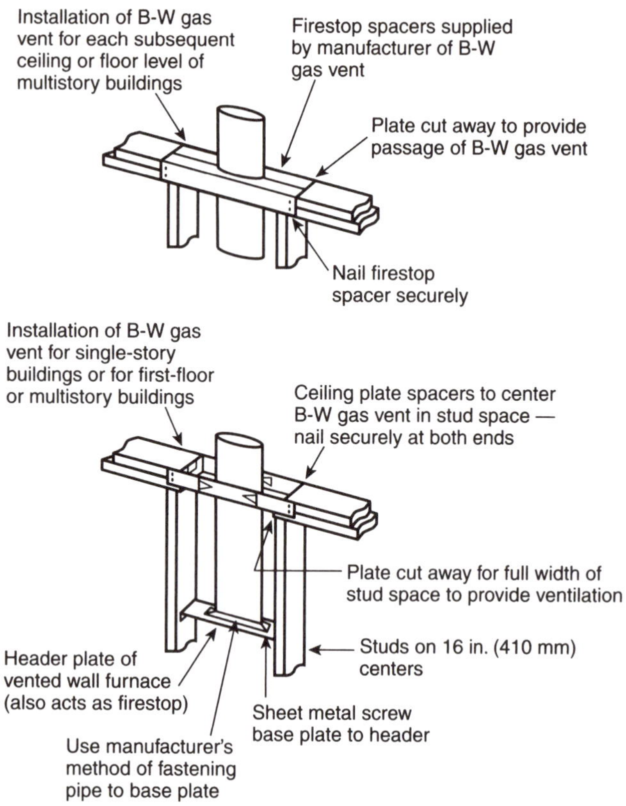


***FIGURE 10.27.1.3*** *Installation of Type B-W Gas Vents for Vented Wall Furnaces.*

system. A Type B-W gas vent is oval shaped and needs the fittings sold as part of the system for correct installation. All Type B and Type B-W gas vent systems are listed products and must be installed per the manufacturer's installation instructions.

**10.27.1.4** Direct vent wall furnaces shall be installed with the vent air intake terminal in the outdoors. The thickness of the walls on which the furnace is mounted shall be within the range of wall thickness marked on the furnace and covered in the manufacturers' installation instructions.

**10.27.1.5** Panels, grilles, and access doors that are required to be removed for normal servicing operations shall not be attached to the building. *(For additional information on the venting of wall furnaces, see Chapter 12.)*

### 10.27.2 Location.

Wall furnaces shall be located so as not to cause a hazard to walls, floors, curtains, furniture, or doors. Wall furnaces installed between bathrooms and adjoining rooms shall not circulate air from bathrooms to other parts of the building.

### 10.27.3 Combustion and Circulating Air.

Combustion and circulating air shall be provided in accordance with Section 9.3.

## 10.28 Water Heaters

Section 10.28 covers water heaters, which are appliances that supply hot water for domestic or commercial purposes. Exhibit 10.37 shows an example of a typical water heater. A direct vent water heater is constructed and installed so that all air for combustion is obtained directly from the outside atmosphere and all flue gases are discharged to the outside atmosphere.

### 10.28.1 Location.

Water heater installations in bedrooms and bathrooms shall comply with one of the following:

◀ **FAQ**
**Can a replacement water heater in manufactured housing be installed using the requirements of NFPA 54?**

Section 10.30 addresses the installation and replacement of gas-fired appliances in manufactured housing after the initial sale. Federal law, using regulations written by the U.S. Department of Housing and Urban Development (HUD), covers the construction of new manufactured housing in the United States. NFPA 501, *Standard on Manufactured Housing,* is also available.

Small water heaters [e.g., 75,000 Btu/hr (22 kW) and less] are listed in accordance with ANSI Z21.10.1, *Gas Water Heaters — Volume I — Storage Water Heaters with Input Ratings of 75,000 Btu per Hour or Less,* which contains the requirements and tests for all storage water heaters of this size. Additional requirements are included in ANSI Z21.10.1 for water heaters used in manufactured housing. These requirements include provisions for securing the water heater to the vehicle structure, a requirement that the water heater be convertible for use with natural gas and propane, special marking and installation requirements, and other construction requirements.

Certification that the water heater is suitable for installation in manufactured housing means that the specific requirements for water heater construction in ANSI Z21.10.1 have been met. Both atmospheric vent and direct vent water heaters can be listed for installation in manufactured housing. New manufactured housing is covered by standards of the federal government in the United States. However, replacement of appliances is covered by the codes in effect at the site. The requirements of the *National Fuel Gas Code* now specifically apply to the installation of water heaters in manufactured housing (after consumer sale), and must be followed.

The decision to replace a direct vent water heater in manufactured housing with a non-direct vent type should be made carefully. Compliance with the provisions of Sections 9.3 and 10.30 is necessary. The replacement appliance must be listed or approved (see the definitions in Section 3.2), as required by Section 10.30. The proper clearance to combustibles must also be provided.

In addition to the water heater installation requirements in Section 10.28, the installer should be aware that requirements that prevent the installation of water heaters in certain locations where flammable vapors are or can be present are stated in Section 9.1. (See 9.1.10 through 9.1.12.) The edition of ANSI Z21.10.1 in effect at the time of publication of this code requires that all storage-type water heaters listed under the standard pass a flammable vapor ignition resistance test. Power vent and direct vent–type water heaters will be required to pass the flammable vapor ignition test in the future. (Refer to the commentary following 9.1.9 and 9.1.10.1 for more information.)

***EXHIBIT 10.37*** *Water Heater. (Courtesy of Bradford White)*

NEW

(1) Water heater shall be installed in a closet equipped with a weather-stripped door with no openings and with a self-closing device. All combustion air shall be obtained from the outdoors in accordance with 9.3.3.

Refer to commentary following 10.3.1(2) for an explanation of this revision in the 2009 edition.

**FAQ ▶**
Can an ordinary gas water heater be installed in a bedroom closet?

Paragraph 10.28.1(1) permits, with specific requirements, the installation of water heaters in special equipment rooms or closets opening into bedrooms and bathrooms. The closet must be used only for a water heater, it must have a weather-stripped door with an automatic self-closing device, and all combustion air must be taken from outdoors, with openings sized per 9.3.3.

(2) Water heater shall be of the direct vent type.

Direct vent water heaters, addressed in 10.28.1(2), are vented the same way as other direct vent gas appliances. For specific information, see Chapter 12, Venting of Appliances. Direct vent water heaters installed in confined spaces are not required to have air for combustion as specified in Section 9.3 of this code, but they do need ventilation air to keep the confined space from overheating.

### 10.28.2 Clearance.

**10.28.2.1** The clearances shall not be such as to interfere with combustion air, draft hood clearance and relief, and accessibility for servicing. Listed water heaters shall be installed in accordance with the manufacturer's installation instructions.

**10.28.2.2** Unlisted water heaters shall be installed with a clearance of 12 in. (300 mm) on all sides and rear. Combustible floors under unlisted water heaters shall be protected in an approved manner.

Refer to commentary following 10.2.4 and 10.3.2 for information on the installation of water heaters and similar appliances on combustible floors.

### 10.28.3 Pressure Limiting Devices.

A water heater installation shall be provided with overpressure protection by means of an approved, listed device installed in accordance with the manufacturer's installation instructions. The pressure setting of the device shall exceed the water service pressure and shall not exceed the maximum pressure rating of the water heater.

Water heaters are required to have overpressure protection in order to prevent internal overpressuring and failure of the heater. This protection is accomplished by installing a pressure relief valve on the water heater with a nonadjustable setting of 75 psi to 150 psi (520 kPa to 1030 kPa), with 125 psi (862 kPa) being a commonly used pressure. Water heaters usually are designed and constructed with a 150 psi (1030 kPa) maximum operating pressure. The pressure relief valve setting must be greater than the water service pressure, to prevent unnecessary operation of the valve and to prevent water leakage. Most municipal water supplies have pressures of 80 psi (560 kPa) or less. In most cases, the overpressure protection device is provided and installed in accordance with the water heater manufacturer's installation instructions.

The pressure relief valve frequently is combined with a temperature relief valve that has a nonadjustable setting of 200°F to 210°F (93°C to 99°C). A temperature and pressure relief valve is shown in Exhibit 10.38, which shows the extended temperature probe passing through the inlet of the relief valve. The valve must be installed so that the temperature probe is immersed in water in the top 6 in. (150 mm) of a storage-type water heater.

*EXHIBIT 10.38 Temperature and Pressure Relief Valve. (Courtesy of Watts Regulator Company)*

### 10.28.4 Temperature Limiting Devices.

A water heater installation or a hot water storage vessel installation shall be provided with overtemperature protection by means of an approved, listed device installed in accordance with the manufacturer's installation instructions.

Water heaters are required to have overtemperature protection, which usually is accomplished by an energy cutoff (ECO) device — an independent, high-temperature switch that shuts off the gas valve. Modern units incorporate a switch, built into the thermostat, which can be a replaceable "fusible link" that cannot be reset. This fusible link is replaceable by a service technician. In the case of ECO operation, this fusible link could require the replacement of the thermostat. Many water heater manufacturers require that a combination temperature and pressure relief valve also be installed. (See the commentary following 10.28.3.)

The ECO device, such as the one shown in Exhibit 10.39, is a circular disc of two different metals that are bonded, similar to two coins bonded together. The two metals have different thermal expansion rates. The disc is in the shape of a shallow cup. When the temperature exceeds the predetermined high temperature, the ECO expands sufficiently so that the shape changes by moving the indentation to the other side. The change shuts off the gas flow. The ECO may not revert to its original shape when cooled, in which case it must be replaced for the water heater to operate.

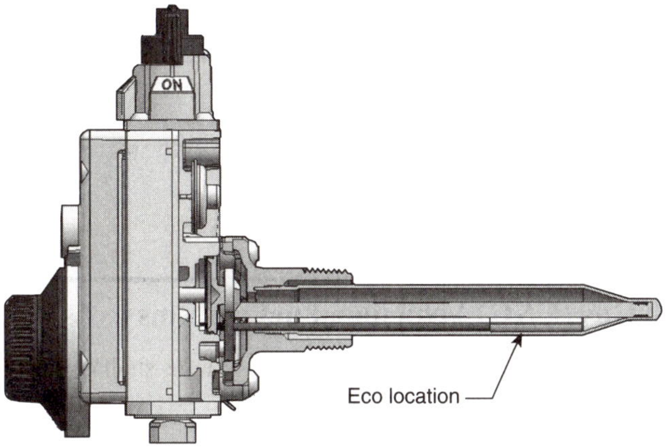


Note: The active trigger mechanism of the thermal cutoff is an electrically non-conductive pellet. Under normal operating temperatures, the solid pellet holds spring-loaded contacts closed. When a predetermined temperature is reached, the pellet melts, allowing the compression spring to relax. The trip spring then slides the contact away from the lead, and the circuit is opened.

***EXHIBIT 10.39*** *Typical ECO Device. (Courtesy of White-Rodgers/Emerson Climate Technologies)*

The temperature limit of the T&P valve is set well below the boiling point of water. When this temperature is reached, the valve releases some water from the heater, allowing cool water to enter the heater. This cool water prevents superheated water from being delivered to the fixture, where it could flash to steam, injuring the user. This process does not control the heater control valve, so water may be released continually until the control valve is adjusted or repaired. The devices described in the preceding commentary paragraphs directly control a malfunctioning water heater gas valve, preventing a "runaway heater."

◄ **FAQ**
Does the temperature and pressure relief valve (T&P valve) serve as an overtemperature protection device?

## 10.28.5 Temperature, Pressure, and Vacuum Relief Devices.

Temperature, pressure, and vacuum relief devices or combinations thereof, and automatic gas shutoff devices, shall be installed in accordance with the manufacturer's installation instructions. A shutoff valve shall not be placed between the relief valve and the water heater or on discharge pipes between such valves and the atmosphere. The hourly Btu discharge capacity or the rated steam relief capacity of the device shall not be less than the input rating of the water heater.

Subsections 10.28.3 through 10.28.5 provide requirements for the installation of safety devices on water heaters. These devices must be installed in accordance with the manufacturer's instructions provided with the devices. Shutoff valves are prohibited in piping to or from relief valves to prevent their being compromised accidentally or intentionally. The discharge line (pipe) from such devices must be full-sized and not be reduced for its full length. Also, the end of the pipe should not be threaded. The lack of threads prevents the installation of a cap on the pipe.

### 10.28.6 Automatic Instantaneous Type: Cold Water Supply.

The water supply to an automatic instantaneous water heater that is equipped with a water flow–actuated control shall be such as to provide sufficient pressure to properly operate the control when water is drawn from the highest faucet served by the heater.

### 10.28.7* Antisiphon Devices.

Means acceptable to the authority having jurisdiction shall be provided to prevent siphoning in any water heater or any tank to which a circulating water heater that incorporates a cold water inlet tube is attached.

**A.10.28.7** A hole near the top of a cold water inlet tube that enters the top of the water heater or tank is commonly accepted for this purpose.

## 10.29 Compressed Natural Gas (CNG) Vehicular Fuel Systems

The installation of compressed natural gas (CNG) fueling (dispensing) systems shall conform to NFPA 52, *Vehicular Fuel Systems Code*.

NFPA 52, *Vehicular Fuel Systems Code*, was revised in 2006 to incorporate liquefied natural gas and hydrogen fuel systems. The reference to NFPA 52 in Section 10.29 refers only to compressed natural gas vehicle fuel systems. NFPA 30A, *Code for Motor Fuel Dispensing Facilities and Repair Garages,* covers dispensing and storage of compressed natural gas at service stations that dispense multiple fuels, and should also be reviewed.

## 10.30 Appliances for Installation in Manufactured Housing

Appliances installed in manufactured housing after the initial sale shall be listed for installation in manufactured housing, or approved, and shall be installed in accordance with the requirements of this code and the manufacturers' installation instructions. Appliances installed in the living space of manufactured housing shall be in accordance with the requirements of Section 9.3.

Section 10.30 provides coverage in the *National Fuel Gas Code* for installation and replacement of appliances in manufactured housing after the initial sale. In the United States, the production of manufactured housing is regulated by the U.S. Department of Housing and Urban Development (HUD). HUD standards governing the production of manufactured housing apply only to the manufacturer and to dealers prior to the first consumer sale. After the first consumer sale, the HUD standard does not apply, and for many years no other codes or standards covered installation and replacement of gas appliances in manufactured housing. The committee recognized this area was lacking, and after reviewing the existing code requirements, the committee determined that the *National Fuel Gas Code* can be used for the safe installation of gas appliances in manufactured housing after the initial consumer sale.

## 10.31 Fuel Cell Power Plants

Fuel cell power plants with a power output of less than 50 kW shall be listed and installed in accordance with the manufacturer's instructions. Fuel cell power plants with a power output of greater than 50 kW shall be installed in accordance with NFPA 853, *Standard for the Installation of Stationary Fuel Cell Power Systems*.

Section 10.31 was added in the 2002 edition to provide requirements for fuel cells, which produce electricity directly from fuel gas. Fuel cells are not new, having been used in satellites for many years. The cost of fuel cells has come down to the point where they are closer to becoming economically viable, especially in areas with high electricity costs.

Fuel cell units must meet the requirements of NFPA 853, *Standard for the Installation of Stationary Fuel Cell Power Systems.*

## 10.32 Outdoor Open Flame Decorative Appliances

Permanently fixed in place outdoor open flame decorative appliances shall be installed in accordance with 10.32.1 through 10.32.3.

### 10.32.1 Listed Units.

Listed outdoor open flame decorative appliances shall be installed in accordance with the manufacturer's installation instructions.

### 10.32.2 Unlisted Units.

Unlisted outdoor open flame decorative appliances shall be installed outdoors in accordance with the manufacturer's installation instructions and with clearances to combustible material of not less than 36 in. (910 mm) from the sides. In no case shall the appliance be located under overhead combustible construction.

### 10.32.3 Connection to Piping System.

The connection to the gas piping system shall be in accordance with 9.6.1(1), (2), (4), or (5).

Section 10.32.3 was added in the 2009 edition to provide minimum code coverage for permanently fixed outdoor open flame decorative appliances. These appliances may be used for decorative lighting and for heat. The major concern is clearance to combustibles to prevent these appliances from causing building fires.

## REFERENCES CITED IN COMMENTARY

The following publications are available from the National Fire Protection Association, 1 Batterymarch Park, Quincy, MA 02169-7471, 1-800-344-3555, www.nfpa.org.

NFPA 20, *Standard for the Installation of Stationary Pumps for Fire Protection,* 2007 edition.
NFPA 30, *Flammable and Combustible Liquids Code*, 2008 edition.
NFPA 30A, *Code for Motor Fuel Dispensing Facilities and Repair Garages,* 2008 edition.
NFPA 31, *Standard for the Installation of Oil-Burning Equipment,* 2006 edition.
NFPA 37, *Standard for the Installation and Use of Stationary Combustion Engines and Gas Turbines,* 2006 edition.
NFPA 52, *Vehicular Fuel Systems Code,* 2006 edition.
NFPA 58, *Liquefied Petroleum Gas Code,* 2008 edition.
*NFPA 70®, National Electrical Code®*, 2008 edition.

NFPA 90A, *Standard for the Installation of Air-Conditioning and Ventilating Systems,* 2009 edition.

NFPA 90B, *Standard for the Installation of Warm Air Heating and Air-Conditioning Systems,* 2009 edition.

NFPA 96, *Standard for Ventilation Control and Fire Protection of Commercial Cooking Operations,* 2008 edition.

NFPA 99, *Standard for Health Care Facilities,* 2005 edition.

NFPA 211, *Standard for Chimneys, Fireplaces, Vents, and Solid Fuel-Burning Appliances,* 2006 edition.

NFPA 501, *Standard on Manufactured Housing,* 2005 edition.

NFPA 853, *Standard for the Installation of Stationary Fuel Cell Power Systems,* 2007 edition.

ANSI A119.2/NFPA 1192, *Standard on Recreational Vehicles,* 2008 edition.

John R. Hall, Jr., *Home Cooking Fire Patterns and Trends,* NFPA Fire Analysis and Research Division, Quincy, MA, Jan. 2005.

John R. Hall, Jr., *Home Heating Fire Patterns and Trends,* NFPA Fire Analysis and Research Division, Quincy, MA, Nov. 2004.

The following publication is available from Engineering and Safety Service (ISO), 545 Washington Blvd., Jersey City, NJ 07310, 1-800-888-4476, www.iso.com.

*National Building Code,* 1976 edition.

The following publications are available from CSA America, Inc., 8501 East Pleasant Valley Road, Cleveland, OH 44131-5575, 216-524-4990, www.csa-america.org.

ANSI Z21.5.1/CSA 7.1, *Gas Clothes Dryers — Volume I, Type 1 Clothes Dryers,* 2006.

ANSI Z21.10.1, *Gas Water Heaters — Volume I — Storage Water Heaters with Input Ratings of 75,000 Btu per Hour or Less,* 2006.

ANSI Z21.11.2, *Gas-Fired Room Heaters — Volume II — Unvented Room Heaters,* 2008.

ANSI Z21.47, *Gas-Fired Central Furnaces,* 2007.

ANSI Z21.50/CSA 2.22, *Vented Gas Fireplaces,* 2008.

ANSI Z21.60/CSA 2.26, *Decorative Gas Appliances for Installation in Solid-Fuel Burning Fireplaces,* 2004.

ANSI Z21.86, *Gas-Fired Gravity Space Heating Equipment,* 2008.

ANSI Z83.4/CSA 3.7, *Non-Recirculating Direct Gas-Fired Industrial Air Heaters,* 2004.

ANSI Z83.18, *Recirculating Direct Gas-Fired Industrial Air Heaters,* 2005.

The following publication is available from McGraw-Hill, Inc., 1221 Avenue of the Americas, New York, NY 10020, 212-512-2000, www.mcgraw-hill.com.

Perry, Robert H., and Cecil H. Chilton. *Chemical Engineers' Handbook,* fifth edition, 1973.

The following publication is available from the American Society of Mechanical Engineers, Three Park Avenue, New York, NY 10016-5990, 1-800-843-2763, www.asme.org.

ASME B31.3, *Process Piping,* 2006.

The following publication is available from the Sheet Metal and Air Conditioning Contractors' National Association, 4201 Lafayette Center Drive, Chantilly, VA 20151-1209, 703-803-2980, www.smacna.org.

*HVAC Duct Construction Standards — Metal and Flexible,* 2005.

The following publications are available from Underwriters Laboratories Inc., 333 Pfingsten Road, Northbrook, IL 60062-2096, 847-272-8800, www.ul.com.

UL 181, *Standard for Factory-Made Air Ducts and Air Connectors,* 2005.

UL 263, *Standard for Fire Tests of Building Construction and Materials,* 2003.

# Procedures to Be Followed to Place Appliance in Operation

CHAPTER 11

Chapter 11 of the code provides requirements for the final part of appliance installation. This chapter describes the procedures that must be followed after an appliance is installed in place, piped, and connected to its venting system. The chapter covers the following specific areas:

- Adjusting the burner input, by pressure adjustment or orifice change, and making adjustments needed due to high altitude [over 2000 ft (600 m)] (See Section 11.1.)
- Making air adjustments (See Section 11.2.)
- Verifying operation of safety shutoffs, automatic ignition, and protective devices (See Section 11.3 through Section 11.5.)
- Checking draft for vent-connected appliances (See Section 11.6.)
- Operating instructions for the consumer (See Section 11.7.)

## 11.1 Adjusting the Burner Input

Section 11.1 covers making adjustments to the burner input. Burner input for each gas appliance is shown on the appliance-rating nameplate. For units with multiple burners, the total input is divided by the number of burners to determine the input rating per burner.

### 11.1.1* Adjusting Input.

The input rate of the burner shall be adjusted to the proper value in accordance with the appliance manufacturer's instructions. Firing at a rate in excess of the nameplate rating shall be prohibited.

Gas appliances are furnished with low-pressure gas regulators that are specific to the type of gas (i.e., LP-Gas or natural gas) to be used. Each burner has an orifice that is rated for the proper input. Overfiring of appliances is prohibited because it may change the flame shape and burning characteristics, which can result in a number of undesirable consequences, including loss of efficiency, overtemperature of the appliance flue, or change in the products of combustion.

In accordance with instructions furnished by the manufacturer, minor adjustments to the appliance input rating can be made with the appliance regulator. The manufacturer should provide upper and lower limits, and the manifold pressure must be checked using either a "U gauge" manometer or a pressure gauge. The appliance instructions contain the location of the tap to be used for measuring manifold pressure. The tap will either be part of an appliance control or be located on the gas manifold itself. In any case, the flame must be inspected for quality (color) and length.

Additional good current practice warrants the use of a combustion analyzer anytime the input to an appliance is adjusted. A properly calibrated combustion analyzer will provide an accurate representation of the appliance operation.

◀ **FAQ**
I would like to increase the input to an appliance by adjusting the appliance regulator to increase the manifold pressure. Is there an upper limit to the pressure?

**A.11.1.1** For most burners, the input rate can be changed only slightly by changing the input pressure. Burner input should be checked in accordance with the appliance manufacturer's

NEW

installation instructions. If no appliance instructions are provided, burner input rate can be checked as follows:

(1) *Checking Burner Input Using a Meter (Clocking).* To check the Btu/hr input rate, the test hand on the gas meter should be timed for at least one revolution and the input determined from this timing. Test dials are generally marked $^1/_2$, 1, 2, or 5 ft$^3$/revolution depending on the size of the meter. Instructions for converting the test hand readings to cubic feet per hour are given in Table A.11.1.1. This table is provided for specific gas

NEW

***TABLE A.11.1.1*** *Gas Flow Rate to Burner in Cubic Feet per Hour at Standard Temperature and Pressure*

| ***Meter Pressure:*** | ***7.0 in. w.c. or 0.25 psi*** | | | | ***11.0 in. w.c. or 0.40 psi*** | | | | ***55.4 in. w.c. or 2 psi*** | | | |
|---|---|---|---|---|---|---|---|---|---|---|---|---|
| ***Seconds for One Revolution*** | ***Size of Test Meter Dial*** | | | | | | | | | | | |
| | $^1/_2$ ***ft$^3$*** | ***1 ft$^3$*** | ***2 ft$^3$*** | ***5 ft$^3$*** | $^1/_2$ ***ft$^3$*** | ***1 ft$^3$*** | ***2 ft$^3$*** | ***5 ft$^3$*** | $^1/_2$ ***ft$^3$*** | ***1 ft$^3$*** | ***2 ft$^3$*** | ***5 ft$^3$*** |
| 10 | 183 | 366 | 732 | 1831 | 185 | 370 | 739 | 1849 | 204 | 409 | 818 | 2044 |
| 11 | 166 | 333 | 666 | 1664 | 168 | 336 | 672 | 1680 | 186 | 372 | 743 | 1859 |
| 12 | 153 | 305 | 610 | 1526 | 154 | 308 | 616 | 1540 | 170 | 341 | 681 | 1704 |
| 13 | 141 | 282 | 563 | 1408 | 142 | 284 | 569 | 1422 | 157 | 315 | 629 | 1573 |
| 14 | 131 | 262 | 523 | 1308 | 132 | 264 | 528 | 1320 | 146 | 292 | 584 | 1460 |
| 15 | 122 | 244 | 488 | 1221 | 123 | 246 | 493 | 1232 | 136 | 273 | 545 | 1363 |
| 16 | 114 | 229 | 458 | 1144 | 116 | 231 | 462 | 1155 | 128 | 256 | 511 | 1278 |
| 17 | 108 | 215 | 431 | 1077 | 109 | 217 | 435 | 1087 | 120 | 241 | 481 | 1203 |
| 18 | 102 | 203 | 407 | 1017 | 103 | 205 | 411 | 1027 | 114 | 227 | 454 | 1136 |
| 19 | 96 | 193 | 385 | 964 | 97 | 195 | 389 | 973 | 108 | 215 | 430 | 1076 |
| 20 | 92 | 183 | 366 | 915 | 92 | 185 | 370 | 924 | 102 | 204 | 409 | 1022 |
| 21 | 87 | 174 | 349 | 872 | 88 | 176 | 352 | 880 | 97 | 195 | 389 | 974 |
| 22 | 83 | 166 | 333 | 832 | 84 | 168 | 336 | 840 | 93 | 186 | 372 | 929 |
| 23 | 80 | 159 | 318 | 796 | 80 | 161 | 321 | 804 | 89 | 178 | 356 | 889 |
| 24 | 76 | 153 | 305 | 763 | 77 | 154 | 308 | 770 | 85 | 170 | 341 | 852 |
| 25 | 73 | 146 | 293 | 732 | 74 | 148 | 296 | 739 | 82 | 164 | 327 | 818 |
| 26 | 70 | 141 | 282 | 704 | 71 | 142 | 284 | 711 | 79 | 157 | 315 | 786 |
| 27 | 68 | 136 | 271 | 678 | 68 | 137 | 274 | 685 | 76 | 151 | 303 | 757 |
| 28 | 65 | 131 | 262 | 654 | 66 | 132 | 264 | 660 | 73 | 146 | 292 | 730 |
| 29 | 63 | 126 | 253 | 631 | 64 | 127 | 255 | 637 | 70 | 141 | 282 | 705 |
| 30 | 61 | 122 | 244 | 610 | 62 | 123 | 246 | 616 | 68 | 136 | 273 | 681 |
| 31 | 59 | 118 | 236 | 591 | 60 | 119 | 239 | 596 | 66 | 132 | 264 | 660 |
| 32 | 57 | 114 | 229 | 572 | 58 | 116 | 231 | 578 | 64 | 128 | 256 | 639 |
| 33 | 55 | 111 | 222 | 555 | 56 | 112 | 224 | 560 | 62 | 124 | 248 | 620 |
| 34 | 54 | 108 | 215 | 538 | 54 | 109 | 217 | 544 | 60 | 120 | 241 | 601 |
| 35 | 52 | 105 | 209 | 523 | 53 | 106 | 211 | 528 | 58 | 117 | 234 | 584 |
| 36 | 51 | 102 | 203 | 509 | 51 | 103 | 205 | 513 | 57 | 114 | 227 | 568 |
| 37 | 49 | 99 | 198 | 495 | 50 | 100 | 200 | 500 | 55 | 111 | 221 | 553 |
| 38 | 48 | 96 | 193 | 482 | 49 | 97 | 195 | 486 | 54 | 108 | 215 | 538 |
| 39 | 47 | 94 | 188 | 469 | 47 | 95 | 190 | 474 | 52 | 105 | 210 | 524 |
| 40 | 46 | 92 | 183 | 458 | 46 | 92 | 185 | 462 | 51 | 102 | 204 | 511 |
| 41 | 45 | 89 | 179 | 447 | 45 | 90 | 180 | 451 | 50 | 100 | 199 | 499 |
| 42 | 44 | 87 | 174 | 436 | 44 | 88 | 176 | 440 | 49 | 97 | 195 | 487 |
| 43 | 43 | 85 | 170 | 426 | 43 | 86 | 172 | 430 | 48 | 95 | 190 | 475 |
| 44 | 42 | 83 | 166 | 416 | 42 | 84 | 168 | 420 | 46 | 93 | 186 | 465 |
| 45 | 41 | 81 | 163 | 407 | 41 | 82 | 164 | 411 | 45 | 91 | 182 | 454 |
| 46 | 40 | 80 | 159 | 398 | 40 | 80 | 161 | 402 | 44 | 89 | 178 | 444 |

*TABLE A.11.1.1 Continued*

| ***Meter Pressure:*** | ***7.0 in. w.c. or 0.25 psi*** | | | | ***11.0 in. w.c. or 0.40 psi*** | | | | ***55.4 in. w.c. or 2 psi*** | | | |
|---|---|---|---|---|---|---|---|---|---|---|---|---|
| ***Seconds for One Revolution*** | ***Size of Test Meter Dial*** | | | | | | | | | | | |
| | ***½ ft³*** | ***1 ft³*** | ***2 ft³*** | ***5 ft³*** | ***½ ft³*** | ***1 ft³*** | ***2 ft³*** | ***5 ft³*** | ***½ ft³*** | ***1 ft³*** | ***2 ft³*** | ***5 ft³*** |
| 47 | 39 | 78 | 156 | 390 | 39 | 79 | 157 | 393 | 43 | 87 | 174 | 435 |
| 48 | 38 | 76 | 153 | 381 | 39 | 77 | 154 | 385 | 43 | 85 | 170 | 426 |
| 49 | 37 | 75 | 149 | 374 | 38 | 75 | 151 | 377 | 42 | 83 | 167 | 417 |
| 50 | 37 | 73 | 146 | 366 | 37 | 74 | 148 | 370 | 41 | 82 | 164 | 409 |
| 51 | 36 | 72 | 144 | 359 | 36 | 72 | 145 | 362 | 40 | 80 | 160 | 401 |
| 52 | 35 | 70 | 141 | 352 | 36 | 71 | 142 | 355 | 39 | 79 | 157 | 393 |
| 53 | 35 | 69 | 138 | 345 | 35 | 70 | 140 | 349 | 39 | 77 | 154 | 386 |
| 54 | 34 | 68 | 136 | 339 | 34 | 68 | 137 | 342 | 38 | 76 | 151 | 379 |
| 55 | 33 | 67 | 133 | 333 | 34 | 67 | 134 | 336 | 37 | 74 | 149 | 372 |
| 56 | 33 | 65 | 131 | 327 | 33 | 66 | 132 | 330 | 37 | 73 | 146 | 365 |
| 57 | 32 | 64 | 128 | 321 | 32 | 65 | 130 | 324 | 36 | 72 | 143 | 359 |
| 58 | 32 | 63 | 126 | 316 | 32 | 64 | 127 | 319 | 35 | 70 | 141 | 352 |
| 59 | 31 | 62 | 124 | 310 | 31 | 63 | 125 | 313 | 35 | 69 | 139 | 347 |
| 60 | 31 | 61 | 122 | 305 | 31 | 62 | 123 | 308 | 34 | 68 | 136 | 341 |
| 62 | 30 | 59 | 118 | 295 | 30 | 60 | 119 | 298 | 33 | 66 | 132 | 330 |
| 64 | 29 | 57 | 114 | 286 | 29 | 58 | 116 | 289 | 32 | 64 | 128 | 319 |
| 66 | 28 | 55 | 111 | 277 | 28 | 56 | 112 | 280 | 31 | 62 | 124 | 310 |
| 68 | 27 | 54 | 108 | 269 | 27 | 54 | 109 | 272 | 30 | 60 | 120 | 301 |
| 70 | 26 | 52 | 105 | 262 | 26 | 53 | 106 | 264 | 29 | 58 | 117 | 292 |
| 72 | 25 | 51 | 102 | 254 | 26 | 51 | 103 | 257 | 28 | 57 | 114 | 284 |
| 74 | 25 | 49 | 99 | 247 | 25 | 50 | 100 | 250 | 28 | 55 | 111 | 276 |
| 76 | 24 | 48 | 96 | 241 | 24 | 49 | 97 | 243 | 27 | 54 | 108 | 269 |
| 78 | 23 | 47 | 94 | 235 | 24 | 47 | 95 | 237 | 26 | 52 | 105 | 262 |
| 80 | 23 | 46 | 92 | 229 | 23 | 46 | 92 | 231 | 26 | 51 | 102 | 256 |
| 82 | 22 | 45 | 89 | 223 | 23 | 45 | 90 | 225 | 25 | 50 | 100 | 249 |
| 84 | 22 | 44 | 87 | 218 | 22 | 44 | 88 | 220 | 24 | 49 | 97 | 243 |
| 86 | 21 | 43 | 85 | 213 | 21 | 43 | 86 | 215 | 24 | 48 | 95 | 238 |
| 88 | 21 | 42 | 83 | 208 | 21 | 42 | 84 | 210 | 23 | 46 | 93 | 232 |
| 90 | 20 | 41 | 81 | 203 | 21 | 41 | 82 | 205 | 23 | 45 | 91 | 227 |
| 94 | 19 | 39 | 78 | 195 | 20 | 39 | 79 | 197 | 22 | 43 | 87 | 217 |
| 98 | 19 | 37 | 75 | 187 | 19 | 38 | 75 | 189 | 21 | 42 | 83 | 209 |
| 100 | 18 | 37 | 73 | 183 | 18 | 37 | 74 | 185 | 20 | 41 | 82 | 204 |
| 104 | 18 | 35 | 70 | 176 | 18 | 36 | 71 | 178 | 20 | 39 | 79 | 197 |
| 108 | 17 | 34 | 68 | 170 | 17 | 34 | 68 | 171 | 19 | 38 | 76 | 189 |
| 112 | 16 | 33 | 65 | 163 | 17 | 33 | 66 | 165 | 18 | 37 | 73 | 183 |
| 116 | 16 | 32 | 63 | 158 | 16 | 32 | 64 | 159 | 18 | 35 | 70 | 176 |
| 120 | 15 | 31 | 61 | 153 | 15 | 31 | 62 | 154 | 17 | 34 | 68 | 170 |
| 130 | 14 | 28 | 56 | 141 | 14 | 28 | 57 | 142 | 16 | 31 | 63 | 157 |
| 140 | 13 | 26 | 52 | 131 | 13 | 26 | 53 | 132 | 15 | 29 | 58 | 146 |
| 150 | 12 | 24 | 49 | 122 | 12 | 25 | 49 | 123 | 14 | 27 | 55 | 136 |
| 160 | 11 | 23 | 46 | 114 | 12 | 23 | 46 | 116 | 13 | 26 | 51 | 128 |
| 170 | 11 | 22 | 43 | 108 | 11 | 22 | 43 | 109 | 12 | 24 | 48 | 120 |
| 180 | 10 | 20 | 41 | 102 | 10 | 21 | 41 | 103 | 11 | 23 | 45 | 114 |
| 190 | 10 | 19 | 39 | 96 | 10 | 19 | 39 | 97 | 11 | 22 | 43 | 108 |
| 200 | 9 | 18 | 37 | 92 | 9 | 18 | 37 | 92 | 10 | 20 | 41 | 102 |

Note: To convert to Btu per hour, multiply the cubic feet per hour of gas by the Btu per cubic foot heating value of the gas used.

pressures within the meters and gives gas flow rate (corrected to standard conditions) in cubic feet of gas per hour. Standard temperature is 60°F (16°C), and standard pressure is 30.00 in. of mercury. Measure the time for at least one revolution of a dial. Look up the gas flow rate in Table A.11.1.1. Gas flow rates can be calculated for meter pressures other than in these tables in the following manner. A pressure adjustment factor **F** should be determined for use in the gas input calculation for the gas pressure difference $\Delta P$ between the meter inlet and the atmosphere. The gas supplier can provide the pressure at the meter inlet. The pressure adjustment factor **F** is calculated with the following formula. Table A.11.1.1 was calculated using this formula.

$$\mathbf{F} = \frac{\Delta P + (B + 13.596)}{30.00 \times 13.596}$$

where:

$\mathbf{F}$ = pressure correction factor

$\Delta P$ = meter inlet pressure (in. w.c.)

$B$ = barometric pressure, unadjusted to sea level (in. of mercury)

NOAA weather reports barometric pressure in inches of mercury, adjusted to sea level. The sea level adjustment must be subtracted from the barometric pressure reported by NOAA weather. The local sea level adjustment can be obtained from NOAA.

For example, NOAA reported barometric pressure to be 30.12 in. of mercury for a city at 250 ft elevation. The barometric pressure adjustment for 250 ft is 0.27 in. of mercury. Subtract the local sea level adjustment from the NOAA barometric pressure to get the unadjusted barometric pressure.

$$30.12 - 0.27 = 29.85$$

The gas flow rate $Q$ is calculated using the following formula:

$$Q = \mathbf{F} \times C$$

where:

$Q$ = gas flow rate at standard conditions ($ft^3/hr$)

$\mathbf{F}$ = pressure adjustment factor

$C$ = timed gas flow rate ($ft^3/hr$)

The gas input rate $I$ is calculated with the following formula:

$$I = Q \times HHV$$

where:

$I$ = gas input rate (Btu/hr)

$Q$ = gas flow rate at standard conditions ($ft^3/hr$)

$HHV$ = average higher heat value of the gas at standard temperature and pressure conditions ($Btu/ft^3$), which can be obtained from the gas supplier

Appliances can be seriously overfired if the timed meter gas flow rate used to set input rate is not adjusted for meter pressure. At 2 psi meter pressure, an appliance would be 13 percent overfired if the gas flow rate is not adjusted for meter pressure.

(2) *Checking Burner Input by Using Orifice Pressure Drop and Orifice Size.* The fixed orifice size for each burner can be determined in accordance with Table F.1.1(a) for utility gases and Table F.1.1(b) for undiluted LP-Gases.

In the 2009 edition, Table A.11.1.1 was revised by correcting a factor in the formula used to calculate the numbers in the table. The changes are small and should not require resizing of orifices selected from the table in previous editions.

The data in Table A.11.1.1 can be used to check the burner input, after timing the meter for at least one full revolution. Table A.11.1.1 shows values for natural gas (7 in. w.c.), propane (11 in. w.c.), and for both gases at 2 psi. The table shows the input in cubic feet per hour. To convert to Btu/hr for natural gas, multiply by 1000 (or the exact heating value of the natural gas if known). To convert to Btu/hr for propane, multiply by 2500.

**11.1.1.1** The input rate can be adjusted by either changing the size of a fixed orifice, changing the adjustment of an adjustable orifice, or readjusting the appliance's gas pressure regulator outlet pressure (where a regulator is provided in the appliance).

**11.1.1.2** Input rate can be determined by either one of the following:

(1) Checking burner input by using a gas meter
(2) Checking burner input by using orifice pressure drop and orifice size

As noted in 11.1.1.2, the gas flow can be determined by using a meter. Note that any type of meter can be used, not just a traditional gas meter. If a meter is not available, the gas flow can be determined by installing an orifice plate and reading the pressure drop across the orifice. The gas flow can be calculated using standard engineering methods. See Exhibit 11.1 for an example of an orifice plate meter.

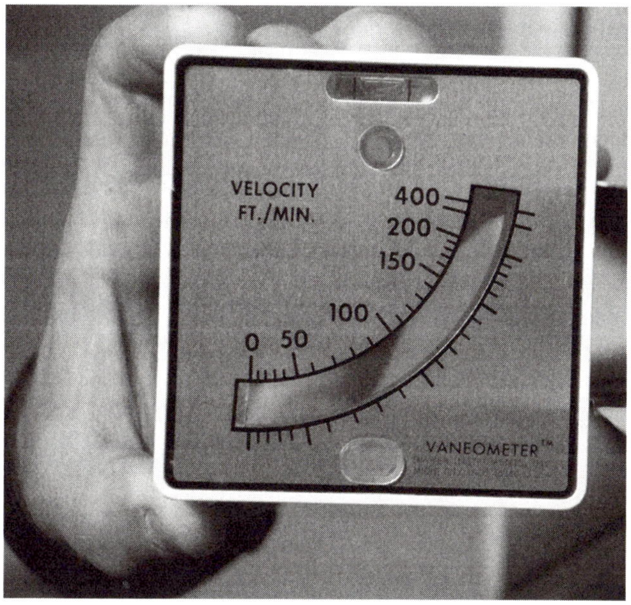


***EXHIBIT 11.1*** *Orifice Plate Meter. (Courtesy of Crane Engineering)*

**11.1.1.3** Overfiring shall be prohibited.

## 11.1.2 High Altitude.

Gas input ratings of appliances shall be used for elevations up to 2000 ft (600 m). The input ratings of appliances operating at elevations above 2000 ft (600 m) shall be reduced in accordance with one of the following methods:

Subsection 11.1.2 addresses derating of the gas supply to an appliance installed at high altitude. To determine the proper orifice size at the prescribed manifold pressure for the appliance, refer to Annex F. See Table F.1.1(a) for utility gases and Table F.1.1(b) for undiluted LP-Gases. For additional information on orifice sizing, refer to the Annex F commentary.

(1) At the rate of 4 percent for each 1000 ft (300 m) above sea level before selecting appropriately sized appliance
(2) As permitted by the authority having jurisdiction
(3) In accordance with the manufacturer's installation instructions

To calculate the derating of an orifice for high altitudes [e.g., any installation above 2000 ft (600 m)], multiply the altitude (in thousands of feet) by 4 percent and then deduct that rating from the sea level rating. For example, a 100,000 Btu (29 kW) furnace installed at an elevation of 5000 ft (1500 m) would have to be reduced by 20 percent to a rating of 80,000 Btu (23 kW) (5 × 4 percent = 20 percent reduction). For additional information on altitude derating, refer to the Annex F commentary.

Subsection 11.1.2(3) allows the appliances to be derated according to the manufacturer's listing, which is more accurate because it involves testing that is specific to the appliance. Paragraph 11.1.2(3) applies primarily to fan-assisted combustion appliances. The induced draft blower, depending on its design and motor characteristics, can provide sufficient air to compensate for the lower air density at higher elevations.

## 11.2* Primary Air Adjustment

The primary air for injection (Bunsen)-type burners shall be adjusted for proper flame characteristics in accordance with the appliance manufacturers' instructions. After setting the primary air, the adjustment means shall be secured in position.

**A.11.2** Normally, the primary air adjustment should first be set to give a soft blue flame having luminous tips and then increased to a point where the yellow tips just disappear. If the burner cannot be so adjusted, the manufacturer or serving gas supplier should be contacted.

A typical primary air adjustment for burners, as covered in Section 11.2, is an air shutter at the inlet of the burner, which can be rotated to open or close the openings, thus varying the amount of primary air. Exhibit 11.2 shows examples of burner primary air shutter adjustments.

Other burners may have an interrupter device (or spoiler) in the venturi of the burner. This device can be turned in accordance with the manufacturer's instructions to change the flame characteristics. The burner should always be adjusted for a proper blue flame with no yellow tips. With LP-Gas, a slight yellow tip may be normal.

As with the adjustment of burner impact, the proper use of a modern combustion analyzer will enhance the air adjustment because there are many situations when the full burner flame may not be visible or easily discernible to the servicer.

## 11.3 Safety Shutoff Devices

Where a safety shutoff device is provided, it shall be checked for proper operation and adjustment in accordance with the appliance manufacturer's instructions. Where the device does

*EXHIBIT 11.2 Examples of Burner Primary Air Shutter Adjustments. (Courtesy of the American Gas Association)*

not function properly to turn off the gas supply in the event of pilot outage or other improper operation, it shall be properly serviced or replaced with a new device.

Section 11.3 covers safety shutoff devices, which shut off the supply of gas either to the main burner or to both the main burner and the pilot burner. The manufacturer's instructions should be consulted to determine the type of safety shutoff device that should be used. The equipment should be tested for proper operation of the shutoff device.

## 11.4 Automatic Ignition

Appliances supplied with means for automatic ignition shall be checked for proper operation. If necessary, proper adjustments shall be made.

Automatic ignition systems are covered in Section 11.4. Most gas appliances are equipped with an automatic ignition system that incorporates either a constant burning pilot and safety shutoff device or one of several different types of spark or glow coil ignition devices. The manufacturer's instructions should be followed when checking these devices for proper operation.

## 11.5 Protective Devices

All protective devices furnished with the appliance, such as a limit control, fan control to blower, temperature and pressure relief valve, low-water cutoff device, or manual operating features, shall be checked for proper operation.

## 11.6* Checking the Draft

Vent-connected appliances shall be operated for several minutes and checked to see that the combustion products are going up the chimney or gas vent properly, by passing a lighted match or taper around the edge of the relief opening of the draft hood. Where the chimney or gas vent is drawing properly, the match flame is drawn into the draft hood. Where not drawing properly, the combustion products tend to extinguish this flame. Where the combustion products are escaping from the relief opening of the draft hood, the appliance shall not be operated until proper adjustments or repairs are made to provide adequate draft through the chimney or gas vent.

Section 11.6 covers checking the draft for vent-connected appliances. In addition to the match or taper method of checking the draft, which is covered in Section 11.6, the draft of a vent-connected appliance can also be verified with a draft gauge placed at the draft hood. This method will allow evaluation of the quality of the appliance draft under difficult operating conditions such as the operation of other combustion appliances and the operation of an exhaust fan.

**A.11.6** A procedure for checking draft can be found in Annex H, Steps 7, 8, and 10 through 14.

In reference to Section 11.6, if the appliance is installed in a closet with insufficient room for service personnel to be in the closet when the door is closed, then using a manometer or gauge with a probe inserted in the vent connector may be necessary to make sure that the burner, draft diverter, and vent are performing satisfactorily.

## 11.7 Operating Instructions

Operating instructions shall be furnished and shall be left in a prominent position near the appliance for the use of the consumer.

Section 11.7 covers operating instructions, which are required for each gas appliance and usually are included in the packet containing the installation instructions. The installer should be sure that the owner/user is aware of the proper operating procedures and, if necessary, should explain the function of the appliance. If the unit is equipped with a remote thermostat, the owner should be shown how to operate the appliance from the thermostat location.

# Venting of Appliances

# CHAPTER 12

Chapter 12 covers venting equipment, including the construction and installation of venting equipment, but not the sizing of the vents. This chapter and Chapter 13, Sizing of Category I Venting Systems, comprise the venting portion of the *National Fuel Gas Code*. Note that Chapter 12 covers installation requirements. However, safe installation of venting systems requires engineering calculations to ensure that sufficient draft is present. The tables in Chapter 13 perform these calculations. These tables do not cover all vent sizes and configurations, and other methods can be used in lieu of using the tables.

The sections in Chapter 12 generally follow the procedures an installer would use to design or evaluate a venting system for use with a Category I, mid-efficiency furnace or boiler. The venting system is roughly divided into two parts. The main, vertical section of the venting system is generally a chimney or Type B vent. The main vent is attached to the appliance(s) by the vent connector, which often has a horizontal component. Each portion of the venting system has different requirements, which are detailed in each portion's corresponding section in Chapter 12. The first five sections of Chapter 12 cover the following:

- Minimum safe performance, which establishes a performance requirement for all venting systems (See Section 12.1.)
- General requirements (See Section 12.2.)
- Specification for venting, which identifies appliances that must be vented and those that do not need to be vented (See Section 12.3.)
- Design and construction of venting systems, including some general design and construction requirements and some more detailed design and construction requirements for mechanical draft systems, ventilating hoods, and exhaust systems (See Section 12.4.)
- The type of venting system to be used for different types of appliances, provided in Table 12.5.1 (See Section 12.5.)

The next three sections of the code cover materials and installation requirements for the three types of venting systems used for conventional venting systems. Conventional venting systems (for Category I) rely on the heat of the exhaust gases as the driving force for venting. These sections cover the following:

- Masonry, metal, and factory-built chimneys, including requirements for construction, termination, sizing (by reference to Chapter 13), inspection, and cleaning of chimneys. Also included are detailed diagrams that specify proper chimney termination requirements, making these requirements easily understood. Chimneys can be listed or unlisted. (See Section 12.6.)
- Gas vents, including requirements for sizing, termination, multistory venting, and gas vent support and marking. Gas vents are always listed. (See Section 12.7.)
- Single-wall metal pipe, including requirements for construction, termination, installation, sizing, support and marking of single-wall metal pipe. Single-wall metal pipe is not listed. (See Section 12.8.)

The appliance category for most listed appliances can be found on the appliance nameplate. Since few visual differences exist between Category I, II, III, and IV appliances, installers should always check the appliance nameplate to determine the appliance venting category. Some appliances equipped with draft hoods might not have a category on the nameplate. These appliances should be installed as Category I appliances.

The next two sections of the code cover the termination of venting systems used for nonconventional venting systems. These systems use a fan as the driving force for venting. These sections cover the following:

- Venting system location, which provides requirements for locating the termination of vent systems exiting through the sides of buildings, relative to other building openings (See Section 12.9.)
- Condensation drain (See Section 12.10.)

The next six sections of the code cover material and installation requirements for the two types of connectors used in venting systems serving Category I appliances. Vent connectors connect gas appliances to vents. Vents are usually vertical, and connectors are usually horizontal. These sections cover the following:

- Vent connectors for Category I appliances, including requirements for installation, materials, sizing, and clearance to combustibles for vent connectors used with conventionally vented appliances (See Section 12.11.)
- Vent connectors for Category II, Category III, and Category IV appliances (Venting systems for these types of appliances are specified in Section 12.4.) (See Section 12.12.)
- Draft hoods and draft controls (This section defines which gas appliances require draft hoods, gives requirements for their installation, and covers other draft control devices and their installation.) (See Section 12.13.)
- Manually operated dampers, which are prohibited (Fixed baffles are not prohibited.) (See Section 12.14.)
- Automatically operated vent dampers, which must be listed (See Section 12.15.)
- Obstructions, which are not permitted except for (1) approved installations of listed draft regulators and safety controls, and (2) heat recovery systems, which are generally used with larger equipment, such as industrial boilers (See Section 12.16.)

When installing venting systems serving Category I appliances, the requirements of Chapter 12 apply for most installations, as does the sizing information in Chapter 13. Both sets of requirements must be used, except where the alternate sizing method provided in 12.6.3.1 and 12.7.3.1 is used for up to two appliances.

Although gas is a clean-burning fuel, the products of combustion must not be allowed to accumulate inside of a building. (See Exhibit 12.1 for the products of combustion formed

***EXHIBIT 12.1*** *Combustion of Natural Gas.*

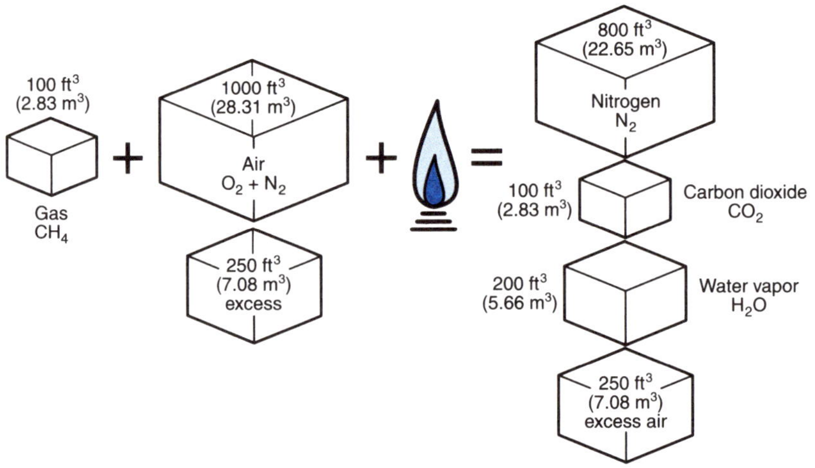

under ideal conditions.) Therefore, venting of most gas appliances is required. (See 12.3.2 for a list of appliances not requiring venting.) To provide for proper appliance function and the safety of building occupants, a properly installed and maintained venting system will perform the following functions:

1. Convey all of the combustion products to the outside atmosphere
2. Prevent damage to the gas equipment, vent, building, and furnishings from water vapor condensation in the flue gases
3. Prevent overheating of walls, building structure, and other combustible materials that are installed with required clearance to the appliance and venting system
4. Provide fast priming of natural draft venting systems to minimize spillage of combustion products into the building

The operation of a venting system may appear complicated, but natural draft venting works because hot air rises. When gas is burned, the products of combustion are hotter than the ambient air and therefore lighter and able to rise. A natural draft venting system's job is to channel these combustion products out of the building. To do so, the venting system must keep the combustion products as warm as possible in the vent. Heat retention maximizes the draft produced.

Prior to the 1990s, when the Department of Energy's minimum appliance efficiency regulation mandated higher-efficiency appliances, gas furnaces had a seasonal efficiency of about 60 percent. This lower efficiency resulted in vent gases that were hotter than they are today, providing ample heat for venting and for controlling condensation. With higher-efficiency space-heating appliances, the old safety margin no longer applies, and extra care must be taken so that the vents operate properly. Higher-efficiency appliances have the benefit of consuming less gas to produce a given amount of heat, which is a significant offsetting factor.

The Category I (mid-efficiency) appliances available today have a seasonal efficiency of about 80 percent. Some use a fan to assist the flow of combustion products through the appliance and are not equipped with a draft hood. Other appliances are equipped with a draft hood but reduce their total flue losses by using an automatic damper to cut off the flow in the vent during the off-cycle. Problems with early models and vent system failures were traced to condensation. As a result, new venting tables were developed. These tables, shown in Chapter 13, were added in the 1992 edition of the code to provide proper sizing of vents for these appliances. For more details on these considerations, see Supplement 1 of this handbook.

Water heaters must also meet minimum efficiency requirements, but they are still typically draft hood–equipped. Vent sizing for draft hood–equipped water heaters (and other appliances) connected to a dedicated venting system did not require an extensive reevaluation when other appliance efficiencies were increased. However, the size of common vents serving water heaters and Category I fan-assisted appliances has changed. Sizing must now be done in accordance with the venting tables in Chapter 13.

High-efficiency condensing appliances have a seasonal efficiency of 90 percent or higher, which reduces vent gas temperatures to a point where the water vapor produced as a product of combustion condenses to liquid water in the appliance or in the vent. These condensing appliances carry a vented appliance category of Category IV. This type of appliance produces much cooler vent gases, resulting in water condensing in the vent. Venting must be accomplished with a fan, because the vent gases are not hot enough to operate the natural draft vent. Water will condense in the vent and will dissolve some of the gases produced during combustion, which are slightly acidic. The vent materials used with these appliances must be able to resist the acidic condensate. For many of these Category IV appliances, plastic vent material is acceptable and preferred for corrosion reasons. The appliance manufacturer specifies the vent material for use with Category IV appliances.

The advantage of high-efficiency appliances is that they significantly reduce the amount of gas consumed with no loss in output. A mid-efficiency appliance uses one-third less gas than a conventional appliance, and a condensing appliance uses only one-half of the gas of a

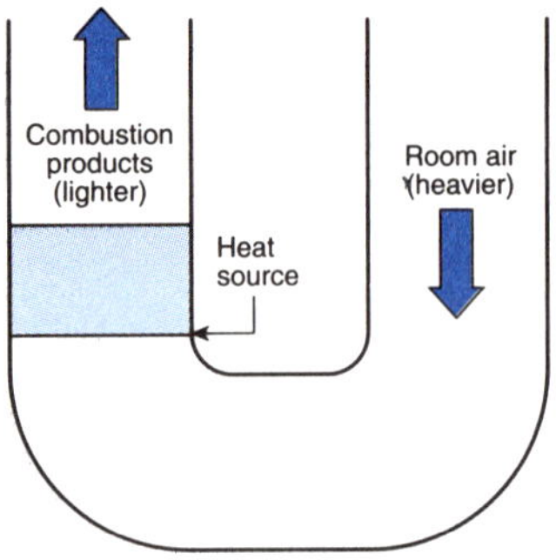


*EXHIBIT 12.2 Principles of Vent Operation.*

conventional appliance. The savings in fuel are offset by higher first cost and the higher maintenance requirements of high-efficiency appliances, as well as the added cost of the electricity to operate the fan.

## 12.1 Minimum Safe Performance

A venting system shall be designed and constructed so as to develop a positive flow adequate to convey flue or vent gases to the outdoors.

This requirement is the essence of safe venting — that is, the products of combustion must be removed from the building! While the rest of Chapter 12 and all of Chapter 13 provide requirements for venting, the end result must be achieved, or the venting system is not in compliance with the code. See Exhibit 12.2.

## 12.2 General

### 12.2.1 Materials.

This chapter recognizes that the choice of venting materials and the methods of installation of venting systems are dependent on the operating characteristics of the appliance.

### 12.2.2 Categories.

The operating characteristics of vented appliances can be categorized with respect to the following:

(1) Positive or negative pressure within the venting system
(2) Whether or not the appliance generates flue or vent gases that can condense in the venting system *(See Section 3.3 for the definition of these vented appliance categories.)*

### 12.2.3 Installation.

Listed vents shall be installed in accordance with Chapter 12 and the manufacturers' installation instructions.

Subsection 12.2.3 points out that the installation of listed venting systems must be in accordance with the manufacturers' instructions *and* the requirements of this chapter (Chapter 12) of the *National Fuel Gas Code.* There should be no conflicts between the code and the manufacturer's instructions. If conflicts occur, it is recommended that the manufacturer or the listing agency be contacted, because installation instructions for listed products are checked as part of the listing process to prevent conflicts with the code.

## 12.3 Specification for Venting

### 12.3.1 Connection to Venting Systems.

Except as permitted in 12.3.2 through 12.3.6, all appliances shall be connected to venting systems.

### 12.3.2 Appliances Not Required to Be Vented.

The following appliances shall not be required to be vented:

(1) Listed ranges
(2) Built-in domestic cooking units listed and marked for optional venting

(3) Listed hot plates and listed laundry stoves
(4) Listed Type 1 clothes dryers exhausted in accordance with Section 10.4
(5) A single listed booster-type (automatic instantaneous) water heater, when designed and used solely for the sanitizing rinse requirements of a dishwashing machine, provided that the appliance is installed with the draft hood in place and unaltered, if a draft hood is required, in a commercial kitchen having a mechanical exhaust system [Where installed in this manner, the draft hood outlet shall not be less than 36 in. (910 mm) vertically and 6 in. (150 mm) horizontally from any surface other than the appliance.]
(6) Listed refrigerators
(7) Counter appliances
(8) Room heaters listed for unvented use *(see 10.23.1 and 10.23.3)*
(9) Direct gas-fired make-up air heaters
(10) Other appliances listed for unvented use and not provided with flue collars
(11) Specialized appliances of limited input such as laboratory burners or gas lights

The appliances listed in 12.3.2 do not require venting. Note that some of the appliances [listed in 12.3.2(5) through 12.3.2(11)] must be vented if the total input of all the appliances in this class exceeds 20 Btu/hr per $ft^3$ (207 W per $m^3$) of room volume. This ratio is the same as the ratio provided by the standard method of calculation of the room volume needed to provide adequate air for combustion and ventilation. (See 9.3.2.1.) Further information can be found in the commentary following 9.3.2.1.

Paragraph 12.3.2(4) references Section 10.4, Clothes Dryers, highlighting that clothes dryers are exhausted, rather than vented. Exhausting means that the products of combustion and the moisture removed from the clothing are combined and removed together. This exhaust is powered with a fan to ensure proper operation. Clothes dryer exhausts and appliance vents must never be combined. Clothes dryer exhausts are under positive pressure, and Category I vents are under negative pressure. The two are incompatible.

Listed ranges, domestic cooking appliances with provision for optional venting, listed hot plates and laundry stoves, and listed clothes dryers can be installed in rooms with a smaller volume than is required by 9.3.2.1. These appliances have one of the following characteristics:

- Limited input (e.g., hot plates and laundry stoves)
- Short operating period (e.g., cooking appliances and dryers)
- Removal of products of combustion (e.g., dryers)
- Locations that have additional air supply (e.g., kitchens)

Listed, unvented appliances have been tested by the listing agency to determine that the burners, when properly adjusted, burn cleanly without producing unacceptable levels of dangerous combustion products. However, unvented appliances should be operated in rooms with ample air for combustion, and the burners should be maintained in proper adjustment. Except in decorative appliances, a yellow flame is indicative of improper combustion. Condensation or mildew in a room is an indication of insufficient air supply. Soot or carbon buildup is another indication. Usually, this fault can be corrected by one of the following:

- Properly adjusting the burner
- Cleaning the burner
- Providing adequate combustion air into the room
- Proper setup and positioning of unvented gas logs

Hot plates, laundry stoves, ranges, other cooking units, and so forth must never be used for room heating. Direct gas-fired make-up heaters are intended for industrial or commercial applications and are designed so that the products of combustion are diluted with large quantities of fresh air.

Unvented room heaters and fireplaces should be listed, installed, and operated strictly in accordance with the code and the manufacturer's instructions, which are required to be furnished with each heater. Listed, unvented room heaters and fireplaces are equipped with an oxygen depletion safety shutoff system (ODS) designed to shut off the gas supply to the heater before the oxygen in the surrounding air is reduced below a safe level. If a specially designed ODS pilot assembly is used, a reduced oxygen level will cause the pilot flame to lift above the thermocouple, which will cool, thereby shutting down the burner. However, in all cases, and as specified in the manufacturer's instructions, the room in which the heater is located must be provided with means for supplying fresh air for combustion and ventilation if the space is smaller than the required volume. (See combustion air requirements in Section 9.3.)

Listed ranges and ovens are not required to be vented when used as designed. They must not be used for space heating. With the oven door closed, oven burners normally cycle on and off to maintain oven temperature. Using ranges and ovens for space heating results in the burners being on constantly, which could affect the oxygen supply to the burners and lead to incomplete combustion.

**12.3.2.1** Where any or all of these appliances in 12.3.2(5) through (11) are installed so the aggregate input rating exceeds 20 Btu/hr/ft$^3$ (207 W/m$^3$) of room or space in which it is installed, one or more shall be provided with venting systems or other approved means for conveying the vent gases to the outdoors so the aggregate input rating of the remaining unvented appliances does not exceed 20 Btu/hr/ft$^3$ (207 W/m$^3$).

**12.3.2.2** Where the calculation includes the volume of an adjacent room or space, the room or space in which the appliances are installed shall be directly connected to the adjacent room or space by a doorway, archway, or other opening of comparable size that cannot be closed.

### 12.3.3* Ventilating Hoods.

Ventilating hoods and exhaust systems shall be permitted to be used to vent appliances installed in commercial applications *(see 12.4.4)* and to vent industrial appliances, particularly where the process itself requires fume disposal. *(See 9.1.6 and 9.1.9.)*

The provision in 12.3.3 is commonly applied in restaurant kitchens, where, for sanitation reasons, very hot water is required for dishwashing. The gas supply to the booster is interlocked with the hood, usually with a normally closed solenoid valve in the gas line that is opened only when the hood is running. This interlock is needed to meet the requirements of 12.4.4.2 and 12.4.4. The commentary following 12.4.4.2 provides more information on this application.

**A.12.3.3** Information on the construction and installation of ventilating hoods can be obtained from NFPA 96, *Standard for Ventilation Control and Fire Protection of Commercial Cooking Operations.*

### 12.3.4 Well-Ventilated Spaces.

The operation of industrial appliances such that its flue gases are discharged directly into a large and well-ventilated space shall be permitted.

The unusual provision in 12.3.4 is another exception to the rule stated in 12.3.1 that all gas utilization equipment be connected to a venting system. Paragraph 12.3.4 applies only to appliances installed in industrial buildings and recognizes common, safe practice in metal-treating and other industries. Some of the equipment normally installed under the requirements of this paragraph are heat-treating furnaces, radiant tube burners, and pot heaters in foundries.

Building ventilation must be designed to dilute flue gases to a safe level by means of natural draft or mechanical ventilation. In many of these installations, ventilation for fume removal or process reasons exceeds the ventilation level needed for safe removal of the products of combustion.

### 12.3.5 Direct Vent Appliances.

Listed direct vent appliances shall be installed in accordance with the manufacturer's installation instructions and 12.9.3.

The venting system is part of a direct vent appliance and is considered to be part of its listing. Listed appliances are tested with inlet and outlet configurations taken from their installation directions. Therefore, the installation instructions must be followed carefully. Exhibit 12.3 shows one type of a direct vent appliance that uses a concentric vent and air supply. Some direct vent appliances use independent ducts for the inlet and outlets.

Note that 12.3.5 applies to listed direct vent appliances, which are the most likely appliances to be encountered. Section 12.9 provides requirements for the location of the vent terminations of direct vent appliances.

Direct vent appliances do not use room air for combustion and therefore can be installed in locations where other types of appliances cannot be installed, such as bedrooms and bathrooms. Direct vent appliances are sometimes called "sealed combustion" appliances, which can be a misleading term. Direct vent appliances do draw combustion air from the outdoors and deliver the products of combustion to the outdoors, but they are not airtight and a small amount of air and combustion products can enter the room. That air and flammable vapors (if present) can be drawn into the direct vent appliances. This means that direct vent appliances can ignite flammable vapors nearby. Listed direct vent water heaters, furnaces, and boilers cannot be installed on the floor of a residential garage unless they are also listed as resistant to the ignition of flammable vapors.

### 12.3.6 Appliances with Integral Vents.

Appliances incorporating integral venting means shall be considered properly vented where installed in accordance with the manufacturer's installation instructions and 12.9.1 and 12.9.2.

An integral vent, addressed in 12.3.6, is a vent that is supplied with a gas appliance by the manufacturer and is part of the appliance.

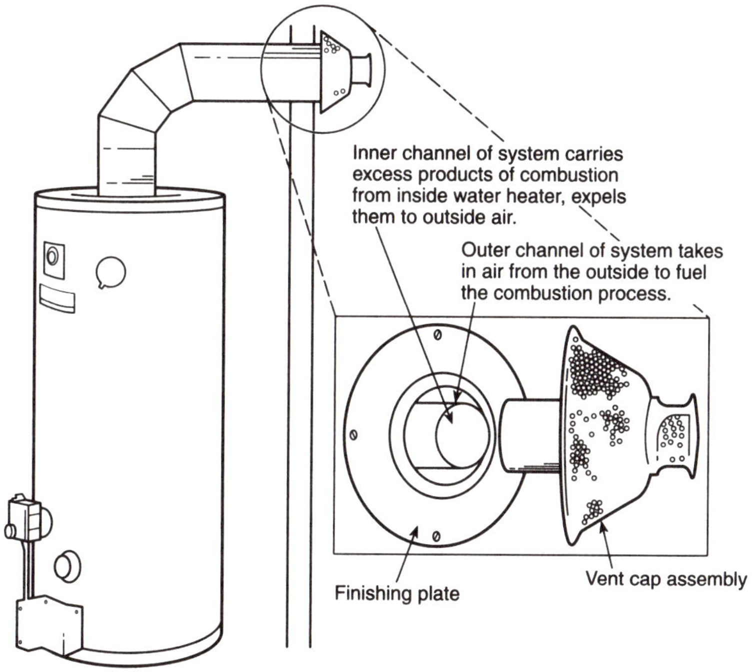


*EXHIBIT 12.3 Direct Vent Water Heater Using a Concentric Vent and Supply.*

## 12.4 Design and Construction

### 12.4.1 Appliance Draft Requirements.

A venting system shall satisfy the draft requirements of the appliance in accordance with the manufacturer's instructions.

Subsection 12.4.1 provides an overall performance requirement for a natural draft vent system. The vent system must remove the products of combustion from gas appliances from a building.

The principle on which natural draft vents operate is simple. Heat is the power that operates a natural draft vent or chimney, as shown in Exhibit 12.2. Combustion gases rise in the chimney or vent only because they are hotter, and therefore lighter, than the surrounding air. The hotter the gases and the higher the vent, the more swiftly and powerfully the gases will rise. Conversely, the cooler the gases and the shorter the vent, the more sluggish the gases' movement will be. The flue gases in the vent must remain hot enough over the length of the vent to provide a strong draft. If cooled enough, the upward motion of the gases stops altogether, and combustion gases can spill into the building through the relief opening of the appliance draft hood.

When the combustion products and dilution air rise in the vent, this volume must be replaced by air from outside the building. This replacement air can be supplied through normal air infiltration (through small openings in the building walls), through outdoor openings purposely installed in outside walls, or by a mechanical air system (usually a fan). Thus, to ensure proper vent operation, the building's airtightness and other devices exhausting air from the building must be taken into account.

Fan-assisted combustion appliances cause some confusion because they are listed as Category I appliances, meaning that their vents operate by natural draft due to the heat of the vent gas. However, these appliances have a fan to assist the flow of the combustion product through the appliance. The fan is necessary because higher-efficiency appliances have more resistance to air movement (or higher pressure drop) through their heat exchangers. The pressure provided by the induced draft blower is matched carefully to the resistance of the heat exchanger. Once the combustion products exit the appliance, the natural buoyancy takes over in the vent. If the vent is designed using the tables of Chapter 13, the pressure in the vent system will be negative.

Some furnaces are listed as Category I or III appliances, depending on the method used to vent the appliance. The fan in these furnaces is sized to result in negative vent pressure when connected to a properly sized vertical vent (Category I) or in positive vent pressure when vented horizontally (Category III). Category III installation requires using venting materials recommended by the manufacturer's installation instructions.

**FAQ ▶** How do I conduct a simple spillage test on a natural draft appliance?

Blockage of a natural draft vent may cause flue gases to spill into the building through the relief opening of the draft hood. Spillage can also be caused if the pressure in the vicinity of the appliance is much lower than the pressure outside the building (depressurization). Mechanical exhausts, fireplaces, and wind — anything that removes air from the building — can cause depressurization. Consequently, periodically verifying the performance of appliance venting systems is wise. Verification can be done readily with a natural draft system as follows (Annex H, Recommended Procedure for Safety Inspection of an Existing Appliance Installation, also provides information on this subject.):

1. Operate the appliance for at least five minutes to allow the flue gases to heat the vent.
2. Move a lighted match across the entire width of the draft hood relief opening. (See Exhibit 12.4.)
3. If the match flame is blown downward or extinguished, the venting system or chimney should be checked by a qualified agency.

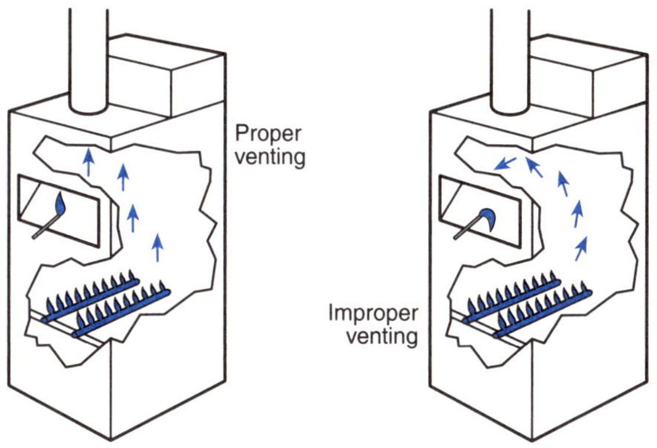


***EXHIBIT 12.4*** *Match Test to Determine Proper and Improper Venting.*

4. If the appliance is not equipped with a draft hood, performing the match test may be impossible, in which case the appliance instruction manual should be consulted.

Good air circulation in adequate amounts is also vital for good venting and efficient appliance operation. Signs of improper operation include a yellow or wavering flame; discoloration around access doors; a pungent odor in the building; excessive humidity, condensation, or mold; corrosion of the vent material; and soot near the burner or vent area. If any of these conditions is observed, a qualified agency should be contacted to have the installation inspected. Where these conditions exist, building occupants — especially occupants who spend most of the day in the building affected — have been reported to have headaches and flu-like symptoms.

Under extremely adverse conditions, carbon monoxide can be produced as a result of either improper venting of combustion products, insufficient fresh air to support the proper burning of gas, or improperly adjusted appliances. If symptoms of carbon monoxide poisoning are experienced — headache, yawning, ringing in ears, weariness, vomiting, or heart fluttering or throbbing — fresh air is needed promptly. Occupants should open windows or go outdoors. Any gas appliance suspected of operating improperly should be shut off and immediately checked by a qualified agency.

### 12.4.2 Design and Construction.

Appliances required to be vented shall be connected to a venting system designed and installed in accordance with the provisions of Sections 12.5 through 12.16.

### 12.4.3 Mechanical Draft Systems.

Mechanical draft systems, addressed in 12.4.3, use the mechanical force, usually from a fan or blower, to vent the combustion products. Natural buoyancy is not needed or significant.

There are two types of mechanical draft systems: forced draft and induced draft. The difference between these types is as follows:

◀ **FAQ**
What is the difference between forced draft and induced draft?

1. Forced draft exists where the fan is located at the entrance to the vent or chimney, resulting in pressure in the vent or chimney that is greater than atmospheric pressure.
2. Induced draft exists where the fan is located at the exit of the vent or chimney, resulting in pressure that is less than atmospheric pressure.

Typically, any section of vent downstream of the blower will be under a positive gauge pressure, as a result of forced draft. Therefore, care must be taken to ensure that the vent pipe used is appropriate for positive-pressure combustion gases. Otherwise, leaks of the combustion products may occur. A mechanical draft water heater is shown in Exhibit 12.5.

**12.4.3.1** Mechanical draft systems shall be listed and shall be installed in accordance with both the appliance and the mechanical draft system manufacturer's installation instructions.

**12.4.3.2** Appliances requiring venting shall be permitted to be vented by means of mechanical draft systems of either forced or induced draft design.

*Exception: Incinerators.*

**12.4.3.3** Forced draft systems and all portions of induced draft systems under positive pressure during operation shall be designed and installed so as to prevent leakage of flue or vent gases into a building.

**12.4.3.4** Vent connectors serving appliances vented by natural draft shall not be connected into any portion of mechanical draft systems operating under positive pressure.

The key term in 12.4.3.3 and 12.4.3.4 is *positive pressure*. Vents serving fan-assisted appliances operate under negative pressure when vented into a chimney, even though the appliance uses a mechanical fan. Fan-assisted appliances can be common-vented with other fan-assisted appliances, draft hood–equipped appliances, and all other Category I vented appliances.

**12.4.3.5** Where a mechanical draft system is employed, provision shall be made to prevent the flow of gas to the main burners when the draft system is not performing so as to satisfy the operating requirements of the appliance for safe performance.

**FAQ ►**
How do you prevent main burner operation when the mechanical draft system is not operating?

An interlock between the mechanical draft system and the appliance gas valve will fulfill the requirement in 12.4.3.5 that gas does not flow to the main burners when the draft system is not performing (creating sufficient air movement to remove the products of combustion). Controls must be configured so that power to open the gas valve is only supplied if the mechanical draft system is operating and producing the required level of draft.

NEW ► **12.4.3.6** The exit terminals of mechanical draft systems shall be not less than 7 ft (2.1 m) above finished ground level where located adjacent to public walkways and shall be located as specified in 12.9.1 and 12.9.2.

Paragraph 12.4.3.6 provides the installer with guidance on the location of exit terminals for mechanical draft systems. This information is needed because of the increased use of power vent kits to vent appliances. The manufacturer of a listed power vent kit provides sizing information in the instructions provided with the kit.

A minimum elevation of 7 ft (2.1 m) above finished ground level is required in 12.4.3.6 so that combustion gases do not create a nuisance for pedestrians. Potential nuisances include the high temperature of the vent and the possibility of condensation of water in the vent gases forming a pool of water, ice, or snow on the pavement. The definition of a "public walkway" is not specified in the code. The dictionary definition and common sense should be used. Note that the requirement was revised in the 2009 edition by changing the reference point for the 7 ft (2.1 m) elevation from "grade" to "finished ground level," as has been done in other places in the code. Refer to the commentary following 9.3.8.8 for the explanation of this change.

Subsections 12.9.1 and 12.9.2 specify the location of vent terminals in relation to air inlets, doors, and windows.

### 12.4.4* Ventilating Hoods and Exhaust Systems.

**A.12.4.4** See A.12.3.3.

Subsection 12.4.4 recognizes a safe, alternate method for venting of gas appliances in commercial applications. For example, in restaurant kitchens a "booster" water heater is used to provide very hot water for sanitation. In this application, the booster water heater is operated only while the restaurant kitchen is in operation, and the range hood is used to vent both the range and the water heater.

Range hoods are covered under NFPA 96, *Standard for Ventilation Control and Fire Protection of Commercial Cooking Operations*, which requires an interlock that operates when the airflow falls below a predetermined value. This circuit can be interlocked with the main gas valve of the water heater (or other gas appliance vented via the range hood) so that gas will not flow to the main burners unless a minimum airflow is achieved in the vent system.

***EXHIBIT 12.5*** *Mechanically Vented Water Heater. (Courtesy of A. O. Smith Water Products)*

**12.4.4.1** Ventilating hoods and exhaust systems shall be permitted to be used to vent appliances installed in commercial applications.

**12.4.4.2** Where automatically operated appliances, other than commercial cooking appliances, are vented through a ventilating hood or exhaust system equipped with a damper or with a power means of exhaust, provisions shall be made to allow the flow of gas to the main burners only when the damper is open to a position to properly vent the appliance and when the power means of exhaust is in operation.

Paragraph 12.4.4.2 was revised in the 2009 edition following a Formal Interpretation (FI) of the 2006 edition to clarify the committee's intent, to allow automatically controlled appliances, such as water heaters, to be vented via a vent hood or exhaust system. This provision has been in NFPA 54 for about 50 years, and it allows appliances in rooms with ventilating hood exhaust systems, usually kitchens, to use the ventilating hood exhaust systems to vent the appliance. The FI to the 2006 edition was issued after the committee became aware that some officials were requiring the interlock for commercial cooking ranges located under an exhausted hood, a common practice. The interpretation was that the commercial cooking appliance oven had a thermostat and was therefore an "automatically operated appliance," and an interlock was required. As this is not the intent of the committee, the FI was issued to so state.

The requirement in 12.4.4.2 recognizes an unusual method of venting appliances in buildings: using ventilation hoods. This method, which is used frequently in restaurant kitchens, is explained in the commentary following 12.3.3. Information on the design and installation of ventilating hoods in industrial plants can be obtained from NFPA 91, *Standard for Exhaust Systems for Air Conveying of Vapors, Gases, Mists, and Noncombustible Particulate Solids*.

### 12.4.5 Circulating Air Ducts, Above-Ceiling Air-Handling Spaces, and Furnace Plenums.

**12.4.5.1** Venting systems shall not extend into or pass through any fabricated air duct or furnace plenum.

The requirement in 12.4.5.1 was revised in the 2009 edition to replace the term *circulating air duct or furnace plenum* with *fabricated air duct or furnace plenum*. It was noted that a circulating air duct was not defined, and it is not consistent with terms used in mechanical codes for ducts.

**12.4.5.2** Where a venting system passes through an above-ceiling air space or other nonducted portion of an air-handling system, it shall conform to one of the following requirements:

(1) The venting system shall be a listed special gas vent, other system serving a Category III or Category IV appliance, or other positive pressure vent, with joints sealed in accordance with the appliance or vent manufacturer's instructions.
(2) The vent system shall be installed such that no fittings or joints between sections are installed in the above-ceiling space.
(3) The venting system shall be installed in a conduit or enclosure with joints between the interior of the enclosure and the ceiling space sealed.

**FAQ ▶**
**Is it acceptable to install a venting system for a natural draft appliance in an above-ceiling space used as a return air plenum?**

The code includes requirements in 12.4.5.2 that provide guidance for the installation of venting systems in above-ceiling spaces or other non-ducted spaces used for transfer of conditioned building air. In commercial buildings, the space above the ceiling is often used as the return air plenum for the heating and air-conditioning system. Special consideration for the installation of vents that pass through these spaces is necessary to prevent flue gases from leaking into the space. The issue of airtightness of that portion of the venting system that passes through the air-handling space is covered in 12.4.5.2(1), (2), and (3).

Paragraph 12.4.5.2(1) allows the installation of a special gas vent or other approved venting system for a Category III or IV appliance. Category III and IV appliances have positive pressure in their venting systems, so the venting system is already designed to be airtight to prevent the outward leakage of flue gases.

Paragraph 12.4.5.2(2) allows the installation of standard venting systems, such as a Type B vent, to pass through the air-handling space when it is installed with no joints or fittings within the air-handling space. Consider as an example the installation of a Type B vent in a 3 ft (0.9 m) high ceiling space where a 5 ft (1.5 m) section of pipe could be used as long as it is installed such that the joints were below and above the ceiling space.

Paragraph 12.4.5.2(3) allows the installation of a standard venting system when it is installed within a chase or conduit that is sealed from the air-handling space. This option essentially removes the venting system from the air-handling space.

## 12.5 Type of Venting System to Be Used

**12.5.1** The type of venting system to be used shall be in accordance with Table 12.5.1.

To use Table 12.5.1, locate the type of gas appliance to be vented in the left-hand column and read across the row to find the type(s) of venting system(s) that is permitted. For example, a listed Category I appliance can be vented using a Type B gas vent, chimney, single-wall metal pipe, chimney lining system that is listed for gas venting, or a special gas vent listed for the appliance.

Table 12.5.1 refers to appliance Category I through Category IV. The categories are based on vent temperature and pressure. (See Commentary Table 12.1, Appliance Vent Categories, and the definitions of *vented appliance — Category I* through *Category IV* in 3.3.6.11. Note that the definitions reference "a vent gas temperature that may cause [or avoid] excessive condensate production in the vent.") A specific temperature is not provided, because it is not the same for all appliances. The ANSI Z21 standards for appliance categorization can be referenced for this information. Annex M contains a complete list of the ANSI Z21 standards. Note that the installer should not need this information, because the appliance vent category is included in the appliance manufacturer's installation instructions and on the appliance nameplate.

The criteria in the ANSI Z21 standards are based on a flue loss of 17 percent of total energy. The 17 percent flue loss is the same flue loss built into the vent sizing tables for fan-assisted appliances. In this way, the standards ensure that the appliance will work properly with the vent system.

In Commentary Table 12.1, the term *nonpositive vent pressure* means that the pressure in the vent will be lower than the surrounding atmosphere if the vent system meets the requirements of Chapter 12 and Chapter 13. The incorporation of a fan into the appliance does not always mean that the vent pressure is positive. If unsure, check the appliance nameplate or manufacturer's instructions for the venting category, or check the vent pressure with a manometer or other pressure gauge when the appliance is operating.

### 12.5.2 Plastic Piping.

Plastic piping used for venting appliances listed for use with such venting materials shall be approved.

***TABLE 12.5.1*** *Type of Venting System to Be Used*

| ***Appliances*** | ***Type of Venting System*** |
|---|---|
| Listed Category I appliances<br>Listed appliances equipped with draft hood<br>Appliances listed for use with Type B gas vent | Type B gas vent *(see Section 12.7)*<br>Chimney *(see Section 12.6)*<br>Single-wall metal pipe *(see Section 12.8)*<br>Listed chimney lining system for gas venting *(see 12.6.1.3)*<br>Special gas vent listed for these appliances *(see 12.5.3)* |
| Listed vented wall furnaces | Type B-W gas vent *(see Section 12.7, Section 10.27)* |
| Category II appliances<br>Category III appliances<br>Category IV appliances | As specified or furnished by manufacturers of listed appliances *(see 12.5.2, 12.5.3)* |
| Incinerators, outdoors | Single-wall metal pipe *[see Section 12.8, 12.8.3(3)]* |
| Incinerators, indoors | |
| Appliances that can be converted to use solid fuel<br>Unlisted combination gas- and oil-burning appliances<br>Combination gas- and solid fuel–burning appliances<br>Appliances listed for use with chimneys only<br>Unlisted appliances | Chimney *(see Section 12.6)* |
| Listed combination gas- and oil-burning appliances | Type L vent *(see Section 12.7)* or chimney *(see Section 12.6)* |
| Decorative appliance in vented fireplace | Chimney *(see 10.6.2)* |
| Gas-fired toilets | Single-wall metal pipe *(see Section 12.8, 10.25.3)* |
| Direct vent appliances | See 12.3.5 |
| Appliances with integral vents | See 12.3.6 |

Before the introduction of high-efficiency (90+ percent efficiency) gas utilization equipment, plastic piping was prohibited as a vent material. High-efficiency (Category IV) appliances reduce vent temperatures, resulting in condensate formation. As accumulation of condensate became a source of corrosion of metal vents, plastic piping became the preferred material. Paragraph 12.5.2 requires that plastic vent materials be used for listed gas utilization equipment only when specified in the manufacturer's instructions.

Exhibit 12.6 shows an example of a direct vent furnace using plastic vent and plastic combustion air supply.

### 12.5.3 Plastic Vent Joints.

NEW

Plastic pipe and fittings used to vent appliances shall be installed in accordance with the appliance manufacturer's installation instructions. Where primer is required, it shall be of a contrasting color.

*COMMENTARY TABLE 12.1 Appliance Vent Categories*

| Appliance Category[1] | Vent Pressure | Temperature Above/Below Defined in Relevant ANSI Z21 Standard | Comment |
|---|---|---|---|
| I | Nonpositive[2] | Above | Natural draft venting<br>Materials must be corrosion resistant. |
| II | Nonpositive[2] | Below | Condensate must be drained. |
| III | Positive[3] | Above | Vent must be gastight.<br>Vent must be liquidtight and gastight. |
| IV | Positive[3] | Below | Condensate must be drained. |

[1]Appliances are identified as Category I, II, III, or IV on the nameplate on the appliance and are stated as such in the manufacturer's installation instructions.

[2]The term *nonpositive vent pressure* means that even if fans or blowers are used in the appliance or vent systems, venting is accomplished by natural draft. (The vent pressure is lower than the atmospheric pressure.)

[3]The term *positive vent pressure* means that fans, blowers, or other means are used to propel vent gases through the vent at above atmospheric pressure.

**EXHIBIT 12.6** *Direct Vent Furnace with Plastic Pipe. (Courtesy of Magic Sweep Corporation)*

Paragraph 12.5.3 was added in the 2009 edition to recognize concerns of enforcement officials over proper plastic vent joint installation. PVC pipe used as a vent requires a primer that is colored. Inspectors can easily verify that the primer has been used by the residual color on the pipe near a joint. The choice of color is left to the primer manufacturer. To the editor's knowledge, all primers as of the date of publication of this handbook are purple, but the code allows other colors if desired or needed to distinguish between different primers in the future.

### 12.5.4 Special Gas Vent.

Special gas vent shall be listed and installed in accordance with the special gas vent manufacturer's installation instructions.

All special gas vents are listed vent materials. Special gas vents, which are covered in 12.5.4, are listed in accordance with UL 1738, *Standard for Venting Systems for Gas-Burning Appliances, Categories II, III and IV*. Installation instructions for special gas vents include limitations on operating temperature, categories of appliance to be used with each vent, clearance to combustible materials, types of fittings and joint sealant to be used, and vent termination requirements.

Special attention should be given to the following areas:

1. Proper support for the special gas vent to prevent sagging and to allow for expansion, contraction, and condensate drainage
2. Proper cutting and cleaning of joints and fittings, and the use of recommended joint sealants (substitutes not usually permitted)
3. Construction of a condensate trap (See the appliance manufacturer's instructions for special requirements.)
4. Wall penetrations (The pipe should not be secured at a thimble, because the pipe must be allowed to move to accommodate expansion and contraction.)
5. Insulation [The vent pipe or the fittings of the inside of a wall thimble must not be insulated when polymeric (nonmetallic) vent materials are used.]

More than 15 years ago, a class of special gas vent known as high temperature plastic vent (HTPV) was introduced to the market for use with mid-efficiency appliances. Field experience has shown that these vent systems are prone to failure. The failure may occur because of improper installation practice and/or corrosion from acidic condensate.

The location of those vent pipes should also be checked. For furnaces, only HTPV systems that have vent pipes that go through the side walls of structures (horizontal systems) are subject to this program. Other plastic vent pipes, such as white PVC or CPVC, are not involved in this program.

## 12.6 Masonry, Metal, and Factory-Built Chimneys

Section 12.6 applies to chimneys of all types, but not to gas vents (including Type B vents), which are covered in Section 12.7, or single-wall metal pipe, which is covered in Section 12.8.

### 12.6.1 Listing or Construction.

**12.6.1.1** Factory-built chimneys shall be installed in accordance with the manufacturer's installation instructions. Factory-built chimneys used to vent appliances that operate at positive vent pressure shall be listed for such application.

Sketches of factory-built chimneys, addressed in 12.6.1.1 and shown in Exhibit 12.7 through Exhibit 12.9, consist entirely of factory-built parts, such as chimney sections, supports, thimbles, flashings, caps, and other parts required to complete a particular installation. The parts are designed to be assembled with other parts of the same model without requiring field alteration or construction. All of the parts for a particular model, as described in the chimney manufacturer's instructions, are to be used. Exhibit 12.10 shows the hardware for a positive-pressure factory-built chimney.

**12.6.1.2** Metal chimneys shall be built and installed in accordance with NFPA 211, *Standard for Chimneys, Fireplaces, Vents, and Solid Fuel–Burning Appliances.*

In 12.6.1.2, the term *metal chimneys* defines field-constructed chimneys made of sheet steel and intended for industrial applications. These chimneys differ greatly from factory-built chimneys in that they are made of metal. The term *metal chimneys* is sometimes misapplied to describe factory-built chimneys. The construction and installation of metal chimneys is covered in NFPA 211, *Standard for Chimneys, Fireplaces, Vents, and Solid Fuel–Burning Appliances,* which prohibits the installation of metal chimneys in residential structures.

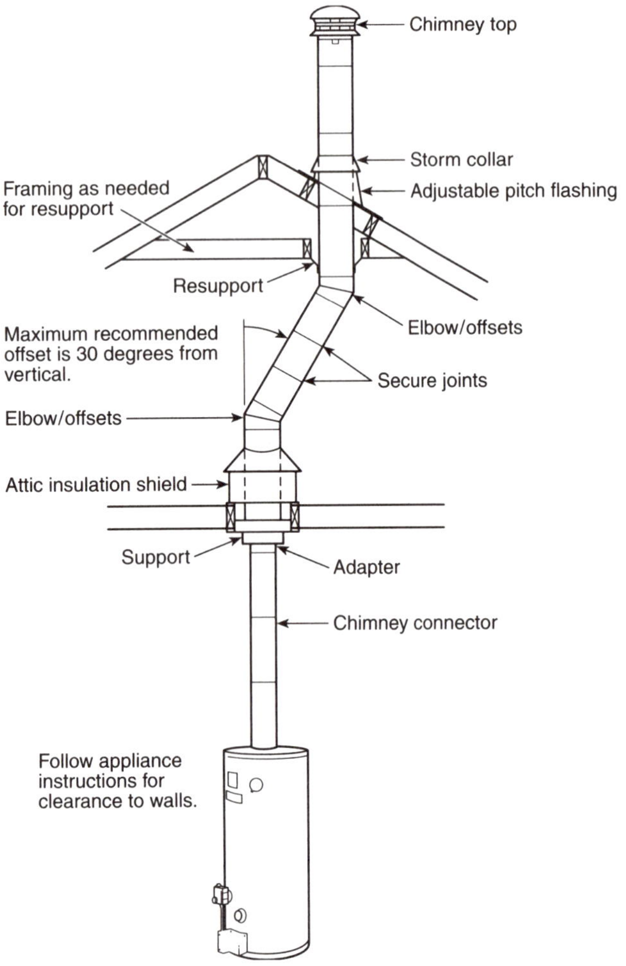


***EXHIBIT 12.7*** *Typical Installation of Residential Factory-Built Chimney.*

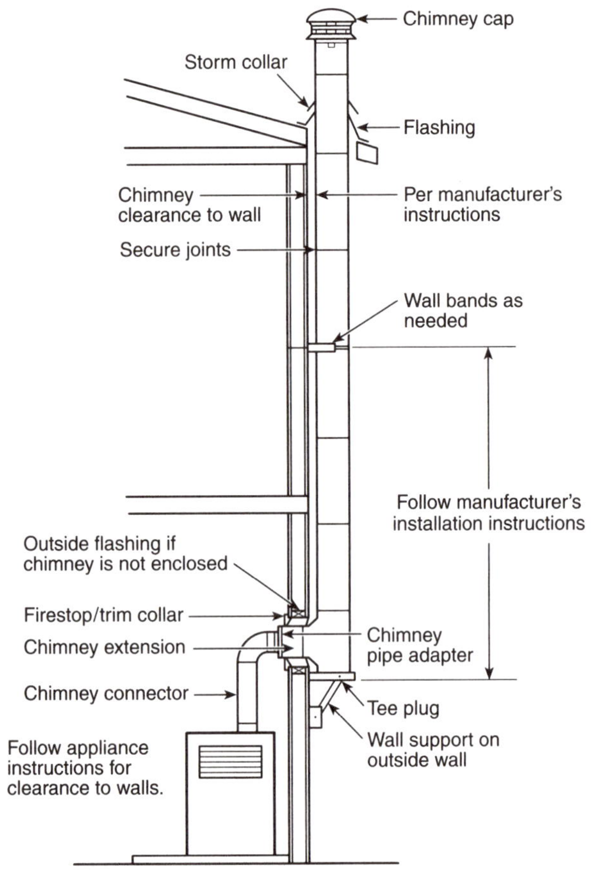


***EXHIBIT 12.8*** *Alternate Installation of Residential Factory-Built Chimney Outside a Building.*

The following three types of factory-built chimneys are available:

1. Residential-type and building heating appliance chimneys that are intended for venting flue gases at a temperature not exceeding 1000°F (538°C) under continuous operating conditions
2. 1400°F chimneys for venting flue gases at a temperature not exceeding 1400°F (760°C) under continuous operating conditions
3. Medium-heat appliance chimneys for venting flue gases at a temperature not exceeding 1800°F (982°C)

The installation of chimneys serving appliances with vent temperatures over 1000°F (538°C) is not covered in this code. See NFPA 211, which also describes the various types of heating appliances mentioned here.

Factory-built chimneys must be installed in accordance with the terms of their listing and the manufacturer's instructions, which are furnished with each chimney assembly and have been verified by the listing agency. Chimneys suitable for positive pressure are now available. The product standard for chimneys, UL 103, *Standard for Factory-Built Chimneys for Residential Type and Building Heating Appliances,* includes separate tests to verify the proper operation of factory-built chimneys that are designed for positive combustion-product pressure.

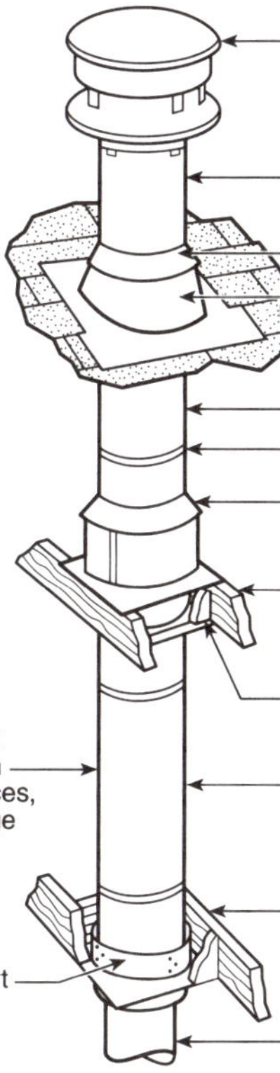


***EXHIBIT 12.9*** *Alternate Installation of Residential Factory-Built Chimney Inside a Building Highlighting Roof Penetration.*

***EXHIBIT 12.10*** *Hardware for a Positive-Pressure Factory-Built Chimney. (Courtesy of Selkirk Metalbestos)*

Category II, III, and IV appliances are not suitable for venting by the conventional method of natural draft venting. These appliances must not be connected to chimneys. These appliances should be vented in accordance with the appliance manufacturer's instructions, which cover the type, size, and length of material to be used; how the venting system is to be installed; and the minimum clearance to combustible material. These instructions have been verified by the listing agency.

**12.6.1.3*** Masonry chimneys shall be built and installed in accordance with NFPA 211, *Standard for Chimneys, Fireplaces, Vents, and Solid Fuel–Burning Appliances*, and lined with approved clay flue lining, a listed chimney lining system, or other approved material that resists corrosion, erosion, softening, or cracking from vent gases at temperatures up to 1800°F (982°C).

The requirements in 12.6.1.3 apply to masonry chimneys. Masonry chimneys can be used for all appliances permitted in 12.5.1 that can be vented using chimneys. The code user must be aware that Section 13.1, Additional Requirements to Single Appliance Vent, includes 13.1.11 and that Section 13.2, Additional Requirements to Multiple-Appliance Vent, includes 13.2.22. Both 13.1.11 and 13.2.22 place restrictions on the use of the sizing tables for chimneys that vent fan-assisted combustion appliances.

Masonry chimneys are rarely suitable to vent fan-assisted appliances without the addition of a properly sized chimney liner, especially when they are not common-vented with draft hood–equipped appliances. Exterior masonry chimneys pose additional challenges and restrictions, and the addition of the exterior masonry chimney venting tables in Chapter 13 makes it clear that exterior masonry chimneys could not be used to vent a fan-assisted appliance unless common-vented with draft hood–equipped appliances. Additionally, the exterior masonry chimney venting tables severely limit installation options in cold climates, due to the high heat loss from exterior chimneys.

*Exception: Masonry chimney flues lined with a chimney lining system specifically listed for use with listed appliances with draft hoods, Category I appliances, and other appliances listed for use with Type B vents shall be permitted. The liner shall be installed in accordance with the liner manufacturer's installation instructions. A permanent identifying label shall be attached at the point where the connection is to be made to the liner. The label shall read "This chimney liner is for appliances that burn gas only. Do not connect to solid or liquid fuel–burning appliances or incinerators."*

The exception to 12.6.1.3 permits installation of liners that may not be covered in NFPA 211 in a masonry chimney. The labeling requirement notifies future installers of the vent liner's limitations. Notification is important if a change in appliance fuels is considered.

The intent of this requirement is to warn installers that a chimney liner is not suitable to vent appliances with higher vent temperatures than gas appliances. If the liner is constructed of a material that can tolerate higher vent temperatures, then its listing should reflect such information. The labeling should state that the chimney liner can be used for listed appliances with draft hoods and other types of appliances. If the liner is suitable for high temperatures, the label is not required.

**A.12.6.1.3** For information on the installation of gas vents in existing masonry chimneys, see Section 12.7.

### 12.6.2 Termination.

**12.6.2.1** A chimney for residential-type or low-heat appliances shall extend at least 3 ft (0.9 m) above the highest point where it passes through a roof of a building and at least 2 ft (0.6 m) higher than any portion of a building within a horizontal distance of 10 ft (3 m). *(See Figure 12.6.2.1.)*

**12.6.2.2** A chimney for medium-heat appliances shall extend at least 10 ft (3 m) higher than any portion of any building within 25 ft (7.6 m).

**12.6.2.3** A chimney shall extend at least 5 ft (1.5 m) above the highest connected appliance draft hood outlet or flue collar.

**12.6.2.4** Decorative shrouds shall not be installed at the termination of factory-built chimneys except where such shrouds are listed and labeled for use with the specific factory-built chimney system and are installed in accordance with manufacturers' installation instructions.

**FAQ ►** What are the hazards associated with installing decorative shrouds around chimneys?

The requirement in 12.6.2.4 prohibits the installation of unlisted shrouds around factory-built chimneys, which would affect the flow of flue gases and trap hot flue gases in close proximity to the combustible roof. Shrouds, as shown in Exhibit 12.11, are used to hide chimneys. Shrouds can interfere with vent operation if they extend above the top of the vent. Hazards include overheating due to reduced airflow around the chimney top and obstruction of chimneys by animal nesting that is not easily seen from the ground. Nesting material can also lead to fires inside the shroud.

### 12.6.3 Size of Chimneys.

Information on the sizing of chimneys, addressed in 12.6.3, is provided in Chapter 13 in the form of tables and additional requirements.

The Chapter 13 tables and additional requirements recognize that there is a significant difference between the operation of interior and exterior chimneys. An exterior chimney is one in which one or more sides are exposed to the outdoors below the roofline. An interior

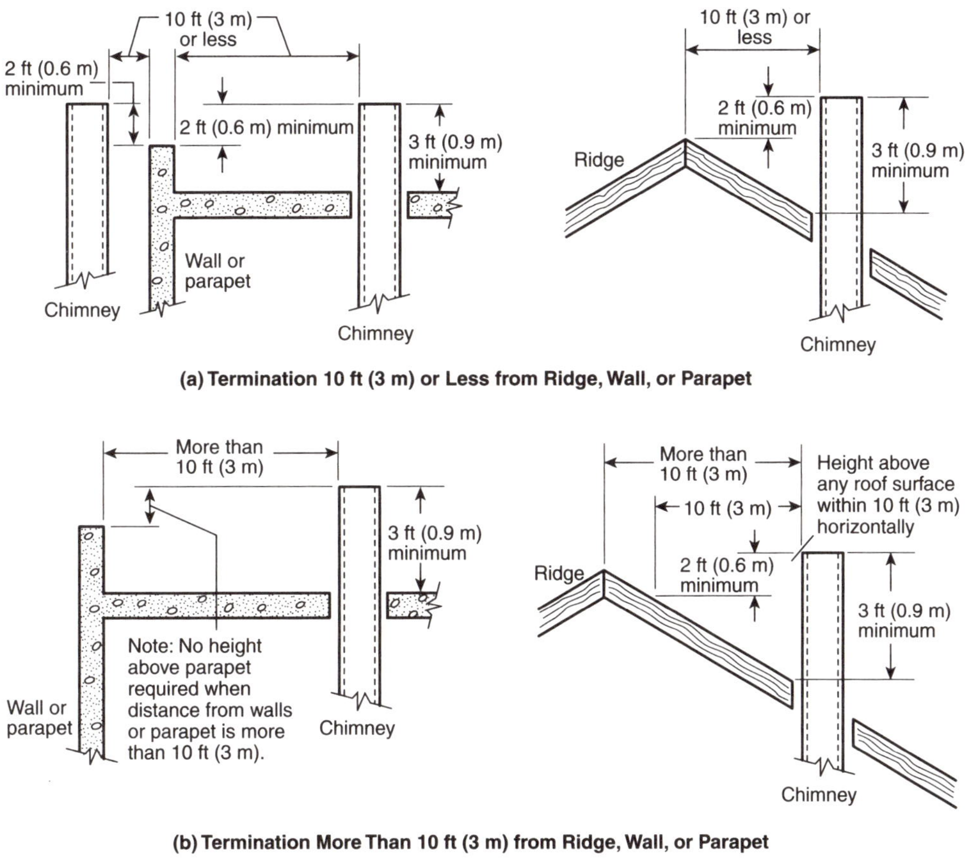


***FIGURE 12.6.2.1*** *Typical Termination Locations for Chimneys and Single-Wall Metal Pipes Serving Residential-Type and Low-Heat Appliances.*

***EXHIBIT 12.11*** *Decorative Vent Shroud. (Courtesy of Magic Sweep Corporation)*

chimney is not exposed to the outdoors below the building's roofline. (Chimneys that pass through unheated attics or garages are not considered to be outdoors.) An interior chimney is largely isolated from weather changes. Outside ambient temperatures affect an exterior chimney. Therefore, the requirements for exterior chimneys are keyed to the lowest temperature expected in the installation location.

The requirements for sizing chimneys were developed using a flue loss of 17 percent for the heating appliance. Most real furnaces and boilers operate at a higher flue loss. Therefore, the code requirements are conservative. Several manufacturers offer customized vent kits, which are keyed to the exact performance of their appliances. In this case, their recommendations should be used.

Exhibit 12.12 shows an example of a custom vent kit. The exhibit illustrates a furnace with a field-installed modification utilizing customized venting instructions for venting using a masonry chimney. Note the draft hood to the right of the induced draft blower, which has been added. The kit includes the hardware and venting instructions.

**12.6.3.1** The effective area of a chimney venting system serving listed appliances with draft hoods, Category I appliances, and other appliances listed for use with Type B vents shall be in accordance with one of the following methods:

(1) Those listed in Chapter 13.
(2) For sizing an individual chimney venting system for a single appliance with a draft hood, the effective areas of the vent connector and chimney flue shall be not less than the area

*EXHIBIT 12.12 Customized Vent Kit. (Courtesy of Carrier Corporation)*

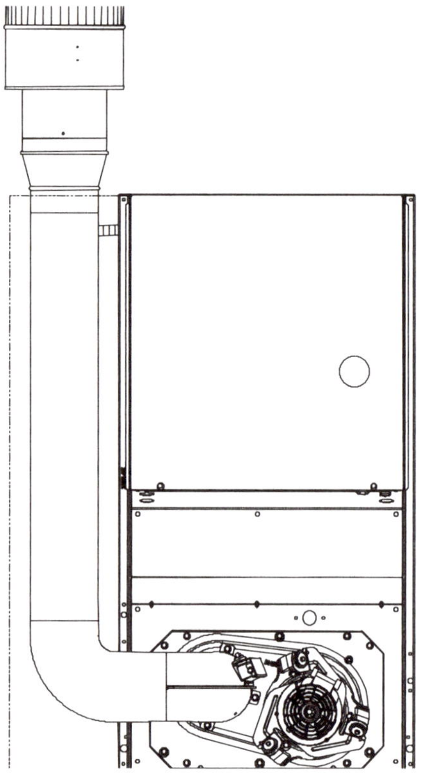

of the appliance flue collar or draft hood outlet or greater than seven times the draft hood outlet area.

(3) For sizing a chimney venting system connected to two appliances with draft hoods, the effective area of the chimney flue shall be not less than the area of the larger draft hood outlet plus 50 percent of the area of the smaller draft hood outlet or greater than seven times the smaller draft hood outlet area.

(4) Chimney venting systems using mechanical draft shall be sized in accordance with approved engineering methods.

(5) Other approved engineering methods.

**FAQ ▶ Are there simple alternative sizing methods that can be used in place of the vent system sizing tables in Chapter 13?**

The sizing methods in 12.6.3.1(2) and 12.6.3.1(3) permit simple, alternate chimney sizing methods that limit the vent to a maximum size of seven times the smallest draft hood outlet area or a minimum size based on the draft hood outlet areas. These methods, one used where one appliance is vented [12.6.3.1(2)] and the other used where two appliances are common-vented [12.6.3.1(3)], limit the maximum size of the vent connector and gas vent to minimize condensation in the gas vent caused by insufficient flow of hot vent gases to heat an excessively large gas vent. Note that the alternate sizing methods are applicable only to appliances equipped with draft hoods. Excessive condensation can cause premature failure of a chimney.

Note that 12.6.3.1(3) limits the use of these alternatives to chimney venting systems serving only two appliances. Sample calculations using the venting tables in Chapter 13 and the alternate method showed that the alternate method results in vent sizes too small when four or more appliances are involved and in the case of some tall vents. The alternate method always provides acceptable vent sizes for two appliances. Venting systems serving more than two appliances must use the vent sizing tables in Chapter 13 or other engineering methods.

In the 2009 edition, former paragraph 12.6.3.2, covering incinerators vented in chimneys serving other appliances, was deleted as it was in conflict with 12.6.5.1, which does not allow solid fuel–burning appliances to be vented in a chimney serving gas appliances. As incinerators in most buildings are no longer installed for environmental reasons, the subject is no longer relevant.

## 12.6.4 Inspection of Chimneys.

**12.6.4.1** Before replacing an existing appliance or connecting a vent connector to a chimney, the chimney passageway shall be examined to ascertain that it is clear and free of obstructions and shall be cleaned if previously used for venting solid or liquid fuel–burning appliances or fireplaces.

Paragraph 12.6.4.1 covers inspection prior to the connection of a new appliance to an existing chimney. It is important that the chimney be cleaned prior to connecting a gas appliance to a chimney that was previously used for oil- or solid fuel–burning appliances because the products of gas combustion can soften a buildup of soot or creosote that may have been caused by oil- or solid fuel–burning appliances. These deposits, if softened, can break away and block the chimney.

Oil- and solid fuel–burning appliances can deposit sulfur residue in the chimney. When this sulfur is exposed to water vapor from gas appliances, it can form sulfuric acid, which leads to rapid chimney deterioration. The requirement to clean the chimney will help minimize, but not eliminate, this problem.

**12.6.4.2** Chimneys shall be lined in accordance with NFPA 211, *Standard for Chimneys, Fireplaces, Vents, and Solid Fuel–Burning Appliances.*

*Exception: Existing chimneys shall be permitted to have their use continued when an appliance is replaced by an appliance of similar type, input rating, and efficiency, where the chimney complies with 12.6.4 and the sizing of the chimney is in accordance with 12.6.3.* NEW

The exception to 12.6.4.2 recognizes that not all appliances have been converted to the higher-efficiency types at this time. Water heaters and space heaters are examples of appliances that are not yet available in high-efficiency types.

Note that the exception was revised in the 2009 edition with the addition of "where the chimney complies with 12.6.4 and the sizing of the chimney is in accordance with 12.6.3," to clarify that the exception has limits. Every chimney must be lined for any appliance to be connected to it, unless the replacement appliance is of the same type, input rating, and efficiency as the one it is replacing. Fan-assisted combustion appliances have been developed to meet the minimum energy efficiency requirements of PL 100-12, National Appliance Energy Conservation Act of 1987. These appliances obtain a higher efficiency by extracting heat that formerly rose with the vent gases. Therefore, the vent gases are cooler, and condensation of water (a product of combustion) occurs during the initial period of appliance operation, making the lining of chimneys mandatory to prevent corrosion. This energy-saving feature resulted in revising Tables 13.1(a) through 13.1(d) and Tables 13.2(a) through 13.2(d), and adding Table 13.1(f) and Tables 13.2(f) through 13.2(i) in the 1992 and subsequent editions of the code. These tables provide a method to properly size vents and chimneys for these new appliances, and they provide a method to size masonry chimneys that are exposed to the outdoors below the roofline.

Also note that the exception to 12.6.4.2 does not remove or eliminate the need to verify other code requirements, such as inspection or proper sizing of the venting system. When venting into unlined masonry chimneys, proper sizing is critical to prevent condensation and deterioration in these old chimneys. Paragraph 12.6.4.4 requires that an unlined chimney be safe for continued use. When new appliances are installed, they must be installed according to the requirements of this chapter and the manufacturer's installation instructions. Most, if not all, manufacturers will prohibit the installation of their product in unlined chimneys.

For additional information on the venting of fan-assisted combustion appliances, see Supplement 1, Development of Revised Venting Guidelines, in this handbook, which describes the work that went into the development of the new tables and the code requirements.

**12.6.4.3** Cleanouts shall be examined to determine that they remain tightly closed when not in use.

It is important that cleanout openings remain closed during operation of the venting system. When air enters the chimney through the cleanout, available draft for the connected appliance(s) is reduced. An open cleanout can cause flue gases to enter the building.

**12.6.4.4** When inspection reveals that an existing chimney is not safe for the intended application, it shall be repaired, rebuilt, lined, relined, or replaced with a vent or chimney to conform to NFPA 211, *Standard for Chimneys, Fireplaces, Vents, and Solid Fuel–Burning Appliances*, and shall be suitable for the appliances to be attached.

Some common chimney troubles and ways to detect and remedy them are shown in Exhibit 12.13. It shows a chimney with two flues, one serving a wood-burning fireplace, and the other serving two gas appliances. Included is a list of common chimney problems. This information provides a relatively simple, practical guide for installers who are not chimney experts to assist in diagnosing chimney problems. If in doubt, a chimney sweep should be contacted for a thorough inspection.

### 12.6.5 Chimney Serving Appliances Burning Other Fuels.

**12.6.5.1** An appliance shall not be connected to a chimney flue serving a separate appliance designed to burn solid fuel.

The common venting of gas and solid fuel–burning appliances is prohibited in 12.6.5.1 for several reasons. Solid fuel–burning appliances increase the amount of deposits in the venting system and can lead to blockages of the venting system. Blocked venting systems serving gas appliances can be dangerous, and the danger may not be apparent immediately.

The possibility of chimney fires is greater in chimneys serving solid fuel–burning appliances. A chimney fire could burn uncontrolled due to air entering the chimney through the draft hoods of gas appliances.

**12.6.5.2** Where one chimney serves gas appliances and liquid fuel–burning appliances, the appliances shall be connected through separate openings or shall be connected through a single opening where joined by a suitable fitting located as close as practical to the chimney. Where two or more openings are provided into one chimney flue, they shall be at different levels. Where the gas appliance is automatically controlled, it shall be equipped with a safety shutoff device.

Placing openings into the chimney at different levels, as required in 12.6.5.2, helps prevent pressurization of one vent or chimney connector by the flue gases of the opposite appliance.

**12.6.5.3*** A listed combination gas- and solid fuel–burning appliance connected to a single chimney flue shall be equipped with a manual reset device to shut off gas to the main burner in the event of sustained backdraft or flue gas spillage. The chimney flue shall be sized to properly vent the appliance.

**A.12.6.5.3** Reference can also be made to the chapter on chimney, gas vent, and fireplace systems of the *ASHRAE Handbook — HVAC Systems and Equipment.*

**12.6.5.4** A single chimney flue serving a listed combination gas- and oil-burning appliance shall be sized to properly vent the appliance.

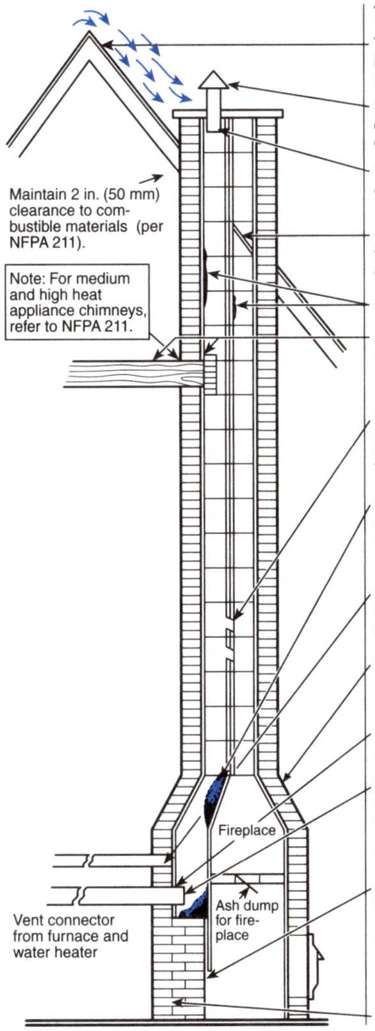


| Trouble | Disclosed by | Remedy |
|---|---|---|
| Top of chimney lower than surrounding objects | Observation | Extend chimney. |
| Undersized chimney cap or ventilator | Observation | Remove or replace. |
| Restricted opening | Measurement | Make opening as large as inside of chimney. |
| Piece of broken tile wedged in chimney | Lowering a light or weight down chimney | Break tile with a rod or weight on string or wire. |
| Creosote buildup | Observation | Clean chimney. |
| Joist protruding into chimney | Lowering a light or weight down chimney | Change support for joist so that chimney will be clear. |
| Leakage between loosely joined tiles | Start smoke fire with top of chimney closed; look for leakage from chimney | Rebuild chimney with a course of brick between flue tiles. |
| Debris accumulated in offset | Lowering a light or weight down chimney | Breakout with rod or weight may be necessary to open chimney. |
| Solid fuel-burning appliance connection | Observation | Remove and seal opening. |
| Offset | Lowering a light or weight down chimney | Change to straight or to long offset. |
| Loosely fitted smoke pipe | Smoke test | Clean leaks with cement. |
| Smoke pipe extends into chimney | Lowering a light or weight down chimney | Make end flush with inside of chimney. |
| Chimney too short | Measurement | Extend. |
| Opening between flues | Smoke test | Close opening permanently. |
| Chimney too small | Measurement | Rebuild. |
| Chimney too large | Measurement | Rebuild or insert liner. |
| No cleanout | Observation | Add cleanout. |

*EXHIBIT 12.13 Common Chimney Troubles.*

## 12.6.6 Support of Chimneys.

All portions of chimneys shall be supported for the design and weight of the materials employed. Listed factory-built chimneys shall be supported and spaced in accordance with the manufacturer's installation instructions.

## 12.6.7 Cleanouts.

Where a chimney that formerly carried flue products from liquid or solid fuel–burning appliances is used with an appliance using fuel gas, an accessible cleanout shall be provided. The cleanout shall have a tight-fitting cover and be installed so its upper edge is at least 6 in. (150 mm) below the lower edge of the lowest chimney inlet opening.

When a gas heating unit or conversion burner is connected to a chimney that previously served to vent an appliance that used another fuel, after a period of operation on natural gas or propane, carbon deposits or mortar materials can become dislodged from the chimney wall and drop to the base of the chimney. The accumulation of this debris, if great enough, can result in blockage of the venting system, sometimes with fatal results.

A 6 in. (150 mm) drop leg and inspection/maintenance cleanout is required in 12.6.7 of the code. The cleanout provides a sump to collect a considerable amount of debris before a potential accident occurs. The cleanout also enables homeowners and service personnel to clean debris easily and to inspect conditions.

### 12.6.8 Space Surrounding Lining or Vent.

**FAQ ►** Can a gas vent be installed in an unused, unlined masonry chimney?

Subsection 12.6.8 covers the installation of gas vents and chimney liners in an abandoned chimney. The use of a masonry chimney as a chase for vents is a practical solution to an existing unlined, oversized, or failed chimney. When installing vents in a chimney space, the vent material manufacturer's installation requirements must be followed thoroughly. If required by the vent manufacturer, penetrating the masonry chimney to install supports might be necessary.

When using an abandoned chimney as a chase for multiple vents, care should be taken to ensure that the vent materials are compatible and that vent failure does not result inadvertently. Category I or III gas appliance vent gas temperatures can exceed 300°F (150°C). Contact between the high-temperature Category I or III vent material and the PVC vent pipe can cause the PVC pipe to degrade or soften to the extent of failure. With the chimney opening sealed around the vent pipes, the following two potential hazards are created:

1. The products of combustion, which can escape from the failed PVC pipe, will be trapped in the chimney. These gases could eventually leak back into the living space or corrode the other nearby vents.
2. Heat leakage from a vent will not be dissipated as quickly in a chase as in the open. This leakage results in increased temperature in the chase, which can cause failure of some plastic vent materials.

**12.6.8.1** The remaining space surrounding a chimney liner, gas vent, special gas vent, or plastic piping installed within a masonry chimney shall not be used to vent another appliance.

*Exception: The insertion of another liner or vent within the chimney as provided in this code and the liner or vent manufacturer's instructions.*

The intent of 12.6.8.1 is to prohibit the venting of an appliance into the original chimney flue (or cavity) around a chimney liner or gas vent serving another appliance. Problems may include an inadequate venting area in the masonry chimney cavity, the effects of flue gases and condensate on the exterior surface of the chimney liner or gas vent, and the likelihood of a top support plate for the chimney liner or gas vent blocking the top of the chimney.

**12.6.8.2** The remaining space surrounding a chimney liner, gas vent, special gas vent, or plastic piping installed within a masonry, metal, or factory-built chimney flue shall not be used to supply combustion air.

*Exception: Direct vent appliances designed for installation in a solid fuel–burning fireplace where installed in accordance with the manufacturer's installation instructions.*

**FAQ ►** Can the area around a chimney liner inside a masonry chimney be used to supply combustion air?

Paragraph 12.6.8.2 prohibits the use of the original chimney cavity as a passageway or duct for combustion air when a chimney liner or gas vent is installed within the chimney. The exception allows the original chimney cavity to supply combustion air to a direct vent insert in a fireplace. In this limited case, a chimney liner is installed within the original chimney to vent the flue gases from the direct vent appliance, while the combustion air is drawn through the original cavity around the exhaust pipe. The exception only applies when this installation method is specifically described in the manufacturer's instructions.

See also the commentary following 12.7.1(3) covering labeling requirements where gas vents are installed in abandoned masonry chimneys.

## 12.7 Gas Vents

See 3.3.107.2.

Section 12.7 applies to gas vents of all types. Not included are chimneys, which are covered in Section 12.6, and single-wall metal pipe, which is covered in Section 12.8. Note that the term *gas vent* is defined as follows:

> **3.3.107.2 Gas Vent.** A passageway composed of *listed* factory-built components assembled in accordance with the manufacturer's installation instructions for conveying vent gases from appliances or their vent connectors to the outdoors.

The word "listed" is highlighted to emphasize that unlisted products are not gas vents and cannot be installed using Section 12.7.

Type B and Type B-W gas vents that currently are listed are of double-wall construction: basically a pipe within a pipe, with airspace between the two walls. This construction reduces the heat loss from the vent gases in much the same way that an insulated coffee mug works.

Usually, the inner wall is made of aluminum to resist corrosion, and the outer wall is made of galvanized steel for strength and corrosion resistance. The parts of the vent system — pipe sections, supports, spacers, caps, and flashings — are furnished for field erection to form a continuous passageway from the gas appliance to the vent termination above the roof, including the cap or roof assembly. Several companies manufacture the vent pipe and other related parts needed to erect a complete vent. The piping and parts produced by the different manufacturers are not interchangeable without the use of special, listed adapters.

Type B gas vents can be round or oval (see Exhibit 12.14). Type B-W gas vents, as shown in Exhibit 12.15, are always oval, because they are designed to be installed within a wall between two studs. Type L vents are similar in construction to Type B gas vents, except that the inner pipe of a Type L vent is stainless steel. Type L vents are intended for venting flue gases at higher temperatures than are normally produced by gas appliances. Some oil furnaces, for instance, are listed for use with Type L vents.

Typical Type B and Type B-W gas vent installations are diagrammed in Exhibit 12.16 and Exhibit 12.17, respectively.

***EXHIBIT 12.14*** *Construction of Double-Wall Type B Gas Vent. (Courtesy of Selkirk Metalbestos)*

***EXHIBIT 12.15*** *Type B and Type B-W Double-Wall Gas Vent Fittings. (Courtesy of Selkirk Metalbestos)*

***EXHIBIT 12.16*** *Typical Type B Gas Vent — Single Appliance Installation.*

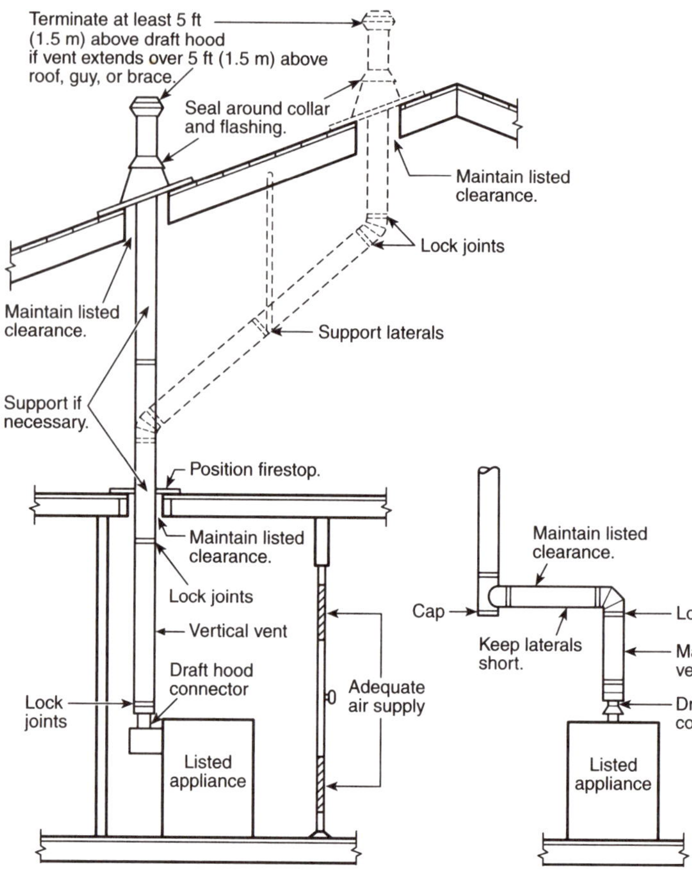


## 12.7.1 Installation.

The installation of gas vents shall meet the following requirements:

Note that the title of 12.7.1 has been changed in the 2009 edition from "Application" to better convey the contents of the requirement.

(1) Gas vents shall be installed in accordance with the manufacturer's installation instructions.

The requirement of 12.7.1(1) is a reminder that gas vents are a listed product and, as such, must be installed in compliance with the installation instructions provided by the manufacturer and in accordance with this code.

(2) A Type B-W gas vent shall have a listed capacity not less than that of the listed vented wall furnace to which it is connected.

Paragraph 12.7.1(2) references a Type B-W gas vent that is installed within the stud space of a wall and intended for venting listed wall furnaces. The installation instructions for the Type B-W vent should provide the maximum capacity of the vent and will usually refer to the code.

Note that in the 2009 edition, the former paragraph 12.7.1(3), which required Category I fan-assisted appliances to be installed in accordance with the manufacturer's instructions and Chapter 13 or other approved engineering methods, was deleted because it repeated the sizing requirements of 12.7.3.

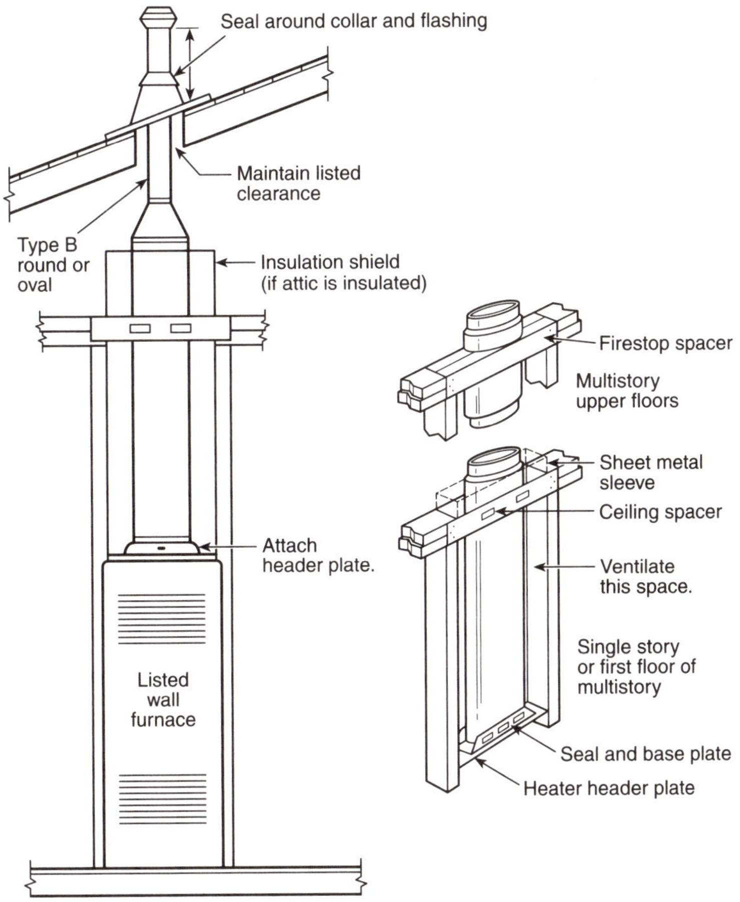


***EXHIBIT 12.17*** *Typical Type B-W Gas Vent Installation.*

(3) Gas vents installed within masonry chimneys shall be installed in accordance with the manufacturer's installation instructions. Gas vents installed within masonry chimneys shall be identified with a permanent label installed at the point where the vent enters the chimney. The label shall contain the following language: "This gas vent is for appliances that burn gas. Do not connect to solid or liquid fuel–burning appliances or incinerators."

The provision in 12.7.1(3) requires masonry chimneys in which gas vents are installed to be identified with a permanent label, and the requirement provides the specific text to be used. Additional text can be added. Because a vent installed in a masonry chimney will probably not be visible, the label is intended to warn future installers not to connect wood- or oil-fired equipment to the gas vent. Exhibit 12.18 shows a Type B vent installed in a masonry chimney.

(4) Screws, rivets, and other fasteners shall not penetrate the inner wall of double wall gas vents, except at the transition from the appliance draft hood outlet, flue collar, or single wall metal connector to a double wall vent. NEW

In the 2009 edition, 12.7.1(4) was added to recognize that screws and other fasteners that penetrate the inner wall of double wall gas vents can result in the following problems:

1. Corrosion of the screw or fastener, which can leave a hole in the vent wall, allowing air to enter the vent. Air entering a vent can negatively affect vent performance.
2. The screw or fastener can cause turbulence in the vent, with reduced vent flow.

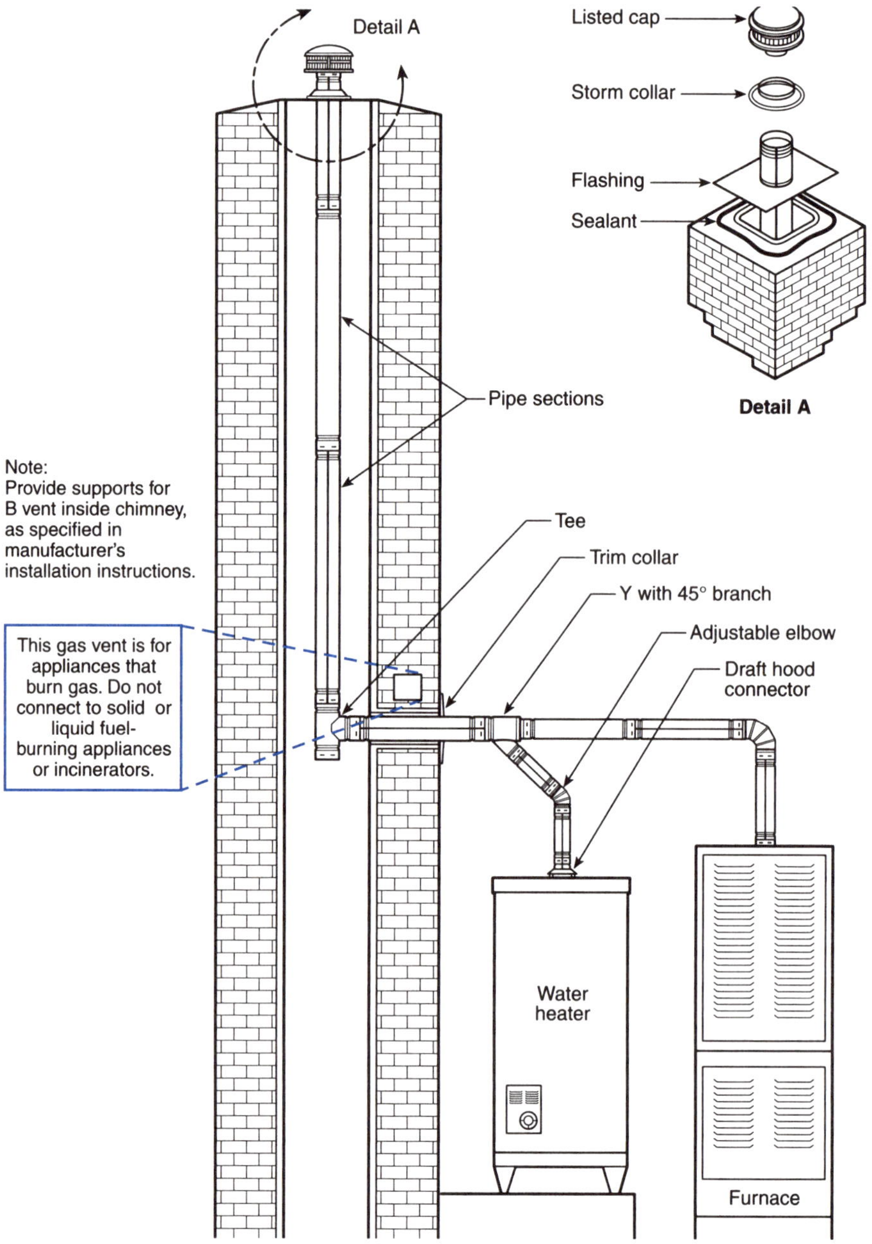


***EXHIBIT 12.18*** *Typical Installation Using Type B Vent to Line a Chimney.*

3. The screw or fastener can conduct heat out of the vent, resulting in reduced vent performance.

## 12.7.2 Gas Vent Termination.

The termination of gas vents shall comply with the following requirements:

(1) A gas vent shall terminate in accordance with one of the following:

(a) Gas vents that are 12 in. (300 mm) or less in size and located not less than 8 ft (2.4 m) from a vertical wall or similar obstruction shall terminate above the roof in accordance with Figure 12.7.2 and Table 12.7.2.

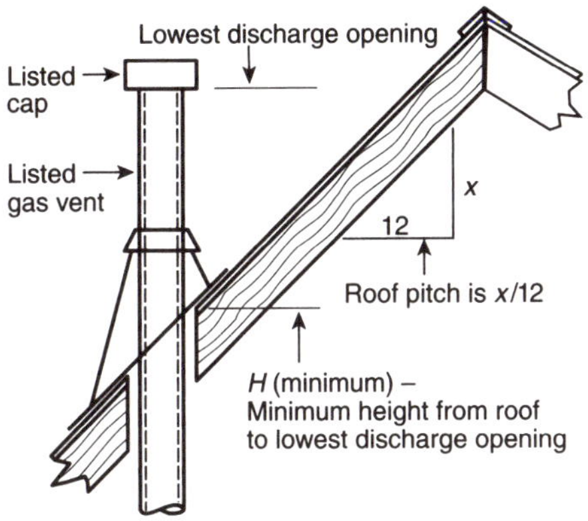


*FIGURE 12.7.2 Termination Locations for Gas Vents with Listed Caps 12 in. (300 mm) or Less in Size at Least 8 ft (2.4 m) from a Vertical Wall.*

*TABLE 12.7.2 Roof Slope Heights*

| | H (minimum) | |
|---|---|---|
| **Roof Slope** | ***ft*** | ***m*** |
| Flat to 6/12 | 1.0 | 0.30 |
| Over 6/12 to 7/12 | 1.25 | 0.38 |
| Over 7/12 to 8/12 | 1.5 | 0.46 |
| Over 8/12 to 9/12 | 2.0 | 0.61 |
| Over 9/12 to 10/12 | 2.5 | 0.76 |
| Over 10/12 to 11/12 | 3.25 | 0.99 |
| Over 11/12 to 12/12 | 4.0 | 1.22 |
| Over 12/12 to 14/12 | 5.0 | 1.52 |
| Over 14/12 to 16/12 | 6.0 | 1.83 |
| Over 16/12 to 18/12 | 7.0 | 2.13 |
| Over 18/12 to 20/12 | 7.5 | 2.27 |
| Over 20/12 to 21/12 | 8.0 | 2.44 |

(b) Gas vents that are over 12 in. (300 mm) in size or are located less than 8 ft (2.4 m) from a vertical wall or similar obstruction shall terminate not less than 2 ft (0.6 m) above the highest point where they pass through the roof and not less than 2 ft (0.6 m) above any portion of a building within 10 ft (3.0 m) horizontally.
(c) Industrial appliances as provided in 12.3.4.
(d) Direct vent systems as provided in 12.3.5.
(e) Appliances with integral vents as provided in 12.3.6.
(f) Mechanical draft systems as provided in 12.4.3.
(g) Ventilating hoods and exhaust systems as provided in 12.4.4.

Paragraphs 12.7.2(1)(c) through 12.7.2(1)(g) are alternatives to the normal vent termination described in 12.7.2(1)(a) and (b). Paragraph 12.7.2(1)(c), which references 12.3.4, Well-Ventilated Spaces, is included here to clarify that vent termination may not apply to spaces in which industrial gas equipment is installed if the spaces are well-ventilated.

(2) A Type B or a Type L gas vent shall terminate at least 5 ft (1.5 m) in vertical height above the highest connected appliance draft hood or flue collar.
(3) A Type B-W gas vent shall terminate at least 12 ft (3.7 m) in vertical height above the bottom of the wall furnace.

Paragraphs 12.7.2(2) and (3) are intended to require minimum vent heights to achieve a level of draft that is adequate to remove flue gases safely to the outdoors. It is important to recognize that the addition of sheds and renovation involving roofs, such as the addition of a dormer, may result in an existing installation that is no longer in compliance with this requirement and that has the potential of improper venting and backdrafting.

(4) A gas vent extending through an exterior wall shall not terminate adjacent to the wall or below eaves or parapets, except as provided in 12.3.5 and 12.4.3.
(5) Decorative shrouds shall not be installed at the termination of gas vents except where such shrouds are listed for use with the specific gas venting system and are installed in accordance with manufacturers' installation instructions.

For more information on decorative shrouds, see the commentary following 12.6.2.4.

(6) All gas vents shall extend through the roof flashing, roof jack, or roof thimble and terminate with a listed cap or listed roof assembly.

The use of a listed cap, required in 12.7.2(6), helps ensure that animal nesting and debris will not obstruct the vent. Vent caps for Type B and Type B-W gas vents are designed and tested to serve the following four purposes:

1. Reduce the chance of downdraft caused by wind effect
2. Provide proper venting under wind conditions
3. Permit vent termination at a minimum height above the roof
4. Prevent rain and debris from entering the vent

Notice that a listed vent cap allows the vent termination to be closer to the nearby roofline, because the terminations are tested to operate properly using the clearances shown in Figure 12.7.2 of the code.

(7) A gas vent shall terminate at least 3 ft (0.9 m) above a forced air inlet located within 10 ft (3.0 m).

### 12.7.3 Size of Gas Vents.

Venting systems shall be sized and constructed in accordance with Chapter 13 or other approved engineering methods and the gas vent and the appliance manufacturer's instructions.

**12.7.3.1* Category I Appliances.** The sizing of natural draft venting systems serving one or more listed appliances equipped with a draft hood or appliances listed for use with Type B gas vent, installed in a single story of a building, shall be in accordance with one of the following:

(1) The provisions of Chapter 13.
(2) Vents serving fan-assisted combustion system appliances, or combinations of fan-assisted combustion system and draft hood–equipped appliances, shall be sized in accordance with Chapter 13 or other approved engineering methods.
(3) For sizing an individual gas vent for a single, draft hood–equipped appliance, the effective area of the vent connector and the gas vent shall be not less than the area of the appliance draft hood outlet or greater than seven times the draft hood outlet area.
(4) For sizing a gas vent connected to two appliances, with draft hoods, the effective area of the vent shall be not less than the area of the larger draft hood outlet plus 50 percent of the area of the smaller draft hood outlet or greater than seven times the smaller draft hood outlet area.
(5) Other approved engineering practices.

The tables in Chapter 13 provide a safe method for sizing vents for all Category I and draft hood–equipped appliances. The alternate methods permitted in 12.7.3.1 offer simple techniques for installers to follow for commonly encountered vent arrangements. These methods provide a maximum size of seven times the smallest draft hood outlet area and a minimum size based on the draft hood outlet area. This provision limits the maximum size of the vent connector and gas vent to minimize gas vent condensation caused by insufficient flow of hot vent gases to heat an excessively large gas vent. Excessive condensation can cause premature failure of a gas vent.

Note that these alternate methods cannot be used for fan-assisted combustion appliances. The tables in Chapter 13 (or other engineering methods) must be used for all installations of fan-assisted combustion appliances. Vents for Category II, III, and IV appliances must be in accordance with the equipment manufacturer's instructions. These appliances are listed by definition, and the manufacturer's instructions reflect the test results determined by the listing laboratory.

Can the "rule of thumb" alternate sizing method be used to size systems for fan-assisted appliances?
**FAQ ►**

Paragraphs 12.7.3.1(3) and 12.7.3.1(4) allow the use of an alternate method for sizing a venting system, rather than the tables in Chapter 13. This alternate method applies only to draft hood–equipped appliances. Paragraph 12.7.3.1(3) provides guidance for installations with one appliance, while 12.7.3.1(4) provides guidance for sizing installations with two appliances. Venting systems serving three or more appliances must use the vent sizing tables in Chapter 13 or other engineering methods. Refer to the commentary following 12.6.3.1, which allows a similar alternate method for chimneys.

**A.12.7.3.1** Additional information on sizing venting systems can be found in the following:

(1) Tables in Chapter 13
(2) The appliance manufacturer's instructions
(3) The vent system manufacturer's sizing instructions
(4) Drawings, calculations, and specifications provided by the vent system manufacturer
(5) Drawings, calculations, and specifications provided by a competent person
(6) The chapter on chimney, gas vent, and fireplace systems of the *ASHRAE Handbook — HVAC Systems and Equipment*

Category I appliances may be either draft hood–equipped or a fan-assisted combustion system in design. Different vent design methods are required for draft hood–equipped and fan-assisted combustion system appliances.

**12.7.3.2 Vent Offsets.** Type B and Type L vents sized in accordance with 12.7.3.1(3) or 12.7.3.1(4) shall extend in a generally vertical direction with offsets not exceeding 45 degrees, except that a vent system having not more than one 60 degree offset shall be permitted. Any angle greater than 45 degrees from the vertical is considered horizontal. The total horizontal distance of a vent plus the horizontal vent connector serving draft hood–equipped appliances shall not be greater than 75 percent of the vertical height of the vent.

Paragraph 12.7.3.2 provides limitations on offsets for venting systems that are sized using the alternative sizing method (commonly referred to as "rule of thumb" method) of 12.7.3.1. Installations with only one offset in the vent are limited to a maximum offset angle of 60 degrees (from vertical) or less, while installations involving multiple offsets are limited to 45-degree offsets (from vertical).

Paragraph 12.7.3.2 also states that any angle greater than 45 degrees (from vertical) is considered horizontal and that the total horizontal distance is limited to 75 percent of the total vent height. Horizontal vent offsets reduce venting performance and should be avoided if possible. Where required, horizontal offsets should be kept as short as practical.

Note that the limitations in 12.7.3.2 apply only where the alternate method is used to size a gas vent. Where the tables in Chapter 13 or engineering methods are used to size the vent, 12.7.3.2 does not apply.

**12.7.3.3 Category II, Category III, and Category IV Appliances.** The sizing of gas vents for Category II, Category III, and Category IV appliances shall be in accordance with the appliance manufacturer's instructions.

**12.7.3.4 Sizing.** Chimney venting systems using mechanical draft shall be sized in accordance with approved engineering methods.

### 12.7.4 Gas Vents Serving Appliances on More than One Floor.

**12.7.4.1** A common gas vent shall be permitted in multistory installations to vent Category I appliances located on more than one floor level, provided the venting system is designed and installed in accordance with approved engineering methods. For the purpose of this section, crawl spaces, basements, and attics shall be considered as floor levels.

When sizing a multi-story common vent, crawl spaces, basements, and attics are considered to be individual floor levels. Where appliances are installed on different levels utilizing the same vent, a vent blockage will cause combustion products to spill out of the draft hood of appliances installed below the blockage and possibly into the living space. The risk of this spillage applies to single-family multistory vents, which are subject to the same requirement as vents in larger multistory buildings. Appliances must be separated from the living space to prevent this potential hazardous condition from occurring.

**12.7.4.2** All appliances connected to the common vent shall be located in rooms separated from occupiable space. Each of these rooms shall have provisions for an adequate supply of combustion, ventilation, and dilution air that is not supplied from occupiable space. *(See Figure 12.7.4.2.)*

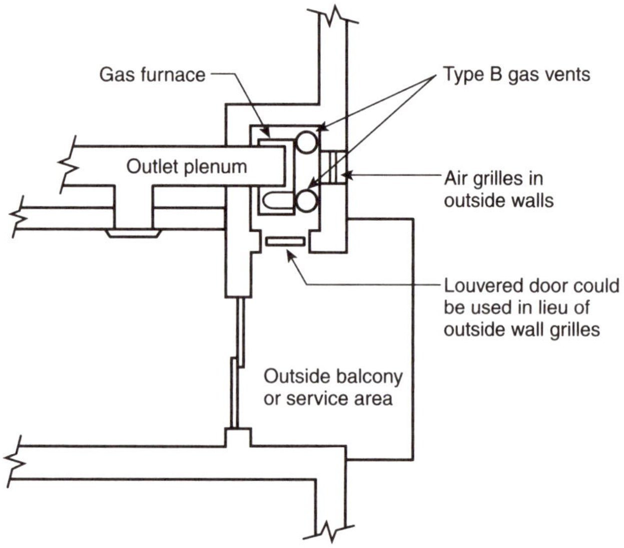


***FIGURE 12.7.4.2** Plan View of Practical Separation Method for Multistory Gas Venting.*

**12.7.4.3** The size of the connectors and common segments of multistory venting systems for appliances listed for use with a Type B double-wall gas vent shall be in accordance with Table 13.2(a), provided all of the following apply:

(1) The available total height *(H)* for each segment of a multistory venting system is the vertical distance between the level of the highest draft hood outlet or flue collar on that floor and the centerline of the next highest interconnection tee. *[See Figure G.1(k).]*
(2) The size of the connector for a segment is determined from the appliance's gas input rate and available connector rise and shall not be smaller than the draft hood outlet or flue collar size.
(3) The size of the common vertical vent segment, and of the interconnection tee at the base of that segment, is based on the total appliance's gas input rate entering that segment and its available total height.

Paragraph 12.7.4.3(3) provides requirements for the separation of appliances from the habitable space (shown in Figure 12.7.4.2 of the code), where multistory venting is used. Further information on sizing vents that serve appliances on more than one floor of a building can be found in 13.2.14 through 13.2.17 and Figure G.1(m) and Figure G.1(n).

FAQ ▶ Under what conditions is it safe to vent gas appliances on different levels of the building into the same venting system?

Safe venting can be achieved in high-rise, multistory appliance installations by separating the appliance from any habitable or occupied space. The separation resolves the question of safety that arises when intercommunication of vents exists between various floors of a building. Separation ensures that, in the event of spillage, no vent gases will enter the occupied space from the appliance room if the common vent becomes obstructed at any level or if the outlet is blocked. When such stoppage occurs, all vent gases from appliances operating below the obstruction will exit through the upper draft hood relief opening rather than through the vent outlet. Large quantities of vent gases will be dumped into the space containing those appliances immediately below the obstruction. At the same time, the appliances located at lower levels will appear to be operating normally.

One practical plan to separate or isolate appliances, if installation can be made adjacent to an outside wall, is shown in Figure 12.7.4.2 of the code and in Exhibit 12.19. Access to the appliance room is through a door that opens onto an outdoor balcony. A panel separates the appliance room from the inside of the building. If the appliance is a central furnace, the cold air return and outlet ducts are attached to the furnace.

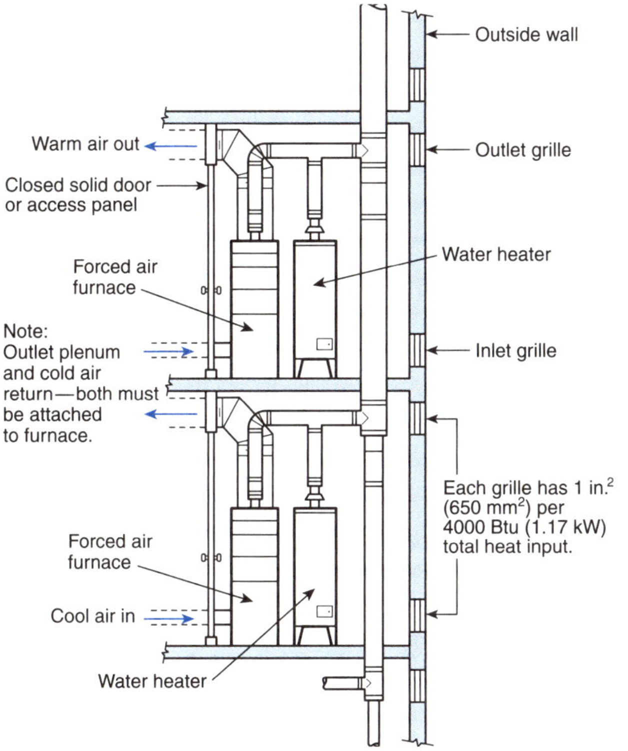


***EXHIBIT 12.19*** *Outside Wall Air Supply in a Multistory Building.*

Previous editions of this handbook suggested that a single combustion air duct could serve multiple floors of a building. In the 1999 edition, the intent was clarified by the addition of 9.3.8, which requires that each room in which an appliance is installed have its own combustion air duct and that there be separate ducts for the top and bottom combustion air openings into the space.

No requirements are given in the code for installation of Category II, III, or IV appliances using a common vent in multistory buildings. A Category II, III, or IV appliance must be installed in accordance with the terms of its listing and the manufacturer's instructions. Category I appliance common vents are not usually suitable for venting Category II, III, or IV appliances.

## 12.7.5 Support of Gas Vents.

Gas vents shall be supported and spaced in accordance with the manufacturer's installation instructions.

## 12.7.6 Marking.

In those localities where solid and liquid fuels are used extensively, gas vents shall be permanently identified by a label attached to the wall or ceiling at a point where the vent connector enters the gas vent. The label shall read: "This gas vent is for appliances that burn gas. Do not connect to solid or liquid fuel–burning appliances or incinerators." The authority having jurisdiction shall determine whether its area constitutes such a locality.

## 12.8 Single-Wall Metal Pipe

The use of single-wall metal pipe (e.g., stove pipe) as a vent for gas appliances is limited by Section 12.8 of the code, especially in dwellings, for the following reasons:

1. The surface of the pipe is usually hot enough to burn a person who accidentally contacts the pipe.
2. The high surface temperature can cause overheating of adjacent combustible material.
3. The heat loss from a long length of pipe in a cool area can result in reduced draft and the condensation of water vapor in the flue gases, which can corrode the single-wall metal pipe.

The use of single-wall metal pipe to vent products of combustion, where permitted in Section 12.8, is limited to Category I appliances. Note that some building codes do not permit single-wall metal chimneys and unlisted metal chimneys inside one- and two-family dwellings.

Single-wall metal pipe is a venting material that has limitations, because it has an inherently higher heat loss than chimneys and gas vents. Exhibit 12.20 shows the limitations of single-wall gas vents when compared with characteristics of double-wall gas vents.

*EXHIBIT 12.20 Comparison of Double-Wall and Single-Wall Gas Vent Characteristics.*

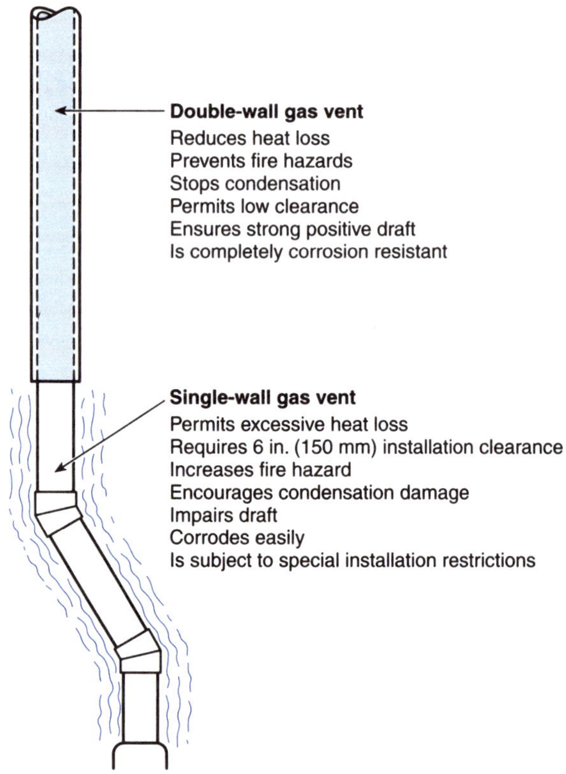


### 12.8.1 Construction.

Single-wall metal pipe shall be constructed of galvanized sheet steel not less than 0.0304 in. (0.7 mm) thick or of other approved, noncombustible, corrosion-resistant material.

### 12.8.2* Cold Climate.

Uninsulated single-wall metal pipe shall not be used outdoors for venting appliances in regions where the 99 percent winter design temperature is below 32°F (0°C).

The outdoor use of single-wall metal pipe in cold climates is restricted to prevent premature corrosion of the vent. In cold climates, the low ambient temperature can cool vent gases to the point at which water vapor condenses in the vent and causes corrosion.

The 99 percent winter design temperature is the temperature that can be expected to be met or exceeded for 99 percent of the time during the winter heating season, which includes the months of December, January, and February. From another perspective, the actual temperature should not be below the design temperature more than 1 percent of the time during the heating season. Annex G contains a map indicating design temperatures for the United States, but a more accurate assessment can be found in the *ASHRAE Handbook — Fundamentals.*

In developing this requirement, the committee recognized that condensation could occur at temperatures above 32°F (0°C), but believed this new requirement would greatly improve the previous requirement, while not being excessively restrictive. In the United States, this new requirement will prohibit single-wall metal pipe outdoors in most areas of the country. Only very warm climates, like the southern portions of California, Texas, and Florida, have winter design temperatures above 32°F (0°C). Where deterioration of single-wall metal pipe has proven to be a problem in those areas, its use should still be discouraged.

**A.12.8.2** Data on winter design temperature can be found in Figure G.2.4 and the 1993 edition of the *ASHRAE Handbook — Fundamentals*.

### 12.8.3 Termination.

The termination of single-wall metal pipe shall meet the following requirements:

(1) Single-wall metal pipe shall terminate at least 5 ft (1.5 m) in vertical height above the highest connected appliance draft hood outlet or flue collar.
(2) Single-wall metal pipe shall extend at least 2 ft (0.6 m) above the highest point where it passes through a roof of a building and at least 2 ft (0.6 m) higher than any portion of a building within a horizontal distance of 10 ft (3 m). *(See Figure 12.6.2.1.)*
(3) An approved cap or roof assembly shall be attached to the terminus of a single-wall metal pipe. *(Also see 12.8.4.4.)*

Paragraphs 12.8.3(1) through 12.8.3(3) provide the limitations under which single-wall metal pipe can be installed. The limitations are based on the fact that single-wall metal pipe can operate at a high temperature that easily can ignite combustible material placed in contact with it. When this type of pipe passes through a combustible roof, protection of the roof is required, and a thimble is specified. Paragraphs 12.8.4.1 and 12.8.4.2 severely limit the application of single-wall venting systems in residential buildings.

### 12.8.4 Installation with Appliances Permitted by 12.5.1.

**12.8.4.1* Prohibited Use.** Single-wall metal pipe shall not be used as a vent in dwellings and residential occupancies. NEW

**A.12.8.4.1** The prohibition only applies to a vent entirely constructed of single-wall metal pipe located in a residential occupancy. The prohibition does not apply to single-wall vent connectors used to connect an appliance to the vent as permitted in Section 12.11 and Chapter 13. NEW

In the 2009 edition, single-wall metal pipe is excluded from residential occupancies. This change reflects the fact that there have been very few, if any, installations where single-wall metal pipe could practically be used in residences. The limitations are due to the following issues:

- **High Surface Temperature.** Single-wall metal pipe is not insulated, and it gets hot enough during appliance operation to cause burns to human skin.
- **Heat Loss.** Single-wall metal pipe has significantly greater heat loss than Type B vent and chimneys. Heat loss reduces vent capacity — larger vents are required, and results in condensation with subsequent corrosion of the pipe.
- **Installation Restrictions.** Single-wall metal pipe is restricted to one floor. If installed in taller buildings it would have to run up alongside the building, increasing heat loss.

**12.8.4.2** Single-wall metal pipe shall be used only for runs directly from the space in which the appliance is located through the roof or exterior wall to the outer air. A pipe passing through a roof shall extend without interruption through the roof flashing, roof jacket, or roof thimble.

**12.8.4.3** Single-wall metal pipe shall not originate in any unoccupied attic or concealed space and shall not pass through any attic, inside wall, concealed space, or floor. *(For the installation of a single-wall metal pipe through an exterior combustible wall, see 12.8.4.7.)*

**12.8.4.4** Single-wall metal pipe used for venting an incinerator shall be exposed and readily examinable for its full length and shall have suitable clearances maintained.

**12.8.4.5** Minimum clearances from single-wall metal pipe to combustible material shall be in accordance with Table 12.8.4.5. Reduced clearances from single-wall metal pipe to combustible material shall be as specified for vent connectors in Table 10.2.3(b).

Paragraph 12.8.4.5 directs the user to Table 12.8.4.4 to determine clearance requirements for single-wall metal pipe based on the appliance type to be vented. Paragraph 12.8.4.5 also permits use of Table 10.2.3(b) to reduce the required clearance between the vent and combustible materials. Depending on the clearance reduction method selected, it may be possible to reduce the required clearance by up to 66 percent.

**12.8.4.6** Where a single-wall metal pipe passes through a roof constructed of combustible material, a noncombustible, nonventilating thimble shall be used at the point of passage. The

***TABLE 12.8.4.5*** *Clearances for Connectors*

| | *Minimum Distance from Combustible Material* | | | |
|---|---|---|---|---|
| *Appliance* | *Listed Type B Gas Vent Material* | *Listed Type L Vent Material* | *Single-Wall Metal Pipe* | *Factory-Built Chimney Sections* |
| Listed appliance with draft hoods and appliance listed for use with Type B gas vents | As listed | As listed | 6 in. | As listed |
| Residential boilers and furnaces with listed gas conversion burner and with draft hood | 6 in. | 6 in. | 9 in. | As listed |
| Residential appliances listed for use with Type L vents | Not permitted | As listed | 9 in. | As listed |
| Listed gas-fired toilets | Not permitted | As listed | As listed | As listed |
| Unlisted residential appliances with draft hood | Not permitted | 6 in. | 9 in. | As listed |
| Residential and low-heat appliances other than those above | Not permitted | 9 in. | 18 in. | As listed |
| Medium-heat appliance | Not permitted | Not permitted | 36 in. | As listed |

For SI units, 1 in. = 25.4 mm.

Note: These clearances shall apply unless the installation instructions of a listed appliance or connector specify different clearances, in which case the listed clearances shall apply.

thimble shall extend at least 18 in. (460 mm) above and 6 in. (150 mm) below the roof with the annular space open at the bottom and closed only at the top. The thimble shall be sized in accordance with 12.8.4.7.

**12.8.4.7** Single-wall metal pipe shall not pass through a combustible exterior wall unless guarded at the point of passage by a ventilated metal thimble not smaller than the following:

NEW

(1) For listed appliances with draft hoods and appliances listed for use with Type B gas vents, the thimble shall be a minimum of 4 in. (100 mm) larger in diameter than the metal pipe. Where there is a run of not less than 6 ft (1.8 m) of metal pipe in the opening between the draft hood outlet and the thimble, the thimble shall be a minimum of 2 in. (50 mm) larger in diameter than the metal pipe.
(2) For unlisted appliances having draft hoods, the thimble shall be a minimum of 6 in. (150 mm) larger in diameter than the metal pipe.
(3) For residential and low-heat appliances, the thimble shall be a minimum of 12 in. (300 mm) larger in diameter than the metal pipe.

*Exception: In lieu of thimble protection, all combustible material in the wall shall be removed a sufficient distance from the metal pipe to provide the specified clearance from such metal pipe to combustible material. Any material used to close up such opening shall be noncombustible.*

In the 2009 edition, this paragraph was relocated from 12.11.14.2, as the Single Wall Metal Pipe (SWMP) section is a more appropriate location for this requirement because it applies to all SWMP, not just vent connectors.

## 12.8.5 Size of Single-Wall Metal Pipe.

Single-wall metal piping shall comply with the following requirements:

(1)* A venting system of a single-wall metal pipe shall be sized in accordance with one of the following methods and the appliance manufacturer's instructions:

(a) For a draft hood–equipped appliance, in accordance with Chapter 13

The sizing tables in Chapter 13 for single-wall metal pipe, which are Table 13.1(e) and Table 13.2(e), may only be used to size single-wall pipe venting draft hood–equipped appliances. The single-wall venting tables do not include information on fan-assisted appliances. The use of single-wall metal pipe to vent fan-assisted appliances is not practical, due to the heat loss in the single-wall metal pipe, which will quickly lead to condensation and corrosion.

◀ **FAQ**
Can the vent sizing tables in Chapter 13 be used to size a single-wall metal pipe venting system for a fan-assisted appliance?

(b) For a venting system for a single appliance with a draft hood, the areas of the connector and the pipe each shall not be less than the area of the appliance flue collar or draft hood outlet, whichever is smaller. The vent area shall not be greater than seven times the draft hood outlet area.

An alternate method to size single-wall metal pipe venting systems is provided in 12.8.5(b) and can only be used for one draft hood–equipped appliance. (Single-wall metal pipe cannot be used to vent appliances not equipped with a draft hood.) The venting tables in Chapter 13 or an approved engineering method must be used to size single-wall vent systems serving two or more draft hood–equipped appliances.

◀ **FAQ**
What are the allowed methods for sizing a single-wall metal pipe venting system that serves a fan-assisted appliance?

(c) Other approved engineering methods

Approved engineering methods can be used to size single-wall metal pipe used to vent gas appliances. If it is desired to vent fan-assisted appliances using single-wall metal pipe, an engineering method must be used because the tables in Chapter 13 and the alternate method cannot be used. The editor is not aware of commercially available software that can do these complex calculations. The VENT-II computer program, discussed in Supplement 1 to this handbook, can be used, but it may not be commercially available, and it requires training to provide useful results.

**A.12.8.5(1)** Reference can also be made to the chapter on chimney, gas vent, and fireplace systems of the *ASHRAE Handbook — HVAC Systems and Equipment.*

(2) Where a single-wall metal pipe is used and has a shape other than round, it shall have an equivalent effective area equal to the effective area of the round pipe for which it is substituted and the minimum internal dimension of the pipe shall be 2 in. (50 mm).

**FAQ ►** Does the area of a round pipe decrease when the pipe is formed into an oval shape?

When a round pipe is formed into an oval shape, the internal area of the pipe decreases. For a simple visual demonstration of this principle, fill a paper cup to the top with water, then squeeze the sides of the cup to form an oval shape. You will see the water spill out of the cup as soon as you begin to squeeze, because the area of the cup is reduced.

(3) The vent cap or a roof assembly shall have a venting capacity not less than that of the pipe to which it is attached.

### 12.8.6 Support of Single-Wall Metal Pipe.

All portions of single-wall metal pipe shall be supported for the design and weight of the material employed.

### 12.8.7 Marking.

Single-wall metal pipe shall comply with the marking provisions of 12.7.6.

## 12.9* Through-the-Wall Vent Termination

Section 12.9 provides requirements for separation of the termination point of venting systems from the building openings for venting systems that terminate through the side of a building.

**A.12.9** See Figure A.12.9.

**12.9.1** A mechanical draft venting system shall terminate at least 3 ft (0.9 m) above any forced air inlet located within 10 ft (3 m).

*Exception No. 1: This provision shall not apply to the combustion air intake of a direct vent appliance.*

The intent of 12.9.1 is to prevent gases from being drawn back into a building. This requirement recognizes that vent gases are lighter than air. Exception No. 1 recognizes that direct vent appliance inlets do not communicate with the air in a building.

*Exception No. 2: This provision shall not apply to the separation of the integral outdoor air inlet and flue gas discharge of listed outdoor appliances.*

Exception No. 2 to 12.9.1 prevents confusion in the installation of outdoor gas appliances. Some authorities have misinterpreted the code to prohibit such appliances or to require them to be modified in the field, which is not the intent of 12.9.1. An example of this type of appliance is a packaged rooftop air conditioner, which incorporates a gas vent and a circulation air inlet used for the building air supply.

NEW **12.9.2** A mechanical draft venting system of other than direct vent type shall terminate at least 4 ft (1.2 m) below, 4 ft (1.2 m) horizontally from, or 1 ft (300 mm) above any door, operable window, or gravity air inlet into any building. The bottom of the vent terminal shall be located at least 12 in. (300 mm) above finished ground level.

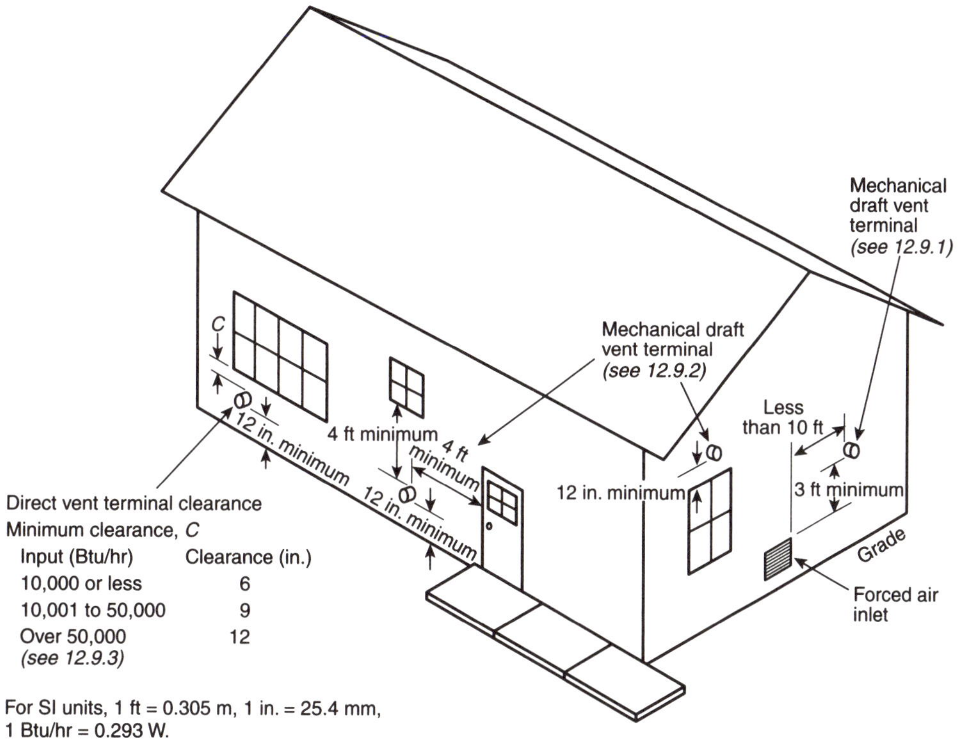


***FIGURE A.12.9*** *Exit Terminals of Mechanical Draft and Direct Vent Venting Systems.*

Paragraph 12.9.2 was revised in the 2009 edition by substituting "finished ground level" for "grade," as has been done throughout the code. Refer to commentary following 9.3.8.8 for the reason for this revision.

Note that the requirement refers to operable windows so that it is clear that the distances in the requirement are not applicable to windows that cannot be opened.

◀ **FAQ**
**Do the separation requirements of 12.9.2 for exit terminals apply to windows that do not open?**

NEW

**12.9.3** The vent terminal of a direct vent appliance with an input of 10,000 Btu/hr (3 kW) or less shall be located at least 6 in. (150 mm) from any air opening into a building, an appliance with an input over 10,000 Btu/hr (3 kW) but not over 50,000 Btu/hr (14.7 kW) shall be installed with a 9 in. (230 mm) vent termination clearance, and an appliance with an input over 50,000 Btu/hr (14.7 kW) shall have at least a 12 in. (300 mm) vent termination clearance. The bottom of the vent terminal and the air intake shall be located at least 12 in. (300 mm) above finished ground level.

Subsections 12.9.1 and 12.9.2 are concerned with preventing appliance combustion products from being drawn into a building through fresh air inlets, including operable windows. The text was changed in the 2009 edition to reference the 12 in. (300 mm) elevation to be measured from "finished ground level," rather than "grade." Refer to the commentary following 9.3.8.8 for an explanation of this change, which was made in several locations in the code.

Subsection 12.9.3 permits the vent terminals of direct vent appliances to be located much closer to air inlets than is permitted in 12.9.1 for non–direct vent appliances. Vent gases from direct vent appliances disperse rapidly upon leaving the vent terminal, even when the terminal is located under an open window. However, a window is unlikely to be open when heat is needed.

◀ **FAQ**
**Why are vent terminals for direct vent appliances allowed to be closer to building openings than exit terminals for other types of appliances?**

Subsection 12.9.3 also specifies the location of the exit terminal of direct vent appliances. All of these locations are shown in Figure A.12.9.

**12.9.4** Through-the-wall vents for Category II and Category IV appliances and noncategorized condensing appliances shall not terminate over public walkways or over an area where condensate or vapor could create a nuisance or hazard or could be detrimental to the operation of regulators, relief valves, or other equipment. Where local experience indicates that condensate is a problem with Category I and Category III appliances, this provision shall also apply. Drains for condensate shall be installed in accordance with the manufacturer's installation instructions.

Subsection 12.9.4 provides requirements for limiting vent termination installations to protect pedestrians and equipment, including gas meters. It places responsibility on the installer to locate vent terminations for Category II and IV appliances away from walkways and gas equipment. See Exhibit 12.21 for an example of correct vent termination placement. Category IV appliances are high efficiency, with low vent temperatures. (See the commentary following 12.5.1.) There are no Category II appliances.

Note that in the 2009 edition the paragraph was revised with the addition of the last sentence, requiring condensate drains be installed using the manufacturer's installation instructions to provide guidance to installers on the installation of drains.

The requirements of 12.9.4 recognize that any appliance can present a condensation problem in a cold climate.

**12.9.5** Where vents, including those for direct-vent appliances or combustion air intake pipes, penetrate outside walls of buildings, the annular spaces around such penetrations shall be permanently sealed using approved materials to prevent entry of combustion products into the building.

As noted in 12.9.5, it is very important that the space around sidewall vent pipes be sealed. Insects must be kept out, and paths for air to travel from the outside near the vent termination to the appliance must be prevented. The latter is based on an incident in Massachusetts that resulted in a carbon monoxide incident.

## 12.10 Condensation Drain

**12.10.1** Provision shall be made to collect and dispose of condensate from venting systems serving Category II and Category IV appliances and noncategorized condensing appliances in accordance with 12.9.4.

**12.10.2** Where local experience indicates that condensation is a problem, provision shall be made to drain off and dispose of condensate from venting systems serving Category I and Category III appliances in accordance with 12.9.4.

The requirement in 12.10.2 is based on the fact that Category II and IV vent systems operate at lower temperatures than do Category I and III systems, and condensation is expected.

***EXHIBIT 12.21*** *Outside Termination of Direct Vent Pipe. (Courtesy of Magic Sweep Corporation)*

## 12.11 Vent Connectors for Category I Appliances

Vent connectors for Category I appliances are relatively short runs of single-wall metal pipe, Type B vent, or other vent materials that are used to connect an appliance to a chimney or vent. When two or more appliances are connected to a chimney or vent, at least one connector is used. When one appliance is connected to a vent with no change in size or venting material, there is no vent connector.

The vent connector has an important effect on the overall operation of the venting system. Gas Research Institute (GRI)–sponsored research has proven that a vent primes better and has a stronger draft if there is a connector rise directly over the appliance. Retaining as much heat as possible in the flue gases as they pass through the connector is also very important. For more information, refer to the Gas Research Institute report, "Vent Oversizing and the 'Seven-Times' Rule: Analysis and Recommendations."

### 12.11.1 Where Required.

A vent connector shall be used to connect an appliance to a gas vent, chimney, or single-wall metal pipe, except where the gas vent, chimney, or single-wall metal pipe is directly connected to the appliance.

### 12.11.2 Materials.

**12.11.2.1** A vent connector shall be made of noncombustible, corrosion-resistant material capable of withstanding the vent gas temperature produced by the appliance and of sufficient thickness to withstand physical damage.

**12.11.2.2** Where the vent connector used for an appliance having a draft hood or a Category I appliance is located in or passes through an unconditioned area, that portion of the vent connector shall be listed Type B, Type L, or listed vent material having equivalent insulation qualities.

*Exception: Single-wall metal pipe located within the exterior walls of the building and located in areas having a local 99 percent winter design temperature of 5°F (−15°C) or higher (see Figure G.2.4).*

Attics and other unconditioned areas of buildings, such as garages and some basements, may experience temperatures approaching those found outdoors. Type B, Type L, or other materials with equivalent insulation are needed to make sure excessive condensation does not occur inside the vent connector. Additionally, areas such as attics, basements, and garages are commonly used for storage, and the requirement for connectors of this type provides protection against accidental ignition of combustible material carelessly placed near or on single-wall connectors.

**12.11.2.3** Where the vent connector used for an appliance having a draft hood or a Category I appliance is located in or passes through attics and crawl spaces, that portion of the vent connector shall be listed Type B, Type L, or listed vent material having equivalent insulation qualities.

See the commentary following 12.11.2.2, Exception.

**12.11.2.4** Vent connectors for residential-type appliances shall comply with the following:

(1) Vent connectors for listed appliances having draft hoods, appliances having draft hoods and equipped with listed conversion burners, and Category I appliances that are not installed in attics, crawl spaces, or other unconditioned areas shall be one of the following:
  (a) Type B or Type L vent material
  (b) Galvanized sheet steel not less than 0.018 in. (0.46 mm) thick
  (c) Aluminum (1100 or 3003 alloy or equivalent) sheet not less than 0.027 in. (0.69 mm) thick
  (d) Stainless steel sheet not less than 0.012 in. (0.31 mm) thick
  (e) Smooth interior wall metal pipe having resistance to heat and corrosion equal to or greater than that of 12.11.2.4(1)(b), (c), or (d)
  (f) A listed vent connector
(2) Vent connectors shall not be covered with insulation.

*Exception: Listed insulated vent connectors shall be installed in accordance with the manufacturer's installation instructions.* NEW

In the 2009 edition, the exception in 12.11.2.4 was revised to substitute the words "manufacturer's installation instructions" for the words "terms of their listing." This change recognizes that installers normally do not have the listing information, but usually have the manufacturer's installation instructions, which include relevant installation items from the listing.

FAQ ▶
Why is it not acceptable to wrap connector pipes with insulation material?

Paragraph 12.11.2.4(2) prohibits wrapping connector pipes with insulating material to retain heat or reduce clearances. Connectors made of single-wall material must be fully visible for inspection and maintenance, and wrapping the connector with insulating material will prevent frequent inspection. Connectors may deteriorate without detection to the point that they collapse, allowing flue gases and heat to enter the building. Further, the application of the insulating material may accelerate the deterioration.

**12.11.2.5** A vent connector for a nonresidential low-heat appliance shall be a factory-built chimney section or steel pipe having resistance to heat and corrosion equivalent to that for the appropriate galvanized pipe as specified in Table 12.11.2.5. Factory-built chimney sections shall be joined together in accordance with the chimney manufacturer's instructions.

***TABLE 12.11.2.5*** *Minimum Thickness for Galvanized Steel Vent Connectors for Low-Heat Appliances*

| *Diameter of Connector (in.)* | *Minimum Thickness (in.)* |
|---|---|
| Less than 6 | 0.019 |
| 6 to less than 10 | 0.023 |
| 10 to 12 inclusive | 0.029 |
| 14 to 16 inclusive | 0.034 |
| Over 16 | 0.056 |

For SI units, 1 in. = 25.4 mm, 1 in.$^2$ = 645 mm$^2$.

Paragraph 12.11.2.5 applies to nonresidential appliances that operate with a flue gas temperature up to 1000°F (538°C). This paragraph identifies the need to properly join sections of factory-built chimneys. Refer to the commentary following 12.6.1.1 for more information on factory-built chimneys.

**12.11.2.6** Vent connectors for medium-heat appliances and commercial and industrial incinerators shall be constructed of factory-built, medium-heat chimney sections or steel of a thickness not less than that specified in Table 12.11.2.6 and shall comply with the following:

(1) A steel vent connector for an appliance with a vent gas temperature in excess of 1000°F (538°C) measured at the entrance to the connector shall be lined with medium-duty fire brick or the equivalent.
(2) The lining shall be at least 2½ in. (64 mm) thick for a vent connector having a diameter or greatest cross-sectional dimension of 18 in. (460 mm) or less.
(3) The lining shall be at least 4½ in. (110 mm) thick laid on the 4½ in. (110 mm) bed for a vent connector having a diameter or greatest cross-sectional dimension greater than 18 in. (460 mm).
(4) Factory-built chimney sections, if employed, shall be joined together in accordance with the chimney manufacturer's instructions.

A medium-heat appliance is intended for nonresidential use and operates with a flue gas temperature of up to 1800°F (982°C). These conditions are more severe than vents from low-temperature appliances, and these more stringent requirements for single-wall metal pipe are needed for a reasonable life for the single-wall metal pipe.

*TABLE 12.11.2.6 Minimum Thickness for Steel Vent Connectors for Medium-Heat Appliances and Commercial and Industrial Incinerators*

| Vent Connector Size | | Minimum Thickness (in.) |
|---|---|---|
| Diameter (in.) | Area (in.$^2$) | |
| Up to 14 | Up to 154 | 0.053 |
| Over 14 to 16 | 154 to 201 | 0.067 |
| Over 16 to 18 | 201 to 254 | 0.093 |
| Over 18 | Larger than 254 | 0.123 |

For SI units, 1 in. = 25.4 mm, 1 in.$^2$ = 645 mm$^2$.

## 12.11.3* Size of Vent Connector.

**A.12.11.3** Reference can also be made to the chapter on chimney, gas vent, and fireplace systems of the *ASHRAE Handbook — HVAC Systems and Equipment.*

Paragraph 12.11.3 provides sizing requirements for vent connectors in Category I venting systems. When sizing connectors serving fan-assisted Category I appliances, the user will note that the selection of single-wall metal pipe connectors is limited. This limitation is not addressed in Chapter 12 but becomes evident when the tables in Chapter 13 are used to size the vent and connector. The limited selection is due to the lower vent-operating temperature of these appliances, which reduces the allowable heat loss in the connector compared to non-fan-assisted combustion appliances. Single-wall connectors are not permitted when Tables 13.1(a) through 13.1(c) are used to size connectors to exterior masonry chimneys.

**12.11.3.1** A vent connector for an appliance with a single draft hood or for a Category I fan-assisted combustion system appliance shall be sized and installed in accordance with Chapter 13 or other approved engineering methods.

**12.11.3.2** For a single appliance having more than one draft hood outlet or flue collar, the manifold shall be constructed according to the instructions of the appliance manufacturer. Where there are no instructions, the manifold shall be designed and constructed in accordance with approved engineering practices. As an alternative method, the effective area of the manifold shall equal the combined area of the flue collars or draft hood outlets, and the vent connectors shall have a minimum 1 ft (0.3 m) rise.

In the absence of specific instructions, it is important to provide an adequate rise above the draft hoods. This rise is needed to ensure proper flue priming on startup. The minimum 1 ft (0.3 m) connector rise required by 12.11.3.2 minimizes spillage. Connector rise is an important factor in venting system performance, and additional connector rise will almost always improve venting system performance.

**12.11.3.3** Where two or more appliances are connected to a common vent or chimney, each vent connector shall be sized in accordance with Chapter 13 or other approved engineering methods.

**12.11.3.4** As an alternative method applicable only where all of the appliances are draft hood–equipped, each vent connector shall have an effective area not less than the area of the draft hood outlet of the appliance to which it is connected.

As stated in 12.11.3.3, connectors serving multiple appliance installations must be sized using the venting tables of Chapter 13 or other approved engineering methods. However, 12.11.3.4 provides an alternate method for sizing connectors in multiple appliance installations serving only draft hood–equipped appliances. Paragraph 12.11.3.4 allows each connector to be sized at least as large as the draft hood outlet.

**12.11.3.5** Where two or more appliances are vented through a common vent connector or vent manifold, the common vent connector or vent manifold shall be located at the highest level consistent with available headroom and clearance to combustible material and shall be sized in accordance with Chapter 13 or other approved engineering methods.

**12.11.3.6** As an alternative method applicable only where there are two draft hood–equipped appliances, the effective area of the common vent connector or vent manifold and all junction fittings shall be not less than the area of the larger vent connector plus 50 percent of the area of the smaller flue collar outlet.

**12.11.3.7** Where the size of a vent connector is increased to overcome installation limitations and obtain connector capacity equal to the appliance input, the size increase shall be made at the appliance draft hood outlet.

The need to increase the vent connector size is common when draft hood appliances with small outlets are common-vented with other appliances. For example, the sizing tables of Chapter 13 often require a 4 in. (100 mm) connector for residential water heaters equipped with 3 in. (80 mm) draft hoods.

### 12.11.4 Two or More Appliances Connected to a Single Vent.

NEW **12.11.4.1** Where two or more openings are provided into one chimney flue or vent, either the openings shall be at different levels or the connectors shall be attached to the vertical portion of the chimney or vent at an angle of 45 degrees or less relative to the vertical.

The requirement in 12.11.4.1 was revised in the 2009 edition to provide guidance on how to safely connect multiple connectors to a common vent. The concern is that if two connectors are installed at the same level, the momentum of the vent gas from one could force some of the vent gases to enter the other appliance and spill out through a draft hood or flue collar.

**12.11.4.2** Where two or more vent connectors enter a common gas vent, chimney flue, or single-wall metal pipe, the smaller connector shall enter at the highest level consistent with the available headroom or clearance to combustible material.

FAQ ► Is it acceptable to connect a Category I appliance to a venting system used for a Category III or IV appliance?

Category I appliances operate with negative pressure in the connectors and vents, while Category III and IV appliances operate with positive pressure. Connecting Category I appliances to a vent system under positive pressure will likely cause flue gases to be forced out of the venting system and into the building.

**12.11.4.3** Vent connectors serving Category I appliances shall not be connected to any portion of a mechanical draft system operating under positive static pressure, such as those serving Category III or Category IV appliances.

### 12.11.5 Clearance.

Minimum clearances from vent connectors to combustible material shall be in accordance with Table 12.8.4.5.

*Exception: The clearance between a vent connector and combustible material shall be permitted to be reduced where the combustible material is protected as specified for vent connectors in Table 10.2.3(b).*

### 12.11.6 Avoid Unnecessary Bends.

A vent connector shall be installed so as to avoid turns or other construction features that create excessive resistance to flow of vent gases.

The requirement in 12.11.6 is intended to limit the number of unnecessary bends or turns in a venting system. Each bend or turn, such as an elbow or tee, in a venting system reduces its capacity, especially in a natural draft vent. The capacities shown in the Chapter 13 tables make an allowance for two 90-degree turns between the draft hood and the top of the venting system.

If the venting system contains more than two 90-degree turns, a capacity reduction of 5 percent should be made for each additional turn or elbow that is less than or equal to 45 degrees. A capacity reduction of 10 percent should be taken for each elbow between 45 and 90 degrees. For example, the capacity of a venting system containing four 90-degree turns is 80 percent of that indicated in the tables, and a system with two 90-degree turns and two 45-degree turns is 90 percent of the table values. (See 13.1.3 and 13.2.6.)

### 12.11.7 Joints.

Joints between sections of connector piping and connections to flue collars or draft hood outlets shall be fastened in accordance with one of the following methods:

(1) Sheet metal screws
(2) Vent connectors of listed vent material assembled and connected to flue collars or draft hood outlets in accordance with the manufacturers' instructions
(3) Other approved means

An appliance must be connected to the vent system, and 12.11.7 provides three methods of connection. The most common are screws to hold the vent material to the appliance. If screws are used, a minimum of three screws is recommended to provide a joint that will not come apart. Listed vent connectors can also be used. Other methods are allowed, but they must be approved.

### 12.11.8 Slope.

A vent connector shall be installed without any dips or sags and shall slope upward toward the vent or chimney at least $^1/_4$ in./ft (20 mm/m).

*Exception: Vent connectors attached to a mechanical draft system installed in accordance with appliance and the draft system manufacturers' instructions.* NEW

The exception in 12.11.8 was revised in the 2009 edition by referencing both the appliance and draft system manufacturer's installation instructions, as both may have requirements for proper operation. Previously, the reference was to manufacturer's installation instructions, which allowed the installer to use the one believed to be appropriate.

### 12.11.9 Length of Vent Connector.

**12.11.9.1** A vent connector shall be as short as practical, and the appliance located as close as practical, to the chimney or vent.

The requirements in 12.11.9 cover the proper length of a vent connector. In new installations, consideration should be given to the location of appliances and venting systems. Routing of systems with additional elbows or lengths of pipe for aesthetic reasons should be avoided. The requirement in 12.11.9.1 is intended to limit the vent connector size as much as possible. Connectors with long horizontal runs can cause a venting system to fail to withdraw the products of combustion from a building. Long connectors also reduce overall venting performance and increase the likelihood of condensation in the venting system.

**12.11.9.2** The maximum horizontal length of a single-wall connector shall be 75 percent of the height of the chimney or vent, except for engineered systems. NEW

The requirement in 12.11.9.2 limits single-wall vent connectors serving a single appliance installation to 75 percent of the vent height unless the system is designed using engineering methods. In the 2009 edition, a second sentence allowing a connector in a venting serving multiple appliances to have a length of 100 percent of the vent height was deleted. The requirement is believed to come from an old "rule of thumb," and could not be confirmed. Note that the venting table notes (Sections 13.1 and 13.2) now limit horizontal length of vent connectors.

**12.11.9.3** The maximum horizontal length of a Type B double-wall connector shall be 100 percent of the height of the chimney or vent, except for engineered systems. The maximum length of an individual connector for a chimney or vent system serving multiple appliances, from the appliance outlet to the junction with the common vent or another connector, shall be 100 percent of the height of the chimney or vent.

Paragraph 12.11.9.3 limits double-wall vent connectors to 100 percent of the vent system height for either single or multiple appliance installations unless the system is designed using engineering methods.

### 12.11.10 Support.

A vent connector shall be supported for the design and weight of the material employed to maintain clearances and prevent physical damage and separation of joints.

### 12.11.11 Chimney Connection.

Where entering a flue in a masonry or metal chimney, the vent connector shall be installed above the extreme bottom to avoid stoppage. Where a thimble or slip joint is used to facilitate removal of the connector, the connector shall be firmly attached to or inserted into the thimble or slip joint to prevent the connector from falling out. Means shall be employed to prevent the connector from entering so far as to restrict the space between its end and the opposite wall of the chimney flue.

### 12.11.12 Inspection.

The entire length of a vent connector shall be readily accessible for inspection, cleaning, and replacement.

Paragraph 12.11.12 requires that the entire vent connector be readily accessible for inspection. "Readily accessible," which is defined in 3.3.1.1, means that there is direct access to the vent connector without the need to remove or move any panel, door, or similar covering. Frequent inspection of connectors is important because connectors, especially single-wall connectors, can deteriorate.

### 12.11.13 Fireplaces.

A vent connector shall not be connected to a chimney flue serving a fireplace unless the fireplace flue opening is permanently sealed.

### 12.11.14 Passage Through Ceilings, Floors, or Walls.

**12.11.14.1** Single-wall metal pipe connectors shall not pass through any wall, floor, or ceiling except as permitted by 12.8.4.2 and 12.8.4.7.

Note that former paragraph 12.11.14.2 has been relocated to 12.8.4.7, in the single-wall metal pipe section — a more appropriate location.

**12.11.14.2** Vent connectors for medium-heat appliances shall not pass through walls or partitions constructed of combustible material.

Note that the term *combustible* is defined in *National Fuel Gas Code* as follows:

> **3.3.67.1 Combustible Material.** As pertaining to materials adjacent to or in contact with heat-producing appliances, vent connectors, gas vents, chimneys, steam and hot water pipes, and warm air ducts, materials made of or surfaced with wood, compressed paper, plant fibers, or other materials that are capable of being ignited and burned. Such material shall be considered combustible even though flame-proofed, fire-retardant treated, or plastered.

Note that the definition is specific and includes wood and other fibrous materials even if they are treated to be flameproof or are covered with plaster.

## 12.12 Vent Connectors for Category II, Category III, and Category IV Appliances

See Section 12.5.

Table 12.5.1 requires that venting materials be furnished or specified by the equipment manufacturer for Category II, III, and IV gas appliances.

## 12.13 Draft Hoods and Draft Controls

◀ **FAQ**
What is the purpose of a draft hood on a gas appliance?

Draft hoods on vent systems perform the following three functions:

1. The negative pressure in the vent system created by the hot exhaust gases draws in dilution air at the draft hood opening. This dilution air is taken from the room in which the draft hood is located. This room air is much cooler than the exhaust gases, thereby lowering the net stack temperature and reducing fire hazards. Dilution air is also much drier than the exhaust gases, thereby lowering the dew point in the venting system and reducing any condensation.
2. The draft hood acts as a break between the vent system and the appliance and eliminates stack action. Appliance manufacturers design their equipment to operate with a specific range of airflow through the appliance. If there were no separation between the appliance and the vent system, excessive drafts created by tall chimneys would affect the combustion process and flame stability, possibly even resulting in pilot outage. Excessive drafts would also lower efficiency by moving the products of combustion through the heat exchanger before optimal heat transfer. Wind effects can also create temporary downdrafts.
3. A draft hood provides a relief opening in the event of a downdraft. Vent systems may temporarily experience poor venting at startup (before the vent heats up) or during windy conditions. Under these conditions, some of the products of combustion may "spill out" at the draft hood. The principal products of combustion from a properly burning appliance, carbon dioxide and water vapor, should cause no immediate harm. Once draft is established (or re-established when the wind dies down), all of the combustion products are vented safely up the vent. During a sustained downdraft, such as in a blockage, all of the combustion products may spill into the living space and may eventually displace the oxygen in the room, potentially leading to incomplete combustion and the formation of carbon monoxide. New central heating appliances equipped with draft hoods must have safety switches, such as spill switches, which will shut off the burner in the event of a sustained downdraft.

Draft hoods are an integral part of the equipment design and should never be altered. The height of the draft hood above the flue collar will affect combustion and airflow through the appliance. If a draft hood were removed entirely, even a temporary downdraft would immediately affect the combustion process, potentially creating carbon monoxide.

Barometric draft regulators perform the same functions as a draft hood but are generally used in connection with power burners and conversion burners. Where power burners are used, the gas input, combustion air, flame pattern, and draft all must be carefully set to match the equipment they serve.

Barometric draft regulators are usually adjustable so that the amount of draft can be set for maximum efficiency and safe burner operation. Barometric draft regulators, when used with gas appliances, are double-acting so that they will act as a relief opening in the event of a downdraft. Safety shutoff devices are required on all conversion burners installed after 1990.

### 12.13.1 Appliances Requiring Draft Hoods.

Vented appliances shall be installed with draft hoods.

*Exception: Dual oven-type combination ranges, incinerators, direct vent appliances; fan-assisted combustion system appliances; appliances requiring chimney draft for operation; single-firebox boilers equipped with conversion burners with inputs greater than 400,000 Btu/hr (117 kW); appliances equipped with blast, power, or pressure burners that are not listed for use with draft hoods; and appliances designed for forced venting.*

### 12.13.2 Installation.

A draft hood supplied with or forming a part of a listed vented appliance shall be installed without alteration, exactly as furnished and specified by the appliance manufacturer.

**12.13.2.1** If a draft hood is not supplied by the appliance manufacturer where one is required, a draft hood shall be installed, be of a listed or approved type, and, in the absence of other instructions, be of the same size as the appliance flue collar. Where a draft hood is required with a conversion burner, it shall be of a listed or approved type.

**12.13.2.2** Where a draft hood of special design is needed or preferable, the installation shall be in accordance with the recommendations of the appliance manufacturer and shall be approved.

### 12.13.3 Draft Control Devices.

Where a draft control device is part of the appliance or is supplied by the appliance manufacturer, it shall be installed in accordance with the manufacturer's instructions. In the absence of manufacturer's instructions, the device shall be attached to the flue collar of the appliance or as near to the appliance as practical.

### 12.13.4* Additional Devices.

Appliances (except incinerators) requiring controlled chimney draft shall be permitted to be equipped with listed double-acting barometric draft regulators installed and adjusted in accordance with the manufacturers' instructions.

**FAQ ►** What is the difference between a single-acting barometric damper and a double-acting barometric damper?

A standard, single-acting barometric draft regulator, such as the one shown in Exhibit 12.22, swings inward toward the chimney or vent to allow room air to enter the vent system. This regulator action reduces and stabilizes the draft through the appliance. Single-acting barometric draft regulators are used with oil- or solid fuel–burning appliances. When a gas appliance is equipped with a barometric draft regulator, it should be of the double-acting type. A double-acting barometric draft regulator, such as the one shown in Exhibit 12.23, swings inward like a single-acting type to allow dilution air to enter the vent, providing the beneficial effects of dilution air in reducing condensation and also reducing the effects of draft or "stack action" on the appliance. Unlike the single-acting type, a double-acting barometric draft regulator can also swing outward to relieve vent overpressurization in the event of a flue blockage or downdraft.

When installing a double-acting barometric draft regulator for a gas appliance, the installer needs to consider the location of the draft regulator in the connector. The draft regulator should be positioned in the connector so that overpressurization "downdrafts" are relieved through the regulator and do not enter the appliance. To accomplish this, the draft regulator should be placed in a bullhead tee, as shown in Figure A.12.13.4. In the three examples marked "correct location" in Figure A.12.13.4, chimney downdrafts would be relieved through the barometric damper instead of flowing into the appliance, because a turn is required from the connector pipe section containing the barometric damper before downdrafts can flow into the appliance.

**A.12.13.4** A device that automatically shuts off gas to the burner in the event of sustained backdraft is recommended if such backdraft might adversely affect burner operation or if flue gas spillage might introduce a hazard. Figure A.12.13.4 shows examples of correct and incorrect locations for barometric draft regulators.

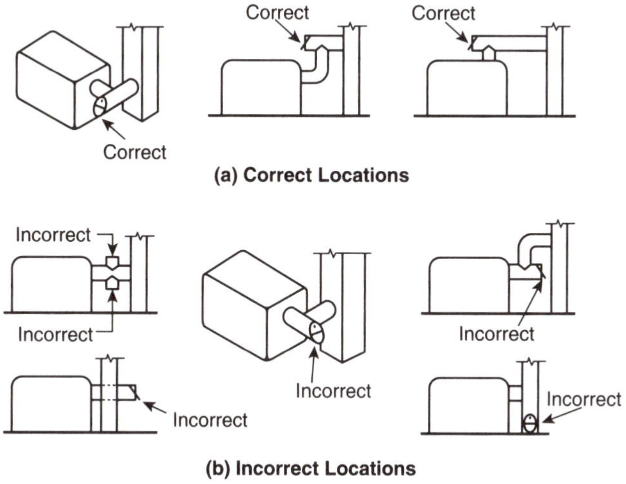


*FIGURE A.12.13.4 Locations for Barometric Draft Regulators.*

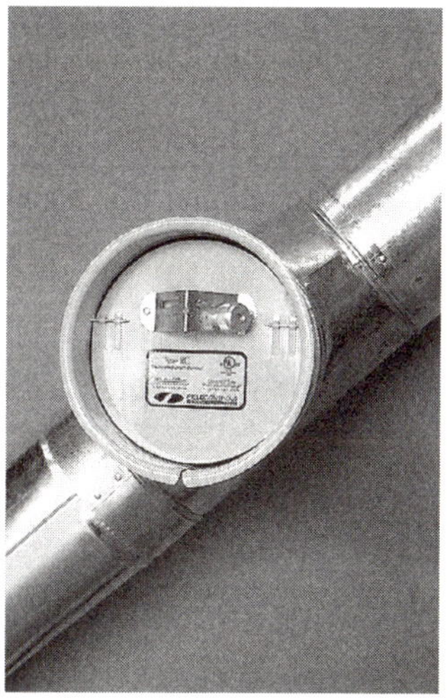

***EXHIBIT 12.22** Standard Single-Acting Barometric Draft Regulator. (Courtesy of Magic Sweep Corporation)*

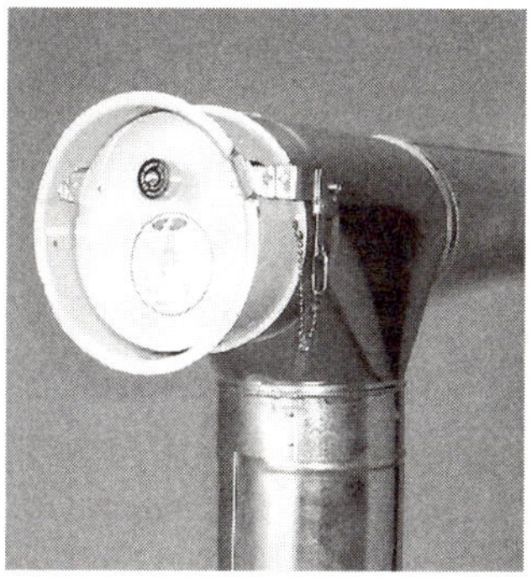

***EXHIBIT 12.23** Standard Double-Acting Barometric Draft Regulator. (Courtesy of Magic Sweep Corporation)*

## 12.13.5 Location.

Draft hoods and barometric draft regulators shall be installed in the same room or enclosure as the appliance in such a manner as to prevent any difference in pressure between the hood or regulator and the combustion air supply.

In the 2002 edition, the title of 12.13.5 was revised editorially to "Location," referring to the location of draft hoods and barometric draft regulators. The former title, "Install in Same Room," might have caused readers to miss the requirement.

## 12.13.6 Positioning.

Draft hoods and draft regulators shall be installed in the position for which they were designed with reference to the horizontal and vertical planes and shall be located so that the relief opening is not obstructed by any part of the appliance or adjacent construction. The appliance and its draft hood shall be located so that the relief opening is accessible for checking vent operation.

## 12.13.7 Clearance.

A draft hood shall be located so that its relief opening is not less than 6 in. (150 mm) from any surface except that of the appliance it serves and the venting system to which the draft hood is connected. Where a greater or lesser clearance is indicated on the appliance label, the clearance shall not be less than that specified on the label. Such clearances shall not be reduced.

## 12.14 Manually Operated Dampers

A manually operated damper shall not be placed in any appliance vent connector. Fixed baffles shall not be classified as manually operated dampers.

Manually operated dampers, as covered in Section 12.14, are those requiring user intervention and are not operated automatically by the appliance. They are prohibited because they could be inadvertently placed in the closed position, causing flue gas spillage into the building.

## 12.15 Automatically Operated Vent Dampers

An automatically operated vent damper shall be of a listed type.

Section 12.15 covers automatically operated vent dampers. Preferably, automatically operated vent dampers, such as the one shown in Exhibit 12.24, should be included as part of the listed gas appliance. Manually operated dampers are not permitted in vent connectors, and the requirement clarifies that fixed baffles, which are sometimes used to balance system draft on startup, are allowed.

***EXHIBIT 12.24*** *Vent Damper. (Courtesy of Magic Sweep Corporation)*

## 12.16 Obstructions

Devices that retard the flow of vent gases shall not be installed in a vent connector, chimney, or vent. The following shall not be considered as obstructions:

(1) Draft regulators and safety controls specifically listed for installation in venting systems and installed in accordance with the manufacturer's installation instructions
(2) Approved draft regulators and safety controls designed and installed in accordance with approved engineering methods
(3) Listed heat reclaimers and automatically operated vent dampers installed in accordance with the manufacturers' installation instructions

(4) Vent dampers serving listed appliances installed in accordance with 13.1.1 and 13.2.1 or other approved engineering methods

Paragraphs 13.1.1 and 13.2.1, referenced in 12.16(4), provide sizing information for vent systems that serve appliances equipped with vent dampers. Special sizing requirements are needed because vent dampers reduce the amount of heat delivered to the venting system during the appliance off-cycle and increase the likelihood of condensation in the vent.

(5) Approved economizers, heat reclaimers, and recuperators installed in venting systems of appliances not required to be equipped with draft hoods, provided the gas utilization appliance manufacturer's instructions cover the installation of such a device in the venting system and performance in accordance with Section 12.1 and 12.4.1 is obtained

Obstructions are not desired in a venting system, because they reduce gas flow and the removal of products of combustion from the building. Five specific types of obstructions are allowed: listed heat reclaimers in 12.16(3); vent dampers in 12.16(4); and economizers, heat reclaimers, and recuperators in 12.16(5) in recognition of the fact that they can be used safely when properly designed and installed. Of the five types, only vent dampers are normally used in residential and commercial venting systems. Note that the use of vent dampers is limited to installations with listed appliances that comply with 13.1.1, Obstructions and Vent Dampers, and 13.1.2, Vent Downsizing, so that these important factors are accommodated where needed. The other four obstructions are normally used with industrial appliances for energy conservation.

## REFERENCES CITED IN COMMENTARY

The following publications are available from the National Fire Protection Association, 1 Batterymarch Park, Quincy, MA 02169-7471, 1-800-344-3555, www.nfpa.org.

NFPA 91, *Standard for Exhaust Systems for Air Conveying of Vapors, Gases, Mists, and Noncombustible Particulate Solids*, 2004 edition.

NFPA 96, *Standard for Ventilation Control and Fire Protection of Commercial Cooking Operations*, 2008 edition.

NFPA 211, *Standard for Chimneys, Fireplaces, Vents, and Solid Fuel–Burning Appliances*, 2006 edition.

The following publication is available from the American Society of Heating, Refrigerating and Air-Conditioning Engineers, Inc., 1791 Tullie Circle, N.E., Atlanta, GA 30329-2305, 404-636-8400, www.ashrae.org.

*ASHRAE Handbook — Fundamentals,* 2005.

The following publication is available from the Gas Technology Institute, 1700 South Mount Prospect Road, Des Plaines, IL 60018-1804, 773-339-8100, www.gastechnology.org.

S. M. Ricci, et al., "Vent Oversizing and the 'Seven-Times' Rule: Analysis and Recommendations," Topical Report, GRI-98/0285, 1998.

The following publication is available from the U.S. Government Printing Office, Washington, DC 20402, www.gpo.gov.

PL 100-12, National Appliance Energy Conservation Act of 1987.

The following publications are available from Underwriters Laboratories Inc., 333 Pfingsten Road, Northbrook, IL 60062-2096, 847-272-8800, www.ul.com.

UL 103, *Standard for Factory-Built Chimneys for Residential Type and Building Heating Appliances,* 2006.

UL 1738, *Standard for Venting Systems for Gas-Burning Appliances, Categories II, III and IV,* 2006.

# Sizing of Category I Venting Systems

## CHAPTER 13

Chapter 13 of the code contains 15 code tables and explanatory text for calculating the required vent size for a variety of Category I venting systems. This chapter was first created for the 1992 edition by relocating the vent sizing tables and related notes from Appendix G and extensively modifying them to cover fan-assisted combustion furnaces designed to meet U.S. energy minimums beginning in 1990. These furnaces provide higher efficiency, which is achieved by reducing vent temperature without jeopardizing vent operation.

The tables in Chapter 13 provide sizes for Category I appliances serving natural draft (NAT) appliances, fan-assisted (FAN) draft appliances, and combinations of both in one venting system. The tables incorporate the FAN Min and FAN Max columns to provide minimum and maximum capacities for each vent connector serving fan-assisted combustion appliances. The "NAT" columns are shaded to make it simpler to use only the shaded columns when sizing a vent serving only a natural draft appliance, such as a water heater. The "FAN Min and FAN Max" columns are not shaded, to distinguish them from the "NAT" columns.

The tables are also used to size the common vent incorporating three terms: FAN + FAN, FAN + NAT, and NAT + NAT, which allow for the three possible combinations of types of appliance in a common vent. These terms are used in the tables as follows:

- *FAN + FAN* columns provide maximum capacities for each connector and vent serving multiple fan-assisted appliances.
- *FAN + NAT* columns provide maximum capacities for each connector and vent serving combinations of natural draft and fan-assisted appliances.
- *NAT + NAT* columns provide maximum capacities for each connector and vent serving multiple natural draft appliances.

*Category I appliances* are defined in 3.3.6.11.1 as appliances that operate with a nonpositive vent static pressure and with a vent gas temperature that avoids excessive condensate production in the vent.

Venting systems serving Category I appliances are conventional venting systems, in which the heat of the flue gases is the force that operates the vent. Chapter 13 of this handbook includes the following:

- Six vent tables for single-appliance installations are listed in Commentary Table 13.1.
- Additional requirements for a single-appliance vent are given. These qualifications must be considered when using single-appliance vent Tables 13.1(a) through 13.1(f) of the code. (See Section 13.1.)
- Nine vent tables for multiple-appliance venting installations are listed in Commentary Table 13.2. The tables listed in Commentary Table 13.2 are used for interior chimney or vent installations and for exterior chimney or vent installations.
- Five tables covering exterior masonry chimneys are listed in Commentary Table 13.3. Table 13.12(a) and Table 13.13(a) show maximum input values for their respective chimneys, calculated to prevent overpressurization of the chimney. Tables 13.11, 13.12(b), and 13.13(b) show the minimum input values and are organized by geographical regions of the United States on the basis of the lowest anticipated winter design temperature. The

***COMMENTARY TABLE 13.1** Vent Tables for Single-Appliance Installations*

| Code Table No. 2002 ed. | Current Code Table No. | Vent | Connector |
|---|---|---|---|
| 13.1 | 13.1(a) | Type B | None |
| 13.2 | 13.1(b) | Type B | Single wall |
| 13.3 | 13.1(c) | Masonry chimney | Type B |
| 13.4 | 13.1(d) | Masonry chimney | Single wall |
| 13.5 | 13.1(e) | Single-wall metal pipe | None |
| 13.11 | 13.1(f) | Exterior masonry chimney | Type B |

***COMMENTARY TABLE 13.2** Vent Tables for Multiple-Appliance Venting Installations*

| Code Table No. 2002 ed. | Current Code Table No. | Vent | Connector |
|---|---|---|---|
| 13.6 | 13.2(a) | Type B | Type B |
| 13.7 | 13.2(b) | Type B | Single wall |
| 13.8 | 13.2(c) | Masonry chimney | Type B |
| 13.9 | 13.2(d) | Masonry chimney | Single wall |
| 13.10 | 13.2(e) | Single wall | Single wall |
| 13.12(a) | 13.2(f) | Exterior masonry chimney Maximum allowed input | Type B |
| 13.12(b) | 13.2(g) | Exterior masonry chimney Minimum allowed input | Type B |
| 13.13(a) | 13.2(h) | Exterior masonry chimney Maximum allowed input | Type B |
| 13.13(b) | 13.2(i) | Exterior masonry chimney Minimum allowed input | Type B |

***COMMENTARY TABLE 13.3** Exterior Masonry Chimney Tables*

| Code Table No. 2002 ed. | Current Code Table No. | Vent | Appliance |
|---|---|---|---|
| 13.11 | 13.1(f) | Exterior masonry chimney Minimum allowed input | Single NAT |
| 13.12(a) | 13.2(f) | Exterior masonry chimney Maximum allowed input | NAT + NAT |
| 13.12(b) | 13.2(g) | Exterior masonry chimney Minimum allowed input | NAT + NAT |
| 13.13(a) | 13.2(h) | Exterior masonry chimney Maximum allowed input | FAN + NAT |
| 13.13(b) | 13.2(i) | Exterior masonry chimney Minimum allowed input | FAN + NAT |

minimum values shown in the tables are calculated to reduce the likelihood of condensation in the chimney.

- Additional requirements for a multiple-appliance vent are provided. These qualifications must be considered when using multiple-appliance vent Tables 13.2(a) through 13.2(i) of the code. (See Section 13.2.)

Table 13.1(f) and Tables 13.2(f) through 13.2(i) of the code provide sizing information for exterior masonry chimneys serving heating appliances. The tables provide sizing options based on local 99 percent winter design temperatures in six temperature ranges. A map showing the boundaries of these temperature ranges in the United States is included in Annex G of the code. (See Figure G.2.4.) Annex G also includes an example that illustrates the use of these tables. Winter design temperature can be calculated using many other sources as well, some of which may be available on the Internet.

Tables 13.1(a) through 13.1(d) and Tables 13.2(a) through 13.2(d) provide maximum capacities for draft hood–equipped appliances while providing both minimum and maximum capacities for fan-assisted appliances. The maximum capacities for both the draft hood and fan-assisted appliances are calculated to prevent a positive pressure in the vent that would lead to spillage from the vent. Notice that the maximum capacities for fan-assisted systems are greater than comparable vent sizes serving draft hood appliances, because vents serving only fan-assisted appliances do not need to exhaust the dilution air drawn in by draft hood appliances.

The minimum capacities shown for fan-assisted appliances are calculated to reduce the likelihood of condensation in the vent by ensuring that an adequate amount of heat is delivered to the venting system. The minimum capacities rise as the vent size increases because larger venting systems need more heat to warm up. The only restriction on vent size for draft hood appliances is the "seven times" rule that limits the largest portion of the venting system to no more than seven times the area of the appliance draft hood outlet.

Further information on the development of these tables is included in Supplement 1 of this handbook. This supplement covers the development of the revised vent tables and the new exterior chimney tables in extensive technical detail and is recommended for those who desire a greater understanding of this important subject.

There are no minimum capacities shown for draft hood appliances because the dilution air drawn in by these appliances provides a margin of safety for minimizing condensation.

◄ **FAQ**
**Why is no "minimum" capacity listed for a draft hood appliance?**

## 13.1 Additional Requirements to Single Appliance Vent

In the 1992 edition of the code, when the tables were first included in the *National Fuel Gas Code*, Section 13.1 and Section 13.2 first appeared as notes to the tables. These notes included necessary modifications to the tables to cover installation conditions and are part of the tables. The tables, which were developed in 1990, were first published by the Gas Appliance Manufacturer's Association and were used prior to the 1992 edition of the code. During the first two years of use, 1990 to 1992, feedback from users led to the development of the notes that were included in the code in the 1992 edition. Additional notes and refinements have been added in subsequent editions.

No Obstructions

Tables 13.1(a) through 13.1(e) of the code were developed with the assumption that there are no restrictions in the path of the vent gas flow. An additional assumption is that there is no deliberate attempt to remove heat. Therefore, devices such as heat economizers may not be used in conjunction with the tables. (If an economizer or other heat recovery device is installed, the vent size must be determined using another method.) Vent dampers installed as part of a listed appliance are to be installed in accordance with the manufacturer's instructions or the general provisions in 13.1.1.

Subsection 13.1.1 treats the listed appliances with vent dampers as a natural draft (NAT) appliance for determining the maximum capacity of the vent system. Because of the reduction of dilution air caused by the damper, the appliance is treated as a fan-assisted (FAN) appliance to determine the minimum capacity of the vent system.

The simplified visual renditions of code requirements shown in the exhibits in this chapter have been provided courtesy of the Gas Research Institute (GRI).

### 13.1.1 Obstructions and Vent Dampers.

Venting Table 13.1(a) through Table 13.1(f) shall not be used where obstructions *(see Section 12.16)* are installed in the venting system. The installation of vents serving listed appliances with vent dampers shall be in accordance with the appliance manufacturer's instructions or in accordance with the following:

(1) The maximum capacity of the vent system shall be determined using the "NAT Max" column.
(2) The minimum capacity shall be determined as though the appliance were a fan-assisted appliance, using the "FAN Min" column to determine the minimum capacity of the vent system. Where the corresponding "Fan Min" is "NA," the vent configuration shall not be permitted and an alternative venting configuration shall be utilized.

Paragraphs 13.1.1(1) and 13.1.1(2) provide guidance for using the venting tables with draft hood appliances that are equipped with a vent damper. An example of such an appliance is a boiler that uses a vent damper to obtain higher efficiencies. The maximum capacity of the vent is found using the NAT Max column in these tables, as stated in 13.1.1(1). This column treats the vent as one serving a natural draft appliance when the appliance is operating. The minimum capacity of the vent is found using the FAN Min column, as stated in 13.1.1(2). This column treats the vent as one serving a fan-assisted (increased efficiency) appliance when the appliance is not operating. The reason for this unusual combination is that the appliance operates as a natural draft appliance, but when the appliance is not operating and the vent damper is closed, it is similar to a fan-assisted (energy-saving) appliance in its propensity to condense water.

FAQ ▶
Does the installation of a vent damper affect the sizing of the venting system?

Draft hood–equipped appliances allow residual heat retained in the appliance along with air from the room to enter the venting system during the appliance "off cycle." This warm air rising into the venting system helps keep the vent warm and helps reduce condensation. When the appliance is equipped with a vent damper, the residual heat in the appliance is not allowed to rise into the venting system during the off cycle. If the vent damper is located between the draft hood and venting system (which is usually the case), room air is also prevented from entering the venting system while the appliance is off.

### 13.1.2 Vent Downsizing.

Where the vent size determined from the tables is smaller than the appliance draft hood outlet or flue collar, the use of the smaller size shall be permitted, provided that the installation complies with all of the following requirements:

(1) The total vent height *(H)* is at least 10 ft (3 m).
(2) Vents for appliance draft hood outlets or flue collars 12 in. (300 mm) in diameter or smaller are not reduced more than one table size.
(3) Vents for appliance draft hood outlets or flue collars larger than 12 in. (300 mm) in diameter are not reduced more than two table sizes.
(4) The maximum capacity listed in the tables for a fan-assisted appliance is reduced by 10 percent (0.90 × maximum table capacity).
(5) The draft hood outlet is greater than 4 in. (100 mm) in diameter. A 3 in. (80 mm) diameter vent shall not be connected to a 4 in. (100 mm) diameter draft hood outlet. This provision shall not apply to fan-assisted appliances.

Downsizing

FAQ ▶
Is it permissible to use a vent smaller than the flue collar or draft hood outlet of the appliance?

There is an economic incentive to using smaller vents wherever possible. If the vent is smaller than the draft hood or flue collar, venting problems can occur. The restrictions in 13.1.2 recognize and avoid these venting problems. Limits are placed on downsizing. In particular, note that a 4 in. (100 mm) draft hood outlet may not be reduced to 3 in. (80 mm).

NEW

### 13.1.3 Elbows.

Single-appliance venting configurations with zero (0) lateral lengths in Table 13.1(a), Table 13.1(b), and Table 13.1(e) shall not have elbows in the venting system. Single-appliance venting with lateral lengths include two 90 degree elbows. For each additional elbow up to and including 45 degrees, the maximum capacity listed in the venting tables shall be reduced by 5 percent. For each additional elbow greater than 45 degrees up to and including 90 degrees, the maximum capacity listed in the venting tables shall be reduced by 10 percent. Where multiple offsets occur in a vent, the total lateral length of all offsets combined shall not exceed that specified in Table 13.1(a) through Table 13.1(e).

Elbows

In the 2009 edition, the requirement was revised with the addition of the last sentence, which addresses multiple lateral offsets. While these are not common, they are used, and the code now addresses them.

The sizing tables were designed with an assumption that up to two 90-degree turns were part of the venting system, except in the zero lateral length case, addressed in 13.1.3. In the zero lateral length case, the vent is assumed to extend straight up from the appliance outlet to the vent termination. Adding additional elbows to the system is possible.

Derating factors for additional elbows are shown in Commentary Table 13.4. Prior to the 1992 edition, a 10 percent reduction was required for each elbow, regardless of its angle. This revision, which was based on input from the Gas Technology Institute's contractor, Battelle, provides flexibility to installers who are forced to use vent offsets, and encourages the use of elbows of less than 90 degrees.

When sizing a vent system with elbows, please note the following important factors:

- The tables include two 90-degree elbows. The table values should not be derated for venting systems with one or two elbows.
- For each additional elbow over two, derate the table values by 5 percent or 10 percent, depending on the angle of the elbows.

**COMMENTARY TABLE 13.4** *Table Capacity Derating for Elbows of Less Than 90 Degrees*

| Elbow | Vent Table Capacity Reduction, Per Elbow (%) |
|---|---|
| 0–45 | 5 |
| >45–90 | 10 |

## 13.1.4 Zero Lateral.

Zero (0) lateral *(L)* shall apply only to a straight vertical vent attached to a top outlet draft hood or flue collar.

No Elbows

Subsection 13.1.4 does not permit the use of elbows in a venting system where the zero lateral is used. If elbows are needed, the table rows for 2 ft (0.6 m) lateral length must be used. Elbows in a vent system with zero offset are not common but may be needed to route the vent to avoid a building obstruction, such as a beam.

## 13.1.5 High-Altitude Installations.

Sea level input ratings shall be used when determining maximum capacity for high-altitude installation. Actual input (derated for altitude) shall be used for determining minimum capacity for high-altitude installation.

High Altitude Installation

Using the sea level input rating for the maximum capacity, as required by 13.1.5, is a conservative measure, because less draft is produced at high altitudes. The derating process will also make condensation more likely. The reduced input rate should be used for the minimum capacity.

## 13.1.6 Two-Stage/Modulating Appliances.

For appliances with more than one input rate, the minimum vent capacity (FAN Min) determined from the Chapter 13 tables shall be less than the lowest appliance input rating, and the maximum vent capacity (FAN Max/NAT Max) determined from the tables shall be greater than the highest appliance rating input.

Multiple Input Rates

If the appliance has multiple input rates, as addressed in 13.1.6, the minimum capacity is determined with the minimum input rate, and the maximum capacity is determined with the maximum input rate.

### 13.1.7* Corrugated Chimney Liners.

Listed corrugated metallic chimney liner systems in masonry chimneys shall be sized by using Table 13.1(a) or Table 13.1(b) for Type B vents, with the maximum capacity reduced by 20 percent (0.80 × maximum capacity) and the minimum capacity as shown in Table 13.1(a) or Table 13.1(b). Corrugated metallic liner systems installed with bends or offsets shall have their maximum capacity further reduced in accordance with 13.1.3. The 20 percent reduction for corrugated metallic chimney liner systems includes an allowance for one long radius 90 degree turn at the bottom of the liner.

Does a corrugated chimney liner have the same capacity as a similar sized smooth wall liner or vent?

**A.13.1.7** A long radius turn is a turn where the centerline radius is equal to or greater than 1.5 times the vent diameter.

FAQ ►

Chimney Liner

Because properly installed corrugated chimney liners have heat loss similar to that of a Type B vent, they are sized using Table 13.1(a) or Table 13.1(b), as stated in 13.1.7. However, such liners' corrugations and their tendency to spiral in the chimney require a 20 percent maximum capacity reduction.

FAQ ►

Does the turn or bend at the bottom of a flexible chimney liner count as an elbow?

Many liners begin at the opening in the chimney wall, then bend up vertically. This 90-degree elbow at the beginning of the liner is included in the table capacity. The code clearly states that the 20 percent reduction includes one 90-degree elbow, which is formed when the liner is bent to exit the chimney at the bottom. This capacity reduction was developed with the assistance of the researchers who developed the tables.

### 13.1.8 Connection to Chimney Liners.

Connections between chimney liners and listed double-wall connectors shall be made with listed adapters designed for such purpose.

The requirement in 13.1.8 recognizes that chimney liners are provided with adapters to allow connection to a Type B vent. These adapters must be used to provide a secure connection between chimney liners and double-wall connectors.

### 13.1.9 Vertical Vent Upsizing/7 × Rule.

Where the vertical vent has a larger diameter than the vent connector, the vertical vent diameter shall be used to determine the minimum vent capacity, and the connector diameter shall be used to determine the maximum vent capacity. The flow area of the vertical vent shall not exceed seven times the flow area of the listed appliance categorized vent area, flue collar area, or draft hood outlet area unless designed in accordance with approved engineering methods.

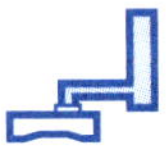

Upsizing a Vertical Vent

FAQ ►

Can the venting system be larger than the appliance flue collar or draft hood outlet?

In a vent system with a larger vent than the vent connector, the draft produced is limited by the small diameter of the vent connector. Condensation will begin in the larger diameter vent or connector. The maximum and minimum capacities must be determined accordingly. Practical limits exist as to how large the vertical vent may be relative to its source of vent gas flow. Therefore, the flow area of the vent may not be more than seven times the flow area of the outlet of the appliance or draft hood, as stated in 13.1.9.

### 13.1.10 Draft Hood Conversion Accessories.

Draft hood conversion accessories for use with masonry chimneys venting listed Category I fan-assisted appliances shall be listed and installed in accordance with the listed accessory manufacturers' installation instructions.

Subsection 13.1.10 covers draft hood conversion accessories, which are add-on kits that convert a fan-assisted appliance to a draft hood appliance by providing an opening for dilution air from the room to enter the venting system. A draft hood conversion accessory is a listed component that is usually provided by the appliance manufacturer. The accessory must be installed according to the instructions included with the kit. The addition of the draft hood kit allows the appliance to be sized as a draft hood appliance, which may be beneficial when venting into masonry chimneys, especially exterior masonry chimneys.

### 13.1.11 Chimneys and Vent Locations.

Table 13.1(a) through Table 13.1(d) shall be used only for chimneys and vents not exposed to the outdoors below the roof line. A Type B vent or listed chimney lining system passing through an unused masonry chimney flue shall not be considered to be exposed to the outdoors. A Type B vent passing through an unventilated enclosure or chase insulated to a value of not less than R8 shall not be considered to be exposed to the outdoors. Table 13.1(b) in combination with Table 13.1(e) shall be used for clay tile lined exterior masonry chimneys, provided all of the following requirements are met:

(1) The vent connector is Type B double wall.
(2) The vent connector length is limited to 18 in./in. (18 mm/mm) of vent connector diameter.
(3) The appliance is draft hood equipped.
(4) The input rating is less than the maximum capacity given in Table 13.1(b).
(5) For a water heater, the outdoor design temperature shall not be less than 5°F (−15°C).
(6) For a space-heating appliance, the input rating is greater than the minimum capacity given by Table 13.1(e).

In the 2009 edition, item (7) and the exception covering alternate designs were deleted here, as they are covered elsewhere in the code.

Subsection 13.1.11 provides guidance for using exterior masonry chimneys for heating appliances. Restrictions are needed because of the high heat loss in chimneys exposed to the outdoors below the roofline in cold climates. Note that these restrictions do not apply to chimneys and vents that are not exposed to the outdoors below the roofline, because these chimneys and vents are considered to be interior chimneys.

The following items are required when sizing exterior masonry chimneys:

- To size chimneys and vents serving only water heaters, the outdoor design temperature must be greater than 5°F (–15°C). See Figure G.2.4 in Annex G for a map of the United States showing the temperature zones. In colder areas, either the chimney must be lined with a metal liner or another vent must be used.
- To size chimneys and vents serving space heating appliances or all appliances providing heat to the building, Table 13.1(f) for the minimum vent capacity, and Table 13.2(h) for the maximum vent capacity, should be used.

Note that only Type B double-wall vent connectors are allowed to minimize heat loss in the vent connector.

Water heaters are treated differently from heating appliances because their total operating hours are not sufficient to keep a chimney warm in cold periods, resulting in condensation of water in the chimney or vent.

### 13.1.12 Corrugated Vent Connector Size.

Corrugated vent connectors shall not be smaller than the listed appliance categorized vent diameter, flue collar diameter, or draft hood outlet diameter.

The requirement in 13.1.12 assists in sizing corrugated vent connectors. Note that the term *appliance categorized vent diameter/area* is defined in 3.3.7 as follows:

> **3.3.7 Appliance Categorized Vent Diameter/Area.** The minimum vent diameter/area permissible for Category I appliances to maintain a nonpositive vent static pressure when tested in accordance with nationally recognized standards.

There is no capacity reduction required when using corrugated connectors as there is when using corrugated chimney liners.

### 13.1.13 Upsizing.

Vent connectors shall not be upsized more than two sizes greater than the listed appliance categorized vent diameter, flue collar diameter, or draft hood outlet diameter.

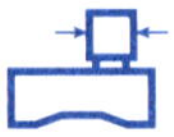
Upsizing

Upsizing, which is a sudden, large expansion of the vent connector diameter and is covered in 13.1.13, creates a pressure drop that may limit the draft and encourage condensation. Therefore, a limit is placed on upsizing.

### 13.1.14 Multiple Vertical Vent Sizes.

In a single run of vent or vent connector, more than one diameter and type shall be permitted to be used, provided that all the sizes and types are permitted by the tables.

Multiple Vertical Vent Sizes

Subsection 13.1.14 requires the installer to check the minimum and maximum capacities for the vent section for each vent type as if the entire vent section were made of that type. For example, if a vent connector is half single wall and half Type B, the installer must show that a vent connector of that length would be allowed if it were all single wall and also if it were all Type B.

### 13.1.15 Interpolation.

Interpolation shall be permitted in calculating capacities for vent dimensions that fall between table entries. *(See Example 3, Annex G.)*

Interpolation

If the installation dimensions fall between two table entries for which there are numerical values, the designer may calculate the "in between" value. This calculation is called interpolation and is addressed in 13.1.15. Interpolation is illustrated in Example 13.1.

***EXAMPLE 13.1***

Table 13.1(a) provides the capacity of a 3 in. diameter, 15 ft high vent with zero lateral serving a natural draft appliance of 58,000 Btu/hr, and of a 3 in. diameter, 20 ft high vent with zero lateral of 61,000 Btu/hr. See Commentary Table 13.5, which is excerpted from Table 13.1(a).

***COMMENTARY TABLE 13.5** Excerpt from Table 13.1(a)*

| Height H (ft) | Lateral L (ft) | Vent Diameter — D (in.) | | |
|---|---|---|---|---|
| | | 3 | | |
| | | FAN | | NAT |
| | | Min | Max | Max |
| 15 | 0 | 0 | 94 | 58 |
| | 2 | 11 | 69 | 48 |
| | 5 | 22 | 65 | 45 |
| | 10 | 29 | 59 | 41 |
| | 15 | 35 | 53 | 37 |

The capacity of a 3 in. diameter, 18 ft high vent with zero lateral can be interpolated by taking $^3/_5$ of the difference between 58,000 and 61,000 and adding it to the capacity of the 15 ft high vent. In this case the difference is 3000 Btu/hr; $^3/_5$ of 3000 is 1800, and the capacity of the 18 ft vent is 58,800 Btu/hr.

### 13.1.16 Extrapolation.

Extrapolation beyond the table entries shall not be permitted.

Extrapolation, which is addressed in 13.1.16, is estimating a value outside of a table. For example, Table 13.1(a) of the code provides vent capacities for vents up to 100 ft. The code does not permit one to use the numbers in Table 13.1(a) to estimate the capacity of a vent that is 110 ft high.

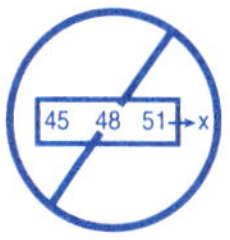

Extrapolation

### 13.1.17 Sizing Vents Not Covered by Tables.

For vent heights lower than 6 ft (1.8 m) and higher than shown in the Chapter 13 tables, engineering methods shall be used to calculate vent capacities.

Subsection 13.1.17 reinforces the requirement that vent heights outside the parameters of the tables in Chapter 13 must be calculated and that the tables cannot be used.

***TABLE 13.1(a)*** *Type B Double-Wall Gas Vent*

| | | | | | | | | | | | | | | | | | | | | | | |
|---|---|---|---|---|---|---|---|---|---|---|---|---|---|---|---|---|---|---|---|---|---|---|
| | | | | | | | | | | *Number of Appliances:* | | | | | *Single* | | | | | | | |
| | | | | | | | | | | *Appliance Type:* | | | | | *Category I* | | | | | | | |
| | | | | | | | | | | *Appliance Vent Connection:* | | | | | *Connected Directly to Vent* | | | | | | | |
| | | *Vent Diameter — D (in.)* | | | | | | | | | | | | | | | | | | | | |
| | | *3* | | | *4* | | | *5* | | | *6* | | | *7* | | | *8* | | | *9* | | |
| | | *Appliance Input Rating in Thousands of Btu per Hour* | | | | | | | | | | | | | | | | | | | | |
| *Height H* | *Lateral L* | *FAN* | | *NAT* | *FAN* | | *NAT* | *FAN* | | *NAT* | *FAN* | | *NAT* | *FAN* | | *NAT* | *FAN* | | *NAT* | *FAN* | | *NAT* |
| *(ft)* | *(ft)* | *Min* | *Max* | *Max* | *Min* | *Max* | *Max* | *Min* | *Max* | *Max* | *Min* | *Max* | *Max* | *Min* | *Max* | *Max* | *Min* | *Max* | *Max* | *Min* | *Max* | *Max* |
| 6 | 0 | 0 | 78 | 46 | 0 | 152 | 86 | 0 | 251 | 141 | 0 | 375 | 205 | 0 | 524 | 285 | 0 | 698 | 370 | 0 | 897 | 470 |
| | 2 | 13 | 51 | 36 | 18 | 97 | 67 | 27 | 157 | 105 | 32 | 232 | 157 | 44 | 321 | 217 | 53 | 425 | 285 | 63 | 543 | 370 |
| | 4 | 21 | 49 | 34 | 30 | 94 | 64 | 39 | 153 | 103 | 50 | 227 | 153 | 66 | 316 | 211 | 79 | 419 | 279 | 93 | 536 | 362 |
| | 6 | 25 | 46 | 32 | 36 | 91 | 61 | 47 | 149 | 100 | 59 | 223 | 149 | 78 | 310 | 205 | 93 | 413 | 273 | 110 | 530 | 354 |
| 8 | 0 | 0 | 84 | 50 | 0 | 165 | 94 | 0 | 276 | 155 | 0 | 415 | 235 | 0 | 583 | 320 | 0 | 780 | 415 | 0 | 1006 | 537 |
| | 2 | 12 | 57 | 40 | 16 | 109 | 75 | 25 | 178 | 120 | 28 | 263 | 180 | 42 | 365 | 247 | 50 | 483 | 322 | 60 | 619 | 418 |
| | 5 | 23 | 53 | 38 | 32 | 103 | 71 | 42 | 171 | 115 | 53 | 255 | 173 | 70 | 356 | 237 | 83 | 473 | 313 | 99 | 607 | 407 |
| | 8 | 28 | 49 | 35 | 39 | 98 | 66 | 51 | 164 | 109 | 64 | 247 | 165 | 84 | 347 | 227 | 99 | 463 | 303 | 117 | 596 | 396 |
| 10 | 0 | 0 | 88 | 53 | 0 | 175 | 100 | 0 | 295 | 166 | 0 | 447 | 255 | 0 | 631 | 345 | 0 | 847 | 450 | 0 | 1096 | 585 |
| | 2 | 12 | 61 | 42 | 17 | 118 | 81 | 23 | 194 | 129 | 26 | 289 | 195 | 40 | 402 | 273 | 48 | 533 | 355 | 57 | 684 | 457 |
| | 5 | 23 | 57 | 40 | 32 | 113 | 77 | 41 | 187 | 124 | 52 | 280 | 188 | 68 | 392 | 263 | 81 | 522 | 346 | 95 | 671 | 446 |
| | 10 | 30 | 51 | 36 | 41 | 104 | 70 | 54 | 176 | 115 | 67 | 267 | 175 | 88 | 376 | 245 | 104 | 504 | 330 | 122 | 651 | 427 |
| 15 | 0 | 0 | 94 | 58 | 0 | 191 | 112 | 0 | 327 | 187 | 0 | 502 | 285 | 0 | 716 | 390 | 0 | 970 | 525 | 0 | 1263 | 682 |
| | 2 | 11 | 69 | 48 | 15 | 136 | 93 | 20 | 226 | 150 | 22 | 339 | 225 | 38 | 475 | 316 | 45 | 633 | 414 | 53 | 815 | 544 |
| | 5 | 22 | 65 | 45 | 30 | 130 | 87 | 39 | 219 | 142 | 49 | 330 | 217 | 64 | 463 | 300 | 76 | 620 | 403 | 90 | 800 | 529 |
| | 10 | 29 | 59 | 41 | 40 | 121 | 82 | 51 | 206 | 135 | 64 | 315 | 208 | 84 | 445 | 288 | 99 | 600 | 386 | 116 | 777 | 507 |
| | 15 | 35 | 53 | 37 | 48 | 112 | 76 | 61 | 195 | 128 | 76 | 301 | 198 | 98 | 429 | 275 | 115 | 580 | 373 | 134 | 755 | 491 |
| 20 | 0 | 0 | 97 | 61 | 0 | 202 | 119 | 0 | 349 | 202 | 0 | 540 | 307 | 0 | 776 | 430 | 0 | 1057 | 575 | 0 | 1384 | 752 |
| | 2 | 10 | 75 | 51 | 14 | 149 | 100 | 18 | 250 | 166 | 20 | 377 | 249 | 33 | 531 | 346 | 41 | 711 | 470 | 50 | 917 | 612 |
| | 5 | 21 | 71 | 48 | 29 | 143 | 96 | 38 | 242 | 160 | 47 | 367 | 241 | 62 | 519 | 337 | 73 | 697 | 460 | 86 | 902 | 599 |
| | 10 | 28 | 64 | 44 | 38 | 133 | 89 | 50 | 229 | 150 | 62 | 351 | 228 | 81 | 499 | 321 | 95 | 675 | 443 | 112 | 877 | 576 |
| | 15 | 34 | 58 | 40 | 46 | 124 | 84 | 59 | 217 | 142 | 73 | 337 | 217 | 94 | 481 | 308 | 111 | 654 | 427 | 129 | 853 | 557 |
| | 20 | 48 | 52 | 35 | 55 | 116 | 78 | 69 | 206 | 134 | 84 | 322 | 206 | 107 | 464 | 295 | 125 | 634 | 410 | 145 | 830 | 537 |
| 30 | 0 | 0 | 100 | 64 | 0 | 213 | 128 | 0 | 374 | 220 | 0 | 587 | 336 | 0 | 853 | 475 | 0 | 1173 | 650 | 0 | 1548 | 855 |
| | 2 | 9 | 81 | 56 | 13 | 166 | 112 | 14 | 283 | 185 | 18 | 432 | 280 | 27 | 613 | 394 | 33 | 826 | 535 | 42 | 1072 | 700 |
| | 5 | 21 | 77 | 54 | 28 | 160 | 108 | 36 | 275 | 176 | 45 | 421 | 273 | 58 | 600 | 385 | 69 | 811 | 524 | 82 | 1055 | 688 |
| | 10 | 27 | 70 | 50 | 37 | 150 | 102 | 48 | 262 | 171 | 59 | 405 | 261 | 77 | 580 | 371 | 91 | 788 | 507 | 107 | 1028 | 668 |
| | 15 | 33 | 64 | NA | 44 | 141 | 96 | 57 | 249 | 163 | 70 | 389 | 249 | 90 | 560 | 357 | 105 | 765 | 490 | 124 | 1002 | 648 |
| | 20 | 56 | 58 | NA | 53 | 132 | 90 | 66 | 237 | 154 | 80 | 374 | 237 | 102 | 542 | 343 | 119 | 743 | 473 | 139 | 977 | 628 |
| | 30 | NA | NA | NA | 73 | 113 | NA | 88 | 214 | NA | 104 | 346 | 219 | 131 | 507 | 321 | 149 | 702 | 444 | 171 | 929 | 594 |
| 50 | 0 | 0 | 101 | 67 | 0 | 216 | 134 | 0 | 397 | 232 | 0 | 633 | 363 | 0 | 932 | 518 | 0 | 1297 | 708 | 0 | 1730 | 952 |
| | 2 | 8 | 86 | 61 | 11 | 183 | 122 | 14 | 320 | 206 | 15 | 497 | 314 | 22 | 715 | 445 | 26 | 975 | 615 | 33 | 1276 | 813 |
| | 5 | 20 | 82 | NA | 27 | 177 | 119 | 35 | 312 | 200 | 43 | 487 | 308 | 55 | 702 | 438 | 65 | 960 | 605 | 77 | 1259 | 798 |
| | 10 | 26 | 76 | NA | 35 | 168 | 114 | 45 | 299 | 190 | 56 | 471 | 298 | 73 | 681 | 426 | 86 | 935 | 589 | 101 | 1230 | 773 |
| | 15 | 59 | 70 | NA | 42 | 158 | NA | 54 | 287 | 180 | 66 | 455 | 288 | 85 | 662 | 413 | 100 | 911 | 572 | 117 | 1203 | 747 |
| | 20 | NA | NA | NA | 50 | 149 | NA | 63 | 275 | 169 | 76 | 440 | 278 | 97 | 642 | 401 | 113 | 888 | 556 | 131 | 1176 | 722 |
| | 30 | NA | NA | NA | 69 | 131 | NA | 84 | 250 | NA | 99 | 410 | 259 | 123 | 605 | 376 | 141 | 844 | 522 | 161 | 1125 | 670 |
| 100 | 0 | NA | NA | NA | 0 | 218 | NA | 0 | 407 | NA | 0 | 665 | 400 | 0 | 997 | 560 | 0 | 1411 | 770 | 0 | 1908 | 1040 |
| | 2 | NA | NA | NA | 10 | 194 | NA | 12 | 354 | NA | 13 | 566 | 375 | 18 | 831 | 510 | 21 | 1155 | 700 | 25 | 1536 | 935 |
| | 5 | NA | NA | NA | 26 | 189 | NA | 33 | 347 | NA | 40 | 557 | 369 | 52 | 820 | 504 | 60 | 1141 | 692 | 71 | 1519 | 926 |
| | 10 | NA | NA | NA | 33 | 182 | NA | 43 | 335 | NA | 53 | 542 | 361 | 68 | 801 | 493 | 80 | 1118 | 679 | 94 | 1492 | 910 |
| | 15 | NA | NA | NA | 40 | 174 | NA | 50 | 321 | NA | 62 | 528 | 353 | 80 | 782 | 482 | 93 | 1095 | 666 | 109 | 1465 | 895 |
| | 20 | NA | NA | NA | 47 | 166 | NA | 59 | 311 | NA | 71 | 513 | 344 | 90 | 763 | 471 | 105 | 1073 | 653 | 122 | 1438 | 880 |
| | 30 | NA | NA | NA | NA | NA | NA | 78 | 290 | NA | 92 | 483 | NA | 115 | 726 | 449 | 131 | 1029 | 627 | 149 | 1387 | 849 |
| | 50 | NA | NA | NA | NA | NA | NA | NA | NA | NA | 147 | 428 | NA | 180 | 651 | 405 | 197 | 944 | 575 | 217 | 1288 | 787 |

*TABLE 13.1(a) Continued*

| | | | | | | | | | | | | | | | | | | | | | | | | | |
|---|---|---|---|---|---|---|---|---|---|---|---|---|---|---|---|---|---|---|---|---|---|---|---|---|---|
| | | | | | | | | | | | | | | *Number of Appliances:* | | | | *Single* | | | | | | | |
| | | | | | | | | | | | | | | *Appliance Type:* | | | | *Category I* | | | | | | | |
| | | | | | | | | | | | | | | *Appliance Vent Connection:* | | | | *Connected Directly to Vent* | | | | | | | |
| | | *Vent Diameter — D (in.)* | | | | | | | | | | | | | | | | | | | | | | | |
| | | *10* | | | *12* | | | *14* | | | *16* | | | *18* | | | *20* | | | *22* | | | *24* | | |
| | | *Appliance Input Rating in Thousands of Btu per Hour* | | | | | | | | | | | | | | | | | | | | | | | |
| *Height H* | *Lateral L* | *FAN* | | *NAT* | *FAN* | | *NAT* | *FAN* | | *NAT* | *FAN* | | *NAT* | *FAN* | | *NAT* | *FAN* | | *NAT* | *FAN* | | *NAT* | *FAN* | | *NAT* |
| *(ft)* | *(ft)* | *Min* | *Max* | *Max* | *Min* | *Max* | *Max* | *Min* | *Max* | *Max* | *Min* | *Max* | *Max* | *Min* | *Max* | *Max* | *Min* | *Max* | *Max* | *Min* | *Max* | *Max* | *Min* | *Max* | *Max* |
| 6 | 0 | 0 | 1121 | 570 | 0 | 1645 | 850 | 0 | 2267 | 1170 | 0 | 2983 | 1530 | 0 | 3802 | 1960 | 0 | 4721 | 2430 | 0 | 5737 | 2950 | 0 | 6853 | 3520 |
| | 2 | 75 | 675 | 455 | 103 | 982 | 650 | 138 | 1346 | 890 | 178 | 1769 | 1170 | 225 | 2250 | 1480 | 296 | 2782 | 1850 | 360 | 3377 | 2220 | 426 | 4030 | 2670 |
| | 4 | 110 | 668 | 445 | 147 | 975 | 640 | 191 | 1338 | 880 | 242 | 1761 | 1160 | 300 | 2242 | 1475 | 390 | 2774 | 1835 | 469 | 3370 | 2215 | 555 | 4023 | 2660 |
| | 6 | 128 | 661 | 435 | 171 | 967 | 630 | 219 | 1330 | 870 | 276 | 1753 | 1150 | 341 | 2235 | 1470 | 437 | 2767 | 1820 | 523 | 3363 | 2210 | 618 | 4017 | 2650 |
| 8 | 0 | 0 | 1261 | 660 | 0 | 1858 | 970 | 0 | 2571 | 1320 | 0 | 3399 | 1740 | 0 | 4333 | 2220 | 0 | 5387 | 2750 | 0 | 6555 | 3360 | 0 | 7838 | 4010 |
| | 2 | 71 | 770 | 515 | 98 | 1124 | 745 | 130 | 1543 | 1020 | 168 | 2030 | 1340 | 212 | 2584 | 1700 | 278 | 3196 | 2110 | 336 | 3882 | 2560 | 401 | 4634 | 3050 |
| | 5 | 115 | 758 | 503 | 154 | 1110 | 733 | 199 | 1528 | 1010 | 251 | 2013 | 1330 | 311 | 2563 | 1685 | 398 | 3180 | 2090 | 476 | 3863 | 2545 | 562 | 4612 | 3040 |
| | 8 | 137 | 746 | 490 | 180 | 1097 | 720 | 231 | 1514 | 1000 | 289 | 2000 | 1320 | 354 | 2552 | 1670 | 450 | 3163 | 2070 | 537 | 3850 | 2530 | 630 | 4602 | 3030 |
| 10 | 0 | 0 | 1377 | 720 | 0 | 2036 | 1060 | 0 | 2825 | 1450 | 0 | 3742 | 1925 | 0 | 4782 | 2450 | 0 | 5955 | 3050 | 0 | 7254 | 3710 | 0 | 8682 | 4450 |
| | 2 | 68 | 852 | 560 | 93 | 1244 | 850 | 124 | 1713 | 1130 | 161 | 2256 | 1480 | 202 | 2868 | 1890 | 264 | 3556 | 2340 | 319 | 4322 | 2840 | 378 | 5153 | 3390 |
| | 5 | 112 | 839 | 547 | 149 | 1229 | 829 | 192 | 1696 | 1105 | 243 | 2238 | 1461 | 300 | 2849 | 1871 | 382 | 3536 | 2318 | 458 | 4301 | 2818 | 540 | 5132 | 3371 |
| | 10 | 142 | 817 | 525 | 187 | 1204 | 795 | 238 | 1669 | 1080 | 298 | 2209 | 1430 | 364 | 2818 | 1840 | 459 | 3504 | 2280 | 546 | 4268 | 2780 | 641 | 5099 | 3340 |
| 15 | 0 | 0 | 1596 | 840 | 0 | 2380 | 1240 | 0 | 3323 | 1720 | 0 | 4423 | 2270 | 0 | 5678 | 2900 | 0 | 7099 | 3620 | 0 | 8665 | 4410 | 0 | 10,393 | 5300 |
| | 2 | 63 | 1019 | 675 | 86 | 1495 | 985 | 114 | 2062 | 1350 | 147 | 2719 | 1770 | 186 | 3467 | 2260 | 239 | 4304 | 2800 | 290 | 5232 | 3410 | 346 | 6251 | 4080 |
| | 5 | 105 | 1003 | 660 | 140 | 1476 | 967 | 182 | 2041 | 1327 | 229 | 2696 | 1748 | 283 | 3442 | 2235 | 355 | 4278 | 2777 | 426 | 5204 | 3385 | 501 | 6222 | 4057 |
| | 10 | 135 | 977 | 635 | 177 | 1446 | 936 | 227 | 2009 | 1289 | 283 | 2659 | 1712 | 346 | 3402 | 2193 | 432 | 4234 | 2739 | 510 | 5159 | 3343 | 599 | 6175 | 4019 |
| | 15 | 155 | 953 | 610 | 202 | 1418 | 905 | 257 | 1976 | 1250 | 318 | 2623 | 1675 | 385 | 3363 | 2150 | 479 | 4192 | 2700 | 564 | 5115 | 3300 | 665 | 6129 | 3980 |
| 20 | 0 | 0 | 1756 | 930 | 0 | 2637 | 1350 | 0 | 3701 | 1900 | 0 | 4948 | 2520 | 0 | 6376 | 3250 | 0 | 7988 | 4060 | 0 | 9785 | 4980 | 0 | 11,753 | 6000 |
| | 2 | 59 | 1150 | 755 | 81 | 1694 | 1100 | 107 | 2343 | 1520 | 139 | 3097 | 2000 | 175 | 3955 | 2570 | 220 | 4916 | 3200 | 269 | 5983 | 3910 | 321 | 7154 | 4700 |
| | 5 | 101 | 1133 | 738 | 135 | 1674 | 1079 | 174 | 2320 | 1498 | 219 | 3071 | 1978 | 270 | 3926 | 2544 | 337 | 4885 | 3174 | 403 | 5950 | 3880 | 475 | 7119 | 4662 |
| | 10 | 130 | 1105 | 710 | 172 | 1641 | 1045 | 220 | 2282 | 1460 | 273 | 3029 | 1940 | 334 | 3880 | 2500 | 413 | 4835 | 3130 | 489 | 5896 | 3830 | 573 | 7063 | 4600 |
| | 15 | 150 | 1078 | 688 | 195 | 1609 | 1018 | 248 | 2245 | 1425 | 306 | 2988 | 1910 | 372 | 3835 | 2465 | 459 | 4786 | 3090 | 541 | 5844 | 3795 | 631 | 7007 | 4575 |
| | 20 | 167 | 1052 | 665 | 217 | 1578 | 990 | 273 | 2210 | 1390 | 335 | 2948 | 1880 | 404 | 3791 | 2430 | 495 | 4737 | 3050 | 585 | 5792 | 3760 | 689 | 6953 | 4550 |
| 30 | 0 | 0 | 1977 | 1060 | 0 | 3004 | 1550 | 0 | 4252 | 2170 | 0 | 5725 | 2920 | 0 | 7420 | 3770 | 0 | 9341 | 4750 | 0 | 11,483 | 5850 | 0 | 13,848 | 7060 |
| | 2 | 54 | 1351 | 865 | 74 | 2004 | 1310 | 98 | 2786 | 1800 | 127 | 3696 | 2380 | 159 | 4734 | 3050 | 199 | 5900 | 3810 | 241 | 7194 | 4650 | 285 | 8617 | 5600 |
| | 5 | 96 | 1332 | 851 | 127 | 1981 | 1289 | 164 | 2759 | 1775 | 206 | 3666 | 2350 | 252 | 4701 | 3020 | 312 | 5863 | 3783 | 373 | 7155 | 4622 | 439 | 8574 | 5552 |
| | 10 | 125 | 1301 | 829 | 164 | 1944 | 1254 | 209 | 2716 | 1733 | 259 | 3617 | 2300 | 316 | 4647 | 2970 | 386 | 5803 | 3739 | 456 | 7090 | 4574 | 535 | 8505 | 5471 |
| | 15 | 143 | 1272 | 807 | 187 | 1908 | 1220 | 237 | 2674 | 1692 | 292 | 3570 | 2250 | 354 | 4594 | 2920 | 431 | 5744 | 3695 | 507 | 7026 | 4527 | 590 | 8437 | 5391 |
| | 20 | 160 | 1243 | 784 | 207 | 1873 | 1185 | 260 | 2633 | 1650 | 319 | 3523 | 2200 | 384 | 4542 | 2870 | 467 | 5686 | 3650 | 548 | 6964 | 4480 | 639 | 8370 | 5310 |
| | 30 | 195 | 1189 | 745 | 246 | 1807 | 1130 | 305 | 2555 | 1585 | 369 | 3433 | 2130 | 440 | 4442 | 2785 | 540 | 5574 | 3565 | 635 | 6842 | 4375 | 739 | 8239 | 5225 |
| 50 | 0 | 0 | 2231 | 1195 | 0 | 3441 | 1825 | 0 | 4934 | 2550 | 0 | 6711 | 3440 | 0 | 8774 | 4460 | 0 | 11,129 | 5635 | 0 | 13,767 | 6940 | 0 | 16,694 | 8430 |
| | 2 | 41 | 1620 | 1010 | 66 | 2431 | 1513 | 86 | 3409 | 2125 | 113 | 4554 | 2840 | 141 | 5864 | 3670 | 171 | 7339 | 4630 | 209 | 8980 | 5695 | 251 | 10,788 | 6860 |
| | 5 | 90 | 1600 | 996 | 118 | 2406 | 1495 | 151 | 3380 | 2102 | 191 | 4520 | 2813 | 234 | 5826 | 3639 | 283 | 7295 | 4597 | 336 | 8933 | 5654 | 394 | 10,737 | 6818 |
| | 10 | 118 | 1567 | 972 | 154 | 2366 | 1466 | 196 | 3332 | 2064 | 243 | 4464 | 2767 | 295 | 5763 | 3585 | 355 | 7224 | 4542 | 419 | 8855 | 5585 | 491 | 10,652 | 6749 |
| | 15 | 136 | 1536 | 948 | 177 | 2327 | 1437 | 222 | 3285 | 2026 | 274 | 4409 | 2721 | 330 | 5701 | 3534 | 396 | 7155 | 4511 | 465 | 8779 | 5546 | 542 | 10,570 | 6710 |
| | 20 | 151 | 1505 | 924 | 195 | 2288 | 1408 | 244 | 3239 | 1987 | 300 | 4356 | 2675 | 361 | 5641 | 3481 | 433 | 7086 | 4479 | 506 | 8704 | 5506 | 586 | 10,488 | 6670 |
| | 30 | 183 | 1446 | 876 | 232 | 2214 | 1349 | 287 | 3150 | 1910 | 347 | 4253 | 2631 | 412 | 5523 | 3431 | 494 | 6953 | 4421 | 577 | 8557 | 5444 | 672 | 10,328 | 6603 |
| 100 | 0 | 0 | 2491 | 1310 | 0 | 3925 | 2050 | 0 | 5729 | 2950 | 0 | 7914 | 4050 | 0 | 10,485 | 5300 | 0 | 13,454 | 6700 | 0 | 16,817 | 8600 | 0 | 20,578 | 10,300 |
| | 2 | 30 | 1975 | 1170 | 44 | 3027 | 1820 | 72 | 4313 | 2550 | 95 | 5834 | 3500 | 120 | 7591 | 4600 | 138 | 9577 | 5800 | 169 | 11,803 | 7200 | 204 | 14,264 | 8800 |
| | 5 | 82 | 1955 | 1159 | 107 | 3002 | 1803 | 136 | 4282 | 2531 | 172 | 5797 | 3475 | 208 | 7548 | 4566 | 245 | 9528 | 5769 | 293 | 11,748 | 7162 | 341 | 14,204 | 8756 |
| | 10 | 108 | 1923 | 1142 | 142 | 2961 | 1775 | 180 | 4231 | 2500 | 223 | 5737 | 3434 | 268 | 7478 | 4509 | 318 | 9447 | 5717 | 374 | 11,658 | 7100 | 436 | 14,105 | 8683 |
| | 15 | 126 | 1892 | 1124 | 163 | 2920 | 1747 | 206 | 4182 | 2469 | 252 | 5678 | 3392 | 304 | 7409 | 4451 | 358 | 9367 | 5665 | 418 | 11,569 | 7037 | 487 | 14,007 | 8610 |
| | 20 | 141 | 1861 | 1107 | 181 | 2880 | 1719 | 226 | 4133 | 2438 | 277 | 5619 | 3351 | 330 | 7341 | 4394 | 387 | 9289 | 5613 | 452 | 11,482 | 6975 | 523 | 13,910 | 8537 |
| | 30 | 170 | 1802 | 1071 | 215 | 2803 | 1663 | 265 | 4037 | 2375 | 319 | 5505 | 3267 | 378 | 7209 | 4279 | 446 | 9136 | 5509 | 514 | 11,310 | 6850 | 592 | 13,720 | 8391 |
| | 50 | 241 | 1688 | 1000 | 292 | 2657 | 1550 | 350 | 3856 | 2250 | 415 | 5289 | 3100 | 486 | 6956 | 4050 | 572 | 8841 | 5300 | 659 | 10,979 | 6600 | 752 | 13,354 | 8100 |

For SI units, 1 in. = 25.4 mm, 1 ft = 0.305 m, 1000 Btu/hr = 0.293 kW, 1 in.$^2$ = 645 mm$^2$.

NA: Not applicable.

*TABLE 13.1(b) Type B Double-Wall Vent*

| | |
|---|---|
| *Number of Appliances:* | *Single* |
| *Appliance Type:* | *Category I* |
| *Appliance Vent Connection:* | *Single-Wall Metal Connector* |

| | | *Vent Diameter — D (in.)* | | | | | | | | | | | | | | | | | | | | | | | | | | |
|---|---|---|---|---|---|---|---|---|---|---|---|---|---|---|---|---|---|---|---|---|---|---|---|---|---|---|---|---|
| | | *3* | | | *4* | | | *5* | | | *6* | | | *7* | | | *8* | | | *9* | | | *10* | | | *12* | | |
| | | *Appliance Input Rating in Thousands of Btu per Hour* | | | | | | | | | | | | | | | | | | | | | | | | | | |
| *Height H* | *Lateral L* | *FAN* | | *NAT* | *FAN* | | *NAT* | *FAN* | | *NAT* | *FAN* | | *NAT* | *FAN* | | *NAT* | *FAN* | | *NAT* | *FAN* | | *NAT* | *FAN* | | *NAT* | *FAN* | | *NAT* |
| *(ft)* | *(ft)* | *Min* | *Max* | *Max* | *Min* | *Max* | *Max* | *Min* | *Max* | *Max* | *Min* | *Max* | *Max* | *Min* | *Max* | *Max* | *Min* | *Max* | *Max* | *Min* | *Max* | *Max* | *Min* | *Max* | *Max* | *Min* | *Max* | *Max* |
| 6 | 0 | 38 | 77 | 45 | 59 | 151 | 85 | 85 | 249 | 140 | 126 | 373 | 204 | 165 | 522 | 284 | 211 | 695 | 369 | 267 | 894 | 469 | 371 | 1118 | 569 | 537 | 1639 | 849 |
| | 2 | 39 | 51 | 36 | 60 | 96 | 66 | 85 | 156 | 104 | 123 | 231 | 156 | 159 | 320 | 213 | 201 | 423 | 284 | 251 | 541 | 368 | 347 | 673 | 453 | 498 | 979 | 648 |
| | 4 | NA | NA | 33 | 74 | 92 | 63 | 102 | 152 | 102 | 146 | 225 | 152 | 187 | 313 | 208 | 237 | 416 | 277 | 295 | 533 | 360 | 409 | 664 | 443 | 584 | 971 | 638 |
| | 6 | NA | NA | 31 | 83 | 89 | 60 | 114 | 147 | 99 | 163 | 220 | 148 | 207 | 307 | 203 | 263 | 409 | 271 | 327 | 526 | 352 | 449 | 656 | 433 | 638 | 962 | 627 |
| 8 | 0 | 37 | 83 | 50 | 58 | 164 | 93 | 83 | 273 | 154 | 123 | 412 | 234 | 161 | 580 | 319 | 206 | 777 | 414 | 258 | 1002 | 536 | 360 | 1257 | 658 | 521 | 1852 | 967 |
| | 2 | 39 | 56 | 39 | 59 | 108 | 75 | 83 | 176 | 119 | 121 | 261 | 179 | 155 | 363 | 246 | 197 | 482 | 321 | 246 | 617 | 417 | 339 | 768 | 513 | 486 | 1120 | 743 |
| | 5 | NA | NA | 37 | 77 | 102 | 69 | 107 | 168 | 114 | 151 | 252 | 171 | 193 | 352 | 235 | 245 | 470 | 311 | 305 | 604 | 404 | 418 | 754 | 500 | 598 | 1104 | 730 |
| | 8 | NA | NA | 33 | 90 | 95 | 64 | 122 | 161 | 107 | 175 | 243 | 163 | 223 | 342 | 225 | 280 | 458 | 300 | 344 | 591 | 392 | 470 | 740 | 486 | 665 | 1089 | 715 |
| 10 | 0 | 37 | 87 | 53 | 57 | 174 | 99 | 82 | 293 | 165 | 120 | 444 | 254 | 158 | 628 | 344 | 202 | 844 | 449 | 253 | 1093 | 584 | 351 | 1373 | 718 | 507 | 2031 | 1057 |
| | 2 | 39 | 61 | 41 | 59 | 117 | 80 | 82 | 193 | 128 | 119 | 287 | 194 | 153 | 400 | 272 | 193 | 531 | 354 | 242 | 681 | 456 | 332 | 849 | 559 | 475 | 1242 | 848 |
| | 5 | 52 | 56 | 39 | 76 | 111 | 76 | 105 | 185 | 122 | 148 | 277 | 186 | 190 | 388 | 261 | 241 | 518 | 344 | 299 | 667 | 443 | 409 | 834 | 544 | 584 | 1224 | 825 |
| | 10 | NA | NA | 34 | 97 | 100 | 68 | 132 | 171 | 112 | 188 | 261 | 171 | 237 | 369 | 241 | 296 | 497 | 325 | 363 | 643 | 423 | 492 | 808 | 520 | 688 | 1194 | 788 |
| 15 | 0 | 36 | 93 | 57 | 56 | 190 | 111 | 80 | 325 | 186 | 116 | 499 | 283 | 153 | 713 | 388 | 195 | 966 | 523 | 244 | 1259 | 681 | 336 | 1591 | 838 | 488 | 2374 | 1237 |
| | 2 | 38 | 69 | 47 | 57 | 136 | 93 | 80 | 225 | 149 | 115 | 337 | 224 | 148 | 473 | 314 | 187 | 631 | 413 | 232 | 812 | 543 | 319 | 1015 | 673 | 457 | 1491 | 983 |
| | 5 | 51 | 63 | 44 | 75 | 128 | 86 | 102 | 216 | 140 | 144 | 326 | 217 | 182 | 459 | 298 | 231 | 616 | 400 | 287 | 795 | 526 | 392 | 997 | 657 | 562 | 1469 | 963 |
| | 10 | NA | NA | 39 | 95 | 116 | 79 | 128 | 201 | 131 | 182 | 308 | 203 | 228 | 438 | 284 | 284 | 592 | 381 | 349 | 768 | 501 | 470 | 966 | 628 | 664 | 1433 | 928 |
| | 15 | NA | NA | NA | NA | NA | 72 | 158 | 186 | 124 | 220 | 290 | 192 | 272 | 418 | 269 | 334 | 568 | 367 | 404 | 742 | 484 | 540 | 937 | 601 | 750 | 1399 | 894 |
| 20 | 0 | 35 | 96 | 60 | 54 | 200 | 118 | 78 | 346 | 201 | 114 | 537 | 306 | 149 | 772 | 428 | 190 | 1053 | 573 | 238 | 1379 | 750 | 326 | 1751 | 927 | 473 | 2631 | 1346 |
| | 2 | 37 | 74 | 50 | 56 | 148 | 99 | 78 | 248 | 165 | 113 | 375 | 248 | 144 | 528 | 344 | 182 | 708 | 468 | 227 | 914 | 611 | 309 | 1146 | 754 | 443 | 1689 | 1098 |
| | 5 | 50 | 68 | 47 | 73 | 140 | 94 | 100 | 239 | 158 | 141 | 363 | 239 | 178 | 514 | 334 | 224 | 692 | 457 | 279 | 896 | 596 | 381 | 1126 | 734 | 547 | 1665 | 1074 |
| | 10 | NA | NA | 41 | 93 | 129 | 86 | 125 | 223 | 146 | 177 | 344 | 224 | 222 | 491 | 316 | 277 | 666 | 437 | 339 | 866 | 570 | 457 | 1092 | 702 | 646 | 1626 | 1037 |
| | 15 | NA | NA | NA | NA | NA | 80 | 155 | 208 | 136 | 216 | 325 | 210 | 264 | 469 | 301 | 325 | 640 | 419 | 393 | 838 | 549 | 526 | 1060 | 677 | 730 | 1587 | 1005 |
| | 20 | NA | NA | NA | NA | NA | NA | 186 | 192 | 126 | 254 | 306 | 196 | 309 | 448 | 285 | 374 | 616 | 400 | 448 | 810 | 526 | 592 | 1028 | 651 | 808 | 1550 | 973 |
| 30 | 0 | 34 | 99 | 63 | 53 | 211 | 127 | 76 | 372 | 219 | 110 | 584 | 334 | 144 | 849 | 472 | 184 | 1168 | 647 | 229 | 1542 | 852 | 312 | 1971 | 1056 | 454 | 2996 | 1545 |
| | 2 | 37 | 80 | 56 | 55 | 164 | 111 | 76 | 281 | 183 | 109 | 429 | 279 | 139 | 610 | 392 | 175 | 823 | 533 | 219 | 1069 | 698 | 296 | 1346 | 863 | 424 | 1999 | 1308 |
| | 5 | 49 | 74 | 52 | 72 | 157 | 106 | 98 | 271 | 173 | 136 | 417 | 271 | 171 | 595 | 382 | 215 | 806 | 521 | 269 | 1049 | 684 | 366 | 1324 | 846 | 524 | 1971 | 1283 |
| | 10 | NA | NA | NA | 91 | 144 | 98 | 122 | 255 | 168 | 171 | 397 | 257 | 213 | 570 | 367 | 265 | 777 | 501 | 327 | 1017 | 662 | 440 | 1287 | 821 | 620 | 1927 | 1243 |
| | 15 | NA | NA | NA | 115 | 131 | NA | 151 | 239 | 157 | 208 | 377 | 242 | 255 | 547 | 349 | 312 | 750 | 481 | 379 | 985 | 638 | 507 | 1251 | 794 | 702 | 1884 | 1205 |
| | 20 | NA | NA | NA | NA | NA | NA | 181 | 223 | NA | 246 | 357 | 228 | 298 | 524 | 333 | 360 | 723 | 461 | 433 | 955 | 615 | 570 | 1216 | 768 | 780 | 1841 | 1166 |
| | 30 | NA | NA | NA | NA | NA | NA | NA | NA | NA | NA | NA | NA | 389 | 477 | 305 | 461 | 670 | 426 | 541 | 895 | 574 | 704 | 1147 | 720 | 937 | 1759 | 1101 |
| 50 | 0 | 33 | 99 | 66 | 51 | 213 | 133 | 73 | 394 | 230 | 105 | 629 | 361 | 138 | 928 | 515 | 176 | 1292 | 704 | 220 | 1724 | 948 | 295 | 2223 | 1189 | 428 | 3432 | 1818 |
| | 2 | 36 | 84 | 61 | 53 | 181 | 121 | 73 | 318 | 205 | 104 | 495 | 312 | 133 | 712 | 443 | 168 | 971 | 613 | 209 | 1273 | 811 | 280 | 1615 | 1007 | 401 | 2426 | 1509 |
| | 5 | 48 | 80 | NA | 70 | 174 | 117 | 94 | 308 | 198 | 131 | 482 | 305 | 164 | 696 | 435 | 204 | 953 | 602 | 257 | 1252 | 795 | 347 | 1591 | 991 | 496 | 2396 | 1490 |
| | 10 | NA | NA | NA | 89 | 160 | NA | 118 | 292 | 186 | 162 | 461 | 292 | 203 | 671 | 420 | 253 | 923 | 583 | 313 | 1217 | 765 | 418 | 1551 | 963 | 589 | 2347 | 1455 |
| | 15 | NA | NA | NA | 112 | 148 | NA | 145 | 275 | 174 | 199 | 441 | 280 | 244 | 646 | 405 | 299 | 894 | 562 | 363 | 1183 | 736 | 481 | 1512 | 934 | 668 | 2299 | 1421 |
| | 20 | NA | NA | NA | NA | NA | NA | 176 | 257 | NA | 236 | 420 | 267 | 285 | 622 | 389 | 345 | 866 | 543 | 415 | 1150 | 708 | 544 | 1473 | 906 | 741 | 2251 | 1387 |
| | 30 | NA | NA | NA | NA | NA | NA | NA | NA | NA | 315 | 376 | NA | 373 | 573 | NA | 442 | 809 | 502 | 521 | 1086 | 649 | 674 | 1399 | 848 | 892 | 2159 | 1318 |
| 100 | 0 | NA | NA | NA | 49 | 214 | NA | 69 | 403 | NA | 100 | 659 | 395 | 131 | 991 | 555 | 166 | 1404 | 765 | 207 | 1900 | 1033 | 273 | 2479 | 1300 | 395 | 3912 | 2042 |
| | 2 | NA | NA | NA | 51 | 192 | NA | 70 | 351 | NA | 98 | 563 | 373 | 125 | 828 | 508 | 158 | 1152 | 698 | 196 | 1532 | 933 | 259 | 1970 | 1168 | 371 | 3021 | 1817 |
| | 5 | NA | NA | NA | 67 | 186 | NA | 90 | 342 | NA | 125 | 551 | 366 | 156 | 813 | 501 | 194 | 1134 | 688 | 240 | 1511 | 921 | 322 | 1945 | 1153 | 460 | 2990 | 1796 |
| | 10 | NA | NA | NA | 85 | 175 | NA | 113 | 324 | NA | 153 | 532 | 354 | 191 | 789 | 486 | 238 | 1104 | 672 | 293 | 1477 | 902 | 389 | 1905 | 1133 | 547 | 2938 | 1763 |
| | 15 | NA | NA | NA | 132 | 162 | NA | 138 | 310 | NA | 188 | 511 | 343 | 230 | 764 | 473 | 281 | 1075 | 656 | 342 | 1443 | 884 | 447 | 1865 | 1110 | 618 | 2888 | 1730 |
| | 20 | NA | NA | NA | NA | NA | NA | 168 | 295 | NA | 224 | 487 | NA | 270 | 739 | 458 | 325 | 1046 | 639 | 391 | 1410 | 864 | 507 | 1825 | 1087 | 690 | 2838 | 1696 |
| | 30 | NA | NA | NA | NA | NA | NA | 231 | 264 | NA | 301 | 448 | NA | 355 | 685 | NA | 418 | 988 | NA | 491 | 1343 | 824 | 631 | 1747 | 1041 | 834 | 2739 | 1627 |
| | 50 | NA | NA | NA | NA | NA | NA | NA | NA | NA | NA | NA | NA | 540 | 584 | NA | 617 | 866 | NA | 711 | 1205 | NA | 895 | 1591 | NA | 1138 | 2547 | 1489 |

For SI units, 1 in. = 25.4 mm, 1 ft = 0.305 m, 1000 Btu/hr = 0.293 kW, 1 in.$^2$ = 645 mm$^2$.

NA: Not applicable.

*TABLE 13.1(c) Masonry Chimney*

| | | | | | | | | | | | | | | | | | | | | | | | | | | | | |
|---|---|---|---|---|---|---|---|---|---|---|---|---|---|---|---|---|---|---|---|---|---|---|---|---|---|---|---|---|
| | | | | | | | | | | | | | | *Number of Appliances:* | | | | | | *Single* | | | | | | | | |
| | | | | | | | | | | | | | | *Appliance Type:* | | | | | | *Category I* | | | | | | | | |
| | | | | | | | | | | | | | | *Appliance Vent Connection:* | | | | | | *Type B Double-Wall Connector* | | | | | | | | |
| | | *Type B Double-Wall Connector Diameter — D (in.) To be used with chimney areas within the size limits at bottom* | | | | | | | | | | | | | | | | | | | | | | | | | | |
| | | *3* | | | *4* | | | *5* | | | *6* | | | *7* | | | *8* | | | *9* | | | *10* | | | *12* | | |
| | | *Appliance Input Rating in Thousands of Btu per Hour* | | | | | | | | | | | | | | | | | | | | | | | | | | |
| *Height H* | *Lateral L* | *FAN* | | *NAT* | *FAN* | | *NAT* | *FAN* | | *NAT* | *FAN* | | *NAT* | *FAN* | | *NAT* | *FAN* | | *NAT* | *FAN* | | *NAT* | *FAN* | | *NAT* | *FAN* | | *NAT* |
| *(ft)* | *(ft)* | *Min* | *Max* | *Max* | *Min* | *Max* | *Max* | *Min* | *Max* | *Max* | *Min* | *Max* | *Max* | *Min* | *Max* | *Max* | *Min* | *Max* | *Max* | *Min* | *Max* | *Max* | *Min* | *Max* | *Max* | *Min* | *Max* | *Max* |
| 6 | 2 | NA | NA | 28 | NA | NA | 52 | NA | NA | 86 | NA | NA | 130 | NA | NA | 180 | NA | NA | 247 | NA | NA | 320 | NA | NA | 401 | NA | NA | 581 |
| | 5 | NA | NA | 25 | NA | NA | 49 | NA | NA | 82 | NA | NA | 117 | NA | NA | 165 | NA | NA | 231 | NA | NA | 298 | NA | NA | 376 | NA | NA | 561 |
| 8 | 2 | NA | NA | 29 | NA | NA | 55 | NA | NA | 93 | NA | NA | 145 | NA | NA | 198 | NA | NA | 266 | 84 | 590 | 350 | 100 | 728 | 446 | 139 | 1024 | 651 |
| | 5 | NA | NA | 26 | NA | NA | 52 | NA | NA | 88 | NA | NA | 134 | NA | NA | 183 | NA | NA | 247 | NA | NA | 328 | 149 | 711 | 423 | 201 | 1007 | 640 |
| | 8 | NA | NA | 24 | NA | NA | 48 | NA | NA | 83 | NA | NA | 127 | NA | NA | 175 | NA | NA | 239 | NA | NA | 318 | 173 | 695 | 410 | 231 | 990 | 623 |
| 10 | 2 | NA | NA | 31 | NA | NA | 61 | NA | NA | 103 | NA | NA | 162 | NA | NA | 221 | 68 | 519 | 298 | 82 | 655 | 388 | 98 | 810 | 491 | 136 | 1144 | 724 |
| | 5 | NA | NA | 28 | NA | NA | 57 | NA | NA | 96 | NA | NA | 148 | NA | NA | 204 | NA | NA | 277 | 124 | 638 | 365 | 146 | 791 | 466 | 196 | 1124 | 712 |
| | 10 | NA | NA | 25 | NA | NA | 50 | NA | NA | 87 | NA | NA | 139 | NA | NA | 191 | NA | NA | 263 | 155 | 610 | 347 | 182 | 762 | 444 | 240 | 1093 | 668 |
| 15 | 2 | NA | NA | 35 | NA | NA | 67 | NA | NA | 114 | NA | NA | 179 | 53 | 475 | 250 | 64 | 613 | 336 | 77 | 779 | 441 | 92 | 968 | 562 | 127 | 1376 | 841 |
| | 5 | NA | NA | 35 | NA | NA | 62 | NA | NA | 107 | NA | NA | 164 | NA | NA | 231 | 99 | 594 | 313 | 118 | 759 | 416 | 139 | 946 | 533 | 186 | 1352 | 828 |
| | 10 | NA | NA | 28 | NA | NA | 55 | NA | NA | 97 | NA | NA | 153 | NA | NA | 216 | 126 | 565 | 296 | 148 | 727 | 394 | 173 | 912 | 567 | 229 | 1315 | 777 |
| | 15 | NA | NA | NA | NA | NA | 48 | NA | NA | 89 | NA | NA | 141 | NA | NA | 201 | NA | NA | 281 | 171 | 698 | 375 | 198 | 880 | 485 | 259 | 1280 | 742 |
| 20 | 2 | NA | NA | 38 | NA | NA | 74 | NA | NA | 124 | NA | NA | 201 | 51 | 522 | 274 | 61 | 678 | 375 | 73 | 867 | 491 | 87 | 1083 | 627 | 121 | 1548 | 953 |
| | 5 | NA | NA | 36 | NA | NA | 68 | NA | NA | 116 | NA | NA | 184 | 80 | 503 | 254 | 95 | 658 | 350 | 113 | 845 | 463 | 133 | 1059 | 597 | 179 | 1523 | 933 |
| | 10 | NA | NA | NA | NA | NA | 60 | NA | NA | 107 | NA | NA | 172 | NA | NA | 237 | 122 | 627 | 332 | 143 | 811 | 440 | 167 | 1022 | 566 | 221 | 1482 | 879 |
| | 15 | NA | NA | NA | NA | NA | NA | NA | NA | 97 | NA | NA | 159 | NA | NA | 220 | NA | NA | 314 | 165 | 780 | 418 | 191 | 987 | 541 | 251 | 1443 | 840 |
| | 20 | NA | NA | NA | NA | NA | NA | NA | NA | 83 | NA | NA | 148 | NA | NA | 206 | NA | NA | 296 | 186 | 750 | 397 | 214 | 955 | 513 | 277 | 1406 | 807 |
| 30 | 2 | NA | NA | 41 | NA | NA | 82 | NA | NA | 137 | NA | NA | 216 | 47 | 581 | 303 | 57 | 762 | 421 | 68 | 985 | 558 | 81 | 1240 | 717 | 111 | 1793 | 1112 |
| | 5 | NA | NA | NA | NA | NA | 76 | NA | NA | 128 | NA | NA | 198 | 75 | 561 | 281 | 90 | 741 | 393 | 106 | 962 | 526 | 125 | 1216 | 683 | 169 | 1766 | 1094 |
| | 10 | NA | NA | NA | NA | NA | 67 | NA | NA | 115 | NA | NA | 184 | NA | NA | 263 | 115 | 709 | 373 | 135 | 927 | 500 | 158 | 1176 | 648 | 210 | 1721 | 1025 |
| | 15 | NA | NA | NA | NA | NA | NA | NA | NA | 107 | NA | NA | 171 | NA | NA | 243 | NA | NA | 353 | 156 | 893 | 476 | 181 | 1139 | 621 | 239 | 1679 | 981 |
| | 20 | NA | NA | NA | NA | NA | NA | NA | NA | 91 | NA | NA | 159 | NA | NA | 227 | NA | NA | 332 | 176 | 860 | 450 | 203 | 1103 | 592 | 264 | 1638 | 940 |
| | 30 | NA | NA | NA | NA | NA | NA | NA | NA | NA | NA | NA | NA | NA | NA | 188 | NA | NA | 288 | NA | NA | 416 | 249 | 1035 | 555 | 318 | 1560 | 877 |
| 50 | 2 | NA | NA | NA | NA | NA | 92 | NA | NA | 161 | NA | NA | 251 | NA | NA | 351 | 51 | 840 | 477 | 61 | 1106 | 633 | 72 | 1413 | 812 | 99 | 2080 | 1243 |
| | 5 | NA | NA | NA | NA | NA | NA | NA | NA | 151 | NA | NA | 230 | NA | NA | 323 | 83 | 819 | 445 | 98 | 1083 | 596 | 116 | 1387 | 774 | 155 | 2052 | 1225 |
| | 10 | NA | NA | NA | NA | NA | NA | NA | NA | 138 | NA | NA | 215 | NA | NA | 304 | NA | NA | 424 | 126 | 1047 | 567 | 147 | 1347 | 733 | 195 | 2006 | 1147 |
| | 15 | NA | NA | NA | NA | NA | NA | NA | NA | 127 | NA | NA | 199 | NA | NA | 282 | NA | NA | 400 | 146 | 1010 | 539 | 170 | 1307 | 702 | 222 | 1961 | 1099 |
| | 20 | NA | NA | NA | NA | NA | NA | NA | NA | NA | NA | NA | 185 | NA | NA | 264 | NA | NA | 376 | 165 | 977 | 511 | 190 | 1269 | 669 | 246 | 1916 | 1050 |
| | 30 | NA | NA | NA | NA | NA | NA | NA | NA | NA | NA | NA | NA | NA | NA | NA | NA | NA | 327 | NA | NA | 468 | 233 | 1196 | 623 | 295 | 1832 | 984 |
| Minimum internal area of chimney (in.$^2$) | | 12 | | | 19 | | | 28 | | | 38 | | | 50 | | | 63 | | | 78 | | | 95 | | | 132 | | |
| Maximum internal area of chimney (in.$^2$) | | Seven times the listed appliance categorized vent area, flue collar area, or draft hood outlet areas. | | | | | | | | | | | | | | | | | | | | | | | | | | |

For SI units, 1 in. = 25.4 mm, 1 ft = 0.305 m, 1000 Btu/hr = 0.293 kW, 1 in.$^2$ = 645 mm$^2$.

NA: Not applicable.

*TABLE 13.1(d) Masonry Chimney*

| | |
|---|---|
| *Number of Appliances:* | *Single* |
| *Appliance Type:* | *Category I* |
| *Appliance Vent Connection:* | *Single-Wall Metal Connector* |

| | | *Single-Wall Metal Connector Diameter — D (in.) To be used with chimney areas within the size limits at bottom* | | | | | | | | | | | | | | | | | | | | | | | | | | |
|---|---|---|---|---|---|---|---|---|---|---|---|---|---|---|---|---|---|---|---|---|---|---|---|---|---|---|---|---|
| | | *3* | | | *4* | | | *5* | | | *6* | | | *7* | | | *8* | | | *9* | | | *10* | | | *12* | | |
| | | *Appliance Input Rating in Thousands of Btu per Hour* | | | | | | | | | | | | | | | | | | | | | | | | | | |
| *Height H (ft)* | *Lateral L (ft)* | *FAN Min* | *FAN Max* | *NAT Max* | *FAN Min* | *FAN Max* | *NAT Max* | *FAN Min* | *FAN Max* | *NAT Max* | *FAN Min* | *FAN Max* | *NAT Max* | *FAN Min* | *FAN Max* | *NAT Max* | *FAN Min* | *FAN Max* | *NAT Max* | *FAN Min* | *FAN Max* | *NAT Max* | *FAN Min* | *FAN Max* | *NAT Max* | *FAN Min* | *FAN Max* | *NAT Max* |
| 6 | 2 | NA | NA | 28 | NA | NA | 52 | NA | NA | 86 | NA | NA | 130 | NA | NA | 180 | NA | NA | 247 | NA | NA | 319 | NA | NA | 400 | NA | NA | 580 |
| | 5 | NA | NA | 25 | NA | NA | 48 | NA | NA | 81 | NA | NA | 116 | NA | NA | 164 | NA | NA | 230 | NA | NA | 297 | NA | NA | 375 | NA | NA | 560 |
| 8 | 2 | NA | NA | 29 | NA | NA | 55 | NA | NA | 93 | NA | NA | 145 | NA | NA | 197 | NA | NA | 265 | NA | NA | 349 | 382 | 725 | 445 | 549 | 1021 | 650 |
| | 5 | NA | NA | 26 | NA | NA | 51 | NA | NA | 87 | NA | NA | 133 | NA | NA | 182 | NA | NA | 246 | NA | NA | 327 | NA | NA | 422 | 673 | 1003 | 638 |
| | 8 | NA | NA | 23 | NA | NA | 47 | NA | NA | 82 | NA | NA | 126 | NA | NA | 174 | NA | NA | 237 | NA | NA | 317 | NA | NA | 408 | 747 | 985 | 621 |
| 10 | 2 | NA | NA | 31 | NA | NA | 61 | NA | NA | 102 | NA | NA | 161 | NA | NA | 220 | 216 | 518 | 297 | 271 | 654 | 387 | 373 | 808 | 490 | 536 | 1142 | 722 |
| | 5 | NA | NA | 28 | NA | NA | 56 | NA | NA | 95 | NA | NA | 147 | NA | NA | 203 | NA | NA | 276 | 334 | 635 | 364 | 459 | 789 | 465 | 657 | 1121 | 710 |
| | 10 | NA | NA | 24 | NA | NA | 49 | NA | NA | 86 | NA | NA | 137 | NA | NA | 189 | NA | NA | 261 | NA | NA | 345 | 547 | 758 | 441 | 771 | 1088 | 665 |
| 15 | 2 | NA | NA | 35 | NA | NA | 67 | NA | NA | 113 | NA | NA | 178 | 166 | 473 | 249 | 211 | 611 | 335 | 264 | 776 | 440 | 362 | 965 | 560 | 520 | 1373 | 840 |
| | 5 | NA | NA | 32 | NA | NA | 61 | NA | NA | 106 | NA | NA | 163 | NA | NA | 230 | 261 | 591 | 312 | 325 | 755 | 414 | 444 | 942 | 531 | 637 | 1348 | 825 |
| | 10 | NA | NA | 27 | NA | NA | 54 | NA | NA | 96 | NA | NA | 151 | NA | NA | 214 | NA | NA | 294 | 392 | 722 | 392 | 531 | 907 | 504 | 749 | 1309 | 774 |
| | 15 | NA | NA | NA | NA | NA | 46 | NA | NA | 87 | NA | NA | 138 | NA | NA | 198 | NA | NA | 278 | 452 | 692 | 372 | 606 | 873 | 481 | 841 | 1272 | 738 |
| 20 | 2 | NA | NA | 38 | NA | NA | 73 | NA | NA | 123 | NA | NA | 200 | 163 | 520 | 273 | 206 | 675 | 374 | 258 | 864 | 490 | 252 | 1079 | 625 | 508 | 1544 | 950 |
| | 5 | NA | NA | 35 | NA | NA | 67 | NA | NA | 115 | NA | NA | 183 | NA | NA | 252 | 255 | 655 | 348 | 317 | 842 | 461 | 433 | 1055 | 594 | 623 | 1518 | 930 |
| | 10 | NA | NA | NA | NA | NA | 59 | NA | NA | 105 | NA | NA | 170 | NA | NA | 235 | 312 | 622 | 330 | 382 | 806 | 437 | 517 | 1016 | 562 | 733 | 1475 | 875 |
| | 15 | NA | NA | NA | NA | NA | NA | NA | NA | 95 | NA | NA | 156 | NA | NA | 217 | NA | NA | 311 | 442 | 773 | 414 | 591 | 979 | 539 | 823 | 1434 | 835 |
| | 20 | NA | NA | NA | NA | NA | NA | NA | NA | 80 | NA | NA | 144 | NA | NA | 202 | NA | NA | 292 | NA | NA | 392 | 663 | 944 | 510 | 911 | 1394 | 800 |
| 30 | 2 | NA | NA | 41 | NA | NA | 81 | NA | NA | 136 | NA | NA | 215 | 158 | 578 | 302 | 200 | 759 | 420 | 249 | 982 | 556 | 340 | 1237 | 715 | 489 | 1789 | 1110 |
| | 5 | NA | NA | NA | NA | NA | 75 | NA | NA | 127 | NA | NA | 196 | NA | NA | 279 | 245 | 737 | 391 | 306 | 958 | 524 | 417 | 1210 | 680 | 600 | 1760 | 1090 |
| | 10 | NA | NA | NA | NA | NA | 66 | NA | NA | 113 | NA | NA | 182 | NA | NA | 260 | 300 | 703 | 370 | 370 | 920 | 496 | 500 | 1168 | 644 | 708 | 1713 | 1020 |
| | 15 | NA | NA | NA | NA | NA | NA | NA | NA | 105 | NA | NA | 168 | NA | NA | 240 | NA | NA | 349 | 428 | 884 | 471 | 572 | 1128 | 615 | 798 | 1668 | 975 |
| | 20 | NA | NA | NA | NA | NA | NA | NA | NA | 88 | NA | NA | 155 | NA | NA | 223 | NA | NA | 327 | NA | NA | 445 | 643 | 1089 | 585 | 883 | 1624 | 932 |
| | 30 | NA | NA | NA | NA | NA | NA | NA | NA | NA | NA | NA | NA | NA | NA | 182 | NA | NA | 281 | NA | NA | 408 | NA | NA | 544 | 1055 | 1539 | 865 |
| 50 | 2 | NA | NA | NA | NA | NA | 91 | NA | NA | 160 | NA | NA | 250 | NA | NA | 350 | 191 | 837 | 475 | 238 | 1103 | 631 | 323 | 1408 | 810 | 463 | 2076 | 1240 |
| | 5 | NA | NA | NA | NA | NA | NA | NA | NA | 149 | NA | NA | 228 | NA | NA | 321 | NA | NA | 442 | 293 | 1078 | 593 | 398 | 1381 | 770 | 571 | 2044 | 1220 |
| | 10 | NA | NA | NA | NA | NA | NA | NA | NA | 136 | NA | NA | 212 | NA | NA | 301 | NA | NA | 420 | 355 | 1038 | 562 | 447 | 1337 | 728 | 674 | 1994 | 1140 |
| | 15 | NA | NA | NA | NA | NA | NA | NA | NA | 124 | NA | NA | 195 | NA | NA | 278 | NA | NA | 395 | NA | NA | 533 | 546 | 1294 | 695 | 761 | 1945 | 1090 |
| | 20 | NA | NA | NA | NA | NA | NA | NA | NA | NA | NA | NA | 180 | NA | NA | 258 | NA | NA | 370 | NA | NA | 504 | 616 | 1251 | 660 | 844 | 1898 | 1040 |
| | 30 | NA | NA | NA | NA | NA | NA | NA | NA | NA | NA | NA | NA | NA | NA | NA | NA | NA | 318 | NA | NA | 458 | NA | NA | 610 | 1009 | 1805 | 970 |
| Minimum internal area of chimney (in.$^2$) | | 12 | | | 19 | | | 28 | | | 38 | | | 50 | | | 6 | | | 78 | | | 95 | | | 132 | | |
| Maximum internal area of chimney (in.$^2$) | | Seven times the listed appliance categorized vent area, flue collar area, or draft hood outlet areas. | | | | | | | | | | | | | | | | | | | | | | | | | | |

NEW

For SI units, 1 in. = 25.4 mm, 1 ft = 0.305 m, 1000 Btu/hr = 0.293 kW, 1 in.$^2$ = 645 mm$^2$.

NA: Not applicable.

*TABLE 13.1(e) Single-Wall Metal Pipe or Type B Asbestos Cement Vent*

| | | | | | *Number of Appliances:* | *Single* | | | |
|---|---|---|---|---|---|---|---|---|---|
| | | | | | *Appliance Type:* | *Draft Hood–Equipped* | | | |
| | | | | | *Appliance Vent Connection:* | *Connected Directly to Pipe or Vent* | | | |
| | | *Diameter — D (in.) To be used with chimney areas within the size limits at bottom* | | | | | | | |
| | | *3* | *4* | *5* | *6* | *7* | *8* | *10* | *12* |
| *Height H (ft)* | *Lateral L (ft)* | *Appliance Input Rating in Thousands of Btu per Hour* | | | | | | | |
| | | *Maximum Appliance Input Rating in Thousands of Btu per Hour* | | | | | | | |
| 6 | 0 | 39 | 70 | 116 | 170 | 232 | 312 | 500 | 750 |
| | 2 | 31 | 55 | 94 | 141 | 194 | 260 | 415 | 620 |
| | 5 | 28 | 51 | 88 | 128 | 177 | 242 | 390 | 600 |
| 8 | 0 | 42 | 76 | 126 | 185 | 252 | 340 | 542 | 815 |
| | 2 | 32 | 61 | 102 | 154 | 210 | 284 | 451 | 680 |
| | 5 | 29 | 56 | 95 | 141 | 194 | 264 | 430 | 648 |
| | 10 | 24 | 49 | 86 | 131 | 180 | 250 | 406 | 625 |
| 10 | 0 | 45 | 84 | 138 | 202 | 279 | 372 | 606 | 912 |
| | 2 | 35 | 67 | 111 | 168 | 233 | 311 | 505 | 760 |
| | 5 | 32 | 61 | 104 | 153 | 215 | 289 | 480 | 724 |
| | 10 | 27 | 54 | 94 | 143 | 200 | 274 | 455 | 700 |
| | 15 | NA | 46 | 84 | 130 | 186 | 258 | 432 | 666 |
| 15 | 0 | 49 | 91 | 151 | 223 | 312 | 420 | 684 | 1040 |
| | 2 | 39 | 72 | 122 | 186 | 260 | 350 | 570 | 865 |
| | 5 | 35 | 67 | 110 | 170 | 240 | 325 | 540 | 825 |
| | 10 | 30 | 58 | 103 | 158 | 223 | 308 | 514 | 795 |
| | 15 | NA | 50 | 93 | 144 | 207 | 291 | 488 | 760 |
| | 20 | NA | NA | 82 | 132 | 195 | 273 | 466 | 726 |
| 20 | 0 | 53 | 101 | 163 | 252 | 342 | 470 | 770 | 1190 |
| | 2 | 42 | 80 | 136 | 210 | 286 | 392 | 641 | 990 |
| | 5 | 38 | 74 | 123 | 192 | 264 | 364 | 610 | 945 |
| | 10 | 32 | 65 | 115 | 178 | 246 | 345 | 571 | 910 |
| | 15 | NA | 55 | 104 | 163 | 228 | 326 | 550 | 870 |
| | 20 | NA | NA | 91 | 149 | 214 | 306 | 525 | 832 |
| 30 | 0 | 56 | 108 | 183 | 276 | 384 | 529 | 878 | 1370 |
| | 2 | 44 | 84 | 148 | 230 | 320 | 441 | 730 | 1140 |
| | 5 | NA | 78 | 137 | 210 | 296 | 410 | 694 | 1080 |
| | 10 | NA | 68 | 125 | 196 | 274 | 388 | 656 | 1050 |
| | 15 | NA | NA | 113 | 177 | 258 | 366 | 625 | 1000 |
| | 20 | NA | NA | 99 | 163 | 240 | 344 | 596 | 960 |
| | 30 | NA | NA | NA | NA | 192 | 295 | 540 | 890 |
| 50 | 0 | NA | 120 | 210 | 310 | 443 | 590 | 980 | 1550 |
| | 2 | NA | 95 | 171 | 260 | 370 | 492 | 820 | 1290 |
| | 5 | NA | NA | 159 | 234 | 342 | 474 | 780 | 1230 |
| | 10 | NA | NA | 146 | 221 | 318 | 456 | 730 | 1190 |
| | 15 | NA | NA | NA | 200 | 292 | 407 | 705 | 1130 |
| | 20 | NA | NA | NA | 185 | 276 | 384 | 670 | 1080 |
| | 30 | NA | NA | NA | NA | 222 | 330 | 605 | 1010 |

For SI units, 1 in. = 25.4 mm, 1 ft = 0.305 m, 1000 Btu/hr = 0.293 kW, 1 in.$^2$ = 645 mm$^2$.

NA: Not applicable.

***TABLE 13.1(f)*** *Exterior Masonry Chimney*

| | *Number of Appliances:* | *Single* |
|---|---|---|
| | *Appliance Type:* | *NAT* |
| | *Appliance Vent Connection:* | *Type B Double-Wall Connector* |

***SPECIAL USE: Minimum Allowable Input Rating of Space-Heating Appliance in Thousands of Btu per Hour***

| *Vent Height H (ft)* | *Internal Area of Chimney (in.$^2$)* | | | | | | | |
|---|---|---|---|---|---|---|---|---|
| | *12* | *19* | *28* | *38* | *50* | *63* | *78* | *113* |
| | Local 99% winter design temperature: 37°F or greater | | | | | | | |
| 6 | 0 | 0 | 0 | 0 | 0 | 0 | 0 | 0 |
| 8 | 0 | 0 | 0 | 0 | 0 | 0 | 0 | 0 |
| 10 | 0 | 0 | 0 | 0 | 0 | 0 | 0 | 0 |
| 15 | NA | 0 | 0 | 0 | 0 | 0 | 0 | 0 |
| 20 | NA | NA | 123 | 190 | 249 | 184 | 0 | 0 |
| 30 | NA | NA | NA | NA | NA | 393 | 334 | 0 |
| 50 | NA | NA | NA | NA | NA | NA | NA | 579 |
| | Local 99% winter design temperature: 27°F to 36°F | | | | | | | |
| 6 | 0 | 0 | 68 | 116 | 156 | 180 | 212 | 266 |
| 8 | 0 | 0 | 82 | 127 | 167 | 187 | 214 | 263 |
| 10 | 0 | 51 | 97 | 141 | 183 | 201 | 225 | 265 |
| 15 | NA | NA | NA | NA | 233 | 253 | 274 | 305 |
| 20 | NA | NA | NA | NA | NA | 307 | 330 | 362 |
| 30 | NA | NA | NA | NA | NA | 419 | 445 | 485 |
| 50 | NA | NA | NA | NA | NA | NA | NA | 763 |
| | Local 99% winter design temperature: 17°F to 26°F | | | | | | | |
| 6 | NA | NA | NA | NA | NA | 215 | 259 | 349 |
| 8 | NA | NA | NA | NA | 197 | 226 | 264 | 352 |
| 10 | NA | NA | NA | NA | 214 | 245 | 278 | 358 |
| 15 | NA | NA | NA | NA | NA | 296 | 331 | 398 |
| 20 | NA | NA | NA | NA | NA | 352 | 387 | 457 |
| 30 | NA | NA | NA | NA | NA | NA | 507 | 581 |
| 50 | NA | NA | NA | NA | NA | NA | NA | NA |
| | Local 99% winter design temperature: 5°F to 16°F | | | | | | | |
| 6 | NA | NA | NA | NA | NA | NA | NA | 416 |
| 8 | NA | NA | NA | NA | NA | NA | 312 | 423 |
| 10 | NA | NA | NA | NA | NA | 289 | 331 | 430 |
| 15 | NA | NA | NA | NA | NA | NA | 393 | 485 |
| 20 | NA | NA | NA | NA | NA | NA | 450 | 547 |
| 30 | NA | NA | NA | NA | NA | NA | NA | 682 |
| 50 | NA | NA | NA | NA | NA | NA | NA | 972 |
| | Local 99% winter design temperature: −10°F to 4°F | | | | | | | |
| 6 | NA | NA | NA | NA | NA | NA | NA | 484 |
| 8 | NA | NA | NA | NA | NA | NA | NA | 494 |
| 10 | NA | NA | NA | NA | NA | NA | NA | 513 |
| 15 | NA | NA | NA | NA | NA | NA | NA | 586 |
| 20 | NA | NA | NA | NA | NA | NA | NA | 650 |
| 30 | NA | NA | NA | NA | NA | NA | NA | 805 |
| 50 | NA | NA | NA | NA | NA | NA | NA | 1003 |
| | Local 99% winter design temperature: −11°F or lower<br>Not recommended for any vent configurations | | | | | | | |

For SI units, 1 in. = 25.4 mm, 1 in.$^2$ = 645 mm$^2$, 1 ft = 0.305 m, 1000 Btu/hr = 0.293 kW, °C = (°F − 32)/1.8.

Note: See Figure G.2.4 for a map showing local 99 percent winter design temperatures in the United States.

NA: Not applicable.

## 13.2 Additional Requirements to Multiple-Appliance Vent

Tables 13.2(a) through 13.2(i) were developed with the assumption that there are no restrictions in the path of the vent gas flow. An additional assumption is that there is no deliberate attempt to remove heat. Therefore, devices such as heat economizers may not be used in conjunction with the tables. Vent dampers installed as part of a listed appliance are allowed and must be installed in accordance with the manufacturer's instructions or the general provisions in 13.2.1.

No Obstructions

### 13.2.1 Obstructions and Vent Dampers.

Venting Table 13.2(a) through Table 13.2(i) shall not be used where obstructions *(see Section 12.16)* are installed in the venting system. The installation of vents serving listed appliances with vent dampers shall be in accordance with the appliance manufacturer's instructions, or in accordance with the following:

(1) The maximum capacity of the vent connector shall be determined using the NAT Max column.
(2) The maximum capacity of the vertical vent or chimney shall be determined using the FAN+NAT column when the second appliance is a fan-assisted appliance, or the NAT+NAT column when the second appliance is equipped with a draft hood.
(3) The minimum capacity shall be determined as if the appliance were a fan-assisted appliance, as follows:
   (a) The minimum capacity of the vent connector shall be determined using the FAN Min column.
   (b) The FAN+FAN column shall be used when the second appliance is a fan-assisted appliance, and the FAN+NAT column shall be used when the second appliance is equipped with a draft hood, to determine whether the vertical vent or chimney configuration is not permitted (NA). Where the vent configuration is NA, the vent configuration shall not be permitted and an alternative venting configuration shall be utilized.

Paragraphs 13.2.1(1), 13.2.1(2), and 13.2.1(3) provide guidance for using the venting tables with draft hood appliances that are equipped with a vent damper. Vent dampers are commonly installed on boilers to increase efficiency. A vent damper can increase efficiency by eliminating off-cycle loss of heat through the vent from the boiler and the room in which it is installed. In other words, the vent is a hole in a building that can leak warm air when the boiler is not operating. The vent damper prevents this leakage by closing the hole. However, the vent damper prevents heated air from entering the venting system while the boiler is not firing, and the vent can cool, increasing the likelihood of condensation in the vent. For this reason, the following special rules apply when sizing venting systems for appliances equipped with vent dampers:

- Paragraph 13.2.1(1) addresses sizing of the vent connector and requires that the maximum capacity of the vent connector be determined using the NAT Max column (shaded).
- Paragraph 13.2.1(2) addresses sizing of the common vent and directs the user to treat the appliance as a draft hood appliance for determining the common vent capacity. Where the appliance is common-vented with other draft hood appliances, use the NAT + NAT column of the common venting tables. Where the appliance is common-vented with fan-assisted appliances, use the FAN + NAT column.

Both of these paragraphs above and the explanatory paragraphs above recognize that when the appliance is operating and the damper is open, the damper has no effect on venting performance and the appliance behaves like a normal draft hood–equipped appliance. Stated differently, determining the *maximum* capacity for a draft hood appliance is the same whether or not it is equipped with a vent damper.

- Paragraph 13.2.1(3) directs the user to determine the *minimum* capacity of the venting system when the draft hood appliance has a vent damper installed. This section essentially treats the appliance as a fan-assisted appliance for determining the minimum capacity of the venting system.
- Paragraph 13.2.1(3)(a) directs the user to determine the minimum capacity of the vent connector using the FAN Min column of the venting table.
- Paragraph 13.2.1(3)(b) requires the user to determine if the venting system configuration is not applicable (NA) by treating the appliance as a fan-assisted appliance and referring to the common vent capacity portion of the table. If the other appliance is a fan-assisted appliance, then the FAN + FAN column is used. If the other appliance is a draft hood appliance, then the FAN + NAT column is used.

Where "NA" appears in the table for a given installation, that configuration is not allowed and an alternative venting arrangement must be used. These requirements in 13.2.1 recognize that when the appliance is off and the damper is closed, the appliance behaves like a fan-assisted appliance by stopping the flow of heated air (off-cycle loss) into the venting system. This unusual combination of rules (using the unshaded FAN Min and shaded NAT Max columns) has the following simple explanations:

- When the appliance is firing (on), the vent damper is open. It operates as a natural draft appliance.
- When the appliance is not firing (off), the vent damper is closed, conserving energy. It is similar to a fan-assisted appliance in this energy savings, and condensation in the vent must be minimized.

### 13.2.2 Vent Connector Maximum Length.

The maximum vent connector horizontal length shall be 18 in./in. (18 mm/mm) of connector diameter as shown in Table 13.2.2.

***TABLE 13.2.2*** *Vent Connector Maximum Length*

| Connector Diameter Maximum (in.) | Connector Horizontal Length (ft) | Connector Diameter Maximum (in.) | Connector Horizontal Length (ft) |
|---|---|---|---|
| 3 | 4½ | 12 | 18 |
| 4 | 6 | 14 | 21 |
| 5 | 7½ | 16 | 24 |
| 6 | 9 | 18 | 27 |
| 7 | 10½ | 20 | 30 |
| 8 | 12 | 22 | 33 |
| 9 | 13½ | 24 | 36 |
| 10 | 15 | | |

For SI units, 1 in. = 25.4 mm, 1 ft = 0.305 m.

Since the multiple appliance tables do not contain rows showing lateral or horizontal length, is there any limit on the connector pipe length?

FAQ ▶ The multiple appliance venting tables do not contain lateral length rows like those found in the single appliance tables. Subsection 13.2.2 limits vent connector length to 18 in. per inch of connector diameter as shown in Table 13.2.2. If these lengths are exceeded, 13.2.3 provides guidance on derating the tables accordingly.

### 13.2.3 Vent Connector Exceeding Maximum Length.

The vent connector shall be routed to the vent utilizing the shortest possible route. Connectors with longer horizontal lengths than those listed in Table 13.2.2 are permitted under the following conditions:

(1) The maximum capacity (FAN Max or NAT Max) of the vent connector shall be reduced 10 percent for each additional multiple of the length listed in Table 13.2.2. For example, the maximum length listed for a 4 in. (100 mm) connector is 6 ft (1.8 m). With a connector length greater than 6 ft (1.8 m) but not exceeding 12 ft (3.7 m), the maximum capacity must be reduced by 10 percent (0.90 × maximum vent connector capacity). With a connector length greater than 12 ft (3.7 m) but not exceeding 18 ft (5.5 m), the maximum capacity must be reduced by 20 percent (0.80 × maximum vent capacity).
(2) For a connector serving a fan-assisted appliance, the minimum capacity (FAN Min) of the connector shall be determined by referring to the corresponding single appliance table. For Type B double-wall connectors, Table 13.1(a) shall be used. For single-wall connectors, Table 13.1(b) shall be used. The height *(H)* and lateral *(L)* shall be measured according to the procedures for a single appliance vent, as if the other appliances were not present.

For brevity, the common venting tables [the lower sections of Tables 13.2(a) through 13.2(e) of the code] are designed with the assumption that the vent connector is no more than 18 in. long for each inch of diameter. Subsections 13.2.2 and 13.2.3, summarized as follows, provide guidance on how to handle connectors that are longer than 18 in.:

Vent Connector Length

- The maximum capacity for vent connectors that exceed 18 in. per inch of diameter is reduced by 10 percent for each extra length of 18 in. per inch of vent diameter.
- The minimum capacity for connectors serving fan-assisted appliances is determined differently for single-wall and Type B double-wall connectors, as described in 13.2.3(2).

The minimum connector capacities in the multiple appliance vent tables do not consider connector length, as the single appliance tables do. Therefore, to determine the minimum capacity of long connectors serving fan-assisted appliances in multiple appliance installations, the user is directed to the single appliance tables. To determine the minimum capacity of the connector, consult the proper row for a given combination of height and lateral, as if the fan-assisted appliance were operating alone. Use Table 13.1(a) if the fan-assisted appliance has a B vent connector and Table 13.1(b) if using a single-wall connector.

### 13.2.4 Vent Connector Manifolds.

Where the vent connectors are combined prior to entering the vertical portion of the common vent to form a common vent manifold, the size of the common vent manifold and the common vent shall be determined by applying a 10 percent reduction (0.90 × maximum common vent capacity) to the common vent capacity part of the common vent tables. The length of the common vent manifold *(LM)* shall not exceed 18 in./in. (18 mm/mm) of common vent diameter *(D)*.

Subsection 13.2.4 considers the "common vent" to be the vertical vent or chimney only. A combination of vent connections prior to entering a vertical chimney or vent is named a "vent manifold," and guidance for sizing vent manifolds is located in this section. Note that a chimney liner system that comes out through the breaching, as many do, should be considered a manifold.

Vent Manifolds

The capacity of the vent manifold is reduced by 10 percent from the table value. Also note that the length of a vent manifold is limited to 18 in. per inch of vent manifold diameter.

### 13.2.5 Vent Offsets.

Where the common vertical vent is offset, the maximum capacity of the common vent shall be reduced in accordance with 13.2.6 and the horizontal length of the common vent offset shall not exceed 18 in./in. (18 mm/mm) of common vent diameter *(D)*. Where multiple offsets occur in a common vent, the total horizontal length of all offsets combined shall not exceed 18 in./in. (18 mm/mm) of the common vent diameter.

Vent Offsets

In the 2009 edition, the requirement was revised with the addition of the last sentence, which addresses multiple lateral offsets. While these are not common, they are used, and the code now addresses them.

**FAQ ▶** Are offsets in the vertical common vent permitted?

Vent offsets, addressed in 13.2.5, typically occur high in the venting system and reduce the draft produced because they slow the flow of vent gases. This reduction in draft is offset by the requirement to derate the capacity of the vent in accordance with 13.2.6. Additionally, the horizontal length of the offset is limited to 18 in. per inch of vent diameter, just as the connectors are limited in 13.2.2. Though permitted, the use of offsets in the common vent should be avoided wherever possible.

### 13.2.6 Elbows in Vents.

For each elbow up to and including 45 degrees in the common vent, the maximum common vent capacity listed in the venting tables shall be reduced by 5 percent. For each elbow greater than 45 degrees up to and including 90 degrees, the maximum common vent capacity listed in the venting tables shall be reduced by 10 percent.

The requirements in 13.2.6 for derating the table capacities for elbows recognize that the flow restriction of all elbows is not the same. The derating factors are shown in Commentary Table 13.6. Prior to the 1992 edition, a 10 percent reduction was required for each elbow, regardless of its angle. The derating factors were based on input from the Gas Technology Institute's contractor, Battelle. The derating factors provide flexibility to installers forced to use vent offsets and encourage the use of elbows of less than 90 degrees.

***COMMENTARY TABLE 13.6*** *Table Capacity Derating for Elbows of Less Than 90 Degrees*

| *Elbow* | *Vent Table Capacity Reduction, Per Elbow (%)* |
|---|---|
| 0–45 | 5 |
| >45–90 | 10 |

### 13.2.7 Elbows in Connectors.

The vent connector capacities listed in the common vent sizing tables include allowance for two 90 degree elbows. For each additional elbow up to and including 45 degrees, the maximum vent connector capacity listed in the venting tables shall be reduced by 5 percent. For each elbow greater than 45 degrees up to and including 90 degrees, the maximum vent connector capacity listed in the venting tables shall be reduced by 10 percent.

Elbows in Connectors

The tables in Section 13.2 were developed using the assumption that up to two 90-degree turns were used in the vent connector. In the 2002 edition, the tables were revised to recognize additional elbows and the lower draft reduction in elbows of less than 45 degrees compared to elbows of 90 degrees. Refer to Commentary Table 13.6 for the capacity reductions.

### 13.2.8 Common Vent Minimum Size.

The cross-sectional area of the common vent shall be equal to or greater than the cross-sectional area of the largest connector.

Subsection 13.2.8 highlights that in multiple appliance installations, no part of the common vent is allowed to be smaller than the largest connector, even if the common vent capacity tables would permit a smaller common vent.

### 13.2.9 Tee and Wye Fittings.

Tee and wye fittings connected to a common gas vent shall be considered as part of the common gas vent and constructed of materials consistent with that of the common gas vent.

Subsection 13.2.9 clarifies that tee and wye fittings, which are shown in Exhibit 13.1, are considered a part of the common vent. The last sentence was added in the 2009 edition to recognize that these fittings cannot be constructed of masonry. The term *gas vent* is defined as follows:

> **3.3.107.2 Gas Vent.** A passageway composed of listed factory-built components assembled in accordance with the manufacturer's installation instructions for conveying vent gases from appliances or their vent connectors to the outdoors.

(a) Tee (T) Fitting (b) Wye (Y) Fitting

***EXHIBIT 13.1*** *Tee and Wye Fittings. (Courtesy of Simpson Dura-Vent Company, Inc.)*

## 13.2.10 Tee and Wye Sizing.

At the point where tee or wye fittings connect to a common gas vent, the opening size of the fitting shall be equal to the size of the common vent. Such fittings shall not be prohibited from having reduced size openings at the point of connection of appliance gas vent connectors.

The tee and wye size must be the same as the common vent size. For example, if 3 in. × 4 in. connectors are joined to a 5 in. common vent, a 3 in. × 4 in. × 5 in. tee may be used.

## 13.2.11 High-Altitude Installations.

High Altitude Installation

Sea level input ratings shall be used when determining maximum capacity for high-altitude installation. Actual input (derated for altitude) shall be used for determining minimum capacity for high-altitude installation.

Using the sea level input rating required by 13.2.11 for the maximum capacity is a conservative measure because less draft is produced at high altitudes. The derating process will also make condensation more likely. The reduced input rate should be used for the minimum capacity.

## 13.2.12 Connector Rise.

The connector rise *(R)* for each appliance connector shall be measured from the draft hood outlet or flue collar to the centerline where the vent gas streams come together.

The requirement in 13.2.12 explains how to measure the connector rise. Providing as much rise as the installation will permit, while still maintaining required clearance, is beneficial to overall vent performance. Additional rise helps establish and maintain flow through the venting system. Total rise is measured from the outlet of the draft hood or flue collar up to the (centerline) point where the highest connector joins the common vent. [See Figures G.1(f) through G.1(l) in Annex G for specific examples.] For a given appliance, this distance may include a portion of the common vent. In addition to providing rise, it is also beneficial to provide as much vertical height between the outlet of the appliance and the first elbow in the connector.

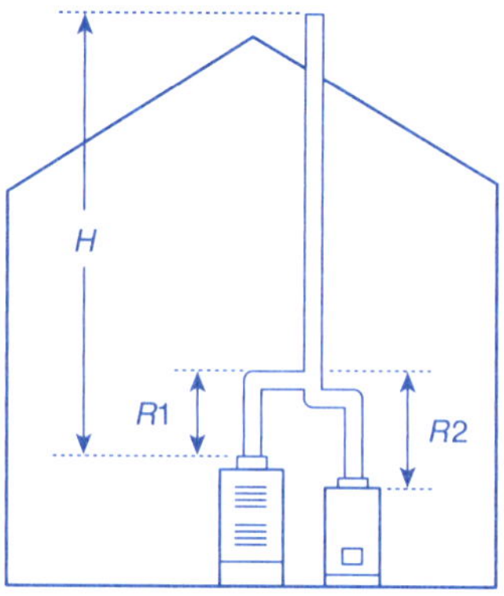


*EXHIBIT 13.2* *Connector Rise and Vent Height.*

### 13.2.13 Vent Height.

For multiple appliances all located on one floor, available total height *(H)* shall be measured from the highest draft hood outlet or flue collar up to the level of the outlet of the common vent.

The requirement in 13.2.13 explains how to measure the vent height. See Exhibit 13.2 for measurement of connector rise, *R*, and total vent height, *H*. Connector rise is described in 13.2.12.

### 13.2.14 Multistory Vent Height.

For multistory installations, available total height *(H)* for each segment of the system shall be the vertical distance between the highest draft hood outlet or flue collar entering that segment and the centerline of the next higher interconnection tee. *[See Figure G.1(m).]*

**FAQ ▶** How is vent height in a multistory venting system determined?

As venting system height increases, so does draft and vent capacity. However, the effect of increased height is negated when additional appliances are connected to the venting system. As dilution air enters the vent through the upper appliances, draft to lower appliances is decreased. Therefore, the vent height for a given segment is based on the distance between the appliances of that segment and the next higher connection tee (usually on the next level of the building).

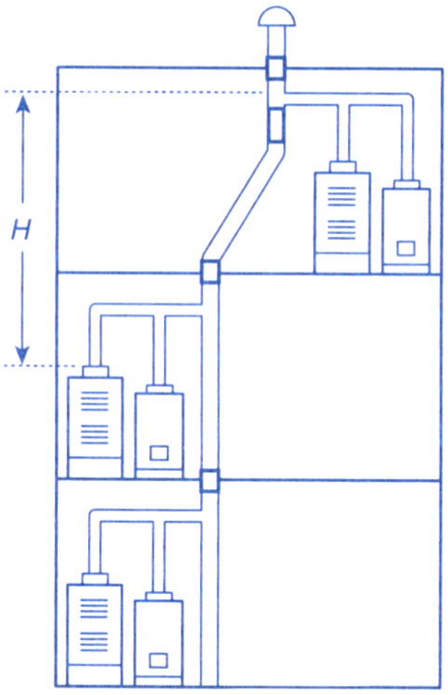


*EXHIBIT 13.3* *Multistory Vents.*

### 13.2.15 Multistory Lowest Vent and Vent Connector Sizing.

The size of the lowest connector and of the vertical vent leading to the lowest interconnection of a multistory system shall be in accordance with Table 13.1(a) or Table 13.1(b) for available total height *(H)* up to the lowest interconnection. *[See Figure G.1(n).]*

See Exhibit 13.3 and the examples in Annex G for more information on the multistory vents described in 13.2.14, 13.2.15, and 13.2.16. The vent on the first floor is treated like a normal venting system whose height ends at the next floor. The higher floors are treated as common vents whose height ends at the floor above. This tends to result in large diameter vertical vents. It may be more cost effective to use separate venting systems for groups of floors.

### 13.2.16 Multistory B Vents Required.

Where used in multistory systems, vertical common vents shall be Type B double-wall and shall be installed with a listed vent cap.

Subsection 13.2.16 states that the common vent of multistory venting systems sized using Chapter 13 must be constructed only of Type B double-wall pipe and that the common vent must be terminated with a listed vent cap.

### 13.2.17 Multistory Vent Offsets and Capacity.

Offsets in multistory common vent systems shall be limited to a single offset in each system, and systems with an offset shall comply with all of the following:

(1) The offset angle shall not exceed 45 degrees from vertical.
(2) The horizontal length of the offset shall not exceed 18 in./in. (18 mm/mm) of common vent diameter of the segment in which the offset is located.

(3) For the segment of the common vertical vent containing the offset, the common vent capacity listed in the common venting tables shall be reduced by 20 percent (0.80 × maximum common vent capacity).
(4) A multistory common vent shall not be reduced in size above the offset.

### 13.2.18 Vertical Vent Size Limitation.

Where two or more appliances are connected to a vertical vent or chimney, the flow area of the largest section of vertical vent or chimney shall not exceed seven times the smallest listed appliance categorized vent areas, flue collar area, or draft hood outlet area unless designed in accordance with approved engineering methods.

The "seven times" rule covered in 13.2.18 limits the common vent diameter to seven times the diameter of the smallest connector in the vent system. For example, if a water heater with a 4 in. vent connector was the only appliance connected to a venting system, the maximum vent area would be 7 × the 4 in. vent connector area. A 4 in. diameter vent connector has an area of about 12.5 in.$^2$ [The area of a circle is $\pi r^2$. Therefore, we multiply $\pi$, or 3.14, by the radius ($^1/_2$ the diameter squared), or 3.14 × 2 × 2 = about 12.5 in$^2$.] Similarly, 7 × 12.5 is about 88 in$^2$. A 10 in. diameter circle has an area of about 79 in.$^2$ and is the largest diameter round vent that could be used with a water heater with a 4 in. vent connector.

### 13.2.19 Two-Stage/Modulating Appliances.

For appliances with more than one input rate, the minimum vent connector capacity (FAN Min) determined from the tables shall be less than the lowest appliance input rating, and the maximum vent connector capacity (FAN Max or NAT Max) determined from the tables shall be greater than the highest appliance input rating.

If an appliance has multiple input rates, as covered in 13.2.19, the minimum capacity is determined with the minimum input rate and the maximum capacity is determined with the maximum input rate.

Multiple Input Rates

### 13.2.20* Corrugated Chimney Liners.

Listed corrugated metallic chimney liner systems in masonry chimneys shall be sized by using Table 13.2(a) or Table 13.2(b) for Type B vents, with the maximum capacity reduced by 20 percent (0.80 × maximum capacity) and the minimum capacity as shown in Table 13.2(a) or Table 13.2(b). Corrugated metallic liner systems installed with bends or offsets shall have their maximum capacity further reduced in accordance with 13.2.5 and 13.2.6. The 20 percent reduction for corrugated metallic chimney liner systems includes an allowance for one long radius 90 degree turn at the bottom of the liner.

Does a corrugated chimney liner have the same capacity as a similar sized smooth wall liner or vent?
◀ **FAQ**

Properly installed corrugated chimney liners have heat loss similar to that of a Type B vent, so they are sized using Table 13.2(a) or Table 13.2(b), as stated in 13.2.20. However, such liners' corrugations and their tendency to spiral in the chimney require a 20 percent maximum capacity reduction.

Many liners begin at the opening in the chimney wall and then bend up vertically. This 90-degree elbow at the beginning of the liner is included in the 20 percent reduction. The inclusion of one 90-degree elbow was clarified in the 2002 edition with the assistance of the researchers who developed the tables.

◀ **FAQ**
Does the turn or bend at the bottom of a flexible chimney liner count as an elbow?

**A.13.2.20** A long radius turn is a turn where the centerline radius is equal to or greater than 1.5 times the vent diameter.

### 13.2.21 Connections to Chimney Liners.

Where double-wall connectors are required, tee and wye fittings used to connect to the common vent chimney liner shall be listed double-wall fittings. Connections between chimney liners and listed double-wall fittings shall be made with listed adapter fittings designed for such purpose.

Subsection 13.2.21 was added to the 2006 edition of the code to clarify that if tee and wye fittings are used to connect double-wall connectors to chimney liners, the tee and wye fittings must also be double wall. Paragraph 13.2.21 also requires the use of listed adapters to connect the double-wall fittings to the chimney liner.

### 13.2.22 Chimneys and Vent Locations.

Table 13.2(a) and Table 13.2(b) shall be used only for chimneys and vents not exposed to the outdoors below the roof line. A Type B vent passing through an unventilated enclosure or chase insulated to a value of not less than R8 shall not be considered to be exposed to the outdoors. Table 13.2(f), Table 13.2(g), Table 13.2(h), and Table 13.2(i) shall be used for clay tile lined exterior masonry chimneys, provided all the following conditions are met:

Subsection 13.2.22 states that Tables 13.2(a) and 13.2(b) are only to be used for venting systems that are not exposed to the outdoors below the roofline. This section aims to clarify that Type B vents installed in unventilated enclosures or chases insulated to a value of R8 could be sized as if they were installed indoors. This requirement permits these installations to be sized using Tables 13.2(a) or 13.2(b).

Exterior Chimney

Table 13.2(c) is used to determine the capacity of chimney connectors attached to chimneys exposed to the outdoors below the roofline, whereas Table 13.2(f) or Table 13.2(h) is used to determine the maximum capacity of the common vent. Tables 13.2(g) and 13.2(i), in turn, have different minimum capacities for the common vent based on the ambient temperatures expected. The vent connectors are sized with Table 13.2(c), calculating both a minimum and maximum connector capacity for fan-assisted appliances and only the maximum connector capacity for draft hood appliances.

The exception to 13.2.22 recognizes that manufacturers may produce products customized for exterior masonry chimneys and use their own instructions.

Requirements are provided in 13.2.22 for using exterior masonry chimneys for heating appliances. These restrictions are needed because of the high heat loss in chimneys exposed to the outdoors below the roofline in cold climates. Note that these restrictions do not apply to chimneys and vents that are not exposed to the outdoors below the roofline, because these chimneys and vents are considered to be interior chimneys. The following items are required when using 13.2.22:

- To size chimneys and vents serving space heating appliances or all appliances providing heat to the building, Table 13.2(f) and Table 13.2(g) or Table 13.2(h) and Table 13.2(i) should be used as appropriate for the type of appliances connected to the chimney or vent exposed to the outdoors below the roofline.
- Vent connectors are sized using Table 13.2(c), the masonry chimney sizing table.

Note that only Type B double-wall vent connectors are required to minimize heat loss in the vent connector. At least one appliance, such as a water heater, must be draft hood–equipped.

(1) The vent connector is Type B double-wall.
(2) At least one appliance is draft hood–equipped.
(3) The combined appliance input rating is less than the maximum capacity given by Table 13.2(f) (for NAT+NAT) or Table 13.2(h) (for FAN+NAT).
(4) The input rating of each space-heating appliance is greater than the minimum input rating given by Table 13.2(g) (for NAT+NAT) or Table 13.2(i) (for FAN+NAT).
(5) The vent connector sizing is in accordance with Table 13.2(c).

In the 2009 edition, item (6) and the exception covering alternate designs were deleted here, as they are covered elsewhere in the code.

### 13.2.23 Draft Hood Conversion Accessories.

Draft hood conversion accessories for use with masonry chimney venting listed Category I fan-assisted appliances shall be listed and installed in accordance with the listed accessory manufacturer's installation instructions.

Draft hood conversion accessories, which are covered in 13.2.23, are add-on kits that convert a fan-assisted appliance to a draft hood appliance by providing an opening for dilution air from the room to enter the venting system. As listed components, these accessories are provided by the appliance manufacturer and must be installed according to the instructions included with the accessory. The addition of the draft hood kit allows the appliance to be sized as a draft hood appliance, which may be beneficial when venting into masonry chimneys, especially exterior masonry chimneys.

### 13.2.24 Vent Connector Sizing.

Vent connectors shall not be increased more than two sizes greater than the listed appliance categorized vent diameter, flue collar diameter, or draft hood outlet diameter. Vent connectors for draft hood–equipped appliances shall not be smaller than the draft hood outlet diameter. Where a vent connector size(s) determined from the tables for a fan-assisted appliance(s) is smaller than the flue collar diameter, the use of the smaller size(s) shall be permitted, provided that the installation complies with all of the following conditions:

(1) Vent connectors for fan-assisted appliance flue collars 12 in. (300 mm) in diameter or smaller are not reduced by more than one table size [e.g., 12 in. to 10 in. (300 mm to 250 mm) is a one-size reduction], and those larger than 12 in. (300 mm) in diameter are not reduced more than two table sizes [e.g., 24 in. to 20 in. (610 mm to 510 mm) is a two-size reduction].
(2) The fan-assisted appliance(s) is common vented with a draft hood–equipped appliance(s).
(3) The vent connector has a smooth interior wall.

A sudden, large expansion of the vent connector diameter creates a pressure drop that may limit the draft and encourage condensation. Therefore, 13.2.24 places a limit on upsizing, limiting the vent connector diameter to two sizes larger than one of the following, whichever is applicable:

1. Appliance categorized vent diameter
2. Flue collar diameter
3. Draft hood outlet diameter

Note that an appliance includes either a draft hood or flue collar for connection to the vent or vent connector. The term *appliance categorized vent diameter/area* is defined in 3.3.7 as follows:

> **3.3.7 Appliance Categorized Vent Diameter/Area.** The minimum vent diameter/area permissible for Category I appliances to maintain a nonpositive vent static pressure when tested in accordance with nationally recognized standards.

Therefore, where connecting an appliance to a vent connector, the vent connector must be sized using Tables 13.2(a) through 13.2(e), as appropriate, to determine the minimum size connector permitted for the vent system configuration.

Note that 13.2.24(3) was added in the 2002 edition to recognize the use of corrugated vent connectors, which cannot be downsized. Only smooth wall connectors may be downsized.

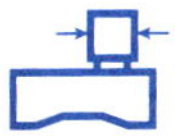
Upsizing

### 13.2.25 Multiple Vent and Connector Sizes.

All combinations of pipe sizes, single-wall metal pipe, and double-wall metal pipe shall be allowed within any connector run(s) or within the common vent, provided ALL of the appropriate tables permit ALL of the desired sizes and types of pipe, as if they were used for the entire length of the subject connector or vent. Where single-wall and Type B double-wall metal pipes are used for vent connectors within the same venting system, the common vent shall be sized using Table 13.2(b) or Table 13.2(d) as appropriate.

The requirements in 13.2.25 cover multiple vent and connector sizes. Table 13.2(b) is the B vent single-wall connector table, and Table 13.2(d) is the masonry chimney single-wall connector table. Using these single-wall tables where combinations of single wall and B vent connectors exist in the same installation provides a conservative factor for the entire installation.

### 13.2.26 Multiple Vent and Connector Sizes Permitted.

Where a Chapter 13 table permits more than one diameter of pipe to be used for a connector or vent, all the permitted sizes shall be permitted to be used.

Multiple Vertical Vent Sizes

Subsections 13.2.25 and 13.2.26 require the installer to check the minimum and maximum capacities for the vent section for each vent type as if the entire vent section were of that size. For example, for a vent connector that is half single-wall and half Type B, the installer must show that a vent connector of that length would be allowed if it were all single-wall and also if it were all Type B.

### 13.2.27 Interpolation.

Interpolation

Interpolation shall be permitted in calculating capacities for vent dimensions that fall between table entries. *(See Example 3, Annex G.)*

If the installation dimensions fall between two table entries for which there are defined values, the installer may calculate the "in between" value. This is called interpolation and is addressed in 13.1.15 and 13.2.27. An example of interpolation is included in the commentary following 13.1.15.

### 13.2.28 Extrapolation.

Extrapolation beyond the table entries shall not be permitted.

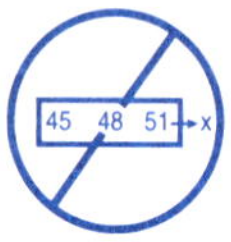

Extrapolation

Extrapolation, which is addressed in 13.1.16, is estimating a value outside of a table. For example, Table 13.2(a) provides common vent capacities for vents up to 100 ft (30 m). The reader is not permitted to use the numbers in the table to estimate the capacity of a common vent that is 110 ft (34 m) high.

### 13.2.29 Sizing Vents Not Covered by Tables.

For vent heights lower than 6 ft (1.8 m) and higher than shown in the tables, engineering methods shall be used to calculate vent capacities.

Subsection 13.2.29 requires that vent heights outside the parameters of the tables must be calculated and that the tables cannot be used for those vent heights.

## *TABLE 13.2(a)* *Type B Double-Wall Vent*

| | |
|---|---|
| *Number of Appliances:* | *Two or More* |
| *Appliance Type:* | *Category I* |
| *Appliance Vent Connection:* | *Type B Double-Wall Connector* |

*Vent Connector Capacity*

| | | *Type B Double-Wall Vent and Connector Diameter — D (in.)* | | | | | | | | | | | | | | | | | | | | | | | |
|---|---|---|---|---|---|---|---|---|---|---|---|---|---|---|---|---|---|---|---|---|---|---|---|---|---|
| | | *3* | | | *4* | | | *5* | | | *6* | | | *7* | | | *8* | | | *9* | | | *10* | | |
| | | *Appliance Input Rating Limits in Thousands of Btu per Hour* | | | | | | | | | | | | | | | | | | | | | | | |
| *Vent Height H* | *Connector Rise R* | *FAN* | | *NAT* | *FAN* | | *NAT* | *FAN* | | *NAT* | *FAN* | | *NAT* | *FAN* | | *NAT* | *FAN* | | *NAT* | *FAN* | | *NAT* | *FAN* | | *NAT* |
| *(ft)* | *(ft)* | *Min* | *Max* | *Max* | *Min* | *Max* | *Max* | *Min* | *Max* | *Max* | *Min* | *Max* | *Max* | *Min* | *Max* | *Max* | *Min* | *Max* | *Max* | *Min* | *Max* | *Max* | *Min* | *Max* | *Max* |
| 6 | 1 | 22 | 37 | 26 | 35 | 66 | 46 | 46 | 106 | 72 | 58 | 164 | 104 | 77 | 225 | 142 | 92 | 296 | 185 | 109 | 376 | 237 | 128 | 466 | 289 |
| | 2 | 23 | 41 | 31 | 37 | 75 | 55 | 48 | 121 | 86 | 60 | 183 | 124 | 79 | 253 | 168 | 95 | 333 | 220 | 112 | 424 | 282 | 131 | 526 | 345 |
| | 3 | 24 | 44 | 35 | 38 | 81 | 62 | 49 | 132 | 96 | 62 | 199 | 139 | 82 | 275 | 189 | 97 | 363 | 248 | 114 | 463 | 317 | 134 | 575 | 386 |
| 8 | 1 | 22 | 40 | 27 | 35 | 72 | 48 | 49 | 114 | 76 | 64 | 176 | 109 | 84 | 243 | 148 | 100 | 320 | 194 | 118 | 408 | 248 | 138 | 507 | 303 |
| | 2 | 23 | 44 | 32 | 36 | 80 | 57 | 51 | 128 | 90 | 66 | 195 | 129 | 86 | 269 | 175 | 103 | 356 | 230 | 121 | 454 | 294 | 141 | 564 | 358 |
| | 3 | 24 | 47 | 36 | 37 | 87 | 64 | 53 | 139 | 101 | 67 | 210 | 145 | 88 | 290 | 198 | 105 | 384 | 258 | 123 | 492 | 330 | 143 | 612 | 402 |
| 10 | 1 | 22 | 43 | 28 | 34 | 78 | 50 | 49 | 123 | 78 | 65 | 189 | 113 | 89 | 257 | 154 | 106 | 341 | 200 | 125 | 436 | 257 | 146 | 542 | 314 |
| | 2 | 23 | 47 | 33 | 36 | 86 | 59 | 51 | 136 | 93 | 67 | 206 | 134 | 91 | 282 | 182 | 109 | 374 | 238 | 128 | 479 | 305 | 149 | 596 | 372 |
| | 3 | 24 | 50 | 37 | 37 | 92 | 67 | 52 | 146 | 104 | 69 | 220 | 150 | 94 | 303 | 205 | 111 | 402 | 268 | 131 | 515 | 342 | 152 | 642 | 417 |
| 15 | 1 | 21 | 50 | 30 | 33 | 89 | 53 | 47 | 142 | 83 | 64 | 220 | 120 | 88 | 298 | 163 | 110 | 389 | 214 | 134 | 493 | 273 | 162 | 609 | 333 |
| | 2 | 22 | 53 | 35 | 35 | 96 | 63 | 49 | 153 | 99 | 66 | 235 | 142 | 91 | 320 | 193 | 112 | 419 | 253 | 137 | 532 | 323 | 165 | 658 | 394 |
| | 3 | 24 | 55 | 40 | 36 | 102 | 71 | 51 | 163 | 111 | 68 | 248 | 160 | 93 | 339 | 218 | 115 | 445 | 286 | 140 | 565 | 365 | 167 | 700 | 444 |
| 20 | 1 | 21 | 54 | 31 | 33 | 99 | 56 | 46 | 157 | 87 | 62 | 246 | 125 | 86 | 334 | 171 | 107 | 436 | 224 | 131 | 552 | 285 | 158 | 681 | 347 |
| | 2 | 22 | 57 | 37 | 34 | 105 | 66 | 48 | 167 | 104 | 64 | 259 | 149 | 89 | 354 | 202 | 110 | 463 | 265 | 134 | 587 | 339 | 161 | 725 | 414 |
| | 3 | 23 | 60 | 42 | 35 | 110 | 74 | 50 | 176 | 116 | 66 | 271 | 168 | 91 | 371 | 228 | 113 | 486 | 300 | 137 | 618 | 383 | 164 | 764 | 466 |
| 30 | 1 | 20 | 62 | 33 | 31 | 113 | 59 | 45 | 181 | 93 | 60 | 288 | 134 | 83 | 391 | 182 | 103 | 512 | 238 | 125 | 649 | 305 | 151 | 802 | 372 |
| | 2 | 21 | 64 | 39 | 33 | 118 | 70 | 47 | 190 | 110 | 62 | 299 | 158 | 85 | 408 | 215 | 105 | 535 | 282 | 129 | 679 | 360 | 155 | 840 | 439 |
| | 3 | 22 | 66 | 44 | 34 | 123 | 79 | 48 | 198 | 124 | 64 | 309 | 178 | 88 | 423 | 242 | 108 | 555 | 317 | 132 | 706 | 405 | 158 | 874 | 494 |
| 50 | 1 | 19 | 71 | 36 | 30 | 133 | 64 | 43 | 216 | 101 | 57 | 349 | 145 | 78 | 477 | 197 | 97 | 627 | 257 | 120 | 797 | 330 | 144 | 984 | 403 |
| | 2 | 21 | 73 | 43 | 32 | 137 | 76 | 45 | 223 | 119 | 59 | 358 | 172 | 81 | 490 | 234 | 100 | 645 | 306 | 123 | 820 | 392 | 148 | 1014 | 478 |
| | 3 | 22 | 75 | 48 | 33 | 141 | 86 | 46 | 229 | 134 | 61 | 366 | 194 | 83 | 502 | 263 | 103 | 661 | 343 | 126 | 842 | 441 | 151 | 1043 | 538 |
| 100 | 1 | 18 | 82 | 37 | 28 | 158 | 66 | 40 | 262 | 104 | 53 | 442 | 150 | 73 | 611 | 204 | 91 | 810 | 266 | 112 | 1038 | 341 | 135 | 1285 | 417 |
| | 2 | 19 | 83 | 44 | 30 | 161 | 79 | 42 | 267 | 123 | 55 | 447 | 178 | 75 | 619 | 242 | 94 | 822 | 316 | 115 | 1054 | 405 | 139 | 1306 | 494 |
| | 3 | 20 | 84 | 50 | 31 | 163 | 89 | 44 | 272 | 138 | 57 | 452 | 200 | 78 | 627 | 272 | 97 | 834 | 355 | 118 | 1069 | 455 | 142 | 1327 | 555 |

*Common Vent Capacity*

| | *Type B Double-Wall Common Vent Diameter — D (in.)* | | | | | | | | | | | | | | | | | | | | |
|---|---|---|---|---|---|---|---|---|---|---|---|---|---|---|---|---|---|---|---|---|---|
| | *4* | | | *5* | | | *6* | | | *7* | | | *8* | | | *9* | | | *10* | | |
| | *Combined Appliance Input Rating in Thousands of Btu per Hour* | | | | | | | | | | | | | | | | | | | | |
| *Vent Height H (ft)* | *FAN +FAN* | *FAN +NAT* | *NAT +NAT* | *FAN +FAN* | *FAN +NAT* | *NAT +NAT* | *FAN +FAN* | *FAN +NAT* | *NAT +NAT* | *FAN +FAN* | *FAN +NAT* | *NAT +NAT* | *FAN +FAN* | *FAN +NAT* | *NAT +NAT* | *FAN +FAN* | *FAN +NAT* | *NAT +NAT* | *FAN +FAN* | *FAN +NAT* | *NAT +NAT* |
| 6 | 92 | 81 | 65 | 140 | 116 | 103 | 204 | 161 | 147 | 309 | 248 | 200 | 404 | 314 | 260 | 547 | 434 | 335 | 672 | 520 | 410 |
| 8 | 101 | 90 | 73 | 155 | 129 | 114 | 224 | 178 | 163 | 339 | 275 | 223 | 444 | 348 | 290 | 602 | 480 | 378 | 740 | 577 | 465 |
| 10 | 110 | 97 | 79 | 169 | 141 | 124 | 243 | 194 | 178 | 367 | 299 | 242 | 477 | 377 | 315 | 649 | 522 | 405 | 800 | 627 | 495 |
| 15 | 125 | 112 | 91 | 195 | 164 | 144 | 283 | 228 | 206 | 427 | 352 | 280 | 556 | 444 | 365 | 753 | 612 | 465 | 924 | 733 | 565 |
| 20 | 136 | 123 | 102 | 215 | 183 | 160 | 314 | 255 | 229 | 475 | 394 | 310 | 621 | 499 | 405 | 842 | 688 | 523 | 1035 | 826 | 640 |
| 30 | 152 | 138 | 118 | 244 | 210 | 185 | 361 | 297 | 266 | 547 | 459 | 360 | 720 | 585 | 470 | 979 | 808 | 605 | 1209 | 975 | 740 |
| 50 | 167 | 153 | 134 | 279 | 244 | 214 | 421 | 353 | 310 | 641 | 547 | 423 | 854 | 706 | 550 | 1164 | 977 | 705 | 1451 | 1188 | 860 |
| 100 | 175 | 163 | NA | 311 | 277 | NA | 489 | 421 | NA | 751 | 658 | 479 | 1025 | 873 | 625 | 1408 | 1215 | 800 | 1784 | 1502 | 975 |

*(continues)*

*TABLE 13.2(a) Continued*

| | | | | | | | | | | | | | | | | | | | | | | |
|---|---|---|---|---|---|---|---|---|---|---|---|---|---|---|---|---|---|---|---|---|---|---|
| | | | | | | | | | | | | *Number of Appliances:* | | | | *Two or More* | | | | | | |
| | | | | | | | | | | | | *Appliance Type:* | | | | *Category I* | | | | | | |
| | | | | | | | | | | | | *Appliance Vent Connection:* | | | | *Type B Double-Wall Connector* | | | | | | |
| | | *Type B Double-Wall Vent and Connector Diameter — D (in.)* | | | | | | | | | | | | | | | | | | | | |
| | | *12* | | | *14* | | | *16* | | | *18* | | | *20* | | | *22* | | | *24* | | |
| *Vent Height H (ft)* | *Connector Rise R (ft)* | *Appliance Input Rating Limits in Thousands of Btu per Hour* | | | | | | | | | | | | | | | | | | | | |
| | | *FAN* | | *NAT* | *FAN* | | *NAT* | *FAN* | | *NAT* | *FAN* | | *NAT* | *FAN* | | *NAT* | *FAN* | | *NAT* | *FAN* | | *NAT* |
| | | *Min* | *Max* | *Max* | *Min* | *Max* | *Max* | *Min* | *Max* | *Max* | *Min* | *Max* | *Max* | *Min* | *Max* | *Max* | *Min* | *Max* | *Max* | *Min* | *Max* | *Max* |
| 6 | 2 | 174 | 764 | 496 | 223 | 1046 | 653 | 281 | 1371 | 853 | 346 | 1772 | 1080 | NA | NA | NA | NA | NA | NA | NA | NA | NA |
| | 4 | 180 | 897 | 616 | 230 | 1231 | 827 | 287 | 1617 | 1081 | 352 | 2069 | 1370 | NA | NA | NA | NA | NA | NA | NA | NA | NA |
| | 6 | NA | NA | NA | NA | NA | NA | NA | NA | NA | NA | NA | NA | NA | NA | NA | NA | NA | NA | NA | NA | NA |
| 8 | 2 | 186 | 822 | 516 | 238 | 1126 | 696 | 298 | 1478 | 910 | 365 | 1920 | 1150 | NA | NA | NA | NA | NA | NA | NA | NA | NA |
| | 4 | 192 | 952 | 644 | 244 | 1307 | 884 | 305 | 1719 | 1150 | 372 | 2211 | 1460 | 471 | 2737 | 1800 | 560 | 3319 | 2180 | 662 | 3957 | 2590 |
| | 6 | 198 | 1050 | 772 | 252 | 1445 | 1072 | 313 | 1902 | 1390 | 380 | 2434 | 1770 | 478 | 3018 | 2180 | 568 | 3665 | 2640 | 669 | 4373 | 3130 |
| 10 | 2 | 196 | 870 | 536 | 249 | 1195 | 730 | 311 | 1570 | 955 | 379 | 2049 | 1205 | NA | NA | NA | NA | NA | NA | NA | NA | NA |
| | 4 | 201 | 997 | 664 | 256 | 1371 | 924 | 318 | 1804 | 1205 | 387 | 2332 | 1535 | 486 | 2887 | 1890 | 581 | 3502 | 2280 | 686 | 4175 | 2710 |
| | 6 | 207 | 1095 | 792 | 263 | 1509 | 1118 | 325 | 1989 | 1455 | 395 | 2556 | 1865 | 494 | 3169 | 2290 | 589 | 3849 | 2760 | 694 | 4593 | 3270 |
| 15 | 2 | 214 | 967 | 568 | 272 | 1334 | 790 | 336 | 1760 | 1030 | 408 | 2317 | 1305 | NA | NA | NA | NA | NA | NA | NA | NA | NA |
| | 4 | 221 | 1085 | 712 | 279 | 1499 | 1006 | 344 | 1978 | 1320 | 416 | 2579 | 1665 | 523 | 3197 | 2060 | 624 | 3881 | 2490 | 734 | 4631 | 2960 |
| | 6 | 228 | 1181 | 856 | 286 | 1632 | 1222 | 351 | 2157 | 1610 | 424 | 2796 | 2025 | 533 | 3470 | 2510 | 634 | 4216 | 3030 | 743 | 5035 | 3600 |
| 20 | 2 | 223 | 1051 | 596 | 291 | 1443 | 840 | 357 | 1911 | 1095 | 430 | 2533 | 1385 | NA | NA | NA | NA | NA | NA | NA | NA | NA |
| | 4 | 230 | 1162 | 748 | 298 | 1597 | 1064 | 365 | 2116 | 1395 | 438 | 2778 | 1765 | 554 | 3447 | 2180 | 661 | 4190 | 2630 | 772 | 5005 | 3130 |
| | 6 | 237 | 1253 | 900 | 307 | 1726 | 1288 | 373 | 2287 | 1695 | 450 | 2984 | 2145 | 567 | 3708 | 2650 | 671 | 4511 | 3190 | 785 | 5392 | 3790 |
| 30 | 2 | 216 | 1217 | 632 | 286 | 1664 | 910 | 367 | 2183 | 1190 | 461 | 2891 | 1540 | NA | NA | NA | NA | NA | NA | NA | NA | NA |
| | 4 | 223 | 1316 | 792 | 294 | 1802 | 1160 | 376 | 2366 | 1510 | 474 | 3110 | 1920 | 619 | 3840 | 2365 | 728 | 4861 | 2860 | 847 | 5606 | 3410 |
| | 6 | 231 | 1400 | 952 | 303 | 1920 | 1410 | 384 | 2524 | 1830 | 485 | 3299 | 2340 | 632 | 4080 | 2875 | 741 | 4976 | 3480 | 860 | 5961 | 4150 |
| 50 | 2 | 206 | 1479 | 689 | 273 | 2023 | 1007 | 350 | 2659 | 1315 | 435 | 3548 | 1665 | NA | NA | NA | NA | NA | NA | NA | NA | NA |
| | 4 | 213 | 1561 | 860 | 281 | 2139 | 1291 | 359 | 2814 | 1685 | 447 | 3730 | 2135 | 580 | 4601 | 2633 | 709 | 5569 | 3185 | 851 | 6633 | 3790 |
| | 6 | 221 | 1631 | 1031 | 290 | 2242 | 1575 | 369 | 2951 | 2055 | 461 | 3893 | 2605 | 594 | 4808 | 3208 | 724 | 5826 | 3885 | 867 | 6943 | 4620 |
| 100 | 2 | 192 | 1923 | 712 | 254 | 2644 | 1050 | 326 | 3490 | 1370 | 402 | 4707 | 1740 | NA | NA | NA | NA | NA | NA | NA | NA | NA |
| | 4 | 200 | 1984 | 888 | 263 | 2731 | 1346 | 336 | 3606 | 1760 | 414 | 4842 | 2220 | 523 | 5982 | 2750 | 639 | 7254 | 3330 | 769 | 8650 | 3950 |
| | 6 | 208 | 2035 | 1064 | 272 | 2811 | 1642 | 346 | 3714 | 2150 | 426 | 4968 | 2700 | 539 | 6143 | 3350 | 654 | 7453 | 4070 | 786 | 8892 | 4810 |

*Common Vent Capacity*

| | | | | | | | | | | | | | | | | | | | | | |
|---|---|---|---|---|---|---|---|---|---|---|---|---|---|---|---|---|---|---|---|---|---|
| | *Type B Double-Wall Common Vent Diameter — D (in.)* | | | | | | | | | | | | | | | | | | | | |
| | *12* | | | *14* | | | *16* | | | *18* | | | *20* | | | *22* | | | *24* | | |
| *Vent Height* | *Combined Appliance Input Rating in Thousands of Btu per Hour* | | | | | | | | | | | | | | | | | | | | |
| *H (ft)* | *FAN +FAN* | *FAN +NAT* | *NAT +NAT* | *FAN +FAN* | *FAN +NAT* | *NAT +NAT* | *FAN +FAN* | *FAN +NAT* | *NAT +NAT* | *FAN +FAN* | *FAN +NAT* | *NAT +NAT* | *FAN +FAN* | *FAN +NAT* | *NAT +NAT* | *FAN +FAN* | *FAN +NAT* | *NAT +NAT* | *FAN +FAN* | *FAN +NAT* | *NAT +NAT* |
| 6 | 900 | 696 | 588 | 1284 | 990 | 815 | 1735 | 1336 | 1065 | 2253 | 1732 | 1345 | 2838 | 2180 | 1660 | 3488 | 2677 | 1970 | 4206 | 3226 | 2390 |
| 8 | 994 | 773 | 652 | 1423 | 1103 | 912 | 1927 | 1491 | 1190 | 2507 | 1936 | 1510 | 3162 | 2439 | 1860 | 3890 | 2998 | 2200 | 4695 | 3616 | 2680 |
| 10 | 1076 | 841 | 712 | 1542 | 1200 | 995 | 2093 | 1625 | 1300 | 2727 | 2113 | 1645 | 3444 | 2665 | 2030 | 4241 | 3278 | 2400 | 5123 | 3957 | 2920 |
| 15 | 1247 | 986 | 825 | 1794 | 1410 | 1158 | 2440 | 1910 | 1510 | 3184 | 2484 | 1910 | 4026 | 3133 | 2360 | 4971 | 3862 | 2790 | 6016 | 4670 | 3400 |
| 20 | 1405 | 1116 | 916 | 2006 | 1588 | 1290 | 2722 | 2147 | 1690 | 3561 | 2798 | 2140 | 4548 | 3552 | 2640 | 5573 | 4352 | 3120 | 6749 | 5261 | 3800 |
| 30 | 1658 | 1327 | 1025 | 2373 | 1892 | 1525 | 3220 | 2558 | 1990 | 4197 | 3326 | 2520 | 5303 | 4193 | 3110 | 6539 | 5157 | 3680 | 7940 | 6247 | 4480 |
| 50 | 2024 | 1640 | 1280 | 2911 | 2347 | 1863 | 3964 | 3183 | 2430 | 5184 | 4149 | 3075 | 6567 | 5240 | 3800 | 8116 | 6458 | 4500 | 9837 | 7813 | 5475 |
| 100 | 2569 | 2131 | 1670 | 3732 | 3076 | 2450 | 5125 | 4202 | 3200 | 6749 | 5509 | 4050 | 8597 | 6986 | 5000 | 10,681 | 8648 | 5920 | 13,004 | 10,499 | 7200 |

For SI units, 1 in. = 25.4 mm, 1 in.$^2$ = 645 mm$^2$, 1 ft = 0.305 m, 1000 Btu/hr = 0.293 kW.

## *TABLE 13.2(b) Type B Double-Wall Vent*

| | |
|---|---|
| *Number of Appliances:* | *Two or More* |
| *Appliance Type:* | *Category I* |
| *Appliance Vent Connection:* | *Single-Wall Metal Connector* |

*Vent Connector Capacity*

| Vent Height H (ft) | Connector Rise R (ft) | 3 FAN Min | 3 FAN Max | 3 NAT Max | 4 FAN Min | 4 FAN Max | 4 NAT Max | 5 FAN Min | 5 FAN Max | 5 NAT Max | 6 FAN Min | 6 FAN Max | 6 NAT Max | 7 FAN Min | 7 FAN Max | 7 NAT Max | 8 FAN Min | 8 FAN Max | 8 NAT Max | 9 FAN Min | 9 FAN Max | 9 NAT Max | 10 FAN Min | 10 FAN Max | 10 NAT Max |
|---|---|---|---|---|---|---|---|---|---|---|---|---|---|---|---|---|---|---|---|---|---|---|---|---|---|
| | | *Single-Wall Metal Vent Connector Diameter — D (in.); Appliance Input Rating Limits in Thousands of Btu per Hour* | | | | | | | | | | | | | | | | | | | | | | | |
| 6 | 1 | NA | NA | 26 | NA | NA | 46 | NA | NA | 71 | NA | NA | 102 | 207 | 223 | 140 | 262 | 293 | 183 | 325 | 373 | 234 | 447 | 463 | 286 |
| | 2 | NA | NA | 31 | NA | NA | 55 | NA | NA | 85 | 168 | 182 | 123 | 215 | 251 | 167 | 271 | 331 | 219 | 334 | 422 | 281 | 458 | 524 | 344 |
| | 3 | NA | NA | 34 | NA | NA | 62 | 121 | 131 | 95 | 175 | 198 | 138 | 222 | 273 | 188 | 279 | 361 | 247 | 344 | 462 | 316 | 468 | 574 | 385 |
| 8 | 1 | NA | NA | 27 | NA | NA | 48 | NA | NA | 75 | NA | NA | 106 | 226 | 240 | 145 | 285 | 316 | 191 | 352 | 403 | 244 | 481 | 502 | 299 |
| | 2 | NA | NA | 32 | NA | NA | 57 | 125 | 126 | 89 | 184 | 193 | 127 | 234 | 266 | 173 | 293 | 353 | 228 | 360 | 450 | 292 | 492 | 560 | 355 |
| | 3 | NA | NA | 35 | NA | NA | 64 | 130 | 138 | 100 | 191 | 208 | 144 | 241 | 287 | 197 | 302 | 381 | 256 | 370 | 489 | 328 | 501 | 609 | 400 |
| 10 | 1 | NA | NA | 28 | NA | NA | 50 | 119 | 121 | 77 | 182 | 186 | 110 | 240 | 253 | 150 | 302 | 335 | 196 | 372 | 429 | 252 | 506 | 534 | 308 |
| | 2 | NA | NA | 33 | 84 | 85 | 59 | 124 | 134 | 91 | 189 | 203 | 132 | 248 | 278 | 183 | 311 | 369 | 235 | 381 | 473 | 302 | 517 | 589 | 368 |
| | 3 | NA | NA | 36 | 89 | 91 | 67 | 129 | 144 | 102 | 197 | 217 | 148 | 257 | 299 | 203 | 320 | 398 | 265 | 391 | 511 | 339 | 528 | 637 | 413 |
| 15 | 1 | NA | NA | 29 | 79 | 87 | 52 | 116 | 138 | 81 | 177 | 214 | 116 | 238 | 291 | 158 | 312 | 380 | 208 | 397 | 482 | 266 | 556 | 596 | 324 |
| | 2 | NA | NA | 34 | 83 | 94 | 62 | 121 | 150 | 97 | 185 | 230 | 138 | 246 | 314 | 189 | 321 | 411 | 248 | 407 | 522 | 317 | 568 | 646 | 387 |
| | 3 | NA | NA | 39 | 87 | 100 | 70 | 127 | 160 | 109 | 193 | 243 | 157 | 255 | 333 | 215 | 331 | 438 | 281 | 418 | 557 | 360 | 579 | 690 | 437 |
| 20 | 1 | 49 | 56 | 30 | 78 | 97 | 54 | 115 | 152 | 84 | 175 | 238 | 120 | 233 | 325 | 165 | 306 | 425 | 217 | 390 | 538 | 276 | 546 | 664 | 336 |
| | 2 | 52 | 59 | 36 | 82 | 103 | 64 | 120 | 163 | 101 | 182 | 252 | 144 | 243 | 346 | 197 | 317 | 453 | 259 | 400 | 574 | 331 | 558 | 709 | 403 |
| | 3 | 55 | 62 | 40 | 87 | 107 | 72 | 125 | 172 | 113 | 190 | 264 | 164 | 252 | 363 | 223 | 326 | 476 | 294 | 412 | 607 | 375 | 570 | 750 | 457 |
| 30 | 1 | 47 | 60 | 31 | 77 | 110 | 57 | 112 | 175 | 89 | 169 | 278 | 129 | 226 | 380 | 175 | 296 | 497 | 230 | 378 | 630 | 294 | 528 | 779 | 358 |
| | 2 | 51 | 62 | 37 | 81 | 115 | 67 | 117 | 185 | 106 | 177 | 290 | 152 | 236 | 397 | 208 | 307 | 521 | 274 | 389 | 662 | 349 | 541 | 819 | 425 |
| | 3 | 54 | 64 | 42 | 85 | 119 | 76 | 122 | 193 | 120 | 185 | 300 | 172 | 244 | 412 | 235 | 316 | 542 | 309 | 400 | 690 | 394 | 555 | 855 | 482 |
| 50 | 1 | 46 | 69 | 34 | 75 | 128 | 60 | 109 | 207 | 96 | 162 | 336 | 137 | 217 | 460 | 188 | 284 | 604 | 245 | 364 | 768 | 314 | 507 | 951 | 384 |
| | 2 | 49 | 71 | 40 | 79 | 132 | 72 | 114 | 215 | 113 | 170 | 345 | 164 | 226 | 473 | 223 | 294 | 623 | 293 | 376 | 793 | 375 | 520 | 983 | 458 |
| | 3 | 52 | 72 | 45 | 83 | 136 | 82 | 119 | 221 | 123 | 178 | 353 | 186 | 235 | 486 | 252 | 304 | 640 | 331 | 387 | 816 | 423 | 535 | 1013 | 518 |
| 100 | 1 | 45 | 79 | 34 | 71 | 150 | 61 | 104 | 249 | 98 | 153 | 424 | 140 | 205 | 585 | 192 | 269 | 774 | 249 | 345 | 993 | 321 | 476 | 1236 | 393 |
| | 2 | 48 | 80 | 41 | 75 | 153 | 73 | 110 | 255 | 115 | 160 | 428 | 167 | 212 | 593 | 228 | 279 | 788 | 299 | 358 | 1011 | 383 | 490 | 1259 | 469 |
| | 3 | 51 | 81 | 46 | 79 | 157 | 85 | 114 | 260 | 129 | 168 | 433 | 190 | 222 | 603 | 256 | 289 | 801 | 339 | 368 | 1027 | 431 | 506 | 1280 | 527 |

*Common Vent Capacity*

| Vent Height H (ft) | 4 FAN +FAN | 4 FAN +NAT | 4 NAT +NAT | 5 FAN +FAN | 5 FAN +NAT | 5 NAT +NAT | 6 FAN +FAN | 6 FAN +NAT | 6 NAT +NAT | 7 FAN +FAN | 7 FAN +NAT | 7 NAT +NAT | 8 FAN +FAN | 8 FAN +NAT | 8 NAT +NAT | 9 FAN +FAN | 9 FAN +NAT | 9 NAT +NAT | 10 FAN +FAN | 10 FAN +NAT | 10 NAT +NAT |
|---|---|---|---|---|---|---|---|---|---|---|---|---|---|---|---|---|---|---|---|---|---|
| | *Type B Double-Wall Vent Diameter — D (in.); Combined Appliance Input Rating in Thousands of Btu per Hour* | | | | | | | | | | | | | | | | | | | | |
| 6 | NA | 78 | 64 | NA | 113 | 99 | 200 | 158 | 144 | 304 | 244 | 196 | 398 | 310 | 257 | 541 | 429 | 332 | 665 | 515 | 407 |
| 8 | NA | 87 | 71 | NA | 126 | 111 | 218 | 173 | 159 | 331 | 269 | 218 | 436 | 342 | 285 | 592 | 473 | 373 | 730 | 569 | 460 |
| 10 | NA | 94 | 76 | 163 | 137 | 120 | 237 | 189 | 174 | 357 | 292 | 236 | 467 | 369 | 309 | 638 | 512 | 398 | 787 | 617 | 487 |
| 15 | 121 | 108 | 88 | 189 | 159 | 140 | 275 | 221 | 200 | 416 | 343 | 274 | 544 | 434 | 357 | 738 | 599 | 456 | 905 | 718 | 553 |
| 20 | 131 | 118 | 98 | 208 | 177 | 156 | 305 | 247 | 223 | 463 | 383 | 302 | 606 | 487 | 395 | 824 | 673 | 512 | 1013 | 808 | 626 |
| 30 | 145 | 132 | 113 | 236 | 202 | 180 | 350 | 286 | 257 | 533 | 446 | 349 | 703 | 570 | 459 | 958 | 790 | 593 | 1183 | 952 | 723 |
| 50 | 159 | 145 | 128 | 268 | 233 | 208 | 406 | 337 | 296 | 622 | 529 | 410 | 833 | 686 | 535 | 1139 | 954 | 689 | 1418 | 1157 | 838 |
| 100 | 166 | 153 | NA | 297 | 263 | NA | 469 | 398 | NA | 726 | 633 | 464 | 999 | 846 | 606 | 1378 | 1185 | 780 | 1741 | 1459 | 948 |

For SI units, 1 in. = 25.4 mm, 1 in.$^2$ = 645 mm$^2$, 1 ft = 0.305 m, 1000 Btu/hr = 0.293 kW.

## *TABLE 13.2(c) Masonry Chimney*

| | |
|---|---|
| *Number of Appliances:* | *Two or More* |
| *Appliance Type:* | *Category I* |
| *Appliance Vent Connection:* | *Type B Double-Wall Connector* |

*Vent Connector Capacity*

| Vent Height H (ft) | Connector Rise R (ft) | *Type B Double-Wall Vent Connector Diameter — D (in.)* | | | | | | | | | | | | | | | | | | | | | | | |
|---|---|---|---|---|---|---|---|---|---|---|---|---|---|---|---|---|---|---|---|---|---|---|---|---|---|
| | | 3 | | | 4 | | | 5 | | | 6 | | | 7 | | | 8 | | | 9 | | | 10 | | |
| | | *Appliance Input Rating Limits in Thousands of Btu per Hour* | | | | | | | | | | | | | | | | | | | | | | | |
| | | FAN | | NAT | FAN | | NAT | FAN | | NAT | FAN | | NAT | FAN | | NAT | FAN | | NAT | FAN | | NAT | FAN | | NAT |
| | | Min | Max | Max | Min | Max | Max | Min | Max | Max | Min | Max | Max | Min | Max | Max | Min | Max | Max | Min | Max | Max | Min | Max | Max |
| 6 | 1 | 24 | 33 | 21 | 39 | 62 | 40 | 52 | 106 | 67 | 65 | 194 | 101 | 87 | 274 | 141 | 104 | 370 | 201 | 124 | 479 | 253 | 145 | 599 | 319 |
| | 2 | 26 | 43 | 28 | 41 | 79 | 52 | 53 | 133 | 85 | 67 | 230 | 124 | 89 | 324 | 173 | 107 | 436 | 232 | 127 | 562 | 300 | 148 | 694 | 378 |
| | 3 | 27 | 49 | 34 | 42 | 92 | 61 | 55 | 155 | 97 | 69 | 262 | 143 | 91 | 369 | 203 | 109 | 491 | 270 | 129 | 633 | 349 | 151 | 795 | 439 |
| 8 | 1 | 24 | 39 | 22 | 39 | 72 | 41 | 55 | 117 | 69 | 71 | 213 | 105 | 94 | 304 | 148 | 113 | 414 | 210 | 134 | 539 | 267 | 156 | 682 | 335 |
| | 2 | 26 | 47 | 29 | 40 | 87 | 53 | 57 | 140 | 86 | 73 | 246 | 127 | 97 | 350 | 179 | 116 | 473 | 240 | 137 | 615 | 311 | 160 | 776 | 394 |
| | 3 | 27 | 52 | 34 | 42 | 97 | 62 | 59 | 159 | 98 | 75 | 269 | 145 | 99 | 383 | 206 | 119 | 517 | 276 | 139 | 672 | 358 | 163 | 848 | 452 |
| 10 | 1 | 24 | 42 | 22 | 38 | 80 | 42 | 55 | 130 | 71 | 74 | 232 | 108 | 101 | 324 | 153 | 120 | 444 | 216 | 142 | 582 | 277 | 165 | 739 | 348 |
| | 2 | 26 | 50 | 29 | 40 | 93 | 54 | 57 | 153 | 87 | 76 | 261 | 129 | 103 | 366 | 184 | 123 | 498 | 247 | 145 | 652 | 321 | 168 | 825 | 407 |
| | 3 | 27 | 55 | 35 | 41 | 105 | 63 | 58 | 170 | 100 | 78 | 284 | 148 | 106 | 397 | 209 | 126 | 540 | 281 | 147 | 705 | 366 | 171 | 893 | 463 |
| 15 | 1 | 24 | 48 | 23 | 38 | 93 | 44 | 54 | 154 | 74 | 72 | 277 | 114 | 100 | 384 | 164 | 125 | 511 | 229 | 153 | 658 | 297 | 184 | 824 | 375 |
| | 2 | 25 | 55 | 31 | 39 | 105 | 55 | 56 | 174 | 89 | 74 | 299 | 134 | 103 | 419 | 192 | 128 | 558 | 260 | 156 | 718 | 339 | 187 | 900 | 432 |
| | 3 | 26 | 59 | 35 | 41 | 115 | 64 | 57 | 189 | 102 | 76 | 319 | 153 | 105 | 448 | 215 | 131 | 597 | 292 | 159 | 760 | 382 | 190 | 960 | 486 |
| 20 | 1 | 24 | 52 | 24 | 37 | 102 | 46 | 53 | 172 | 77 | 71 | 313 | 119 | 98 | 437 | 173 | 123 | 584 | 239 | 150 | 752 | 312 | 180 | 943 | 397 |
| | 2 | 25 | 58 | 31 | 39 | 114 | 56 | 55 | 190 | 91 | 73 | 335 | 138 | 101 | 467 | 199 | 126 | 625 | 270 | 153 | 805 | 354 | 184 | 1011 | 452 |
| | 3 | 26 | 63 | 35 | 40 | 123 | 65 | 57 | 204 | 104 | 75 | 353 | 157 | 104 | 493 | 222 | 129 | 661 | 301 | 156 | 851 | 396 | 187 | 1067 | 505 |
| 30 | 1 | 24 | 54 | 25 | 37 | 111 | 48 | 52 | 192 | 82 | 69 | 357 | 127 | 96 | 504 | 187 | 119 | 680 | 255 | 145 | 883 | 337 | 175 | 1115 | 432 |
| | 2 | 25 | 60 | 32 | 38 | 122 | 58 | 54 | 208 | 95 | 72 | 376 | 145 | 99 | 531 | 209 | 122 | 715 | 287 | 149 | 928 | 378 | 179 | 1171 | 484 |
| | 3 | 26 | 64 | 36 | 40 | 131 | 66 | 56 | 221 | 107 | 74 | 392 | 163 | 101 | 554 | 233 | 125 | 746 | 317 | 152 | 968 | 418 | 182 | 1220 | 535 |
| 50 | 1 | 23 | 51 | 25 | 36 | 116 | 51 | 51 | 209 | 89 | 67 | 405 | 143 | 92 | 582 | 213 | 115 | 798 | 294 | 140 | 1049 | 392 | 168 | 1334 | 506 |
| | 2 | 24 | 59 | 32 | 37 | 127 | 61 | 53 | 225 | 102 | 70 | 421 | 161 | 95 | 604 | 235 | 118 | 827 | 326 | 143 | 1085 | 433 | 172 | 1379 | 558 |
| | 3 | 26 | 64 | 36 | 39 | 135 | 69 | 55 | 237 | 115 | 72 | 435 | 180 | 98 | 624 | 260 | 121 | 854 | 357 | 147 | 1118 | 474 | 176 | 1421 | 611 |
| 100 | 1 | 23 | 46 | 24 | 35 | 108 | 50 | 49 | 208 | 92 | 65 | 428 | 155 | 88 | 640 | 237 | 109 | 907 | 334 | 134 | 1222 | 454 | 161 | 1589 | 596 |
| | 2 | 24 | 53 | 31 | 37 | 120 | 60 | 51 | 224 | 105 | 67 | 444 | 174 | 92 | 660 | 260 | 113 | 933 | 368 | 138 | 1253 | 497 | 165 | 1626 | 651 |
| | 3 | 25 | 59 | 35 | 38 | 130 | 68 | 53 | 237 | 118 | 69 | 458 | 193 | 94 | 679 | 285 | 116 | 956 | 399 | 141 | 1282 | 540 | 169 | 1661 | 705 |

*Common Vent Capacity*

| Vent Height H (ft) | *Minimum Internal Area of Masonry Chimney Flue (in.$^2$)* | | | | | | | | | | | | | | | | | | | | | | | |
|---|---|---|---|---|---|---|---|---|---|---|---|---|---|---|---|---|---|---|---|---|---|---|---|---|
| | 12 | | | 19 | | | 28 | | | 38 | | | 50 | | | 63 | | | 78 | | | 113 | | |
| | *Combined Appliance Input Rating in Thousands of Btu per Hour* | | | | | | | | | | | | | | | | | | | | | | | |
| | FAN +FAN | FAN +NAT | NAT +NAT | FAN +FAN | FAN +NAT | NAT +NAT | FAN +FAN | FAN +NAT | NAT +NAT | FAN +FAN | FAN +NAT | NAT +NAT | FAN +FAN | FAN +NAT | NAT +NAT | FAN +FAN | FAN +NAT | NAT +NAT | FAN +FAN | FAN +NAT | NAT +NAT | FAN +FAN | FAN +FAN | NAT +NAT |
| 6 | NA | 74 | 25 | NA | 119 | 46 | NA | 178 | 71 | NA | 257 | 103 | NA | 351 | 143 | NA | 458 | 188 | NA | 582 | 246 | 1041 | 853 | NA |
| 8 | NA | 80 | 28 | NA | 130 | 53 | NA | 193 | 82 | NA | 279 | 119 | NA | 384 | 163 | NA | 501 | 218 | 724 | 636 | 278 | 1144 | 937 | 408 |
| 10 | NA | 84 | 31 | NA | 138 | 56 | NA | 207 | 90 | NA | 299 | 131 | NA | 409 | 177 | 606 | 538 | 236 | 776 | 686 | 302 | 1226 | 1010 | 454 |
| 15 | NA | NA | 36 | NA | 152 | 67 | NA | 233 | 106 | NA | 334 | 152 | 523 | 467 | 212 | 682 | 611 | 283 | 874 | 781 | 365 | 1374 | 1156 | 546 |
| 20 | NA | NA | 41 | NA | NA | 75 | NA | 250 | 122 | NA | 368 | 172 | 565 | 508 | 243 | 742 | 668 | 325 | 955 | 858 | 419 | 1513 | 1286 | 648 |
| 30 | NA | NA | NA | NA | NA | NA | NA | 270 | 137 | NA | 404 | 198 | 615 | 564 | 278 | 816 | 747 | 381 | 1062 | 969 | 496 | 1702 | 1473 | 749 |
| 50 | NA | NA | NA | NA | NA | NA | NA | NA | NA | NA | NA | NA | NA | 620 | 328 | 879 | 831 | 461 | 1165 | 1089 | 606 | 1905 | 1692 | 922 |
| 100 | NA | NA | NA | NA | NA | NA | NA | NA | NA | NA | NA | NA | NA | NA | 348 | NA | NA | 499 | NA | NA | 669 | 2053 | 1921 | 1058 |

## *TABLE 13.2(d) Masonry Chimney*

| | |
|---|---|
| *Number of Appliances:* | *Two or More* |
| *Appliance Type:* | *Category I* |
| *Appliance Vent Connection:* | *Single-Wall Metal Connector* |

***Vent Connector Capacity***

| | | *Single-Wall Metal Vent Connector Diameter — D (in.)* | | | | | | | | | | | | | | | | | | | | | | | |
|---|---|---|---|---|---|---|---|---|---|---|---|---|---|---|---|---|---|---|---|---|---|---|---|---|---|
| | | *3* | | | *4* | | | *5* | | | *6* | | | *7* | | | *8* | | | *9* | | | *10* | | |
| *Vent Height H (ft)* | *Connector Rise R (ft)* | *Appliance Input Rating Limits in Thousands of Btu per Hour* | | | | | | | | | | | | | | | | | | | | | | | |
| | | *FAN* | | *NAT* | *FAN* | | *NAT* | *FAN* | | *NAT* | *FAN* | | *NAT* | *FAN* | | *NAT* | *FAN* | | *NAT* | *FAN* | | *NAT* | *FAN* | | *NAT* |
| | | *Min* | *Max* | *Max* | *Min* | *Max* | *Max* | *Min* | *Max* | *Max* | *Min* | *Max* | *Max* | *Min* | *Max* | *Max* | *Min* | *Max* | *Max* | *Min* | *Max* | *Max* | *Min* | *Max* | *Max* |
| 6 | 1 | NA | NA | 21 | NA | NA | 39 | NA | NA | 66 | 179 | 191 | 100 | 231 | 271 | 140 | 292 | 366 | 200 | 362 | 474 | 252 | 499 | 594 | 316 |
| | 2 | NA | NA | 28 | NA | NA | 52 | NA | NA | 84 | 186 | 227 | 123 | 239 | 321 | 172 | 301 | 432 | 231 | 373 | 557 | 299 | 509 | 696 | 376 |
| | 3 | NA | NA | 34 | NA | NA | 61 | 134 | 153 | 97 | 193 | 258 | 142 | 247 | 365 | 202 | 309 | 491 | 269 | 381 | 634 | 348 | 519 | 793 | 437 |
| 8 | 1 | NA | NA | 21 | NA | NA | 40 | NA | NA | 68 | 195 | 208 | 103 | 250 | 298 | 146 | 313 | 407 | 207 | 387 | 530 | 263 | 529 | 672 | 331 |
| | 2 | NA | NA | 28 | NA | NA | 52 | 137 | 139 | 85 | 202 | 240 | 125 | 258 | 343 | 177 | 323 | 465 | 238 | 397 | 607 | 309 | 540 | 766 | 391 |
| | 3 | NA | NA | 34 | NA | NA | 62 | 143 | 156 | 98 | 210 | 264 | 145 | 266 | 376 | 205 | 332 | 509 | 274 | 407 | 663 | 356 | 551 | 838 | 450 |
| 10 | 1 | NA | NA | 22 | NA | NA | 41 | 130 | 151 | 70 | 202 | 225 | 106 | 267 | 316 | 151 | 333 | 434 | 213 | 410 | 571 | 273 | 558 | 727 | 343 |
| | 2 | NA | NA | 29 | NA | NA | 53 | 136 | 150 | 86 | 210 | 255 | 128 | 276 | 358 | 181 | 343 | 489 | 244 | 420 | 640 | 317 | 569 | 813 | 403 |
| | 3 | NA | NA | 34 | 97 | 102 | 62 | 143 | 166 | 99 | 217 | 277 | 147 | 284 | 389 | 207 | 352 | 530 | 279 | 430 | 694 | 363 | 580 | 880 | 459 |
| 15 | 1 | NA | NA | 23 | NA | NA | 43 | 129 | 151 | 73 | 199 | 271 | 112 | 268 | 376 | 161 | 349 | 502 | 225 | 445 | 646 | 291 | 623 | 808 | 366 |
| | 2 | NA | NA | 30 | 92 | 103 | 54 | 135 | 170 | 88 | 207 | 295 | 132 | 277 | 411 | 189 | 359 | 548 | 256 | 456 | 706 | 334 | 634 | 884 | 424 |
| | 3 | NA | NA | 34 | 96 | 112 | 63 | 141 | 185 | 101 | 215 | 315 | 151 | 286 | 439 | 213 | 368 | 586 | 289 | 466 | 755 | 378 | 646 | 945 | 479 |
| 20 | 1 | NA | NA | 23 | 87 | 99 | 45 | 128 | 167 | 76 | 197 | 303 | 117 | 265 | 425 | 169 | 345 | 569 | 235 | 439 | 734 | 306 | 614 | 921 | 387 |
| | 2 | NA | NA | 30 | 91 | 111 | 55 | 134 | 185 | 90 | 205 | 325 | 136 | 274 | 455 | 195 | 355 | 610 | 266 | 450 | 787 | 348 | 627 | 986 | 443 |
| | 3 | NA | NA | 35 | 96 | 119 | 64 | 140 | 199 | 103 | 213 | 343 | 154 | 282 | 481 | 219 | 365 | 644 | 298 | 461 | 831 | 391 | 639 | 1042 | 496 |
| 30 | 1 | NA | NA | 24 | 86 | 108 | 47 | 126 | 187 | 80 | 193 | 347 | 124 | 259 | 492 | 183 | 338 | 665 | 250 | 430 | 864 | 330 | 600 | 1089 | 421 |
| | 2 | NA | NA | 31 | 91 | 119 | 57 | 132 | 203 | 93 | 201 | 366 | 142 | 269 | 518 | 205 | 348 | 699 | 282 | 442 | 908 | 372 | 613 | 1145 | 473 |
| | 3 | NA | NA | 35 | 95 | 127 | 65 | 138 | 216 | 105 | 209 | 381 | 160 | 277 | 540 | 229 | 358 | 729 | 312 | 452 | 946 | 412 | 626 | 1193 | 524 |
| 50 | 1 | NA | NA | 24 | 85 | 113 | 50 | 124 | 204 | 87 | 188 | 392 | 139 | 252 | 567 | 208 | 328 | 778 | 287 | 417 | 1022 | 383 | 582 | 1302 | 492 |
| | 2 | NA | NA | 31 | 89 | 123 | 60 | 130 | 218 | 100 | 196 | 408 | 158 | 262 | 588 | 230 | 339 | 806 | 320 | 429 | 1058 | 425 | 596 | 1346 | 545 |
| | 3 | NA | NA | 35 | 94 | 131 | 68 | 136 | 231 | 112 | 205 | 422 | 176 | 271 | 607 | 255 | 349 | 831 | 351 | 440 | 1090 | 466 | 610 | 1386 | 597 |
| 100 | 1 | NA | NA | 23 | 84 | 104 | 49 | 122 | 200 | 89 | 182 | 410 | 151 | 243 | 617 | 232 | 315 | 875 | 328 | 402 | 1181 | 444 | 560 | 1537 | 580 |
| | 2 | NA | NA | 30 | 88 | 115 | 59 | 127 | 215 | 102 | 190 | 425 | 169 | 253 | 636 | 254 | 326 | 899 | 361 | 415 | 1210 | 488 | 575 | 1570 | 634 |
| | 3 | NA | NA | 34 | 93 | 124 | 67 | 133 | 228 | 115 | 199 | 438 | 188 | 262 | 654 | 279 | 337 | 921 | 392 | 427 | 1238 | 529 | 589 | 1604 | 687 |

***Common Vent Capacity***

| | *Minimum Internal Area of Masonry Chimney Flue (in.²)* | | | | | | | | | | | | | | | | | | | | | | | |
|---|---|---|---|---|---|---|---|---|---|---|---|---|---|---|---|---|---|---|---|---|---|---|---|---|
| | *12* | | | *19* | | | *28* | | | *38* | | | *50* | | | *63* | | | *78* | | | *113* | | |
| *Vent Height H (ft)* | *Combined Appliance Input Rating in Thousands of Btu per Hour* | | | | | | | | | | | | | | | | | | | | | | | |
| | *FAN +FAN* | *FAN +NAT* | *NAT +NAT* | *FAN +FAN* | *FAN +NAT* | *NAT +NAT* | *FAN +FAN* | *FAN +NAT* | *NAT +NAT* | *FAN +FAN* | *FAN +NAT* | *NAT +NAT* | *FAN +FAN* | *FAN +NAT* | *NAT +NAT* | *FAN +FAN* | *FAN +NAT* | *NAT +NAT* | *FAN +FAN* | *FAN +NAT* | *NAT +NAT* | *FAN +FAN* | *FAN +FAN* | *NAT +NAT* |
| 6 | NA | NA | 25 | NA | 118 | 45 | NA | 176 | 71 | NA | 255 | 102 | NA | 348 | 142 | NA | 455 | 187 | NA | 579 | 245 | NA | 846 | NA |
| 8 | NA | NA | 28 | NA | 128 | 52 | NA | 190 | 81 | NA | 276 | 118 | NA | 380 | 162 | NA | 497 | 217 | NA | 633 | 277 | 1136 | 928 | 405 |
| 10 | NA | NA | 31 | NA | 136 | 56 | NA | 205 | 89 | NA | 295 | 129 | NA | 405 | 175 | NA | 532 | 234 | 771 | 680 | 300 | 1216 | 1000 | 450 |
| 15 | NA | NA | 36 | NA | NA | 66 | NA | 230 | 105 | NA | 335 | 150 | NA | 400 | 210 | 677 | 602 | 280 | 866 | 772 | 360 | 1359 | 1139 | 540 |
| 20 | NA | NA | NA | NA | NA | 74 | NA | 247 | 120 | NA | 362 | 170 | NA | 503 | 240 | 765 | 661 | 321 | 947 | 849 | 415 | 1495 | 1264 | 640 |
| 30 | NA | NA | NA | NA | NA | NA | NA | NA | 135 | NA | 398 | 195 | NA | 558 | 275 | 808 | 739 | 377 | 1052 | 957 | 490 | 1682 | 1447 | 740 |
| 50 | NA | NA | NA | NA | NA | NA | NA | NA | NA | NA | NA | NA | NA | 612 | 325 | NA | 821 | 456 | 1152 | 1076 | 600 | 1879 | 1672 | 910 |
| 100 | NA | NA | NA | NA | NA | NA | NA | NA | NA | NA | NA | NA | NA | NA | NA | NA | NA | 494 | NA | NA | 663 | 2006 | 1885 | 1046 |

For SI units, 1 in. = 25.4 mm, 1 in.² = 645 mm², 1 ft = 0.305 m, 1000 Btu/hr = 0.293 kW.

***TABLE 13.2(e)*** *Single-Wall Metal Pipe or Type B Asbestos Cement Vent*

| | |
|---|---|
| *Number of Appliances:* | *Two or More* |
| *Appliance Type:* | *Draft Hood–Equipped* |
| *Appliance Vent Connection:* | *Direct to Pipe or Vent* |

*Vent Connector Capacity*

| *Total Vent Height H (ft)* | *Connector Rise R (ft)* | *Vent Connector Diameter — D (in.)* | | | | | |
|---|---|---|---|---|---|---|---|
| | | *3* | *4* | *5* | *6* | *7* | *8* |
| | | *Maximum Appliance Input Rating in Thousands of Btu per Hour* | | | | | |
| 6–8 | 1 | 21 | 40 | 68 | 102 | 146 | 205 |
| | 2 | 28 | 53 | 86 | 124 | 178 | 235 |
| | 3 | 34 | 61 | 98 | 147 | 204 | 275 |
| 15 | 1 | 23 | 44 | 77 | 117 | 179 | 240 |
| | 2 | 30 | 56 | 92 | 134 | 194 | 265 |
| | 3 | 35 | 64 | 102 | 155 | 216 | 298 |
| 30 and up | 1 | 25 | 49 | 84 | 129 | 190 | 270 |
| | 2 | 31 | 58 | 97 | 145 | 211 | 295 |
| | 3 | 36 | 68 | 107 | 164 | 232 | 321 |

*Common Vent Capacity*

| *Total Vent Height H (ft)* | *Common Vent Diameter — D (in.)* | | | | | | |
|---|---|---|---|---|---|---|---|
| | *4* | *5* | *6* | *7* | *8* | *10* | *12* |
| | *Combined Appliance Input Rating in Thousands of Btu per Hour* | | | | | | |
| 6 | 48 | 78 | 111 | 155 | 205 | 320 | NA |
| 8 | 55 | 89 | 128 | 175 | 234 | 365 | 505 |
| 10 | 59 | 95 | 136 | 190 | 250 | 395 | 560 |
| 15 | 71 | 115 | 168 | 228 | 305 | 480 | 690 |
| 20 | 80 | 129 | 186 | 260 | 340 | 550 | 790 |
| 30 | NA | 147 | 215 | 300 | 400 | 650 | 940 |
| 50 | NA | NA | NA | 360 | 490 | 810 | 1190 |

For SI units, 1 in. = 25.4 mm, 1 in.$^2$ = 645 mm$^2$, 1 ft = 0.305 m, 1000 Btu/hr = 0.293 kW.

Note: See Figure G.1(f) and Section 13.2.

***TABLE 13.2(f)*** *Exterior Masonry Chimney*

| | |
|---|---|
| *Number of Appliances:* | *Two or More* |
| *Appliance Type:* | *NAT + NAT* |
| *Appliance Vent Connection:* | *Type B Double-Wall Connector* |

***SPECIAL USE: Combined Appliance Maximum Input Rating in Thousands of Btu per Hour***

| *Vent Height H (ft)* | *Internal Area of Chimney (in.$^2$)* | | | | | | | |
|---|---|---|---|---|---|---|---|---|
| | *12* | *19* | *28* | *38* | *50* | *63* | *78* | *113* |
| 6 | 25 | 46 | 71 | 103 | 143 | 188 | 246 | NA |
| 8 | 28 | 53 | 82 | 119 | 163 | 218 | 278 | 408 |
| 10 | 31 | 56 | °90 | 131 | 177 | 236 | 302 | 454 |
| 15 | NA | 67 | 106 | 152 | 212 | 283 | 365 | 546 |
| 20 | NA | NA | NA | NA | NA | 325 | 419 | 648 |
| 30 | NA | NA | NA | NA | NA | NA | 496 | 749 |
| 50 | NA | NA | NA | NA | NA | NA | NA | 922 |
| 100 | NA | NA | NA | NA | NA | NA | NA | NA |

For SI units, 1 in. = 25.4 mm, 1 in.$^2$ = 645 mm$^2$, 1 ft = 0.305 m, 1000 Btu/hr = 0.293 kW.

***TABLE 13.2(g)*** *Exterior Masonry Chimney*

| | |
|---|---|
| *Number of Appliances:* | *Two or More* |
| *Appliance Type:* | *NAT + NAT* |
| *Appliance Vent Connection:* | *Type B Double-Wall Connector* |

***SPECIAL USE: Minimum Allowable Input Rating of Space-Heating Appliance in Thousands of Btu per Hour***

| *Vent Height H (ft)* | *Internal Area of Chimney (in.$^2$)* | | | | | | | |
|---|---|---|---|---|---|---|---|---|
| | *12* | *19* | *28* | *38* | *50* | *63* | *78* | *113* |
| | Local 99% winter design temperature: 37°F or greater | | | | | | | |
| 6 | 0 | 0 | 0 | 0 | 0 | 0 | 0 | NA |
| 8 | 0 | 0 | 0 | 0 | 0 | 0 | 0 | 0 |
| 10 | 0 | 0 | 0 | 0 | 0 | 0 | 0 | 0 |
| 15 | NA | 0 | 0 | 0 | 0 | 0 | 0 | 0 |
| 20 | NA | NA | NA | NA | NA | 184 | 0 | 0 |
| 30 | NA | NA | NA | NA | NA | 393 | 334 | 0 |
| 50 | NA | NA | NA | NA | NA | NA | NA | 579 |
| 100 | NA | NA | NA | NA | NA | NA | NA | NA |
| | Local 99% winter design temperature: 27°F to 36°F | | | | | | | |
| 6 | 0 | 0 | 68 | NA | NA | 180 | 212 | NA |
| 8 | 0 | 0 | 82 | NA | NA | 187 | 214 | 263 |
| 10 | 0 | 51 | NA | NA | NA | 201 | 225 | 265 |
| 15 | NA | NA | NA | NA | NA | 253 | 274 | 305 |
| 20 | NA | NA | NA | NA | NA | 307 | 330 | 362 |
| 30 | NA | NA | NA | NA | NA | NA | 445 | 485 |
| 50 | NA | NA | NA | NA | NA | NA | NA | 763 |
| 100 | NA | NA | NA | NA | NA | NA | NA | NA |
| | Local 99% winter design temperature: 17°F to 26°F | | | | | | | |
| 6 | NA | NA | NA | NA | NA | NA | NA | NA |
| 8 | NA | NA | NA | NA | NA | NA | 264 | 352 |
| 10 | NA | NA | NA | NA | NA | NA | 278 | 358 |
| 15 | NA | NA | NA | NA | NA | NA | 331 | 398 |
| 20 | NA | NA | NA | NA | NA | NA | 387 | 457 |
| 30 | NA | NA | NA | NA | NA | NA | NA | 581 |
| 50 | NA | NA | NA | NA | NA | NA | NA | 862 |
| 100 | NA | NA | NA | NA | NA | NA | NA | NA |
| | Local 99% winter design temperature: 5°F to 16°F | | | | | | | |
| 6 | NA | NA | NA | NA | NA | NA | NA | NA |
| 8 | NA | NA | NA | NA | NA | NA | NA | NA |
| 10 | NA | NA | NA | NA | NA | NA | NA | 430 |
| 15 | NA | NA | NA | NA | NA | NA | NA | 485 |
| 20 | NA | NA | NA | NA | NA | NA | NA | 547 |
| 30 | NA | NA | NA | NA | NA | NA | NA | 682 |
| 50 | NA | NA | NA | NA | NA | NA | NA | NA |
| 100 | NA | NA | NA | NA | NA | NA | NA | NA |
| | Local 99% winter design temperature: 4°F or lower<br>Not recommended for any vent configurations | | | | | | | |

For SI units, 1 in. = 25.4 mm, 1 in.$^2$ = 645 mm$^2$, 1 ft = 0.305 m, 1000 Btu/hr = 0.293 kW, °C = (°F − 32)/1.8.

Note: See Figure G.2.4 for a map showing local 99 percent winter design temperatures in the United States.

*TABLE 13.2(h) Exterior Masonry Chimney*

| | | | | | Number of Appliances: | Two or More | | |
|---|---|---|---|---|---|---|---|---|
| | | | | | Appliance Type: | FAN + NAT | | |
| | | | | | Appliance Vent Connection: | Type B Double-Wall Connector | | |
| SPECIAL USE: Combined Appliance Maximum Input Rating in Thousands of Btu per Hour | | | | | | | | |
| Vent Height H (ft) | Internal Area of Chimney (in.$^2$) | | | | | | | |
| | 12 | 19 | 28 | 38 | 50 | 63 | 78 | 113 |
| 6 | 74 | 119 | 178 | 257 | 351 | 458 | 582 | 853 |
| 8 | 80 | 130 | 193 | 279 | 384 | 501 | 636 | 937 |
| 10 | 84 | 138 | 207 | 299 | 409 | 538 | 686 | 1010 |
| 15 | NA | 152 | 233 | 334 | 467 | 611 | 781 | 1156 |
| 20 | NA | NA | 250 | 368 | 508 | 668 | 858 | 1286 |
| 30 | NA | NA | NA | 404 | 564 | 747 | 969 | 1473 |
| 50 | NA | NA | NA | NA | NA | 831 | 1089 | 1692 |
| 100 | NA | NA | NA | NA | NA | NA | NA | 1921 |

For SI units, 1 in. = 25.4 mm, 1 in.$^2$ = 645 mm$^2$, 1 ft = 0.305 m, 1000 Btu/hr = 0.293 kW.

*TABLE 13.2(i) Exterior Masonry Chimney*

| | |
|---|---|
| *Number of Appliances:* | *Two or More* |
| *Appliance Type:* | *FAN + NAT* |
| *Appliance Vent Connection:* | *Type B Double-Wall Connector* |

*SPECIAL USE: Minimum Allowable Input Rating of Space-Heating Appliance in Thousands of Btu per Hour*

| *Vent Height H (ft)* | *Internal Area of Chimney (in.²)* | | | | | | | |
|---|---|---|---|---|---|---|---|---|
| | *12* | *19* | *28* | *38* | *50* | *63* | *78* | *113* |
| | Local 99% winter design temperature: 37°F or greater | | | | | | | |
| 6 | 0 | 0 | 0 | 0 | 0 | 0 | 0 | 0 |
| 8 | 0 | 0 | 0 | 0 | 0 | 0 | 0 | 0 |
| 10 | 0 | 0 | 0 | 0 | 0 | 0 | 0 | 0 |
| 15 | NA | 0 | 0 | 0 | 0 | 0 | 0 | 0 |
| 20 | NA | NA | 123 | 190 | 249 | 184 | 0 | 0 |
| 30 | NA | NA | NA | 334 | 398 | 393 | 334 | 0 |
| 50 | NA | NA | NA | NA | NA | 714 | 707 | 579 |
| 100 | NA | NA | NA | NA | NA | NA | NA | 1600 |
| | Local 99% winter design temperature: 27°F to 36°F | | | | | | | |
| 6 | 0 | 0 | 68 | 116 | 156 | 180 | 212 | 266 |
| 8 | 0 | 0 | 82 | 127 | 167 | 187 | 214 | 263 |
| 10 | 0 | 51 | 97 | 141 | 183 | 210 | 225 | 265 |
| 15 | NA | 111 | 142 | 183 | 233 | 253 | 274 | 305 |
| 20 | NA | NA | 187 | 230 | 284 | 307 | 330 | 362 |
| 30 | NA | NA | NA | 330 | 319 | 419 | 445 | 485 |
| 50 | NA | NA | NA | NA | NA | 672 | 705 | 763 |
| 100 | NA | NA | NA | NA | NA | NA | NA | 1554 |
| | Local 99% winter design temperature: 17°F to 26°F | | | | | | | |
| 6 | 0 | 55 | 99 | 141 | 182 | 215 | 259 | 349 |
| 8 | 52 | 74 | 111 | 154 | 197 | 226 | 264 | 352 |
| 10 | NA | 90 | 125 | 169 | 214 | 245 | 278 | 358 |
| 15 | NA | NA | 167 | 212 | 263 | 296 | 331 | 398 |
| 20 | NA | NA | 212 | 258 | 316 | 352 | 387 | 457 |
| 30 | NA | NA | NA | 362 | 429 | 470 | 507 | 581 |
| 50 | NA | NA | NA | NA | NA | 723 | 766 | 862 |
| 100 | NA | NA | NA | NA | NA | NA | NA | 1669 |
| | Local 99% winter design temperature: 5°F to 16°F | | | | | | | |
| 6 | NA | 78 | 121 | 166 | 214 | 252 | 301 | 416 |
| 8 | NA | 94 | 135 | 182 | 230 | 269 | 312 | 423 |
| 10 | NA | 111 | 149 | 198 | 250 | 289 | 331 | 430 |
| 15 | NA | NA | 193 | 247 | 305 | 346 | 393 | 485 |
| 20 | NA | NA | NA | 293 | 360 | 408 | 450 | 547 |
| 30 | NA | NA | NA | 377 | 450 | 531 | 580 | 682 |
| 50 | NA | NA | NA | NA | NA | 797 | 853 | 972 |
| 100 | NA | NA | NA | NA | NA | NA | NA | 1833 |
| | Local 99% winter design temperature: −10°F to 4°F | | | | | | | |
| 6 | NA | NA | 145 | 196 | 249 | 296 | 349 | 484 |
| 8 | NA | NA | 159 | 213 | 269 | 320 | 371 | 494 |
| 10 | NA | NA | 175 | 231 | 292 | 339 | 397 | 513 |
| 15 | NA | NA | NA | 283 | 351 | 404 | 457 | 586 |
| 20 | NA | NA | NA | 333 | 408 | 468 | 528 | 650 |
| 30 | NA | NA | NA | NA | NA | 603 | 667 | 805 |
| 50 | NA | NA | NA | NA | NA | NA | 955 | 1003 |
| 100 | NA | NA | NA | NA | NA | NA | NA | NA |
| | Local 99% winter design temperature: −11°F or lower<br>Not recommended for any vent configurations | | | | | | | |

For SI units, 1 in. = 25.4 mm, 1 in.$^2$ = 645 mm$^2$, 1 ft = 0.305 m, 1000 Btu/hr = 0.293 kW.

Note: See Figure G.2.4 for a map showing local 99 percent winter design temperatures in the United States.

# Explanatory Material

## ANNEX A

*Annex A is not a part of the requirements of this NFPA document but is included for informational purposes only. This annex contains explanatory material, numbered to correspond with the applicable text paragraphs.*

The material contained in Annex A of NFPA 54 is not part of the requirements of this code but is included with the code for informational purposes only. For the convenience of readers of this handbook, Annex A text is interspersed among the text of Chapter 1 through Chapter 13 and, therefore, is not repeated here.

# Coordination of Appliance and Equipment Design, Construction, and Maintenance

ANNEX B

*This annex is not a part of the requirements of this NFPA document but is included for informational purposes only.*

Annex B provides a mechanism for coordination of the design, construction, and maintenance of industrial gas appliances. Annex B text applies to specialized gas equipment that is not mass produced. This annex coordinates the work of the designer of the equipment, the installer, the gas supplier, and the appliance manufacturer. A checklist is provided, and provisions for maintenance are addressed. Although this annex is not used often, it is a valuable tool when new, customized gas appliances are installed, which usually occurs in industrial applications.

## B.1 Coordination

**B.1.1** Because industrial gas applications are so varied in nature, many agencies are jointly involved with their safe and satisfactory use. Prior to installation, the specific assignments should be agreed upon by the parties concerned. A typical, but not mandatory, delineation of assignments is given in B.1.2 through B.1.5, and a detailed checklist is given in Section B.2.

**B.1.2** The person or agency planning an installation of appliances and equipment does the following:

(1) Verifies the adequacy of the gas supply, volume, pressure, and meter location
(2) Determines suitability of gas for the process
(3) Notifies gas suppliers of significant changes in requirements

**B.1.3** Upon request, the gas supplier furnishes the user complete information on the following:

(1) Combustion characteristics and physical or chemical properties such as specific gravity, heating value, pressure, and the approximate analysis of the gas
(2) Conditions under which an adequate supply of gas at suitable pressure can be brought to the site
(3) Continuity of the gas supply

**B.1.4** The appliance or equipment manufacturer or builder provides the following:

(1) Design and construction of all appliances or equipment or assemblies shipped from its plant
(2) Design and construction of all appliances or equipment fabricated, erected, or assembled by the appliances or equipment manufacturer or builder in the field
(3) A statement of the maximum hourly Btu input, type of gas, and design pressure range
(4) Written installation and operating instructions for the user

**B.1.5** The person or agency installing the appliances or equipment and the person or agency authorizing the installation of appliances or equipment (purchaser) jointly should do the following:

(1) Select, erect, or assemble appliances and equipment, components, or designs purchased or developed by that person or agency
(2) Ensure conformance to codes, ordinances, or regulations applicable to the installation
(3) Provide adequate means of disposal of products of combustion
(4) Initially operate the appliances or equipment in a safe manner

## B.2 Appliance and Equipment Design and Construction Checklist

**B.2.1** The basic design and installation should consider the following:

(1) Suitability of appliance and equipment for process requirements
(2) Adequate structural strength and stability
(3) Reasonable life expectancy
(4) Conformance to existing safety standards
(5) Adequate combustion space and venting
(6) Means for observation and inspection of combustion

**B.2.2** Materials of construction used, other than pipe, fittings, and valves, should provide reasonable life expectancy for the service intended and should be capable of satisfactorily withstanding the following:

(1) Operating temperatures
(2) Chemical action
(3) Thermal shock
(4) Load stresses

**B.2.3** Combustion systems should be selected for the characteristics of the available gas so that they operate properly at the elevation at point of use and produce the following:

(1) Proper heat distribution
(2) Adequate operating temperature range
(3) Suitable flame geometry
(4) Flame stability
(5) Operating flexibility
(6) Desired heating chamber atmosphere

**B.2.4** Pipe, fittings, and valves should conform to applicable American National Standards as indicated in Section 5.6. Piping, bushings, and material in fittings should not be selected or used until the following factors have been considered:

(1) Correct size to handle required volume (consideration of pressure drop in controls and manifolds is particularly important in low-pressure systems)
(2) Material specifications suitable for pressures and temperatures encountered
(3) Adequate supports and protection against physical damage
(4) Tight assembly and thorough leak inspection
(5) Use of sufficient unions and flanges, where permitted, for convenient field replacement or repair
(6) Arrangement of piping to provide accessibility for appliance and equipment adjustments and freedom from thermal damage

**B.2.5** Information concerning the characteristics of the gas and electricity available at the point of utilization should be specific and complete. Gas controls and electrical systems should be selected to conform to these characteristics, which include the following:

(1) Gas characteristics: Heat content, pressure, specific gravity, and approximate analysis
(2) Electrical characteristics: Voltages, number of phases, and frequencies for both control and power circuits
(3) Location of electrical equipment and wiring to avoid thermal damage and excessive concentrations of dust, dirt, or foreign material
(4) Requirements of applicable electrical codes and standards, with particular reference to *NFPA 70, National Electrical Code*

*NFPA 70®*, *National Electrical Code®*, is referenced in B.2.5(4). *NFPA 70®*, particularly in its Article 500, Hazardous (Classified) Locations, Classes I, II, and III, Divisions 1 and 2, provides requirements for electrical equipment and wiring in locations where fire or explosion hazards could exist due to flammable gases or vapors, flammable liquids, combustible dust, or ignitible fibers or flyings.

**B.2.6** Temperature controls, if used, should be selected, with consideration of the following:

(1) Range and type of instruments and sensing elements
(2) Type of control action
(3) Suitability for service required
(4) Correlation of control instruments with operating equipment

**B.2.7** In enclosed chambers, the accumulation of gas–air or solvent–air mixtures that can be accidentally ignited constitutes a potential hazard to life and property. For this reason, consideration should be given to the selection and installation of suitable protective equipment. The selection of a satisfactory protective system and components not otherwise covered by existing codes or standards should be based on the requirements of each individual installation after consultation with the various interested parties, including the user, designer, insurance company, and local authorities having jurisdiction. Factors and considerations involved in the selection of protective equipment include the following:

(1) Feasibility of its installation
(2) Its adaptability to process and control requirements
(3) Conformance to existing standards, ordinances, requirements, and other regulations that apply *(See Annex M for the listing of standards and specifications.)*

Boilers, ovens, and furnaces are examples of equipment in which accumulations of gas–air mixtures could ignite, as discussed in B.2.7. Consult the following standards for additional information on this topic:

- NFPA 85, *Boiler and Combustion Systems Hazards Code*
- NFPA 86, *Standard for Ovens and Furnaces*

## B.3 Maintenance of Appliances and Equipment

**B.3.1** These recommendations are prepared for maintenance of appliances and equipment. Special types of appliances and equipment demand special attention.

**B.3.2** Burners and pilots should be kept clean and in proper operating condition. Burner refractory parts should be examined at frequent regular intervals to ensure good condition.

Subsection B.3.2 recommends that refractory linings be thoroughly inspected for spalling, cracking, and voids. Annual inspections are common. Cracks, spalls, and voids are imperfections in refractory linings. If any defects are observed, a qualified inspector should inspect the linings. The appliance manufacturer should be consulted for repair procedures if any defects are found upon inspection.

**B.3.3** Where automatic flame safeguards are used, a complete shutdown and restart should be made at frequent intervals to check the components for proper operation.

### B.3.4 Other Safeguard Equipment.

**B.3.4.1** Accessory safeguard equipment, such as manual reset valves with pressure or vacuum switches, high-temperature limit switches, draft controls, shutoff valves, airflow switches, door switches, and gas valves, should be operated at frequent regular intervals to ensure proper functioning. If inoperative, they should be repaired or replaced promptly.

**B.3.4.2** Where firechecks are installed in gas–air mixture piping to prevent flashbacks from traveling farther upstream, the pressure loss across the firechecks should be measured at regular intervals. When excessive pressure loss is found, screens should be removed and cleaned. Water-type backfire checks should be inspected at frequent regular intervals and liquid level maintained.

**B.3.4.3** All safety shutoff valves should be checked for leakage and proper operation at frequent regular intervals.

### B.3.5 Auxiliary Devices.

**B.3.5.1** A necessary part of the appliance or equipment maintenance is the proper maintenance of auxiliary devices. Maintenance instructions as supplied by the manufacturers of these devices should be followed.

**B.3.5.2** Gas combustion systems, including blowers, mechanical mixers, control valves, temperature control instruments, air valves, and air filters, should be kept clean and should be examined at frequent regular intervals.

**B.3.5.3** Necessary repairs and replacements should be made promptly.

**B.3.6** Regulator and zero governor vents and impulse or control piping and tubing should be kept clear. Regulator valves that operate improperly should be cleaned, repaired, or replaced promptly.

**B.3.7** A necessary part of the appliance or equipment maintenance is the proper maintenance of the gas piping system. It is recommended that gas piping be inspected and tested for leakage at regular intervals in accordance with the provisions of 8.1.5. Air piping should be kept internally clean to prevent accumulation of dust, lint, and grease in air jets and valves. Where conditions warrant, filters should be installed at the intake to the fans.

**B.3.8** Standby or substitute fuel equipment and systems for appliances or equipment should be kept in good operating condition and tested periodically.

**B.3.9** An adequate supply of repair parts should be maintained.

## REFERENCES CITED IN COMMENTARY

The following publications are available from the National Fire Protection Association, 1 Batterymarch Park, Quincy, MA 02169-7471, 1-800-344-3555, www.nfpa.org.

*NFPA 70®*, *National Electrical Code®*, 2008 edition.
NFPA 85, *Boiler and Combustion Systems Hazards Code*, 2007 edition.
NFPA 86, *Standard for Ovens and Furnaces*, 2007 edition.

# Sizing and Capacities of Gas Piping

## ANNEX C

*This annex is not a part of the requirements of this NFPA document but is included for informational purposes only.*

Annex C provides additional information on the sizing of gas piping systems beyond that provided in Chapter 5 and Chapter 6. Instructions are provided for sizing using both the longest length method (C.4.1) and the branch length method (C.4.2).

In addition, the hybrid pressure method, used for systems with two or more pressures (and a pressure regulator), is discussed in C.4.3. Pipe sizing examples also are provided for these methods in C.8.1, C.8.2, and C.8.3 and for sizing existing systems that have been modified through the addition of new corrugated stainless steel tubing (CSST) branches.

## C.1 Sizing Factors

The first goal of determining the pipe sizing of a fuel gas piping system is to be assured that the gas pressure at the inlet to each appliance is sufficient. The majority of systems are residential, and the appliances all have the same, or nearly the same, requirement for minimum gas pressure at the appliance inlet. This pressure is about 5 in. (1.2 kPa) w.c., which is enough for proper operation of the appliance regulator to deliver about 3.5 in. (0.87 kPa) w.c. to the burner itself. The pressure drop in the piping is subtracted from the source delivery pressure to verify that the minimum is available at the appliance.

There are other systems, however, where the required inlet pressure to the different appliances may be quite varied. In such cases, the greatest inlet pressure required must be satisfied, as well as the farthest appliance, which is almost always the critical appliance in small systems.

There is an additional requirement to be observed besides the capacity of the system at 100 percent flow. That requirement is that at minimum flow, the pressure at the inlet to any appliance does not exceed the pressure rating of the appliance regulator. This factor would seldom be of concern in small systems if the source pressure is $^1/_2$ psi (14 in. w.c.) (3.4 kPa) or less, but it should be verified for systems with greater gas pressure at the point of supply.

Section C.1 recognizes that each appliance must be supplied with the minimum gas pressure required by the appliance for proper operation at its inlet. The farthest appliance and the highest appliance inlet demands must be met.

## C.2 General Pipe Sizing Considerations

To determine the size of piping used in a gas piping system, the following factors must be considered:

(1) Allowable loss in pressure from point of delivery to appliance
(2) Maximum gas demand
(3) Length of piping and number of fittings

(4) Specific gravity of the gas
(5) Diversity factor

For any gas piping system, or special appliance, or for conditions other than those covered by the tables provided in this code, such as longer runs, greater gas demands, or greater pressure drops, the size of each gas piping system should be determined by standard engineering practices acceptable to the authority having jurisdiction.

The "standard engineering practices" addressed in Section C.2 include calculations using flow equations recognized in engineering textbooks.

## C.3 Description of Tables

The tables in Chapter 6 provide the designer with information for piping flow capacity for gas of 0.60 specific gravity (i.e., natural gas) and for propane. The tables are organized in the following order:

1. Gas covered (natural gas, propane)
2. Piping material (steel, copper, CSST, polyethylene)
3. Pressure (lowest to highest)

Tables 6.2(a) through 6.2(v) are used for natural gas (0.60 specific gravity). Tables 6.3(a) through 6.3(m) are used for propane. For each gas, the tables for metallic tubing follow the tables for steel pipe and tables for CSST and polyethylene tubing follow the tables for metallic tubing.

Table 5.4.2.1 in Chapter 5 of the code shows approximate gas input for various appliances. Table C.3.2 shows equivalent lengths for fittings. Note, however, that if the system contains "typical" fittings, no allowance for fittings needs to be made to use the pipe sizing tables.

The capacities of Tables 6.2(a) through 6.2(v) are given in standard cubic feet per hour, so the heating capacity can be obtained by multiplying the tabular value by the heating value of the gas. Because the heating value of natural gas is approximately 1000 Btu/ft$^3$ (37.5 MJ/m$^3$), the tabular values can be taken as the heating value in MBH (thousands of Btu per hour) for natural gas.

Tables 6.3(a) through 6.3(m) for propane capacities are in thousands of Btu per hour, which is commonly used for propane system calculations. Older editions of the code included a table in Chapter 6 to convert to gases of different specific gravities.

The pipe sizing tables include flow capacities for several initial pressures and for various pressure losses, so most smaller jobs will be covered by one of the tables. The explanation in Section C.4 and the examples in Section C.8 illustrate how to use the tables.

### C.3.1 General.

The quantity of gas to be provided at each outlet should be determined, whenever possible, directly from the manufacturer's gas input Btu/hr rating of the appliance to be installed. In case the ratings of the appliances to be installed are not known, Table 5.4.2.1 shows the approximate consumption (in Btu per hour) of certain types of typical household appliances.

To obtain the cubic feet per hour of gas required, divide the total Btu/hr input of all appliances by the average Btu heating value per cubic foot of the gas. The average Btu per cubic foot of the gas in the area of the installation can be obtained from the serving gas supplier.

### C.3.2 Low-Pressure Natural Gas Tables.

Capacities for gas at low pressure [2.0 psi (14 kPa gauge) or less] in cubic feet per hour of 0.60 specific gravity gas for different sizes and lengths are shown in Table 6.2(a) and Table

6.2(b) for iron pipe or equivalent rigid pipe, in Table 6.2(f) through Table 6.2(h) for smooth wall semirigid tubing, and in Table 6.2(m) and Table 6.2(o) for corrugated stainless steel tubing. Table 6.2(a) and Table 6.2(f) are based on a pressure drop of 0.3 in. w.c. (75 Pa), whereas Table 6.2(b), Table 6.2(g), and Table 6.2(m) are based on a pressure drop of 0.5 in. w.c. (125 Pa). Table 6.2(h), Table 6.2(n), and Table 6.2(o) are special low-pressure applications based on pressure drops greater than 0.5 in. w.c. (125 Pa). In using Table 6.2(h), Table 6.2(n), or Table 6.2(o), an allowance (in equivalent length of pipe) should be considered for any piping run with four or more fittings *(see Table C.3.2)*.

### C.3.3 Undiluted LP-Gas Tables.

Capacities in thousands of Btu per hour of undiluted LP-Gases based on a pressure drop of 0.5 in. w.c. (125 Pa) for different sizes and lengths are shown in Table 6.3(d) for iron pipe or equivalent rigid pipe, in Table 6.3(f) for smooth wall semirigid tubing, in Table 6.3(h) for corrugated stainless steel tubing, and in Table 6.3(k) and Table 6.3(m) for polyethylene plastic pipe and tubing. Table 6.3(i) and Table 6.3(j) for corrugated stainless steel tubing and Table 6.3(l) for polyethylene plastic pipe are based on operating pressures greater than 0.5 psi (3.5 kPa) and pressure drops greater than 0.5 in. w.c. (125 Pa). In using these tables, an allowance (in equivalent length of pipe) should be considered for any piping run with four or more fittings *(see Table C.3.2)*.

### C.3.4 Natural Gas Specific Gravity.

Gas piping systems that are to be supplied with gas of a specific gravity of 0.70 or less can be sized directly from the tables provided in this code, unless the authority having jurisdiction specifies that a gravity factor be applied. Where the specific gravity of the gas is greater than 0.70, the gravity factor should be applied.

Application of the gravity factor converts the figures given in the tables provided in this code to capacities for another gas of different specific gravity. Such application is accomplished by multiplying the capacities given in the tables by the multipliers shown in Table C.3.4. In case the exact specific gravity does not appear in the table, choose the next higher value specific gravity shown.

### C.3.5 Higher Pressure Natural Gas Tables.

Capacities for gas at pressures of 2 psi and greater in cubic feet per hour of 0.60 specific gravity gas for different sizes and lengths are shown in Table 6.2(c) and Table 6.2(d) for iron pipe or equivalent rigid pipe, Table 6.2(j) through Table 6.2(l) for semirigid tubing, Table 6.2(p) and Table 6.2(q) for corrugated stainless steel tubing, and Table 6.2(s) and Table 6.2(t) for polyethylene plastic pipe.

## C.4 Use of Capacity Tables

There are several methods that can be used to size pipes. These methods include the Longest Length Method, the Branch Length Method, the Hybrid Pressure Method, and the Pressure Drop per 100 ft Method. The pros and cons of using each method are shown in Commentary Table C.1, and the sizing methods are explained in C.4.1 through C.4.4.

### C.4.1 The Longest Length Method.

This sizing method is conservative in its approach by applying the maximum operating conditions in the system as the norm for the system and by setting the length of pipe used to size any given part of the piping system to the maximum value.

*TABLE C.3.2 Equivalent Lengths of Pipe Fittings and Valves*

| | | *Screwed Fittings*[1] | | | | *90° Welding Elbows and Smooth Bends*[2] | | | | | |
|---|---|---|---|---|---|---|---|---|---|---|---|
| | | *45°/Ell* | *90°/Ell* | *180° Close Return Bends* | *Tee* | *R/d = 1* | *R/d = 1⅓* | *R/d = 2* | *R/d = 4* | *R/d = 6* | *R/d = 8* |
| | *k factor =* | *0.42* | *0.90* | *2.00* | *1.80* | *0.48* | *0.36* | *0.27* | *0.21* | *0.27* | *0.36* |
| | *L/d′ratio*[4]*n =* | *14* | *30* | *67* | *60* | *16* | *12* | *9* | *7* | *9* | *12* |
| *Nominal Pipe Size (in.)* | *Inside Diam. d (in.), Sched. 40*[6] | | | | | | | | | | |
| | | *L = Equivalent Length in Feet of Schedule 40 (Standard Weight) Straight Pipe*[6] | | | | | | | | | |
| ½ | 0.622 | 0.73 | 1.55 | 3.47 | 3.10 | 0.83 | 0.62 | 0.47 | 0.36 | 0.47 | 0.62 |
| ¾ | 0.824 | 0.96 | 2.06 | 4.60 | 4.12 | 1.10 | 0.82 | 0.62 | 0.48 | 0.62 | 0.82 |
| 1 | 1.049 | 1.22 | 2.62 | 5.82 | 5.24 | 1.40 | 1.05 | 0.79 | 0.61 | 0.79 | 1.05 |
| 1¼ | 1.380 | 1.61 | 3.45 | 7.66 | 6.90 | 1.84 | 1.38 | 1.03 | 0.81 | 1.03 | 1.38 |
| 1½ | 1.610 | 1.88 | 4.02 | 8.95 | 8.04 | 2.14 | 1.61 | 1.21 | 0.94 | 1.21 | 1.61 |
| 2 | 2.067 | 2.41 | 5.17 | 11.5 | 10.3 | 2.76 | 2.07 | 1.55 | 1.21 | 1.55 | 2.07 |
| 2½ | 2.469 | 2.88 | 6.16 | 13.7 | 12.3 | 3.29 | 2.47 | 1.85 | 1.44 | 1.85 | 2.47 |
| 3 | 3.068 | 3.58 | 7.67 | 17.1 | 15.3 | 4.09 | 3.07 | 2.30 | 1.79 | 2.30 | 3.07 |
| 4 | 4.026 | 4.70 | 10.1 | 22.4 | 20.2 | 5.37 | 4.03 | 3.02 | 2.35 | 3.02 | 4.03 |
| 5 | 5.047 | 5.88 | 12.6 | 28.0 | 25.2 | 6.72 | 5.05 | 3.78 | 2.94 | 3.78 | 5.05 |
| 6 | 6.065 | 7.07 | 15.2 | 33.8 | 30.4 | 8.09 | 6.07 | 4.55 | 3.54 | 4.55 | 6.07 |
| 8 | 7.981 | 9.31 | 20.0 | 44.6 | 40.0 | 10.6 | 7.98 | 5.98 | 4.65 | 5.98 | 7.98 |
| 10 | 10.02 | 11.7 | 25.0 | 55.7 | 50.0 | 13.3 | 10.0 | 7.51 | 5.85 | 7.51 | 10.0 |
| 12 | 11.94 | 13.9 | 29.8 | 66.3 | 59.6 | 15.9 | 11.9 | 8.95 | 6.96 | 8.95 | 11.9 |
| 14 | 13.13 | 15.3 | 32.8 | 73.0 | 65.6 | 17.5 | 13.1 | 9.85 | 7.65 | 9.85 | 13.1 |
| 16 | 15.00 | 17.5 | 37.5 | 83.5 | 75.0 | 20.0 | 15.0 | 11.2 | 8.75 | 11.2 | 15.0 |
| 18 | 16.88 | 19.7 | 42.1 | 93.8 | 84.2 | 22.5 | 16.9 | 12.7 | 9.85 | 12.7 | 16.9 |
| 20 | 18.81 | 22.0 | 47.0 | 105 | 94.0 | 25.1 | 18.8 | 14.1 | 11.0 | 14.1 | 18.8 |
| 24 | 22.63 | 26.4 | 56.6 | 126 | 113 | 30.2 | 22.6 | 17.0 | 13.2 | 17.0 | 22.6 |

*TABLE C.3.2 Continued*

| *Miter Elbows[3] (No. of Miters)* | | | | | *Welding Tees* | | *Valves (Screwed, Flanged, or Welded)* | | | |
|---|---|---|---|---|---|---|---|---|---|---|
| ***1–45°*** | ***1–60°*** | ***1–90°*** | ***2–90°*** | ***3–90°*** | ***Forged*** | ***Miter[3]*** | ***Gate*** | ***Globe*** | ***Angle*** | ***Swing Check*** |
| ***0.45*** | ***0.90*** | ***1.80*** | ***0.60*** | ***0.45*** | ***1.35*** | ***1.80*** | ***0.21*** | ***10*** | ***5.0*** | ***2.5*** |
| ***15*** | ***30*** | ***60*** | ***20*** | ***15*** | ***45*** | ***60*** | ***7*** | ***333*** | ***167*** | ***83*** |
|  |  |  | 5  | 5  |  |  |  | 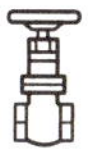 |  |  |
| ***L = Equivalent Length in Feet of Schedule 40 (Standard Weight) Straight Pipe[6]*** | | | | | | | | | | |
| 0.78 | 1.55 | 3.10 | 1.04 | 0.78 | 2.33 | 3.10 | 0.36 | 17.3 | 8.65 | 4.32 |
| 1.03 | 2.06 | 4.12 | 1.37 | 1.03 | 3.09 | 4.12 | 0.48 | 22.9 | 11.4 | 5.7 2 |
| 1.31 | 2.62 | 5.24 | 1.75 | 1.31 | 3.93 | 5.24 | 0.61 | 29.1 | 14.6 | 7.27 |
| 1.72 | 3.45 | 6.90 | 2.30 | 1.72 | 5.17 | 6.90 | 0.81 | 38.3 | 19.1 | 9.58 |
| 2.01 | 4.02 | 8.04 | 2.68 | 2.01 | 6.04 | 8.04 | 0.94 | 44.7 | 22.4 | 11.2 |
| 2.58 | 5.17 | 10.3 | 3.45 | 2.58 | 7.75 | 10.3 | 1.21 | 57.4 | 28.7 | 14.4 |
| 3.08 | 6.16 | 12.3 | 4.11 | 3.08 | 9.25 | 12.3 | 1.44 | 68.5 | 34.3 | 17.1 |
| 3.84 | 7.67 | 15.3 | 5.11 | 3.84 | 11.5 | 15.3 | 1.79 | 85.2 | 42.6 | 21.3 |
| 5.04 | 10.1 | 20.2 | 6.71 | 5.04 | 15.1 | 20.2 | 2.35 | 112 | 56.0 | 28.0 |
| 6.30 | 12.6 | 25.2 | 8.40 | 6.30 | 18.9 | 25.2 | 2.94 | 140 | 70.0 | 35.0 |
| 7.58 | 15.2 | 30.4 | 10.1 | 7.58 | 22.8 | 30.4 | 3.54 | 168 | 84.1 | 42.1 |
| 9.97 | 20.0 | 40.0 | 13.3 | 9.97 | 29.9 | 40.0 | 4.65 | 222 | 111 | 55.5 |
| 12.5 | 25.0 | 50.0 | 16.7 | 12.5 | 37.6 | 50.0 | 5.85 | 278 | 139 | 69.5 |
| 14.9 | 29.8 | 59.6 | 19.9 | 14.9 | 44.8 | 59.6 | 6.96 | 332 | 166 | 83.0 |
| 16.4 | 32.8 | 65.6 | 21.9 | 16.4 | 49.2 | 65.6 | 7.65 | 364 | 182 | 91.0 |
| 18.8 | 37.5 | 75.0 | 25.0 | 18.8 | 56.2 | 75.0 | 8.75 | 417 | 208 | 104 |
| 21.1 | 42.1 | 84.2 | 28.1 | 21.1 | 63.2 | 84.2 | 9.85 | 469 | 234 | 117 |
| 23.5 | 47.0 | 94.0 | 31.4 | 23.5 | 70.6 | 94.0 | 11.0 | 522 | 261 | 131 |
| 28.3 | 56.6 | 113 | 37.8 | 28.3 | 85.0 | 113 | 13.2 | 629 | 314 | 157 |

For SI units, 1 ft = 0.305 m.

Note: Values for welded fittings are for conditions where bore is not obstructed by weld spatter or backing rings. If appreciably obstructed, use values for "Screwed Fittings."

[1]Flanged fittings have three-fourths the resistance of screwed elbows and tees.

[2]Tabular figures give the extra resistance due to curvature alone to which should be added the full length of travel.

[3]Small size socket-welding fittings are equivalent to miter elbows and miter tees.

[4]Equivalent resistance in number of diameters of straight pipe computed for a value of $f - 0.0075$ from the relation $n - k/4f$.

[5]For condition of minimum resistance where the centerline length of each miter is between $d$ and $2\frac{1}{2}d$.

[6]For pipe having other inside diameters, the equivalent resistance may be computed from the above $n$ values.

Source: From *Piping Handbook*, Table XIV, pp. 100–101. Used by permission of McGraw-Hill Book Company.

***TABLE C.3.4*** *SPECIAL USE: Multipliers to Be Used with Tables 6.2(a) Through 6.2(v) When the Specific Gravity of the Gas Is Other than 0.60*

| Specific Gravity | Multiplier | Specific Gravity | Multiplier |
|---|---|---|---|
| 0.35 | 1.31 | 1.00 | 0.78 |
| 0.40 | 1.23 | 1.10 | 0.74 |
| 0.45 | 1.16 | 1.20 | 0.71 |
| 0.50 | 1.10 | 1.30 | 0.68 |
| 0.55 | 1.04 | 1.40 | 0.66 |
| 0.60 | 1.00 | 1.50 | 0.63 |
| 0.65 | 0.96 | 1.60 | 0.61 |
| 0.70 | 0.93 | 1.70 | 0.59 |
| 0.75 | 0.90 | 1.80 | 0.58 |
| 0.80 | 0.87 | 1.90 | 0.56 |
| 0.85 | 0.84 | 2.00 | 0.55 |
| 0.90 | 0.82 | 2.10 | 0.54 |

***COMMENTARY TABLE C.1*** *Pros and Cons of Sizing Methods*

| Sizing Method | Pros | Cons |
|---|---|---|
| Longest Length Method | • Simplest pipe sizing method to use<br>• Always works<br>• Provides some capacity for future system expansion | • May provide larger than necessary pipe sizing for closer outlets<br>• May result in a costlier installation |
| Branch Length Method | • May provide smaller yet adequate pipe sizes for branches<br>• More simple than engineered methods (although not as simple as the longest length method) | • Not as easy to use as the longest length method<br>• Requires training |
| Hybrid Pressure Method | • May provide a smaller line from the meter to the regulator<br>• More simple than engineered methods<br>• May allow a manifold system of smaller branches (CSST) and semirigid copper tubing | • May require a vent on the regulator if the regulator is not outdoors |

(The Pressure Drop per 100 ft Method is a method to optimize existing piping systems and cannot be compared with the other methods in this table.)

The longest length method, which is described in C.4.1, is not new to the code, but it did not have a name in previous editions. The longest length method is conservative and sizes all branches using the longest length of piping from the point of delivery in the system.

To determine the size of each section of gas piping in a system within the range of the capacity tables, proceed as follows *(also see sample calculations included in this annex)*:

(1) Divide the piping system into appropriate segments consistent with the presence of tees, branch lines, and main runs. For each segment, determine the gas load (assuming all appliances operate simultaneously) and its overall length. An allowance (in equivalent length of pipe) as determined from Table C.3.2 shall be considered for piping segments that include four or more fittings.
(2) Determine the gas demand of each appliance to be attached to the piping system. Where Table 6.2(a) through Table 6.2(v) are to be used to select the piping size, calculate the gas demand in terms of cubic feet per hour for each piping system outlet. Where Table 6.3(a) through Table 6.3(m) are to be used to select the piping size, calculate the gas demand in terms of thousands of Btu per hour for each piping system outlet.
(3) Where the piping system is for use with other than undiluted LP-Gases, determine the design system pressure, the allowable loss in pressure (pressure drop), and specific gravity of the gas to be used in the piping system.
(4) Determine the length of piping from the point of delivery to the most remote outlet in the building/piping system.
(5) In the appropriate capacity table, select the row showing the measured length or the next longer length if the table does not give the exact length. This length is the only length used in determining the size of any section of gas piping. If the gravity factor is to be applied, the values in the selected row of the table are multiplied by the appropriate multiplier from Table C.3.4.
(6) Use this horizontal row to locate ALL gas demand figures for this particular system of piping.
(7) Starting at the most remote outlet, find the gas demand for that outlet in the horizontal row just selected. If the exact figure of demand is not shown, choose the next larger figure left in the row.
(8) Opposite this demand figure, in the first row at the top, the correct size of gas piping will be found.
(9) Proceed in a similar manner for each outlet and each section of gas piping. For each section of piping, determine the total gas demand supplied by that section.

When a large number of piping components (such as elbows, tees, and valves) are installed in a pipe run, additional pressure loss can be accounted for by the use of equivalent lengths. Pressure loss across any piping component can be equated to the pressure drop through a length of pipe. The equivalent length of a combination of only four elbows/tees can result in a jump to the next larger length row, resulting in a significant reduction in capacity. The equivalent lengths in feet shown in Table C.3.2 have been computed on a basis that the inside diameter corresponds to that of Schedule 40 (standard weight) steel pipe, which is close enough for most purposes involving other schedules of pipe. Where a more specific solution for equivalent length is desired, this may be made by multiplying the actual inside diameter of the pipe in inches by $n/12$, or the actual inside diameter in feet by $n$. $N$ can be read from the table heading. The equivalent length values can be used with reasonable accuracy for copper or brass fittings and bends, although the resistance per foot of copper or brass pipe is less than that of steel. For copper or brass valves, however, the equivalent length of pipe should be taken as 45 percent longer than the values in the table, which are for steel pipe.

## C.4.2 The Branch Length Method.

This sizing method reduces the amount of conservatism built into the traditional Longest Length Method. The longest length as measured from the meter to the farthest remote appliance is used only to size the initial parts of the overall piping system. The Branch Length Method is applied in the following manner:

(1) Determine the gas load for each of the connected appliances.
(2) Starting from the meter, divide the piping system into a number of connected segments, and determine the length and amount of gas that each segment would carry assuming that

all appliances were operated simultaneously. An allowance (in equivalent length of pipe) as determined from Table C.3.2 should be considered for piping segments that include four or more fittings.

(3) Determine the distance from the outlet of the gas meter to the appliance farthest removed from the meter.

(4) Using the longest distance (found in Step 3), size each piping segment from the meter to the most remote appliance outlet.

(5) For each of these piping segments, use the longest length and the calculated gas load for all of the connected appliances for the segment and begin the sizing process in Steps 6 through 8.

(6) Referring to the appropriate sizing table (based on operating conditions and piping material), find the longest length distance in the first column or the next larger distance if the exact distance is not listed. The use of alternative operating pressures and/or pressure drops requires the use of a different sizing table but does not alter the sizing methodology. In many cases, the use of alternative operating pressures and/or pressure drops requires the approval of both the authority having jurisdiction and the local gas serving utility.

(7) Trace across this row until the gas load is found or the closest larger capacity if the exact capacity is not listed.

(8) Read up the table column and select the appropriate pipe size in the top row. Repeat Steps 6, 7, and 8 for each pipe segment in the longest run.

(9) Size each remaining section of branch piping not previously sized by measuring the distance from the gas meter location to the most remote outlet in that branch, using the gas load of attached appliances, and follow the procedures of Steps 2 through 8.

The branch length method, which is described in C.4.2, was added to the code in the 2002 edition and uses the tables in Chapter 6, as the longest length method does. Refer to Figure C.8.3 to see an example of how this method is used. The branch length method allows branches to be sized using the length of each branch from the point of delivery.

The longest pipe path from the point of delivery to the most remote appliance is sized at its length. The branches serving other, closer appliances are sized at the length of each outlet from the point of delivery. There will be no difference in the pipe size serving the most remote outlet, whether the longest length method or the branch length method is used. However, outlets located closer to the point of delivery may have smaller pipe sizes when the branch length method is used.

## C.4.3 Hybrid Pressure Method.

The sizing of a 2 psi (14 kPa) gas piping system is performed using the traditional Longest Length Method but with modifications. The 2 psi (14 kPa) system consists of two independent pressure zones, and each zone is sized separately. The Hybrid Pressure Method is applied using the following steps.

The hybrid pressure method is described in C.4.3. This method is used to size piping systems that use more than one pressure, such as piping systems containing a line pressure regulator to provide two pressures in the system. An example of such a system is a 2 psi (13.8 kPa) system in which the piping from the point of delivery to the line pressure regulator is 2 psi (13.8 kPa), and the pressure from the line pressure regulator to the appliances is 7 in. w.c. (1.75 kPa) for natural gas, and 11 in. w.c. (2.75 kPa) for propane systems. First, the higher-pressure system is sized using the longest length method (as shown in C.4.1) or by another method. Second, the lower-pressure system is sized using the longest length method (as shown in C.4.1), by the branch length method (as shown in C.4.2), or by another method. If more than one low-pressure system exists, as might be the case in an apartment building, each low-pressure system should be sized separately.

The 2 psi (14 kPa) section (from the meter to the line regulator) is sized as follows:

(1) Calculate the gas load (by adding up the nameplate ratings) from all connected appliances. (In certain circumstances the installed gas load may be increased up to 50 percent to accommodate future addition of appliances.) Ensure that the line regulator capacity is adequate for the calculated gas load and that the required pressure drop (across the regulator) for that capacity does not exceed $^3/_4$ psi (5.2 kPa) for a 2 psi (14 kPa) system. If the pressure drop across the regulator is too high (for the connected gas load), select a larger regulator.
(2) Measure the distance from the meter to the line regulator located inside the building.
(3) If multiple line regulators are used, measure the distance from the meter to the regulator farthest removed from the meter.
(4) The maximum allowable pressure drop for the 2 psi (14 kPa) section is 1 psi (7 kPa).
(5) Referring to the appropriate sizing table (based on piping material) for 2 psi (14 kPa) systems with a 1 psi (7 kPa) pressure drop, find this distance in the first column, or the closest larger distance if the exact distance is not listed.
(6) Trace across this row until the gas load is found or the closest larger capacity if the exact capacity is not listed.
(7) Read up the table column to the top row and select the appropriate pipe size.
(8) If multiple regulators are used in this portion of the piping system, each line segment must be sized for its actual gas load, using the longest length previously determined.

The low-pressure section (all piping downstream of the line regulator) is sized as follows:

(1) Determine the gas load for each of the connected appliances.
(2) Starting from the line regulator, divide the piping system into a number of connected segments and/or independent parallel piping segments and determine the amount of gas that each segment would carry assuming that all appliances were operated simultaneously. An allowance (in equivalent length of pipe) as determined from Table C.3.2 should be considered for piping segments that include four or more fittings.
(3) For each piping segment, use the actual length or longest length (if there are sub-branch lines) and the calculated gas load for that segment and begin the sizing process as follows:
    (a) Referring to the appropriate sizing table (based on operating pressure and piping material), find the longest length distance in the first column or the closest larger distance if the exact distance is not listed. The use of alternative operating pressures and/or pressure drops requires the use of a different sizing table but does not alter the sizing methodology. In many cases, the use of alternative operating pressures and/or pressure drops may require the approval of the authority having jurisdiction.
    (b) Trace across this row until the appliance gas load is found or the closest larger capacity if the exact capacity is not listed.
    (c) Read up the table column to the top row and select the appropriate pipe size.
    (d) Repeat this process for each segment of the piping system.

## C.4.4 Pressure Drop per 100 ft Method.

This sizing method is less conservative than the others, but it allows the designer to immediately see where the largest pressure drop occurs in the system. With this information, modifications can be made to bring the total drop to the critical appliance within the limitations that are presented to the designer.

The pressure drop per 100 ft method, which is discussed in C.4.4, requires the designer to know the lengths, fittings used, and pipe sizes of every pipe in the run. This method allows the designer to accommodate specific appliance inlet pressures by accurately calculating the pressure loss of the piping to that appliance.

Follow the procedures described in the Longest Length Method for steps (1) through (4) and step (9).

For each piping segment, calculate the pressure drop based on pipe size, length as a percentage of 100 ft, and gas flow. Table C.4.4 shows pressure drop per 100 ft for pipe sizes from $^1/_2$ in. through 2 in. The sum of pressure drops to the critical appliance is subtracted from the supply pressure to verify that sufficient pressure is available. If not, the layout can be examined to find the high drop section(s), and sizing selections modified.

***TABLE C.4.4*** *Thousands of Btu/hr of Natural Gas per 100 ft of Pipe at Various Pressure Drops and Pipe Diameters*

| ***Press. Drop/ 100 ft (in. w.c.)*** | ***Pipe Sizes (in.)*** | | | | | |
|---|---|---|---|---|---|---|
| | $^1/_2$ | $^3/_4$ | 1 | $1^1/_4$ | $1^1/_2$ | 2 |
| 0.2 | 31 | 64 | 121 | 248 | 372 | 716 |
| 0.3 | 38 | 79 | 148 | 304 | 455 | 877 |
| 0.5 | 50 | 104 | 195 | 400 | 600 | 1160 |
| 1.0 | 71 | 147 | 276 | 566 | 848 | 1640 |

Note: Other values can be obtained using the following equation:

$$\text{Desired Value} = \text{thousand of Btu/hr} \times \sqrt{\frac{\text{Desired Drop}}{\text{Table Drop}}}$$

For example, if it is desired to get flow through $^3/_4$ in. pipe at 2 in. w.c./100 ft, multiply the capacity of $^3/_4$ in. pipe at 1 in./100 ft by the square root of the pressure ratio:

$$147{,}000 \text{ Btu/hr} \times \sqrt{\frac{2 \text{ in. w.c.}}{1 \text{ in. w.c.}}} = 147{,}000 \times 1.414 = 208{,}000 \text{ Btu/hr}$$

## C.5 Use of Sizing Equations

Capacities of smooth wall pipe or tubing can also be determined by using the following formulas:

The two equations shown in C.5(1) and C.5(2) have been in Annex C for many editions. In the 2002 edition, these equations were reconfigured to solve for pipe diameter, *D*, and the reconfigured versions were moved to Chapter 6. System designers are interested in the pipe size, and the equations are useful in calculating pipe size. The equations in their original form are retained in Annex C because some users need to determine the actual flow of pipe of a given size. A table of values of *Cr* and *Y* was also added to Chapter 6 in the 2002 edition.

(1) *High Pressure* [1.5 psi (10.3 kPa) and above]:

$$Q = 181.6\sqrt{\frac{D^5 \cdot (P_1^2 - P_2^2) \cdot Y}{Cr \cdot fba \cdot L}}$$

$$= 2237\, D^{2.623}\left[\frac{(P_1^2 - P_2^2) \cdot Y}{Cr \cdot L}\right]^{0.541}$$

(2) *Low Pressure* [less than 1.5 psi (10.3 kPa)]:

$$Q = 187.3\sqrt{\frac{D^5 \cdot \Delta H}{Cr \cdot fba \cdot L}}$$

$$= 2313\, D^{2.623}\left(\frac{\Delta H}{Cr \cdot L}\right)^{0.541}$$

where:

$Q$ = rate (cubic feet per hour at 60°F and 30 in. mercury column)
$D$ = inside diameter of pipe (in.)
$P_1$ = upstream pressure (psia)
$P_2$ = downstream pressure (psia)
$Y$ = superexpansibility factor = 1/supercompressibility factor
$Cr$ = factor for viscosity, density, and temperature
$fba$ = base friction factor for air at 60°F (CF = 1)
$L$ = length of pipe (ft)
$H$ = pressure drop [in. w.c. (27.7 in. $H_2O$ = 1 psi) = $0.00354\ ST(Z/S)^{0.152}$]
$S$ = specific gravity of gas at 60°F and 30 in. mercury column (0.60 for natural gas, 1.53 for propane)
$T$ = absolute temperature (°F or = $t$ + 460)
$Z$ = viscosity of gas [centipoise (0.012 for natural gas, 0.008 for propane), or = 1488]
$t$ = temperature (°F)

See Table 6.4.2 for values of $Cr$ and $Y$ for natural gas and propane.

## C.6 Pipe and Tube Diameters

Where the internal diameter is determined by the formulas in Section 6.4, Table C.6(a) and Table C.6(b) can be used to select the nominal or standard pipe size based on the calculated internal diameter.

Information on the size of steel and copper pipe and tubing, in Section C.6, is especially helpful when considering changing piping materials, because it allows comparison of the actual inside diameter of different piping materials.

## C.7 Use of Sizing Charts

A third method of sizing gas piping is detailed here as a useful option when large quantities of piping are involved in a job (e.g., an apartment house) and material costs are of concern. If the user is not completely familiar with this method, the resulting pipe sizing should be checked by a knowledgeable gas engineer. The sizing charts are applied as follows:

(1) With the layout developed according to Section 5.1 of the code, indicate in each section the *design gas flow* under maximum operation conditions. For many layouts, the maximum design flow is the sum of all connected loads. However, in some cases, certain combinations of appliances do not occur simultaneously (e.g., gas heating and air conditioning). For these cases, the design flow is the greatest gas flow that can occur at any one time.
(2) Determine the *inlet gas pressure* for the system being designed. In most cases, the point of inlet is the gas meter or service regulator, but in the case of a system addition, it could be the point of connection to the existing system.
(3) Determine the *minimum pressure* required at the inlet to the critical appliance. Usually, the critical item is the appliance with the highest required pressure for satisfactory operation. If several items have the same required pressure, it is the one with the greatest length of piping from the system inlet.
(4) The difference between the inlet pressure and critical item pressure is the *allowable system pressure drop*. Figure C.7(a) and Figure C.7(b) show the relationship between gas flow, pipe size, and pipe length for natural gas with 0.60 specific gravity.

*TABLE C.6(a) Schedule 40 Steel Pipe Standard Sizes*

| Nominal Size (in.) | Internal Diameter (in.) | Nominal Size (in.) | Internal Diameter (in.) |
|---|---|---|---|
| $^1/_4$ | 0.364 | $1^1/_2$ | 1.610 |
| $^3/_8$ | 0.493 | 2 | 2.067 |
| $^1/_2$ | 0.622 | $2^1/_2$ | 2.469 |
| $^3/_4$ | 0.824 | 3 | 3.068 |
| 1 | 1.049 | $3^1/_2$ | 3.548 |
| $1^1/_4$ | 1.380 | 4 | 4.026 |

*TABLE C.6(b) Copper Tube Standard Sizes*

| Tube Type | Nominal or Standard Size (in.) | Internal Diameter (in.) | Tube Type | Nominal or Standard Size (in.) | Internal Diameter (in.) |
|---|---|---|---|---|---|
| K | $^1/_4$ | 0.305 | K | 1 | 0.995 |
| L | $^1/_4$ | 0.315 | L | 1 | 1.025 |
| ACR (D) | $^3/_8$ | 0.315 | ACR (D,A) | $1^1/_8$ | 1.025 |
| ACR (A) | $^3/_8$ | 0.311 | K | $1^1/_4$ | 1.245 |
| K | $^3/_8$ | 0.402 | L | $1^1/_4$ | 1.265 |
| L | $^3/_8$ | 0.430 | ACR (D,A) | $1^3/_8$ | 1.265 |
| ACR (D) | $^1/_2$ | 0.430 | K | $1^1/_2$ | 1.481 |
| ACR (A) | $^1/_2$ | 0.436 | L | $1^1/_2$ | 1.505 |
| K | $^1/_2$ | 0.527 | ACR (D,A) | $1^5/_8$ | 1.505 |
| L | $^1/_2$ | 0.545 | K | 2 | 1.959 |
| ACR (D) | $^5/_8$ | 0.545 | L | 2 | 1.985 |
| ACR (A) | $^5/_8$ | 0.555 | ACR (D,A) | $2^1/_8$ | 1.985 |
| K | $^5/_8$ | 0.652 | K | $2^1/_2$ | 2.435 |
| L | $^5/_8$ | 0.666 | L | $2^1/_2$ | 2.465 |
| ACR (D) | $^3/_4$ | 0.666 | ACR (D,A) | $2^5/_8$ | 2.465 |
| ACR (A) | $^3/_4$ | 0.680 | K | 3 | 2.907 |
| K | $^3/_4$ | 0.745 | L | 3 | 2.945 |
| L | $^3/_4$ | 0.785 | ACR (D,A) | $3^1/_8$ | 2.945 |
| ACR (D,A) | $^7/_8$ | 0.785 | | | |

(5) To use Figure C.7(a) (low-pressure applications), calculate the piping length from the inlet to the critical appliance. Increase this length by 50 percent to allow for fittings. Divide the allowable pressure drop by the equivalent length (in hundreds of feet) to determine the allowable pressure drop per hundred feet. Select the pipe size from Figure C.7(a) for the required volume of flow.

(6) To use Figure C.7(b) (high-pressure applications), calculate the equivalent length as in item (5). Calculate the *index number* for Figure C.7(b) by dividing the difference between the squares of the absolute values of inlet and outlet pressures by the equivalent length (in hundreds of feet). Select the pipe size from Figure C.7(b) for the gas volume required.

The sizing graphs in Figure C.7(a) and Figure C.7(b) were added in the 2002 edition. They appeared previously in the commentary of the *National Fuel Gas Code Handbook* but were relocated to this annex of the code in 2002. The sizing graphs are useful for larger piping systems and might result in smaller pipe sizes compared to those derived from using the Longest Length Method and the Branch Length Method.

## C.8 Examples of Piping System Design and Sizing

### C.8.1 Example 1 — Longest Length Method.

Determine the required pipe size of each section and outlet of the piping system shown in Figure C.8.1, with a designated pressure drop of 0.50 in. w.c. (125 Pa), using the Longest Length Method. The gas to be used has 0.60 specific gravity and a heating value of 1000 Btu/ft$^3$ (37.5 MJ/m$^3$).

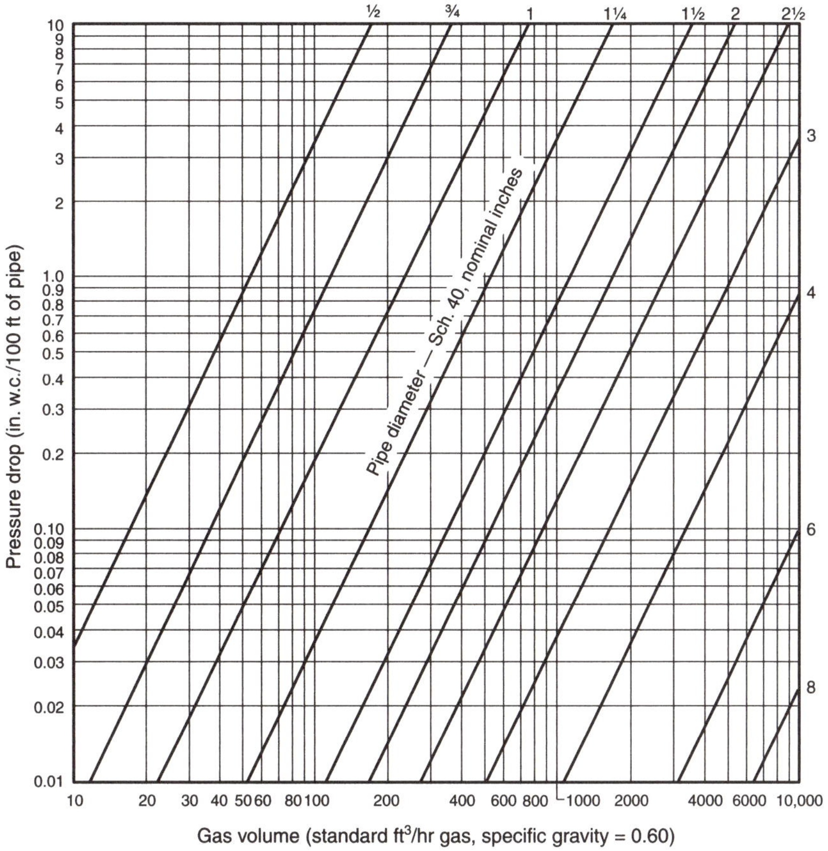


***FIGURE C.7(a)*** *Capacity of Natural Gas Piping, Low Pressure (0.60 in. w.c.).*

*Solution*

(1) Maximum gas demand for outlet A:

$$\frac{\text{Consumption(rating plate input, or Table 5.4.2.1 if necessary)}}{\text{Btu of gas}}$$

$$= \frac{\text{35,000 Btu/hr rating}}{\text{1000 Btu/ft}} = 35 \text{ ft}^3\text{/hr} = 35 \text{ cfh}$$

Maximum gas demand for outlet B:

$$\frac{\text{Consumption}}{\text{Btu of gas}} = \frac{75{,}000}{1000} = 75 \text{ cfh}$$

Maximum gas demand for outlet C:

$$\frac{\text{Consumption}}{\text{Btu of gas}} = \frac{35{,}000}{1000} = 35 \text{ cfh}$$

Maximum gas demand for outlet D:

$$\frac{\text{Consumption}}{\text{Btu of gas}} = \frac{100{,}000}{1000} = 100 \text{ cfh}$$

(2) The length of pipe from the point of delivery to the most remote outlet (A) is 60 ft (18.3 m). This is the only distance used.

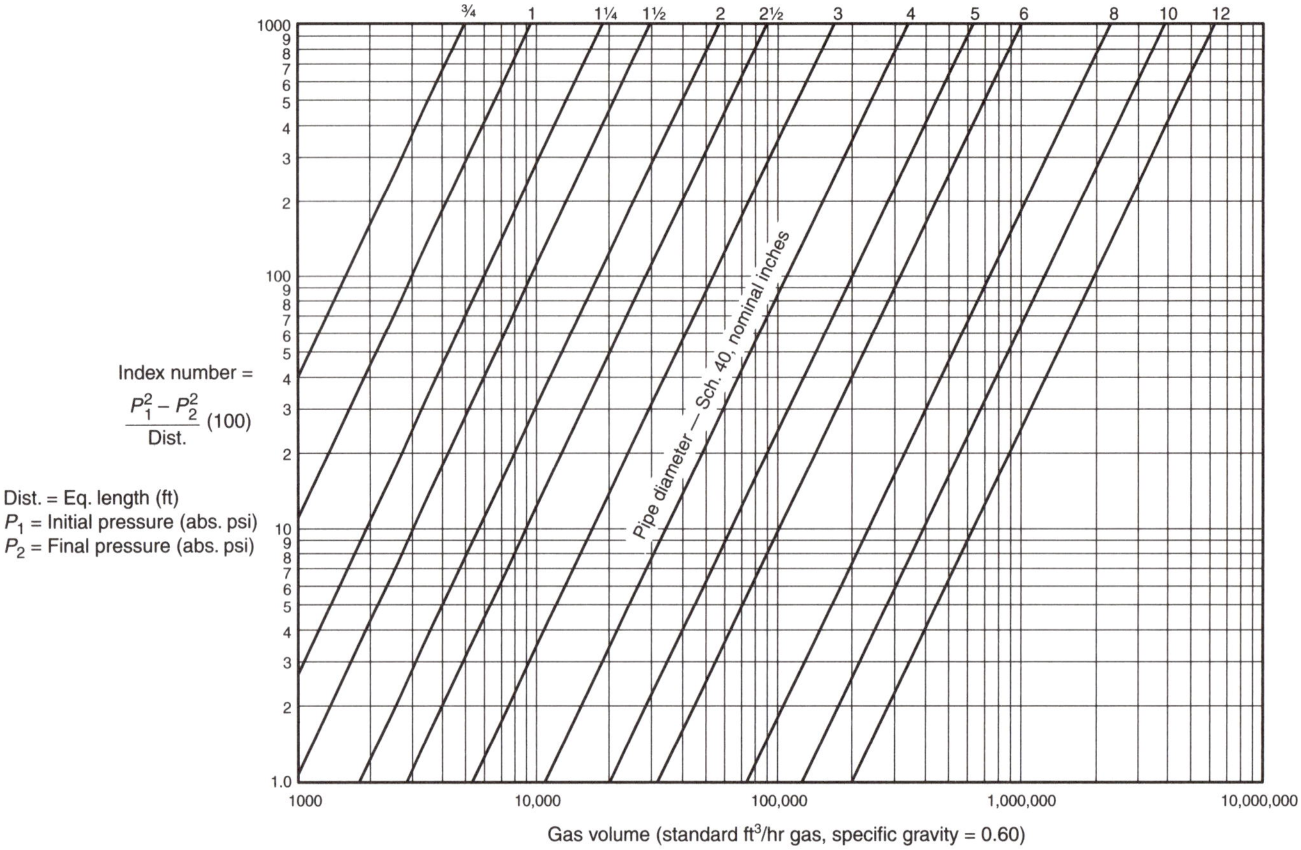


***FIGURE C.7(b)*** *Capacity of Natural Gas Piping, High Pressure (1.5 psi and above).*

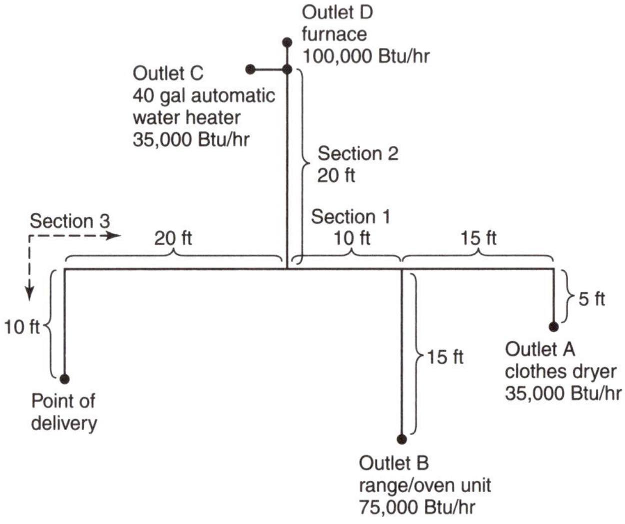


For SI units, 1 ft = 0.305 m, 1 gal = 3.785 L, 1000 Btu/hr = 0.293 kW.

***FIGURE C.8.1*** *Piping Plan Showing a Steel Piping System.*

(3) Using the row marked 60 ft (18.3 m) in Table 6.2(b):

(a) Outlet A, supplying 35 cfh (0.99 $m^3$/hr), requires $^1/_2$ in. pipe.
(b) Outlet B, supplying 75 cfh (2.12 $m^3$/hr), requires $^3/_4$ in. pipe.
(c) Section 1, supplying outlets A and B, or 110 cfh (3.11 $m^3$/hr), requires $^3/_4$ in. pipe.
(d) Section 2, supplying outlets C and D, or 135 cfh (3.82 $m^3$/hr), requires $^3/_4$ in. pipe.
(e) Section 3, supplying outlets A, B, C, and D, or 245 cfh (6.94 $m^3$/hr), requires 1 in. pipe.

(4) If a different gravity factor is applied to this example, the values in the row marked 60 ft (18.3 m) of Table 6.2(b) would be multiplied by the appropriate multiplier from Table C.3.4 and the resulting cubic feet per hour values would be used to size the piping.

Commentary Table C.2 is an excerpt from Table 6.2(b). It shows the portion of the table that is relevant to Example 1 — Longest Length Method. Note that the 60 ft length row is highlighted because this is the only row used in this solution. Read across the row to find a capacity equal to or greater than that needed. For example, outlet A requires 35 cfh (0.99 m/hr), and the smallest pipe size, $^1/_2$ in., delivers more than that, so $^1/_2$ in. pipe is used. Larger sizes are also acceptable if desired for other reasons. Worksheet C.1 details the calculation for Example 1 — Longest Length Method, described in C.8.1. A full-sized, reproducible copy of this form is included in Supplement 6 of this handbook.

***COMMENTARY TABLE C.2** Extract from Table 6.2(b)*

| Length (ft) | Capacity in Cubic Feet of Gas/Hour | | | |
|---|---|---|---|---|
| | $^1/_2$ in. | $^3/_4$ in. | 1 in. | $1^1/_4$ in. |
| 50 | 72 | 151 | 284 | 583 |
| 60 | 65 | 137 | 257 | 528 |
| 70 | 60 | 126 | 237 | 486 |

## C.8.2 Example 2 — Hybrid or Dual Pressure Systems.

Determine the required CSST size of each section of the piping system shown in Figure C.8.2, with a designated pressure drop of 1 psi (7 kPa) for the 2 psi (14 kPa) section and 3 in. w.c. (0.75 kPa) pressure drop for the 10 in. w.c. (2.49 kPa) section. The gas to be used has 0.60 specific gravity and a heating value of 1000 Btu/$ft^3$ (37.5 MJ/$m^3$).

*Solution*

(1) Size 2 psi (14 kPa) line using Table 6.2(p).
(2) Size 10 in. w.c. (2.5 kPa) lines using Table 6.2(n).
(3) Using the following steps, determine if sizing tables can be used:

(a) Total gas load shown in Figure C.8.2 equals 110 chf (3.11 $m^3$/hr).
(b) Determine pressure drop across regulator [see notes in Table 6.2(p)].
(c) If pressure drop across regulator exceeds $^3/_4$ psi (5.2 kPa), Table 6.2(p) cannot be used. Note that if pressure drop exceeds $^3/_4$ psi (5.2 kPa), a larger regulator must be selected or an alternative sizing method must be used.
(d) Pressure drop across the line regulator [for 110 cfh/(3.11 $m^3$/hr)] is 4 in. w.c. (0.99 kPa) based on manufacturer's performance data.
(e) Assume the CSST manufacturer has tubing sizes or EHDs of 13, 18, 23, and 30.

## CALCULATION WORKSHEET: PIPE SIZING, LONGEST LENGTH METHOD

**Step 1:**
- Draw a sketch of a piping system in the space to the right. Use the back of this page or a separate sheet if more space is needed.

**Step 2:**
- Enter the system information. Note that demand is the amount of gas flowing through a section of pipe.
- Use total Btu/hr rating/1000 ($ft^3/hr$) for natural gas.
- Use total Btu/hr for propane.

**Step 3:**
- Determine the gas used and system pressure, and enter it to the right.
- Determine the piping material and enter it to the right.
- Select the appropriate pipe sizing table from Chapter 6 and enter it to the right.

**Step 4:**
- On the sketch, label the section of pipe from the point of delivery (meter or regulator) to the first tee as Section 1.
- Label the section from the first tee to the second tee as Section 2, and label the section from the first tee to the third tee as Section 3. Use similar section numbers for additional sections.

**Step 5:**
- Determine the longest length of piping from the point of delivery to the most remote appliance. Enter this length for all pipe sections in Table 1.

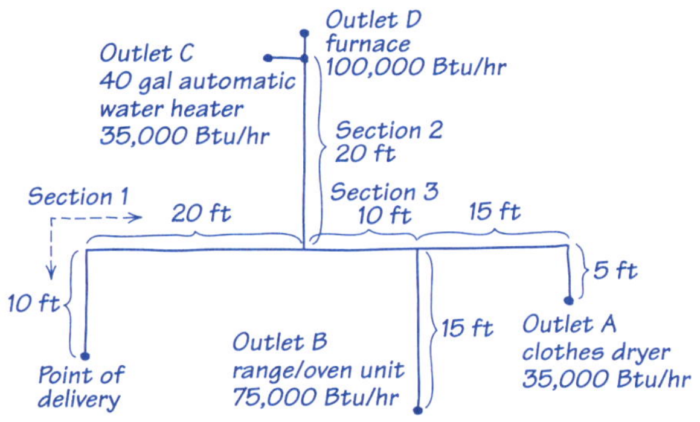


Pipe system sketch

Gas: Natural
System pressure: <0.5 psi
Piping material: Sch 40 steel
Table used: 6.2(b)
Pressure drop: 0.50 in. w.c.

**Table 1 Piping System Table**

| Section | Demand | Section length | Size |
|---|---|---|---|
| 1 | 110 | 60 ft | 3/4 in. |
| 2 | 135 | 60 ft | 3/4 in. |
| 3 | 245 | 60 ft | 1 in. |
| 4 | | | |
| 5 | | | |

**Step 6:**
- Enter the input rating for each appliance in Table 2. For natural gas appliances, enter the input rating in Btu/hr/1000 ($ft^3/hr$). For propane appliances, enter the input rating in Btu/hr.

**Step 7:**
- From the table, determine the length of each pipe section using the appropriate table, using only the row with the longest length. Round up to the lengths in the table. Read across until a capacity equal to or greater than the required demand for the section is found. Read up to find the size. Repeat for each section of piping. Enter this size in Table 2.

**Table 2 Appliances Table**

| Appliance | Demand | Section length | Size |
|---|---|---|---|
| Furnace | 100 | 60 ft | 3/4 in. |
| Furnace | | | |
| Water heater | 35 | 60 ft | 3/8 in. |
| Water heater | | | |
| Range | 75 | 60 ft | 3/4 in. |
| Oven | | | |
| Dryer | 35 | 60 ft | 3/8 in. |
| Other | | | |
| Other | | | |
| Other | | | |
| Other | | | |
| Total | 245 | — | — |

Note that in this example, the 60 ft row in Table 6.2(b) was used. It is okay to interpolate between rows.

**Job:** 25 Main Street **Prepared by:** TL **Date:** 1/1/09



*WORKSHEET C.1 Sample Calculation for Determining Pipe Size Using the Longest Length Method.*

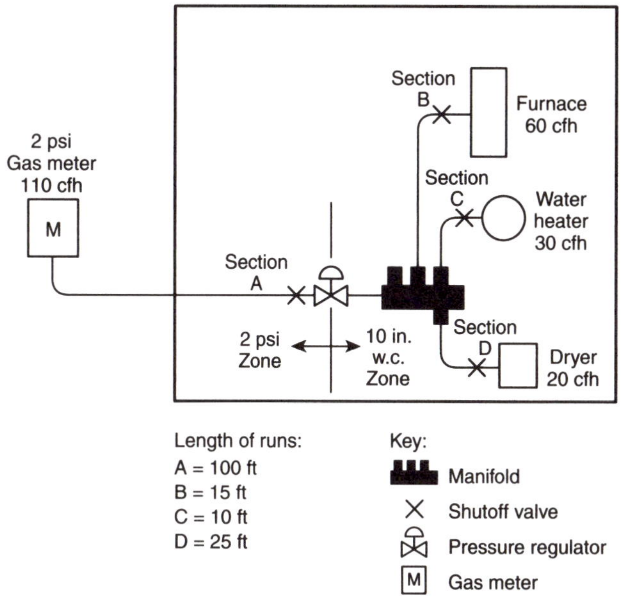


***FIGURE C.8.2*** *Piping Plan Showing a CSST System.*

Note 1 and Note 2 to Table 6.2(p) state the following:

1. Table does not include effect of pressure drop across the line regulator. Where regulator loss exceeds $^3/_4$ psi, do not use this table. Consult with regulator manufacturer for pressure drops and capacity factors. Pressure drops across a regulator may vary with flow rate.
2. CAUTION: Capacities shown in table may exceed maximum capacity for a selected regulator. Consult with regulator or tubing manufacturer for guidance.

Example 2 — Hybrid or Dual Pressure Systems, shown in C.8.2, is based on a specific regulator, a Maxitrol model 325-3L. Refer to Exhibit C.1, the pressure drop chart for the 325-3L model. To use the chart, read across along the bottom scale of the chart, flow rate. This example has a flow rate of 110 cfh (3.11 m/hr). Read up to the angled line for the 325-3L regulator. From there read to the left axis, pressure drop. The capacity of the regulator is 10 mbar, which equals 0.4 psi (3 kPa). Because this value is less than $^3/_4$ psi (5 kPa), the regulator can be used.

Paragraph C.8.2(3)(d) provides information on the pressure drop of the regulator: 4 in. w.c. (0.99 kPa). This information is provided by the regulator manufacturer, usually as part of the product information supplied with the regulator.

(4) From Section A [2 psi (14 kPa) zone]:

(a) Determine distance from meter to regulator = 100 ft (30.48 m).

(b) Determine total load supplied by A = 110 cfh (3.11 $m^3$/hr) (furnace + water heater + dryer).

Commentary Table C.3 is an excerpt from Table 6.2(p). It shows the portion of Table 6.2(p) relevant to this example. Note that the 100 ft length row is highlighted because the length of CSST from the meter to the regulator is 100 ft. Read across the row to find a capacity equal to or greater than that needed. In this example, the load is 110 cfh (3.11 m/hr) and pipe with an EHD of 15 has a capacity of 107 cfh (3.01 m/hr), which is insufficient. EHD 18 has a

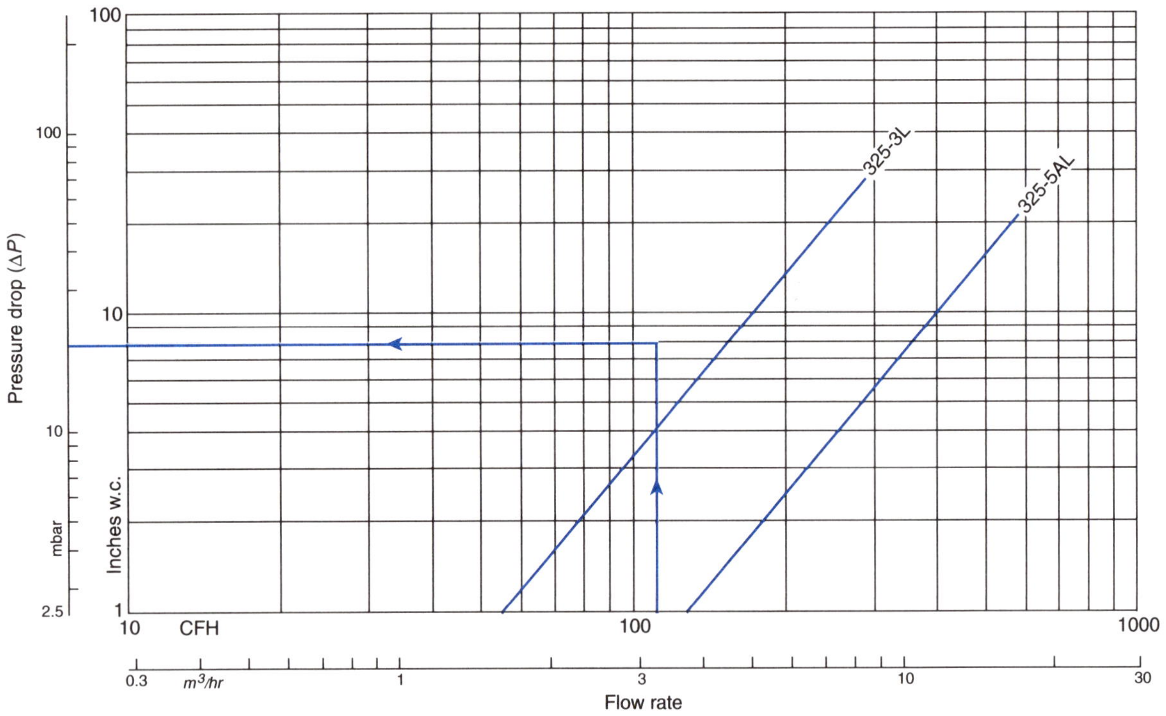


***EXHIBIT C.1*** *Maxitrol Regulator Capacity Chart. (Courtesy of Maxitrol)*

**COMMENTARY TABLE C.3** Extract from Table 6.2(p)

| Flow Designation | 13 | 15 | 18 | 19 | 23 |
|---|---|---|---|---|---|
| **Length (ft)** | | | **Tube Size (EHD)*** | | |
| 10 | 270 | 353 | 587 | 700 | 1,100 |
| 25 | 166 | 220 | 374 | 444 | 709 |
| 30 | 151 | 200 | 342 | 405 | 650 |
| 40 | 129 | 172 | 297 | 351 | 567 |
| 50 | 115 | 154 | 266 | 314 | 510 |
| 75 | 93 | 124 | 218 | 257 | 420 |
| 80 | 89 | 120 | 211 | 249 | 407 |
| 100 | 79 | 107 | 189 | 222 | 366 |

*EHD = Equivalent hydraulic diameter, which is a measure of the relative hydraulic efficiency between different tubing sizes. The greater the value of EHD, the greater the gas capacity of the tubing.

capacity of 189 cfh (5.35 m/hr), which is acceptable, so EHD 18 is selected. Larger sizes are also acceptable if desired for other reasons.

Note that CSST manufacturers do not offer products in all sizes. If the manufacturer does not offer EHD 18 tubing, the next larger size offered by the manufacturer should be used.

(c) Table 6.2(p) shows that EHD size 18 should be used. Note that it is not unusual to oversize the supply line by 25 to 50 percent of the as-installed load. EHD size 18 has a capacity of 189 cfh (5.35 $m^3$/hr).

(5) From Section B (low-pressure zone):

(a) Distance from regulator to furnace is 15 ft (4.57 m).
(b) Load is 60 cfh (1.70 $m^3$/hr).
(c) Table 6.2(n) shows that EHD size 13 should be used.

(6) From Section C (low-pressure zone):

(a) Distance from regulator to water heater is 10 ft (3 m).
(b) Load is 30 cfh (0.85 $m^3$/hr).
(c) Table 6.2(n) shows that EHD size 13 should be used.

(7) From Section D (low-pressure zone):

(a) Distance from regulator to dryer is 25 ft (7.62 m).
(b) Load is 20 cfh (0.57 $m^3$/hr).
(c) Table 6.2(n) shows that EHD size 13 should be used.

Commentary Table C.4 is an excerpt from Table 6.2(n). It shows the portion of Table 6.2(n) relevant to this example. Note that the 100 ft length row is highlighted because the length of CSST from the meter to the regulator is 100 ft. Read across the row to find a capacity equal to or greater than that needed. In this example, the load is 110 cfh (3.11 m/hr) and a pipe with an EHD of 15 has a capacity of 107 cfh (3.01 m/hr), which is not acceptable. EHD 18 has a capacity of 189 cfh (5.35 m/hr), which is acceptable, so EHD 18 is selected. Larger sizes are also acceptable if desired for other reasons. Finally, the individual pipe sections are sized using Table 6.2(o).

***COMMENTARY TABLE C.4 Extract from Table 6.2 (n)***

| Flow Designation | 13 | 15 | 18 | 19 | 23 |
|---|---|---|---|---|---|
| Length (ft) | Tube Size (EHD) | | | | |
| 10 | 83 | 112 | 197 | 231 | 380 |
| 15 | 67 | 90 | 161 | 189 | 313 |
| 20 | 57 | 78 | 140 | 164 | 273 |
| 25 | 51 | 69 | 125 | 147 | 245 |
| 30 | 46 | 63 | 115 | 134 | 225 |
| 40 | 39 | 54 | 100 | 116 | 196 |
| 50 | 35 | 48 | 89 | 104 | 176 |
| 60 | 32 | 44 | 82 | 95 | 161 |
| 70 | 29 | 41 | 76 | 88 | 150 |
| 80 | 27 | 38 | 71 | 82 | 141 |
| 90 | 26 | 36 | 67 | 77 | 133 |
| 100 | 24 | 34 | 63 | 73 | 126 |

**Section B: 15 ft Length.** Because there is no 15 ft length in the table, the next longest length of 25 ft is used. Reading across, we find the smallest CSST size will carry 270 cfh (7.65 m/hr), which is more than the required 60 cfh (1.7 m/hr). EHD 13 is used.

**Section C: 10 ft Length.** Reading across the 10 ft row, we find the smallest CSST size will carry 270 cfh (7.65 m/hr), which is more than the required 30 cfh (0.85 m/hr). EHD 13 is used.

**Section D: 25 ft Length.** Reading across the 25 ft row, we find the smallest CSST size will carry 270 cfh (7.65 m/hr), which is more than the required 20 cfh (0.57 m/hr). EHD 13 is used.

Remember that CSST manufacturers do not offer products in all sizes. If the manufacturer does not offer EHD 13 tubing, the next larger size offered by the manufacturer should be used.

## C.8.3 Example 3 — Branch Length Method.

Determine the required semirigid copper tubing size of each section of the piping system shown in Figure C.8.3, with a designated pressure drop of 1 in. w.c. (250 Pa) (using the Branch Length Method). The gas to be used has 0.60 specific gravity and a heating value of 1000 Btu/ft$^3$ (37.5 MJ/m$^3$).

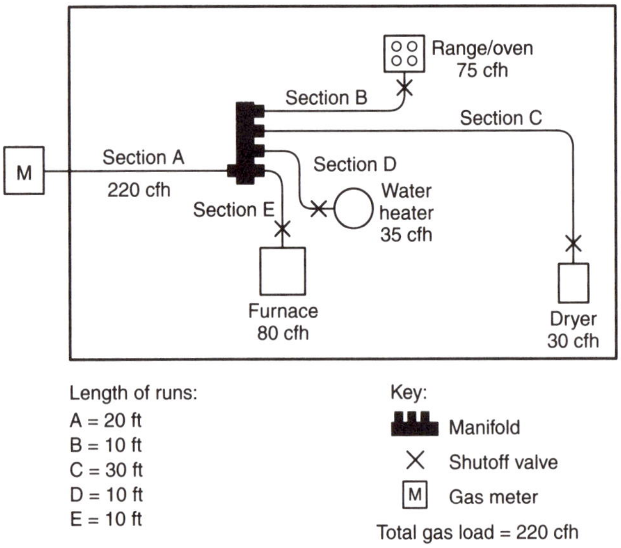


*FIGURE C.8.3 Piping Plan Showing a Copper Tubing System.*

*Solution*

(1) Section A:

(a) The length of tubing from the point of delivery to the most remote appliance is 50 ft (15 m), A + C.

(b) Use this longest length to size Sections A and C.

(c) Using the row marked 50 ft (15 m) in Table 6.2(h), Section A supplying 220 cfh (6.23 m$^3$/hr) for four appliances requires 1 in. tubing.

(2) Section B:

(a) The length of tubing from the point of delivery to the range/oven at the end of Section B is 30 ft (9.14 m), A + B.

(b) Use this branch length to size Section B only.

(c) Using the row marked 30 ft (9.14 m) in Table 6.2(h), Section B supplying 75 cfh (2.12 m$^3$/hr) for the range/oven requires $^1/_2$ in. tubing.

(3) Section C:

(a) The length of tubing from the point of delivery to the dryer at the end of Section C is 50 ft (15 m), A + C.

(b) Use this branch length (which is also the longest length) to size Section C.

(c) Using the row marked 50 ft (15 m) in Table 6.2(h), Section C supplying 30 cfh (0.85 m$^3$/hr) for the dryer requires $^3/_8$ in. tubing.

(4) Section D:

(a) The length of tubing from the point of delivery to the water heater at the end of Section D is 30 ft (9.14 m), A + D.

(b) Use this branch length to size Section D only.
(c) Using the row marked 30 ft (9.14 m) in Table 6.2(h), Section D supplying 35 cfh (34.69 $m^3$/hr) for the water heater requires $^3/_8$ in. tubing.

(5) Section E:

(a) The length of tubing from the point of delivery to the furnace at the end of Section E is 30 ft (9.14 m), A + E.
(b) Use this branch length to size Section E only.
(c) Using the row marked 30 ft (9.14 m) in Table 6.2(h), Section E supplying 80 cfh (0.99 $m^3$/hr) for the furnace requires $^1/_2$ in. tubing.

Example 3 — Branch Length Method of the code — is summarized in Commentary Table C.5, which is an excerpt from Table 6.2(h). Commentary Table C.5 shows the portions of Table 6.2(h) relevant to this example. Note that the 30 ft and 50 ft length rows are highlighted because these are the lengths from the meter to each appliance. Read across to find a capacity equal to or greater than the capacity needed.

***COMMENTARY TABLE C.5*** *Extract from Table 6.2(h)*

| | *Tube Size (in.)* | | | | | |
|---|---|---|---|---|---|---|
| ***Length (ft)*** | $^1/_4$ | $^3/_8$ | $^1/_2$ | $^5/_8$ | $^3/_4$ | 1 |
| 10 | 39 | 80 | 162 | 283 | 402 | 859 |
| 20 | 27 | 55 | 111 | 195 | 276 | 590 |
| 30 | 21 | 44 | 89 | 156 | 222 | 474 |
| 40 | 18 | 38 | 77 | 134 | 190 | 406 |
| 50 | 16 | 33 | 68 | 119 | 168 | 359 |

When using the branch length method, first determine the longest run of piping from the point of delivery to the most remote outlet. In this example, Sections A and C, with a combined length of 50 ft, comprise the longest length of piping. Size the pipe sections individually for the given lengths.

**Section A: 20 ft Length.** Section A is part of the longest length run. Therefore, size Section A using the longest length of 50 ft. Locate the 50 ft row, and read across to find a value equal to or exceeding the required gas flow of 220 cfh (6.23 m/hr). From the table, we find that $^3/_4$ in. tube has a capacity of 168 cfh (4.76 m/hr), which is insufficient, and that 1 in. tube has a capacity of 359 cfh (10.17 m/hr), which is acceptable. Thus, 1 in. tube is used for Section A.

**Section B: 10 ft Length.** To size the 10 ft run from the manifold to the range/oven, use the total length from the point of delivery (marked M in Figure C.8.3) to the range/oven. In this case, add the 10 ft from the manifold to the range/oven to the 20 ft from the meter (point of delivery) to the manifold, for a total length of 30 ft. Reading across the 30 ft row, we find that $^1/_2$ in. tube will carry 89 cfh (2.5 m/hr), which is more than the required 75 cfh (2.1 m/hr). Thus, $^1/_2$ in. tube is used for Section B.

**Section C: 30 ft Length.** To size the 30 ft run from the manifold to the dryer, use the total length from the point of delivery (marked M in Figure C.8.3) to the range/oven. In this case, add the 30 ft from the manifold to the range/oven to the 20 ft from the meter (point of delivery) to the manifold, for a total length of 50 ft. Reading across the 50 ft row, we find that $^3/_8$ in. tube will carry 44 cfh (1.2 m/hr), which is more than the required 30 cfh (0.85 m/hr). Thus, $^3/_8$ in. tube is used for Section C.

**Section D: 10 ft Length.** To size the 10 ft run from the manifold to the water heater, use the total length from the point of delivery (marked M in Figure C.8.3) to the water heater. In this case, add the 10 ft from the manifold to the range/oven to the 20 ft from the point of delivery to the manifold, for a total length of 30 ft. Reading across the 30 ft row, we find that $^3/_8$ in. tube will carry 33 cfh (0.93 m/hr), which is more than the required 30 cfh (0.85 m/hr). Thus, $^3/_8$ in. tube is used for Section D.

**Section E: 10 ft Length.** To size the 10 ft run from the manifold to the furnace, use the total length from the point of delivery (marked M in Figure C.8.3) to the furnace. In this case, add the 10 ft from the manifold to the furnace to the 20 ft from the meter (point of delivery) to the manifold, for a total length of 30 ft. Reading across the 30 ft row, we find that $^1/_2$ in. tube will carry 89 cfh (2.5 m/hr), which is more than the required 80 cfh (2.3 m/hr). Thus, $^1/_2$ in. tube is used for Section E.

The results are shown in Commentary Table C.6.

***COMMENTARY TABLE C.6** Results for Example 3 — Branch Length Method*

| Section | Section Length (ft) | Length from Point of Delivery (ft) | Size (in.) |
|---|---|---|---|
| A | 20 | 50 | 1 |
| B | 10 | 30 | $^1/_2$ |
| C | 30 | 50 | $^3/_8$ |
| D | 10 | 30 | $^3/_8$ |
| E | 10 | 30 | $^1/_2$ |

Worksheet C.2 summarizes the calculation for Example 3 — Branch Length Method. A full-sized, reproducible copy of this form is included in Supplement 6 of this handbook.

### C.8.4 Example 4 — Modification to Existing Piping System.

Determine the required CSST size for Section G (retrofit application) of the piping system shown in Figure C.8.4, with a designated pressure drop of 0.50 in. w.c. (125 Pa) using the Branch Length Method. The gas to be used has 0.60 specific gravity and a heating value of 1000 Btu/ft$^3$ (37.5 MJ/m$^3$).

*Solution*

(1) The length of pipe and CSST from the point of delivery to the retrofit appliance (barbecue) at the end of Section G is 40 ft (12.19 m), A + B + G.
(2) Use this branch length to size Section G.
(3) Assume the CSST manufacturer has tubing sizes or EHDs of 13, 18, 23, and 30.
(4) Using the row marked 40 ft (12.19 m) in Table 6.2(m), Section G supplying 40 cfh (1.13 m$^3$/hr) for the barbecue requires EHD 18 CSST.
(5) The sizing of Sections A, B, F, and E must be checked to ensure adequate gas carrying capacity since an appliance has been added to the piping system. See C.8.1 for details.

Example 4 (Figure C.8.4) shows how to size an additional branch using the branch length method and the CSST sizing chart found in Table 6.2(m). Two existing pipe runs — Section A and Section B — will carry the additional load. These must be recalculated to determine whether they will handle the additional load of the barbeque. The longest length is needed to calculate A and B, which is the sum of A, B, C, and D, or 60 ft.

## CALCULATION WORKSHEET: PIPE SIZING, BRANCH LENGTH METHOD

**Step 1:**

- Draw a sketch of a piping system in the space to the right.

**Step 2:**

- Enter the system information. Note that demand is the amount of gas flowing through a section of pipe.
- Use total Btu/hr rating/1000 ($ft^3$/hr) for natural gas.
- Use total Btu/hr for propane.

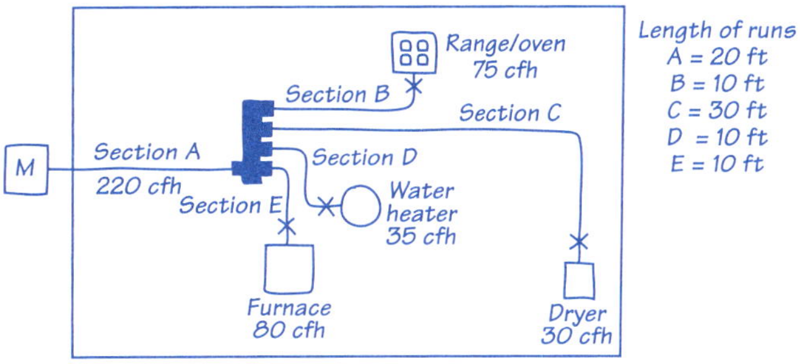


Pipe system sketch

**Step 3:**

- Determine the gas used and system pressure, and enter it to the right.
- Determine the piping material and enter it to the right.
- Select the appropriate pipe sizing table from Chapter 6 and enter it to the right.

Gas: Natural

Pressure: 7 in. w.c.

Piping material: Copper

Table used: 6.2 (h)

Pressure drop: 1.0 in. w.c.

**Step 4:**

- On the sketch above, label the section of pipe from the point of delivery (meter or regulator) to the manifold as Section A.

**Table 1 Piping System Table**

| Section | Demand (cfh) | Section length | Size |
|---|---|---|---|
| A | 220 | 50 ft | 1 in. |
| | | | |
| | | | |
| | | | |
| | | | |

**Step 5:**

- Determine the length of the branch serving each appliance. Enter this length in Table 1.

**Step 6:**

- Enter the input rating for each appliance in Table 2. For natural gas appliances, enter the input rating in $ft^3$/hr. For propane appliances, enter the input rating in thousands of Btu/hr.

**Step 7:**

- From the table, determine the length of each pipe section using the appropriate table, using only the row with the longest length. Round up to the lengths in the table. Read across until a capacity equal to or greater than the required demand for the section is found. Read up to find the size. Repeat for each section of piping. Enter this size in Table 2.

**Table 2 Appliances Table**

| Appliance | Demand | Section length | Size |
|---|---|---|---|
| Furnace | 80 | 30 ft | 1/2 in. |
| Furnace | | | |
| Water heater | 35 | 30 ft | 3/8 in. |
| Water heater | | | |
| Range | 75 | 30 ft | 1/2 in. |
| Oven | | | |
| Dryer | 30 | 50 ft | 3/8 in. |
| Other | | | |
| Other | | | |
| Other | | | |
| Other | | | |
| Total | 220 | — | — |

**Job:** 25 Main Street **Prepared by:** TL **Date:** 1/1/09



***WORKSHEET C.2*** *Sample Calculation for Determining Pipe Size Using the Branch Length Method.*

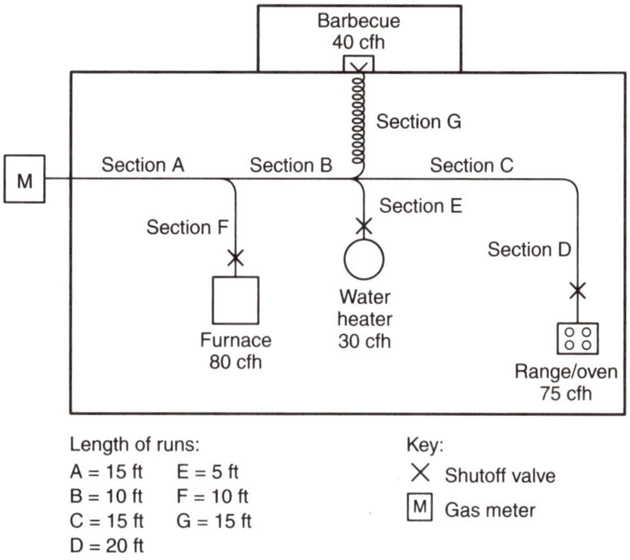


***FIGURE C.8.4*** *Piping Plan Showing Modification to an Existing Piping System.*

**Section A:** This 15 ft length of pipe serves all the appliances, with a total capacity of 225 cfh (6.37 m/hr). Using the sizing Table 6.2(m) on the 60 ft row, the pipe would need to have an EHD of 37.

**Section B:** This 10 ft length of pipe serves all the appliances less the 80 cfh (2.3 m/hr) of the furnace, which is a total of 145 cfh (4.1 m/hr). Using Table 6.2(m) on the 60 ft row, the pipe would need to have an EHD of 31.

**Section G:** This 15 ft pipe is a branch and will be sized based on its distance from the meter (the sum of A, B, and G), which is 40 ft. The load on this pipe is 40 cfh (1.1 m/hr). Using Table 6.2 (m), the row for 40 ft, the pipe would need to have an EHD of 18.

## C.8.5 Example 5 — Calculating Pressure Drops Due to Temperature Changes.

A test piping system is installed on a warm autumn afternoon when the temperature is 70 degrees. In accordance with local custom, the new piping system is subjected to an air pressure test at 20 psi. Overnight, the temperature drops, and when the inspector shows up first thing in the morning the temperature is 40 degrees.

If the volume of the piping system is unchanged, the formula based on Boyle's and Charles' law for determining the new pressure at a reduced temperature is as follows:

(1)
$$\frac{T_1}{T_2} = \frac{P_1}{P_2}$$

where:

$T_1$ = initial temperature [absolute ($T_1$ + 459)]

$T_2$ = final temperature [absolute ($T_2$ + 459)]

$P_1$ = initial pressure [psia ($P_1$ + 14.7)]

$P_2$ = final pressure [psia ($P_2$ + 14.7)]

(2)
$$\frac{(70 + 459)}{(40 + 459)} = \frac{(20 + 14.7)}{(P_2 + 14.7)}$$

(3)
$$\frac{529}{499} = \frac{34.7}{(P_2 + 14.7)}$$

(4)
$$(P_2 + 14.7) = \frac{34.7}{1.06}$$

$$P_2 = 32.7 - 14.7$$
$$P_2 = 18 \text{ psi}$$

Therefore, you could expect the gauge to register 18 psi when the ambient temperature is 40°F.

### C.8.6 Example 6 — Pressure Drop per 100 ft of Pipe Method.

Using the layout shown in Figure C.8.1 and $\Delta H$ = pressure drop, in. w.c. (27.7 in. $H_2O$ = 1 psi), proceed as follows:

(1) Length to A = 20 ft, with 35,000 Btu/hr:

For $^1/_2$ in. pipe:

$$\Delta H = \frac{20 \text{ ft}}{100 \text{ ft}} \times 0.3 \text{ in. w.c.} = 0.06 \text{ in. w.c.}$$

(2) Length to B = 15 ft, with 75,000 Btu/hr:

For $^3/_4$ in. pipe:

$$\Delta H = \frac{15 \text{ ft}}{100 \text{ ft}} \times 0.3 \text{ in. w.c.} = 0.045 \text{ in. w.c.}$$

(3) Section 1 = 10 ft, with 110,000 Btu/hr. Here a choice is available:

For 1 in. pipe:

$$\Delta H = \frac{10 \text{ ft}}{100 \text{ ft}} \times 0.2 \text{ in. w.c.} = 0.02 \text{ in. w.c.}$$

For $^3/_4$ in. pipe:

$$\Delta H = \frac{10 \text{ ft}}{100 \text{ ft}} \times \left[0.5 \text{ in. w.c.} + \left(\frac{110{,}000 \text{ Btu/hr} - 104{,}000 \text{ Btu/hr}}{147{,}000 \text{ Btu/hr} - 104{,}000 \text{ Btu/hr}}\right) \times (1.0 \text{ in. w.c.} - 0.5 \text{ in. w.c.})\right]$$
$$= 0.1 \times 0.57 \text{ in. w.c.} \approx 0.6 \text{ in. w.c.})$$

*Notice that the pressure drop for 110,000 Btu/hr between 104,000 Btu/hr and 147,000 Btu/hr has been interpolated.*

(4) Section 2 = 20 ft, with 135,000 Btu/hr. Here a choice is available:
For 1 in. pipe:

$$\Delta H = \frac{20 \text{ ft}}{100 \text{ ft}} \times \left[0.2 \text{ in. w.c.} + \frac{14{,}000 \text{ Btu/hr}}{27{,}000 \text{ Btu/hr}} \times \Delta 0.1 \text{ in. w.c.}\right] = 0.05 \text{ in. w.c.}$$

For $^3/_4$ in. pipe:

$$H = \frac{20 \text{ ft}}{100 \text{ ft}} \times 1.0 \text{ in. w.c.} = 0.2 \text{ in. w.c.}$$

*Notice that the pressure drop for 135,000 Btu/hr between 121,000 Btu/hr and 148,000 Btu/hr has been interpolated, but interpolation was not used for the $^3/_4$ in. pipe (trivial for 104,000 Btu/hr to 147,000 Btu/hr).*

(5) Section 3 = 30 ft, with 245,000 Btu/hr. Here a choice is available:
For 1 in. pipe:

$$\Delta H = \frac{30\ \text{ft}}{100\ \text{ft}} \times 1.0\ \text{in. w.c.} = 0.3\ \text{in. w.c.}$$

For $1^1/_4$ in. pipe:

$$\Delta H = \frac{30\ \text{ft}}{100\ \text{ft}} \times 0.2\ \text{in. w.c.} = 0.06\ \text{in. w.c.}$$

*Notice that interpolation was not used for these options, since the table values are close to the 245,000 Btu/hr carried by that section.*

(6) The total pressure drop is the sum of the section approaching A, Section 1, and Section 3, or either of the following, depending on whether an absolute minimum is required or the larger drop can be accommodated:
Minimum Pressure Drop to farthest appliance:
$\Delta H$ = 0.06 in. w.c. + 0.02 in. w.c. + 0.06 in. w.c = 0.14 in. w.c.
Larger Pressure Drop to the farthest appliance:
$\Delta H$ = 0.06 in. w.c. + 0.06 in. w.c. + 0.3 in. w.c. = 0.42 in. w.c.
*Notice that Section 2 and the run to B do not enter into this calculation, provided that the appliances have similar input pressure requirements.*
For SI units, 1 Btu/hr = 0.293 W, 1 $ft^3$ = 0.028 $m^3$, 1 ft = 0.305 m, 1 in. w.c. = 249 Pa.

# Suggested Method of Checking for Leakage

ANNEX D

*This annex is not a part of the requirements of this NFPA document but is included for informational purposes only.*

Annex D contains a method for checking for leakage from gas piping systems. This method is widely used, but it is not the only method that can be used to check for leakage. Subsection 8.2.3 requires checking for leakage after the pressure test has been passed or, for a piping system, after an interruption of service. Paragraph A.8.2.3 refers to Annex D. Checking for leakage is needed in addition to the pressure test because the leakage check tests more than the piping itself. This procedure tests the appliance connectors and the appliances up to their operating valves.

Regardless of which method is used, it is strongly recommended that all companies use the leakage check consistently and accurately.

## D.1 Use of Lights

Artificial illumination used in connection with a search for gas leakage should be restricted to battery-operated flashlights (preferably of the safety type) or approved safety lamps. In searching for leaks, electric switches should not be operated. If electric lights are already turned on, they should not be turned off.

## D.2 Leak Check Using the Gas Meter

Immediately prior to the leak check, it should be determined that the meter is in operating condition and has not been bypassed.

The leak check can be done by carefully watching the test dial of the meter to determine whether gas is passing through the meter. To assist in observing any movement of the test hand, wet a small piece of paper and paste its edge directly over the centerline of the hand as soon as the gas is turned on. This observation should be made with the test hand on the upstroke. Table D.2 can be used for determining the length of observation time.

In case careful observation of the test hand for a sufficient length of time reveals no movement, the piping should be purged and a small gas burner turned on and lighted and the hand of the test dial again observed. If the dial hand moves (as it should), it shows that the meter is operating properly. If the test hand does not move or register flow of gas through the meter to the small burner, the meter is defective and the gas should be shut off and the serving gas supplier notified.

***TABLE D.2*** *Observation Times for Various Meter Dials*

| ***Dial Styles ($ft^3$)*** | ***Test Time (minutes)*** |
|---|---|
| 1/4 | 5 |
| 1/2 | 5 |
| 1 | 7 |
| 2 | 10 |
| 5 | 20 |
| 10 | 30 |

For SI units, 1 $ft^3$ = 0.028 $m^3$.

## D.3 Leak Check Not Using a Meter

This test can be done using one of the following methods:

(1) *For Any Gas System.* To an appropriate checkpoint, attach a manometer or pressure gauge between the inlet to the piping system and the first regulator in the piping system,

momentarily turn on the gas supply, and observe the gauging device for pressure drop with the gas supply shut off. No discernible drop in pressure should occur during a period of 3 minutes.

NEW (2) *For Gas Systems Using Undiluted LP-Gas System Preparation for Propane.* A leak check performed on an LP-Gas system being placed back in service may be performed by using one of the following methods:

(a) By inserting a pressure gauge between the container gas shutoff valve and the first regulator in the system, admitting full container pressure to the system and then closing the container shutoff valve. Enough gas should then be released from the system to lower the pressure gauge reading by 10 psi (69 kPa). The system should then be allowed to stand for 3 minutes without showing an increase or a decrease in the pressure gauge reading.

(b) For systems serving appliances that receive gas at pressures of $^1/_2$ psi (3.5 kPa) or less, by inserting a water manometer or pressure gauge into the system downstream of the final system regulator, pressurizing the system with either fuel gas or air to a test pressure of 9 in. w.c. ± $^1/_2$ in. w.c. (2.2 kPa ± 0.1 kPa), and observing the device for a pressure change. If fuel gas is used as a pressure source, it is necessary to pressurize the system to full operating pressure, close the container service valve, and then release enough gas from the system through a range burner valve or other suitable means to drop the system pressure to 9 in. w.c. ± $^1/_2$ in. w.c. (2.2 kPa ± 0.1 kPa). This ensures that all regulators in the system are unlocked and that a leak anywhere in the system is communicated to the gauging device. The gauging device should indicate no loss or gain of pressure for a period of 3 minutes.

(c) By inserting a 30 psi (207 kPa) pressure gauge on the downstream side of the first-stage regulator, admitting normal operating pressure to the system and then closing the container valve. Enough pressure should be released from the system to lower the pressure gauge reading by 5 psi (34.5 kPa). The system should be allowed to stand for 3 minutes without showing an increase or a decrease in pressure gauge reading.

Sections D.2 and D.3 describe methods of checking for leakage. The first method in Section D.2 can be used for both propane and natural gas systems and involves the use of a gas meter with a test hand. The tester must be sure that the gas meter is functioning and that all equipment valves are off. The system is then pressurized, and the test dial is observed carefully for the amount of time indicated in Table D.2. If there is a leak, the test dial will move.

Section D.3(1) can be used for propane and natural gas systems that do not have a meter. A manometer is connected to the piping or appliance orifice and is checked for a drop in pressure while the main supply valve is shut off for 3 minutes. Section D.3(2) contains the following three methods of testing propane systems without the use of a gas meter.

The first procedure requires the temporary installation of a pressure gauge between the propane container shutoff valve and the first stage pressure regulator. The section between the first and second regulator is then charged to normal pressure and the propane tank valve is closed, and then enough gas is drawn off to reduce the pressure to approximately 10 psi (69 kPa) below the initial pressure. The initial pressure will be the tank pressure, which will vary depending on the tank temperature. If the pressure falls, there is a leak in the piping system. If the pressure rises, the supply valve is leaking.

The second method is useful for most residential and commercial systems that serve appliances operating at less than $^1/_2$ psi (3.5 kPa). This method recognizes that some appliances incorporate regulators that can "lock up" and isolate the appliance from the piping system, which makes running a valid leak check impossible. A manometer or pressure gauge is connected downstream of the final stage pressure regulator (usually the second stage pressure regulator). The system is pressurized to its normal operating pressure, and then the pressure is lowered to 9 in. w.c. ± $^1/_2$ in. w.c. (2.2 kPa ± 0.1 kPa). If the manometer or gauge rises or falls after 3 minutes, the system passes the test. This test ensures that regulator lockup does not interfere with the leak check.

Note that D.3(2) was revised in the 2009 edition to delete a recommendation that the leak check using this method should include all regulators, including appliance regulators, and control valves in the system with all appliance shutoff valves open, so that the leak check would check for leakage in the piping system and the appliances. This change recognized that 8.2.3 requires that the piping system be leak checked, and does not require a leak check for the appliances. While Annex D provides a suggested method of checking for leakage, it is not a code requirement and any method can be used.

The third method, added in the 2006 edition, uses a pressure gauge with a full-scale reading of 30 psi (207 kPa), installed between the first and second stage pressure regulators. The system is pressurized, the propane tank container is closed, and the pressure at the meter is reduced by 5 psi (34.5 kPa) and should read about 5 psi. After 3 minutes, a drop in pressure indicates a leak in the system downstream of the first stage pressure regulator, and an increase in pressure indicates a leaking propane container valve.

## D.4 When Leakage Is Indicated

If the meter test hand moves or a pressure drop on the gauge is noted, all appliances and equipment or outlets supplied through the system should be examined to see whether they are shut off and do not leak. If they are found to be tight, the piping system has a leak.

# Suggested Emergency Procedure for Gas Leaks

ANNEX E

*This annex is not a part of the requirements of this NFPA document but is included for informational purposes only.*

Annex E is not mandated in the code but is included to provide information to workers and emergency responders to gas leakage situations. This annex provides a simple, stepwise procedure to deal with gas leakage in buildings and surrounding buildings and to minimize the possibility of ignition of the leaked gas.

## E.1

Where an investigation discloses a concentration of gas inside of a building, it is suggested the following immediate actions be taken:

(1) Clear the room, building, or area of all occupants. Do not re-enter the room, building, or area until the space has been determined to be safe.
(2) Use every practical means to eliminate sources of ignition. Take precautions to prevent smoking, striking matches, operating electrical switches or devices, opening furnace doors, and so on. If possible, cut off all electric circuits at a remote source to eliminate operation of automatic switches in the dangerous area. Safety flashlights designed for use in hazardous atmospheres are recommended for use in such emergencies.
(3) Notify all personnel in the area and the gas supplier from a telephone remote from the area of the leak.
(4) Ventilate the affected portion of the building by opening windows and doors.
(5) Shut off the supply of gas to the areas involved.
(6) Investigate other buildings in the immediate area to determine the presence of escaping gas therein.

Section E.1 offers a list of recommended actions to take after determining that there is a gas concentration inside a building. Following the procedures in this annex could minimize the risk of fire or explosion. The most important point is provided in E.1(1): Clear the building of all people. Occupants should be evacuated completely, and evacuation should be verified before continuing.

All possible ignition sources should be avoided, and the gas utility or supplier should be notified immediately to obtain as much assistance as possible in dealing with the possible leak. Notifying the local fire department might also be advisable. The space(s) should then be ventilated. Safety flashlights listed for Class I, Group D locations should be the only source of illumination.

A check should be made using a gas detector for possible sources of gas outside the space where the gas has accumulated. A leak could be located in an adjacent space or underground, in which case the gas could travel into and accumulate in the space where it was first noticed. Also, underground leaks can be deodorized by adsorption of the odorant by some soils. Therefore, the lack of gas odor does not ensure that no gas is present.

# Flow of Gas Through Fixed Orifices

ANNEX F

*This annex is not a part of the requirements of this NFPA document but is included for informational purposes only.*

Annex F provides orifice sizing tables and procedures for checking burner input. This knowledge enables field adjustment of input rate of appliances. Field adjustment can be applied on startup of large equipment, such as boilers and furnaces, that requires high-altitude adjustment. It must not be used for residential appliances unless specifically allowed by the appliance manufacturer.

## F.1 Use of Orifice Tables

### F.1.1 To Check Burner Input Not Using a Meter.

Gauge the size of the burner orifice and determine flow rate at sea level from Table F.1.1(a), Utility Gases (cubic feet per hour), or from Table F.1.1(b), LP-Gases (Btu per hour). When the specific gravity of the utility gas is other than 0.60, select the multiplier from Table F.1.1(c) for the specific gravity of the utility gas served and apply to the flow rate as determined from Table F.1.1(a). When the altitude is above 2000 ft (600 m), first select the equivalent orifice size at sea level using Table F.1.1(d), then determine the flow rate from Table F.1.1(a) or Table F.1.1(b) as directed. Having determined the flow rate (as adjusted for specific gravity and/or altitude where necessary), check the burner input at sea level with the manufacturer's rated input.

Table F.1.1(a) presents data for gas flow in standard cubic feet per hour at sea level, through orifices, as a function of orifice size and pressure at the orifice inlet. The data apply to gas with 0.60 specific gravity and an orifice coefficient of 0.90. The size is given in terms of numbered drill sizes, from #80 (the smallest) to #1 (the largest). For larger orifice sizes, refer to AGA Report No. 3, "Orifice Metering of Natural Gas."

Table F.1.1(b) presents data for propane and butane in Btu per hour at sea level for orifices from 0.008 in. (0.2 mm) through #18 drill size [0.1695 in. (4.3 mm)]. Because the #18 drill size delivers 233 MBH (68 MW), the sizes in this table cover all but the largest installations. The table covers gases with 11 in. w.c. (2.7 kPa) pressure at the orifice inlet and an orifice coefficient of 0.90 for the Btu per cubic foot and specific gravities given in the table.

Table F.1.1(c) presents the corrections [for Table F.1.1(a)] for gases with specific gravities from 0.45 through 1.40. The volume taken from Table F.1.1(a) must be multiplied by the factor in Table F.1.1(c) to obtain the volume of gas of the new specific gravity that will flow through an orifice of a given size.

Table F.1.1(d) presents data for reducing the input rate of appliances installed at elevations of 2000 ft (600 m) and higher. As altitude increases, the atmosphere becomes less dense. Consequently, less oxygen is available (in the air) for the combustion of fuel. There may also be a slight reduction in the Btu per hour passing through the orifice, caused by a reduction in the density of the fuel gas (presuming the same orifice size and manifold pressure as at sea level) but not enough to compensate for the reduction in available oxygen.

*TABLE F.1.1(a) Utility Gases (cubic feet per hour at sea level)*

| *Orifice or Drill Size* | *Pressure at Orifice (in. w.c.)* | | | | | | | | |
|---|---|---|---|---|---|---|---|---|---|
| | *3* | *3.5* | *4* | *5* | *6* | *7* | *8* | *9* | *10* |
| 80 | 0.48 | 0.52 | 0.55 | 0.63 | 0.69 | 0.73 | 0.79 | 0.83 | 0.88 |
| 79 | 0.55 | 0.59 | 0.64 | 0.72 | 0.80 | 0.84 | 0.90 | 0.97 | 1.01 |
| 78 | 0.70 | 0.76 | 0.78 | 0.88 | 0.97 | 1.04 | 1.10 | 1.17 | 1.24 |
| 77 | 0.88 | 0.95 | 0.99 | 1.11 | 1.23 | 1.31 | 1.38 | 1.47 | 1.55 |
| 76 | 1.05 | 1.13 | 1.21 | 1.37 | 1.52 | 1.61 | 1.72 | 1.83 | 1.92 |
| 75 | 1.16 | 1.25 | 1.34 | 1.52 | 1.64 | 1.79 | 1.91 | 2.04 | 2.14 |
| 74 | 1.33 | 1.44 | 1.55 | 1.74 | 1.91 | 2.05 | 2.18 | 2.32 | 2.44 |
| 73 | 1.51 | 1.63 | 1.76 | 1.99 | 2.17 | 2.32 | 2.48 | 2.64 | 2.78 |
| 72 | 1.64 | 1.77 | 1.90 | 2.15 | 2.40 | 2.52 | 2.69 | 2.86 | 3.00 |
| 71 | 1.82 | 1.97 | 2.06 | 2.33 | 2.54 | 2.73 | 2.91 | 3.11 | 3.26 |
| 70 | 2.06 | 2.22 | 2.39 | 2.70 | 2.97 | 3.16 | 3.38 | 3.59 | 3.78 |
| 69 | 2.25 | 2.43 | 2.61 | 2.96 | 3.23 | 3.47 | 3.68 | 3.94 | 4.14 |
| 68 | 2.52 | 2.72 | 2.93 | 3.26 | 3.58 | 3.88 | 4.14 | 4.41 | 4.64 |
| 67 | 2.69 | 2.91 | 3.12 | 3.52 | 3.87 | 4.13 | 4.41 | 4.69 | 4.94 |
| 66 | 2.86 | 3.09 | 3.32 | 3.75 | 4.11 | 4.39 | 4.68 | 4.98 | 5.24 |
| 65 | 3.14 | 3.39 | 3.72 | 4.28 | 4.62 | 4.84 | 5.16 | 5.50 | 5.78 |
| 64 | 3.41 | 3.68 | 4.14 | 4.48 | 4.91 | 5.23 | 5.59 | 5.95 | 6.26 |
| 63 | 3.63 | 3.92 | 4.19 | 4.75 | 5.19 | 5.55 | 5.92 | 6.30 | 6.63 |
| 62 | 3.78 | 4.08 | 4.39 | 4.96 | 5.42 | 5.81 | 6.20 | 6.59 | 6.94 |
| 61 | 4.02 | 4.34 | 4.66 | 5.27 | 5.77 | 6.15 | 6.57 | 7.00 | 7.37 |
| 60 | 4.21 | 4.55 | 4.89 | 5.52 | 5.95 | 6.47 | 6.91 | 7.35 | 7.74 |
| 59 | 4.41 | 4.76 | 5.11 | 5.78 | 6.35 | 6.78 | 7.25 | 7.71 | 8.11 |
| 58 | 4.66 | 5.03 | 5.39 | 6.10 | 6.68 | 7.13 | 7.62 | 8.11 | 8.53 |
| 57 | 4.84 | 5.23 | 5.63 | 6.36 | 6.96 | 7.44 | 7.94 | 8.46 | 8.90 |
| 56 | 5.68 | 6.13 | 6.58 | 7.35 | 8.03 | 8.73 | 9.32 | 9.92 | 10.44 |
| 55 | 7.11 | 7.68 | 8.22 | 9.30 | 10.18 | 10.85 | 11.59 | 12.34 | 12.98 |
| 54 | 7.95 | 8.59 | 9.23 | 10.45 | 11.39 | 12.25 | 13.08 | 13.93 | 14.65 |
| 53 | 9.30 | 10.04 | 10.80 | 12.20 | 13.32 | 14.29 | 15.27 | 16.25 | 17.09 |
| 52 | 10.61 | 11.46 | 12.31 | 13.86 | 15.26 | 16.34 | 17.44 | 18.57 | 19.53 |
| 51 | 11.82 | 12.77 | 13.69 | 15.47 | 16.97 | 18.16 | 19.40 | 20.64 | 21.71 |
| 50 | 12.89 | 13.92 | 14.94 | 16.86 | 18.48 | 19.77 | 21.12 | 22.48 | 23.65 |
| 49 | 14.07 | 15.20 | 16.28 | 18.37 | 20.20 | 21.60 | 23.06 | 24.56 | 25.83 |
| 48 | 15.15 | 16.36 | 17.62 | 19.88 | 21.81 | 23.31 | 24.90 | 26.51 | 27.89 |
| 47 | 16.22 | 17.52 | 18.80 | 21.27 | 23.21 | 24.93 | 26.62 | 28.34 | 29.81 |
| 46 | 17.19 | 18.57 | 19.98 | 22.57 | 24.72 | 26.43 | 28.23 | 30.05 | 31.61 |
| 45 | 17.73 | 19.15 | 20.52 | 23.10 | 25.36 | 27.18 | 29.03 | 30.90 | 32.51 |
| 44 | 19.45 | 21.01 | 22.57 | 25.57 | 27.93 | 29.87 | 31.89 | 33.96 | 35.72 |
| 43 | 20.73 | 22.39 | 24.18 | 27.29 | 29.87 | 32.02 | 34.19 | 36.41 | 38.30 |
| 42 | 23.10 | 24.95 | 26.50 | 29.50 | 32.50 | 35.24 | 37.63 | 40.07 | 42.14 |
| 41 | 24.06 | 25.98 | 28.15 | 31.69 | 34.81 | 37.17 | 39.70 | 42.27 | 44.46 |
| 40 | 25.03 | 27.03 | 29.23 | 33.09 | 36.20 | 38.79 | 41.42 | 44.10 | 46.38 |
| 39 | 26.11 | 28.20 | 30.20 | 34.05 | 37.38 | 39.97 | 42.68 | 45.44 | 47.80 |
| 38 | 27.08 | 29.25 | 31.38 | 35.46 | 38.89 | 41.58 | 44.40 | 47.27 | 49.73 |
| 37 | 28.36 | 30.63 | 32.99 | 37.07 | 40.83 | 43.62 | 46.59 | 49.60 | 52.17 |
| 36 | 29.76 | 32.14 | 34.59 | 39.11 | 42.76 | 45.77 | 48.88 | 52.04 | 54.74 |

*TABLE F.1.1(a) Continued*

| *Orifice or Drill Size* | *Pressure at Orifice (in. w.c.)* | | | | | | | | |
|---|---|---|---|---|---|---|---|---|---|
| | *3* | *3.5* | *4* | *5* | *6* | *7* | *8* | *9* | *10* |
| 35 | 32.36 | 34.95 | 36.86 | 41.68 | 45.66 | 48.78 | 52.10 | 55.46 | 58.34 |
| 34 | 32.45 | 35.05 | 37.50 | 42.44 | 46.52 | 49.75 | 53.12 | 56.55 | 59.49 |
| 33 | 33.41 | 36.08 | 38.79 | 43.83 | 48.03 | 51.46 | 54.96 | 58.62 | 61.55 |
| 32 | 35.46 | 38.30 | 40.94 | 46.52 | 50.82 | 54.26 | 57.95 | 61.70 | 64.89 |
| 31 | 37.82 | 40.85 | 43.83 | 49.64 | 54.36 | 58.01 | 61.96 | 65.97 | 69.39 |
| 30 | 43.40 | 46.87 | 50.39 | 57.05 | 62.09 | 66.72 | 71.22 | 75.86 | 79.80 |
| 29 | 48.45 | 52.33 | 56.19 | 63.61 | 69.62 | 74.45 | 79.52 | 84.66 | 89.04 |
| 28 | 51.78 | 55.92 | 59.50 | 67.00 | 73.50 | 79.50 | 84.92 | 90.39 | 95.09 |
| 27 | 54.47 | 58.83 | 63.17 | 71.55 | 78.32 | 83.59 | 89.27 | 95.04 | 99.97 |
| 26 | 56.73 | 61.27 | 65.86 | 74.57 | 81.65 | 87.24 | 93.17 | 99.19 | 104.57 |
| 25 | 58.87 | 63.58 | 68.22 | 77.14 | 84.67 | 90.36 | 96.50 | 102.74 | 108.07 |
| 24 | 60.81 | 65.67 | 70.58 | 79.83 | 87.56 | 93.47 | 99.83 | 106.28 | 111.79 |
| 23 | 62.10 | 67.07 | 72.20 | 81.65 | 89.39 | 94.55 | 100.98 | 107.49 | 113.07 |
| 22 | 64.89 | 70.08 | 75.21 | 85.10 | 93.25 | 99.60 | 106.39 | 113.24 | 119.12 |
| 21 | 66.51 | 71.83 | 77.14 | 87.35 | 95.63 | 102.29 | 109.24 | 116.29 | 122.33 |
| 20 | 68.22 | 73.68 | 79.08 | 89.49 | 97.99 | 104.75 | 111.87 | 119.10 | 125.28 |
| 19 | 72.20 | 77.98 | 83.69 | 94.76 | 103.89 | 110.67 | 118.55 | 125.82 | 132.36 |
| 18 | 75.53 | 81.57 | 87.56 | 97.50 | 108.52 | 116.03 | 123.92 | 131.93 | 138.78 |
| 17 | 78.54 | 84.82 | 91.10 | 103.14 | 112.81 | 120.33 | 128.52 | 136.82 | 143.91 |
| 16 | 82.19 | 88.77 | 95.40 | 107.98 | 118.18 | 126.78 | 135.39 | 144.15 | 151.63 |
| 15 | 85.20 | 92.02 | 98.84 | 111.74 | 122.48 | 131.07 | 139.98 | 149.03 | 156.77 |
| 14 | 87.10 | 94.40 | 100.78 | 114.21 | 124.44 | 133.22 | 142.28 | 151.47 | 159.33 |
| 13 | 89.92 | 97.11 | 104.32 | 118.18 | 128.93 | 138.60 | 148.02 | 157.58 | 165.76 |
| 12 | 93.90 | 101.41 | 108.52 | 123.56 | 135.37 | 143.97 | 153.75 | 163.69 | 172.13 |
| 11 | 95.94 | 103.62 | 111.31 | 126.02 | 137.52 | 147.20 | 157.20 | 167.36 | 176.03 |
| 10 | 98.30 | 106.16 | 114.21 | 129.25 | 141.82 | 151.50 | 161.81 | 172.26 | 181.13 |
| 9 | 100.99 | 109.07 | 117.11 | 132.58 | 145.05 | 154.71 | 165.23 | 175.91 | 185.03 |
| 8 | 103.89 | 112.20 | 120.65 | 136.44 | 149.33 | 160.08 | 170.96 | 182.00 | 191.44 |
| 7 | 105.93 | 114.40 | 123.01 | 139.23 | 152.56 | 163.31 | 174.38 | 185.68 | 195.30 |
| 6 | 109.15 | 117.88 | 126.78 | 142.88 | 156.83 | 167.51 | 178.88 | 190.46 | 200.36 |
| 5 | 111.08 | 119.97 | 128.93 | 145.79 | 160.08 | 170.82 | 182.48 | 194.22 | 204.30 |
| 4 | 114.75 | 123.93 | 133.22 | 150.41 | 164.36 | 176.18 | 188.16 | 200.25 | 210.71 |
| 3 | 119.25 | 128.79 | 137.52 | 156.26 | 170.78 | 182.64 | 195.08 | 207.66 | 218.44 |
| 2 | 128.48 | 138.76 | 148.61 | 168.64 | 184.79 | 197.66 | 211.05 | 224.74 | 235.58 |
| 1 | 136.35 | 147.26 | 158.25 | 179.33 | 194.63 | 209.48 | 223.65 | 238.16 | 250.54 |

For SI units, 1 Btu/hr = 0.293 W, 1 $ft^3$ = 0.028 $m^3$, 1 ft = 0.305 m, 1 in. w.c. = 249 Pa.

Notes:

(1) Specific gravity = 0.60; orifice coefficient = 0.90.

(2) For utility gases of another specific gravity, select multiplier from Table F.1.1(c). For altitudes above 2000 ft, first select the equivalent orifice size at sea level from Table F.1.1(d).

*TABLE F.1.1(b) LP-Gases (Btu per hour at sea level)*

| *Orifice or Drill Size* | *Propane* | *Butane* | *Orifice or Drill Size* | *Propane* | *Butane* |
|---|---|---|---|---|---|
| 0.008 | 519 | 589 | 50 | 39,842 | 45,168 |
| 0.009 | 656 | 744 | 49 | 43,361 | 49,157 |
| 0.010 | 812 | 921 | 48 | 46,983 | 53,263 |
| 0.011 | 981 | 1,112 | 47 | 50,088 | 56,783 |
| 0.012 | 1,169 | 1,326 | 46 | 53,296 | 60,420 |
| 80 | 1,480 | 1,678 | 45 | 54,641 | 61,944 |
| 79 | 1,708 | 1,936 | 44 | 60,229 | 68,280 |
| 78 | 2,080 | 2,358 | 43 | 64,369 | 72,973 |
| 77 | 2,629 | 2,980 | 42 | 71,095 | 80,599 |
| 76 | 3,249 | 3,684 | 41 | 74,924 | 84,940 |
| 75 | 3,581 | 4,059 | 40 | 78,029 | 88,459 |
| 74 | 4,119 | 4,669 | 39 | 80,513 | 91,215 |
| 73 | 4,678 | 5,303 | 38 | 83,721 | 94,912 |
| 72 | 5,081 | 5,760 | 37 | 87,860 | 99,605 |
| 71 | 5,495 | 6,230 | 36 | 92,207 | 104,532 |
| 70 | 6,375 | 7,227 | 35 | 98,312 | 111,454 |
| 69 | 6,934 | 7,860 | 34 | 100,175 | 113,566 |
| 68 | 7,813 | 8,858 | 33 | 103,797 | 117,672 |
| 67 | 8,320 | 9,433 | 32 | 109,385 | 124,007 |
| 66 | 8,848 | 10,031 | 31 | 117,043 | 132,689 |
| 65 | 9,955 | 11,286 | 30 | 134,119 | 152,046 |
| 64 | 10,535 | 11,943 | 29 | 150,366 | 170,466 |
| 63 | 11,125 | 12,612 | 28 | 160,301 | 181,728 |
| 62 | 11,735 | 13,304 | 27 | 168,580 | 191,114 |
| 61 | 12,367 | 14,020 | 26 | 175,617 | 199,092 |
| 60 | 13,008 | 14,747 | 25 | 181,619 | 205,896 |
| 59 | 13,660 | 15,486 | 24 | 187,828 | 212,935 |
| 58 | 14,333 | 16,249 | 23 | 192,796 | 218,567 |
| 57 | 15,026 | 17,035 | 22 | 200,350 | 227,131 |
| 56 | 17,572 | 19,921 | 21 | 205,525 | 232,997 |
| 55 | 21,939 | 24,872 | 20 | 210,699 | 238,863 |
| 54 | 24,630 | 27,922 | 19 | 223,945 | 253,880 |
| 53 | 28,769 | 32,615 | 18 | 233,466 | 264,673 |
| 52 | 32,805 | 37,190 | | | |
| 51 | 36,531 | 41,414 | | | |

Notes:

| | Propane | Butane |
|---|---|---|
| (1) Btu per cubic foot | 2516 | 3280 |
| (2) Specific gravity | 1.52 | 2.01 |
| (3) Pressure at orifice (in. w.c.) | 11 | 11 |
| (4) Orifice coefficient | 0.9 | 0.9 |

(5) For altitudes above 2000 ft (610 m), first select the equivalent orifice size at sea level from Table F.1.1(d).

*TABLE F.1.1(c) Multipliers for Utility Gases of Another Specific Gravity*

| Specific Gravity | Multiplier | Specific Gravity | Multiplier |
|---|---|---|---|
| 0.45 | 1.155 | 0.95 | 0.795 |
| 0.50 | 1.095 | 1.00 | 0.775 |
| 0.55 | 1.045 | 1.05 | 0.756 |
| 0.60 | 1.000 | 1.10 | 0.739 |
| 0.65 | 0.961 | 1.15 | 0.722 |
| 0.70 | 0.926 | 1.20 | 0.707 |
| 0.75 | 0.894 | 1.25 | 0.693 |
| 0.80 | 0.866 | 1.30 | 0.679 |
| 0.85 | 0.840 | 1.35 | 0.667 |
| 0.90 | 0.817 | 1.40 | 0.655 |

*TABLE F.1.1(d) Equivalent Orifice Sizes at High Altitudes (includes 4% input reduction for each 1000 ft above sea level)*

| Orifice Size at Sea Level | Orifice Size Required at Other Elevations (ft) | | | | | | | | |
|---|---|---|---|---|---|---|---|---|---|
| | 2000 | 3000 | 4000 | 5000 | 6000 | 7000 | 8000 | 9000 | 10,000 |
| 1 | 2 | 2 | 3 | 3 | 4 | 5 | 7 | 8 | 10 |
| 2 | 3 | 3 | 4 | 5 | 6 | 7 | 9 | 10 | 12 |
| 3 | 4 | 5 | 7 | 8 | 9 | 10 | 12 | 13 | 15 |
| 4 | 6 | 7 | 8 | 9 | 11 | 12 | 13 | 14 | 16 |
| 5 | 7 | 8 | 9 | 10 | 12 | 13 | 14 | 15 | 17 |
| 6 | 8 | 9 | 10 | 11 | 12 | 13 | 14 | 16 | 17 |
| 7 | 9 | 10 | 11 | 12 | 13 | 14 | 15 | 16 | 18 |
| 8 | 10 | 11 | 12 | 13 | 13 | 15 | 16 | 17 | 18 |
| 9 | 11 | 12 | 12 | 13 | 14 | 16 | 17 | 18 | 19 |
| 10 | 12 | 13 | 13 | 14 | 15 | 16 | 17 | 18 | 19 |
| 11 | 13 | 13 | 14 | 15 | 16 | 17 | 18 | 19 | 20 |
| 12 | 13 | 14 | 15 | 16 | 17 | 17 | 18 | 19 | 20 |
| 13 | 15 | 15 | 16 | 17 | 18 | 18 | 19 | 20 | 22 |
| 14 | 16 | 16 | 17 | 18 | 18 | 19 | 20 | 21 | 23 |
| 15 | 16 | 17 | 17 | 18 | 19 | 20 | 20 | 22 | 24 |
| 16 | 17 | 18 | 18 | 19 | 19 | 20 | 22 | 23 | 25 |
| 17 | 18 | 19 | 19 | 20 | 21 | 22 | 23 | 24 | 26 |
| 18 | 19 | 19 | 20 | 21 | 22 | 23 | 24 | 26 | 27 |
| 19 | 20 | 20 | 21 | 22 | 23 | 25 | 26 | 27 | 28 |
| 20 | 22 | 22 | 23 | 24 | 25 | 26 | 27 | 28 | 29 |
| 21 | 23 | 23 | 24 | 25 | 26 | 27 | 28 | 28 | 29 |
| 22 | 23 | 24 | 25 | 26 | 27 | 27 | 28 | 29 | 29 |
| 23 | 25 | 25 | 26 | 27 | 27 | 28 | 29 | 29 | 30 |
| 24 | 25 | 26 | 27 | 27 | 28 | 28 | 29 | 29 | 30 |
| 25 | 26 | 27 | 27 | 28 | 28 | 29 | 29 | 30 | 30 |
| 26 | 27 | 28 | 28 | 28 | 29 | 29 | 30 | 30 | 30 |
| 27 | 28 | 28 | 29 | 29 | 29 | 30 | 30 | 30 | 31 |
| 28 | 29 | 29 | 29 | 30 | 30 | 30 | 30 | 31 | 31 |
| 29 | 29 | 30 | 30 | 30 | 30 | 31 | 31 | 31 | 32 |
| 30 | 30 | 31 | 31 | 31 | 31 | 32 | 32 | 33 | 35 |

*(continues)*

*TABLE F.1.1(d) Continued*

| Orifice Size at Sea Level | Orifice Size Required at Other Elevations (ft) | | | | | | | | |
|---|---|---|---|---|---|---|---|---|---|
| | 2000 | 3000 | 4000 | 5000 | 6000 | 7000 | 8000 | 9000 | 10,000 |
| 31 | 32 | 32 | 32 | 33 | 34 | 35 | 36 | 37 | 38 |
| 32 | 33 | 34 | 35 | 35 | 36 | 36 | 37 | 38 | 40 |
| 33 | 35 | 35 | 36 | 36 | 37 | 38 | 38 | 40 | 41 |
| 34 | 35 | 36 | 36 | 37 | 37 | 38 | 39 | 40 | 42 |
| 35 | 36 | 36 | 37 | 37 | 38 | 39 | 40 | 41 | 42 |
| 36 | 37 | 38 | 38 | 39 | 40 | 41 | 41 | 42 | 43 |
| 37 | 38 | 39 | 39 | 40 | 41 | 42 | 42 | 43 | 43 |
| 38 | 39 | 40 | 41 | 41 | 42 | 42 | 43 | 43 | 44 |
| 39 | 40 | 41 | 41 | 42 | 42 | 43 | 43 | 44 | 44 |
| 40 | 41 | 42 | 42 | 42 | 43 | 43 | 44 | 44 | 45 |
| 41 | 42 | 42 | 42 | 43 | 43 | 44 | 44 | 45 | 46 |
| 42 | 42 | 43 | 43 | 43 | 44 | 44 | 45 | 46 | 47 |
| 43 | 44 | 44 | 44 | 45 | 45 | 46 | 47 | 47 | 48 |
| 44 | 45 | 45 | 45 | 46 | 47 | 47 | 48 | 48 | 49 |
| 45 | 46 | 47 | 47 | 47 | 48 | 48 | 49 | 49 | 50 |
| 46 | 47 | 47 | 47 | 48 | 48 | 49 | 49 | 50 | 50 |
| 47 | 48 | 48 | 49 | 49 | 49 | 50 | 50 | 51 | 51 |
| 48 | 49 | 49 | 49 | 50 | 50 | 50 | 51 | 51 | 52 |
| 49 | 50 | 50 | 50 | 51 | 51 | 51 | 52 | 52 | 52 |
| 50 | 51 | 51 | 51 | 51 | 52 | 52 | 52 | 53 | 53 |
| 51 | 51 | 52 | 52 | 52 | 52 | 53 | 53 | 53 | 54 |
| 52 | 52 | 53 | 53 | 53 | 53 | 53 | 54 | 54 | 54 |
| 53 | 54 | 54 | 54 | 54 | 54 | 54 | 55 | 55 | 55 |
| 54 | 54 | 55 | 55 | 55 | 55 | 55 | 56 | 56 | 56 |
| 55 | 55 | 55 | 55 | 56 | 56 | 56 | 56 | 56 | 57 |
| 56 | 56 | 56 | 57 | 57 | 57 | 58 | 59 | 59 | 60 |
| 57 | 58 | 59 | 59 | 60 | 60 | 61 | 62 | 63 | 63 |
| 58 | 59 | 60 | 60 | 61 | 62 | 62 | 63 | 63 | 64 |
| 59 | 60 | 61 | 61 | 62 | 62 | 63 | 64 | 64 | 65 |
| 60 | 61 | 61 | 62 | 63 | 63 | 64 | 64 | 65 | 65 |
| 61 | 62 | 62 | 63 | 63 | 64 | 65 | 65 | 66 | 66 |
| 62 | 63 | 63 | 64 | 64 | 65 | 65 | 66 | 66 | 67 |
| 63 | 64 | 64 | 65 | 65 | 65 | 66 | 66 | 67 | 68 |
| 64 | 65 | 65 | 65 | 66 | 66 | 66 | 67 | 67 | 68 |
| 65 | 65 | 66 | 66 | 66 | 67 | 67 | 68 | 68 | 69 |
| 66 | 67 | 67 | 68 | 68 | 68 | 69 | 69 | 69 | 70 |
| 67 | 68 | 68 | 68 | 69 | 69 | 69 | 70 | 70 | 70 |
| 68 | 68 | 69 | 69 | 69 | 70 | 70 | 70 | 71 | 71 |
| 69 | 70 | 70 | 70 | 70 | 71 | 71 | 71 | 72 | 72 |
| 70 | 70 | 71 | 71 | 71 | 71 | 72 | 72 | 73 | 73 |
| 71 | 72 | 72 | 72 | 73 | 73 | 73 | 74 | 74 | 74 |
| 72 | 73 | 73 | 73 | 73 | 74 | 74 | 74 | 74 | 75 |
| 73 | 73 | 74 | 74 | 74 | 74 | 75 | 75 | 75 | 76 |
| 74 | 74 | 75 | 75 | 75 | 75 | 76 | 76 | 76 | 76 |
| 75 | 75 | 76 | 76 | 76 | 76 | 77 | 77 | 77 | 77 |
| 76 | 76 | 76 | 77 | 77 | 77 | 77 | 77 | 77 | 77 |
| 77 | 77 | 77 | 77 | 78 | 78 | 78 | 78 | 78 | 78 |
| 78 | 78 | 78 | 78 | 79 | 79 | 79 | 79 | 80 | 80 |
| 79 | 79 | 80 | 80 | 80 | 80 | 0.013 | 0.012 | 0.012 | 0.01 |
| 80 | 80 | 0.013 | 0.013 | 0.013 | 0.012 | 0.012 | 0.012 | 0.012 | 0.011 |

For SI units, 1 ft = 0.305 m.

The input rate specified on an appliance nameplate applies to elevations up to 2000 ft (600 m) above sea level. For elevations above 2000 ft (600 m), the standard cubic feet per hour input rate must be reduced by 4 percent for each 1000 ft (300 m) above sea level to provide sufficient oxygen for proper combustion of the fuel gas. Table F.1.1(d) also shows the proper burner orifice size to use at elevations of 2000 ft (600 m) and above, based on the orifice size for sea level operation from either Table F.1.1(a) [as corrected by Table F.1.1(c)] or Table F.1.1(b).

The following two examples illustrate the use of these Annex F tables in sizing burner orifices at various altitudes.

### *EXAMPLE 1*

What burner orifices are needed for an appliance, given the following parameters?

- Burner elevation: 4000 ft (1200 m)
- Sea level firing rate: 100 MBH (29 MW) total in two burners at 4 in. w.c. (1.0 kPa) pressure
- Gas specific gravity: 0.80, 850 Btu/ft$^3$ (88 MW/m$^3$)

*Answer:* Each burner is rated at 50,000 Btu/hr (15 MW):

(50,000)/(850) = 58.82 ft$^3$/hour

(58.82)/(0.866) = 67.92 equiv. ft$^3$/hour

[0.866 is obtained from Table F1.1(c)]

Table F.1.1(a) provides the orifice needed at sea level for natural gas. From Table F.1.1(a), 67.92 falls between a #26 and a #25 drill. (Select #26 to provide at least the required flow.)

Table F.1.1(d) corrects for altitude. From Table F.1.1(d) it is seen that a #28 drill is used at 4000 ft (1200 m). Thus, two orifices are required with a #28 drill opening.

### *EXAMPLE 2*

What burner orifices are needed for an appliance, given the following parameters?

- Burner elevation: 5000 ft (1500 m)
- Inshot burner: 75 MBH (22 MW) sea level input rating
- Propane fuel

*Answer:* Table F.1.1(b) shows that a #41 drill is needed (74,924 Btu/hr). Table F.1.1(d) shows that a #43 drill should be used at a 5000 ft (1500 m) elevation. Table F.1.1(b) provides the orifice needed at sea level for liquefied petroleum gases. From Table F.1.1(b), 75,000 falls between a #41 (74,924 Btu/hr) and a #40 (78,029 Btu/hr) drill size for propane. (Select a #41, as it coincides with a pressure of 50,000 Btu/hr when rounded.) From Table F.1.1(d) it is seen that a #43 drill is used at an elevation of 5000 ft (1520 m).

If the orifice is drilled on the job, drilling an orifice opening at least two sizes smaller than called for is good practice. If the orifice blank is held with pliers and the drill is powered by a handheld power tool, getting a straight, round hole is impossible. More gas will pass through the resulting opening than would be expected according to the tables. If, after test firing, increasing the opening is necessary, drilling one or two sizes larger will produce a hole much closer to the true size than the first hole drilled in the blank. Do not plan to compensate for an oversized or undersized orifice by reducing or increasing the appliance manifold pressure at the orifice by more than 10 percent of the manifold pressure specified for the appliance.

Some burners will operate satisfactorily only over a narrow range of orifice inlet pressures. If the pressure is too low, the flame may become soft and lazy. In extreme cases, sooting will occur. If the pressure is too high, the flame can become hard, noisy, and unstable, and lifting off of the burner head could occur.

### F.1.2 To Select Correct Orifice Size for Rated Burner Input.

The selection of a fixed orifice size for any rated burner input is affected by many variables, including orifice coefficient, and it is recommended that the appliance manufacturer be consulted for that purpose. When the correct orifice size cannot be readily determined, the orifice flow rates, as stated in the tables in this annex, can be used to select a fixed orifice size with a flow rate to approximately equal the required rated burner input.

For gases of the specific gravity and pressure conditions stipulated at elevations under 2000 ft (600 m), Table F.1.1(a) (in cubic feet per hour) or Table F.1.1(b) (in Btu per hour) can be used directly.

Where the specific gravity of the gas is other than 0.60, select the multiplier from Table F.1.1(c) for the utility gas served and divide the rated burner input by the selected factor to determine equivalent input at a specific gravity of 0.60, then select orifice size.

Where the appliance is located at an altitude of 2000 ft (600 m) or above, first use the manufacturer's rated input at sea level to select the orifice size as directed, then use Table F.1.1(d) to select the equivalent orifice size for use at the higher altitude.

## REFERENCE CITED IN COMMENTARY

The following publication is available from the American Gas Association, 400 N. Capitol St. NW, Washington, DC 20001, 202-824-7000, www.aga.org.

AGA Report No. 3, Orifice Metering of Natural Gas, Part 1: General Equations & Uncertainty Guidelines, 1990; Part 2: Specification and Installation Requirements, 2000; Part 3: Natural Gas Applications, 1992; and Part 4: Background, Development Implementation Procedure, 1992.

# Sizing of Venting Systems Serving Appliances Equipped with Draft Hoods, Category I Appliances, and Appliances Listed for Use with Type B Vents

## ANNEX G

*This annex is not a part of the requirements of this NFPA document but is included for informational purposes only.*

Annex G provides additional information on the sizing of gas vents beyond what is provided in Chapter 12, Venting of Appliances, and Chapter 13, Sizing of Category I Venting Systems. The annex is useful to the user of the code in the proper sizing of venting systems. The annex consists of two parts:

- Fifteen figures showing typical vent configurations and how to measure the variables needed when using the tables in Chapter 13.
- Seven examples of vent system calculations using the tables in Chapter 13. An example of another vent configuration (multistory vent) is included in the handbook commentary.

The code figures in Annex G illustrate common venting systems and are used in conjunction with the tables in Chapter 13. These figures illustrate the types of systems and the measurement of dimensions so that the tables in Chapter 13 can be used properly. The diagrams illustrate typical installations. Since other configurations that meet the code are possible, the diagrams should not be interpreted as being the only permissible geometry.

A series of examples are given that show how to use the tables and other requirements of Chapter 13. The examples also illustrate which dimensions on a real venting system correspond to the table entries. The examples include excerpts from the tables in Chapter 13 to clarify the calculations.

## G.1 Examples Using Single Appliance Venting Tables

See Figure G.1(a) through Figure G.1(n).

Figure G.1(a) through Figure G.1(n), on pp. 410–413, illustrate how the dimensions called for in the venting tables in Sections 13.1 and 13.2 are measured.

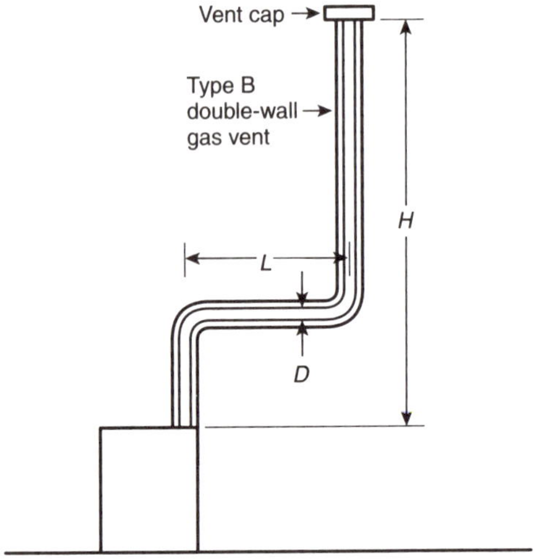


Table 13.1.1(a) is used when sizing Type B double-wall gas vent connected directly to the appliance.

Note: The appliance can be either Category I draft hood–equipped or fan-assisted type.

***FIGURE G.1(a)*** *Type B Double-Wall Vent System Serving a Single Appliance with a Type B Double-Wall Vent.*

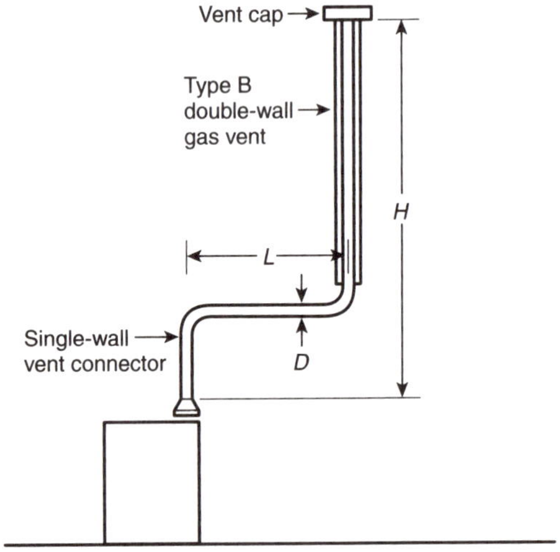


Table 13.1.1(b) is used when sizing a single-wall metal vent connector attached to a Type B double-wall gas vent.

Note: The appliance can be either Category I draft hood–equipped or fan-assisted type.

***FIGURE G.1(b)*** *Type B Double-Wall Vent System Serving a Single Appliance with a Single-Wall Metal Vent Connector.*

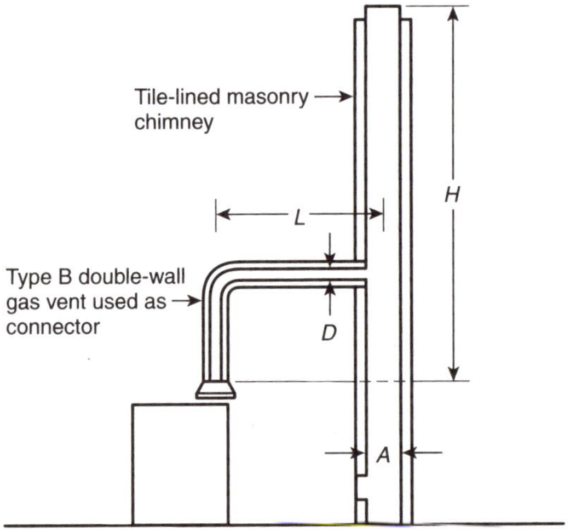


Table 13.1.1(c) is used when sizing a Type B double-wall gas vent connector attached to a tile-lined masonry chimney.

Notes:
(1) *A* is the equivalent cross-sectional area of the tile liner.
(2) The appliance can be either Category I draft hood–equipped or fan-assisted type.

***FIGURE G.1(c)*** *Vent System Serving a Single Appliance with a Masonry Chimney and a Type B Double-Wall Vent Connector.*

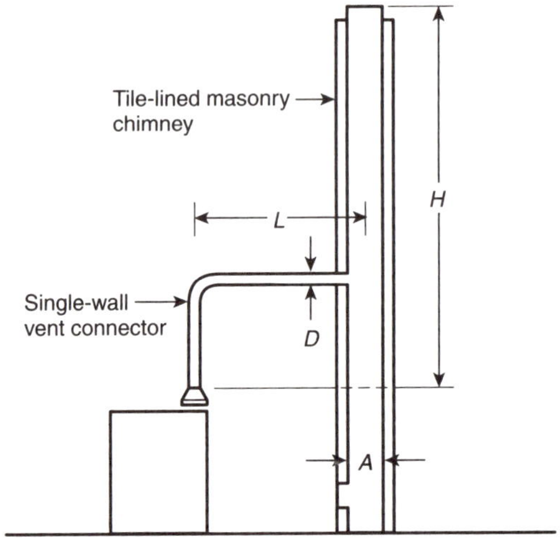


Table 13.1.1(d) is used when sizing a single-wall vent connector attached to a tile-lined masonry chimney.

Notes:
(1) *A* is the equivalent cross-sectional area of the tile liner.
(2) The appliance can be either Category I draft hood–equipped or fan-assisted type.

***FIGURE G.1(d)*** *Vent System Serving a Single Appliance Using a Masonry Chimney and a Single-Wall Metal Vent Connector.*

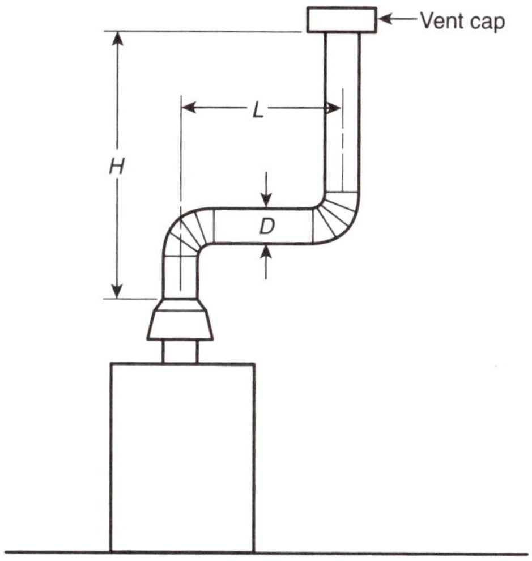


Asbestos cement Type B or single-wall metal vent serving a single draft hood–equipped appliance. *[See Table 13.1.1(e).]*

***FIGURE G.1(e)*** *Asbestos Cement Type B or Single-Wall Metal Vent System Serving a Single Draft Hood–Equipped Appliance.*

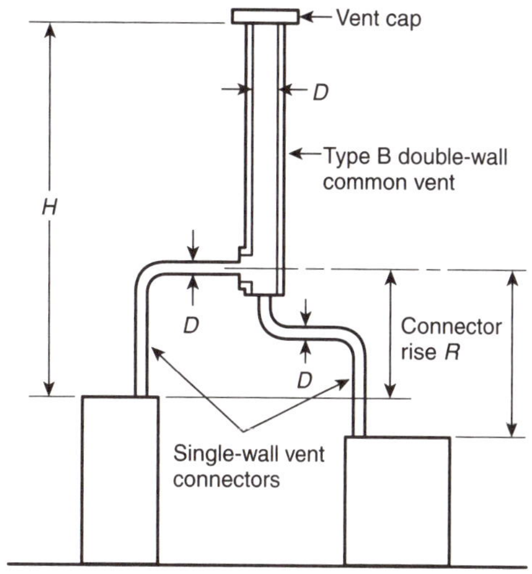


Table 13.2.1(b) is used when sizing single-wall vent connectors attached to a Type B double-wall common vent.

Note: Each appliance can be either Category I draft hood–equipped or fan-assisted type.

***FIGURE G.1(g)*** *Vent System Serving Two or More Appliances with Type B Double-Wall Vent and Single-Wall Metal Vent Connectors.*

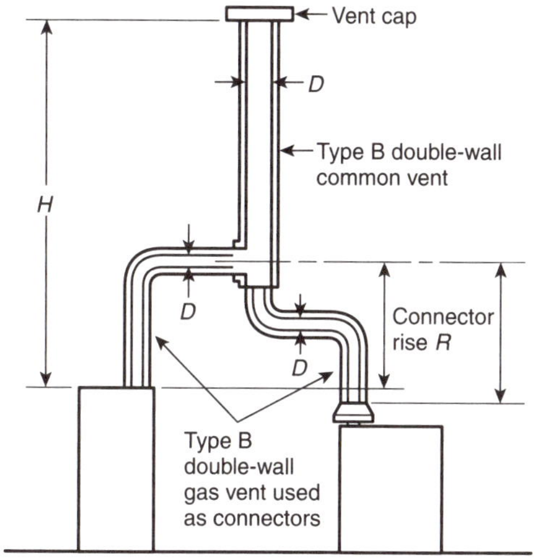


Table 13.2.1(a) is used when sizing Type B double-wall gas vent connectors attached to a Type B double-wall common vent.

Note: Each appliance can be either Category I draft hood–equipped or fan-assisted type.

***FIGURE G.1(f)*** *Vent System Serving Two or More Appliances with Type B Double-Wall Vent and Type B Double-Wall Vent Connectors.*

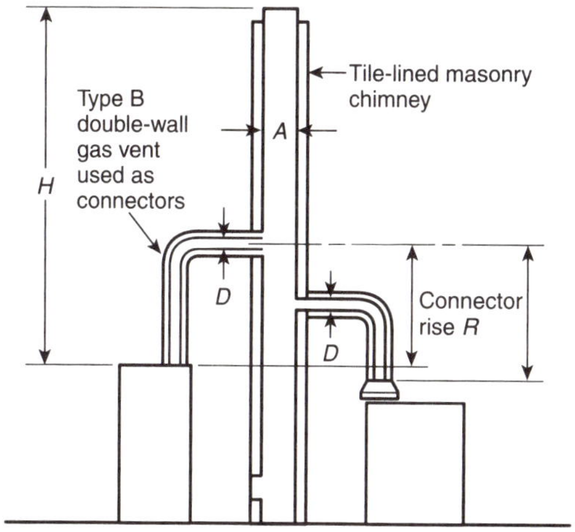


Table 13.2.1(c) is used when sizing Type B double-wall vent connectors attached to a tile-lined masonry chimney.

Notes:
(1) *A* is the equivalent cross-sectional area of the tile liner.
(2) Each appliance can be either Category I draft hood–equipped or fan-assisted type.

***FIGURE G.1(h)*** *Masonry Chimney Serving Two or More Appliances with Type B Double-Wall Vent Connectors.*

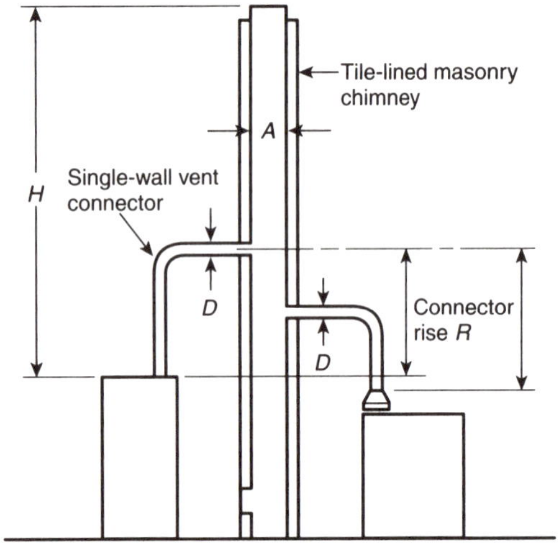


Table 13.2.1(d) is used when sizing single-wall metal vent connectors attached to a tile-lined masonry chimney.

Notes:
(1) *A* is the equivalent cross-sectional area of the tile liner.
(2) Each appliance can be either Category I draft hood–equipped or fan-assisted type.

***FIGURE G.1(i)*** *Masonry Chimney Serving Two or More Appliances with Single-Wall Metal Vent Connectors.*

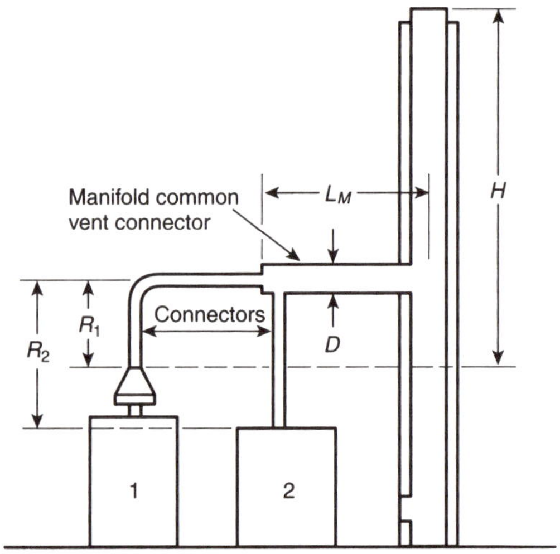


Example: Manifolded common vent connector $L_M$ can be no greater than 18 times the common vent connector manifold inside diameter; that is, a 4 in. (100 mm) inside diameter common vent connector manifold should not exceed 72 in. (1800 mm) in length. *(See 13.2.4.)*

Note: This is an illustration of a typical manifolded vent connector. Different appliance, vent connector, or common vent types are possible. *(See Section 13.2.)*

***FIGURE G.1(k)*** *Use of Manifolded Common Vent Connector.*

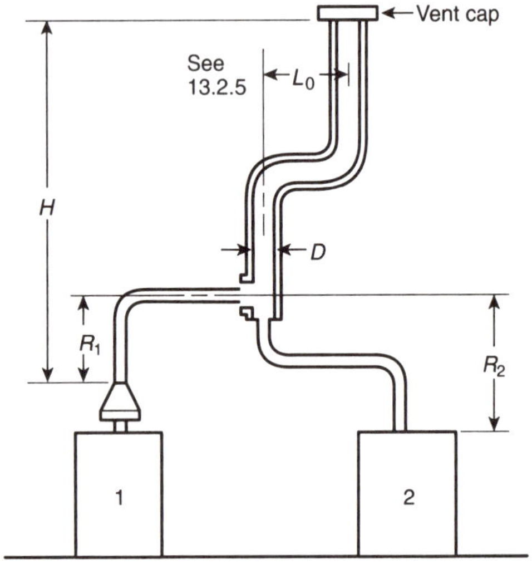


Asbestos cement Type B or single-wall metal pipe vent serving two or more draft hood–equipped appliances. *[See Table 13.2.1(e).]*

***FIGURE G.1(j)*** *Asbestos Cement Type B or Single-Wall Metal Vent System Serving Two or More Draft Hood–Equipped Appliances.*

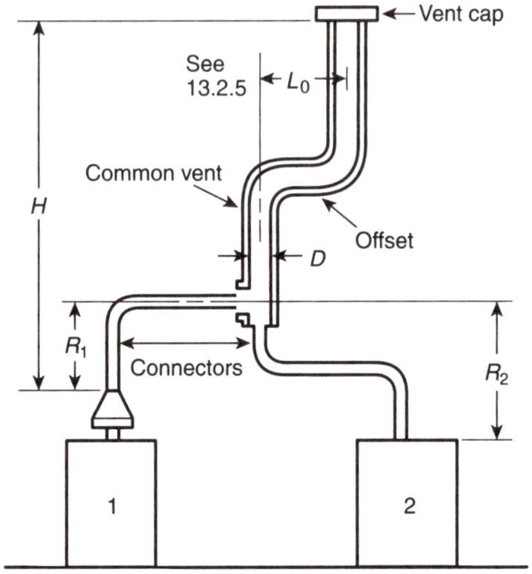


Example: Offset common vent.

Note: This is an illustration of a typical offset vent. Different appliance, vent connector, or vent types are possible. *(See Sections 13.1 and 13.2.)*

***FIGURE G.1(l)*** *Use of Offset Common Vent.*

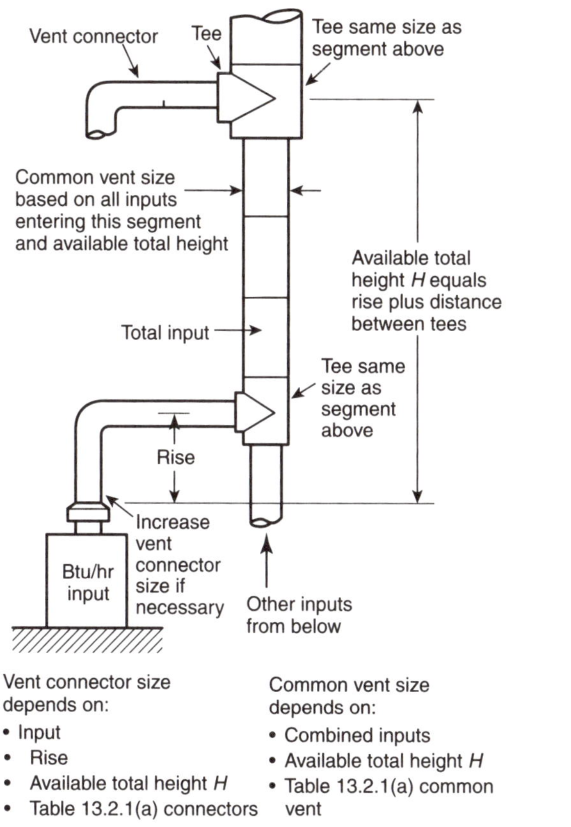


***FIGURE G.1(m)*** *Multistory Gas Vent Design Procedure for Each Segment of System.*

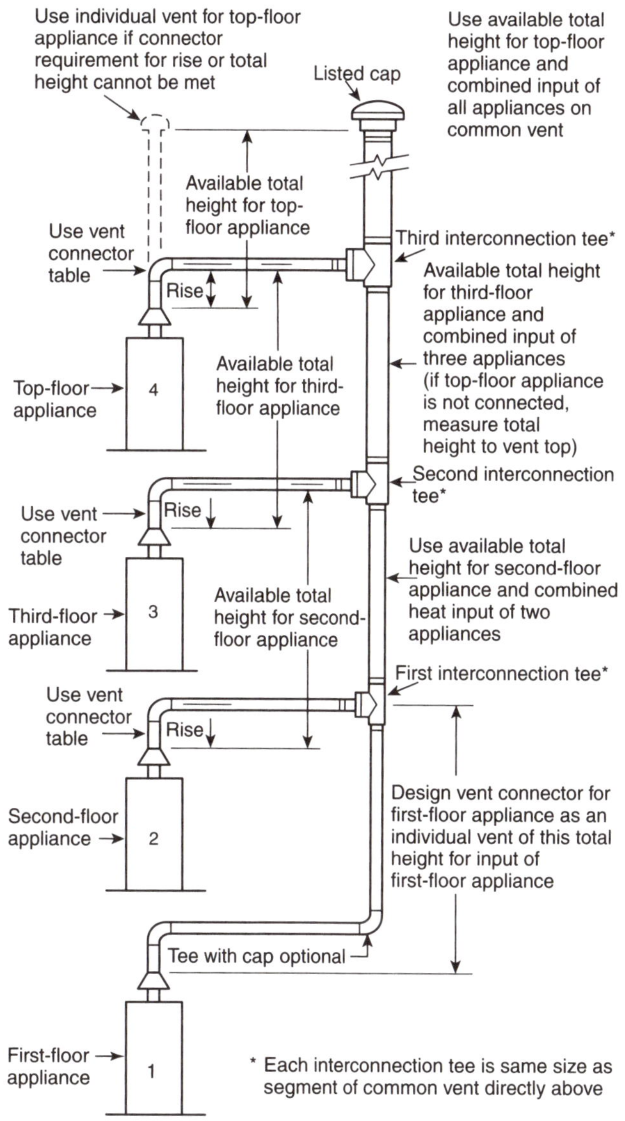


***FIGURE G.1(n)*** *Principles of Design of Multistory Vents Using Vent Connector and Common Vent Design Tables. (See 13.2.14 through 13.2.17.)*

## SIZING MULTISTORY VENTS

The following is an example of the use of the tables in Chapter 13 to size a multistory vent.

Assume that Figure G.1(n) represents a four-story apartment building that has a listed fan-assisted combustion furnace installed on each floor. Each furnace has a 4 in. (100 mm) flue collar and an input of 80,000 Btu/hr (23 MW). All the furnaces are installed outside of the conditioned space (i.e., isolated combustion) and are to be vented into the common vent, which is located 5 ft (1.5 m) from the furnaces. For the purpose of calculation, the overall system is divided into smaller, simple vent systems for each level, as shown in Figure G.1(m).

The structure is such that the vent connector rise is to be 2 ft (0.6 m) and the available total height for each level is 10 ft (3.0 m), except for the top floor, which is 6 ft (1.8 m). The vent connector from the first floor or lowest furnace to the common vent is considered to be an individual vent terminating at the first tee or interconnection, and this connector is sized in accordance with Table 13.1(a). Every other vent connector is sized in accordance with the Vent Connector Capacity section of Table 13.2(a). Each section of the common vent is sized in accordance with the Common Vent Capacity section of Table 13.2(a) to accommodate the accumulated total input of all appliances discharging into it, but sections are never smaller than the largest section below them.

The following example sizes a multistory vent. The vent connectors are sized using the appropriate table.

## PROCEDURE FOR SIZING CONNECTORS

Sizing a multistory vent requires the following three steps: sizing the vent connectors, sizing the sections of the common vent, and checking for oversizing in the common vent.

Vent connectors are sized individually. The vent connector sizing the appliance on the lowest level uses a single appliance vent table, as the vent (looking downward) is serving one appliance. The sizing of the other vent connectors uses a multiple appliance table, as the section of the common vent above the connector is serving multiple appliances.

**First-Floor Connector (Furnace 1).** The input is 80,000 Btu/hr, and the total height is 10 ft. This section of the vent is treated as a single appliance vent with 5 ft of vent connector. Use Table 13.1(a), reading across the row for a 10 ft height (*H*) and a lateral (*L*) of 5 ft. Look under the FAN columns for a range that includes 80,000. The 4 in. column has a range of 32,000 Btu/hr to 113,000 Btu/hr, and the 5 in. column has a range of 41,000 Btu/hr to 187,000 Btu/hr. Both sizes are acceptable, and a 4 in. size is selected because it is smaller and less costly.

Subsection 13.2.2 limits a 4 in. vent connector to a maximum length of 6 ft (1.5 ft length for each inch of diameter), which is not exceeded in this example.

**Second- and Third-Floor Connectors (Furnaces 2 and 3).** The input is 80,000 Btu/hr, the connector rise is 2 ft, and the vent height is 10 ft. Use the Vent Connector Capacity section of Table 13.2(a). Reading across the 10 ft height, 2 ft rise row to the FAN columns shows that a 4 in. connector has a range of 36,000 Btu/hr to 86,000 Btu/hr, a 5 in. connector has a range of 51,000 Btu/hr to 136,000 Btu/hr, and a 6 in. connector has a range of 67,000 Btu/hr to 206,000 Btu/hr. A connector with a diameter of 4 in., 5 in., or 6 in. will work, and a 4 in. connector is selected because it is smaller and less costly.

**Fourth-Floor Connector (Furnace 4).** The input is 80,000 Btu/hr, the connector rise is 2 ft, and the vent height is 6 ft. Again, use the Vent Connector Capacity section of Table 13.2(a). Reading across the 6 ft height, 2 ft rise row shows that a 5 in. connector has a range of 48,000 Btu/hr to 121,000 Btu/hr and a 6 in. connector has a range of 60,000 Btu/hr to 183,000 Btu/hr. Either a 5 in. or 6 in. connector will work, and a 5 in. connector is selected because it is smaller and less costly.

Subsection 13.2.2 limits a 5 in. vent connector to a maximum length of $7\frac{1}{2}$ ft (1.5 ft length for each inch of diameter), which is not exceeded by the connector size selected.

## PROCEDURES FOR SIZING COMMON VENT

The common vent is sized in the sections between appliances and above the uppermost appliance. The lowest section of the common vent, between Appliances 1 and 2, is sized using a single appliance table, because it serves only one appliance. The remaining sections of the common vent are sized using a multiple appliance table.

**Common Vent for Furnaces 1 and 2.** The input is 160,000 Btu/hr, and the vent height is 10 ft. Use the Common Vent Capacity section of Table 13.2(a), FAN+FAN columns only. Reading across the 10 ft row shows that a 5 in. vent has a capacity of 169,000 Btu/hr, which exceeds the 160,000 Btu/hr input. The 5 in. size is selected.

**Common Vent for Third-Floor Furnace (Furnace 3).** The input is 240,000 Btu/ hr, and the height is 10 ft. Use the Common Vent Capacity section of Table 13.2(a). Reading across the

10 ft line in the FAN+FAN columns shows that a 6 in. common vent has a capacity of 243,000 Btu/hr. The 6 in. size is selected.

**Common Vent for Fourth-Floor Furnace (Furnace 4).** The input is 320,000 Btu/hr, and the height is 6 ft. Again, use the FAN+FAN columns in the Common Vent Capacity section of Table 13.2(a). Reading across the 6 ft row shows that an 8 in. vent has a capacity of 404,000 Btu/hr, and this size is selected.

These sizing procedures are summarized in Commentary Tables G.1 and G.2.

***COMMENTARY TABLE G.1*** *Common Vent Serving Appliances on Four Floors*

| *Furnace* | *Total Input to Common Vent (Btu/hr)* | *Available Total Height (ft)* | *Vent Connector Size (in.)* | *Common Vent Size (in.)* |
|---|---|---|---|---|
| 1 | 80,000 | 10 | 4 | 4 |
| 2 | 160,000 | 10 | 4 | 5 |
| 3 | 240,000 | 10 | 4 | 6 |
| 4 | 320,000 | 6 | 5 | 8 |

***COMMENTARY TABLE G.2*** *Common Vent Serving Appliances on Three Floors; Independent Vent Serving Appliances on the Fourth Floor*

| *Furnace* | *Total Input to Common Vent (Btu/hr)* | *Available Total Height (ft)* | *Vent Connector Size (in.)* | *Common Vent Size (in.)* |
|---|---|---|---|---|
| 1 | 80,000 | 10 | 4 | 4 |
| 2 | 160,000 | 10 | 4 | 5 |
| 3 | 240,000 | 10 | 4 | 6 |
| 4 | 80,000 | 6 | 4 | None |

## CHECK FOR EXCESSIVE VENT AREA

Because the vent tables provide only the minimum and not the maximum size for non-fan-assisted appliances (and this example uses non-fan-assisted appliances), the common vent must be checked for oversizing using the "7-times" rule.

The vent connector and common vent have been sized using Table 13.2(a) and Section 13.2, Additional Requirements to Multiple-Appliance Vent tables. To check that the size is correct, refer to 13.2.18, which states:

> **13.2.18 Vertical Vent Size Limitation.** Where two or more appliances are connected to a vertical vent or chimney, the flow area of the largest section of vertical vent or chimney shall not exceed seven times the smallest listed appliance categorized vent areas, flue collar area, or draft hood outlet area unless designed in accordance with approved engineering methods.

In the multistory vent example, the smallest vent connector diameter is 4 in., and the largest common vent diameter is 8 in.

For the 4 in. diameter connector, use the following equation:

$$\text{Area} = \pi r^2 = 3.14(2)^2 = 12.56 \text{ in.}^2$$

where:

$\pi = 3.14$

$r = 2$ ($\frac{1}{2}$ the 4 in. diameter)

For the 8 in. diameter common vent, use the following equation:

$$\text{Area} = \pi r^2 = 3.14(4)^2 = 50.25 \text{ in.}^2$$

where:

$\pi = 3.14$

$r = 4$ ($\frac{1}{2}$ the 8 in. diameter)

Subsection 13.2.18 limits the common vent to 7 times the smallest connector area. The ratio of the area of the smallest connector to the common vent area in this example must be checked using an equation such as the following to be sure it is less than 7:

Ratio = area (largest common vent section)/area (smallest connector)

$$\text{Ratio} = \frac{50.25}{12.56} = 4$$

Since 4 is less than 7, the sizing of the common vent is acceptable.

### G.1.1 Example 1: Single Draft Hood–Equipped Appliance.

An installer has a 120,000 Btu/hr input appliance with a 5 in. diameter draft hood outlet that needs to be vented into a 10 ft high Type B vent system. What size vent should be used assuming (1) a 5 ft lateral single-wall metal vent connector is used with two 90 degree elbows or (2) a 5 ft lateral single-wall metal vent connector is used with three 90 degree elbows in the vent system? See Figure G.1.1.

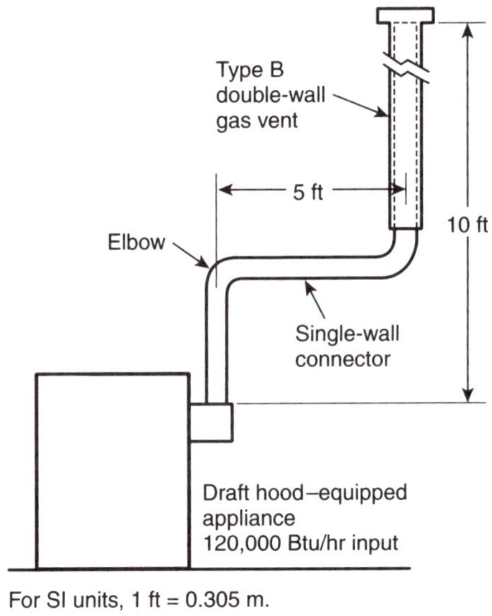


*FIGURE G.1.1 Single Draft Hood–Equipped Appliance — Example 1.*

This example illustrates the sizing of a single appliance using a single-wall connector and B-vent, a common installation. The solution explains the need to be aware of the number of elbows in the vent system, because elbows reduce venting system performance.

*Solution*

Table 13.1(b) should be used to solve this problem, because single-wall metal vent connectors are being used with a Type B vent, as follows:

(1) Read down the first column in Table 13.1(b) until the row associated with a 10 ft height and 5 ft lateral is found. Read across this row until a vent capacity greater than 120,000 Btu/hr is located in the shaded columns labeled NAT Max for draft hood–equipped

appliances. In this case, a 5 in. diameter vent has a capacity of 122,000 Btu/hr and can be used for this application.

(2) If three 90 degree elbows are used in the vent system, the maximum vent capacity listed in the tables must be reduced by 10 percent *(see 13.1.3).* This implies that the 5 in. diameter vent has an adjusted capacity of only 110,000 Btu/hr. In this case, the vent system must be increased to 6 in. in diameter. See the following calculations:

$$122{,}000 \times 0.90 = 110{,}000 \text{ for 5 in. vent}$$

From Table 13.1(b), select 6 in. vent.

$$186{,}000 \times 0.90 = 167{,}000$$

This figure is greater than the required 120,000. Therefore, use a 6 in. vent and connector where three elbows are used.

Refer to Commentary Table G.3, which is an extract from code Table 13.1(b). This extract illustrates the use of the vent tables as described in G.1.1.

***COMMENTARY TABLE G.3 Extract from Code Table 13.1(b) for Example 1***

| | | Vent Diameter D | | | | | | |
|---|---|---|---|---|---|---|---|---|
| | | 4 in. | | | 5 in. | | | 6 in. |
| | | Appliance Input Rating in Thousands of Btu per Hour | | | | | | |
| | | FAN | | NAT | FAN | | NAT | FAN |
| Height H (ft) | Lateral L (ft) | Min | Max | Max | Min | Max | Max | Min |
| (10) | 0 | 57 | 174 | 99 | 82 | 293 | 165 | 120 |
| | 2 | 59 | 117 | 80 | 82 | 193 | 128 | 119 |
| | (5) | 76 | 111 | 76 | 105 | 185 | (122) | 148 |
| | 10 | 97 | 100 | 68 | 132 | 171 | 112 | 188 |

The solution to Example 1 is summarized in Worksheet G.1, Sizing Vents for Single Draft Hood–Equipped Appliances.

## G.1.2 Example 2: Single Fan-Assisted Appliance.

An installer has an 80,000 Btu/hr input fan-assisted appliance that must be installed using 10 ft of lateral connector attached to a 30 ft high Type B vent. Two 90 degree elbows are needed for the installation. Can a single-wall metal vent connector be used for this application? See Figure G.1.2.

Example 2 in G.1.2 illustrates a very important point. The sizing tables for fan-assisted appliances are designed to reduce the allowable length of single-wall galvanized vent connector runs, as compared to those for Type B connectors. The reason for this limitation is that single-wall connectors lose more heat in the vent gases than double-wall B vent connectors, which adversely affects the performance of the entire venting system. Type B connectors are of double-wall construction with an air space. This design provides insulation and reduces heat loss. Remember that fan-assisted combustion appliances operate at a lower vent temperature than do draft hood–equipped appliances, so further heat loss in the venting system must be minimized to prevent condensation and corrosion of the vent.

There is no hard and fast rule of thumb for the maximum length of single-wall connectors. However, use of Table 13.1(b) in practical situations shows there are not many instances where a connector of more than 5 ft (1.5 m) long is allowed.

## CALCULATION WORKSHEET: VENT SIZING, SINGLE APPLIANCE
### (DRAFT HOOD–EQUIPPED APPLIANCE)

**Step 1:**

- Determine the type of chimney or vent used and enter it to the right.
- If used, determine the type of vent connector material and enter it to the right.
- Select the appropriate table and enter it to the right.

Type of vent: *B vent*
Type of connector: *Single wall*
Height: *10 ft*
Lateral: *5 ft*
Table used: *13.1(b)*

**Step 2:**

- Draw a sketch of the proposed design in the space to the right. Use the back of this page or a separate sheet if more space is needed.

Note: Factors such as combining connectors prior to entering the common vent will affect capacity.

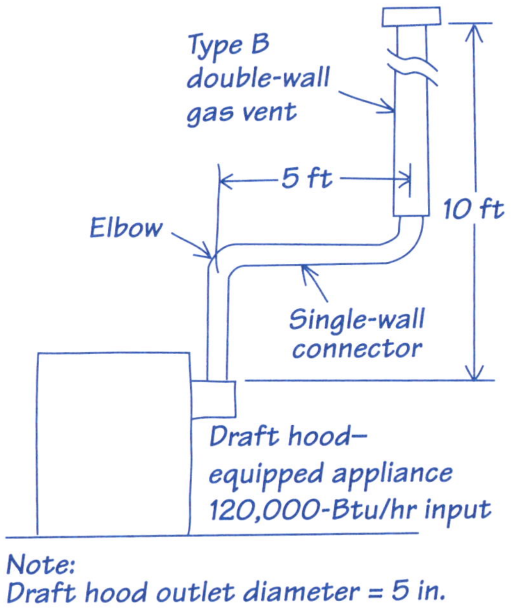


Proposed design sketch

**Step 3:**

- Determine the common vent size using the single appliance vent table. Read across the row with the correct height and lateral offset. Use the NAT Min. column for draft hood appliances and the FAN Min and Max columns for draft hood appliances.

Vent size: *5 in.*

**Step 4:**

- Check for vent downsizing.
- If the vent is smaller than the draft hood outlet or flue collar, see 13.1.2 for restrictions.

Draft hood outlet/ flue collar diameter: *5 in.*

**Step 5:**

- Check for excessive elbows.
- If more than two elbows are used, derate the single appliance vent table values by 10% for each elbow above 2 (see 13.1.3).

Number of elbows: *2*

☑ Derating not required.
☐ Derating required. Show calculation on a separate page.

**Job:** *25 Main Street* **Prepared by:** *TL* **Date:** *1/1/09*



*WORKSHEET G.1 Sizing Vents for Single Draft Hood–Equipped Appliances.*

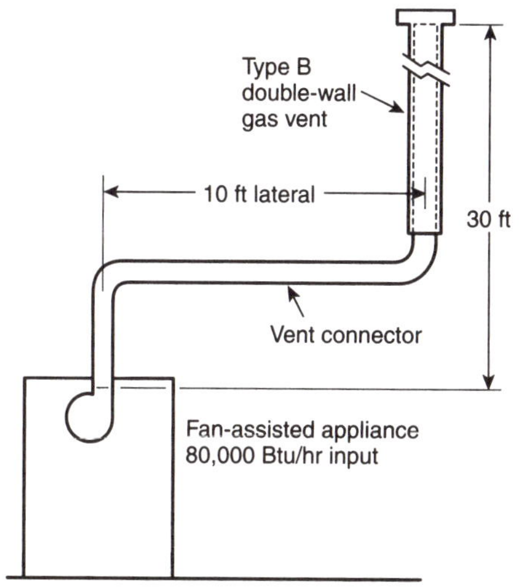


*FIGURE G.1.2 Single Fan-Assisted Appliance — Example 2.*

*Solution*

Table 13.1(b) refers to the use of single-wall metal vent connectors with Type B vent. In the first column find the row associated with a 30 ft height and a 10 ft lateral. Read across this row, looking at the FAN Min and FAN Max columns, to find that a 3 in. diameter single-wall metal vent connector is not recommended. Moving to the next larger size single-wall connector (4 in.), we find that a 4 in. diameter single-wall metal connector has a recommended minimum vent capacity of 91,000 Btu/hr and a recommended maximum vent capacity of 144,000 Btu/hr. The 80,000 Btu/hr fan-assisted appliance is outside this range, so the conclusion is that a single-wall metal vent connector cannot be used to vent this appliance using 10 ft of lateral for the connector.

However, if the 80,000 Btu/hr input appliance could be moved to within 5 ft of the vertical vent, a 4 in. single-wall metal connector could be used to vent the appliance. Table 13.1(b) shows the acceptable range of vent capacities for a 4 in. vent with 5 ft of lateral to be between 72,000 Btu/hr and 157,000 Btu/hr.

Refer to Commentary Table G.4, which is an extract from code Table 13.1(b). This commentary table illustrates the use of the vent tables as described in the solution listed in G.1.2. The solution to this example is summarized in a calculation worksheet, Worksheet G.2. Blank worksheets are located in Supplement 6 of this handbook.

If the appliance cannot be moved closer to the vertical vent, a Type B vent could be used as the connector material. In this case, Table 13.1(a) shows that, for a 30 ft high vent with 10 ft of lateral, the acceptable range of vent capacities for a 4 in. diameter vent attached to a fan-assisted appliance is between 37,000 Btu/hr and 150,000 Btu/hr.

Commentary Table G.5 is an extract from code Table 13.1(b).

Commentary Table G.6 is an extract from code Table 13.1(a).

### G.1.3 Example 3: Interpolating Between Table Values.

An installer has an 80,000 Btu/hr input appliance with a 4 in. diameter draft hood outlet that needs to be vented into a 12 ft high Type B vent. The vent connector has a 5 ft lateral length and is also Type B. Can this appliance be vented using a 4 in. diameter vent?

**COMMENTARY TABLE G.4** *Extract from Code Table 13.1(b) for Example 2 with a 30 ft Height and a 10 ft Lateral Using Single-Wall Metal Connector*

| | | Vent Diameter D | | | | | |
|---|---|---|---|---|---|---|---|
| | | 3 in. | | | 4 in. | | |
| | | Appliance Input Rating in Thousands of Btu per Hour | | | | | |
| | | FAN | | NAT | FAN | | NAT |
| Height H (ft) | Lateral L (ft) | Min | Max | Max | Min | Max | Max |
| (30) | 0 | 34 | 99 | 63 | 53 | 211 | 127 |
| | 2 | 37 | 80 | 56 | 55 | 164 | 111 |
| | 5 | 49 | 74 | 52 | 72 | 157 | 106 |
| | (10) | NA | NA | NA | (91 | 144) | 98 |
| | 15 | NA | NA | NA | 115 | 131 | NA |
| | 20 | NA | NA | NA | NA | NA | NA |
| | 30 | NA | NA | NA | NA | NA | NA |

**COMMENTARY TABLE G.5** *Extract from Code Table 13.1(b) for Example 2 with a 30 ft Height and a 5 ft Lateral Using Single-Wall Metal Connector*

| | | Vent Diameter D | | | | | |
|---|---|---|---|---|---|---|---|
| | | 3 in. | | | 4 in. | | |
| | | Appliance Input Rating in Thousands of Btu per Hour | | | | | |
| | | FAN | | NAT | FAN | | NAT |
| Height H (ft) | Lateral L (ft) | Min | Max | Max | Min | Max | Max |
| (30) | 0 | 34 | 99 | 63 | 53 | 211 | 127 |
| | 2 | 37 | 80 | 56 | 55 | 164 | 111 |
| | (5) | 49 | 74 | 52 | (72 | 157) | 106 |
| | 10 | NA | NA | NA | 91 | 144 | 98 |
| | 15 | NA | NA | NA | 115 | 131 | NA |
| | 20 | NA | NA | NA | NA | NA | NA |
| | 30 | NA | NA | NA | NA | NA | NA |

**COMMENTARY TABLE G.6** *Extract from Code Table 13.1(a) for Example 2 with a 30 ft Height and a 10 ft Lateral Using Type B Double-Wall Connector*

| | | Vent Diameter D | | | | | |
|---|---|---|---|---|---|---|---|
| | | 3 in. | | | 4 in. | | |
| | | Appliance Input Rating in Thousands of Btu per Hour | | | | | |
| | | FAN | | NAT | FAN | | NAT |
| Height H (ft) | Lateral L (ft) | Min | Max | Max | Min | Max | Max |
| (30) | 0 | 0 | 100 | 64 | 0 | 213 | 128 |
| | 2 | 9 | 81 | 56 | 13 | 166 | 112 |
| | 5 | 21 | 77 | 54 | 28 | 160 | 108 |
| | (10) | 27 | 70 | 50 | (37 | 150) | 102 |
| | 15 | 33 | 64 | NA | 44 | 141 | 96 |
| | 20 | 56 | 58 | NA | 53 | 132 | 90 |
| | 30 | NA | NA | NA | 73 | 113 | NA |

## CALCULATION WORKSHEET: VENT SIZING, SINGLE APPLIANCE
### (FAN-ASSISTED APPLIANCE)

**Step 1:**

- Determine the type of chimney or vent used and enter it to the right.
- If used, determine the type of vent connector material and enter it to the right.
- Select the appropriate table and enter it to the right.

Type of vent: B vent/B vent
Type of connector: Single wall/Single wall
Height: 30 ft /30 ft
Lateral: 10 ft/5 ft
Table used: 13.1 (b)/13.1 (b)

**Step 2:**

- Draw a sketch of the proposed design in the space to the right. Use the back of this page or a separate sheet if more space is needed.

Note: Factors such as combining connectors prior to entering the common vent will affect capacity.

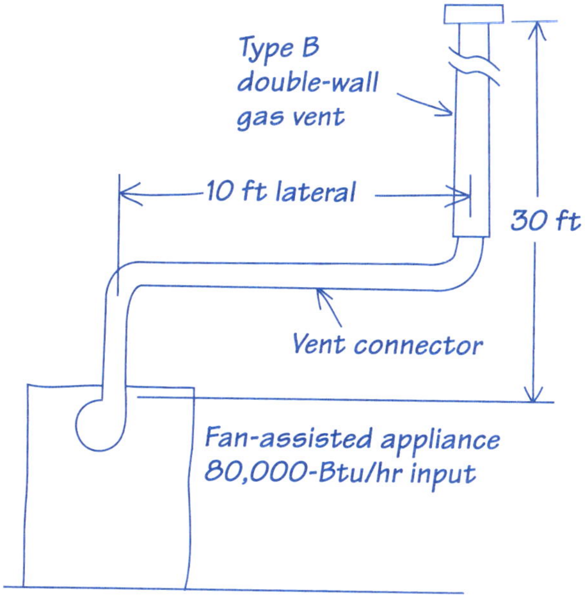


Proposed design sketch

**Step 3:**

- Determine the common vent size using the single appliance vent table. Read across the row with the correct height and lateral offset. Use the NAT Min column for draft hood appliances and the FAN Min and Max columns for draft hood appliances.

Vent size: None/4 in.

**Step 4:**

- Check for vent downsizing.
- If the vent is smaller than the draft hood outlet or flue collar, see 13.1.2 for restrictions.

Draft hood outlet/ flue collar diameter: N/A

**Step 5:**

- Check for excessive elbows.
- If more than two elbows are used, derate the single appliance vent table values by 10% for each elbow above 2 (see 13.1.3).

Number of elbows: 2

☑ Derating not required.
❑ Derating required. Show calculation on a separate page.

**Job:** 25 Main Street **Prepared by:** TL **Date:** 1/1/09



***WORKSHEET G.2*** *Calculating Size of Vents for Single Fan-Assisted Appliance.*

Example 3 in G.1.3 shows how to interpolate between existing table values (to find an intermediate value between two table values). However, it is not permissible to extrapolate out to values beyond the outer bounds of the tables (to infer a value beyond the values in the table). In a case where values exceed those shown in the tables, some other approved engineering method must be used.

*Solution*

Table 13.1(a) is used in the case of an all Type B vent system. However, Table 13.1(a) does not have an entry for a height of 12 ft, and interpolation must be used. Read down the 4 in. diameter NAT Max column to the row associated with 10 ft height and 5 ft lateral to find the capacity value of 77,000 Btu/hr. Read further down to the 15 ft height, 5 ft lateral row to find the capacity value of 87,000 Btu/hr. The difference between the 15 ft height capacity value and the 10 ft height capacity value is 10,000 Btu/hr. The capacity for a vent system with a 12 ft height is equal to the capacity for a 10 ft height plus $^2/_5$ of the difference between the 10 ft and 15 ft height values, or $77{,}000 + {}^2/_5 \times 10{,}000 = 81{,}000$ Btu/hr. Therefore, a 4 in. diameter vent can be used in the installation.

Commentary Table G.7 is an extract from code Table 13.1(a).

***COMMENTARY TABLE G.7** Extract from Code Table 13.1(a) for Example 3*

| | | Vent Diameter D | | |
|---|---|---|---|---|
| | | 4 in. | | |
| | | FAN | | |
| Height H (ft) | Lateral L (ft) | Min | Max | NAT Max |
| 10 | 0 | 0 | 175 | 100 |
| | 2 | 17 | 118 | 81 |
| | 5 | 32 | 113 | 77 |
| | 10 | 41 | 104 | 70 |
| 15 | 0 | 0 | 191 | 112 |
| | 2 | 15 | 136 | 93 |
| | 5 | 30 | 130 | 87 |
| | 10 | 40 | 121 | 82 |
| | 15 | 48 | 112 | 76 |

## G.2 Examples Using Common Venting Tables

### G.2.1 Example 4: Common Venting Two Draft Hood–Equipped Appliances.

A 35,000-Btu/hr water heater is to be common vented with a 150,000 Btu/hr furnace, using a common vent with a total height of 30 ft. The connector rise is 2 ft for the water heater with a horizontal length of 4 ft. The connector rise for the furnace is 3 ft with a horizontal length of 8 ft. Assume single-wall metal connectors will be used with Type B vent. What size connectors and combined vent should be used in this installation? See Figure G.2.1.

*Solution*

Table 13.2(b) should be used to size single-wall metal vent connectors attached to Type B vertical vents. In the vent connector capacity portion of Table 13.2(b), find the row associated with a 30 ft vent height. For a 2 ft rise on the vent connector for the water heater, read the shaded columns for draft hood–equipped appliances to find that a 3 in. diameter vent connector has a capacity of 37,000 Btu/hr. Therefore, a 3 in. single-wall metal vent connector can

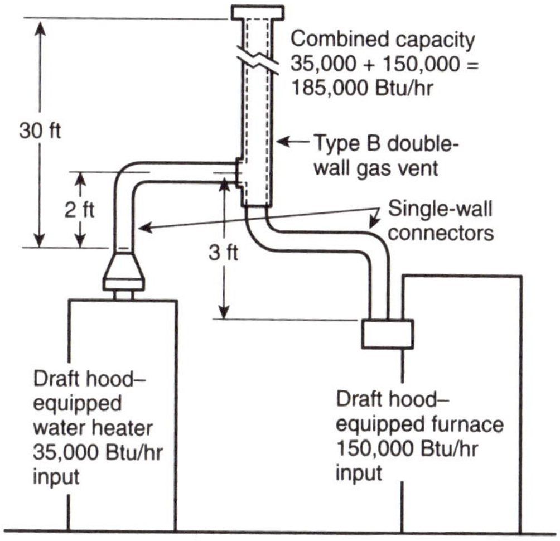


***FIGURE G.2.1** Common Venting Two Draft Hood–Equipped Appliances — Example 4.*

be used with the water heater. For a draft hood–equipped furnace with a 3 ft rise, read across the appropriate row to find that a 5 in. diameter vent connector has a maximum capacity of 120,000 Btu/hr (which is too small for the furnace) and a 6 in. diameter vent connector has a maximum vent capacity of 172,000 Btu/hr. Therefore, a 6 in. diameter vent connector should be used with the 150,000 Btu/hr furnace. Because both vent connector horizontal lengths are less than the maximum lengths listed in 13.2.2, the table values can be used without adjustments.

In the common vent capacity portion of Table 13.2(b), find the row associated with a 30 ft vent height and read over to the NAT + NAT portion of the 6 in. diameter column to find a maximum combined capacity of 257,000 Btu/hr. Since the two appliances total only 185,000 Btu/hr, a 6 in. common vent can be used.

Commentary Table G.8 is an extract from code Table 13.2(b). The vent connector capacity is shown in the upper portion of the table. The common vent capacity is shown in the lower portion of the table. The solution to Example 4 is summarized in the calculation worksheet, Worksheet G.3. Blank worksheets are located in Supplement 6 of this book.

## G.2.2 Example 5(a): Common Venting a Draft Hood–Equipped Water Heater with a Fan-Assisted Furnace into a Type B Vent.

In this case, a 35,000 Btu/hr input draft hood–equipped water heater with a 4 in. diameter draft hood outlet, 2 ft of connector rise, and 4 ft of horizontal length is to be common vented with a 100,000 Btu/hr fan-assisted furnace with a 4 in. diameter flue collar, 3 ft of connector rise, and 6 ft of horizontal length. The common vent consists of a 30 ft height of Type B vent. What are the recommended vent diameters for each connector and the common vent? The installer would like to use a single-wall metal vent connector. See Figure G.2.2.

*Solution [See Table 13.2(b).]*

See Commentary Table G.9, which is an extract from code Table 13.2(b), for calculation of the water heater vent connector diameter for Example 5(a).

*Water Heater Vent Connector Diameter.* Since the water heater vent connector horizontal length of 4 ft is less than the maximum value listed in Table 13.2(b), the venting table values can be used without adjustments. Using the Vent Connector Capacity portion of Table

*COMMENTARY TABLE G.8 Extract from Code Table 13.2(b) for Example 4 — Common Vent Sizing*

*Vent Connector Capacity*

| | | Single-Wall Metal Vent Connector Diameter D | | | | | | | | | | | |
|---|---|---|---|---|---|---|---|---|---|---|---|---|---|
| | | 3 in. | | | 4 in. | | | 5 in. | | | 6 in. | | |
| | | Appliance Input Rating Limits in Thousands of Btu per Hour | | | | | | | | | | | |
| | | FAN | | NAT | FAN | | NAT | FAN | | NAT | FAN | | NAT |
| Height H (ft) | Rise R (ft) | Min | Max | Max | Min | Max | Max | Min | Max | Max | Min | Max | Max |
| 30 | 1 | 47 | 60 | 31 | 77 | 110 | 57 | 112 | 175 | 89 | 169 | 278 | 129 |
| | 2 | 51 | 62 | 37 | 81 | 115 | 67 | 117 | 185 | 106 | 177 | 290 | 152 |
| | 3 | 54 | 64 | 42 | 85 | 119 | 76 | 122 | 193 | 120 | 185 | 300 | 172 |

*Common Vent Capacity*

| | Type B Double-Wall Vent Diameter D | | | | | | | | |
|---|---|---|---|---|---|---|---|---|---|
| | 4 in. | | | 5 in. | | | 6 in. | | |
| | Combined Appliance Input Rating in Thousands of Btu per Hour | | | | | | | | |
| Vent Height H (ft) | FAN + FAN | FAN + NAT | NAT + NAT | FAN + FAN | FAN + NAT | NAT + NAT | FAN + FAN | FAN + NAT | NAT + NAT |
| 15 | 121 | 108 | 88 | 189 | 159 | 140 | 275 | 221 | 200 |
| 20 | 131 | 118 | 98 | 208 | 177 | 156 | 305 | 247 | 223 |
| 30 | 145 | 132 | 113 | 236 | 202 | 180 | 350 | 286 | 257 |
| 50 | 159 | 145 | 128 | 268 | 233 | 208 | 406 | 337 | 296 |

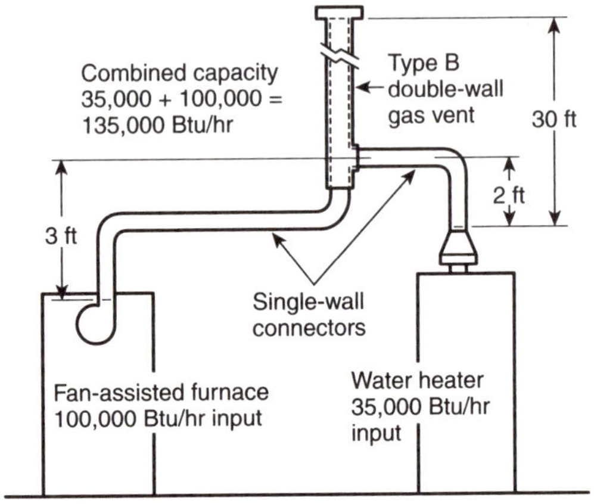


For SI units, 1000 Btu/hr = 0.293 kW, 1 ft = 0.305 m.

***FIGURE G.2.2*** *Common Venting a Draft Hood–Equipped Water Heater with a Fan-Assisted Furnace into a Type B Double-Wall Common Vent — Example 5(a).*

## CALCULATION WORKSHEET: VENT SIZING, TWO APPLIANCES
### (COMMON VENTING OF TWO DRAFT HOOD–EQUIPPED APPLIANCES)

**Step 1:**
- Determine the type of chimney or vent used and enter it to the right.
- Determine the type of vent connector material to be used and enter it to the right.
- Determine the number of appliances and enter it to the right.
- Select the appropriate table and enter it to the right.

Type of vent: *B vent*
Type of connector: *Single wall*
Number of appliances: *2*
Table used: *13.2 (b)*

**Step 2:**
- Draw a sketch of the proposed design in the space to the right. Use the back of this page or a separate sheet if more space is needed.

Note: Factors such as combining connectors prior to entering the common vent will affect capacity.

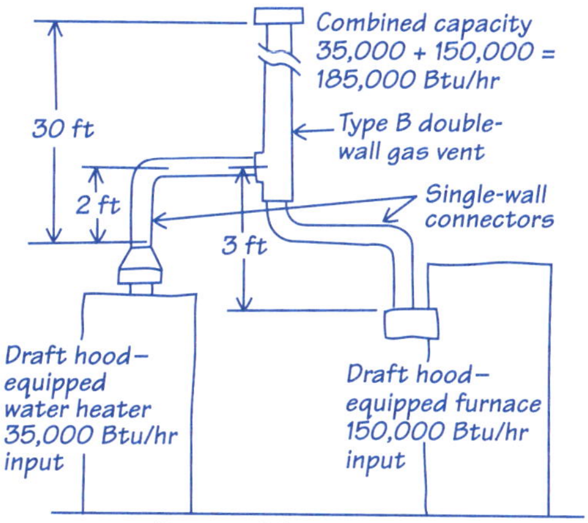

Proposed design sketch

**Step 3:**
- Determine the vent connector size using the top portion of the multiple appliance vent table, as shown in Table 1.

**Step 4:**
- Check for vent connector size limits, 13.2.24.

Note that Table 13.2(b) allows a 3 in. connector for the water heater, but 13.2.24 does not allow the vent conector to be smaller than the draft hood.

**Table 1 Top of Multiple Appliance Vent Table**

| Appliance | Connector rise | Vent height | Input rating | FAN or NAT | Connector size | Table used |
|---|---|---|---|---|---|---|
| Furnace | 3 ft | 30 | 150,000 | NAT | 6 in. | 13.2(b) |
| Water heater | 2 ft | 30 | 35,000 | NAT | 6 in. | 13.2(b) |
| | | | | | | |
| | | | | | | |
| Total | | | 185,000 | | | |

**Step 5:**
- Check for excessive connector horizontal length using Table 2.
- If any connector horizontal length in feet exceeds 1.5 times the diameter in inches, derate the table by 10% for each length unit and recalculate connector length (see 13.2.2).

**Table 2 Connector Horizontal Length**

| Connector | Diameter | Max length | Length | OK? |
|---|---|---|---|---|
| Furnace | 4 in. | 6 ft | 6 ft | ✔ |
| Water heater | 4 in. | 6 ft | 4 ft | ✔ |
| | | | | |
| | | | | |

**Step 6:**
- Determine the common vent size using the bottom portion of the multiple appliance vent table, as shown in Table 3.

**Table 3 Bottom of Multiple Appliance Vent Table**

| Appliance types | Vent height | Combined input | Common vent size | Table used |
|---|---|---|---|---|
| NAT + NAT | 30 ft | 185,000 | 6 in. | 13.2(b) |

**Job:** *25 Main Street* **Prepared by:** *TL* **Date:** *1/1/09*



***WORKSHEET G.3*** *Vent Sizing, Common Venting of Two Draft Hood–Equipped Appliances.*

**COMMENTARY TABLE G.9** *Extract from Code Table 13.2(b) for Example 5(a) — Water Heater Sizing*

*Vent Connector Capacity*

| | | Single-Wall Metal Vent Connector Diameter D | | | | | |
|---|---|---|---|---|---|---|---|
| | | 3 in. | | | 4 in. | | |
| | | Appliance Input Rating Limits in Thousands of Btu per Hour | | | | | |
| Vent Height H (ft) | Connector Rise R (ft) | FAN Min | FAN Max | NAT Max | FAN Min | FAN Max | NAT Max |
| (30) | 1 | 47 | 60 | 31 | 77 | 110 | 57 |
| | (2) | 51 | 62 | (37) | 81 | 115 | (67) |
| | 3 | 54 | 64 | 42 | 85 | 119 | 76 |

13.2(b), read down the Total Vent Height *(H)* column to 30 ft and read across the 2 ft Connector Rise *(R)* row to the first Btu/hr rating in the NAT Max column that is equal to or greater than the water heater input rating. The table shows that a 3 in. vent connector has a maximum input rating of 37,000 Btu/hr. Although this rating is greater than the water heater input rating, a 3 in. vent connector is prohibited by 13.2.22. A 4 in. vent connector has a maximum input rating of 67,000 Btu/hr and is equal to the draft hood outlet diameter. A 4 in. vent connector is selected. Since the water heater is equipped with a draft hood, there are no minimum input rating restrictions.

See Commentary Table G.10, which is an extract from code Table 13.2(b).

**COMMENTARY TABLE G.10** *Extract from Code Table 13.2(b) for Example 5(a) — Furnace Sizing*

*Vent Connector Capacity*

| | | Single-Wall Metal Vent Connector Diameter D | | | | | | | | |
|---|---|---|---|---|---|---|---|---|---|---|
| | | 3 in. | | | 4 in. | | | 5 in. | | |
| | | Appliance Input Rating Limits in Thousands of Btu per Hour | | | | | | | | |
| Vent Height H (ft) | Connector Rise R (ft) | FAN Min | FAN Max | NAT Max | FAN Min | FAN Max | NAT Max | FAN Min | FAN Max | NAT Max |
| (30) | 1 | 47 | 60 | 31 | 77 | 110 | 57 | 112 | 175 | 89 |
| | 2 | 51 | 62 | 37 | 81 | 115 | 67 | 117 | 185 | 106 |
| | (3) | 54 | 64 | 42 | (85 | 119) | 76 | 122 | 193 | 120 |

*Furnace Vent Connector Diameter.* Using the Vent Connector Capacity portion of Table 13.2(b), read down the Total Vent Height *(H)* column to 30 ft and across the 3 ft Connector Rise *(R)* row. Because the furnace has a fan-assisted combustion system, find the first FAN Max column with a Btu/hr rating greater than the furnace input rating. The 4 in. vent connector has a maximum input rating of 119,000 Btu/hr and a minimum input rating of 85,000 Btu/hr.

The 100,000 Btu/hr furnace in this example falls within this range, so a 4 in. connector is adequate. Because the furnace vent connector horizontal length of 6 ft is less than the maximum value listed in 13.2.2, the venting table values can be used without adjustment. If the furnace had an input rating of 80,000 Btu/hr, a Type B vent connector *[see Table 13.2(a)]* would be needed in order to meet the minimum capacity limit.

*Common Vent Diameter.* The total input to the common vent is 135,000 Btu/hr. Using the Common Vent Capacity portion of Table 13.2(b), read down the Total Vent Height *(H)* column to 30 ft and across this row to find the smallest vent diameter in the FAN + NAT column that has a Btu/hr rating equal to or greater than 135,000 Btu/hr. The 4 in. common vent has a capacity of 132,000 Btu/hr, and the 5 in. common vent has a capacity of 202,000 Btu/hr. Therefore, the 5 in. common vent should be used in this example.

*Summary.* In this example, the installer can use a 4 in. diameter, single-wall metal vent connector for the water heater and a 4 in. diameter, single-wall metal vent connector for the furnace. The common vent should be a 5 in. diameter Type B vent.

See Commentary Table G.11, which is an extract from code Table 13.2(b).

***COMMENTARY TABLE G.11** Extract from Code Table 13.2(b) for Example 5(a) — Common Vent Sizing*

*Common Vent Capacity*

| Vent Height H (ft) | Type B Double-Wall Vent Diameter D | | | | | | | | |
|---|---|---|---|---|---|---|---|---|---|
| | 4 in. | | | 5 in. | | | 6 in. | | |
| | Combined Appliance Input Rating in Thousands of Btu per Hour | | | | | | | | |
| | FAN + FAN | FAN + NAT | NAT + NAT | FAN + FAN | FAN + NAT | NAT + NAT | FAN + FAN | FAN + NAT | NAT + NAT |
| 15 | 121 | 108 | 88 | 189 | 159 | 140 | 275 | 221 | 200 |
| 20 | 131 | 118 | 98 | 208 | 177 | 156 | 305 | 247 | 223 |
| 30 | 145 | 132 | 113 | 236 | 202 | 180 | 350 | 286 | 257 |
| 50 | 159 | 145 | 128 | 268 | 233 | 208 | 406 | 337 | 296 |

The solution to Example 5(a) is summarized in Worksheet G.4. Blank worksheets are located in Supplement 6 of this handbook.

## G.2.3 Example 5(b): Common Venting into an Interior Masonry Chimney.

In this case, the water heater and fan-assisted furnace of Example 5(a) are to be common-vented into a clay-tile-lined masonry chimney with a 30 ft height. The chimney is not exposed to the outdoors below the roof line. The internal dimensions of the clay tile liner are nominally 8 in. × 12 in. Assuming the same vent connector heights, laterals, and materials found in Example 5(a), what are the recommended vent connector diameters, and is this an acceptable installation?

*Solution*

Table 13.2(d) is used to size common venting installations involving single-wall connectors into masonry chimneys.

*Water Heater Vent Connector Diameter.* Using Table 13.2(d), Vent Connector Capacity, read down the Total Vent Height *(H)* column to 30 ft, and read across the 2 ft Connector Rise *(R)* row to the first Btu/hr rating in the NAT Max column that is equal to or greater than the water heater input rating. The table shows that a 3 in. vent connector has a maximum input of only 31,000 Btu/hr, while a 4 in. vent connector has a maximum input of 57,000 Btu/hr. A 4 in. vent connector must therefore be used.

## CALCULATION WORKSHEET: VENT SIZING, TWO APPLIANCES
### (COMMON VENTING OF A DRAFT HOOD–EQUIPPED WATER HEATER WITH A FAN-ASSISTED FURNACE INTO A TYPE B DOUBLE-WALL COMMON VENT)

**Step 1:**
- Determine the type of chimney or vent used and enter it to the right.
- Determine the type of vent connector material to be used and enter it to the right.
- Determine the number of appliances and enter it to the right.
- Select the appropriate table and enter it to the right.

Type of vent: *B vent*
Type of connector: *Single wall*
Number of Appliances: *2*
Table used: *13.2 (b)*

**Step 2:**
- Draw a sketch of the proposed design in the space to the right. Use the back of this page or a separate sheet if more space is needed.

Note: Factors such as combining connectors prior to entering the common vent will affect capacity.

*Note:*
*Water heater connector horizontal length is 4 ft.*
*Furnace connector horizontal length is 6 ft.*

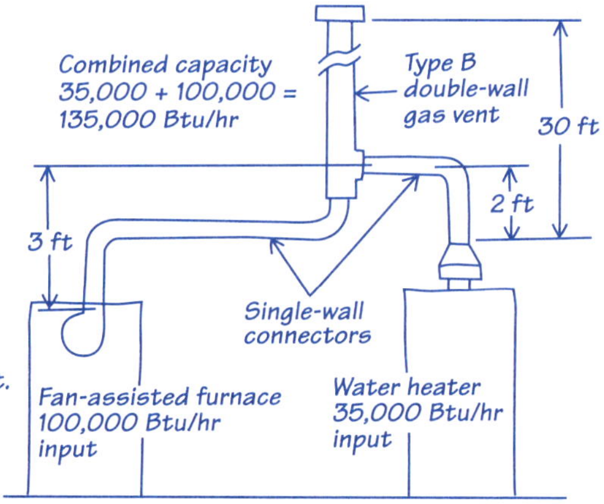


Proposed design sketch

**Step 3:**
- Determine the vent connector size using the top portion of the multiple appliance vent table, as shown in Table 1.

**Step 4:**
- Check for vent connector size limits, 13.2.21.

Note that Table 13.2(b) allows a 3 in. connector for the water heater, but 13.2.21 does not allow the vent connector to be smaller than the draft hood.

**Table 1 Top of Multiple Appliance Vent Table**

| Appliance | Connector rise | Vent height | Input rating | FAN or NAT | Connector size | Table used |
|---|---|---|---|---|---|---|
| Furnace | *3 ft* | *30 ft* | *100,000* | *FAN* | *4 in.* | *13.2(b)* |
| Water heater | *2 ft* | *30 ft* | *35,000* | *NAT* | *4 in.* | *13.2(b)* |
| | | | | | | |
| Total | | | *135,000* | | | |

**Step 5:**
- Check for excessive connector horizontal length using Table 2.
- If any connector horizontal length in feet exceeds 1.5 times the diameter in inches, derate the table by 10% for each length unit and recalculate connector length (see 13.2.2).

**Table 2 Connector Horizontal Length**

| Connector | Diameter | Max length | Length | OK? |
|---|---|---|---|---|
| *Furnace* | *4 in.* | *6 ft* | *6 ft* | ✓ |
| *Water heater* | *4 in.* | *6 ft* | *4 ft* | ✓ |
| | | | | |
| | | | | |

**Step 6:**
- Determine the common vent size using the bottom portion of the multiple appliance vent table, as shown in Table 3.

**Table 3 Bottom of Multiple Appliance Vent Table**

| Appliance types | Vent height | Combined input | Common vent size | Table used |
|---|---|---|---|---|
| *FAN + NAT* | *30 ft* | *135,000* | *5 in.* | *13.2(b)* |

**Job:** *25 Main Street* **Prepared by:** *TL* **Date:** *1/1/09*



***WORKSHEET G.4*** *Vent Sizing, Common Venting a Draft Hood–Equipped Water Heater with a Fan-Assisted Furnace into a Type B Double-Wall Common Vent.*

See Commentary Table G.12, which is an extract from code Table 13.2(d).

***COMMENTARY TABLE G.12 Extract from Code Table 13.2(d) for Example 5(b) — Water Heater Sizing***

***Vent Connector Capacity***

| | | Single-Wall Metal Vent Connector Diameter D | | | | | |
|---|---|---|---|---|---|---|---|
| | | 3 in. | | | 4 in. | | |
| | | Appliance Input Rating Limits in Thousands of Btu per Hour | | | | | |
| Vent Height H (ft) | Connector Rise R (ft) | FAN Min | FAN Max | NAT Max | FAN Min | FAN Max | NAT Max |
| (30) | 1 | NA | NA | 24 | 86 | 108 | 47 |
| | (2) | NA | NA | 31 | 91 | 119 | (57) |
| | 3 | NA | NA | 35 | 95 | 127 | 65 |

*Furnace Vent Connector Diameter.* Using the Vent Connector Capacity portion of Table 13.2(d), read down the Total Vent Height *(H)* column to 30 ft and across the 3 ft Connector Rise *(R)* row. Because the furnace has a fan-assisted combustion system, find the first FAN Max column with a Btu/hr rating greater than the furnace input rating. The 4 in. vent connector has a maximum input rating of 127,000 Btu/hr and a minimum input rating of 95,000 Btu/hr. The 100,000 Btu/hr furnace in this example falls within this range, so a 4 in. connector is adequate.

Commentary Table G.13, which is extracted from code Table 13.2(d), is used to calculate the diameter of the furnace vent connector.

***COMMENTARY TABLE G.13 Extract from Code Table 13.2(d) for Example 5(b) — Furnace Vent Sizing***

***Vent Connector Capacity***

| | | Single-Wall Metal Vent Connector Diameter D | | | | | |
|---|---|---|---|---|---|---|---|
| | | 3 in. | | | 4 in. | | |
| | | Appliance Input Rating Limits in Thousands of Btu per Hour | | | | | |
| Vent Height H (ft) | Connector Rise R (ft) | FAN Min | FAN Max | NAT Max | FAN Min | FAN Max | NAT Max |
| (30) | 1 | NA | NA | 24 | 86 | 108 | 47 |
| | 2 | NA | NA | 31 | 91 | 119 | 57 |
| | (3) | NA | NA | 35 | (95 | 127) | 65 |

*Masonry Chimney.* From Table G.2.3, the Equivalent Area for a Nominal Liner size of 8 in. × 12 in. is 63.6 in.$^2$. Using Table 13.2(d), Common Vent Capacity, read down the FAN + NAT column under the Minimum Internal Area of Chimney value of 63 to the row for 30 ft height to find a capacity value of 739,000 Btu/hr. The combined input rating of the furnace and water heater, 135,000 Btu/hr, is less than the table value, so this is an acceptable installation.

Commentary Tables G.14 and G.15 are used to answer the final portion of Example 5(b). They are extracted from code Table G.2.3 and code Table 13.2(d), respectively.

***COMMENTARY TABLE G.14** Extract from Code Table G.2.3 for Example 5(b)*

| Nominal Liner Size (in.) | Inside Dimensions of Liner (in.) | Inside Diameter or Equivalent Diameter (in.) | Equivalent Area (in.$^2$) |
|---|---|---|---|
| 8 × 12 | $6\frac{1}{2} \times 10\frac{1}{2}$ | 9 | 63.6 |

***COMMENTARY TABLE G.15** Extract from Code Table 13.2(d) for Example 5(b) — Common Venting*

*Common Vent Capacity*

| Vent Height H (ft) | Minimum Internal Area of Masonry Chimney Flue (in.$^2$) | | | | | | | | |
|---|---|---|---|---|---|---|---|---|---|
| | 50 in. | | | (63 in.) | | | 78 in. | | |
| | Combined Appliance Input Rating in Thousands of Btu per Hour | | | | | | | | |
| | FAN + FAN | FAN + NAT | NAT + NAT | FAN + FAN | FAN + NAT | NAT + NAT | FAN + FAN | FAN + NAT | NAT + NAT |
| 20 | NA | 503 | 240 | 765 | 661 | 321 | 947 | 849 | 415 |
| (30) | NA | 558 | 275 | 808 | (739) | 377 | 1052 | 957 | 490 |
| 50 | NA | 612 | 325 | NA | 821 | 456 | 152 | 1076 | 600 |

Subsection 13.2.16 requires the common vent area to be no greater than seven times the smallest listed appliance categorized vent area, flue collar area, or draft hood outlet area. Both appliances in this installation have 4 in. diameter outlets. From Table G.2.3, the equivalent area for an inside diameter of 4 in. is 12.2 in.$^2$. Seven times 12.2 equals 85.4, which is greater than 63.6, so this configuration is acceptable.

The solution to Example 5(b) is summarized in the calculation worksheet, Worksheet G.5. Blank worksheets are located in Supplement 6 of this handbook.

Table G.2.3 is included in the code as a convenience. The Equivalent Diameter and Equivalent Area columns tabulate conversion between diameter and area of a circle for each whole inch diameter (8 in., 9 in., etc.). The table also offers suggestions for appropriate free areas to use for common clay tile sizes.

In practice, the dimensions of clay tile vary because of different manufacturing practices and tolerances. In addition, the internal corners of tile are not square. For this reason, the equivalent areas listed for the tiles are somewhat smaller than what would be calculated by the Inside Dimensions column.

Note that code Table G.2.3 contains informational material and is not mandated. In the field, it may be advantageous to carefully measure the actual free area of the tile.

## G.2.4 Example 5(c): Common Venting into an Exterior Masonry Chimney.

In this case, the water heater and fan-assisted furnace of Examples 5(a) and 5(b) are to be common-vented into an exterior masonry chimney. The chimney height, clay-tile-liner dimensions, and vent connector heights and laterals are the same as in Example 5(b). This system is being installed in Charlotte, North Carolina. Does this exterior masonry chimney need to be relined? If so, what corrugated metallic liner size is recommended? What vent connector diameters are recommended? See Table G.2.3 and Figure G.2.4.

*Solution*

According to 13.2.25, Type B vent connectors are required to be used with exterior masonry chimneys. Use Table 13.2(h) and Table 13.2(i) to size FAN+NAT common venting installations involving Type B double-wall connectors into exterior masonry chimneys.

## CALCULATION WORKSHEET: VENT SIZING, TWO APPLIANCES
### (TWO APPLIANCES WITH SINGLE-WALL CONNECTORS, MASONRY CHIMNEY)

**Step 1:**

- Determine the type of chimney or vent used and enter it to the right.
- Determine the type of vent connector material to be used and enter it to the right.
- Determine the number of appliances and enter it to the right.
- Select the appropriate table and enter it to the right.

Type of vent: *Masonry, 8 in. x 12 in.*

Type of connector: *Single + B vent*

Number of appliances: *2*

Table used: *13.2 (c) and 13.2 (d)*

**Step 2:**

- Draw a sketch of the proposed design in the space to the right. Use the back of this page or a separate sheet if more space is needed.

Note: Factors such as combining connectors prior to entering the common vent will affect capacity.

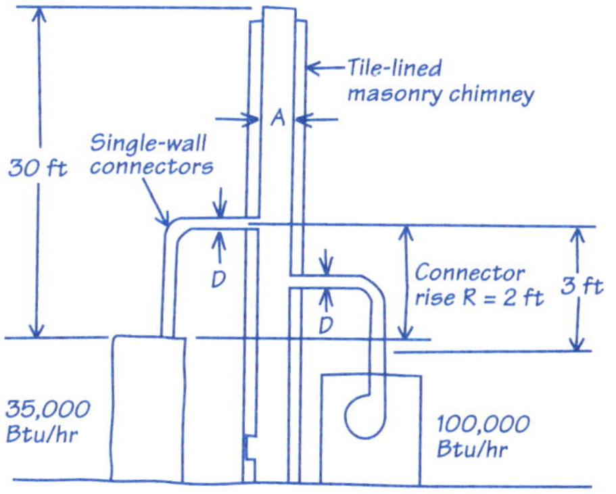


Proposed design sketch

*Notes:*
*A = Equivalent cross-sectional area of the tile liner.*
*Lateral distance from chimney to water heater is 4 ft, and from chimney to furnace is 6 ft.*

*Chimney not exposed to the outdoors below the roofline.*

**Step 3:**

- Determine the vent connector size using the top portion of the multiple appliance vent table, as shown in Table 1.

**Step 4:**

- Check for vent connector size limits, 13.2.21.

**Table 1 Top of Multiple Appliance Vent Table**

| Appliance | Connector rise | Vent height | Input rating | FAN or NAT | Connector size | Table used |
|---|---|---|---|---|---|---|
| Furnace | *3 ft* | *30 ft* | *100,000* | *FAN* | *4 in.* | *13.2(c)* |
| Water heater | *2 ft* | *30 ft* | *35,000* | *NAT* | *4 in.* | *13.2(d)* |
| | | | | | | |
| Total | | | *135,000* | | | |

**Step 5:**

- Check for excessive connector horizontal length using Table 2.
- If any connector horizontal length in feet exceeds 1.5 times the diameter in inches, derate the table by 10% for each length unit and recalculate connector length (see 13.2.2).

**Table 2 Connector Horizontal Length**

| Connector | Diameter | Max. length | Length | OK? |
|---|---|---|---|---|
| *Furnace* | *4 in.* | *6 ft* | *6 ft* | |
| *Water heater* | *4 in.* | *6 ft* | *4 ft* | |
| | | | | |
| | | | | |

**Step 6:**

- Determine the common vent size using the bottom portion of the multiple appliance vent table, as shown in Table 3.

**Table 3 Bottom of Multiple Appliance Vent Table**

| Appliance types | Vent height | Combined input | Common vent size | Table used |
|---|---|---|---|---|
| *FAN + NAT* | *30 ft* | *135,000* | *8 x 12 is OK* | *13.2(d)* |

Note that Table G.2.3 shows that an 8 in. × 12 in. chimney with a 9 in. nominal liner has an equivalent area of 63.6 in.$^2$.

Check for "7 × rule": A 4 in. diameter connector has an area of 2 × 2 × 3.14 = 12.6 in.$^2$. Seven times 12.6 = 88.2 in.$^2$. The total of 88.2 is greater than 63.6. Sizing of common vent is okay.

**Job:** *25 Main Street* **Prepared by:** *TL* **Date:** *1/1/09*



***WORKSHEET G.5*** *Vent Sizing, Two Appliances with Single-Wall Connectors, Masonry Chimney.*

***TABLE G.2.3** Masonry Chimney Liner Dimensions with Circular Equivalents*

| *Nominal Liner Size (in.)* | *Inside Dimensions of Liner (in.)* | *Inside Diameter or Equivalent Diameter (in.)* | *Equivalent Area (in.²)* |
|---|---|---|---|
| 4 × 8 | $2\frac{1}{2} \times 6\frac{1}{2}$ | 4.0 | 12.2 |
| | | 5.0 | 19.6 |
| | | 6.0 | 28.3 |
| | | 7.0 | 38.3 |
| 8 × 8 | $6\frac{3}{4} \times 6\frac{3}{4}$ | 7.4 | 42.7 |
| | | 8.0 | 50.3 |
| 8 × 12 | $6\frac{1}{2} \times 10\frac{1}{2}$ | 9.0 | 63.6 |
| | | 10.0 | 78.5 |
| 12 × 12 | $9\frac{3}{4} \times 9\frac{3}{4}$ | 10.4 | 83.3 |
| | | 11.0 | 95.0 |
| 12 × 16 | $9\frac{1}{2} \times 13\frac{1}{2}$ | 11.8 | 107.5 |
| | | 12.0 | 113.0 |
| | | 14.0 | 153.9 |
| 16 × 16 | $13\frac{1}{4} \times 13\frac{1}{4}$ | 14.5 | 162.9 |
| | | 15.0 | 176.7 |
| 16 × 20 | 13 × 17 | 16.2 | 206.1 |
| | | 18.0 | 254.4 |
| 20 × 20 | $16\frac{1}{2} \times 16\frac{3}{4}$ | 18.2 | 260.2 |
| | | 20.0 | 314.1 |
| 20 × 24 | $16\frac{1}{2} \times 20\frac{1}{2}$ | 20.1 | 314.2 |
| | | 22.0 | 380.1 |
| 24 × 24 | $20\frac{1}{4} \times 20\frac{1}{4}$ | 22.1 | 380.1 |
| | | 24.0 | 452.3 |
| 24 × 28 | $20\frac{1}{4} \times 24\frac{1}{4}$ | 24.1 | 456.2 |
| 28 × 28 | $24\frac{1}{4} \times 24\frac{1}{4}$ | 26.4 | 543.3 |
| | | 27.0 | 572.5 |
| 30 × 30 | $25\frac{1}{2} \times 25\frac{1}{2}$ | 27.9 | 607.0 |
| | | 30.0 | 706.8 |
| 30 × 36 | $25\frac{1}{2} \times 31\frac{1}{2}$ | 30.9 | 749.9 |
| | | 33.0 | 855.3 |
| 36 × 36 | $31\frac{1}{2} \times 31\frac{1}{2}$ | 34.4 | 929.4 |
| | | 36.0 | 1017.9 |

For SI units, 1 in. = 25.4 mm, 1 in.$^2$ = 645 mm$^2$.

Note: When liner sizes differ dimensionally from those shown in this table, equivalent diameters can be determined from published tables for square and rectangular ducts of equivalent carrying capacity or by other engineering methods.

The local 99 percent winter design temperature needed to use Table 13.2(h) and Table 13.2(i) can be found in *ASHRAE Handbook — Fundamentals.* For Charlotte, North Carolina, this design temperature is 19°F.

*Chimney Liner Requirement.* As in Example 5(b), use the 63 in.$^2$ Internal Area columns for this size clay tile liner. Read down the 63 in.$^2$ column of Table 13.2(h) to the 30 ft height row to find that the Combined Appliance Maximum Input is 747,000 Btu/hr. The combined input rating of the appliances in this installation, 135,000 Btu/hr, is less than the maximum value, so this criterion is satisfied. Table 13.2(i), at a 19°F Design Temperature, and at the

**99% Winter Design Temperatures for the Contiguous United States**

This map is a necessarily generalized guide to temperatures in the contiguous United States. Temperatures shown for areas such as mountainous regions and large urban centers are not necessarily accurate. The data used to develop this map are from the 1993 *ASHRAE Handbook—Fundamentals* (Chapter 24, Table 1: Climate Conditions for the United States).

For 99% winter design temperatures in Alaska, consult the *ASHRAE Handbook—Fundamentals.*

99% winter design temperatures for Hawaii are greater than 37°F.

***FIGURE G.2.4*** *Range of Winter Design Temperatures Used in Analyzing Exterior Masonry Chimneys in the United States.*

same Vent Height and Internal Area used earlier, shows that the minimum allowable input rating of a space-heating appliance is 470,000 Btu/hr. The furnace input rating of 100,000 Btu/hr is less than this minimum value. So this criterion is not satisfied, and an alternative venting design needs to be used, such as a Type B vent shown in Example 5(a) or a listed chimney liner system shown in the remainder of the example.

According to 13.2.19, Table 13.2(a) or Table 13.2(b) is used for sizing corrugated metallic liners in masonry chimneys, with the maximum common vent capacities reduced by 20 percent. This example will be continued assuming Type B vent connectors.

See Commentary Table G.16, which is extracted from code Table 13.2(h).

*Water Heater Vent Connector Diameter.* Using Table 13.2(a), Vent Connector Capacity, read down the Total Vent Height *(H)* column to 30 ft, and read across the 2 ft Connector Rise *(R)* row to the first Btu/hour rating in the NAT Max column that is equal to or greater than the water heater input rating. The table shows that a 3 in. vent connector has a maximum capacity of 39,000 Btu/hr. Although this rating is greater than the water heater input rating, a 3 in. vent connector is prohibited by 13.2.20. A 4 in. vent connector has a maximum input rating of 70,000 Btu/hr and is equal to the draft hood outlet diameter. A 4 in. vent connector is selected.

**COMMENTARY TABLE G.16** *Extract from Code Table 13.2(h) for Example 5(c) — Chimney Liner Sizing*

| Vent Height H (ft) | Internal Area of Chimney (in.$^2$) | | | |
|---|---|---|---|---|
| | 38 in. | 50 in. | 63 in. | 78 in. |
| 15 | 334 | 467 | 611 | 781 |
| 20 | 368 | 508 | 668 | 858 |
| 30 | 404 | 564 | 747 | 969 |
| 50 | NA | NA | 831 | 1089 |

*Furnace Vent Connector Diameter.* Using Table 13.2(a), Vent Connector Capacity, read down the Total Vent Height *(H)* column to 30 ft, and read across the 3 ft Connector Rise *(R)* row to the first Btu/hr rating in the FAN Max column that is equal to or greater than the furnace input rating. The 100,000 Btu/hr furnace in this example falls within this range, so a 4 in. connector is adequate.

Commentary Table G.17, which is extracted from code Table 13.2(a), is used to calculate the diameter of a single-wall vent connector for Example 5(c).

**COMMENTARY TABLE G.17** *Extract from Code Table 13.2(a) for Example 5(c) — Water Heater Sizing*

*Vent Connector Capacity*

| Vent Height H (ft) | Connector Rise R (ft) | Single-Wall Metal Vent Connector Diameter D | | | | | |
|---|---|---|---|---|---|---|---|
| | | 3 in. | | | 4 in. | | |
| | | Appliance Input Rating Limits in Thousands of Btu per Hour | | | | | |
| | | FAN | | NAT | FAN | | NAT |
| | | Min | Max | Max | Min | Max | Max |
| 30 | 1 | 20 | 62 | 33 | 31 | 113 | 59 |
| | 2 | 21 | 64 | 39 | 33 | 118 | 70 |
| | 3 | 22 | 66 | 44 | 34 | 123 | 79 |

Assume that a corrugated chimney liner has been installed. Subsection 13.2.24 then applies, which states the following:

**13.2.24 Vent Connector Sizing.** Vent connectors shall not be increased more than two sizes greater than the listed appliance categorized vent diameter, flue collar diameter, or draft hood outlet diameter. ***Vent connectors for draft hood–equipped appliances shall not be smaller than the draft hood outlet diameter.*** Where a vent connector size(s) determined from the tables for a fan-assisted appliance(s) is smaller than the flue collar diameter, the use of the smaller size(s) shall be permitted, provided that the installation complies with all of the following conditions:

(1) Vent connectors for fan-assisted appliance flue collars 12 in. (300 mm) in diameter or smaller are not reduced by more than one table size [e.g., 12 in. to 10 in. (300 mm to 250 mm) is a one-size reduction], and those larger than 12 in. (300 mm) in diameter are not reduced more than two table sizes [e.g., 24 in. to 20 in. (610 mm to 510 mm) is a two-size reduction].
(2) The fan-assisted appliance(s) is common vented with a draft hood–equipped appliance(s).
(3) The vent connector has a smooth interior wall.

Note that the bold text is the relevant part of 13.2.24 to this example. The vent connector must have a diameter of at least the draft hood outlet, which in this example is 4 in.

*Chimney Liner Diameter.* The total input to the common vent is 135,000 Btu/hr. Using the Common Vent Capacity portion of Table 13.2(a), read down the Total Vent Height *(H)* column to 30 ft and across this row to find the smallest vent diameter in the FAN+NAT column that has a Btu/hr rating greater than 135,000 Btu/hr. The 4 in. common vent has a capacity of 138,000 Btu/hr. Reducing the maximum capacity by 20 percent *(see 13.2.20)* results in a maximum capacity for a 4 in. corrugated liner of 110,000 Btu/hr, less than the total input of 135,000 Btu/hr. So a larger liner is needed. The 5 in. common vent capacity listed in Table 13.2(a) is 210,000 Btu/hr, and after reducing by 20 percent is 168,000 Btu/hr. Therefore, a 5 in. corrugated metal liner should be used in this example.

Commentary Table G.18, which is extracted from code Table 13.2(a), is used to calculate the diameter of the single-wall furnace vent connector for Example 5(c).

***COMMENTARY TABLE G.18** Extract from Code Table 13.2(a) for Example 5(c) — Furnace Sizing*

*Vent Connector Capacity*

| | | Single-Wall Metal Vent Connector Diameter D | | | | | |
|---|---|---|---|---|---|---|---|
| | | 3 in. | | | 4 in. | | |
| | | Appliance Input Rating Limits in Thousands of Btu per Hour | | | | | |
| | | FAN | | NAT | FAN | | NAT |
| Vent Height H (ft) | Connector Rise R (ft) | Min | Max | Max | Min | Max | Max |
| (30) | 1 | 20 | 62 | 33 | 31 | 113 | 59 |
| | 2 | 21 | 64 | 39 | 33 | 118 | 70 |
| | (3) | 22 | 66 | 44 | (34 | 123) | 79 |

*Single-Wall Connectors.* Once it has been established that relining the chimney is necessary, Type B double-wall vent connectors are not specifically required. This example could be redone using Table 13.2(b) for single-wall vent connectors. For this case, the vent connector and liner diameters would be the same as found for Type B double-wall connectors.

Commentary Table G.19, which is extracted from code Table 13.2(a), is used to calculate the diameter of the chimney liner for Example 5(c).

***COMMENTARY TABLE G.19** Extract from Code Table 13.2(a) for Example 5(c) — Chimney Liner Diameter*

*Common Vent Capacity*

| | Type B Double-Wall Vent Diameter D | | | | | | | | |
|---|---|---|---|---|---|---|---|---|---|
| | 4 in. | | | 5 in. | | | 6 in. | | |
| | Combined Appliance Input Rating in Thousands of Btu per Hour | | | | | | | | |
| Vent Height H (ft) | FAN + FAN | FAN + NAT | NAT + NAT | FAN + FAN | FAN + NAT | NAT + NAT | FAN + FAN | FAN + NAT | NAT + NAT |
| 15 | 125 | 112 | 91 | 195 | 164 | 144 | 283 | 228 | 206 |
| 20 | 136 | 123 | 102 | 215 | 183 | 160 | 314 | 255 | 229 |
| (30) | 152 | 138 | 118 | 244 | (210*) | 185 | 361 | 297 | 266 |
| 50 | 167 | 153 | 134 | 279 | 244 | 214 | 421 | 353 | 310 |

*Derate by 20% Capacity = 168,000 Btu/hr

# Recommended Procedure for Safety Inspection of an Existing Appliance Installation

## ANNEX H

*This annex is not a part of the requirements of this NFPA document but is included for informational purposes only.*

## H.1 General

The following procedure is intended as a guide to aid in determining that an appliance is properly installed and is in a safe condition for continuing use.

This procedure is intended for central furnace and boiler installations, and may not be applicable to all installations.

This procedure should be performed prior to any attempt to modify the appliance or the installation.

If it is determined a condition that could result in unsafe operation exists, shut off the appliance and advise the owner of the unsafe condition.

The following steps should be followed in making the safety inspection:

(1) Conduct a test for gas leakage. *(See Section 8.2.)*
(2) Visually inspect the venting system for proper size and horizontal pitch, and determine that there is no blockage, restriction, leakage, corrosion, or other deficiencies that could cause an unsafe condition.

When inspecting the venting system, it is important to check the top of horizontal single-wall vent connectors. The top of these connectors is the area where corrosive condensate tends to accumulate early in the operation of the appliance. A single-wall vent connector is especially prone to corrosion because it cools off much faster (and condenses water vapor more easily) than Type B vent connectors, and holes may not be visible when the observer is looking up. Vent systems require a vent cap designed as part of the system. Simple rain-cap-type replacement vent caps may not be sufficient for proper operation of the vent system.

(3) Shut off all gas to the appliance, and shut off any other fuel gas burning appliance within the same room. Use the shutoff valve in the supply line to each appliance.
(4) Inspect burners and crossovers for blockage and corrosion.
(5) *Furnace Installations:* Inspect the heat exchanger for cracks, openings, or excessive corrosion.
(6) *Boiler Installations:* Inspect for evidence of water or combustion product leaks.
(7) Close all building doors and windows and all doors between the space in which the appliance is located and other spaces of the building that can be closed. Turn on any clothes dryers. Turn on any exhaust fans, such as range hoods and bathroom exhausts, so they operate at maximum speed. Do not operate a summer exhaust fan. Close fireplace dampers. If, after completing Steps 8 through 13, it is believed sufficient combustion air is not available, refer to Section 9.3 of this code.
(8) Place the appliance being inspected in operation. Follow the lighting instructions. Adjust the thermostat so the appliance operates continuously.
(9) Determine that the pilot, where provided, is burning properly and that the main burner ignition is satisfactory by interrupting and re-establishing the electrical supply to the

appliance in any convenient manner. If the appliance is equipped with a continuous pilot, test all pilot safety devices to determine whether it is operating properly by extinguishing the pilot when the main burner is off and determining, after 3 minutes, that the main burner gas does not flow upon a call for heat. If the appliance is not provided with a pilot, test for proper operation of the ignition system in accordance with the appliance manufacturer's lighting and operating instructions.

(10) Visually determine that the main burner gas is burning properly (i.e., no floating, lifting, or flashback). Adjust the primary air shutters as required. If the appliance is equipped with high and low flame controls or flame modulation, check for proper main burner operation at low flame.

Exhibit H.1 shows an example of lifting burner flames, which can be seen at the left and right. Lifting may occur on cold burner assemblies and stop once the burner is warmed up. "Flashback" is when the flame ignites the fuel–air mixture in the burner head. Flashback sometimes makes a popping sound. Increasing the primary air to the burner jet may alleviate this problem.

(11) Test for spillage at the draft hood relief opening after 5 minutes of main burner operation. Use the flame of a match or candle or smoke. *(See Section 11.6.)*

A smoking match held near the draft hood should have the smoke sucked into the hood. This indicates a positive flow out of the building. Exhibit H.2 illustrates a typical draft hood test.

(12) Turn on all other fuel gas burning appliances within the same room so they operate at their full inputs. Follow lighting instructions for each appliance.

(13) Repeat Steps 10 and 11 on the appliance being inspected.

(14) Return doors, windows, exhaust fans, fireplace dampers, and any other fuel gas burning appliance to their previous conditions of use.

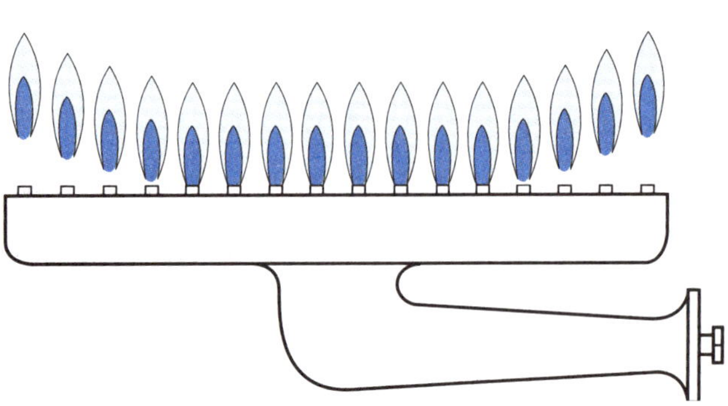

**EXHIBIT H.1** *Lifting Burner Flames.*

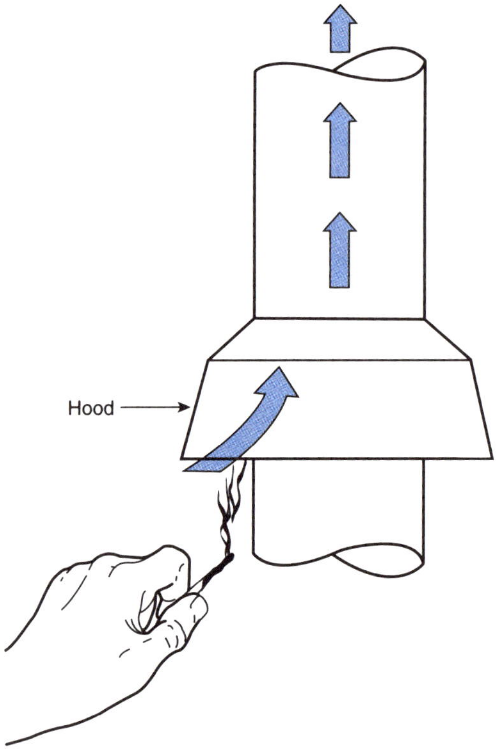


**EXHIBIT H.2** *Draft Hood Test.*

(15) *Furnace Installations:* Check both the limit control and the fan control for proper operation. Limit control operation can be checked by blocking the circulating air inlet or temporarily disconnecting the electrical supply to the blower motor and determining that the limit control acts to shut off the main burner gas. NEW

(16) *Boiler Installations:* Verify that the water pumps are in operating condition. Test low-water cutoffs, automatic feed controls, pressure and temperature limit controls, and relief valves in accordance with the manufacturer's recommendations to determine that they are in operating condition. NEW

# Indoor Combustion Air Calculation Examples

# ANNEX I

*This annex is not a part of the requirements of this NFPA document but is included for informational purposes only.*

Annex I provides a calculation example for a method of supplying air for combustion and ventilation to confined spaces. The example is illustrated in Worksheet I.1 by calculating combustion air using the standard method. A full-sized, reproducible copy of this form, which can be copied for use, is included in Supplement 6 of this handbook.

## I.1 New Installation

Determine if the indoor volume is sufficient to supply combustion air for the following new installation example.

*Example Installation 1:* A 100,000 Btu/hr fan-assisted furnace and a 40,000 Btu/hr draft hood–equipped water heater are being installed in a basement of a new single-family home. The basement measures 25 ft × 40 ft with an 8 ft ceiling.

*Solution*

(1) *Determine the total required volume:* Since the air infiltration rate is unknown, the standard method to determine combustion air is used to calculate the required volume.

- (a) The combined input for the appliances located in the basement is calculated as follows: 100,000 Btu/hr + 40,000 Btu/hr = 140,000 Btu/hr
- (b) The Standard Method requires that the required volume be determined based on 50 cubic feet per 1000 Btu/hour.
- (c) Using Table A.9.3.2.1, the required volume for a 140,000 Btu/hr combined input is 7000 $ft^3$

(2) *Determine available volume:* The available volume is the total basement volume:

Available Volume: 25 ft × 40 ft × 8 ft ceiling = 8000 $ft^3$

*Conclusion:* The installation can use indoor air because the available volume of 8000 $ft^3$ exceeds the total required volume of 7000 $ft^3$. No outdoor air openings are required.

## I.2 New Installation, Known Air Infiltration Rate Method

The solution to Example Installation 2 is also illustrated in Worksheet I.2, a form for calculating combustion air using the KAIR Method. A full-sized, reproducible copy of this form is provided in Supplement 6 of this handbook.

Determine if the indoor volume is sufficient to supply combustion air for the following replacement installation example.

## CALCULATION WORKSHEET: COMBUSTION AIR, STANDARD METHOD

**Step 1:**

- Calculate the room volume.

Room volume: Room length: 40 ft
Room width: 25 ft
Room height: 8 ft

Room volume: = Length × width × height
= $8000\ ft^3$

**Step 2:**

- Calculate the total input of all appliances in the room.
- Enter the input rating of all appliances in Table 1. (Per 9.3.1.1, Exception 2, dryers are not included.)
- Total the column.
- Divide the total by 1000 (of Btu/hr).

**Table 1 Appliances Table**

| Appliance | Input rating (Btu/hr) |
|---|---|
| Furnace | 100,000 |
| Water heater | 40,000 |
| Space heater | |
| Range | |
| Other | |
| Total | 140,000 |
| Total/1000 | 140 |

**Step 3:**

- Calculate the required volume. Divide room volume (Step 1) by total/1000 (Step 2).
- If less than 50, additional air is needed.
- If greater than or equal to 50, no additional air is needed.

Required volume:
= 8000/140
= 57.1

Additional air needed? (Check one) Yes ❑ No ☑

## ALTERNATE CALCULATION METHOD

**Step 1:**

- Calculate the room volume.

Room volume = $8000\ ft^3$ (from Step 1 above)

**Step 2:**

- Calculate the maximum appliance input.

Maximum appliance input: = Room volume × 20
= 8000 × 20
= 160,000 Btu/hr

**Step 3:**

- Determine if additional air is needed.
- If less than max., no additional air is needed.
- If greater than or equal to max., additional air is needed.

Total appliance input: = 140,000 Btu/hr

Additional air needed? (Check one) Yes ❑ No ☑

**Job:** 25 Main Street **Prepared by:** TL **Date:** 1/1/09



***WORKSHEET I.1*** *Sample Calculation for Determining Combustion Air Using the Standard Method (Example Installation 1).*

## CALCULATION WORKSHEET: COMBUSTION AIR, KAIR METHOD, EXAMPLE INSTALLATION 2

**Step 1:**
- Enter the input ratings of all non-fan-assisted appliances in Table 1.
- Total the column.

**Table 1 Ratings for Non-Fan-Assisted Appliances**

| Appliance | Input rating (Btu/hr) |
|---|---|
| Furnace | |
| Water heater | 40,000 |
| Space heater | |
| Refrigerator | |
| Total ($I_{other}$) | 40,000 |

**Step 2:**
- Enter the air infiltration rate of the space. This can be determined by measurement or estimation. See Supplement 5 for an estimation method.
- If the air infiltration rate exceeds 0.6, use 0.6.

Air infiltration rate: 0.65 ACH (max. 0.6)
Note: Use 0.6 in the calculation.

**Step 3:**
- Calculate the required volume for non-fan-assisted appliances using the equation.
- Enter the required volume here and in Step 6.

Required volume:

$$\frac{21\ \text{ft}^3}{\text{ACH}}\left(\frac{I_{other}}{100{,}000\ \text{Btu/hr}}\right) = \frac{21\ \text{ft}^3}{0.6}\left(\frac{40{,}000\ \text{Btu/hr}}{100{,}000\ \text{Btu/hr}}\right)$$

$$35\ (40) = 1400\ \text{ft}^3$$

**Step 4:**
- Enter the input ratings of all fan-assisted appliances in Table 2.
- Total the column.

**Table 2 Ratings for Fan-Assisted Appliances**

| Appliance | Input rating (Btu/hr) |
|---|---|
| Furnace | 100,000 |
| Furnace | |
| Space heater | |
| Other | |
| Other | |
| Other | |
| Total | 100,000 |

**Step 5:**
- Calculate the required volume for fan-assisted appliances using the equation.
- Enter the required volume here and below.

Required volume:

$$\frac{15\ \text{ft}^3}{\text{ACH}}\left(\frac{I_{fan}}{100{,}000\ \text{Btu/hr}}\right) = \frac{15\ \text{ft}^3}{0.6}\left(\frac{100{,}000\ \text{Btu/hr}}{100{,}000\ \text{Btu/hr}}\right)$$

$$25\ (100) = 2500\ \text{ft}^3$$

**Step 6:**
- Add the two required volumes calculated to find the total required volume for all appliances.

Required volume (Non-fan assisted) = 1400 ft$^3$
Required volume (Non-fan assisted) = 2500 ft$^3$
Total required volume = 3900 ft$^3$

**Step 7:**
- Enter the available room volume.

Available volume = Room volume = 20 x 35 x 8 = 5600 ft$^3$

**Step 8:**
- If the total required volume is greater than the available room volume, air openings are required. Check Yes and determine the opening size and location. If the total required volume is less than the available room volume, no air openings are required. Check No.

Air openings required? (Check one)

Yes ☐ No ☑

**Job:** 25 Main Street **Prepared by:** TL **Date:** 1/1/09



***WORKSHEET I.2*** *Sample Calculation for Determining Combustion Air Using the KAIR Method (Example Installation 2).*

*Example Installation 2:* A 100,000 Btu/hr fan-assisted furnace and a 40,000 Btu/hr draft hood–equipped water heater are being installed in a new single-family house. It was determined (either by use of the ASHRAE calculation method or blower door test) that the house has 0.65 air changes per hour (ACH). The furnace and water heater are being installed in a 20 ft × 35 ft basement with an 8 ft ceiling height.

*Solution*

(1) *Determine the required volume:* Because two types of appliances are located in the space — a fan-assisted furnace and a draft hood–equipped water heater — the required volume must be determined for each appliance and then combined to determine the total required volume:

(a) *Fan-assisted furnace:* For structures for which the air infiltration rate is known, method 9.3.2.2 permits the use of the equation in 9.3.2.2(2) to determine the required volume for a fan-assisted appliance. Paragraph 9.3.2.2(3) limits the use of the equation to air change rates equal to or less than 0.60 *ACH*. While the house was determined to have a 0.65 *ACH*, 0.60 is used to calculate the required volume. Using the equation in 9.3.2.2(2), the required volume for a 100,000 Btu/hr fan-assisted furnace is calculated as follows:

$$= \frac{15\ \text{ft}^3}{0.60}\left(\frac{100{,}000\ \text{Btu/hr}}{1000\ \text{Btu/hr}}\right)$$

$$= 2308\ \text{ft}^3$$

Paragraph 9.3.2.2 specifies a lower required volume limitation for fan-assisted appliances at no smaller than 25 $ft^3$ per 1000 Btu/hr. From Table A.9.3.2.2(b), the lower limit is

$$2500\ \text{ft}^3$$

Because the calculated required volume of 2308 $ft^3$ falls below the lower required volume limit, the lower limit of 2500 $ft^3$ must be used as the minimum required volume.

(b) *Draft hood–equipped water heater:* For structures for which the air infiltration rate is known, method 9.3.2.2 permits the use of the equation in 9.3.2.2(1) to determine the required volume for a draft hood–equipped appliance. Paragraph 9.3.2.2(3) limits the use of the equation to air change rates equal to or less than 0.60 *ACH*. While the house was determined to have a 0.65 *ACH*, 0.60 is used to calculate the required volume. Using the equation in 9.3.2.2(1), the required volume for the 40,000 Btu/hr water heater is calculated as follows:

$$= \frac{21\ \text{ft}^3}{0.60}\left(\frac{40{,}000\ \text{Btu/hr}}{1000\ \text{Btu/hr}}\right)$$

$$= 1400\ \text{ft}^3$$

Paragraph 9.3.2.2 specifies a lower required volume limitation for appliances other than fan-assisted at no smaller than 35 $ft^3$ per 1000 Btu/hr. From Table A.9.3.2.2(a), the lower limit is

$$1400\ \text{ft}^3$$

Because the calculated required volume of 1292 $ft^3$ falls below the lower required volume limit, the lower limit of 1400 $ft^3$ must be used as the minimum required volume.

(c) *Total required volume:* Subsection 9.3.2 states that the total required volume of indoor air is the sum of the required volumes for all appliances located in the space:

$$\text{Total Required} = 2500\ \text{ft}^3 + 1400\ \text{ft}^3 = 3900\ \text{ft}^3$$

(2) *Determine available volume:* The available volume is determined as follows:

$$(20 \text{ ft} \times 35 \text{ ft}) \times 8 \text{ ft} = 5600 \text{ ft}^3$$

*Conclusion:* The installation can use indoor air because the available volume of 5600 $ft^3$ exceeds the total required volume of 3900 $ft^3$. No outdoor air openings are required.

## I.3 New Installation, Known Air Infiltration Rate Method

The solution to Example Installation 3 is also illustrated in Worksheet I.3, a form for calculating combustion air using the KAIR Method. A full-sized, reproducible copy is provided in Supplement 6 of this handbook.

Determine if the indoor volume is sufficient to supply combustion air for the following replacement installation example.

*Example Installation 3:* A 100,000 Btu/hr fan-assisted furnace and a 40,000 Btu/hr draft hood–equipped water heater are being installed in a new single-family house. It was determined (either by use of the ASHRAE calculation method or blower door test) that the house has 0.30 air changes per hour (ACH). The furnace and water heater are being installed in a 20 ft × 35 ft basement with an 8 ft ceiling height.

*Solution*

(1) *Determine the required volume:* Because two types of appliances are located in the space — a fan-assisted furnace and a draft hood–equipped water heater — the required volume must be determined for each appliance and then combined to determine the total required volume:

(a) *Fan-assisted furnace:* For structures for which the air infiltration rate is known, method 9.3.2.2 permits the use of the equation in 9.3.2.2(2) to determine the required volume for a fan-assisted appliance. Paragraph 9.3.2.3 limits the use of the equation to air change rates equal to or less than 0.60 *ACH*. Because 0.30 *ACH* is less than 0.60 *ACH*, 0.30 can be used to calculate the required volume. Using the equation in 9.3.2.2(2), the required volume for a 100,000 Btu/hr fan-assisted furnace is calculated as follows:

$$= \frac{15 \text{ ft}^3}{0.30}\left(\frac{100{,}000 \text{ Btu/hr}}{1000 \text{ Btu/hr}}\right)$$
$$= 5000 \text{ ft}^3$$

Paragraph 9.3.2.2 specifies a lower required volume limitation for fan-assisted appliances at no smaller than 25 $ft^3$ per 1000 Btu/hr. From Table A.9.3.2.2(b), the lower limit is

$$2500 \text{ ft}^3$$

Because the calculated required volume of 5000 $ft^3$ is above the lower required volume limit, use this amount as the minimum required volume.

(b) *Draft hood–equipped water heater:* For structures for which the air infiltration rate is known, method 9.3.2.2 permits the use of the equation in 9.3.2.2(1) to determine the required volume for a draft hood–equipped appliance. Paragraph 9.3.2.2(3) limits the use of the equation to air change rates equal to or less than 0.60 *ACH*. While the house was determined to have a 0.65 *ACH*, 0.30 *ACH* is used to calculate the required volume. Using the equation in 9.3.2.2(1), the required volume for the 40,000 Btu/hr water heater is calculated as follows:

$$= \frac{21 \text{ ft}^3}{0.30}\left(\frac{40{,}000 \text{ Btu/hr}}{1000 \text{ Btu/hr}}\right)$$
$$= 2800 \text{ ft}^3$$

## CALCULATION WORKSHEET: COMBUSTION AIR, KAIR METHOD, EXAMPLE INSTALLATION 3

**Step 1:**
- Enter the input ratings of all non-fan-assisted appliances in Table 1.
- Total the column.

**Table 1 Ratings for Non-Fan Assisted Appliances**

| Appliance | Input rating (Btu/hr) |
|---|---|
| Furnace | |
| Water heater | 40,000 |
| Space heater | |
| Refrigerator | |
| Total ($I_{other}$) | 40,000 |

**Step 2:**
- Enter the air infiltration rate of the space. This can be determined by measurement or estimation. See Supplement 5 for an estimation method.
- If the air infiltration rate exceeds 0.6, use 0.6.

Air infiltration rate: 0.3 ACH (max. 0.6)

**Step 3:**
- Calculate the required volume for non-fan-assisted appliances using the equation.
- Enter the required volume here and in Step 6.

Required volume:

$$\frac{21\text{ ft}^3}{ACH}\left(\frac{I_{other}}{100{,}000\text{ Btu/hr}}\right) = \frac{21\text{ ft}^3}{0.3}\left(\frac{40{,}000\text{ Btu/hr}}{100{,}000\text{ Btu/hr}}\right)$$

$$70\,(40) = 2800\text{ ft}^3$$

**Step 4:**
- Enter the input ratings of all fan-assisted appliances in Table 2.
- Total the column.

**Table 2 Ratings for Fan-Assisted Appliances**

| Appliance | Input rating (Btu/hr) |
|---|---|
| Furnace | 100,000 |
| Furnace | |
| Space heater | |
| Other | |
| Other | |
| Other | |
| Total | 100,000 |

**Step 5:**
- Calculate the required volume for fan-assisted appliances using the equation.
- Enter the required volume here and below.

Required volume:

$$\frac{15\text{ ft}^3}{ACH}\left(\frac{I_{fan}}{100{,}000\text{ Btu/hr}}\right) = \frac{15\text{ ft}^3}{0.3}\left(\frac{100{,}000\text{ Btu/hr}}{100{,}000\text{ Btu/hr}}\right)$$

$$50\,(100) = 5000\text{ ft}^3$$

**Step 6:**
- Add the two required volumes calculated to find the total required volume for all appliances.

Required volume (Non-fan assisted) = 2800 ft³
Required volume (Non-fan assisted) = 5000 ft³
Total required volume = 7800 ft³

**Step 7:**
- Enter the available room volume.

Available volume = Room volume = 20 x 35 x 8 = 5600 ft³

**Step 8:**
- If the total required volume is greater than the available room volume, air openings are required. Check Yes and determine the opening size and location. If the total required volume is less than the available room volume, no air openings are required. Check No.

Air openings required? (Check one)

Yes ☑ No ☐

**Job:** 25 Main Street **Prepared by:** TL **Date:** 1/1/09



***WORKSHEET I.3*** *Sample Calculation for Determining Combustion Air Using the KAIR Method (Example Installation 3).*

Paragraph 9.3.2.2 specifies a lower required volume limitation for appliances other than fan-assisted at no smaller than 35 $ft^3$ per 1000 Btu/hr. From Table A.9.3.2.2(a), the lower limit is

$$1400\ ft^3$$

Because the calculated required volume of 2800 $ft^3$ is above the lower required volume limit, use this amount as the minimum required volume.

(c) *Total required volume:* Subsection 9.3.2 states that the total required volume to use indoor air is the sum of the required volumes for all appliances located in the space:

$$\text{Total Required} = 5000\ ft^3 + 2800\ ft^3 = 7800\ ft^3$$

(2) *Determine available volume:* The available volume is determined as follows:

$$(20\ ft \times 35\ ft) \times 8\ ft = 5600\ ft^3$$

*Conclusion:* The installation cannot use indoor air alone, because the available volume of 5600 $ft^3$ is less than the total required volume of 7800 $ft^3$. Outdoor air openings can be sized in accordance with all air from the outdoors *(see 9.3.3)* or by use of the combination of indoor/outdoor air method *(see 9.3.4)*.

# Example of Combination of Indoor and Outdoor Combustion and Ventilation Opening Design

## ANNEX J

*This annex is not a part of the requirements of this NFPA document but is included for informational purposes only.*

Annex J provides an example of the use of 9.3.4, which lists the requirements for using a combination of indoor and outdoor air to provide the necessary air for combustion and ventilation for gas appliances.

### J.1 Example of Combination Indoor and Outdoor Combustion Air Opening Design

Determine the required combination of indoor and outdoor combustion air opening sizes for the following appliance installation example.

*Example Installation:* A fan-assisted furnace and a draft hood–equipped water heater with the following inputs are located in a 15 ft × 30 ft basement with an 8 ft ceiling. No additional indoor spaces can be used to help meet the appliance combustion air needs.

Fan-Assisted Furnace Input: 100,000 Btu/hr

Draft Hood–Equipped Water Heater Input: 40,000 Btu/hr

*Solution*

(1) *Determine the total available room volume:* Appliance room volume: 15 ft × 30 ft with an 8 ft ceiling = 3600 $ft^3$

(2) *Determine the total required volume:* The Standard Method to determine combustion air is used to calculate the required volume. The combined input for the appliances located in the basement is calculated as follows:

$$100{,}000 \text{ Btu/hr} + 40{,}000 \text{ Btu/hr} = 140{,}000 \text{ Btu/hr}$$

The Standard Method requires that the required volume be determined based on 50 $ft^3$ per 1000 Btu/hr. Using Table A.9.3.2.1, the required volume for a 140,000 Btu/hr water heater is

$$7000 \text{ ft}^3$$

*Conclusion:* The indoor volume is insufficient to supply combustion air since the total of 3600 $ft^3$ does not meet the required volume of 7000 $ft^3$. Therefore, additional combustion air must be provided from the outdoors.

(3) Determine ratio of the available volume to the required volume:

$$\frac{3600 \text{ ft}^3}{7000 \text{ ft}^3} = 0.51$$

(4) Determine the reduction factor to be used to reduce the full outdoor air opening size to the minimum required based on ratio of indoor spaces:

$$1.00 - 0.51 \text{ (from Step 3)} = 0.49$$

(5) Determine the single outdoor combustion air opening size as though all combustion air is to come from outdoors. In this example, the combustion air opening directly communicates with the outdoors:

$$\frac{140{,}000 \text{ Btu/hr}}{3000 \text{ Btu/in.}^2} = 47 \text{ in.}^2$$

(6) Determine the minimum outdoor combustion air opening area:

$$\text{Outdoor opening area} = 0.49 \text{ (from Step 4)} \times 47 \text{ in.}^2 = 23 \text{ in.}^2$$

Paragraph 9.3.4(3)(c) requires the minimum dimension of the air opening should not be less than 3 in.

# Other Useful Definitions

## ANNEX K

*This annex is not a part of the requirements of this NFPA document but is included for informational purposes only.*

Annex K contains definitions of terms that are no longer used in the code. These terms apply to gas appliances and parts of gas appliances. Because the definitions are useful, they have been retained in this handbook for reference.

### K.1 Useful Terms

The following terms are not used in the code. They are used in appliance standards and by manufacturers of products covered by the code.

**K.1.1 Ambient Temperature.** The temperature of the surrounding medium; usually used to refer to the temperature of the air in which a structure is situated or a device operates.

**K.1.2 Automatic Damper Regulator.** A mechanically or electrically actuated device designed to maintain a constant draft on combustion appliances.

**K.1.3 Burner, Induced-Draft.** A burner that depends on draft induced by a fan that is an integral part of the appliance and is located downstream from the burner.

**K.1.4 Burner, Injection (Atmospheric).** A burner in which the air at atmospheric pressure is injected into the burner by a jet of gas.

**K.1.5 Burner, Power, Premixing.** A power burner in which all or nearly all of the air for combustion is mixed with the gas as primary air.

**K.1.6 Conversion Burner, Gas, Firing Door Type.** A conversion burner specifically for boiler or furnace firing door installation.

**K.1.7 Conversion Burner, Gas, Inshot Type.** A conversion burner normally for boiler or furnace ash pit installation and fired in a horizontal position.

**K.1.8 Conversion Burner, Gas, Upshot Type.** A conversion burner normally for boiler or furnace ash pit installation and fired in a vertical position at approximately grate level.

**K.1.9 Decorative Appliance for Installation in a Vented Fireplace, Coal Basket.** An open-flame-type appliance consisting of a metal basket that is filled with simulated coals and gives the appearance of a coal fire when in operation.

**K.1.10 Decorative Appliance for Installation in a Vented Fireplace, Fireplace Insert.** Consists of an open-flame, radiant-type appliance mounted in a decorative metal panel to cover the fireplace or mantel opening and having provisions for venting into the fireplace chimney.

**K.1.11 Decorative Appliance for Installation in a Vented Fireplace, Gas Log.** An open-flame-type appliance consisting of a metal frame or base supporting simulated logs.

**K.1.12 Decorative Appliance for Installation in a Vented Fireplace, Radiant Appliance.** An open-front appliance designed primarily to convert the energy in fuel gas to radiant heat by means of refractory radiants or similar radiating materials.

**K.1.13 Fireplace, Factory-Built.** A fireplace composed of listed factory-built components assembled in accordance with the manufacturer's installation instructions to form the completed fireplace.

**K.1.14 Fireplace, Masonry.** A hearth and fire chamber of solid masonry units such as bricks, stones, listed masonry units, or reinforced concrete, provided with a suitable chimney.

**K.1.15 Floor Furnace, Fan-Type.** A floor furnace equipped with a fan that provides the primary means for circulation of air.

**K.1.16 Floor Furnace, Gravity-Type.** A floor furnace depending primarily on circulation of air by gravity. This classification also includes floor furnaces equipped with booster-type fans that do not materially restrict free circulation of air by gravity flow when such fans are not in operation.

**K.1.17 Furnace, Direct Vent Central.** A system consisting of (1) a central furnace for indoor installation, (2) combustion air connections between the central furnace and the outdoor atmosphere, (3) flue gas connections between the central furnace and the vent cap, and (4) a vent cap for installation outdoors, supplied by the manufacturer and constructed so that all air for combustion is obtained from the outdoor atmosphere and all flue gases are discharged to the outdoor atmosphere.

**K.1.18 Furnace, Downflow.** A furnace designed with airflow discharge vertically downward at or near the bottom of the furnace.

**K.1.19 Furnace, Forced Air, with Cooling Unit.** A single-package unit, consisting of a gas-fired, forced-air furnace of the downflow, horizontal, or upflow type combined with an electrically or gas-operated summer air-conditioning system, contained in a common casing.

**K.1.20 Furnace, Gravity.** A furnace depending primarily on circulation of air by gravity.

**K.1.21 Furnace, Gravity, with Booster Fan.** A furnace equipped with a booster fan that does not materially restrict free circulation of air by gravity flow when the fan is not in operation.

**K.1.22 Furnace, Gravity, with Integral Fan.** A furnace equipped with a fan or blower as an integral part of its construction and operable on gravity systems only. The fan or blower is used only to overcome the internal furnace resistance to airflow.

**K.1.23 Furnace, Horizontal.** A furnace designed for low headroom installation with airflow across the heating element essentially in a horizontal path.

**K.1.24 Furnace, Upflow.** A furnace designed with airflow discharge vertically upward at or near the top of the furnace. This classification includes "highboy" furnaces with the blower mounted below the heating element and "lowboy" furnaces with the blower mounted beside the heating element.

**K.1.25 Gas Main or Distribution Main.** A pipe installed in a community to convey gas to individual services or other mains.

**K.1.26 Household Cooking Appliance, Floor-Supported Unit.** A self-contained cooking appliance for installation directly on the floor. It has a top section and an oven section. It may have additional sections.

**K.1.27 Indirect Oven.** An oven in which the flue gases do not flow through the oven compartment.

**K.1.28 Joint, Adhesive.** A joint made in plastic piping by the use of an adhesive substance that forms a continuous bond between the mating surfaces without dissolving either one of them.

**K.1.29 Joint, Solvent Cement.** A joint made in thermoplastic piping by the use of a solvent or solvent cement that forms a continuous bond between the mating surfaces.

**K.1.30 Leak Detector.** An instrument for determining concentration of gas in air.

**K.1.31 Loads, Connected.** Sum of the rated Btu/hr gas input to individual appliances connected to a piping system. May also be expressed in cubic feet per hour.

**K.1.32 Orifice Cap (Hood).** A movable fitting having an orifice that permits adjustment of the flow of gas by the changing of its position with respect to a fixed needle or other device.

**K.1.33 Orifice Spud.** A removable plug or cap containing an orifice that permits adjustment of the flow of gas either by substitution of a spud with a different sized orifice or by the motion of a needle with respect to it.

**K.1.34 Pressure Control.** Manual or automatic maintenance of pressure, in all or part of a system, at a predetermined level, or within a selected range.

**K.1.35 Regulator, Appliance, Adjustable.** (1) Spring type, limited adjustment: a regulator in which the regulating force acting upon the diaphragm is derived principally from a spring, the loading of which is adjustable over a range of not more than ±15 percent of the outlet pressure at the midpoint of the adjustment range; (2) spring type, standard adjustment: a regulator in which the regulating force acting on the diaphragm is derived principally from a spring, the loading of which is adjustable.

**K.1.36 Regulator, Appliance, Multistage.** A regulator for use with a single gas whose adjustment means can be positioned manually or automatically to two or more predetermined outlet pressure settings.

**K.1.37 Regulator, Appliance, Nonadjustable.** (1) Spring type, nonadjustable: a regulator in which the regulating force acting on the diaphragm is derived principally from a spring, the loading of which is not field adjustable; (2) weight type: a regulator in which the regulating force acting upon the diaphragm is derived from a weight or combination of weights.

**K.1.38 Room Heater, Unvented Circulator.** A room heater designed to convert the energy in fuel gas to convected and radiant heat by direct mixing of air to be heated with the combustion products and excess air inside the jacket.

**K.1.39 Room Heater, Vented.** A vented, self-contained, freestanding, nonrecessed, fuel gas burning appliance for furnishing warm air to the space in which installed, directly from the heater without duct connections.

**K.1.40 Room Heater, Vented Circulator.** A room heater designed to convert the energy in fuel gas to convected and radiant heat, by transfer of heat from flue gases to a heat exchanger surface, without mixing of flue gases with circulating heated air.

**K.1.41 Room Heater, Vented Circulator, Fan Type.** A vented circulator equipped with an integral circulating air fan, the operation of which is necessary for satisfactory appliance performance.

**K.1.42 Room Heater, Vented Overhead Heater.** A room heater designed for suspension from or attachment to or adjacent to the ceiling of the room being heated and transferring the energy of the fuel gas to the space being heated primarily by radiation downward from a hot surface, and in which there is no mixing of flue gases with the air of the space being heated.

**K.1.43 Room Heater, Wall Heater, Unvented Closed Front.** An unvented circulator having a closed front, for insertion in or attachment to a wall or partition.

**K.1.44 Valve, Automatic Gas Shutoff.** A valve used in conjunction with an automatic gas shutoff device to shut off the gas supply to a fuel gas burning water heating system.

**K.1.45 Valve, Individual Main Burner.** A valve that controls the gas supply to an individual main burner.

**K.1.46 Valve, Main Burner Control.** A valve that controls the gas supply to the main burner manifold.

**K.1.47 Valve, Manual Main Gas Control.** A manually operated valve in the gas line for the purpose of completely turning on or shutting off the gas supply to the appliance, except to a pilot or pilots that are provided with independent shutoff.

**K.1.48 Vented Wall Furnace, Fan-Type.** A wall furnace that is equipped with a fan.

**K.1.49 Vented Wall Furnace, Gravity-Type.** A wall furnace that depends on circulation of air by gravity.

**K.1.50 Venting System, Mechanical Draft, Induced.** A portion of a venting system using a fan or other mechanical means to cause the removal of flue or vent gases under nonpositive static vent pressure.

**K.1.51 Venting System, Mechanical Draft, Power.** *See 3.3.100.7.1*, Forced Mechanical Draft Venting System.

**K.1.52 Water Heater, Automatic Instantaneous.** A water heater that has a rated input of at least 4000 Btu/hr/gal (5 kW/L) of self-stored water. Automatic control is obtained by water-actuated control, thermostatic control, or a combination of water-actuated control and thermostatic control. This classification includes faucet-type water heaters designed to deliver water through a single faucet integral with or directly adjacent to the appliance.

**K.1.53 Water Heater, Commercial Storage.** A water heater that heats and stores water at a thermostatically controlled temperature for delivery on demand. Input rating is 75,000 Btu/hr (21,980 W) or more.

**K.1.54 Water Heater, Domestic Storage.** A water heater that heats and stores water at a thermostatically controlled temperature for delivery on demand. Input rating may not exceed 75,000 Btu/hr (21,980 W).

# Enforcement

# ANNEX L

*This annex is not a part of the requirements of this NFPA document but is included for informational purposes only.*

The model adoption text in L.1 is intended for government entities adopting the *National Fuel Gas Code*. Note that this is "model" legislation and may be modified or replaced with other legislation or official notice as required by local laws.

## L.1

The following sample ordinance is provided to assist a jurisdiction in the adoption of this code and is not part of this code.

ORDINANCE NO.__________________

An ordinance of the *[jurisdiction]* adopting the 2009 edition of NFPA 54/ANSI Z223.1, *National Fuel Gas Code*, documents listed in Chapter 2 of that code; prescribing regulations governing conditions hazardous to life and property from fire or explosion; providing for the issuance of permits and collection of fees; repealing Ordinance No.______ of the *[jurisdiction]* and all other ordinances and parts of ordinances in conflict therewith; providing a penalty; providing a severability clause; and providing for publication; and providing an effective date.

BE IT ORDAINED BY THE *[governing body]* OF THE *[jurisdiction]*:

SECTION 1 That the *National Fuel Gas Code* and documents adopted by Chapter 2, three (3) copies of which are on file and are open to inspection by the public in the office of the *[jurisdiction's keeper of records]* of the *[jurisdiction]*, are hereby adopted and incorporated into this ordinance as fully as if set out at length herein, and from the date on which this ordinance shall take effect, the provisions thereof shall be controlling within the limits of the *[jurisdiction]*. The same are hereby adopted as the code of the *[jurisdiction]* for the purpose of prescribing regulations governing conditions hazardous to life and property from fire or explosion and providing for issuance of permits and collection of fees.

SECTION 2 Any person who shall violate any provision of this code or standard hereby adopted or fail to comply therewith; or who shall violate or fail to comply with any order made thereunder; or who shall build in violation of any detailed statement of specifications or plans submitted and approved thereunder; or failed to operate in accordance with any certificate or permit issued thereunder; and from which no appeal has been taken; or who shall fail to comply with such an order as affirmed or modified by or by a court of competent jurisdiction, within the time fixed herein, shall severally for each and every such violation and noncompliance, respectively, be guilty of a misdemeanor, punishable by a fine of not less than $ _________ nor more than $_________ or by imprisonment for not less than_________ days nor more than__________ days or by both such fine and imprisonment. The imposition of one penalty for any violation shall not excuse the violation or permit it to continue; and all such persons shall be required to correct or remedy such violations or defects within a reasonable time; and when not otherwise specified the application of the above penalty shall not be held to prevent the enforced removal of prohibited conditions. Each day that prohibited conditions are maintained shall constitute a separate offense.

SECTION 3 Additions, insertions, and changes — that the 2009 edition of NFPA 54/ANSI Z223.1, *National Fuel Gas Code*, is amended and changed in the following respects:

List Amendments

SECTION 4 That ordinance No.____________ of *[jurisdiction]* entitled *[fill in the title of the ordinance or ordinances in effect at the present time]* and all other ordinances or parts of ordinances in conflict herewith are hereby repealed.

SECTION 5 That if any section, subsection, sentence, clause, or phrase of this ordinance is, for any reason, held to be invalid or unconstitutional, such decision shall not affect the validity or constitutionality of the remaining portions of this ordinance. The *[governing body]* hereby declares that it would have passed this ordinance, and each section, subsection, clause, or phrase hereof, irrespective of the fact that any one or more sections, subsections, sentences, clauses, and phrases be declared unconstitutional.

SECTION 6 That the *[jurisdiction's keeper of records]* is hereby ordered and directed to cause this ordinance to be published. [NOTE: An additional provision may be required to direct the number of times the ordinance is to be published and to specify that it is to be in a newspaper in general circulation. Posting may also be required.]

SECTION 7 That this ordinance and the rules, regulations, provisions, requirements, orders, and matters established and adopted hereby shall take effect and be in full force and effect *[time period]* from and after the date of its final passage and adoption.

# Informational References

ANNEX M

## M.1 Referenced Publications

The documents or portions thereof listed in this annex are referenced within the informational sections of this code and are not part of the requirements of this document unless also listed in Chapter 2 for other reasons.

### M.1.1 NFPA Publications.

National Fire Protection Association, 1 Batterymarch Park, Quincy, MA 02169-7471.

NFPA 58, *Liquefied Petroleum Gas Code*, 2008 edition.
NFPA 68, *Standard on Explosion Protection by Deflagration Venting*, 2007 edition.
*NFPA 70*®, *National Electrical Code*®, 2008 edition.
NFPA 90A, *Standard for the Installation of Air-Conditioning and Ventilating Systems*, 2009 edition.
NFPA 90B, *Standard for the Installation of Warm Air Heating and Air-Conditioning Systems*, 2009 edition.
NFPA 96, *Standard for Ventilation Control and Fire Protection of Commercial Cooking Operations*, 2008 edition.
NFPA 780, *Standard for the Installation of Lightning Protection Systems*, 2008 edition. NEW

### M.1.2 Other Publications.

**M.1.2.1 API Publications.** American Petroleum Institute, 1220 L Street, N.W., Washington, DC 20005-4070.

API 1104, *Welding of Pipelines and Related Facilities,* 2005.

**M.1.2.2 ASHRAE Publications.** American Society of Heating, Refrigerating and Air Conditioning Engineers, Inc., 1791 Tullie Circle, N.E., Atlanta, GA 30329-2305, (404)636-8400, www.ashrae.org.

*ASHRAE Handbook — Fundamentals,* 2005.
*ASHRAE Handbook — HVAC Systems and Equipment,* 2004.

**M.1.2.3 ASME Publications.** American Society of Mechanical Engineers, Three Park Avenue, New York, NY 10016-5990, (800)843-2763, www.asme.org.

ASME *Boiler and Pressure Vessel Code,* Section IX and Section IV, 1998.
ASME B1.20.1, *Pipe Threads, General Purpose, Inch,* 1983.
ASME B16.1, *Gray Iron Pipe Flanges and Flanged Fittings: (Classes 25, 125, and 250),* 2005.
ASME B36.10M, *Welded and Seamless Wrought Steel Pipe*, 2004.

**M.1.2.4 ASTM Publications.** ASTM International, 100 Barr Harbor Drive, P.O. Box C700, West Conshohocken, PA 19428-2959, (610)833-9585, www.astm.org.

ANSI/ASTM D 2385, *Method of Test for Hydrogen Sulfide and Mercaptan Sulfur in Natural Gas (Cadmium Sulfate — Iodometric Titration Method),* 1981.

ANSI/ASTM D 2420, *Method of Test for Hydrogen Sulfide in Liquefied Petroleum (LP) Gases (Lead Acetate Method),* 1991 (Reaffirmed 1996).

ASTM A 53, *Standard Specification for Pipe, Steel, Black and Hot-Dipped, Zinc-Coated, Welded and Seamless*, 2007.

ASTM A 106, *Standard Specification for Seamless Carbon Steel Pipe for High-Temperature Service*, 2006.

ASTM A 254, *Standard Specification for Copper Brazed Steel Tubing,* 2002.

ASTM A 539, *Standard Specification for Electric-Resistance-Welded Coiled Steel Tubing for Gas and Fuel Oil Lines*, 1999.

ASTM B 88, *Standard Specification for Seamless Copper Water Tube*, 2003.

ASTM B 210, *Specification for Aluminum and Aluminum-Alloy Drawn Seamless Tubes*, 2004.

ASTM B 241, *Standard Specification for Aluminum and Aluminum Alloy Seamless Pipe and Seamless Extruded Tube*, 2002.

ASTM B 280, *Standard Specification for Seamless Copper Tube for Air Conditioning and Refrigeration Field Service*, 2003.

ASTM D 2513, *Standard Specification for Thermoplastic Gas Pressure Pipe, Tubing, and Fittings*, 2007b

ASTM F 1973, *Standard Specification for Factory Assembled Anodeless Risers and Transition Fittings in Polyethylene (PE) and Polyamide 11 (PA11) Fuel Gas Distribution Systems,* 2002.

ASTM F 2509, *Standard Specification for Field-Assembled Anodeless Riser Kits for Use on Outside Diameter Controlled Polyethylene Gas Distribution Pipe and Tubing,* 2006.

**M.1.2.5 AWS Publications.** American Welding Society, 550 N.W. LeJeune Road, Miami, FL 33126, (800)443-9353, www.aws.org.

AWS B2.1, *Standard for Welding Procedure and Performance Qualification,* 1998.

AWS B2.2, *Standard for Brazing Procedure and Performance Qualification,* 1991.

**M.1.2.6 CSA America Publications.** Canadian Standards Association, 8501 East Pleasant Valley Road, Cleveland, OH 44131-5575, (216)524-4990, www.csa-america.org.

ANSI LC 1/CSA 6.26, *Fuel Gas Piping Systems Using Corrugated Stainless Steel Tubing (CSST)*, 2005.

ANSI LC-4, *Press-Connect Copper and Copper Alloy Fittings for Use in Fuel Gas Distribution Systems*, 2006.

ANSI Z21.50/CSA 2.22, *Vented Gas Fireplaces,* 2000.

ANSI Z21.60/CGA 2.26, *Decorative Gas Appliances for Installation in Solid-Fuel Burning Fireplaces,* 1996.

ANSI Z21.90, *Gas Convenience Outlets and Optional Enclosures,* 2003.

**M.1.2.7 NACE Publications.** NACE International, 1440 South Creek Drive, Houston, TX 77084-4906, www.nace.org.

NACE RP 0169, *Control of External Corrosion on Underground or Submerged Metallic Piping Systems,* 1996.

**M.1.2.8 SAE Publications.** Society of Automotive Engineers, 400 Commonwealth Drive, Warrendale, PA 15096, www.sae.org.

SAE J533, *Flares for Tubing*, 2007.

**M.1.2.9 UL Publications.** Underwriters Laboratories Inc., 333 Pfingsten Road, Northbrook, IL 60062-2096, www.ul.com.

ANSI/UL 651, *Schedule 40 and 80 Rigid PVC Conduit and Fittings*, 2007.

**M.1.2.10 Other Publications.**

*Piping Handbook*, 2000, New York: McGraw-Hill Book Company.

## M.2 Informational References

### M.2.1 NFPA Publications.

National Fire Protection Association, 1 Batterymarch Park, Quincy, MA 02169-7471.

NFPA 30, *Flammable and Combustible Liquids Code*, 2008 edition.
NFPA 59, *Utility LP-Gas Plant Code*, 2008 edition.
NFPA 61, *Standard for the Prevention of Fires and Dust Explosions in Agricultural and Food Processing Facilities*, 2008 edition.
NFPA 86, *Standard for Ovens and Furnaces*, 2007 edition.
NFPA 211, *Standard for Chimneys, Fireplaces, Vents, and Solid Fuel–Burning Appliances*, 2006 edition.
NFPA 501A, *Standard for Fire Safety Criteria for Manufactured Home Installations, Sites, and Communities*, 2005 edition.

### M.2.2 CSA America Publications.

Canadian Standards Association, 8501 East Pleasant Valley Road, Cleveland, OH 44131-5575, (216)524-4990, www.csa-america.org.

AGA/CGA NGV 1, *Compressed Natural Gas Vehicle (NGV) Fueling Connection Devices,* 2006.
AGA/CGA NGV 3.1, *Fuel System Components for Natural Gas Powered Vehicles,* 2007.
ANSI/IAS LC 2A, *Agricultural Heaters,* 1998.
ANSI Z21.1, *Household Cooking Gas Appliances,* 2007.
ANSI Z21.5.1/CSA 7.1, *Gas Clothes Dryers — Volume I — Type 1 Clothes Dryers,* 2006.
ANSI Z21.5.2/CSA 7.2, *Gas Clothes Dryers — Volume II — Type 2 Clothes Dryers,* 2006.
ANSI Z21.10.1/CSA 4.1, *Gas Water Heaters — Volume I — Storage, Water Heaters with Input Ratings of 75,000 Btu per Hour or Less,* 2004.
ANSI Z21.10.3/CSA 4.3, *Gas Water Heaters — Volume III — Storage, Water with Input Ratings above 75,000 Btu per Hour, Circulating and Instantaneous,* 2007.
ANSI Z21.11.2, *Gas-Fired Room Heaters — Volume II — Unvented Room Heaters,* 2007.
ANSI Z21.12, *Draft Hoods,* 1990 (Reaffirmed 2000).
ANSI Z21.13/CSA 4.9, *Gas-Fired Low-Pressure Steam and Hot Water Boilers,* 2004.
ANSI Z21.15/CGA 9.1, *Manually Operated Gas Valves for Appliances, Appliance Connector Valves, and Hose End Valves,* 1997.
ANSI Z21.17/CSA 2.7, *Domestic Gas Conversion Burners,* 2004.
ANSI Z21.18/CSA 6.3, *Gas Appliance Pressure Regulators,* 2007.
ANSI Z21.19/CSA 1.4, *Refrigerators Using Gas Fuel,* 2007.
ANSI Z21.20, *Automatic Gas Ignition Systems and Components,* 2007.
ANSI Z21.21/CGA 6.5, *Automatic Valves for Gas Appliances,* 2005.
ANSI Z21.22/CSA 4.4, *Relief Valves and Automatic Gas Shutoff Devices for Hot Water Supply Systems,* 1999.
ANSI Z21.23, *Gas Appliance Thermostats,* 2005.
ANSI Z21.24/CSA 6.10, *Metal Connectors for Gas Appliances,* 2006.
ANSI Z21.35/CGA 6.8, *Pilot Gas Filters,* 2005.
ANSI Z21.40.1/CGA 2.91-M96, *Gas-Fired Absorption Summer Air Conditioning Appliances,* 2002.
ANSI Z21.40.2/CGA 2.92, *Gas-Fired Work Activated Air-Conditioning and Heat Pump Appliances (Internal Combustion),* 2002.
ANSI Z21.40.4/CGA 2.94, *Performance Testing and Rating of Gas Fired, Air-Conditioning and Heat Pump Appliances,* 2002.
ANSI Z21.42, *Gas-Fired Illuminating Appliances,* 2002.

ANSI Z21.47/CSA 2.3, *Gas-Fired Central Furnaces,* 2006.

ANSI Z21.54/CSA 8.4, *Gas Hose Connectors for Portable Outdoor Gas-Fired Appliances,* 2002.

ANSI Z21.56/CSA 4.7, *Gas-Fired Pool Heaters,* 2006.

ANSI Z21.57, *Recreational Vehicle Cooking Gas Appliances,* 2005.

ANSI Z21.58/CGA 1.6, *Outdoor Cooking Gas Appliances,* 2007.

ANSI Z21.61, *Gas-Fired Toilets,* 2004.

ANSI Z21.66/CGA 6.14, *Automatic Vent Damper Devices for Use with Gas-Fired Appliances,* 2001.

ANSI Z21.69/CSA 6.16, *Connectors for Movable Gas Appliances,* 2007.

ANSI Z21.71, *Automatic Intermittent Pilot Ignition Systems for Field Installations,* 2007.

ANSI Z21.77/CGA 6.23, *Manually-Operated Piezo-Electric Spark Gas Ignition Systems and Components,* 2005.

ANSI Z21.78/CGA 6.20, *Combination Gas Controls for Gas Appliances,* 2005.

ANSI Z21.83, *Fuel Cell Power Plants*, 2007.

ANSI Z21.84, *Manually Lighted, Natural Gas, Decorative Gas Appliances for Installation in Solid-Fuel Burning Appliances*, 2002.

ANSI Z21.86/CSA 2.32, *Vented Gas-Fired Space Heating Equipment,* 2008.

ANSI Z21.87/CSA 4.6, *Automatic Gas Shutoff Devices for Hot Water Supply Systems*, 2007.

ANSI Z21.88/CSA 2.33, *Vented Gas-Fireplace Heaters,* 2007.

ANSI Z21.91, *Ventless Firebox Enclosures for Gas-Fired Unvented Decorative Room Heaters*, 2007.

ANSI Z83.3, *Gas Utilization Equipment in Large Boilers,* 1971 (Reaffirmed 1995).

ANSI Z83.4/CSA 3.7, *Non-Recirculating Direct Gas-Fired Industrial Air Heaters,* 2003.

ANSI Z83.6, *Gas-Fired Infrared Heaters,* 1990 (Reaffirmed 1998).

ANSI Z83.8/CGA 2.6, *Gas Fired Duct Furnaces and Unit Heaters,* 2006.

ANSI Z83.11/CGA 1.8, *Food Service Equipment,* 2006.

IAS U.S. 7, *Requirements for Gas Convenience Outlets and Optional Enclosures,* 1990.

IAS U.S. 9, *Requirements for Gas-Fired, Desiccant Type Dehumidifiers and Central Air Conditioners,* 1990.

IAS U.S. 42, *Requirement for Gas Fired Commercial Dishwashers,* 1992.

IAS NGV2, *Basic Requirements for Compressed Natural Gas Vehicle (NGV) Containers,* 2007.

### M.2.3 MSS Publications.

Manufacturers Standardization Society of the Valve and Fittings Industry, 127 Park Street, NE, Vienna, VA 22180-6671, www.mss-hq.com.

MSS SP-6, *Standard Finishes for Contact Faces of Pipe Flanges and Connecting-End Flanges of Valves and Fittings,* 2007.

ANSI/MSS SP-58, *Pipe Hangers and Supports — Materials, Design and Manufacture,* 2002.

### M.2.4 UL Publications.

Underwriters Laboratories Inc., 333 Pfingsten Road, Northbrook, IL 60062-2096, www.ul.com.

UL 103, *Chimneys, Factory-Built, Residential Type and Building Heating Appliances,* 1995.

ANSI/UL 441, *Gas Vents,* 2006.

NEW ANSI/UL 641, *Type L Low-Temperature Venting Systems,* 2005.

UL 1738, *Venting Systems for Gas Burning Appliances, Categories II, III and IV,* 2006.

UL 1777, *Chimney Liners,* 2004.

### M.2.5 U.S. Government Publications.

U.S. Government Printing Office, Washington, DC 20402, www.access.gpo.gov.

*Manufactured Home Construction and Safety Standard,* Title 24, *Code of Federal Regulations,* Part 3280.

## M.3 References for Extracts in Informational Sections. (Reserved)

# PART TWO

# Supplements

In addition to the 2009 edition of NFPA 54, *National Fuel Gas Code*, and commentary presented in Part One, the *National Fuel Gas Code Handbook* includes supplements.

The eight supplements in Part Two explore the background of selected topics related to NFPA 54 in more detail than does the commentary. These supplements are not part of the code; they are included as additional information for handbook users.

The eight supplements in Part Two are as follows:

1. Development of Revised Venting Guidelines
2. Update on the Design and Installation Requirements for CSST Gas Piping Systems
3. Carbon Monoxide
4. Fuel Gas Odorization
5. Procedure to Estimate Infiltration Rate for Residential Structures
6. Calculation Worksheets
7. Clearance Distance for Gas Appliance Sidewall Venting
8. Technical/Substantive Changes from the 2006 Edition to the 2009 Edition of NFPA 54

# SUPPLEMENT 1

# Development of Revised Venting Guidelines

***Editor's Note:*** *The 1992 edition of the National Fuel Gas Code incorporated extensive changes to the tables used to size vents serving Category I appliances. The 1996 edition added three new tables that provide a method of sizing masonry chimneys exposed to the outdoors below the roofline. These changes were significant and resulted from research conducted by Battelle Laboratories, with the American Gas Association (AGA) Laboratories, and funded by the Gas Research Institute (GRI), a nonprofit institute that sponsors projects promoting the use of natural gas. The need for this work was identified after problems arose in the field with these newer appliances. The exterior masonry chimney tables provide specific, scientifically based guidance for these chimneys that was not included in the research that led to the 1992 tables.*

*Since the introduction of the revised tables, their use has become the norm as appliance efficiencies have changed. The 2002 edition made only minor changes to the code, to resolve issues that had arisen from the use of these tables. In the 2006 edition, minor changes were made to the code, and again no new tables were added, but the tables were renumbered. This supplement is as useful now as in 1992 to those who seek to understand the development of the tables.*

*This supplement, written by the researchers at Battelle and AGA, describes in detail the problem that led to the research and the technical aspects of the work. Many users of the code will find this explanation helpful in understanding the reasons for the new, more complex tables.*

*The editor thanks the GRI and its successor, the Gas Technology Institute, for its cooperation in making this supplement possible and thanks the researchers who worked with the National Fuel Gas Code Committee to make the tables possible. Specifically, special thanks are extended to Darrel D. Paul and Allen L. Rutz of Battelle Laboratories, Robert A. Borgeson of AGA Research, and Douglas W. DeWerth of the International Approval Services — U.S., Inc. This supplement was taken from a topical report of research conducted as part of GRI's venting program and was revised in 1996 by the principal investigator, Allen L. Rutz.*

## OVERVIEW

Conventional, atmospheric gas heating appliances, such as furnaces, boilers, and water heaters, have had an enviable field service record. These systems have provided the consumer with safe, economical space or water heating that requires minimal maintenance over a long service life. Single-wall galvanized vent connectors and either a masonry chimney or Type B double-wall metal vent pipe have been used to vent flue gases from these appliances. Here, too, the design principles, which were developed over the years, have resulted in trouble-free venting. The widespread use of higher-efficiency appliances, however, has made altering the design of vent systems necessary.

The requirements appear in Chapter 12 and Chapter

13, and guidance on their use appears in Annex G. These were specifically designed for Category I fan-assisted furnaces and boilers. A Category I appliance is one whose vent is expected to operate under negative static pressure with a limited amount of condensation occurring. Category I appliances include most traditional, draft hood–equipped units, as well as many mid-efficiency models designed for vertical venting. A more complete description of the categories is found on the following pages.

These requirements apply to both new construction and retrofitted applications. When a new appliance is retrofit into an existing installation or an existing appliance is removed from a common vent, the entire venting system, which may include an existing masonry chimney, should conform to current codes. The existing venting system may need to be modified to match the new or retrofitted appliance installation.

## Background of the Vent Sizing Tables

The venting recommendations found in the code prior to 1992 were developed in the 1950s for atmospheric appliances that were equipped with a draft hood. The efficiency of these appliances was low compared to the level required by the National Appliance Energy Conservation Act (NAECA) of 1987. About 35 percent of the annual energy input to a conventional appliance could have been lost from the building through the vent. This loss resulted in appliances that operated with an annual fuel utilization efficiency (AFUE) of approximately 65 percent to 70 percent and with high flue gas temperatures, which made venting relatively simple. NAECA required these losses to be reduced to no more than approximately 20 percent.

Today's mid-efficiency, Category I appliance (with an AFUE of approximately 80 percent) has lower flue gas temperatures and reduced off-cycle losses. Elimination of the draft hood or diverter in some new furnace designs also alters the vent gas temperature, dew point, flow rate, and the amount of dilution air in the vent. Overall, the potential for condensation increases, and vent system design for the modern furnace requires more care.

Modern lifestyles and energy-efficient buildings that incorporate tighter construction practices also have increased the potential for indoor combustion air contamination. This fact, coupled with increased condensate, can increase the potential for vent system corrosion. Updated venting practice must take these factors into account for both ventilation and venting of appliances.

The primary factors affecting modern venting practice are as follows:

1. The reduced level of vent system dilution air increases the potential for more condensate to form in the vent.
2. Higher appliance efficiencies and lower vent gas temperatures mean that oversized vents will not heat up as quickly. The tendency for condensation also will increase.
3. Limiting condensate formation reduces the chance of vent corrosion.
4. The reduction of dilution air level increases the maximum capacity of the vent. In other words, a given vent can handle a larger input rate.

The results were reviewed and the researchers determined that revised vent sizing tables were needed. The following changes were made to the existing tables:

1. *Minimum* vent system capacities (i.e., appliance input rates) were added for fan-assisted appliances to limit condensation.
2. *Maximum* vent system capacities, which were designed to avoid positive vent pressures for fan-assisted appliances and are higher than those for draft hood–equipped models, also were added.

The maximum vent system capacities for draft hood–equipped appliances were left unchanged.

## Conventional Venting

The traditional gas-heating appliance is equipped with a draft hood or draft diverter and an atmospheric burner (see Exhibit S1.1). The products of combustion are driven by the heat in the flue gases through the heat exchanger and the vent. The tendency of hot gases to rise is called buoy-

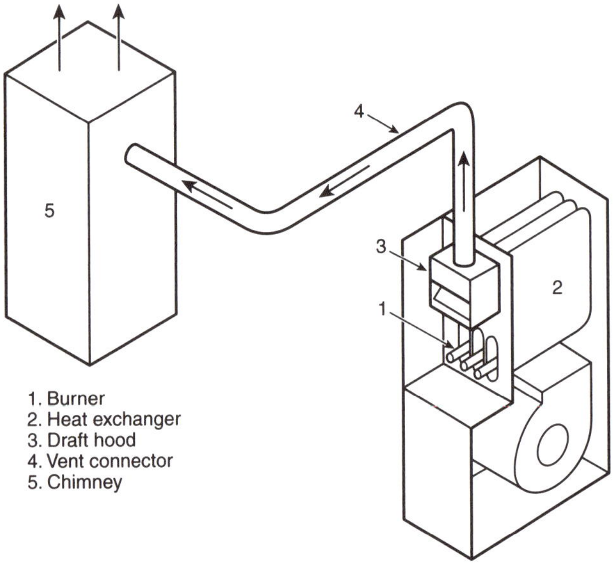


*EXHIBIT S1.1 Draft Hood–Equipped Furnace. (Courtesy of Gas Technology Institute)*

ancy. In the vent, buoyancy creates what is usually called draft or stack action.

The purpose of the draft hood is to separate the appliance from the vent. The draft hood relief opening protects the appliance from wind or too much draft. It also allows additional air, called dilution air, to enter the vent and mix with the combustion products. This dilution air reduces the humidity or dew point of the vent gases, which in turn helps to reduce condensation.

Many people do not realize that products of combustion contain water. In fact, a 100,000 Btu/hr (29 kW) appliance, operating continuously, will produce over 1 gallon (4 L) of water per hour in the flue gas. Of course, the water in the flue gas is vapor and cannot be seen unless it condenses into a liquid.

The amount of water vapor in the vent gases is usually measured as the dew point. The dew point of combustion products in a gas vent is generally about 90°F (32°C) to 130°F (54°C). If the vent gases contact a surface that has a temperature below the dew point, water will condense out of the vent gases onto the surface. As an analogy, this effect is seen on grass in the morning, when the grass temperature is below the dew point of the air. This type of condensation in the vent must be limited unless the vent has been designed for handling condensate.

## Fan-Assisted Combustion Systems

The energy crisis of the 1970s caused consumers to demand appliances with higher AFUEs. Since then, appliances have been designed to reduce their off-cycle losses. In a few cases, a power burner or a combustion air or built-in flue damper is used for this purpose.

However, most of these appliances use fan-assisted combustion systems with an induced-draft blower to assist the combustion products through the heat exchanger (see Exhibit S1.2). Both mid-efficiency (AFUE = 78 percent to 83 percent) and high-efficiency (AFUE > 90 percent) condensing appliances can use fan-assisted systems.

A fan-assisted system does not use a draft hood, thus reducing the dilution air in the vent. The elimination of dilution air has the following important effects on vent performance:

1. The vent gas dew point (or humidity) increases.
2. The total amount of gases flowing in the vent decreases.
3. When the appliance is off, there is much less airflow through the vent.

The higher dew point means that the vent must warm up more to stop condensation. At the same time, the lower flow rate makes warming up the vent more difficult. All other things being equal, a mid-efficiency appliance will produce more condensate in the vent system than a conventional-efficiency draft hood–equipped model.

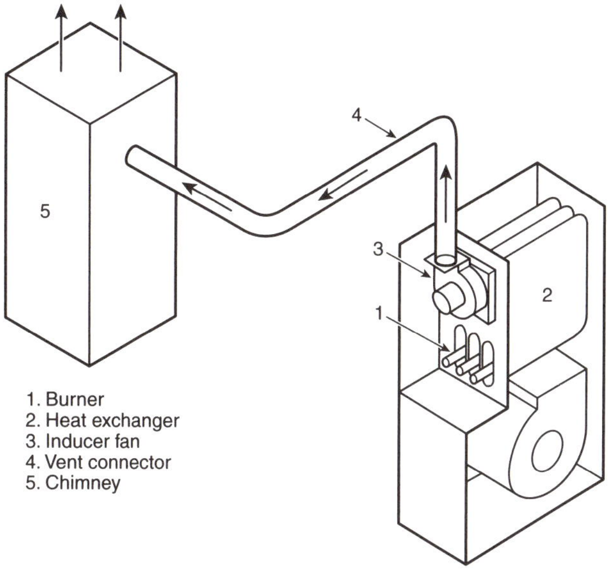


***EXHIBIT S1.2*** *Fan-Assisted Furnace. (Courtesy of Gas Technology Institute)*

Exhibit S1.3 and Exhibit S1.4 show how condensation during startup lasts longer in a vent serving a fan-assisted appliance than in a traditional, draft hood–equipped appliance. In Exhibit S1.3, the average vent gas, vent wall, and dew points were measured at the same point in a Type B vent serving a conventional, draft hood–equipped appliance. The point when the wall temperature falls below the dew point is called the wet time. At this point, condensation will be present on the vent wall surface. The dew point decreases with time, because the amount of dilution air increases as the draft becomes stronger.

Exhibit S1.4 presents the same temperature measurements found in Exhibit S1.3, but for a mid-efficiency, fan-assisted appliance instead of a draft hood–equipped appliance. Notice that the dew point no longer declines because there is no source of dilution air. The higher dew point nearly doubles the wet time. However, many mid-efficiency, fan-assisted appliances can successfully use traditional vents that are designed properly.

## Appliance Categorization

Overall, the development of more efficient appliances has resulted in many new product designs with different venting requirements. Space-heating appliances were organized into four categories, based on the pressure produced in a special test rig and the difference between the actual temperature and dew point of the flue gas. Exhibit S1.5 shows the criteria for each appliance category, as well as

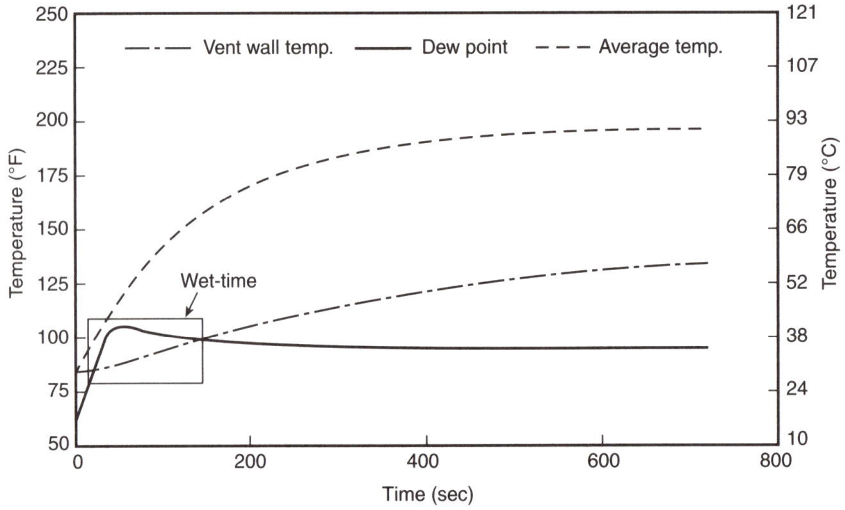


**EXHIBIT S1.3** *Typical Vent Temperatures for a Draft Hood–Equipped Appliance. (Courtesy of American Gas Association)*

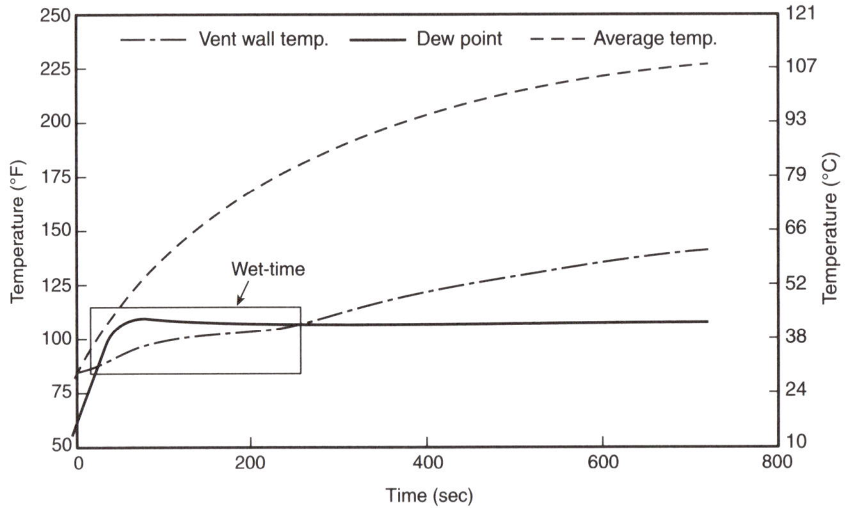


**EXHIBIT S1.4** *Typical Vent Temperatures for a Fan-Assisted Appliance. (Courtesy of American Gas Association)*

the requirements that a particular vent system needs to meet.

Installers and code officials should understand the definitions of the different categories, because each space-heating appliance has a rating plate containing its category number. This rating plate tells field personnel, in a generic sense, the venting requirements of each appliance. The category of an appliance is determined exclusively by the equipment design. Third-party testing laboratories verify that the appliance operates satisfactorily according to the category marked on the rating plate. There is no further need to check the category in the field.

Exhibit S1.5 illustrates the organization of the four categories. Category I appliances operate with a negative static vent pressure and have vent gases that are relatively warm. They operate with only a limited amount of condensation. The traditional, draft hood–equipped appliance is considered Category I. However, many mid-efficiency, fan-assisted appliances are also Category I. How is this possible? First, the vent gas temperature of these appliances is in the same range as the traditional, draft hood appliances (after dilution). Second, the fan-assisted appliance is designed so that the vent system will remain at a negative pressure. If the vent is designed correctly, the vent pressure in a fan-assisted appliance is negative because the draft action should be stronger than the fan pressure rise.

All Category I appliances should be vented using the manufacturer's installation instructions, which usually refer to the *National Fuel Gas Code*. Appliances in Categories II, III, or IV should also be vented according to the manufacturer's installation instructions. However, the *National Fuel Gas Code* does not contain any additional venting information for appliances in these categories.

If a special vent system is used, certain Category I ap-

***EXHIBIT S1.5*** *ANSI Categorization Criteria. (Courtesy of American Gas Association)*

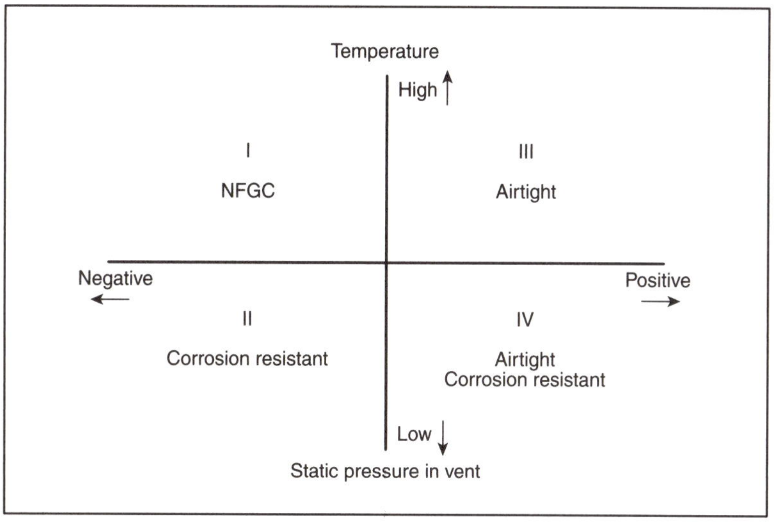


pliances may also be vented in unconventional ways, including horizontal, through-the-wall arrangements. The manufacturer must supply the special vent system with the appliance. The installer should follow the manufacturer's installation instructions exactly.

For residential installations, Category I appliances should never be sidewall-vented through the wall using traditional venting products, such as Type B or single-wall galvanized vent pipe, because the pressure in the vent may be positive. These products should be used only for venting systems that are primarily vertical and sized according to the manufacturer's instructions.

Category II appliances also operate with negative vent pressure but with a vent gas temperature that could produce excessive condensation. These appliances require a corrosion-resistant vent. There are few, if any, Category II appliances on the market.

Category III appliances operate with a positive vent pressure and with a vent gas temperature that will produce limited condensation. These appliances require an airtight vent to avoid leakage. An example of a Category III appliance is a mid-efficiency furnace that is vented horizontally through the wall.

Category IV includes furnaces, boilers, and other appliances that operate with a positive vent pressure and with a low vent gas temperature. These appliances need an airtight, corrosion-resistant vent that includes a method of condensate disposal. Category IV appliances are usually high-efficiency, condensing models.

The category of an appliance determines the type, size, materials, and installation requirements of the venting system for that specific appliance. For example, a Category IV furnace requires a vent system built of corrosion-resistant materials because condensate is corrosive. Gastight vent systems are essential for Category III or IV appliances because the pressure within their vent systems exceeds the surrounding atmospheric pressure. On the other hand, Category I appliances can use traditional venting products, such as Type B vent pipe or masonry chimneys, within recommended limits.

## Vent Sizing Tables

The pre-1992 vent sizing tables for Category I appliances in the *National Fuel Gas Code* were developed in the 1950s for draft hood–equipped models. Therefore, the tables assumed an appliance efficiency lower than that of our modern, mid-efficiency appliances. The tables also assumed that a large amount of dilution air would be in the vent. The research, sponsored by the Gas Research Institute, found that the tables indeed could be improved. A summary of the reasons for these changes follows:

1. Fan-assisted appliances are likely to produce more condensate in a vent than do draft hood–equipped models. Therefore, a way to limit wet time is needed.
2. The reduction of dilution air increases the maximum capacity of the vent.
3. The assumed efficiency for fan-assisted appliances used for determining the capacities in the tables should be raised to reflect modern appliances.

The potential to produce condensate must be controlled to limit corrosion. The corrosion can be accelerated by contamination of the combustion air by household chemicals. If contamination occurs, the condensate can become highly acidic.

Traditionally, the vent sizing tables have contained the maximum vent capacities that could be vented using a given vent diameter and length. These values were designed to prevent spillage.

The updated tables introduced the concept of minimum vent capacity. The minimum capacity is provided to prevent corrosion by limiting the wet time. Minimum vent capacity is the smallest appliance input rating recommended for a given vent. A vent that is operating below its minimum capacity can experience excessive condensation and, possibly, corrosion.

The tables also contain separate, maximum vent capacities for fan-assisted appliances. These values are meant to prevent positive vent pressure. Therefore, a given vent for a fan-assisted appliance has both a minimum and a maximum capacity. Together, these capacity limitations form a range of acceptable appliance inputs and a range of acceptable vent sizes. The maximum capacity prevents vents from being too small; the minimum capacity prevents vents from being too large.

## VENT-II COMPUTER PROGRAM

The venting tables for Category I fan-assisted appliances were generated using the VENT-II computer program to calculate the minimum and maximum vent capacities for each configuration. The VENT-II program uses a transient mathematical model to solve the mass, momentum, and energy balance equations for a user-specified appliance(s) connected to a user-specified vent system. The user specifies the number of appliances connected to the vent system (one or two), the appliance cycling times, and the number of consecutive appliance on/off cycles in the transient analysis. VENT-II then calculates the temperatures, pressures, flow rates, and the amount of condensate formation for all of the user-specified sections of the venting system as a function of time.

### Benefits of Using Computer Model

The primary benefit of using a computer model is the laboratory time saved by not having to determine experimentally the minimum and maximum vent capacities at the precisely controlled, ambient temperatures specified for each of the over 3600 unique vent systems listed in the venting tables in Chapter 13. These vent systems consist of either single-wall metal vent connectors or Type B double-wall metal vent connectors attached to either a Type B double-wall vertical vent or to a tile-lined masonry chimney, meeting the requirements of NFPA 211, *Standard for Chimneys, Fireplaces, Vents, and Solid Fuel–Burning Appliances*. The analyzed vent systems range in vertical height from 6 ft to 100 ft (2 m to 30 m) and have horizontal, lateral lengths from 0 ft to 50 ft (0 m to 15 m).

Over the years, the original VENT-II computer program has been updated, as new venting capabilities were needed. For instance, VENT-II, Version 3.0, allowed the program to calculate wet times for each section of the vent system. VENT-II, Version 4.0, added the capability of simulating the vent system performance when two appliances were connected to the vent. VENT-II, Version 4.1, added additional features that allowed the program to accurately predict the performance of a masonry chimney when one or two appliances were vented. In all cases, the later versions of VENT-II retained the primary features of the earlier versions while improving the mathematical models.

The venting tables were calculated over a period of approximately five years, using the most up-to-date version of VENT-II available at the time that each table was generated. In chronological order, the single-appliance venting tables for Type B double-wall vent systems were generated first. Next, the multiple-appliance venting tables for Type B double-wall vent systems were generated. The single- and multiple-appliance venting tables for interior masonry chimneys followed. Finally, the single- and multiple-appliance venting tables for exterior chimneys were generated.

### Validation of VENT-II

The origins of VENT-II date back to a 1983 GRI study in which Battelle developed a computer model that was validated experimentally by the AGA Laboratories. That study compared the computer program predictions of vent static pressures and temperatures to measurements on 20 unique combinations of appliance and vent configurations. In addition, the study compared computer program predictions to natural draft design data then contained in Table G.1 of the 1988 *National Fuel Gas Code*; in Chapter 26 of the 1979 American Society of Heating, Refrigerating, and Air-Conditioning Engineers (ASHRAE) *Equipment Handbook*; and in an updated, hand-calculator, magnetic tape model developed by Richard Stone. Overall, the study found close agreement between the vent simulation model predictions, the experimental data, and the existing information in the handbooks and codes.

Since 1983, VENT-II has been converted for use on an IBM® personal computer (Version 1.0), the user-interface and condensation models were refined (Version 2.0), a transient analysis capability was added (Version 3.0), a common vent model was included (Version 4.0), and an experimentally validated masonry chimney model was developed and refined (Version 4.1). Each step in the development of VENT-II compared the new models against the previously developed sources of data and conducted additional validation experiments.

In addition to experimental validation in the laboratory, VENT-II has undergone field validation. Battelle has used VENT-II to predict the venting performance of field installations in Newark, New Jersey; Hartford, Connecticut; and Columbus, Ohio. Battelle has also received numerous phone calls and letters from users of VENT-II who have successfully used the program to either diagnose problems or design vent systems. Chapter 30 of the 2000 *ASHRAE HVAC Systems and Equipment Handbook* also recognizes VENT-II as a design tool for gas appliance venting systems.

## BASELINE CONDITIONS FOR CATEGORY I FAN-ASSISTED APPLIANCE VENTING TABLES

Before venting tables could be generated, baseline conditions needed to be established to define a set of typical, appliance-operating parameters and ambient conditions. Where appropriate, the same conditions used to generate the existing *National Fuel Gas Code* tables were assumed for this analysis.

The appliance-operating parameters were set after consultations with and review by the research project's Technical Advisory Group, which included appliance manufacturers. These operating parameters were chosen so that the resulting venting tables would be applicable to the majority of Category I fan-assisted appliances being manufactured. However, ensuring that the venting guidelines are applicable to a particular appliance model is the responsibility of the appliance manufacturer, especially if the operation of that model deviates significantly from the baseline conditions assumed here. Table S1.1 summarizes the baseline appliance conditions used to generate the venting tables for Category I fan-assisted appliances.

***TABLE S1.1*** *Baseline Appliance Conditions Used to Generate Venting Tables for Category I Fan-Assisted Appliances*

| *Appliance Attribute* | *Baseline Values* |
|---|---|
| Thermal efficiency | 83% |
| Excess combustion air for natural gas | 65% (7% $CO_2$) |
| Off-cycle pressure loss coefficient | 30 |

### Appliance Thermal Efficiency

One important appliance parameter that affects venting is the steady-state thermal efficiency of the appliance. In this case, steady-state thermal efficiency is defined as follows:

$$\text{Thermal efficiency (at steady state)} = \frac{\begin{bmatrix}\text{Appliance heat} \\ \text{input rate}\end{bmatrix} - \begin{bmatrix}\text{Appliance flue} \\ \text{heat loss rate}\end{bmatrix}}{\text{Appliance heat input rate}} \times 100$$

The thermal efficiency is the percentage of heat generated by the gas burner (i.e., appliance input rate) that is used for space or water heating. For natural draft venting systems (i.e., Category I appliances), a high appliance thermal efficiency means proportionately lower vent gas temperatures, and therefore less buoyancy is available to exhaust the gases through the venting system. Lower vent gas temperatures can lead to an increased potential for condensate formation in the vent.

Category I fan-assisted appliances typically operate at thermal efficiencies between 80 percent and 83 percent, resulting in AFUEs greater than 78 percent. The venting tables for Category I fan-assisted appliances were generated assuming an appliance thermal efficiency of 83 percent. If all other parameters were held constant, the minimum and maximum vent capacity range, based on an efficiency of 83 percent, would be a subset of the range for lower appliance thermal efficiencies. Therefore, the venting tables generated at an 83 percent thermal efficiency are also applicable (and conservative) for Category I fan-assisted appliances with lower thermal efficiencies.

### Appliance Excess Combustion Air

Appliance excess combustion air is another parameter that has a strong influence on venting. The excess combustion air level directly affects the quantity of vent gas to be vented; in turn, the maximum capacity of the vent is affected. At the same time, the amount of excess air also changes the dew point of the vent gas, influencing the amount of condensate formation in the vent and the minimum capacity of the vent.

Category I fan-assisted appliances typically operate with excess combustion air levels between 30 percent and 65 percent (9 percent to 7 percent $CO_2$). The venting tables for Category I fan-assisted appliances were generated assuming 65 percent excess combustion air (7 percent $CO_2$ level). If all other parameters are held constant, then a decrease in the excess combustion air generally will result in an increase in the minimum vent capacities and a decrease in the maximum vent capacities.

### Appliance Cycle Times

For space-conditioning equipment, appliance cycle times depend on the thermostat characteristics and setting, the outdoor temperature, the building heat loss characteristics, and the oversizing factor used at design load to determine the appliance input rating. Fortunately, the U.S. Department of Energy has done extensive research on typical appliance cycle times as part of their methodology for determining AFUEs of furnaces. The selection of cycle times that resulted in the revised venting tables relies heavily on the results of the Department of Energy's analyses.

The generic, Category I fan-assisted appliance used to generate the venting tables was assumed to be installed in a typical Midwestern residence with an average outdoor temperature during the heating season of 42°F (5.6°C). Under these circumstances, the appliance cycles on for 3.87 minutes and off for 13.3 minutes in accordance with the Department of Energy AFUE test.

Table S1.2 lists, as a function of time, the flue gas temperatures exiting the Category I fan-assisted appliance that were used to generate the venting tables. These temperatures are based on typical heat-up and cool-down response times for a Category I fan-assisted appliance with a thermal efficiency of 83 percent and an excess combustion air level of 65 percent (7 percent $CO_2$).

***TABLE S1.2 Category I Fan-Assisted Appliance Outlet Temperatures Used to Generate Venting Tables***

| Time (min:sec) | Fan-Assisted Appliances (83% Eff., 7% $CO_2$) (°F) | (°C) |
|---|---|---|
| 00:00 | 131.8 | 55.4 |
| 0:23 | 181.6 | 83.1 |
| 0:46 | 215.4 | 101.9 |
| 1:10 | 238.0 | 114.4 |
| 1:33 | 254.0 | 123.3 |
| 1:56 | 264.6 | 129.2 |
| 2:19 | 271.8 | 133.2 |
| 2:43 | 276.7 | 135.9 |
| 3:06 | 280.0 | 137.8 |
| 3:29 | 282.2 | 139.0 |
| 3:52 | 283.8 | 139.9 |
| 4:46 | 244.4 | 118.0 |
| 5:38 | 215.1 | 101.7 |
| 6:32 | 193.4 | 89.7 |
| 7:25 | 177.1 | 80.6 |
| 8:18 | 165.1 | 73.9 |
| 9:11 | 156.1 | 68.9 |
| 10:05 | 149.4 | 65.2 |
| 10:58 | 144.5 | 62.5 |
| 11:51 | 140.8 | 60.4 |
| 12:44 | 138.0 | 58.9 |
| 13:37 | 136.0 | 57.8 |
| 14:31 | 134.4 | 56.9 |
| 15:24 | 133.3 | 56.3 |
| 16:17 | 132.5 | 55.8 |
| 17:15 | 131.8 | 55.4 |

During the off cycle, the flow rate of air entering the vent system was calculated by assuming that the appliance had an off-cycle pressure loss coefficient of 30. The off-cycle pressure loss coefficient is the number of velocity heads ($^1/_2 \rho v^2$) of pressure that is lost between the appliance air intake and the appliance flue gas exit. A value of 30 was found to be representative of appliances with fan-assisted combustion systems as measured in the laboratory.

## Gas Appliance Fuel Type

The four different fuel gases used in gas appliances in the United States were evaluated with respect to the products of combustion and the effect on venting performance. The four fuels and the reason for their inclusion in the comparison are as follows:

1. *Natural gas.* The fuel assumed by VENT-II, Version 4.1, and used to develop the masonry chimney venting tables. The specific composition of natural gas (see Table S1.3) is derived from the *ASHRAE Fundamentals Handbook* and has an equivalent Btu content of 1032 Btu/ft$^3$ (38 kJ/m$^3$) of gas.
2. *Methane.* The fuel assumed by VENT-II prior to Version 4.1. (Natural gas is almost entirely methane.)
3. *Propane.* A fuel used as a substitute for natural gas.
4. *Butane.* A fuel used in the past in southern climates as a substitute for natural gas.

***TABLE S1.3 Natural Gas Composition (Mole Percentages) Used in VENT-II, Version 4.1***

| Component | Mole Percent |
|---|---|
| $N_2$ nitrogen | 0.362 |
| $CO_2$ carbon dioxide | 0.599 |
| $CH_4$ methane | 96.324 |
| $C_2H_6$ ethane | 2.061 |
| $C_3H_8$ propane | 0.367 |
| $C_4H_{10}$ isobutane | 0.150 |
| $C_5H_{12}$ isopentane | 0.056 |
| $C_6H_{14}$ n-hexane | 0.081 |
| Higher heating value, Btu/lb | 23,207 |
| Higher heating value, Btu/SCF | 1,032 |
| Specific gravity | 0.5807 |

The important factors that affect venting performance with respect to the venting tables are the total amount of vent gas generated, the temperature of the vent gas, and the amount of moisture in the vent gas. The total volume and temperature of vent gas generated by an appliance per Btu/hr of energy are determining factors in calculating the maximum capacity of an appliance that is connected to a specified vent system. The amount of moisture in the vent gas affects the vent gas dew point and, in turn, the minimum capacity of a fan-assisted appliance.

Table S1.4 shows a comparison of the products of the stoichiometric combustion of the four gases, as derived

*TABLE S1.4 Products of Stoichiometric Combustion of Four Different Gaseous Fuels*

| *Property* | *Units* | *Methane* $CH_4$ | *Propane* $C_3H_8$ | *Butane* $C_4H_{10}$ | *Natural Gas* $C_{1.04}H_{3.88}$ |
|---|---|---|---|---|---|
| **Combustion products from stoichiometric combustion** | | | | | |
| Volume of combustion product per kBtu | $\left[\frac{ft^3 \text{ comb. product}}{\text{kBtu fuel}}\right]$ | 10.4 | 10.2 | 10.2 | 10.3 |
| Water vapor concentration | $\left[\frac{ft^3 H_2O \text{ vapor}}{ft^3 \text{ comb. product}}\right]$ | 0.190 | 0.155 | 0.149 | 0.184 |
| Flue gas dew point | °F (°C) | 139 (59.4) | 131 (55.0) | 129 (53.9) | 138 (58.9) |
| **Stoichiometric combustion parameters** | | | | | |
| Combustion air required | $\left[\frac{ft^3 \text{ air}}{ft^3 \text{ fuel}}\right]$ | 9.53 | 23.82 | 30.97 | 9.6 |
| Ultimate $CO_2$ | (%) | 11.73 | 13.75 | 14.05 | 12.1 |
| Nitrogen in combustion products | $\left[\frac{ft^3 N_2}{ft^3 \text{ fuel}}\right]$ | 7.53 | 18.82 | 24.47 | 7.58 |
| Carbon dioxide in combustion products | $\left[\frac{ft^3 CO_2}{ft^3 \text{ fuel}}\right]$ | 1.0 | 3.0 | 4.0 | 1.04 |
| Water vapor in combustion products | $\left[\frac{ft^3 H_2O \text{ vapor}}{ft^3 \text{ gas}}\right]$ | 2.0 | 4.0 | 5.0 | 1.94 |
| Total combustion gas volume | $\left[\frac{ft^3 \text{ comb. product}}{ft^3 \text{ fuel}}\right]$ | 10.53 | 25.82 | 33.47 | 10.56 |
| Energy content | $\left[\frac{\text{Btu}}{ft^3 \text{ fuel}}\right]$ | 1012 | 2524 | 3266 | 1030 |

from the *Gas Engineers Handbook* (Segeler, 1965). Note that the volume of combustion products per kBtu fuel values in Table S1.4 are all within 1 percent of the value for natural gas. Also, note that the water vapor concentration in the combustion products is significantly less for both propane and butane than for natural gas, resulting in lower dew points in these fuels.

Although specific appliance designs will vary, certain conditions must be taken into consideration when characterizing the field conversion of an appliance from natural gas to propane. These conditions are the 1 percent to 2 percent increase in carbon dioxide that is measured at the appliance outlet and a slight increase in the efficiency of the appliance. Therefore, an appliance that operates at 65 percent excess air (7 percent $CO_2$) with natural gas can be expected to operate at approximately 48 percent excess air (9 percent $CO_2$) when converted to propane.

Table S1.5 outlines the change in vent gas based on the conversion from natural gas to propane. Note that the total volume of vent gas for propane and butane is less than the volume for natural gas. From laboratory test data of an appliance converted from natural gas to propane, an increase in the vent gas temperature was observed when burning propane. Also, note that the dew point of the vent gas is lower with both propane and butane than with natural gas. Therefore, based on the stoichiometric data presented in Table S1.4 and the example comparison in Table S1.5, the revised venting tables are considered appropriate, without modification, for propane-fired and butane-fired appliances.

## Ambient Conditions — Type B Vents and Interior Masonry Chimneys

In generating the venting tables for Type B vents and interior masonry chimneys, only interior venting configurations were considered, as shown in Exhibit S1.6. By definition, interior venting configurations are those that are not exposed to outdoor temperatures below the roofline. Above the roofline, only 3 ft (1 m) of vent or chimney is assumed to be exposed to outdoor temperatures.

*TABLE S1.5 Rule-of-Thumb Vent Gas Comparison*

| *Parameter* | *Units* | *Propane* | *Butane* | *Natural Gas* |
|---|---|---|---|---|
| Carbon dioxide in vent gas | % | 9.0 | 9.0 | 7.0 |
| Flue product per Btu | $\left[\frac{ft^3 \text{ vent gas}}{kBtu}\right]$ | 14.79 | 15.11 | 16.28 |
| Water vapor concentration | $\left[\frac{ft^3\ H_2O \text{ vapor}}{ft^3 \text{ vent gas}}\right]$ | 0.1071 | 0.1011 | 0.1202 |
| Vent gas dew point | °F | 117.4 | 115.4 | 121.5 |

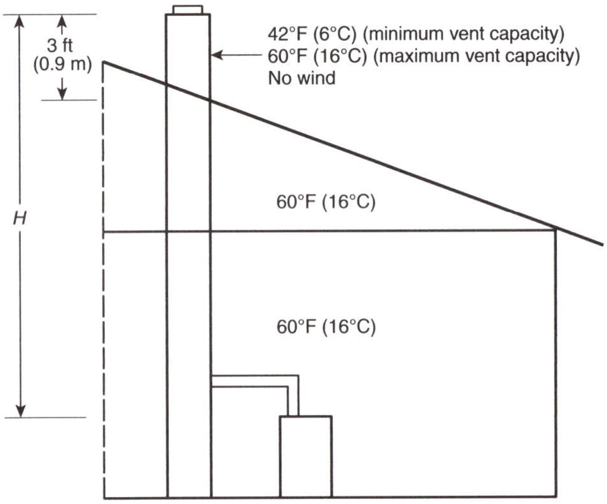


*EXHIBIT S1.6 Ambient Conditions Assumed for Generating Venting Tables for Type B Vents and Interior Masonry Chimneys. (Courtesy of Gas Technology Institute)*

Minimum vent capacities were calculated by assuming a 42°F (5.6°C) outdoor temperature and a 60°F (16°C) indoor temperature. Maximum vent capacities were calculated by assuming both the indoor and outdoor temperatures were 60°F (16°C). In both cases, zero house depressurization and no wind were assumed.

## Maximum Vent Capacities

For Category I fan-assisted appliances, the maximum vent capacity was calculated by VENT-II, Version 4.1, at the appliance input rating that results in a zero static vent pressure 1 ft (0.3 m) from the appliance outlet and 1 minute after the start of the third appliance cycle. These criteria were selected because Category I appliance vent systems are not designed to operate under positive static pressures. However, when the appliance burner first ignites, the vent system undergoes a pressure spike until the vent primes (i.e., becomes filled with warm vent gases). Therefore, a time of 1 minute into the third burner cycle was selected to allow the vent to prime. Additional checks were put into place to ensure that, when fan-assisted appliances are common vented with a draft hood appliance, spillage of vent gas due to vent priming will remain below 1.5 percent of the total vent gas. The Type B double-wall venting tables that were generated using earlier versions of VENT-II (prior to Version 4.1) used a slightly different location and time for determining the maximum vent capacity that corresponded to a zero static pressure. However, essentially the same maximum capacities will be calculated if these criteria are used for Type B double-wall vent systems using VENT-II, Version 4.1.

Maximum vent capacities are relatively insensitive to which appliance cycle is chosen, because gas temperatures in the vent tend to stay approximately the same from one cycle to the next. Therefore, the third appliance cycle was chosen for calculating maximum vent capacities for both Type B double-wall vent systems and masonry chimneys.

## Minimum Vent Capacities

The minimum vent capacities of Category I fan-assisted appliances assume that the vent materials do not exceed the specified wet-time limits. If these wet-time limits are exceeded, a high potential exists for the vent materials to experience excessive corrosion (in the case of metal vents) or for water damage to occur in surrounding structures (in the case of masonry chimneys).

For metal vent systems, the wet-time limits are based on conventional vent conditions that have existed for many years, as well as on the existing data available for the corrosion of condensing heat exchanger materials in the presence of vent gases. For vent connectors made of galvanized steel, aluminum, Type B vent, or 300 series stainless steels, the wet-time limits are set so that these materials can dry out within 3.8 minutes from the start of the third appliance cycle. For vertical Type B double-wall vents, the wet-time limits are set so that these materials can dry out within 12 minutes from the start of the third appliance cycle. The wet-time limits for vent connectors are lower than the wet-time limits for vertical vents, because the vent connectors usually experience a higher concentration of corrosive contaminants (e.g., chlorides, nitrates, sulfates, and fluorides) in the vent gases. By the time the vent gases enter the vertical vent, many of the corrosive contaminants have already been removed from the gas as it passed through the vent connector. Therefore, the vertical vent can accommodate a longer wet time without necessarily experiencing excessive corrosion. In all cases, if the minimum vent capacity exceeds the maximum vent capacity, the specific vent configuration is not allowed (NA) for use with Category I fan-assisted appliances.

### Masonry Chimneys

For vent systems employing masonry chimneys, wet-time limits cannot be set based on the materials' degradation, because of the lack of data available to make this determination. Therefore, the assumption was made that the increased wetness associated with Category I fan-assisted appliances did *not* increase the rate of degradation of the materials used to construct masonry chimneys. This assumption was confirmed through a series of spray chamber tests using masonry chimney material specimens exposed to synthetic vent gas condensate.

The criterion used to determine the adequacy of masonry chimneys is whether or not the chimney is wet continuously for long periods of time. Masonry chimneys built in accordance with NFPA 211, *Standard for Chimneys, Fireplaces, Vents, and Solid Fuel–Burning Appliances*, are not watertight and are not required to have condensate drains. Clay tile liners are porous, and mortar joints between clay tile liners can crack, which can allow moisture and condensate to migrate into surrounding structures, causing water damage. A homeowner will notice this problem when plaster walls adjacent to the chimney become wet and discolored or when paint or wallpaper starts to peel. Condensate can also collect in the bottom of the chimney and seep out, potentially causing water damage to the floor or furnishings in the vicinity. All of these possibilities point to the need for venting guidelines that will limit the amount of condensate formation in existing masonry chimneys when Category I fan-assisted appliances are employed.

### Interior Masonry Chimneys

For interior masonry chimneys, the wet-time limit is checked at the maximum vent capacity and at the maximum liner size. In the VENT-II simulation, the 3 ft (1 m) of chimney that extends through the roof is assumed to be exposed to a 42°F (5.6°C) ambient temperature. The remainder of the chimney is assumed to be located in the interior of the house and exposed to a 60°F (16°C) ambient temperature. The VENT-II simulation begins by allowing the Category I fan-assisted appliance to operate continuously at its input rating for 1 hour to heat up the chimney, followed by an off cycle lasting 13 minutes. Next, the appliance is allowed to cycle ON and OFF (according to the criteria stated earlier in this supplement) for 42 cycles (approximately 12 hours). At the end of this time period, if the clay tile in the chimney is still wet, the table entry is NA (not allowed). If the clay tile has dried out at the end of this time period, then the minimum vent capacity is determined based on the wet-time limit of the vent connector material.

### Exterior Masonry Chimneys

The maximum capacity of an exterior chimney is identical to the maximum capacity of an interior chimney of the same dimensions. However, because exterior masonry chimneys are exposed to the outdoor environment over their entire length, the impact of climate on the potential for excessive condensation is greater than the impact of climate for interior chimneys. Therefore, the tables for exterior chimneys are indexed by outdoor temperature range, becoming increasingly restrictive with increasingly colder climates.

Explicit minimum capacities, based on the chimney dimensions and the climate, are listed for exterior masonry chimneys. These minimum capacities were determined from VENT-II analysis, assuming that the chimney was allowed to be continuously wet for up to 1000 hours during a single heating season and was used to vent an appliance operating at a steady-state thermal efficiency of 83 percent.

The appliance efficiency is the most critical parameter affecting exterior masonry chimney performance. GRI research shows that when the appliance efficiency is about 80.5 percent rather than 83 percent, the guidelines for exterior masonry chimneys can be relaxed significantly. Use Table 13.1(f) or Tables 13.2(e) through 13.2(h) from Chapter 13 unless the installation instructions provide an alternate method.

## SINGLE-APPLIANCE VENTING TABLES

Single-appliance venting tables apply when only one appliance is connected to the vent system. This section describes the rationale used to develop single-appliance venting tables for Category I fan-assisted appliances. Four separate venting tables have been generated that cover the use of either single-wall metal vent connectors or Type B double-wall metal vent connectors with either Type B double-wall metal vertical vents or masonry chimneys.

## Vent Configuration for Single-Appliance Vents

For single-appliance vents, as shown in Exhibit S1.7, the basic configurations for interior vents have been modeled using VENT-II. The following guidelines have been used to model the venting systems:

1. Section A1 is 1 ft (0.3 m) in vertical height.
2. The total lateral length (*L*) is split evenly between Sections A2 and A3.
3. Section C4 is 3 ft (0.9 m) in vertical height.
4. The remaining vertical height [the total height less 4 ft (1.2 m)] is split evenly among Sections C1, C2, and C3.

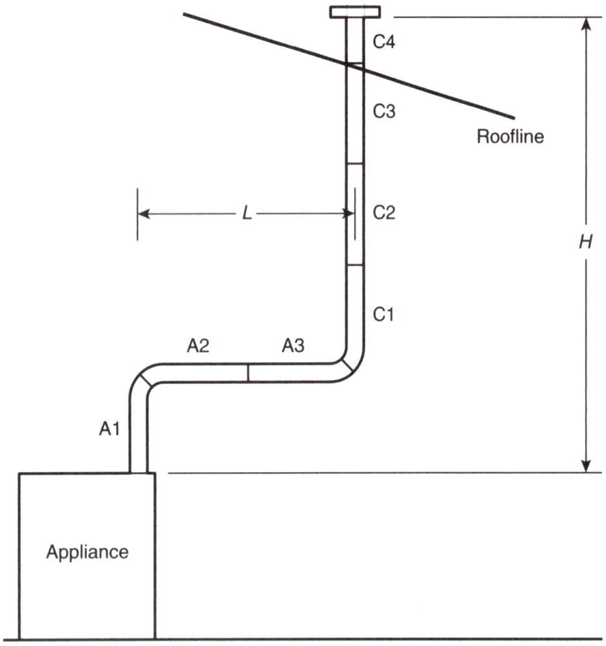


***EXHIBIT S1.7*** *VENT-II Configuration for Single-Appliance Venting Tables. (Courtesy of Gas Technology Institute)*

The A sections are used to designate the vent connector. These sections are modeled as either single-wall metal vent or Type B double-wall metal vent. All A sections are assumed to be exposed to a 60°F (16°C) indoor ambient temperature.

The C sections are used to designate the vertical vent. These sections are modeled as either Type B double-wall metal vent or as a tile-lined masonry chimney meeting NFPA 211 design requirements. The last vertical section extending through the roofline, C4, is assumed to be exposed to the outdoor air temperature, which is 42°F (5.6°C) when minimum vent capacities are calculated and 60°F (16°C) when maximum vent capacities are calculated. All other C sections (C1, C2, and C3) are assumed to be exposed to a 60°F (16°C) indoor ambient temperature.

For exterior chimneys, as shown in Exhibit S1.8, Sections C1, C2, and C3 are exposed on three sides to the outdoor air temperature while Section C4 is exposed on all sides. For exterior chimneys, calculations were made at a range of outdoor air temperatures corresponding to the major climates of the United States. (See Figure G.2.4 of the code.)

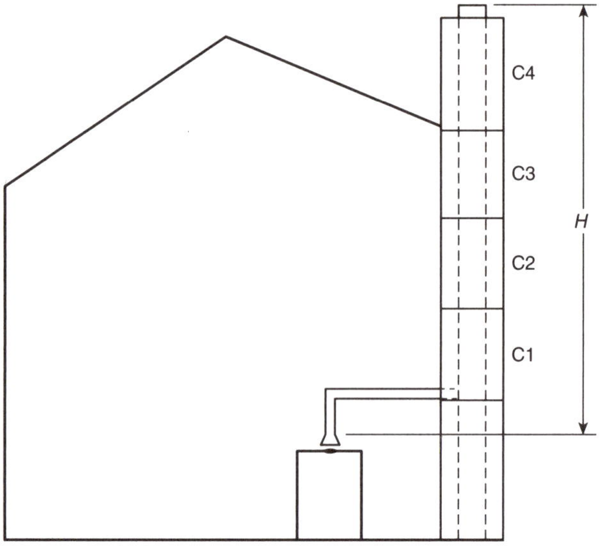


***EXHIBIT S1.8*** *VENT-II Configuration for Exterior Masonry Chimney Venting Tables. (Courtesy of Gas Technology Institute)*

Although the exterior chimney calculations assumed three sides of the chimney were exposed, the potential for excessive condensation can exist even if only one side is exposed. The presence of any surface at a temperature below the dew point of the flue gases creates the potential for condensate formation. Therefore, the definition of an exterior chimney encompasses any chimney with one or more sides exposed to the outdoors below the roofline.

## Format of Single-Appliance Venting Tables

Venting tables for Category I draft hood appliances have been in place for many editions of the code and have withstood the test of time. These tables should be retained for venting conventional draft hood appliances. For easy understanding, the format for the venting tables for Category I fan-assisted appliances was chosen deliberately so that they resemble the format of the tables in editions of the code prior to 1992. At the suggestion of the Gas Research Institute's Technical Advisory Group, the venting tables for Category I fan-assisted appliances and Category I draft hood appliances were merged. However, the table entries for draft hood appliances were not changed from the existing code tables except where interpolation or extrapolation was required for consistency.

The following is an explanation of the terms and abbreviations used in the single-appliance venting tables:

**Fan-Assisted Combustion System.** An appliance equipped with an integral, mechanical means to either draw or force products of combustion through the combustion chamber and heat exchangers.

**FAN Min.** The minimum input rating of a Category I appliance with a fan-assisted combustion system that could be attached to the vent.

**FAN Max.** The maximum input rating of a Category I appliance with a fan-assisted combustion system that could be attached to the vent.

**NAT Max.** The maximum input rating of a Category I appliance equipped with a draft hood that could be attached to the vent. There are no minimum appliance input ratings for draft hood–equipped appliances.

**NA.** Vent configuration is not allowed due to potential for condensate formation or pressurization of the venting system, or not applicable due to physical or geometric restraints.

*** (following number).** Indicates potential for continuous condensation.

In Annex G of the code, Figure G.1(a) shows the configuration used to generate the venting table [Table 13.1(a)] when the entire vent system is composed of Type B double-wall metal vent. Figure G.1(b) shows the configuration used to generate the venting table [Table 13.1(b)] for Type B double-wall metal vents that are used with single-wall metal vent connectors. Figure G.1(c) shows the configuration used to generate the venting tables [Table 13.1(c) and Table 13.1(f)] for masonry chimneys that are used with Type B double-wall metal connectors. Figure G.1(d) shows the configuration used to generate the venting table [Table 13.1(d)] for masonry chimneys that are used with single-wall metal vent connectors.

In all of these tables, the NAT Max values were derived directly from the tables for draft hood appliances in editions of the code prior to 1992. The FAN Min and FAN Max values for fan-assisted Category I appliances were derived using the VENT-II computer program with the criteria described earlier in the Baseline Conditions for Category I Fan-Assisted Appliance Venting Tables section. In Table 13.1(a) and Table 13.1(b), single-appliance venting configurations with zero lateral lengths are assumed to have no elbows in the vent system. For vent configurations with lateral lengths, the venting tables include allowance for two 90-degree turns. For each additional turn up to and including 45 degrees, the maximum capacity listed in the venting tables should be reduced by 5 percent. For each additional turn greater than 45 degrees up to and including 90 degrees, the maximum capacity listed in the venting tables should be reduced by 10 percent (0.90 × maximum table capacity). Two 45-degree turns are equivalent to one 90-degree turn.

Sea level input ratings should be used when determining maximum capacity for high-altitude installations. Actual input (derated for altitude) should be used to determine the minimum capacity for high-altitude installations. For appliances with more than one input firing rate, the minimum vent capacity that is determined from the tables should be less than the lowest appliance input rating, and the maximum vent capacity that is determined from the tables should be greater than the highest appliance input rating.

In generating the masonry chimney venting tables, the minimum flow area of the chimney was assumed to be one size equivalent round larger than the vent connector flow area. The vent connector flow area was assumed to be equal to the appliance vent outlet flow area. The maximum vent capacities for fan-assisted appliances were calculated based on the minimum chimney flow area. This assumption was also used in the existing code table entries for draft hood appliances. For either Type B double-wall vertical vents or masonry chimneys, the vent flow area was assumed to be less than or equal to seven times the appliance outlet flow area.

For a single appliance venting into an exterior masonry chimney, the code specifies the following criteria:

1. The appliance must be draft hood equipped.
2. The vent connector must be Type B double-wall.
3. The vent connector length must be less than $1\frac{1}{2}$ ft per inch of vent connector diameter.
4. The appliance maximum input rate must be less than the rate listed in Table 13.1(c).
5. If the appliance is used for space heating, then its input rate must be greater than the rate listed in Table 13.1(f).

If the appliance is not used for space heating, then consult the appliance manufacturer, local gas supplier, or authority having jurisdiction. Table 13.1(f) lists space-heating appliance minimum input rates for exterior masonry chimneys for six climates (i.e., winter design temperatures).

## COMMON VENTING TABLES FOR TWO APPLIANCES

Common venting tables apply when two appliances are connected to the same vent. This section describes the rationale used to develop common venting tables for

Category I appliances. Venting tables have been generated that cover the use of either single-wall metal vent connectors or Type B double-wall vent connectors with either Type B double-wall metal vertical vents or masonry chimneys (interior and exterior).

## Vent Size Combinations for Common Venting Tables

The common vent tables were generated assuming specific size combinations for the vent connectors on each appliance and for the common vent. For a Type B double-wall common vent, Table S1.6 lists the size combinations assumed for the vent connectors for each appliance and for the common vent. For a masonry chimney, Table S1.7 lists the size combinations used to generate common vent capacities, and Table S1.8 lists the size combinations used to generate vent connector capacities.

## Format of Common Venting Tables

The common venting tables for fan-assisted appliances were combined with the common venting tables for draft hood appliances and put in the same format as the tables in editions of the *National Fuel Gas Code* prior to 1992. The table entries for draft hood appliances were not changed from the pre-1992 edition code tables, except where interpolation or extrapolation was required for consistency.

The following is an explanation of additional terms and abbreviations (see Chapter 3) used in the common venting tables:

**FAN+FAN.** The maximum combined appliance input rating of two or more Category I fan-assisted appliances that are attached to a common vent.

**FAN+NAT.** The maximum combined appliance input rating of one or more Category I fan-assisted appliances and one or more Category I draft hood–equipped appliances that are attached to a common vent.

**NAT+NAT.** The maximum combined appliance input rating of two or more Category I draft hood–equipped appliances that are attached to a common vent.

**NA.** Vent configuration is not applicable due to physical or geometric constraints.

Figures G.1(f) and G.1(g) show the vent configuration used to generate the common venting tables for Type B double-wall vertical vents with either Type B double-wall connectors [Table 13.2(a)] or single-wall metal connectors [Table 13.2(b)]. Figures G.1(h) and G.1(i) show the vent configuration used to generate the common venting tables for masonry chimneys with either Type B double-wall connectors [Tables 13.2(c), 13.2(f), 13.2(g), or 13.2(h)] or single-wall metal connectors [Table 13.2(d)].

*TABLE S1.6 Size Combinations of Type B Vent Used to Generate Common Venting Tables*

| Appliance 1 Vent Connector Diameter | | Appliance 2 Vent Connector Diameter | | Type B Combined Vent Diameter | |
|---|---|---|---|---|---|
| *in.* | *mm* | *in.* | *mm* | *in.* | *mm* |
| 3 | 80 | 3 | 80 | 4 | 100 |
| 3 | 80 | 4 | 100 | 5 | 130 |
| 3 | 80 | 5 | 130 | 6 | 150 |
| 4 | 100 | 7 | 178 | 8 | 200 |
| 5 | 130 | 8 | 200 | 9 | 230 |
| 5 | 130 | 9 | 230 | 10 | 250 |
| 6 | 150 | 10 | 250 | 12 | 310 |
| 7 | 180 | 12 | 310 | 14 | 360 |
| 8 | 200 | 14 | 360 | 16 | 410 |
| 9 | 230 | 16 | 410 | 18 | 457 |
| 10 | 250 | 18 | 460 | 20 | 510 |
| 11 | 280 | 20 | 510 | 22 | 560 |
| 12 | 310 | 22 | 566 | 24 | 610 |
| 14 | 360 | 26 | 660 | 28 | 710 |
| 16 | 410 | 30 | 760 | 32 | 810 |
| 18 | 450 | 32 | 810 | 36 | 910 |
| 20 | 510 | 36 | 910 | 40 | 1020 |
| 22 | 560 | 40 | 1010 | 44 | 1120 |
| 24 | 610 | 44 | 1020 | 48 | 1220 |

***TABLE S1.7*** *Size Combinations of Masonry Chimneys Used to Generate Common Vent Capacities in the Multiple-Appliance Venting Tables*

| *Appliance 1 Vent Connector Diameter* | | *Appliance 2 Vent Connector Diameter* | | *Minimum Masonry Chimney Flow Area* | | *Maximum Masonry Chimney Flow Area* | |
|---|---|---|---|---|---|---|---|
| *in.* | *mm* | *in.* | *mm* | *in.²* | *mm²* | *in.²* | *mm²* |
| 3 | 80 | 3 | 80 | 12 | 7,740 | 49 | 31,600 |
| 3 | 80 | 4 | 100 | 19 | 12,300 | 88 | 55,800 |
| 3 | 80 | 5 | 130 | 28 | 18,100 | 137 | 88,400 |
| 3.5 | 89 | 6 | 150 | 38 | 24,500 | 198 | 128,000 |
| 4 | 100 | 7 | 180 | 50 | 33,300 | 269 | 173,000 |
| 4.5 | 110 | 8 | 200 | 63 | 40,600 | 352 | 550,000 |
| 5 | 130 | 9 | 230 | 78 | 50,300 | 445 | 286,000 |
| 6 | 150 | 10 | 250 | 113 | 72,900 | 550 | 355,000 |

***TABLE S1.8*** *Size Combinations of Masonry Chimneys Used to Generate Vent Connector Capacity in the Multiple-Appliance Venting Tables*

| *Appliance 1 Vent Connector Diameter* | | *Appliance 2 Vent Connector Diameter* | | *Masonry Chimney Flow Area* | |
|---|---|---|---|---|---|
| *in.* | *mm* | *in.* | *mm* | *in.²* | *mm²* |
| 3 | 76 | 5 | 130 | 28 | 18,100 |
| 4 | 100 | 7 | 180 | 50 | 32,300 |
| 5 | 130 | 9 | 230 | 78 | 50,300 |
| 6 | 150 | 10 | 250 | 113 | 73,000 |
| 7 | 180 | 12 | 300 | 153 | 98,700 |
| 8 | 200 | 14 | 360 | 201 | 130,000 |
| 9 | 230 | 16 | 410 | 254 | 164,000 |
| 10 | 250 | 18 | 460 | 314 | 203,000 |

## Vent Connector Capacity

In generating common venting tables, the vent connector capacities for each appliance were calculated by assuming that the appliance attached to the vent connector was on and that the other appliance was off the entire time. The lateral length of each vent connector was assumed to be equal to 1.5 ft per inch of connector diameter (see Table S1.9). The connector rise for appliance 1 was assumed to be equal to the value listed in the appropriate venting table, with the connector rise for appliance 2 set equal to the maximum value (either 3 ft or 6 ft) shown in the venting table for a given vent connector diameter. The maximum and minimum vent connector capacities were calculated as described earlier in this supplement.

The FAN Max column in the vent connector capacity tables refers to the maximum vent connector capacity when a Category I appliance with a fan-assisted combustion system is attached to the vent. The column labeled NAT Max refers to the maximum vent connector capacity for a draft hood–equipped appliance. The FAN Min col-

***TABLE S1.9*** *Vent Connector Horizontal Length Used to Generate Common Venting Tables*

| *Connector Diameter* | | *Maximum Connector Horizontal Length* | |
|---|---|---|---|
| *in.* | *mm* | *ft* | *m* |
| 3 | 76 | 4 | 1.2 |
| 4 | 100 | 6 | 1.8 |
| 5 | 130 | 7.5 | 2.3 |
| 6 | 150 | 9 | 2.7 |
| 7 | 180 | 10.5 | 3.2 |
| 8 | 200 | 12 | 3.7 |
| 9 | 230 | 13.5 | 4.1 |
| 10 | 250 | 15 | 4.6 |
| 12 | 300 | 18 | 5.5 |
| 14 | 360 | 21 | 6.4 |
| 16 | 410 | 24 | 7.3 |
| 18 | 460 | 27 | 8.2 |
| 20 | 510 | 30 | 9.1 |
| 22 | 560 | 33 | 10.1 |
| 24 | 610 | 36 | 11.0 |

umn refers to the minimum vent connector capacity for use with fan-assisted appliances.

### Common Vent Capacities

When determining common vent capacities for the tables, the size combinations assumed for the individual vent connectors affect the calculated capacities. Tables S1.6 and S1.7 list the size combinations used to generate the combined vent capacities. The connector rise for the first appliance is fixed at 1 ft, and the rise for the second appliance is fixed at 3 ft.

The common vent capacities are calculated assuming two Category I appliances are connected to the combined vent with both appliances cycling simultaneously. Only maximum capacity values are required for the common vent portion of the Type B double-wall venting tables. The FAN Min restrictions of the vent connector portion of the tables ensure that condensation in the common vent does not reach unacceptable levels.

In the venting tables, the column labeled FAN+FAN refers to the maximum combined vent capacity when two appliances with fan-assisted combustion systems are attached to the same vent. Both fan-assisted appliances are required to have zero static pressure in the first section of their respective vent connectors. The capacities of the two appliances are totaled to produce the value shown in the table.

The FAN+NAT column refers to the maximum combined vent capacity when one fan-assisted appliance and one draft hood–equipped appliance are connected to the same vent. The fan-assisted appliance is required to have zero static pressure in the first section of its vent connector, while the draft hood–equipped appliance, which is assumed to operate at 78 percent thermal efficiency with an 8 percent $CO_2$ excess combustion air level, is required to have 29 percent dilution air entering through the draft hood. The capacities of the two appliances are added to produce the value shown in the table.

Finally, the column labeled NAT+NAT shows the maximum combined vent capacity when two appliances that are equipped with draft hoods are connected to the same vent. Common vent capacities for two draft hood–equipped appliances were not computed using VENT-II. Instead, the values from pre-1992 editions of the code were retained in the NAT+NAT column.

### Exterior Masonry Chimneys

For two or more appliances, with common venting into an exterior masonry chimney, Tables 13.2(f), 13.2(g), 13.2(h), and 13.2(i) provide minimum and maximum appliance input rates. To use these tables for sizing an exterior masonry chimney, the vent connector must be Type B double-wall, at least one appliance must be draft hood–equipped, and at least one appliance must be used for space heating.

- Draft Hood–Equipped Appliances. Tables 13.2(f) and 13.2(g) are used when all appliances are draft hood–equipped. Table 13.2(f) provides combined appliance maximum input rates, and Table 13.2(g) provides minimum space-heating appliance input rates for five climates.
- Combination of Draft Hood–Equipped and Fan-Assisted Appliances. Tables 13.2(h) and 13.2(i) are used when at least one appliance (but not all) is fan-assisted. Table 13.2(h) provides combined maximum input rates, and Table 13.2(i) provides minimum space-heating appliance input rates for five climates.

### Common Vent Connector Manifolds and Offsets

Connector manifolds in common vent systems result when the vent connectors of the appliances are combined before entering the vertical vent, as depicted in Figure G.1(k). The vertical vent may also be offset when going through the roof, as depicted in Figure G.1(l). These two cases are addressed separately in the following paragraphs. In Figures G.1(j) and G.1(l) the horizontal length is labeled $L_0$.

*Case 1: Manifolding Vent Connectors.* Based on the connector rise, the connector diameter, and the vent height, the vent connectors of both appliances are sized as for any common vent system. However, the common vent capacities listed in Table 13.2(a) through Table 13.2(d) are reduced by 10 percent to account for the additional restriction caused by the connector manifold. The horizontal length of the connector manifold should not exceed 1.5 ft per inch of common vent connector manifold diameter.

*Case 2: Offsets in the Vertical Vent.* Again, the vent connectors of both appliances are sized as for any common vent system. However, the common vent capacity is reduced by 20 percent for each offset. The horizontal length of the offset should not exceed 1.5 ft per inch of combined vent diameter.

### Additional Guidelines for Multiple-Appliance Venting Tables

The multiple-appliance venting tables were generated by assuming a vent connector horizontal length of 1.5 ft (18 in.) (460 mm) for each inch of connector diameter, as shown in Table 13.2.2. The vent connector should be routed to the vent using the shortest possible route. Longer connectors than those listed above are permitted under the following conditions:

1. The maximum capacity (i.e., FAN Max or NAT Max) of the vent connector should be reduced 10 percent for

each additional multiple of the length listed in Table 13.2.2 for Tables 13.2(a) through 13.2(e). For example, the maximum length listed in this table for a 4 in. (100 mm) connector is 6 ft (1.8 m). With a connector length greater than 6 ft (1.8 m) but not exceeding 12 ft (3.7 m), the maximum capacity must be reduced by 10 percent (0.90 × maximum vent connector capacity). With a connector length greater than 12 ft (3.7 m) but not exceeding 18 ft (5.5 m), the maximum capacity must be reduced by 20 percent (0.80 × maximum vent capacity).

**2.** The minimum capacity (i.e., FAN Min) should be determined by referring to the corresponding single-appliance tables [Table 13.1(a) through Table 13.1(d)]. In this case, for each appliance the entire vent connector and common vent from the appliance to the vent termination should be treated as a single appliance vent — that is, as if the other appliance were not present.

When two or more appliances are connected to a vertical vent or chimney, the flow area of the largest section of vertical vent or chimney is assumed to be less than or equal to seven times the vent outlet flow area of the smallest appliance. For appliances with more than one input rate, the minimum vent connector capacity (i.e., FAN Min) determined from the tables should be less than the lowest appliance input rating, and the maximum vent connector capacity (i.e., FAN Max or NAT Max) determined from the tables should be greater than the highest appliance input rating.

Vent connectors should not be increased more than two table sizes greater than the listed appliance categorized vent diameter, flue collar diameter, or draft hood outlet diameter. Corrugated and smooth wall vent connectors for draft hood–equipped appliances should not be smaller than the listed appliance outlet diameter. Smooth wall vent connectors for fan-assisted appliances may be reduced one size in accordance with code-prescribed conditions (13.2.24).

## SUMMARY

In 1987, the National Appliance Energy Conservation Act began to mandate higher minimum efficiency standards for nearly all gas appliances. Effective January 1, 1992, all fan-assisted central furnaces manufactured in the United States must have a minimum AFUE of at least 78 percent. (This percent corresponds to a steady-state thermal efficiency based on flue loss considerations of approximately 80.5 percent.) Also, the Department of Energy is evaluating minimum efficiency standards of gas-fired water heaters.

The furnace manufacturers have responded to these new regulations by developing higher-efficiency Category I gas furnaces with fan-assisted combustion systems. These new Category I furnaces have steady-state thermal efficiencies of up to 83 percent based on flue loss considerations. Revised venting guidelines and tables have been developed for these appliances because of the higher operating efficiencies and the presence of a fan-assisted combustion system.

The majority of gas appliances installed in homes prior to 1992 are draft hood–equipped models with steady-state thermal efficiencies around 75 percent. These appliances can be vented using recommendations found in the *National Fuel Gas Code*, which are also included with the tables in this report.

## SUMMARY OF RESEARCH FINDINGS

Guidelines and tables found in editions of the *National Fuel Gas Code* prior to 1992 are still valid for lower-efficiency, Category I draft hood appliances and remain unchanged. The revised guidelines and tables developed herein apply only to higher-efficiency, Category I fan-assisted appliances currently being produced to meet the new minimum efficiency standards.

The VENT-II computer program has been used to develop venting guidelines and tables for Category I fan-assisted appliances with steady-state thermal efficiencies that are based on flue loss considerations of up to 83 percent. Experimental validation of the VENT-II program was performed using typical venting systems composed of Type B double-wall gas vents meeting Underwriters Laboratories Standard UL 441, *Gas Vents*, and also using an interior masonry chimney constructed to meet NFPA 211 requirements.

Venting guidelines and tables have been developed for venting either a single Category I fan-assisted appliance or for common venting Category I fan-assisted appliances with other Category I appliances. Vent systems covered by these guidelines and tables include Type B double-wall vents and interior and exterior masonry chimneys connected to Category I fan-assisted appliances by using either single-wall metal vent connectors or Type B double-wall vent connectors.

Compared with draft hood–equipped Category I appliances, Category I fan-assisted appliances have a greater potential for producing condensate in the vent system. This condensate can cause excessive corrosion in metal vent systems and water damage to surrounding structures if the condensate should leak from the vent system through joints or cracks. To limit condensate problems in vent systems with Category I fan-assisted appliances, wet-time limits have been established for various vent materials.

For a given vent size and configuration, wet-time limits are used to calculate minimum fan-assisted appliance input ratings, and vent static pressure limits are used to calculate maximum fan-assisted appliance input ratings. This information has been put into the same format as the venting guidelines and tables in previous editions of the code.

For the sake of completeness, the venting tables presented herein contain entries for sizing vents for both fan-assisted appliances and draft hood appliances. However, the table entries for draft hood appliances are identical (or interpolated) to those values in previous editions of the code. The scope of this supplement is limited to establishing venting guidelines and tables for Category I fan-assisted appliances.

## REFERENCES


Agrawal, A. K., R. A. Cudnik, B. Hindin, D. W. Locklin, M. J. Murphy, J. H. Payer, R. Razgaitis, G. H. Stickford, and S. G. Talbert, "Technology Development for Corrosion-Resistant Condensing Heat Exchangers," GRI-85/0282 (NTIS No. PB86-172038), Gas Research Institute, Des Plaines, IL, 1985.

ANSI Z21.47, *Gas-Fired Central Furnaces (Except Direct Vent Central Furnaces)*, American Gas Association (Secretariat), 1990.

ANSI/ASHRAE 103, *Method of Testing for Annual Fuel Utilization Efficiency of Residential Central Furnaces and Boilers*, American National Standards Institute and American Society of Heating, Refrigerating and Air-Conditioning Engineers, Atlanta, GA, 1982.

*ASHRAE Equipment Handbook*, "Chimney, Gas Vent, and Fireplace Systems," Chapter 26, American Society of Heating, Refrigerating and Air-Conditioning Engineers, Atlanta, GA, 1979.

*ASHRAE HVAC Systems and Equipment Handbook*, Chapter 30, American Society of Heating, Refrigerating and Air-Conditioning Engineers, Atlanta, GA, 2000.

*ASHRAE Fundamentals Handbook*, "Combustion and Fuels," Chapter 15, American Society of Heating, Refrigerating and Air-Conditioning Engineers, Atlanta, GA, 1985.

Ball, D. A., and D. W. Locklin, eds., "Symposium on Condensing Heat Exchangers," Conference Proceedings, Battelle Laboratories, GRI-82/0009.3 (NTIS No. PB82-240078), Gas Research Institute, Des Plaines, IL, 1982.

Borgeson, R. A., D. W. DeWerth, V. Kam (A.G.A. Laboratories); and J. J. Crisafulli, A. L. Rutz, and S. G. Talbert (Battelle), "Development of Venting Guidelines for Category I Fan-Assisted Gas Appliances with Fan-Assisted Combustion Systems," GRI Venting Program, American Gas Association Laboratories, and Battelle Laboratories for Gas Research Institute, Gas Research Institute, Des Plaines, IL, 1992.

Connelly, S. M., and D. W. DeWerth (AGA Laboratories); and D. W. Locklin, D. D. Paul, and S. G. Talbert (Battelle), "Venting Requirements for High-Efficiency Gas-Fired Heating Equipment," American Gas Association Laboratories and Battelle Laboratories for Gas Research Institute, GRI Report No.83/0062 (NTIS No. PB84-159912), Gas Research Institute, Des Plaines, IL, 1983.

Crisafulli, J. J., R. D. Fischer, D. D. Paul, A. L. Rutz, S. G. Talbert, and G. R. Whitacre, "User's Manual for VENT-II (Version 3.0) with Diskette: A Dynamic Microcomputer Program for Analyzing Gas Venting Systems," GRI-88/0304, Gas Research Institute, Des Plaines, IL, 1988.

Crisafulli, J. J., R. D. Fischer, D. D. Paul, A. L. Rutz, S. G. Talbert, and G. R. Whitacre, "User's Manual for VENT-II (Version 4.0) with Diskette: A Dynamic Microcomputer Program for Analyzing Common Venting Systems for Two Gas Appliances," GRI-89/0094, Gas Research Institute, Des Plaines, IL, 1989.

Crisafulli, J. J., R. D. Fischer, D. D. Paul, A. L. Rutz, S. G. Talbert, and G. R. Whitacre, "User's Manual for VENT-II (Version 4.1) with Diskette: An Interactive Personal Computer Program for Design and Analysis of Venting Systems for One or Two Gas Appliances," GRI-90/0178 (NTIS No. PB91-509950), Gas Research Institute, Des Plaines, IL, 1990.

Cudnik, R. A., C. A. Farnsworth, D. W. Locklin, J. H. Payer, R. Razgaitis, G. H. Stickford, S. G. Talbert, and E. L. White, "Research on Heat-Exchanger Corrosion," Gas Research Institute for Battelle Laboratories, GRI-84/0157 (NTIS No. PB85-126316), Gas Research Institute, Des Plaines, IL, 1984.

NFPA 211, *Standard for Chimneys, Fireplaces, Vents, and Solid Fuel–Burning Appliances*, National Fire Protection Association, Quincy, MA, 2006 edition.

Paul, D. D., A. L. Rutz, and S. G. Talbert, "User's Manual for VENT-II: A Microcomputer Program for Designing Vent Systems for High-Efficiency Gas Appliances," Gas Research Institute Topical Report, GRI-85/0226 (NTIS No. PB86-20785), Gas Research Institute, Des Plaines, IL, 1985.

Paul, D. D., A. L. Rutz, and S. G. Talbert, "User's Manual for VENT-II (Version 2.0) with Diskette: A Microcomputer Program for Designing Vent Systems for Gas Appliances," GRI-88/0027, Gas Research Institute, Des Plaines, IL, 1988.

Philips, D. B., A. L. Rutz, S. G. Talbert, and G. R. Whitacre, "Venting Gas Appliances into Exterior Masonry Chimneys: Analysis and Recommendations," GRI-94/0193, Gas Research Institute, Des Plaines, IL, 1994.

Private communication with Richard Stone of Metalbestos Systems, Wallace Murray Corp., Belmont, CA, 1983.

Segeler, C. G., ed., *Gas Engineers Handbook*, 1st edition, New York: Industrial Press, 1965.

Talbert, S. G., "Evaluation of Masonry Chimney Materials for Venting Category I Gas Appliances," GRI-94/0338, Gas Research Institute, Des Plaines, IL, 1993.

UL 441, *Gas Vents*. Underwriters Laboratories Inc., Northbrook, IL, 1991.

U.S. Department of Energy, Title 10, Code of Federal Regulations, Part 430, "Energy Conservation Program for Consumer Products: Final Rule Regarding Test Procedures and Energy Conservation Standards for Water Heaters," Federal Register, Part III, 55(201): 42162-42177, 1990.

U.S. Department of Energy, *National Appliance Energy Conservation Act of 1987*. Public Law 100-12, U.S. Government Printing Office, Washington, DC, 1987.

U.S. Department of Energy and Brookhaven National Laboratory for Gas Research Institute, "1987 International Symposium on Condensing Heat Exchangers," Abstracts (GRI-87/0091.1) and Proceedings (GRI-87/0091.2 and /0091.3, BNL-52086), 1987.

## SUPPLEMENT 2

# Update on the Design and Installation Requirements for CSST Gas Piping Systems

**Robert Torbin**
Cutting Edge Solutions LLC

***Editor's Note:*** *Requirements for the corrugated stainless steel tubing (CSST) system of piping for fuel gases in buildings have been included in the National Fuel Gas Code since 1988, and this system continues to be popular with plumbers, mechanical contractors, and other installers of gas piping systems. CSST provides another piping option in addition to steel pipe and copper tubing that has advantages in many installations. This supplement was revised for the 2006 edition to update the user with current information on CSST.*

*Special thanks are extended to the Institute for Gas Technology, formerly Gas Research Institute, which funded the research program that identified CSST and sponsored demonstration projects.*

Corrugated stainless steel tubing (CSST) is commonly used throughout the United States to distribute natural and LP-Gases inside all types of residential and commercial buildings. The design, sizing, and installation of CSST are covered in NFPA 54, *National Fuel Gas Code,* which references ANSI/LC-1, *Standard for Fuel Gas Piping Systems Using Corrugated Stainless Steel Tubing*. In the 2009 edition of the *National Fuel Gas Code,* sizing tables have been updated and new installation practices for the electrical bonding of these systems have been added.

### TUBING TECHNOLOGY

CSST is commercially available in a variety of sizes ranging from 0.375 in. to 2 in. (9.5 mm to 50.8 mm) internal diameter. The tubing is packaged in long coils on spools, which facilitate the dispensing of the tubing, and they vary in length from 100 ft to 500 ft (30 m to 150 m) depending on the tubing diameter. Special mechanical fittings are provided to connect the tubing with other components within the overall gas distribution network. The tubing can be easily cut and field assembled with the mating fittings using standard hand tools. Although each manufacturer's fitting attaches to the tubing in a unique fashion, all fittings are required to incorporate standard pipe threads for connection to conventional plumbing components and appliances.

One of the unusual characteristics of CSST is the unique shape of each manufacturer's tubing profile. No two profiles are alike. Therefore, a special designation was developed to denote the flow capacity of each tubing size rather than use the internal diameter dimension. An Equivalent Hydraulic Diameter (EHD) designator is assigned to each new CSST product by the certifying agency in accordance with the requirements of the ANSI LC-1 standard. Table S2.1 summarizes the currently available CSST products listed by manufacturer and EHD designator. It is important to note that not every manufacturer makes a full range of sizes. The system designer must exercise care

*TABLE S2.1 Availability of CSST Product by Manufacturer*

| *EHD Value* | *13* | *15* | *18* | *19* | *23* | *25* | *30* | *31* | *37* | *39* | *46* | *48* | *60* | *62* |
|---|---|---|---|---|---|---|---|---|---|---|---|---|---|---|
| **Nominal Diameter (inch)** | $^3/_8$ | $^3/_8$ | $^1/_2$ | $^1/_2$ | $^3/_4$ | $^3/_4$ | 1 | 1 | $1^1/_4$ | $1^1/_4$ | $1^1/_2$ | $1^1/_2$ | 2 | 2 |
| **CSST Manufacturer** | | | | | | | | | | | | | | |
| Ward | | X | | X | | X | X | | X | | | X | | X |
| Gastite | X | | X | | X | | | X | X | | X | | X | |
| Omegaflex | | X | | X | | X | | X | X | | X | | | X |
| Parker | X | | X | | X | | | X | | X | | | | X |
| Tru-Flex | | X | X | | | X | | X | X | | | | | |
| Metal-Fab | | X | | X | | X | | X | X | | | | | |

when making a tubing selection using the standard sizing tables to ensure that the size chosen is available from the chosen supplier of the CSST product. If the selected size is not available, then the next higher value of EHD must be used. In addition, CSST is used for retrofit applications, especially for the installation of additional appliances to an existing steel pipe system. This application results in a piping system comprised of two different materials. In these situations, the line sizing for the new tubing must be chosen and the size for the old piping must be recalculated to ensure that it has adequate capacity to carry the new load without excessive pressure reduction.

The lack of a mandatory standard for the tubing shape has created a situation where CSST from different manufacturers are not identical even though they are of the same nominal size. Thus, a fitting from one manufacturer cannot be assembled on the CSST from another. In other words, the fittings are not interchangeable. This discrepancy has raised concerns regarding future modifications and/or repairs to existing CSST systems. These concerns have been addressed in two ways. Although not interchangeable, CSST fittings can be interconnected through standard pipe components, such as couplings, elbows, and tees, because each CSST fitting terminates in an NPT threaded nipple. Should it become necessary to either extend an existing CSST run or replace a damaged section, a CSST fitting from another manufacturer can be used in conjunction with the existing termination fitting that is already installed at the end of the run. The repair is completed using a malleable iron pipe coupling and the two fittings as shown in Exhibit S2.1. The installer sometimes has the option of abandoning the existing line (if practical) and installing a completely new run of CSST and fittings from another manufacturer. Starting from the same port on the manifold (by removing the old fitting), new tubing is installed in parallel with the abandoned appliance run and connected to the appliance with a new fitting. The installer must ensure that the new run of CSST is at least of equal or greater EHD value than the line being replaced or repaired.

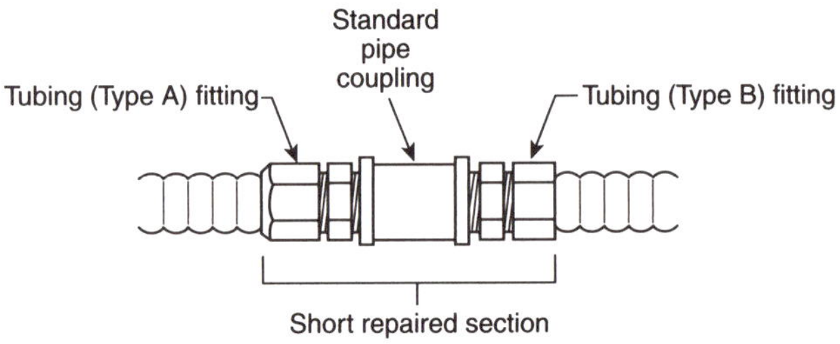


*EXHIBIT S2.1 Repair Method for Damaged CSST Tubing Run. (Courtesy of Foster-Miller, Inc.)*

## LIGHTNING AND CSST

The protection of structures and equipment from both direct and indirect lightning strikes is considered outside the scope of the various building codes, including the electrical, fuel gas, and plumbing codes. However, all gas piping materials and equipment are susceptible to damage from lightning strikes. Based on anecdotal data gathered from plumbing contractors and forensic investigations, it is known that lightning energy will cause pipe dope to flow inside threaded pipe joints, burn perforations through CSST and flexible connectors, and result in seal failures in gas meters, regulators, and equipment controls.

Corrugated stainless steel is more vulnerable to arcing damage than other gas piping materials because of its thinner wall thickness. NFPA 54 was revised in the 2009 edition to address this, based on a number of incidents that became apparent after a class action lawsuit was settled. Section 7.13, Electrical Bonding and Grounding, has been revised to require that all CSST systems be bonded to the electrical service ground where the gas service enters the building using at least a 6 AWG copper conductor or equivalent. This bond is intended to divert much of the lightning energy directly to earth before it can jump to another nearby electrically conductive pathway, such as copper water pipe or electrical wiring. Lowering the voltage level impressed on the CSST and minimizing the difference in electrical potential between conductive pathways will significantly reduce the magnitude and/or the occurrence of arcing.

CSST installation instructions have been revised to include similar requirements. These new practices require the installation of bonding clamps and a bonding conductor at the service entrance between the gas piping and the grounding electrode, grounding electrode conductor, or the service panel enclosure. NFPA 54 requires that CSST be installed in accordance with the manufacturer's installation instructions in 7.2.8.

## BONDING OF GAS PIPING

Effective grounding of the electrical system depends on a low-resistance electrical pathway to the earth. The pathway to ground is affected by the resistance within the electrical system, the grounding electrode(s), and the soil conditions. While *NFPA 70®*, *National Electrical Code®* (*NEC®*), does not require a minimum resistance for the connection of a grounding electrode system that consists of metallic water pipes or concrete-encased electrodes, it does require that the electrical resistance of the connection to ground for ground-rod and plate-type electrodes must be 25 ohms or less. If this resistance is not achieved, an additional grounding electrode must be installed. In addition, Section 250.4(A) (1) of the *NEC* states:

> **(1) Electrical System Grounding.** Electrical systems that are grounded shall be connected to earth in a manner that will limit the voltage imposed by lightning, line surges, or unintentional contact with higher-voltage lines and that will stabilize the voltage to earth during normal operation.

Electrical safety is enhanced if all metallic piping systems, including CSST, are bonded equipotentially to the grounded electrical system so that voltage differences are minimized between these systems. If the piping is inadequately bonded or not bonded at all, gas piping of any type or material is more vulnerable to damage from indirect lightning strikes. Where installed in or attached to a building or structure, metal piping systems (including gas piping) that are likely to become energized are required to be bonded to the supply side of the electrical system in accordance with Article 250 of the *NEC*. Article 250 covers the bonding of different types of electrically conductive metallic systems, including metal water pipe, other metal pipe, including gas piping systems, and structural steel. Section 250.104(B), Other Metal Piping, provides the *NEC* requirements for bonding of gas piping systems.

Traditionally, the bonding conductor is sized in accordance with *NEC* Table 250.122 using the rating of the branch circuit that can energize the piping. Although the typical bonding conductor is of sufficient size to address ground fault conditions, this same conductor has not proven to be of adequate capacity to accommodate the frequency and voltage surge levels associated with lightning energy. The new requirement for bonding of CSST specifies that the bonding conductor be at least a 6 AWG copper conductor or equivalent. A 6 AWG copper conductor is often used as a bonding jumper for other metallic systems.

Direct bonding of metallic structures, such as piping systems, to the grounding electrode system on the power supply side provides a low-resistance path back to earth. This connection is intended to keep each metallic system at or near earth potential (ideally near 0 volts). This type of bonding for gas piping exceeds the requirements of the *NEC*. While the *NEC* requirements intend to maximize life safety protection, the additional bonding is intended to also minimize the impact of indirect lightning strikes on the CSST. This added protection will reduce the likelihood of arc-induced damage of the CSST by significantly reducing the difference in electrical potential between the gas piping, other metallic systems, and the electrical grounding system.

All of the existing CSST manufacturers have revised their installation instructions regarding the bonding of the CSST system to the electrical grounding system. Direct bonding of the CSST gas piping system is required for all new installations regardless of the type of connected gas appliance. The gas piping system is required to be bonded to the electrical earth grounding system of the structure through the use of a bonding clamp and wire. To keep it simple yet cost-effective, the size of the bonding conductor should be at least a 6 AWG copper wire (or equivalent) for most residential applications. However, commercial buildings may require larger size bonding conductors and additional points of attachment. Refer to the *NEC* for specific requirements. In residential applications, the bonding connection must be made to one of the following locations: the service equipment enclosure, the grounded conductor at the service, the grounding electrode conductor (if of sufficient size), or one of the grounding electrodes.

The bonding clamp must be a listed electrical fitting, and it must be attached to the gas piping system on a segment of rigid pipe, as shown on Exhibit S2.2, or a pipe component, such as a coupling. The clamp must always be located on the building side of the gas meter, as shown in Exhibit S2.3, near the entrance to the structure. The bonding connection for propane systems without meters must

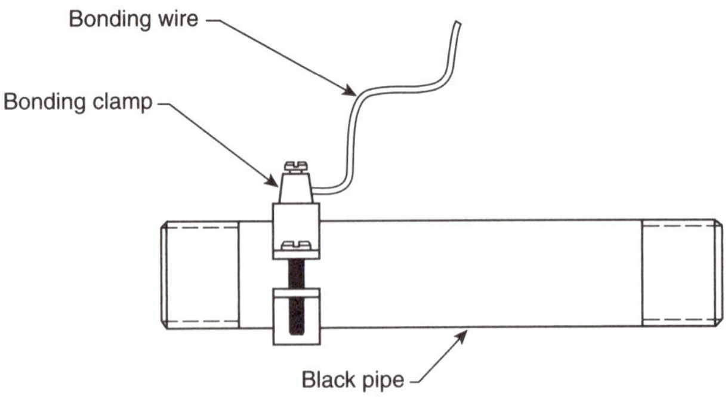


***EXHIBIT S2.2*** *Attachment of Clamp to Steel Pipe. (Courtesy of Cutting Edge Solutions, LLC)*

***EXHIBIT S2.3*** *Location of Bonding Clamp. (Courtesy of Cutting Edge Solutions, LLC)*

be made downstream of the second-stage pressure regulator at the entrance to the building. The clamp can be installed either indoors or outdoors and the bonding jumper must be protected in accordance with the *NEC*, based on its size. The clamp must make metal-to-metal contact with the pipe or component. The corrugated stainless steel tubing portion of the gas piping system must not be used as the point of attachment of the bonding clamp at any location along its length under any circumstances. The points of attachment of the bonding clamp and conductor must be accessible and must remain accessible after construction is completed. The length of the bonding conductor should be as short as practical, consistent within the physical layout and constraints of the installation.

Depending on site conditions specific to the location of the structure, including but not limited to whether the area is prone to frequent lightning strikes, the builder or owner of any structure has the option of installing a lightning protection system. The design of lightning protection systems are covered by NFPA 780, *Standard for the Installation of Lightning Protection Systems*. Bonding required by the lightning protection system standard must be connected to the gas piping system on the building side of the gas meter using one of the connection methods permitted for gas piping. The bonding of the lightning protection system is in addition to the bonding requirements for the CSST piping system.

## REFERENCES

American National Standard, *Standard for Fuel Gas Piping Systems Using Corrugated Stainless Steel Tubing (CSST)*, ANSI LC-1-2005/CSA 6.26-2005.

*NFPA 70®*, *National Electrical Code®*, 2008 edition, National Fire Protection Association, Quincy, MA.

NFPA 780, *Standard for the Installation of Lightning Protection Systems*, 2008 edition, National Fire Protection Association, Quincy, MA.

### Additional Readings

*National Electrical Code® Handbook*, National Fire Protection Association, Quincy, MA, 2008.

*Effectiveness of Direct Bonding of Gas Piping in Mitigating Damage from an Indirect Lightning Flash*, Final Technical Report, Brian Kraft and Robert Torbin, August 2007.

# SUPPLEMENT 3

# Carbon Monoxide

***Editor's Note:*** *Supplement 3 provides technical information on carbon monoxide to code users. This supplement consists of four parts:*

- *Part I, Carbon Monoxide (CO) Information and Data*
- *Part II, Guidelines for Fire and Other Emergency Response Personnel*
- *Part III, Sidewall Vented Appliances*
- *Part IV, Carbon Monoxide Incident Investigation*

*This supplement is a compilation of several existing documents. Jennifer Flynn, NFPA Research Analyst, provided the first part, which references her report* Non-Fire Carbon Monoxide Incidents Reported in 2005 *as well as stats from the CPSC and CDC. The second part is extracted from* Responding to Residential Carbon Monoxide Incidents, *published by the U.S. Consumer Product Safety Commission. (This report is updated periodically, and readers can obtain subsequent issues from NFPA.) The third part of this supplement is provided to inform code users of the complex interaction of factors that are evaluated during the investigation of a carbon monoxide incident. Thomas Crane of Crane Engineering provided the fourth part.*

*The editor gratefully acknowledges all the contributors to this supplement.*

## PART I: CARBON MONOXIDE (CO) INFORMATION AND DATA

Carbon monoxide is a colorless, odorless, poisonous gas that results from the incomplete burning of "common" fuels such as natural or liquefied petroleum gas (LP-Gas), oil, wood, or coal. When CO is inhaled, it enters the bloodstream and reduces the ability of the blood to carry oxygen to vital organs, such as the heart and the brain.

Fire departments and gas suppliers are often called to assist when people suspect that carbon monoxide might be present. Some of these calls are reported after a CO alarm sounds. Some may be in response to symptoms of CO poisoning.

In 2005, U.S. fire departments responded to an estimated 61,100 non-fire carbon monoxide incidents in which carbon monoxide was found, or an average of seven such calls per hour. The number of incidents increased 18 percent from 51,700 incidents reported in 2003 [1].

In the home, heating and cooking equipment that burn natural gas, propane, and other fuels are potential sources of carbon monoxide, which is why installation codes and standards for fueled equipment emphasize arrangements for adjusting venting. Vehicles or generators running in an attached garage can also produce dangerous levels of carbon monoxide.

For example, in March 2004, a carbon monoxide leak from a gas furnace left two families suffering the effects of carbon monoxide poisoning. The leak began in one family's basement, spread throughout the house and into a neighbor's home. The leak resulted in the carbon monoxide poisoning and death of a 14-year-old girl, who lived in the home where the leak originated. Another seven individuals were treated for CO poisoning, including two women, ages 40 and 19 years old, and two males, ages 20 and 4 years old, who lived with the 14-year-old girl, and a 24-year-old woman and two boys, ages 3 years old and 1 month old, who lived next door. These seven victims were treated for carbon monoxide poisoning and survived the incident [2].

In 2005, three children were taken to the hospital and treated for carbon monoxide poisoning when exhaust fumes from a power generator leaked into the house through a crack in a door for several hours. The children included a 7-year-old boy and two girls, ages 4 and 11 [3]. According to the Consumer Product Safety Commission, the amount of exhaust from one generator is equivalent to that of hundreds of idling cars [4]. The three children were lucky to have been removed from the area and treated early enough that they did not die.

## Carbon Monoxide Facts

The following are some basic facts about carbon monoxide incidents:

- In 2005, January and December were the peak months for non-fire carbon monoxide incidents in which CO was found.
- The peak time of day for reported carbon monoxide incidents is between 6:00 p.m. and 9:59 p.m.
- Overall, 75 percent of non-fire carbon monoxide incidents are reported between the hours of 9:00 a.m. and 10:59 p.m.
- Almost 9 out of every 10 (89 percent) reported non-fire carbon monoxide incidents took place in the home. In contrast, homes accounted for 75 percent of the structure fires reported that year [5]. (Homes include one- or two-family dwellings, manufactured homes, and multifamily dwellings, including apartments, condos, town houses, row houses, and tenements.)
- In 2003, there were an estimated 60,600 unintentional carbon monoxide detector activations in which carbon monoxide was not detected; this includes CO detector malfunctions and false alarms. (Due to the increasingly large size of the national database, false alarms and false calls were not included in the publicly released NFIRS data for 2004 and 2005.)
- In 2003, 46 percent of all CO-related non-fire calls reported to fire departments were carbon monoxide incidents in which carbon monoxide was found. Fifty-four percent of all CO-related non-fire calls reported to fire departments were false alarms, or no CO was found.

## Dangers of CO Exposure

The dangers of CO exposure depend on a number of variables, including the victim's health and activity level. Infants, pregnant women, and people with physical conditions that limit their body's ability to use oxygen (i.e., emphysema, asthma, heart disease) can be more severely affected by lower concentrations of CO than healthy adults would be.

Carbon monoxide poisoning can be confused with flu symptoms, food poisoning, and other illnesses. Some symptoms include shortness of breath, nausea, dizziness, lightheadedness, or headaches. High levels of CO can be fatal, causing death within minutes. Although most CO poisoning happens during a single incident, it is possible to suffer from chronic CO poisoning when a person is exposed to low levels of CO over weeks or months and experiences symptoms over that time.

Even single-incident CO exposures can have long-term health consequences, such as cardiovascular manifestations, neurological dysfunction, or brain damage, which can occur days to weeks after the exposure. In 14 percent to 40 percent of serious initial CO poisonings, the individual exposed still faces the prospect of delayed neurologic dysfunction [6].

According to the Centers for Disease Control and Prevention [7], an estimated 15,200 people were treated annually during 2001–2003 in emergency departments for nonfatal, unintentional, non-fire-related carbon monoxide exposure. During 2001–2002, the CDC estimates 480 people died, on average, each year from unintentional non-fire-related carbon monoxide exposure.

The Consumer Product Safety Commission (CPSC) is a federal regulatory agency that works to reduce the risk of injuries and deaths that result specifically from consumer products. According to the CPSC, 166 unintentional non-fire carbon monoxide poisoning deaths were associated with consumer products on average, annually from 2002–2004 [8]. This represents a 34 percent increase in unintentional non-fire carbon monoxide poisoning deaths from 1999–2000, in which an average of 124 deaths were reported annually [9].

Of the CO non-fire deaths related to consumer products in 2003 and 2004, 47 percent were associated with the use of heating systems, most often gas heating systems. Another 35 percent of the CO deaths were associated with engine-driven tools. The CPSC reports that 73 percent of CO deaths occurred in the home, while deaths in tents, campers, and other temporary shelters accounted for an estimated 14 percent of deaths during 2003–2004.

The CPSC examined carbon monoxide incidents associated specifically with engine-driven generators and other engine-driven tools that occurred between 1990 and 2004. During these 15 years, generators resulted in 264 CO potential exposure deaths and were the leading engine-driven consumer products involved in CO exposure incidents [10].

## NFPA 720 and State Legislation

NFPA 720, *Standard for the Installation of Carbon Monoxide (CO) Detection and Warning Equipment,* is the standard for the installation of carbon monoxide (CO) warning equipment in dwelling units [11]. NFPA 720 covers the selection, application, installation, location, testing, and maintenance of carbon monoxide warning equipment within dwelling units that contain fuel-burning appliances or fireplaces, or have attached garages. Beginning with the 2009 edition, the standard also covers nonresidential buildings and structures. The purpose of the standard is to provide a warning of the presence of CO in sufficient time to allow occupants to either escape or take other appropriate action.

NFPA 720 requires a carbon monoxide alarm or detector centrally located outside each separate sleeping area in the immediate vicinity of the bedrooms. Each alarm or detector should be located on the wall, ceiling, or other location as specified in the installation instructions that accompany the unit.

As of 2007, 15 states and more than 40 other authorities having jurisdiction require some form of carbon monoxide detection. Texas, for example, only requires CO detectors in day care and group homes, while Oklahoma only requires detectors in child care facilities. Many states only require carbon monoxide detectors in newly constructed buildings. Of the 15 states that have adopted requirements for carbon monoxide detection, ten require the installation of carbon monoxide detectors in occupancies not covered by the standard, which only applies to dwelling units. The 15 states currently requiring some form of carbon monoxide detection include the following:

- Alaska
- Connecticut
- Florida
- Illinois
- Maryland
- Massachusetts
- Minnesota
- New Jersey
- New York
- Oklahoma
- Rhode Island
- Texas
- Utah
- Vermont
- West Virginia

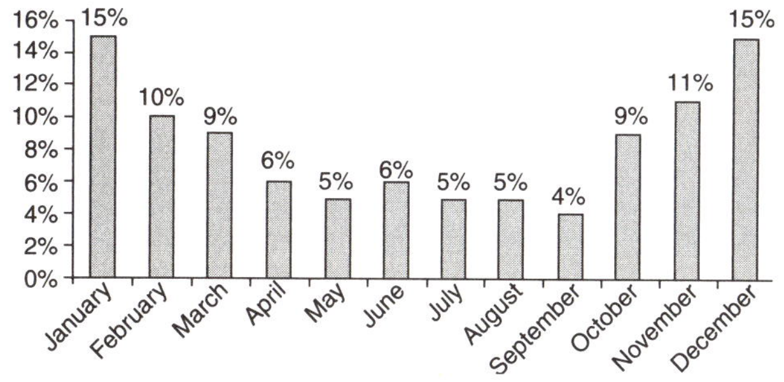


**EXHIBIT S3.1** *Non-Fire Carbon Monoxide Incidents in Which Carbon Monoxide Was Found, Reported by Responding Fire Departments in 2005, by Month. (Source: NFIRS and NFPA Survey)*

See Exhibits S3.1 and S3.2 and Table S3.1. Note that NFPA's national estimates are based on data from the U.S. Fire Administration's (USFA) National Fire Incident Reporting System (NFIRS) and NFPA's sample-based fire department survey. The NFPA survey collects summary data about U.S. fire department activities. Because NFIRS is not a census, the survey results are used to scale up the details from NFIRS into national estimates.

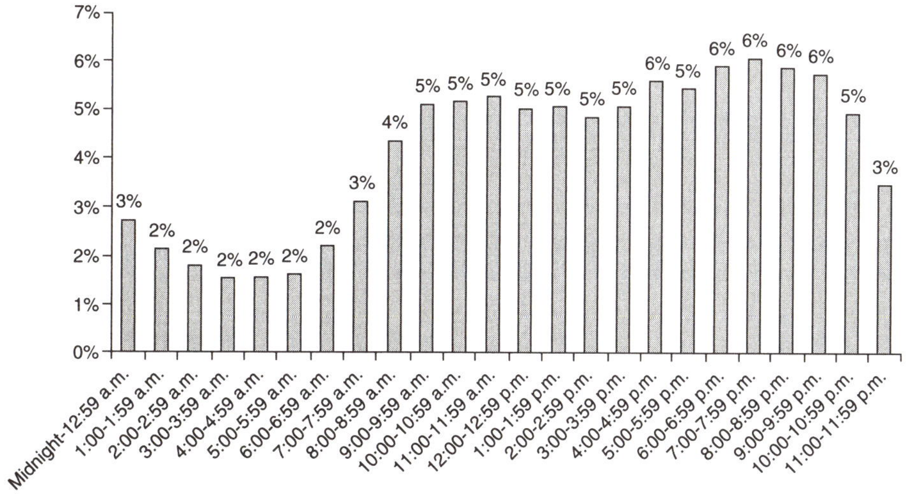


**EXHIBIT S3.2** *Non-Fire Carbon Monoxide Incidents in Which Carbon Monoxide Was Found, Reported by Responding Fire Departments in 2005, by Alarm Hour. (Source: NFIRS and NFPA Survey)*

*TABLE S3.1 Non-Fire Unintentional-Injury Deaths Due to Poisoning by Gases and Vapors*

| Year | Total | Gas from Pipeline | Motor Vehicle Exhaust Gas | Other Utility Gas or Carbon Monoxide | Other | Estimated Carbon Monoxide Poisoning Involving Consumer Products |
|---|---|---|---|---|---|---|
| 1980 | 1,242 | — | — | — | — | 282 |
| 1981 | 1,280 | — | — | — | — | 311 |
| 1982 | 1,259 | 72 | 596 | 426 | 165 | 340 |
| 1983 | 1,251 | 82 | 580 | 414 | 175 | 323 |
| 1984 | 1,103 | 48 | 511 | 354 | 190 | 275 |
| 1985 | 1,079 | 49 | 488 | 392 | 150 | 284 |
| 1986 | 1,009 | 29 | 475 | 341 | 164 | 240 |
| 1987 | 900 | 53 | 402 | 288 | 157 | 232 |
| 1988 | 873 | 36 | 372 | 314 | 151 | 238 |
| 1989 | 921 | 48 | 355 | 353 | 165 | 296 |
| 1990 | 748 | 33 | 293 | 289 | 133 | 243 |
| 1991 | 736 | 20 | 278 | 316 | 122 | 250 |
| 1992 | 633 | 21 | 223 | 281 | 108 | 211 |
| 1993 | 660 | 14 | 245 | 290 | 111 | 214 |
| 1994 | 685 | 24 | 246 | 307 | 108 | 223 |
| 1995 | 611 | 27 | 234 | 272 | 78 | 201 |
| 1996 | 638 | 23 | 219 | 283 | 113 | 217 |
| 1997 | 576 | 13 | 208 | 251 | 104 | 180 |
| 1998 | 546 | 15 | 190 | 254 | 87 | 180 |
| 1999 | 597 | — | — | — | — | 109 |
| 2000 | 631 | — | — | — | — | 138 |
| 2001 | 656 | — | — | — | — | 130 |
| 2002 | 691 | — | — | — | — | 188 |
| 2003 | 690 | — | — | — | — | 154 |
| 2004 | 629 | — | — | — | — | 162 |

Note: Detailed breakdowns were not provided prior to 1982 or after 1998. The last column has estimates from the U.S. Consumer Product Safety Commission.

*Source:* National Safety Council, *Accident Facts* and *Injury Facts*, 1981–2007 editions, 1121 Spring Lake Drive, Itasca, IL 60143; http://www.cdc.gov/nchs; Matthew V. Hnatov, *Non-fire carbon monoxide deaths associated with the use of consumer products*, U.S. Consumer Product Safety Commission, Bethesda, MD, August 8, 2007, and previous reports in the series; http://www.nchs.gov. The national death certificate database, maintained by the U.S. National Center for Health Statistics (NCHS), is coded according to the International Classification of Diseases (ICD), which was substantially revised in 1999. These modifications changed the categories that can be used to analyze trends in deaths involving gases.

## PART II: GUIDELINES FOR FIRE AND OTHER EMERGENCY RESPONSE PERSONNEL

*This guide from the U.S. Consumer Product Safety Commission is intended to help public emergency response personnel act quickly and accurately to a call that may involve carbon monoxide (CO). Such a response can reduce injuries and save lives.*

CO is associated with about 500 unintentional non-fire-related deaths each year. About 60 percent of these deaths are from motor vehicle exhaust, and about 40 percent are associated with consumer products. CO in its pure form is colorless and odorless; however, other gases that do have an odor often accompany it.

Whether you're part of a fire department or a public emergency response organization, you'll find information here that will help you accomplish the following:

- Determine how much assistance occupants need.
- Ensure your own protection from CO when entering a home.
- Confirm whether a CO hazard exists.
- Make a preliminary assessment of the potential sources of elevated CO.

- Give advice to the resident on how to make the home safe for re-entry.

To use this guide most effectively, follow these steps and refer to the following exhibits later in this supplement:

1. Read the section "What to Know Before You Respond to a CO Incident Call."
2. Become familiar with the guidelines for dispatchers (Exhibit S3.3).
3. Follow the steps in the CO incident reporting form (Exhibit S3.4) and the "CO Measurement Form" (Exhibit S3.6), where you record the readings from your CO meter/measuring equipment.
4. Read the section "CO Levels: Advice to Give, Actions to Take."
5. Fill in the form "Advice for Residents" (Exhibit S3.7), and give a copy of this to occupants.

As you use this guide, feel free to adopt all or part of it for your particular organization or community. If you have suggestions for future editions, contact the U.S. Consumer Product Safety Commission by calling 1-800-638-2772 or visiting www.cpsc.gov.

## Agency Disclaimer

This guide has been developed for use by public emergency response organizations such as professional and volunteer fire departments, emergency community service units, and other organizations delegated to respond to carbon monoxide (CO) incident calls; it is not directed toward non-emergency organizations that might receive CO incident calls.

These guidelines consist of procedures that emergency response personnel can use to help residents who call and report a suspected CO incident. The procedures are designed to help responders provide for their own safety when answering a call, determine the level of care needed by the residents, make a preliminary assessment of the CO condition in the residence, and determine when it is safe for the occupants to re-enter the home. These procedures can help in finding a *significant* source of CO in the home; in some circumstances, it is possible that other professionals with technical expertise will need to be called to find the cause of hard-to-trace elevated CO levels in the home.

These guidelines assume that the responder has a basic level of familiarity with CO, including how it is generated, how it can become abnormally elevated, how it is distributed through a home, and what the exposure symptoms are. References for background material are cited.

Finally, these guidelines are not a CPSC standard and are not mandatory requirements. CPSC believes that this document contains the basic information needed to provide a first emergency response to consumers. State and local fire departments and emergency response organizations can choose to adopt all or part of this material to meet their own needs and resources.

## What You Need to Know Before You Respond to a CO Incident Call

- Emergency response procedures for CO incidents are most effective when coordinated with local medical personnel, poison control centers, utilities, and heating and ventilating contractors.
- For your own protection, it is essential to establish a "Safe Entry Procedure" when entering a home where there is a reported CO incident.
- It is important to know what the potential sources of elevated CO are in a home. A list of these sources and a list of background references about CO are provided elsewhere in this document.
- It is important to be familiar with the various types of residential CO alarms and how they work, including the differences among alarm, low-battery, malfunction, and error and warning signals. Manufacturers' product literature provides this information.
- CO incident calls may be made when a CO alarm has sounded or because the occupants suspect possible CO exposure, because of illness or strange odors. A CO alarm signal should never be ignored. CO alarm signals require a different response than smoke alarms. CO alarms are designed to activate at exposures below those at which symptoms occur, while occupants still have time to take action to protect themselves. However, CO levels can build up quickly, so the CO alarm might be responding to a high level. Immediate evacuation is necessary when a CO alarm sounds. CO alarms should not alarm below 30 ppm.
- There have been instances where CO alarms activated and neither a CO elevation nor source could be identified. Changes in the voluntary standard for CO alarms, effective October 1, 1998, have reduced the occurrence of unexplained alarms. If a resident has an alarm manufactured before October 1, 1998, and a CO elevation or CO source cannot be located, the resident should be advised to purchase a new CO alarm. The new alarm should meet one of the following standards:

  Canadian Standards Association (CSA) 6.19-01, 2001

  Underwriters Laboratories Inc. (UL) 2034, Second Edition, October 1998

  International Approval Services (IAS) 6-96, Second Edition, June 1, 1998

- Professional CO detection equipment, and training in its use, is essential in finding the source of elevated CO. Use, maintenance, and periodic calibration procedures should be established. Manufacturers of such equipment generally provide recommendations and instructions for maintenance and re-calibration; follow these instructions carefully.
- In most cases where a CO incident is reported, a professional CO monitor that measures room CO concentrations will help you assess the immediate risk. Finding the reason for the CO alarm can be time-consuming and difficult, particularly if the elevated CO levels are the result of a transient condition such as down drafting from an exhaust flue. Heating, ventilating, and air conditioning (HVAC) contractors may be needed to assess conditions in the residence and provide advice to the resident.

## Guidelines for Dispatchers

Dispatchers are the first contact when residents call for help about a CO incident. A dispatcher can use the guidelines in Exhibit S3.3 to decide what kind of help is needed and what type of equipment should be sent to the residence.

## CO Incident Reporting

The CO measurement procedures can help identify the cause of the resident's request for help and can also help determine when it is safe for a resident to re-enter the home.

When you arrive on the scene to investigate a call about CO, follow the steps listed in Exhibit S3.4. Plan to make multiple copies of this form, so you can fill out separate forms for each separate investigation. Be sure to keep a copy of each completed investigation form.

It is important that residents who show symptoms of CO poisoning be examined on the scene by EMS or appropriately trained personnel. If medical treatment is indicated, the residents should be taken to the appropriate medical facility. Residents should not be allowed to drive themselves to a doctor's office or a hospital. Residents should not go back inside until you have inspected the conditions inside the home and authorized re-entry.

Exhibit S3.5 shows sources of and clues to a possible CO problem. Potential sources of CO include automobiles; motorcycles; trucks; golf carts; RVs; gasoline-, propane-, or diesel-fueled appliances; lawn mowers; power generators; furnaces; water heaters; clothes dryers; natural gas or propane refrigerators; ranges; ovens; space heaters; fireplaces; gas logs; wood and coal stoves; charcoal or gas grills; kerosene heaters; wood stoves; and any other equipment or appliance that burns fuel.

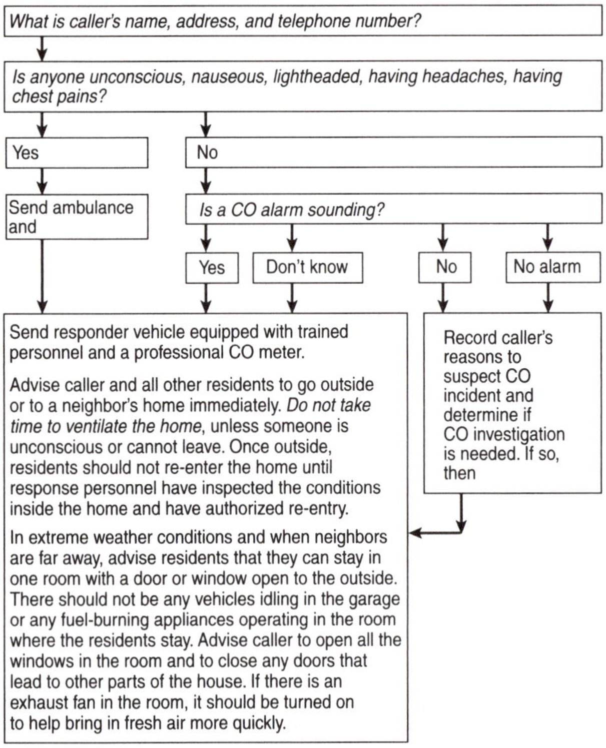


***EXHIBIT S3.3*** *Guidelines for Dispatcher Response.*

These potential sources can be improperly installed, used, or maintained, leading to elevated levels of CO. Improper home remodeling can cause CO problems by interfering with combustion air supplies, for example. It is important to note whether you see any indication of this. For example, gasoline generators improperly used indoors and gas ovens improperly used for heating or lined tightly on the bottom with aluminum foil can increase CO emissions and lead to CO buildup in homes.

When taking indoor CO measurements, you should take one set of CO measurements in areas or rooms of the house where there are potential sources of elevated CO that were in operation at the time of the call or were in use in the 24 hours before the call. Record your measurements on the CO measurement form, shown in Exhibit S3.6.

To complete the entire investigation, you will need to take one set of measurements. These measurements will help you assess whether there are elevated levels of CO in the home and what the sources(s) of those elevated levels might be.

## CO Levels: Advice to Give, Actions to Take

During the investigation process, residents should be evacuated from the home. After the investigation is

## CO INCIDENT REPORTING FORM

### STEP 1: Determine the Level of Care Needed by Household Members

***Part A.*** *If residents are inside the house, go to Part B below. If residents are outside the house, find out the following information:*

Are any members of the household feeling ill? _____ Yes _____ No

Which symptoms are they experiencing?

_____ Headache _____ Nausea _____ Dizziness _____ Confusion _____ Shortness of Breath _____ Chest Pains

_____ Other (describe): ________________________________

*Note: Cardiac patients are the most susceptible to CO poisoning and may experience chest pains.*

*Note: The symptoms of moderate CO poisoning are similar to those of the flu, and CO poisoning can be misdiagnosed.*

***Part B.*** *If residents are inside the house, follow local procedures for safe entry before determining the condition of the residents. If no local safe entry procedures for a CO incident exist, take the following steps:*

1. Have self-contained breathing apparatus (SCBA) available and ready for use. Follow your jurisdiction's guidelines for using this equipment. If there is any doubt about CO levels, use SCBA.
2. Using your CO meter and following the manufacturer's instructions, take a CO reading outside. Make sure you are standing away from any vehicle exhaust or other source of CO.
3. Record the outside reading on the CO measurement form (Exhibit S3.4). This will be the baseline reading.
4. Before entering the home (standing in the doorway), take a second CO reading.
5. Record this reading on the CO measurement form.
6. Make sure you continue to be properly protected until safe CO levels are attained. Leave windows and doors open or closed — just as you found them — until you begin Step 3 of this investigation.
7. Take all residents outside. Occupants should remain outside until you have inspected conditions inside the home and authorized re-entry.
8. Determine the medical condition of all of the residents by following Part A above. This should be done by appropriately trained EMS/fire fighter personnel.

### STEP 2: Identify Potential Sources of Elevated CO in the Home and Attached Garage

***Part A.*** *List the potential sources of elevated CO in the home that were in operation at the time of the call or were in use in the 24 hours before the call. Note their location in the home, and whether they are vented to the outside of the home.*

| Potential Source | Location | Vented or Unvented | |
|---|---|---|---|
| *Example: Kerosene Heater* | *Living Room* | *X* | |
| | | | |
| | | | |
| | | | |
| | | | |
| | | | |

***Part B.*** *Examine flues, vents, and chimneys for blockage by birds' or other animals' nests or by debris, such as chipped or cracked masonry. Note any loose or disconnected vent or chimney connections, any loose or missing furnace panels, and any debris or soot in the chimney. Advise occupants to have flues, vents, and chimneys professionally serviced if problems are indicated.*

________________________________

________________________________

________________________________

(NFPA 54HB, 1 of 2)

***EXHIBIT S3.4*** *Procedure for CO Incident Investigations.*

*(continues)*

***Part C.*** *If there is an attached garage, answer the following questioins:*

Were any vehicles in the garage in the last 12 hours? ______ Yes ______ No

Were any vehicles' engines on in the last 12 hours? ______ Yes ______ No

**STEP 3: Take Indoor CO Measurements**

***Part A.*** *Take indoor measurements by completing the following steps:*

1. If you have not already taken (and recorded) a CO reading outside and at the entry to the home, then return to Step 1, Part B, 1 through 5, and follow those instructions. After taking indoor readings, you will need to subtract the outdoor reading from the indoor reading in order to determine the contribution that indoor sources make to any measured CO indoors. For example, if the outdoor reading is 25 ppm and the indoor reading is 40 ppm, then the contribution from indoor sources is 15 ppm.
2. Close all windows and doors. Turn on the fuel-burning appliances and other CO sources that have been in operation in the past 24 hours and let them reach operating temperature (about 15 minutes). Keep them on during the measurements. Be sure to put pots with water in them on all range-top burners that have been used in the past 24 hours because cold pots on the burner can elevate CO readings.
3. Measure and record indoor CO levels throughout the house. *Important*: Take measurements in the ambient air in the center of each room, standing back approximately 5 feet from any potential CO source. Do *not* take measurements in or near vents or flue pipes.
4. Check for proper draft in a natural draft furnace by using a smoke test. Hold a lit match near the draft hood of each of the appliances that you turned on. Blow the match out. The smoke will be drawn into the hood if there is proper draft.
5. When you have finished taking all measurements and if you measured elevated CO levels, then open all doors and windows of the residence so that the home will be ventilated.

**STEP 4: Evaluate Information**

A. Consider all of the information gathered in the investigation. Review the CO measurements you recorded.
B. Consider whether a vehicle was left idling near the home or in an attached garage (even with the garage door open).
C. Consider whether a portable propane, diesel, or gasoline-powered appliance was in use.
D. Consider whether any ventilation equipment or exhaust fans were used in the 24 hours before the call.

**STEP 5: Complete the Investigation**

Gather the following information:

Resident's Name ______________________________

Address ______________________________

Phone Number ______________________________

Incident Location (if different from address)

______________________________

Type of Residence: ______ Apartment ______ Single-family ______ Townhouse ______ Duplex

______ Commercial establishment ______ Other (describe): ______________________________

Today's Date ______________________________

Date/Time of Call ______________________________

Date/Time Alarm Sounded (if applicable) ______________________________

**STEP 6: Advice and Follow-up**

A. Read the section, "CO Levels: Advice to Give, Actions to Take," elsewhere in this document.
B. Give residents written written advice based on this section and on the information you gathered in your investigation. Use the form in Exhibit S3.5.
C. Record the advice you gave to the resident:

______________________________

______________________________

______________________________

______________________________

D. Tell occupants that regular maintenance of fuel-burning appliances is the first line of defense against CO poisoning.
E. Keep a copy of this report (including the CO measurement form in Exhibit S3.4) for your files.

Your Name (Print): ______________________________

Your Signature: ______________________________

(NFPA 54HB, 2 of 2)

***EXHIBIT S3.4*** *Continued*

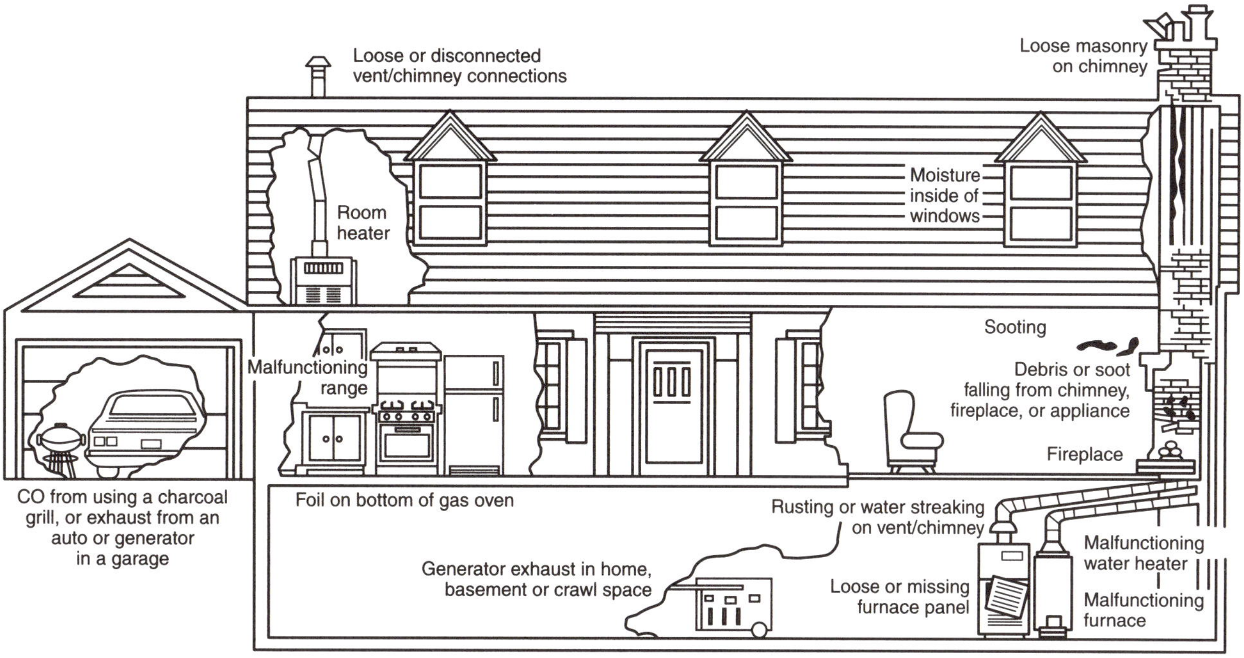


**Carbon monoxide clues you can see...**

- Rusting or water streaking on vent/chimney
- Loose or missing furnace panel
- Sooting
- Debris or soot falling from chimney, fireplace, or appliances
- Loose or disconnected vent/chimney, fireplace, or appliance
- Loose masonry on chimney
- Moisture inside of windows

**Carbon monoxide clues you cannot see...**

- Internal appliance damage or malfunctioning components
- Improper burner adjustments
- Hidden blockage or damage in chimneys

Only a trained service technician can detect hidden problems and correct these conditions!

- CO poisoning symptoms have been experienced when you are home, but they lessen or disappear when you are away from home.

**Warnings...**

- Never leave a car running in a garage even with the garage door open.
- Never run a generator in the home, garage, or crawl space. Opening doors and windows or using fans will NOT prevent CO buildup in the home. When running a generator outdoors, keep it away from open windows and doors.
- Never burn charcoal in homes, tents, vehicles, or garages.
- Never install or service combustion appliances without proper knowledge, skills, and tools.
- Never use a gas range, oven, or dryer for heating.
- Never put foil on the bottom of a gas oven, because it interferes with combustion.
- Never operate an unvented gas-burning appliance in a closed room or in a room in which you are sleeping.

***EXHIBIT S3.5*** *Sources of and Clues to a Possible CO Problem.*

complete, it is time to assess the information you have and provide advice to the residents, including whether they can safely re-enter the home. The final decision on re-entry may need to be made by someone other than the first responder.

The CO levels you measured in various rooms of the residence should have been taken in the center of the room in ambient air, making sure you were about 5 ft away from any appliance. Based on those measurements and your evaluation of the information you gathered in your investigation, identify any areas of the house with measured CO levels of (1) 70 ppm or higher, (2) between 30 ppm and 70 ppm, or (3) less than 30 ppm. Then use the guidelines in this section to give advice to residents using the form shown in Exhibit S3.7.

**CO Levels: 70 ppm or Higher.** *If the source of these CO levels appears to be a vehicle or an appliance, such as a lawn mower or generator, in a garage*

- Turn off the engine if it is still on.
- If you suspect that an appliance or a generator or a vehicle engine that was idling in a garage several hours earlier is the cause of elevated CO levels, advise the consumer that engines should not be left idling in an attached garage, even with the garage door open.

## CO MEASUREMENT FORM

Record levels of CO found throughout the home. Take measurements in the center of each room under the conditions specified in the CO incident reporting form (Exhibit S3.2).

***Measurements (CO ppm):***

Outdoors (baseline) ____________________ Entrance (doorway) ____________________

Measurements (ppm) with recently used potential CO sources turned on and doors and windows closed:

| | | | |
|---|---|---|---|
| Attached Garage | ________ | Basement | ________ |
| Utility Room | ________ | Recreation Room | ________ |
| Den | ________ | Living Room | ________ |
| Dining Room | ________ | Kitchen | ________ |
| Family Room | ________ | Powder Room | ________ |
| Bedroom 1 | ________ | Bathroom 1 | ________ |
| Bedroom 2 | ________ | Bathroom 2 | ________ |
| Bedroom 3 | ________ | Bathroom 3 | ________ |
| Bedroom 4 | ________ | Bathroom 4 | ________ |
| Bedroom 5 | ________ | | |
| Other Rooms | ________ | | |

Draw a diagram of the home showing locations of rooms and possible sources of CO (refer to Exhibit S3.3):

***EXHIBIT S3.6*** *Form for Recording CO Measurements.*

## ADVICE FOR RESIDENTS

Based on my investigation of CO levels in your home, I recommend that the following actions be taken:

____________________________________________________________

____________________________________________________________

____________________________________________________________

____________________________________________________________

____________________________________________________________

____________________________________________________________

____________________________________________________________

____________________________________________________________

____________________________________________________________

____________________________________________________________

Information provided by: ____________________________________________

Name/Organization/Phone/Date: ________________________________________

***EXHIBIT S3.7*** *Form for Advising Residents After a CO Incident Investigation.*

*If the source of these CO levels appears to be a permanently installed appliance*

- Turn off the appliance.
- If the fuel supply can be shut off, do so. Where appropriate, contact the fuel supplier and advise it of the possible need to "red tag" the appliance.
- Advise residents that dangerous levels of CO have been detected and that they should not use the appliance until the cause of the elevated CO has been identified and corrected by a qualified technician. The appliance may not need to be repaired or replaced; it is possible that an adjustment to exhaust fans or some other house situation may need attention.

*If the source of these CO levels appears to be a portable appliance*

- Turn off the appliance.
- Advise residents that dangerous levels of CO have been detected and the appliance must be inspected and repaired or replaced before being used again.
- Advise residents that if the appliance passes inspection, but the CO problem recurs, then a qualified contractor or service person should investigate the appliance while it is being operated in the home. This may help identify why elevated CO levels are in the home and what is causing elevated CO levels.

*If you believe that misuse is the cause of CO (for example, using a charcoal grill or a gas oven to heat a home during a power outage)*

- Educate the resident about the causes and dangers of CO and the proper use of appliances.

*If you cannot find an apparent source of elevated levels of CO*

- Follow your jurisdiction's guidelines or advise the resident to call a heating, ventilating, or air conditioning (HVAC) company immediately to identify and correct the conditions causing the elevated CO levels.
- Advise residents that they can return to the home after it is cleared of CO. When your equipment shows levels of CO below 30 ppm, residents can return to the home. If your equipment shows a CO level below 30 ppm and if a household CO alarm is still sounding, remove the alarm and advise the consumer to replace it as soon as possible.
- Advise residents to have their fuel-burning appliances inspected immediately and maintained once a year by qualified professionals.
- Advise residents to use a CO alarm manufactured to one of the following standards:

  Canadian Standards Association (CSA) 6.19-01, 2001

  Underwriters Laboratories Inc. (UL) 2034, Second Edition, October 1998

  International Approval Services (IAS) 6-96, Second Edition, June 1, 1998

- Advise all residents to call again if their CO alarm sounds — particularly if they have followed the above recommendations.

**CO Levels: Between 30 ppm and 70 ppm.** *If the source of these CO levels appears to be a vehicle or an appliance, such as a lawn mower or generator, in a garage*

- Turn off the engine if it is still on.
- If you suspect that an appliance or a generator or a vehicle engine that had been idling in a garage several hours earlier is the cause of elevated CO levels, advise the consumer that engines should not be left idling in an attached garage, even with the garage door open.

*If the source of these CO levels appears to be a permanently installed appliance or a portable appliance*

- Turn off the appliance.
- Advise residents that potentially dangerous levels of CO have been detected and that they should not use the appliance until the cause of elevated CO levels has been identified and corrected. Advise residents that a condition in the home may be the cause and that a qualified contractor or serviceperson may need to be in the home while the appliance is in operation to identify and repair the CO source.

*If you believe that misuse is the cause of CO (for example, using a charcoal grill or a gas oven to heat a home during a power outage)*

- Educate the resident about the causes and dangers of CO and about the proper use of appliances.

*If you cannot find an apparent source of elevated levels of CO*

- Follow your jurisdiction's guidelines or advise the resident to call a heating, ventilating, and air-conditioning (HVAC) company immediately.
- Advise residents that they can return to the home after it is cleared of CO. When your equipment shows levels of CO below 30 ppm, residents can return to the home. If your equipment shows a CO level below 30 ppm and a household CO alarm is still sounding, then remove the alarm and advise the consumer to replace it as soon as possible.
- Advise residents to have their fuel-burning appliances inspected immediately and maintained once a year by qualified professionals.
- Advise residents to use a CO alarm manufactured to the requirements of one of the following standards:

  Canadian Standards Association (CSA) 6.19-01, 2001

  Underwriters Laboratories Inc. (UL) 2034, Second Edition, October 1998

  International Approval Services (IAS) 6-96, Second Edition, June 1, 1998

- Advise residents to call again if their CO alarm sounds — even if they have followed the above recommendations.

**CO Levels: Less Than 30 ppm.** Advise residents that you did not find high levels of CO.

- If the call was the result of a sounding CO alarm, advise the resident to review the manufacturer's instructions about a sounding CO alarm and to contact the manufacturer if necessary.
- If the alarm was manufactured before October 1, 1998, advise residents that they may want to purchase a new alarm that meets the requirements of one of these standards:

  Canadian Standards Association (CSA) 6.19-01, 2001

  Underwriters Laboratories Inc. (UL) 2034, Second Edition, October 1998

  International Approval Services (IAS) 6-96, Second Edition, June 1, 1998

- Alarms made before that time were subject to nuisance alarming; the new standards reduce the possibility of nuisance alarming. Advise residents that CO alarms have a limited life and that their alarm may need to be replaced. Advise residents never to ignore a sounding CO alarm and to call you again if the alarm activates.

(The following section provides information about why 30 ppm and 70 ppm are used as benchmark levels.)

**A Note About Benchmark CO Levels.** You may have read or heard about standards or recommendations for allowable (or safe) CO levels. Usually, these apply to particular environments, such as outdoors or the workplace. There is no U.S. federal government standard that sets an allowable *residential indoor* level for carbon monoxide; however, there are industry standards for allowable CO emissions from individual indoor appliances, such as furnaces or ovens.

Varying levels of CO exposure may cause varying health effects; there are levels that are not likely to cause adverse health effects in healthy adults, higher exposure levels that can be potentially life-threatening, and levels high enough to cause death. The effects of CO exposure depend on the length of time that a person has been exposed to CO, how high the CO concentration is, how physically active the person is at the time of exposure, and the person's general state of health. When a person breathes in CO, the CO combines with oxygen in the blood to form poisonous carboxyhemoglobin (COHb). The amount of COHb in a person's blood is expressed as a percentage. For example, a person exposed to CO might be described as having 8 percent COHb. Heavy smokers may have baseline levels of 10 percent COHb.

A 10 percent COHb level is a reference value that is widely regarded by physicians as confirmation of CO poisoning, even if there are no symptoms; when used as an upper limit, it is widely recognized as protecting healthy adults against the harmful effects of CO. At this level, a person generally is still able to take action to alleviate a potential life-threatening CO exposure. This benchmark is built into industry standards for CO alarms; a CO alarm MUST sound at exposures predicted to reach 10 percent COHb in heavily exercising individuals. This will occur at concentrations and times of 70 ppm CO for slightly more than 3 hours, a level of 150 ppm for 50 minutes, or a level of 400 ppm for approximately 15 minutes. At these same exposures, less-active individuals will have less than 10 percent COHb. At levels less than 30 ppm, it is unlikely that adverse health effects will occur in healthy adults. This is a second benchmark level in the industry standards for CO alarms; alarms cannot sound at 30 ppm or less. This helps prevent false or nuisance alarms from temporary "spikes" in CO that may occur when there is an outdoor air inversion or when an appliance is turned on and used for a relatively short time.

For the reasons described above, CPSC uses two benchmarks in this document as the basis for advice on what actions a resident should take when CO is present in the home. This first is the false alarm resistance level of 30 ppm. The second is 70 ppm, the lowest level that can cause COHb to exceed 10 percent.

The benchmark levels in this document, and the resulting advice, are suggested guidelines and do not constitute a standard. The benchmark levels are conservative in that they recommend action BEFORE potentially life-threatening CO exposure occurs. You may already be using or have come across other published standard operating procedures (SOPs) that use different CO ppm levels as a benchmark. If your jurisdiction has an SOP that works well for your needs and resources, and if the benchmark levels of the SOP, based on your experience and that of medical personnel in your community, adequately provide for the safety of residents, then you may want to continue using that SOP.

## PART III: SIDEWALL VENTED APPLIANCES

The use of sidewall vented appliances in residential construction is becoming prevalent. Sidewall vented appliances, by design, are less likely to introduce carbon monoxide to the indoor atmosphere. However, they are not immune from problems. An incident with a sidewall vented boiler installed in a two-year-old home in Plymouth, Massachusetts, resulted in the tragic death of a young girl. The incident occurred during an unusually windy snowstorm during which the area had extensive electric power failures, which may have been factors. The causes of this incident are still under investigation at the time of this publication.

It is critical that considerations be made for climate and potential changes to landscaping and construction. Changes that impede the proper operation of the vent can result in appliance shutdown or the production of carbon monoxide. Under certain conditions, such as cold weather, wind, snow accumulation, etc., flue products emitted from a sidewall vent can be drawn into the building through passive openings that exist in most construction. Installers of sidewall vented appliances should always follow code requirements and manufacturers' instructions. Sidewall vent terminations must have proper clearance to grade or decks, windows and doors, combustion and ventilation air openings, regulator vents, etc.

As with all industry codes and standards, improvements are constantly being developed and one should always be aware of the latest changes.

## PART IV: CARBON MONOXIDE INCIDENT INVESTIGATION

Part II of this supplement, "Guidelines for Fire and Other Emergency Response Personnel," is intended to assist public emergency response personnel in fulfilling their duties of rendering medical assistance and public service. Serious carbon monoxide incidents that involve fatalities or serious illness induced by the incident will often be subjected to detailed investigation. The parties involved in such investigations range from public agencies, representatives on behalf of injured or deceased parties, manufacturers of appliances and other fuel gas equipment, installers of fuel gas equipment, contractors, builders, code officials, and others. The discussion that follows is intended to be referenced in incidents that are suspected of having causal factors relating to fuel gas utilization equipment and systems. This presupposes that the initial broad-based investigation of evaluation of non–fuel gas utilization carbon monoxide sources has been conducted in a proper and thorough manner.

In addition to the material presented here, two U.S. Consumer Product Safety Commission publications may also be of interest. They are too large to reprint here, but are available from the CPSC web site:

- CO Investigation Guideline
- CO Investigation Guideline – Furnaces

## Fuel Gas Appliance Fundamentals

The *National Fuel Gas Code* is a critical link to the safe and effective utilization of fuel gas appliances. The code requirements intersect with many other appliance, equipment, and building code documents whose aim is to provide safe and effective utilization of fuel gas energy. In the broadest sense, the event represented by a carbon monoxide incident wherein carbon monoxide produced by a fuel gas appliance enters a habitable space is an indication that one or more elements of these codes and standards failed to exist at the time of the incident.

Conceptually, a fuel gas appliance produces only an inconsequential amount of carbon monoxide in its combustion products during proper operation. If the operation of a particular appliance has deteriorated (for whatever reason), a carbon monoxide–related event will still not occur if the combustion products continue to be vented properly from the subject appliance to the outdoors. The investigation of a carbon monoxide incident must therefore evaluate both the operation of specific appliances as well as the ventilation and venting of the building envelope under consideration.

## First Responder Alterations

As is the case with many categories of accident investigations including automobile, fire, industrial, and so on, first responders to a carbon monoxide incident often provide a critical link for understanding the conditions in the habitable space at the time of the event. Part 1 of this supplement provides an indication of possible data that might be obtained by first responders and any subsequent detailed investigation should necessarily inquire as to whether such data is available. First responder data certainly might also include photographs of the building envelope and any combustion appliances.

Another element of first responder activity is the possible alteration to the existing system as a result of necessary or inadvertent actions on the part of first responders. The degree of alteration to systems by initial responders is known to have varied from minor and inadvertent, such as the removal of portable obstructions found covering combustion air intakes, to major modifications even including "troubleshooting" and repair of appliances by service personnel or others. The consequences to the subsequent carbon monoxide investigation of such scene alterations can be significant if no documentation as to the sequence and methodology of the first responders is known.

## Why Now?

Carbon monoxide incidents can commonly involve complex and interacting system and environmental factors. Whatever the combination of causal factors that comes together to produce carbon monoxide in the habitable space, there is always a significant question to be resolved related to the timing of the occurrence. Clearly, significant factors related to possible deterioration or alteration to appliances or venting systems have to be considered if the event occurs with systems that are many years old as compared to a system that has just been installed with service initiated shortly before the incident.

Also fundamental to the question of the timing of a carbon monoxide incident are the characteristics of the incident itself. Broadly speaking (and there is a substantial range of actual incident type), carbon monoxide incidents can commonly be thought of as either long-term chronic exposure to elevated but nonfatal levels of carbon monoxide or shorter-duration acute exposures to high and possibly fatal levels of carbon monoxide. Understanding the fundamental background of a carbon monoxide incident can assist in evaluating the pertinent system elements, which would be important to either type of case.

Common to both categories of incident is the necessity to understand the history of the fuel gas systems involved, including a history of the building envelope. Pertinent portions of the system history can include the timing and details of the initial fuel gas system installation, service information, additions or alterations to the habitable space that would influence the performance of the fuel gas system, such as the addition of walls, partitions, additional combustion appliances, etc., and evaluation as to whether or not any of these changes had an effect at the time of the incident. The changes to the building envelope can be so broad as to include exterior changes to the building that may include alterations to exterior vents or landscaping, which obscures vents in a pre-existing location or otherwise alters or impairs their operation. Accidental environmental alterations, such as obstruction of vents or ventilation by weather or animal activities, must also be considered.

Gas system service calls can also be an indication as to when the system may have been altered or, possibly, when it may last have been known to be operating safely and efficiently. Changes in the system may often be tied to service records for the system to indicate changes to the appliances or other building envelope alterations.

A significant factor in some carbon monoxide incidents is the influence of adverse weather. Weather effects can produce an aberrant system performance as influenced by extremes of temperature, wind, and precipitation. During the investigation of a carbon monoxide incident, it is understood that various weather conditions may not be reproducible at the time the habitable space is tested for evaluation but the weather data is nonetheless pertinent as an evaluative point for determination of the system performance.

### Incident Scene Investigation

The most effective carbon monoxide incident scene investigation starts well before the physical examination and testing of specific appliances and building systems. As is common and recommended for many types of accident investigation, notification of interested parties, such as those listed at the beginning of this section, can be one of the most useful tools in executing an effective scene investigation. Specifically, the manufacturers of appliances and other fuel gas system components can provide valuable technical input to the scene examination as well as provide an opportunity for the development of a proper test protocol. Similarly, builders, installation contractors, and other persons involved with the appliances and structure may provide valuable information as to the circumstances of the appliance installation and operation and, in many jurisdictions, these parties may have legal rights of notification when they are involved in such an investigation.

A protocol for examination and testing of a building and gas appliance system is extremely useful in providing a thorough and efficient investigation. The most effective investigations benefit when this protocol is made available to participating parties in advance of the joint investigation. Obvious elements during any such investigation include a detailed visual examination of all system components, including the condition of the building envelope. This inspection is commonly expected to include evaluation of the venting systems, combustion air accommodation, and gas pressure regulation system, and a general visual examination of gas utilization equipment without altering the as-found condition.

The testing of fuel gas appliances and the carbon monoxide incident investigation should proceed in a methodical stepwise matter so as to allow documentation of all pertinent data and so as to disclose any operating anomalies that exist or are suspected to exist. Information commonly accepted to be pertinent to such testing includes current weather conditions, gas system delivery pressure, the monitoring of the internal to external building envelope pressure (as influenced by appliance operation, wind effects, or other factors), combustion performance, thermostatic device operation, and, of course, monitoring of carbon monoxide within the habitable space under examination. Such investigations are undertaken to evaluate the cause of harmful or lethal situations. As such, appropriate personal protection measures must be taken that may vary widely from incident to incident but at the least should monitor carbon monoxide levels for personnel safety during all testing. Other scene safety measures may be deemed necessary by criteria established by responsible and knowledgeable safety officials or consultants associated with the investigation.

### General Considerations

Responsibility for conducting carbon monoxide incident scene investigations may fall to one of several categories of possible parties. These certainly include, at a minimum, public officials, representatives of injured or deceased victims of the incident, property owners, and others. Representatives of broad elements of the gas industry are commonly called upon for direct or indirect assistance for investigation of carbon monoxide incidents. Any individual acting in this capacity must remain aware of the complexity of most carbon monoxide incidents. Ordinary troubleshooting techniques used in appliance servicing are inadequate to properly document and memorialize appliance operation, building envelope behavior, the influence of external factors, such as other solid fuel–burning appliances and fan operations. Participants in the investigation must understand the need to evaluate and document these many contributing factors.

## REFERENCES

1. Jennifer D. Flynn, *Non-Fire Carbon Monoxide Incidents Report in 2005*, NFPA Division of Fire Analysis and Research, June 2007.
2. CPSC, National Electronic Injury Surveillance System (NEISS).
3. CPSC, National Electronic Injury Surveillance System (NEISS).
4. CPSC, *CPSC Warns of Dangers After a Hurricane or Power Outage Hits,* News Release #06-169, May 24, 2006.
5. Michael J. Karter, Jr., *Fire Loss In The United States During 2005,* Quincy, MA: NFPA Fire Analysis and Research Division, September 2006.
6. Christian Tomaszewski, MD, Carolinas Medical Center, Charlotte, NC & Department of Emergency Medicine, UNC at Chapel Hill School of Medicine, "Carbon Monoxide Poisoning," *Postgraduate Medicine*, Vol 105, No 1, January 1999.

7. CDC, *Unintentional Non-Fire Related Carbon Monoxide Exposures-United States, 2001–2003,* Morbidity and Mortality Weekly Report, January 21, 2005.
8. Matthew V. Hnatov, *Non-Fire Carbon Monoxide Deaths Associated with the Use of Consumer Products 2003 and 2004 Annual Estimates,* Bethesda, MD: CPSC Division of Hazard Analysis, August 2007.
9. Susan A. Vagts, *Non-Fire Related Carbon Monoxide Deaths Associated with the Use of Consumer Products 1999 and 2000 Annual Estimates,* Bethesda, MD: CPSC Division of Hazard Analysis, July 2003.
10. CPSC, *Incidents, Deaths, and In-Depth Investigations Associated with Carbon Monoxide from Engine-Driven Generators and Other Engine-Driven Tools, 1990-2004*, December 2005.
11. NFPA 720, *Standard for the Installation of Carbon Monoxide (CO) Detection and Warning Equipment*, National Fire Protection Association, Quincy, MA, 2009 edition.

# SUPPLEMENT 4

# Fuel Gas Odorization

**John S. Roberts, Ph.D.**
Principal Chemist
Materials & Chemicals Consulting, LLC
Kingwood, Texas

***Editor's Note:*** *NFPA 54, National Fuel Gas Code, does not cover the subject of odorization of fuel gases; however, it is an important topic to many in the gas business and to users of fuel gases. Concise summaries of the history of odorization and practices are nonexistent; therefore, this supplement is provided to fill that void. A similar supplement appeared in the 1992 and 1996 editions of the National Fuel Gas Code Handbook but was discontinued due to lack of feedback that it was being used. The supplement has been included again, in response to requests from users of the handbook. We are fortunate to have had one of the leading experts in the field of fuel gas odorization agree to write this supplement.*

## INTRODUCTION

Fuel gases, liquefied petroleum gases (LP-Gases), and natural gas are odorized to allow a warning at 20 percent (1/5) of the lower flammable limit (LFL) of the gas being odorized. For natural gas this concentration is 1 percent gas in air (the LFL is 5 percent), and for propane this concentration is 0.43 percent gas in air (the LFL is 2.15 percent). Fuel gases have been odorized for well over 100 years, with much of the early experience being with the odorization of European coal gases. Although the odorants used have changed in this time, both types of gases (LP and natural) have used the same odorants for 50 years or more. The use of odorants in LP-Gas and natural gas has been successful in this time period. Odorization has protected lives and property by giving a warning that an otherwise odorless gas is escaping.

## EARLY HISTORY OF FUEL GAS ODORIZATION

Fuel gases have been odorized since the 1880s [1]. In 1880, ethyl mercaptan was used to odorize water gas (hydrogen/carbon monoxide) using a shallow dish placed in the gas line. As this was unsatisfactory, a wick was finally used [2]. All of the original work in both Europe and the United States was directed at coal gas, as very little natural gas was being used. It should be noted that many coal gases in use in the United States and Europe contained significant amounts of carbon monoxide, which made them much more toxic than natural gas. See Commentary Table 1.1 in Chapter 1 for the composition of several fuel gases, including manufactured gases. The use of odorants as stenches for industrial purposes was suggested in 1919 by Allison [3]; Katz and Allison suggested odorants for use as warning agents for miners [4] and for use in stenching blue water gas (carbon monoxide and hydrogen) and natural gas [5]. An extensive summary of the early history of gas odorization in Europe and the United States is found in the work of Fieldner et al. [1].

In 1920, the American Gas Association (AGA) "requested the United States Bureau of Mines to make a comprehensive search for suitable substances that might be added to fuel gases . . ."[1]. The Bureau of Mines study covered the following areas:

1. Warning properties: intensity, quality, and nature for suitability as a warning to conscious and sleeping persons.
2. Toxic properties: danger to health in concentrations necessary for use of the unburned warning agent and its products of combustion.
3. Corrosion: reaction with the distribution system, appliances, utensils, and household furnishings of the unburned warning agent and its products of combustion.
4. Carrying power: ability to carry through distribution system without loss due to reaction with materials, other components of the gas, oil or water films and drips, oxidation, or polymerization.
5. Cost: probable cost when considering the amount necessary to give the desired warning.

The Bureau of Mines study covered a broad range of odorant candidates. Among the classes of compounds that were evaluated were hydrocarbons (mostly olefins), halogenated hydrocarbons, alcohols, mercaptans, sulfides, halogenated ethers, halogenated sulfides (mustards), aldehydes, ketones, esters, selenium compounds, and arsenic compounds. Many of these types of warning agents were ruled out on the basis that they were unsuited due to corrosivity or toxicity.

The Bureau began testing candidate odorants by building an odorimeter for mixing the odorant candidate with the airstream to give a known concentration of odorant in the air. They also developed an intensity scale, still called the Bureau of Mines scale, which was used to measure the odor intensity. The scale ranged from 0 (no odor) to 5 (very strong), with 2 being a readily recognizable level. They also established scales for irritants (tear gas–type warning agents). They ran odorimeter tests on 57 different candidates. Sleep tests, wherein a candidate material was introduced into a room containing an observer (awake) and a subject (asleep), gave rather uneven results, due in part to the rather unnatural circumstances surrounding the tests. Irritants were found to be better at waking people.

A group of field tests were conducted at four different fuel gas companies. Ethyl mercaptan was tested at three of four companies, while crotonaldehyde was tested at one company. One company tested nine different odorants. Interestingly, one of the odorants tested, pyridine, was described as having a "gassy" odor, which would not be true today. Only one of the gas systems tested was not a coal-derived fuel gas. Ethyl mercaptan gave the lowest concentration to give a warning effect. A final test of two towns in one gas system, where ethyl mercaptan was used to leak survey the towns, resulted in the isolation of many leaks. There were 1800 complaints regarding odor during these two town tests.

The study came to a number of conclusions, but the one that seemed to be of the most interest was that ethyl mercaptan was the most effective compound for causing complaints of leaks. Crotonaldehyde, an irritant, was found to be the next best material tested.

While the Bureau of Mines was working on this study, a gas company had already instituted an odorization program [6, 7]. The product used in their program was called Cal-Odorant #3, a refinery by-product (with a 300°F boiling point range). The criteria for use as odorants, then as now, are that they be harmless and neither toxic nor nauseating, have a penetrating odor similar to the coal gas smell, be noncorrosive, be insoluble in water, be retained by the gas and not adsorbed by mains or meters, burn completely without harmful or odorous products of combustion, not set up chemical reactions, and be readily available [6].

During this same time period, the LP-Gas industry was expending significant effort to come up with a warning agent for propane. Propane, being a liquefied gas, had different requirements for a warning agent than did a gaseous fuel. Any warning agent had to vaporize from the liquid state to the gaseous state, in the presence of propane. Therefore, the effect of the warning agent concentrating in the liquid had to be taken into account. Much of the work resembled the work conducted by the Bureau of Mines, although the effect of a fractional distillation of the odorant was taken into account for each candidate odorant [8]. They concluded that ethyl mercaptan was the best choice for odorizing liquefied petroleum gases.

The explosion of the New London, Texas, high school on March 18, 1937, was the point where odorization became required in fuel gases. In this incident, 239 people lost their lives when unodorized gas from a nearby oil field was used as a fuel in the building. The day after the accident, the state of Texas proposed the first law requiring that fuel gases be odorized [9]. The requirement for odorizing propane was added to NFPA 58, *Liquefied Petroleum (LP) Gas Code*, shortly after [10].

In the period ending in the mid-1960s, the gas and gas odorant industries were developing hand-in-hand. New pipelines were beginning to move gas into the parts of the United States that did not have access to natural gas. During this time, natural gas odorants were moving from refinery by-products and ethyl mercaptan to designed products, containing primarily *t*-butyl mercaptan [9]. The LP-Gas industry stayed with ethyl mercaptan, moving almost completely away from thiophane. In the last 10 to 15 years, the use of thiophane in LP-Gas odorization in the United States has become nonexistent.

## NATURAL GAS

Within the continental United States, natural gas is transported in pipes from the points of origin to the points of use. Natural gas transported in pipelines falls under either federal or state regulations regarding odorization. Transmission pipelines move gas from one area to another. Distribution pipelines move the gas from a receiving point (or points) and to the ultimate consumer of the gas. Any transmission pipeline that moves gas between states is covered by federal regulations. It may be covered by state regulations, if the state opts to regulate interstate pipelines. Each state has different regulations regarding odorization. Federal regulations require that gas in transmission pipelines be odorized when the population along the pipeline reaches a certain level [11]. All distribution systems (consumer systems) must be odorized. All states require that gas be detectable at 1/5 the LFL of gas in air as the minimum requirement, although some states require that this level be lower.

The typical odorant addition level necessary to achieve the 1/5 LFL is usually 0.5 to 1 pound of odorant per million standard cubic feet of gas. This addition level is strongly dependent on the type of odorant being used and may be different due to a multiplicity of factors. This concentration is monitored by smell testing, using an odorimeter. An odorimeter is a mobile device that allows air and gas to be mixed, and then allows an operator to sniff the diluted gas. The gas flow is increased until the odorant is readily recognizable. The gas concentration is then determined from the flow rate of gas. The concentration is compared with regulatory or company requirements. There is an ASTM method [12] describing the correct procedures for using this equipment.

Natural gas, being a flowing, dynamic system, requires an odorization system that can vary with the flow rate. Over the years, a variety of odorizer systems have been utilized. It is outside the scope of this discussion to cover this area, but the AGA *Odorization Manual* covers this topic.

The types of odorants used for natural gas vary significantly. Most of the odorant sold in the United States contains *t*-butyl mercaptan. This mercaptan has a high melting point (–0.5°C) [13], so it is always mixed with other components to avoid handling difficulties. These components have been dimethyl sulfide, thiophane, methyl ethyl sulfide, or light mercaptan blends. Other odorant blends exist that contain either isopropyl mercaptan or thiophane as the primary odorant component.

Isopropyl mercaptan has a good vapor pressure, which makes it useful in situations where there is a significant amount of hydrocarbon liquids condensing in the gas system. This is a situation where it will help, but in the long run, it is still necessary to remove the liquids. Thiophane has been used alone, but is not nearly as effective as when it is blended with *t*-butyl mercaptan. There have been instances where gas systems changing from thiophane to thiophane/*t*-butyl mercaptan blends have been bombarded with odor complaints. Many of these complaints were leaks that had not been found previously.

## LIQUEFIED PETROLEUM GAS

The odorization system for LP-Gases is different, as these fuels are distributed as liquids. Ethyl mercaptan (EM) is added to LP-Gas at a concentration of 1 pound per 10,000 gallons (16 ppmv) by DOT regulation [14]. NFPA 58 [10] contains a requirement that most propane be odorized, and it regulates uses of propane not covered by DOT regulations. Industry practice has been to add EM at a higher rate. Thiophane (tetrahydrothiophene, or THT) had been used in certain areas of the country, but is currently not sold for odorizing propane in the United States. Ethyl mercaptan is added to trucks or rail cars as they leave refineries, gas plants, and pipeline or rail terminals. The original system for adding the odorant to propane was to pour a container full of the odorant into the truck or the line, filling the truck or rail car (this container was oftentimes a soft drink bottle). The current systems use pumps to add EM to the propane as the truck is filled. Many of the new odorization systems are designed to shut off propane flow if there is a positive indication that the odorant is not being added.

The concentration of 1 pound per 10,000 gallons of ethyl mercaptan is higher than required. An industry-sponsored study in 1977, the so-called Bartlesville study [15], demonstrated that lower concentrations would function to warn that portion of the populace that can smell odorants. Individuals who cannot smell odorant (or anything) are termed anosmic. The number of anosmics is not well defined, but is thought to be 0.2 percent of the population [16, 17]. Although this study did not give any real clear-cut conclusions, the industry determined that 1 pound per 10,000 gallons was sufficient for the addition of EM and that THT needed to be added at 6.4 pounds per 10,000 gallons. Because THT had been added at 1 pound per 10,000 gallons and it was priced higher than EM, much of the interest in the use of this odorant disappeared.

The Bartlesville study evaluated the warning effect of ethyl mercaptan and thiophane and also evaluated the effect of temperature on the amount of odorant in the gas phase. As odorized LP-Gas is a mixture of propane [boiling point –44°F (–42°C)] and EM [boiling point 95°F (35°C)], the effect of temperature on vapor composition is quite important. A mixture of liquids with different boiling points will have a vapor composition that is different from the liquid composition. This fact is the basis of fractional distillation, wherein mixtures of different boiling points are separated. This will occur in LP-Gas/odorant systems. The larger the difference in boiling point, the lower the concentration of higher boiling component in the vapor phase of the mixture. The amount of EM in the vapor of odorized propane versus the amount in the liquid propane has been measured at several temperatures [15]. A value, called a K-factor, was determined for EM and THT at three temperatures (see Table S4.1). A K-factor is defined as the concentration of the higher boiling component in the gas phase divided by the concentration of that component in the liquid phase. The K-factor will vary with temperature.

*TABLE S4.1 K-Factors for Odorants in Propane*

| Temperature | | Odorant Propane System | |
|---|---|---|---|
| | | Ethyl Mercaptan | Thiophane |
| (°C) | (°F) | K-Factor | K-Factor |
| −23 | −10 | 0.11 | 0.007 |
| 0 | 32 | 0.18 | 0.029 |
| 35 | 95 | 0.22 | 0.033 |

As noted here, only 11 to 22 percent of the ethyl mercaptan in the liquid is available in the vapor. A higher boiling mercaptan (or odorant) has a smaller amount in the vapor. The K-factor for thiophane indicates that only 0.7 to 3.3 percent of thiophane is available in the vapor. Thiophane has the highest vapor pressure of the odorant materials. All other odorant compounds, although not used for propane, have K-factors between ethyl mercaptan and thiophane. The K-factors for several odorant components and propane have been measured [18].

In the mid-1980s, questions regarding the selection of ethyl mercaptan as the odorant for propane were brought forward by the Consumer Products Safety Commission (CPSC) [19]. The CPSC sponsored research into some of the facets of propane odorization that it felt were problem areas. They sponsored research on the adsorption of ethyl mercaptan on building materials, which showed that in a static system the ethyl mercaptan is depleted [20, 21]. The experimental design in this study magnified the effect. In a dynamic system, the amount of odorant is being continually replaced. The static system would model filling a room with propane and then stopping the propane flow. The dynamic study modeled an ongoing gas leak and noted that the amount of odorant was not as strongly affected as observed in the static study [21].

The CPSC also funded research into the affect of propane tanks on the oxidation of ethyl mercaptan [22, 23]. They found that there was an affect. Further work by other parties indicated that although a tank will oxidize ethyl mercaptan, the odorant will not disappear as rapidly as reported in the previous studies. The primary change created in the industry was that NFPA 58 was modified to require odorant verification upon product transfer. Gas detectors, another modification suggested by the CPSC, have not seen significant use in the United States. As of this writing, there is no standard on the placement and use of gas detectors, although a committee did exist for several years trying to work out the details of utilizing fuel gas detectors.

Research on the issue of propane odorization was conducted by the propane industry and reported at two special conferences (sponsored by the National Propane Gas Association, Gas Processors Association, and Propane Gas Association of Canada). This work was part of an industry-wide task force, which came to the conclusion that ethyl mercaptan was the best available odorant for LP-Gas [25].

## ODORIZATION PROBLEMS

No odorant will ever be able to function to prevent 100 percent of all accidents. A gas could be perfectly odorized, and if no one is present in the area where gas is present, no warning will be given. It is also possible for an individual to have lost the sense of smell completely (anosmia) or to have a cold or some other condition involving the olfactory system. Some individuals have specific anosmia, which is the inability to smell certain substances. Gas odorants can be oxidized when they come in contact with oxidizing agents, such as rust. Gas odorants can be adsorbed onto surfaces and absorbed into liquids.

The gas industry has recognized the problem with oxidation of mercaptan odorants by iron oxides [most particularly ferric oxides, iron (III) oxides] for a long time [26–28]. The inside of pipes and the inside of propane tankage contain rust, which has been identified as $Fe_2O_3$, FeOOH, $Fe_3O_4$, and FeO. $Fe_3O_4$ and FeO are not considered to be active in the oxidation of mercaptans [29, 30].

Sulfides are not oxidized under normal pipeline conditions. Of the three sulfides used in natural gas odorization, only thiophane is a component with odorization value. Dimethyl sulfide and ethyl methyl sulfide are essentially worthless in terms of their odorization value. Experience has shown that any system using thiophane alone will discover significant gas leakage when moving to a blend of *t*-butyl mercaptan and thiophane.

Adsorption is defined as the adherence to a solid surface by a chemical compound. Any chemical with a vapor pressure will be adsorbed. Adsorption is a reversible process. Any compound that is adsorbed can be desorbed, and an equilibrium between adsorbed and desorbed species can be established. This equilibrium is a function of temperature, the presence of chemical reactions occurring between the adsorbent (surface) and the adsorbate (compound adsorbed), and the concentration of the adsorbable species in the vapor space. Adsorption is the first step in any oxidation process.

When gas leaks underground, there may be an odorant loss due to adsorption and/or oxidation, assuming that there are suitable oxidants in the soil. The Bureau of Mines study [1] indicated that soil adsorption may be a difficulty in underground leaks. The only soil constituents that adsorb significant amounts of odorant are dry clays. Wet clay does not adsorb significant amounts of odorant, but all soils will become dry in the presence of a gas leak due to the extreme dryness of most fuel gases. A significant body of work on odorant loss due to soil adsorption has been published over the years [31–35]. The common odorant constituents are adsorbed on clay (montmorillonite) in the following order: *t*-butyl mercaptan (least), ethyl mercaptan, isopropyl mercaptan, thiophane (most). With the exception of *t*-butyl mercaptan, the order of adsorption is in the same order as the vapor pressure, which is the expected order for adsorption. The difference with *t*-butyl mercaptan is due to the nature of clay, where the size and shape of *t*-butyl mercaptan may exclude it from interacting with significant portions of the surface area of clay due to steric hindrance. Many types of clay have a smectite or zeolite structure, which gives a significant amount of internal surface area, which may not be accessible to a bulky compound.

Adsorption on environmental surfaces was investigated as part of the CPSC inquiry regarding propane odorization. A static study, in which odorized gas was inletted into an enclosure, showed that odorant was lost over time to the surfaces inserted into the chamber [20, 21]. A chamber with no air exchange and only a short-term leak is not a realistic situation. If gas is inletted for only a short period of time, the air exchanges within the volume that the gas is in will bring the level of fuel gas below dangerous levels quite rapidly. Because the volume of the test chamber was quite small, when compared to normal areas in which propane leaks occur, the surface-to-volume ratio was too high. In other words, the distance to an adsorbing surface is much higher in a basement than in the test chambers. A second study, in which the odorized gas was continually inletted into a space, showed a lower loss of odor, as odorant that was lost to adsorption was being replaced [21]. The building materials that were studied were unpainted masonry, painted masonry, poured concrete, and painted drywall. The odorant was adsorbed most rapidly in the case of unpainted masonry and decreased in the above order. The author made no attempt to define the relative surface areas of these test specimens.

Absorption is an effect that is important only in natural gas systems [36]. In natural gas systems, there have been occasions when significant quantities of higher-molecular-weight hydrocarbons have condensed out in the pipeline system, creating "puddles." These "puddles" gather at low points in the system and act as a scrubber for the gas. The odorant in the gas will be absorbed by the liquid, significantly reducing the amount of odor in the gas phase. When the line dries out, the odorant will also evaporate, creating an overodorized system. Liquids in the pipeline are generally encountered in gas-producing areas, although anecdotal evidence indicates that these problems can occur in gas that has been transported for significant distances.

Olfactory fatigue or adaptation is another problem with any and all odorants. After a period of time, the human olfactory system saturates with an odor and one stops detecting the odor. Adaptation is discussed at length in the olfactory literature [37, 38], but has not been specifically addressed in any studies in the fuel gas industry. A classic example of this phenomenon is a visit to a house with an unfamiliar background odor. After a short period of time, the odor is forgotten. If one steps outside and refreshes one's nose, the odor will again be readily detectable. In other words, for any odorant-type warning agent, the warning must be heeded early.

Anosmia, the lack of a sense of smell, is a condition that affects a small percentage (0.2 percent) [16, 17] of the population. Odorants cannot protect someone who cannot smell them. The use of scratch-and-sniff brochures or familiarizing a customer with the smell of gas through a stove burner should help to find those individuals who might not be warned. There are gas detectors available for these individuals. Age is also a factor in being able to detect odors. Cain [39, 40] noted that an elderly population required ten times higher concentration to detect the presence of ethyl mercaptan. Amoore estimated that up to 2 percent of the population may not be able to be warned by an odor-borne warning [16].

Specific anosmia is a condition where one has a normal sense of smell, except there are certain compounds that one cannot smell. This is a condition that Amoore [41] was attempting to use to find the "primary odors," in an analogy to the light receptors in the eye, in which different types of receptors are used for certain colors. He had an extensive listing of compounds for which there is some report of specific anosmia [42]. There is some credible evidence for several different compounds, such as androstenone (sweaty, urinous odor) and isovaleric acid (goaty odor). The evidence for mercaptans and sulfides is minimal, with the only actual study showing only 0.1 percent of a test population being unable to detect *n*-butyl mercaptan [17]. The data in Amoore's reference for *t*-butyl mercaptan [43] do not demonstrate this effect. The report on thiophane is a verbal account of one individual who was unable to detect thiophane in natural gas. Interestingly, Wysocki et al. [44] have reported that the ability to smell androstenone can be acquired by those who were unable to detect it. A review of the literature for specific anosmia has been compiled [45].

## SUMMARY

Fuel gases have been odorized throughout the world since the 1800s. Their use as a fuel gas odorant was mandated in the United States following a tragic incident in Texas where an occupied high school suffered an explosion of unodorized natural gas. The use of odorants does not give a perfect warning system, but fuel gas odorization has proved effective for the identification of countless leaks within gas systems since its inception. Odorants provide warning to those with a sense of smell. Most everyone has a sense of smell most of the time, but it can be temporarily lost during an illness. Some people have a reduced or no sense of smell, and other means of warning of gas leaks should be considered for them.

## REFERENCES


1. Fieldner, A. C., R. R. Sayers, W. P. Yant, S. H. Katz, J. B. Shohan, and R. D. Leitch, "Warning Agents for Fuel Gases," U.S. Bureau of Mines, Monograph 4, 1931.
2. Reitmeyer, K., *J. Gasbeleucbtung und Wasserversorgung*, **50** (318), 1907.
3. Allison, V. C., and S. H. Katz, "An investigation of stenches and odors for industrial purposes," *Ind. Eng. Chem.* **11** (336), 1919.
4. Katz, S. H., V. C. Allison, and W. L. Egy, U.S. Bureau of Mines Technical Paper 244, 1920.
5. Katz, S. H., and V. C. Allison, U.S. Bureau of Mines Technical Paper 267, 1920.
6. Holtz, Lee, "The Odorization of Natural Gas," *Gas Age-Record*, **66** (10), 341,9/6/30.
7. Anon., "Odorization of Natural Gas," *Gas Age-Record*, **508**, 4/12/30.
8. Phillips Petroleum Memos.
9. Roberson, Seth, "History of Gas Odorization," *Odorization III*, G. G. Wilson and A. A. Attari, editors, 1, IGT, Chicago, 1993.
10. NFPA 58, *Liquefied Petroleum Gas Code*, 2004 edition.
11. Code of Federal Regulations 49 CFR 192.625.
12. ASTM D2673, "Standard Test Methods for Natural Gas Odor Intensity," 1998.
13. Windholz, Martha, *The Merck Index*, 10th Edition, Merck & Co, Inc., 218, Rahway, NJ, 1983.
14. Code of Federal Regulations 49 CFR 173.315(b)(1).
15. Whisman, M. L., J. W. Goetzinger, F. O. Cotton, D. W. Brinkman, and C. J. Thompson, "A New Look at Odorization Levels for Propane Gas," USERDA Document, BERC/R1-77/1, 1977.
16. Amoore, J. E., "Odor blindness as a problem in odorization," American Gas Association Operating Section Proceedings, Distribution Conference, pp. 242–247, 1968.
17. Patterson, P. M., and B. A. Lauder, "The incidence and probable inheritance of "smell blindness" to Normal Butyl Mercaptan," *J. Heredity* **39**, pp. 295–297, 1948.
18. Ng, Heng-Joo, and Donald B. Robinson, "Vapor Liquid Equilibrium in Propane-Odorant Systems," Research Report RR-113, GPA/NPGA Project 876, 1989.
19. Switzer, Donald W., "CPSC Fuel Gas Study," Proceedings of the Symposium on LP-Gas Odorization, Dallas, Texas, 1989.
20. Beltis, Kevin J., and Daniel J. Ehntholt, "Characterization of LP-Gas Odorant Fade," Proceedings of the Symposium on LP-Gas Odorization, Dallas, Texas, 1989.
21. Kuhlman, Michael R., "LP-Gas Odorant Stability Under Simulated Residential Conditions," Proceedings of the Symposium on LP-Gas Odorization, Dallas, Texas, 1990.
22. Marshall, M. D., and C. A. Palladino, "Ethyl Mercaptan Stability After Refill of New and Air-Exposed Used LP-Gas Storage Tanks," Proceedings of the Symposium on LP-Gas Odorization, Dallas, Texas, 1990.
23. Marshall, M. D., and C. A. Palladino, "Field Testing for Ethyl Mercaptan in LP-Gas Storage Tanks in Four Marketing Areas," Proceedings of the Symposium on LP-Gas Odorization, Dallas, Texas, 1990.
24. Johnson, S. J., "Ethyl Mercaptan Odorant Stability in Stored Liquid Propane," Proceedings of the Symposium on LP-Gas Odorization, Dallas, Texas, 1989.
25. Report of Investigations Odorization of LP Gas, October 1991, available from the Gas Processor Association, Tulsa OK.
26. Phillips Petroleum Company Internal Report.
27. Gill, Charles E., "Gas Odorizing," U. S. Patent 2,430,050, 11/4/47.
28. McClure, John S., "Warning agent and Process for the Odorization of a Hydrocarbon Fuel Gas Therewith,"

U.S. Patent 2,823,104, assigned to California Research Corp., 2/11/58.

**29.** Campbell, Ian D., N. A. Chambers, and John Jacobus, "The Chemical Oxidation of Ethyl Mercaptan in Steel Vessels," Proceedings, International Symposium on Odorization, Chicago, IL, 1994.

**30.** Roberts, John S., and O. John Jacobus, "Odor Fade in Propane Tanks: A Case Study," Proceedings, International Symposium on Odorization, Chicago, IL,1997.

**31.** Williams, Ralph P., "Soil Penetrabilities of Natural Gas Odorants," *Pipeline & Gas Journal*, March 1976.

**32.** Parlman, Robert M., and R. P. Williams, "Penetrabilities of Gas Odorant Compounds in Natural Soils," Proceedings of the American Gas Association Conference, 1979.

**33.** Fraser, W. R., "The Odorization of Natural Gas for City Distribution," Proceedings of the American Gas Association Conference, 729, 1940.

**34.** Sullivan, Frederick, "New Gas Odorants," *Odorization II*, G. G. Wilson and A. A. Attari, editors, **459**, IGT, 1987.

**35.** Tarman, P. B., and H. R. Linden, "Soil Adsorption of Odorant Compounds," IGT Research Bulletin # 33, July 1, 1961.

**36.** Cooper, L. S., A. B. Densham, A. J. DeRose, and B. Juren, "Some Aspects of the Reception and Transmission of North Sea Gas," Proceedings, 34th Autumn Meeting, Institution of Gas Engineers, November 12, 1968.

**37.** Cain, W.S., "Perception of odor intensity and the time course of olfactory adaptation," *ASHRAE Transactions*, **80**(53), 1974.

**38.** Moncrieff, R. W., "Olfactory adaptation and odour intensity," *American Journal of Psychology*, **70**(1), 1957.

**39.** Cain, William S., Joseph C. Stevens, and David E. Weinstein, "Aging Impairs the Ability to Detect Gas Odor," *Fire Technology*, **23**(3) 198, 1987.

**40.** Stevens, Joseph C., and William S. Cain, "Detecting Gas Odor in Old Age," *Annals of the NY Academy of Science*, **510**(644), 1987.

**41.** Amoore, J. E., Molecular Basis of Odor (Charles C. Thomas, Springfield, IL, 1970) for a complete exposition regarding his theories involving specific anosmia.

**42.** Amoore, John E., "A Plan to Identify Most of the Primary Odors," in *Olfaction & Taste*, C. Pfaffman, editor, **158**, Rockefeller Univ. Press, 1969.

**43.** Wilby, Frank V., "Variation in Recognition Odor Threshold of a Panel," *Journal of Air Pollution and Control Association*, **19**(96), 1969.

**44.** Wysocki, C. J., K. M. Domes, and G. K. Beauchamp, "Ability to perceive androstenone can be acquired by ostensibly anosmic people," Proceedings of the National Academy of Sciences USA, **86**(7976), 1989.

**45.** Roberts, John S., "Natural Gas and LP Gas Odorants: A Review of the Basic Science," Natural Gas and LP Odorization Conference, GTI, Chicago, IL, August 2–4, 2004.

# SUPPLEMENT 5

# Procedure to Estimate Infiltration Rate for Residential Structures

**Paul Cabot**
American Gas Association

***Editor's Note:*** *Users of the National Fuel Gas Code may select the Known Air Infiltration Rate (KAIR) Method in order to calculate the required volume of indoor combustion air to meet the needs of installed appliances. This supplement describes a method used by the American Society of Heating, Refrigerating, and Air Conditioning Engineers (ASHRAE) for calculating air changes per hour (ACH) and presents calculation examples.*

*NFPA is grateful to Paul Cabot of the American Gas Association for preparing this supplement. We also appreciate the permission from ASHRAE to reproduce tables from its 2001 and 2005 Fundamentals Handbooks.*

## INTRODUCTION

The 2009 *National Fuel Gas Code* contains two methods, the Standard Method and the Known Air Infiltration Rate (KAIR) Method, for calculating the required volume of indoor air to meet the combustion, dilution, and ventilation needs of appliances. The Standard Method is applicable to structures with an air infiltration rate of 0.40 air change rate per hour (ACH) or greater. The KAIR Method is applicable to all structures, but an upper air infiltration rate limit of 0.60 ACH is established. ACH is based on the effective air leakage area at 0.016 in. water column pressure difference.

A large number of building design and construction features impact the amount of air infiltration of a structure. These inherent features include the amount of window area, amount of exterior wall area, number of envelope penetrations, construction quality, and age. The American Society of Heating, Refrigerating, and Air-Conditioning Engineers (ASHRAE) provides a method to calculate ACH.

## ASHRAE METHOD

ASHRAE is the HVAC industry's principal professional organization. ASHRAE develops and publishes state-of-the-art technical material for the benefit of design professionals and the well-being of the general public. This supplement provides a residential calculation method based on the ASHRAE tables, equations, and calculation methods described in the 2005 *ASHRAE Fundamentals Handbook*, Chapter 27, "Ventilation and Infiltration." Tables S5.1, S5.2, S5.3, and S5.4 and two equations are used to estimate ACH as follows:

$$Q = A_L = \sqrt{C_S \Delta t + C_W U^2}$$

$Q$ = airflow rate, cfm

$A_L$ = effective air leakage area, in.$^2$

$C_S$ = stack coefficient, cfm$^2$/(in.$^4$ · °F)

$\Delta t$ = average indoor-outdoor temperature difference for time interval of calculation, °F

$C_W$ = wind coefficient, cfm$^2$/(in.$^4$ · mph$^2$)

$U$ = average wind speed measured at local weather station for time interval of calculation, mph

**TABLE S5.1** *Effective Air Leakage Areas (Low-Rise Residential Applications Only)*[†]

| | *Units** | *Best Estimate* | *Minimum* | *Maximum* |
|---|---|---|---|---|
| Ceiling | | | | |
| General | in.$^2$/ft$^2$ | 0.026 | 0.011 | 0.04 |
| Drop | in.$^2$/ft$^2$ | 0.0027 | 0.00066 | 0.003 |
| Ceiling penetrations | | | | |
| Whole-house fans | in.$^2$ ea | 3.1 | 0.25 | 3.3 |
| Recessed lights | in.$^2$ ea | 1.6 | 0.23 | 3.3 |
| Ceiling/Flue vent | in.$^2$ ea | 4.8 | 4.3 | 4.8 |
| Surface-mounted lights | in.$^2$ ea | 0.13 | | |
| Chimney | in.$^2$ ea | 4.5 | 3.3 | 5.6 |
| Crawl space | | | | |
| General (area for exposed wall) | in.$^2$/ft$^2$ | 0.144 | 0.1 | 0.24 |
| 8 in. by 16 in. vents | in.$^2$ ea | 20 | | |
| Door frame | | | | |
| General | in.$^2$ ea | 1.9 | 0.37 | 3.9 |
| Masonry, not caulked | in.$^2$/ft$^2$ | 0.07 | 0.024 | 0.07 |
| Masonry, caulked | in.$^2$/ft$^2$ | 0.014 | 0.004 | 0.014 |
| Wood, not caulked | in.$^2$/ft$^2$ | 0.024 | 0.009 | 0.024 |
| Wood, caulked | in.$^2$/ft$^2$ | 0.004 | 0.001 | 0.004 |
| Trim | in.$^2$/lftc | 0.05 | | |
| Jamb | in.$^2$/lftc | 0.4 | 0.3 | 0.5 |
| Threshold | in.$^2$/lftc | 0.1 | 0.06 | 1.1 |
| Doors | | | | |
| Attic/crawl space, not weather stripped | in.$^2$ ea | 4.6 | 1.6 | 5.7 |
| Attic/crawl space, weatherstripped | in.$^2$ ea | 2.8 | 1.2 | 2.9 |
| Attic fold down, not weather stripped | in.$^2$ ea | 6.8 | 3.6 | 13 |
| Attic fold down, weather stripped | in.$^2$ ea | 3.4 | 2.2 | 6.7 |
| Attic fold down, with insulated box | in.$^2$ ea | 0.6 | | |
| Attic from unconditioned garage | in.$^2$ ea | 0 | 0 | 0 |
| Double, not weather stripped | in.$^2$/ft$^2$ | 0.16 | 0.1 | 0.32 |
| Double, weather stripped | in.$^2$/ft$^2$ | 0.12 | 0.04 | 0.33 |
| Elevator (passenger) | in.$^2$ ea | 0.04 | 0.022 | 0.054 |
| General average | in.$^2$/lftc | 0.015 | 0.011 | 0.021 |
| Interior (pocket on top floor) | in.$^2$ ea | 2.2 | | |
| Interior (stairs) | in.$^2$/lftc | 0.04 | 0.012 | 0.070 |
| Mail slot | in.$^2$/lftc | 0.2 | | |
| Sliding exterior glass patio | in.$^2$ ea | 3.4 | 0.46 | 9.3 |
| Sliding exterior glass patio | in$^2$/ft$^2$ | 0.079 | 0.009 | 0.22 |
| Storm (difference between with and without) | in.$^2$ ea | 0.9 | 0.46 | 0.96 |
| Single, not weather stripped | in.$^2$ ea | 3.3 | 1.9 | 8.2 |
| Single, weather stripped | in.$^2$ ea | 1.9 | 0.6 | 4.2 |
| Vestibule (subtract per each location) | in.$^2$ ea | 1.6 | | |
| Electrical outlets/Switches | | | | |
| No gaskets | in.$^2$ ea | 0.38 | 0.08 | 0.96 |
| With gaskets | in.$^2$ ea | 0.023 | 0.012 | 0.54 |
| Furnace | | | | |
| Sealed (or no) combustion | in.$^2$ ea | 0 | 0 | 0 |
| Retention head or stack damper | in.$^2$ ea | 4.6 | 3.1 | 4.6 |
| Retention head and stack damper | in.$^2$ ea | 3.7 | 2.8 | 4.6 |
| Floors over crawl spaces | | | | |
| General | in.$^2$/ft$^2$ | 0.032 | 0.006 | 0.071 |
| Without ductwork in crawl space | in.$^2$/ft$^2$ | 0.0285 | | |
| With ductwork in crawl space | in.$^2$/ft$^2$ | 0.0324 | | |

*TABLE S5.1 Continued*

| | Units* | Best Estimate | Minimum | Maximum |
|---|---|---|---|---|
| Fireplace | | | | |
| With damper closed | in.$^2$/ft$^2$ | 0.62 | 0.14 | 1.3 |
| With damper open | in.$^2$/ft$^2$ | 5.04 | 2.09 | 5.47 |
| With glass doors | in.$^2$/ft$^2$ | 0.58 | 0.06 | 0.58 |
| With insert and damper closed | in.$^2$/ft$^2$ | 0.52 | 0.37 | 0.66 |
| With insert and damper open | in.$^2$/ft$^2$ | 0.94 | 0.58 | 1.3 |
| Gas water heater | in.$^2$ ea | 3.1 | 2.3 | 3.9 |
| Joints | | | | |
| Ceiling-wall | in.$^2$/lftc | 0.070 | 0.0075 | 0.12 |
| Sole plate, floor/wall, uncaulked | in.$^2$/lftc | 0.2 | 0.018 | 0.26 |
| Sole plate, floor/wall, caulked | in.$^2$/lftc | 0.04 | 0.0035 | 0.056 |
| Top plate, band joist | in.$^2$/lftc | 0.005 | 0.0035 | 0.018 |
| Piping/Plumbing/Wiring penetrations | | | | |
| Uncaulked | in.$^2$ ea | 0.9 | 0.31 | 3.7 |
| Caulked | in.$^2$ ea | 0.3 | 0.16 | 0.3 |
| Vents | | | | |
| Bathroom with damper closed | in.$^2$ ea | 1.6 | 0.39 | 3.1 |
| Bathroom with damper open | in.$^2$ ea | 3.1 | 0.95 | 3.4 |
| Dryer with damper | in.$^2$ ea | 0.46 | 0.45 | 1.1 |
| Dryer without damper | in.$^2$ ea | 2.3 | 1.9 | 5.3 |
| Kitchen with damper open | in.$^2$ ea | 6.2 | 2.2 | 11 |
| Kitchen with damper closed | in.$^2$ ea | 0.8 | 0.16 | 1.1 |
| Kitchen with tight gasket | in.$^2$ ea | 0.16 | | |
| Walls (exterior) | | | | |
| Cast-in-place concrete | in.$^2$/ft$^2$ | 0.007 | 0.0007 | 0.026 |
| Clay brick cavity wall, finished | in.$^2$/ft$^2$ | 0.0098 | 0.0007 | 0.033 |
| Precast concrete panel | in.$^2$/ft$^2$ | 0.017 | 0.0004 | 0.024 |
| Lightweight concrete block, unfinished | in.$^2$/ft$^2$ | 0.05 | 0.019 | 0.058 |
| Lightweight concrete block, painted or stucco | in.$^2$/ft$^2$ | 0.016 | 0.0075 | 0.016 |
| Heavyweight concrete block, unfinished | in.$^2$/ft$^2$ | 0.0036 | | |
| Continuous air infiltration barrier | in.$^2$/ft$^2$ | 0.0022 | 0.0008 | 0.003 |
| Rigid sheathing | in.$^2$/ft$^2$ | 0.005 | 0.0042 | 0.006 |
| Window framing | | | | |
| Masonry, uncaulked | in.$^2$/ft$^2$ | 0.094 | 0.082 | 0.148 |
| Masonry, caulked | in.$^2$/ft$^2$ | 0.019 | 0.016 | 0.03 |
| Wood, uncaulked | in.$^2$/ft$^2$ | 0.025 | 0.022 | 0.039 |
| Wood, caulked | in.$^2$/ft$^2$ | 0.004 | 0.004 | 0.007 |
| Windows | | | | |
| Awning, not weather stripped | in.$^2$/ft$^2$ | 0.023 | 0.011 | 0.035 |
| Awning, weather stripped | in.$^2$/ft$^2$ | 0.012 | 0.006 | 0.017 |
| Casement, weather stripped | in.$^2$/lftc | 0.011 | 0.005 | 0.14 |
| Casement, not weather stripped | in.$^2$/lftc | 0.013 | | |
| Double horizontal slider, not weather stripped | in.$^2$/lftc | 0.052 | 0.0009 | 0.16 |
| Double horizontal slider, wood, weather stripped | in.$^2$/lftc | 0.026 | 0.0070 | 0.081 |
| Double horizontal slider, aluminum, weather stripped | in.$^2$/lftc | 0.034 | 0.027 | 0.038 |
| Double-hung, not weather stripped | in.$^2$/lftc | 0.12 | 0.040 | 0.29 |
| Double-hung, weather stripped | in.$^2$/lftc | 0.031 | 0.009 | 0.089 |
| Double-hung with storm, not weather stripped | in.$^2$/lftc | 0.046 | 0.023 | 0.080 |
| Double-hung with storm, weather stripped | in.$^2$/lftc | 0.037 | 0.021 | 0.05 |
| Double-hung with pressurized track, weather stripped | in.$^2$/lftc | 0.023 | 0.018 | 0.026 |
| Jalousie | in.$^2$/louver | 0.524 | | |

*(continues)*

*TABLE S5.1 Effective Air Leakage Areas (Low-Rise Residential Applications Only)*†

| | *Units** | *Best Estimate* | *Minimum* | *Maximum* |
|---|---|---|---|---|
| Windows *(continued)* | | | | |
| Lumped | in.$^2$/lfts | 0.022 | 0.00042 | 0.097 |
| Single horizontal slider, weather-stripped | in.$^2$/lfts | 0.031 | 0.009 | 0.097 |
| Single horizontal slider, aluminum | in.$^2$/lfts | 0.04 | 0.013 | 0.097 |
| Single horizontal slider, wood | in.$^2$/lfts | 0.021 | 0.013 | 0.047 |
| Single horizontal slider, wood clad | in.$^2$/lfts | 0.030 | 0.025 | 0.038 |
| Single-hung, weather stripped | in.$^2$/lfts | 0.041 | 0.029 | 0.058 |
| Sill | in.$^2$/lftc | 0.0099 | 0.0065 | 0.010 |
| Storm inside, heat shrink | in.$^2$/lfts | 0.00085 | 0.00042 | 0.00085 |
| Storm inside, rigid sheet with magnetic seal | in.$^2$/lfts | 0.0056 | 0.00085 | 0.011 |
| Storm inside, flexible sheet with mechanical seal | in.$^2$/lfts | 0.0072 | 0.00085 | 0.039 |
| Storm inside, rigid sheet with mechanical seal | in.$^2$/lfts | 0.019 | 0.0021 | 0.039 |
| Storm outside, pressurized track | in.$^2$/lftc | 0.025 | | |
| Storm outside, 2-track | in.$^2$/lftc | 0.058 | | |
| Storm outside, 3-track | in.$^2$/lftc | 0.116 | | |

*Air leakage areas are based on values found in the literature. The effective air leakage area (in square inches) is based on a pressure difference of 0.016 in. of water and $C_S = 1$.

Abbreviations: ft$^2$ = gross area in square feet, ea = each, lftc = linear foot of crack, lfts = linear foot of sash.

*Source:* 2001 *ASHRAE Handbook — Fundamentals*, Chapter 26, Table 1. © American Society of Heating, Refrigerating and Air-Conditioning Engineers, Inc.

† *This table is no longer published in the current ASHRAE Handbook — Fundamentals. The preferred method to determine the effective air leakage area is through a whole-building pressurization test. However, this table can be used to provide a reasonable estimate where the building has not yet been constructed.*

*TABLE S5.2 Stack Coefficient $C_S$*

| | *House Height (Stories)* | | |
|---|---|---|---|
| | *One* | *Two* | *Three* |
| Stack coefficient | 0.0150 | 0.0299 | 0.0449 |

*Source:* 2005 *ASHRAE Handbook — Fundamentals*, Chapter 27, Table 5. © American Society of Heating, Refrigerating and Air-Conditioning Engineers, Inc.

To convert $Q$ (cfm) to $I$ (ACH):

$$I = \frac{Q \times 60 \text{ min/hr}}{V}$$

$I$ = air exchange rate, ft$^3$/hr

$Q$ = airflow rate, cfm

$V$ = interior volume of space, ft$^3$

## FACTORS IN ESTIMATING AIR INFILTRATION

Two of the most important factors in estimating ACH are the determination of the effective air leakage area ($A_L$) and selection of the appropriate indoor/outdoor temperature difference for time interval of calculation ($\Delta t$).

*TABLE S5.3 Wind Coefficient ($C_W$)*

| | *House Height (Stories)* | | |
|---|---|---|---|
| *Shielding Class* | *One* | *Two* | *Three* |
| 1 | 0.0119 | 0.0157 | 0.0184 |
| 2 | 0.0092 | 0.0121 | 0.0143 |
| 3 | 0.0065 | 0.0086 | 0.0101 |
| 4 | 0.0039 | 0.0051 | 0.0060 |
| 5 | 0.0012 | 0.0016 | 0.0018 |

*Source:* 2005 *ASHRAE Handbook — Fundamentals*, Chapter 27, Table 6. © American Society of Heating, Refrigerating and Air-Conditioning Engineers, Inc.

### Effective Air Leakage Area

Effective Air Leakage Area ($A_L$) can be estimated from a test value from a whole-building pressurization test, or by using the ASHRAE method in conjunction with a building's blueprints or measurements.

**Whole-Building Pressurization Test.** Commonly known as a blower door test, the whole-building pressurization test is the preferred way to estimate $A_L$ for existing structures. A qualified person must conduct this test because it requires skill and experience to set up and interpret. For most code

*TABLE S5.4 Local Shielding Classes*

| *Shelter Class* | *Description* |
|---|---|
| 1 | No obstructions or local shielding |
| 2 | Typical shelter for an isolated rural house |
| 3 | Typical shelter caused by other buildings across the street from the building under study |
| 4 | Typical shelter for urban buildings on larger lots where sheltering obstacles are more than one building height away |
| 5 | Typical shelter produced by buildings or other structures that are immediately adjacent (closer than one house height): e.g., neighboring houses on the same side of the street, trees, bushes, etc. |

*Source:* 2005 *ASHRAE Handbook — Fundamentals*, Chapter 27, Table 4. © American Society of Heating, Refrigerating and Air-Conditioning Engineers, Inc

users, a calculation based on the ASHRAE method using Table S5.1 is a sufficient alternative.

**ASHRAE Method.** A calculated method is likely to be necessary because code users often must make combustion air decisions before a building is constructed. The effective air leakage areas for commonly used construction components for low-rise residential structures can be determined using Table S5.1 and the ASHRAE equations. The two examples in this supplement demonstrate this method.

## Indoor/Outdoor Temperature Difference

The ACH used to determine the required indoor volume is based on an average rate. Therefore, the user must select a time period within which to average the formula's indoor/outdoor temperature difference ($t$). For the purpose of determining combustion air, the time period chosen should be the coldest period of the season in which the building is likely to experience the highest $\Delta t$. This period is a baseline for the calculation and provides an indicator of a building's overall ability to supply combustion air. Its duration can vary and should be based on local weather conditions. Generally, a one-week period provides a conservative evaluation period.

## Calculation Examples

### EXAMPLE 1

### *Two-Story House — Average Temperature, 20°F*

Calculate the ACH for a planned, rectangular two-story house that contains 1600 ft$^2$ of living space, with 8 ft ceilings. The house is located in a new subdivision, and there are adjacent structures or houses. The space will be maintained at 70°F. The average outdoor temperature is 20°F during the coldest, one-week period, and the average wind speed is 7 mph.

*Solution*

**1. Determine the Effective Air Leakage Area ($A_L$).** Estimate the effective air leakage area ($A_L$) of a residential structure based on the construction details shown in Table S5.1. Examine the house blueprints and make a list of the construction details, their area or linear foot of crack, and the appropriate leakage area units. For all construction details in this example, the best estimate number was selected from Table S5.1. The list of calculated air leakage rates for this example is shown in Worksheet S5.1.

**2. Calculate the ACH.** The house has a volume of 10,240 ft$^3$ (1600 ft$^2$ × 8 ft × 0.80) and an effective air leakage area ($A_L$) of 66.4 in.$^2$. Note that for this example, an 80 percent reduction factor is used in the volume calculation to account for interior walls, cabinets, and other structures that reduce the interior air volume available for combustion air. The stack coefficient ($C_S$) for a two-story house is 0.0299 (see Table S5.2), and the local shielding class is #4 (see Table S5.4). The wind coefficient ($C_W$) is 0.0051 (see Table S5.3). The airflow rate calculation is shown in Worksheet S5.1.

### EXAMPLE 2

### *Two-Story House — Average Outdoor Temperature, 30°F*

Estimate the average ACH for a two-story house during a period when the average outdoor temperature is 30°F and the indoor space is maintained at 70°F. The house size is 50 ft × 30 ft, with 9 ft ceilings (volume = 21,600 with 80 percent reduction factor). The average wind speed over the period is 7 mph.

*Solution*

**1. Determining the Effective Air Leakage Area ($A_L$).** The effective air leakage area is determined in a similar manner as shown in Example 1 by making a list based on Table S5.1 and the building's blueprints (list not shown). The calculation worksheet in Supplement 6 can be used. The effective air leakage area is found to be 110 in.$^2$.

**2. Calculate the ACH.** The stack coefficient ($C_S$) for a two-story house is 0.0299 (see Table S5.2), and the local shielding class is #4 (see Table S5.4). The wind coefficient ($C_W$) is 0.0051 (see Table S5.3).

The airflow rate is calculated as follows:

$$Q = 110\sqrt{(0.0299)(70 - 30) + (0.0051)(7^2)}$$

$$Q = 132 \text{ cfm}$$

## CALCULATION WORKSHEET: AIR CHANGES PER HOUR

**Step 1:**

- Enter the building parameters in the spaces to the right.

Calculation parameters:
No. of stories: 2
Area of living space: 1600 $ft^2$
Ceiling height: 8 ft

Interior design temperature: 70°F
Coldest 1-week outdoor temperature: 20°F

**Step 2:**

- Determine the effective air leakage rate, $A_L$, by entering the number, area, or length of construction features in Table 1.
- Multiply by the factor in Table 1 and add to find the total $A_L$.

**Table 1 Effective Air Leakage Area**

| Area | Component | Description | Size/No. | $A_L$/Unit | $A_L$ |
|---|---|---|---|---|---|
| Ceiling | | General | 800 $ft^2$ | 0.026 in.²/ft² | 20.8 |
| Walls | | Cont. air infilt. barrier | 1857 $ft^2$ | 0.0022 in.²/ft² | 4.1 |
| Joints | Top plate | Band joist | 120 $ft^2$ | 0.005 in.²/lftc | 0.6 |
| | Sole plate | Caulked | 120 $ft^2$ | 0.04 in.²/lftc | 4.8 |
| Windows | Exterior | Double hung with pressurized | 420 lftc | 0.023 in.²/lftc | 9.7 |
| | Framing | Wood, caulked | 240 $ft^2$ | 0.004 in.²/ft² | 1 |
| Doors | Front exterior | Double | 21 $ft^2$ | 0.12 in.²/ft² | 2.5 |
| | Patio | Sliding | 42 $ft^2$ | 0.079 in.²/ft² | 3.3 |
| | Framing | Doors, wood | 63 $ft^2$ | 0.004 in.²/ft² | 0.3 |
| | Attic | Fold down | 1 | 3.4 in.² ea. | 3.4 |
| Fireplace | | With damper closed | 9 $ft^2$ | 0.62 in.²/ft² | 5.6 |
| Exhaust vents | Kitchen | With damper closed | 1 | 0.8 in.² ea. | 0.8 |
| | Bathroom | With damper closed | 2 | 1.6 in.² ea. | 3.2 |
| | Dryer | With damper | 1 | 0.46 in.² ea. | 0.5 |
| Penetrations | Plumbing | Vent stacks | 2 | 0.3 in.² ea. | 0.6 |
| | Vent | Furnace/water heater | 1 | 3.1 in.² ea. | 3.1 |
| | Wiring | Caulked or gasketed | 4 | 0.3 in.² ea. | 1.2 |
| | Electrical | Interior outlets on outside walls | 20 | 0.023 in.² ea. | 0.5 |
| | Ceiling lights | Surface mounted | 5 | 0.13 in.² ea. | 0.7 |
| | | | | $A_L$ = | 66.7 |

**Step 3:**

- Calculate the air changes per hour (ACH).
- Calculate the house volume. The 0.8 reduction factor corrects for interior spaces not available for combustion (i.e., closets, wall spaces, cabinets).
- Determine values for $C_S$ and $C_W$ (from Tables S5.2, S5.3, or S5.4 as appropriate), and $U$.

House volume = 1600 $ft^2$ x 8 ft x 0.8 = 10,240 $ft^3$
$C_S$ = 0.0299, stack coefficient, from Table S5.2 for a two-story house
$C_W$ = 0.0051, wind coefficient, from Tables S5.3 and S5.4
$U$ = 7 mph average wind speed over a time interval (1 week in this example)

**Step 4:**

- Calculate the air flow through the house, $Q$.
- Calculate the air changes per hour, $I$.

$Q = A_L / C_S \Delta t + C_W U^2$
$= 66.7/0.0299 \times 0.0051 \times 7^2$
= 88 cfm

$I = Q \times 60$ min/hr/V
$= 88 \times 60/10{,}240$
= 0.52 ACH

**Job:** 25 Main Street **Prepared by:** TL **Date:** 1/1/09



**WORKSHEET S5.1** *Calculating Air Changes per Hour.*

The ACH rate is calculated as follows:

$$I = \frac{(132)(60)}{21{,}600}$$

$$I = 0.37\ ACH$$

## SUMMARY

The calculated method to estimate the air infiltration rate in low-rise residential buildings is relatively straightforward and can be easily accomplished if one knows the basic layout and features of the structure. The selection of "best estimate" effective leakage areas from Table S5.1 offers a realistic estimate for most typical construction.

## REFERENCE

*ASHRAE Handbook — Fundamentals,* "Ventilation and Infiltration," Chapter 27, American Society of Heating, Refrigerating, and Air-Conditioning Engineers, Atlanta, GA, 2001 and 2005.

## SUPPLEMENT 6

# Calculation Worksheets

***Editor's Note:*** *This supplement contains a series of calculation worksheets in a reproducible format for use by the installer of gas piping, gas appliances, and gas venting systems. Duplication and use of these worksheets is encouraged.*

This supplement consists of the following calculation worksheets, which the reader may reproduce and use. These worksheets may not be copied for commercial sale or distribution:

- Worksheet S6.1–Calculating Pipe Size Using the Longest Length Method
- Worksheet S6.2–Calculating Pipe Size Using the Branch Length Method
- Worksheet S6.3–Sizing Combustion Air Openings Using the Standard Method
- Worksheet S6.4–Sizing Combustion Air Openings Using the KAIR Method
- Worksheet S6.5–Calculating the Size of Combustion Air Openings
- Worksheet S6.6–Calculating the Size of Reduced Building Area Openings
- Worksheet S6.7–Calculating Vent Size for Single Appliance
- Worksheet S6.8–Calculating Vent Size for Two Appliances
- Worksheet S6.9–Calculating Air Changes per Hour

## CALCULATION WORKSHEET: PIPE SIZING, LONGEST LENGTH METHOD

**Step 1:**

- Draw a sketch of a piping system in the space to the right. Use the back of this page or a separate sheet if more space is needed.

**Step 2:**

- Enter the system information. Note that demand is the amount of gas flowing through a section of pipe.
- Use total Btu/hr rating/1000 ($ft^3$/hr) for natural gas.
- Use total Btu/hr for propane.

**Step 3:**

- Determine the gas used and system pressure, and enter it to the right.
- Determine the piping material and enter it to the right.
- Select the appropriate pipe sizing table from Chapter 6 and enter it to the right.

**Step 4:**

- On the sketch, label the section of pipe from the point of delivery (meter or regulator) to the first tee as Section 1.
- Label the section from the first tee to the second tee as Section 3. Use similar section numbers for additional sections.

**Step 5:**

- Determine the longest length of piping from the point of delivery to the most remote appliance. Enter this length for all pipe sections in Table 1.

Pipe system sketch

Gas: ____________________

System pressure: ____________________

Piping material: ____________________

Table used: ____________________

Pressure drop: ____________________

**Table 1 Piping System Table**

| Section | Demand | Section length | Size |
|---|---|---|---|
| 1 | | | |
| 2 | | | |
| 3 | | | |
| 4 | | | |
| 5 | | | |

**Step 6:**

- Enter the input rating for each appliance in Table 2. For natural gas appliances, enter the input rating in Btu/hr/1000 ($ft^3$/hr). For propane appliances, enter the input rating in Btu/hr.

**Step 7:**

- From the table, determine the length of each pipe section using the appropriate table, using only the row with the longest length. Round up to the lengths in the table. Read across until a capacity equal to or greater than the required demand for the section is found. Read up to find the size. Repeat for each section of piping. Enter this size in Table 2.

**Table 2 Appliances Table**

| Appliance | Demand | Section length | Size |
|---|---|---|---|
| Furnace | | | |
| Furnace | | | |
| Water heater | | | |
| Water heater | | | |
| Range | | | |
| Oven | | | |
| Dryer | | | |
| Other | | | |
| Other | | | |
| Other | | | |
| Other | | | |
| Total | | | |

**Job:** ____________________ **Prepared by:** ____________________ **Date:** ____________________



***WORKSHEET S6.1*** *Calculating Pipe Size Using the Longest Length Method.*

## CALCULATION WORKSHEET: PIPE SIZING, BRANCH LENGTH METHOD

**Step 1:**
- Draw a sketch of a piping system in the space to the right.

**Step 2:**
- Enter the system information. Note that demand is the amount of gas flowing through a section of pipe.
- Use total Btu/hr rating/1000 ($ft^3$/hr) for natural gas.
- Use total Btu/hr for propane.

**Step 3:**
- Determine the gas used and system pressure, and enter it to the right.
- Determine the piping material and enter it to the right.
- Select the appropriate pipe sizing table from Chapter 6 and enter it to the right.

Pipe system sketch

Gas: ____________________

Pressure: ____________________

Piping material: ____________________

Table used: ____________________

Pressure drop: ____________________

**Step 4:**
- On the sketch above, label the section of pipe from the point of delivery (meter or regulator) to the first tee as Section 1.
- Label the section from the first tee to the second tee as Section 2. Use similar section numbers for additional sections.

**Step 5:**
- Determine the length of the branch serving each appliance. Enter this length in Table 1.

**Table 1 Piping System Table**

| Section | Demand | Section length | Size |
|---|---|---|---|
| 1 | | | |
| 2 | | | |
| 3 | | | |
| 4 | | | |
| 5 | | | |

**Step 6:**
- Enter the input rating for each appliance in Table 2. For natural gas appliances, enter the input rating in $ft^3$/hr. For propane appliances, enter the input rating in thousands of Btu/hr.

**Step 7:**
- From the table, determine the length of each pipe section using the appropriate table, using only the row with the longest length. Round up to the lengths in the table. Read across until a capacity equal to or greater than the required demand for the section is found. Read up to find the size. Repeat for each section of piping. Enter this size in Table 2.

**Table 2 Appliances Table**

| Appliance | Demand | Section length | Size |
|---|---|---|---|
| Furnace | | | |
| Furnace | | | |
| Water heater | | | |
| Water heater | | | |
| Range | | | |
| Oven | | | |
| Dryer | | | |
| Other | | | |
| Other | | | |
| Other | | | |
| Other | | | |
| Total | | | |

**Job:** ____________________ **Prepared by:** ____________________ **Date:** ____________________



***WORKSHEET S6.2*** *Calculating Pipe Size Using the Branch Length Method.*

## CALCULATION WORKSHEET: COMBUSTION AIR, STANDARD METHOD

**Step 1:**

- Calculate the room volume.

Room volume: Room length: ________________

Room width: ________________

Room height: ________________

Room volume: = Length × width × height

= ______________

**Step 2:**

- Calculate the total input of all appliances in the room.
- Enter the input rating of all appliances in Table 1. (Per 9.3.1.1, Exception 2, dryers are not included.)
- Total the column.
- Divide the total by 1000 (of Btu/hr).

**Table 1 Appliances Table**

| Appliance | Input rating (Btu/hr) |
|---|---|
| Furnace | |
| Water heater | |
| Space heater | |
| Range | |
| Other | |
| Total | |
| Total/1000 | |

**Step 3:**

- Calculate the required volume. Divide room volume (Step 1) by total/1000 (Step 2).
- If less than 50, additional air is needed.
- If greater than or equal to 50, no additional air is needed.

Required volume:

Additional air needed? (Check one) Yes ❑ No ❑

## ALTERNATE CALCULATION METHOD

**Step 1:**

- Calculate the room volume.

Room volume = ________________ (from Step 1 above)

**Step 2:**

- Calculate the maximum appliance input.

Maximum appliance input: = Room volume × 20

= ___________ × 20

= ___________

**Step 3:**

- Determine if additional air is needed.
- If less than max., no additional air is needed.
- If greater than or equal to max., additional air is needed.

Total appliance input: = ___________

Additional air needed? (Check one) Yes ❑ No ❑

**Job:** ________________________ **Prepared by:** ________________ **Date:** ________________



***WORKSHEET S6.3*** *Sizing Combustion Air Openings Using the Standard Method.*

## CALCULATION WORKSHEET: COMBUSTION AIR, KAIR METHOD

**Step 1:**
- Enter the input ratings of all non-fan-assisted appliances in Table 1.
- Total the column.

**Table 1 Ratings for Non-Fan-Assisted Appliances**

| Appliance | Input rating (Btu/hr) |
|---|---|
| Furnace | |
| Water heater | |
| Space heater | |
| Refrigerator | |
| Other | |
| Other | |
| Total ($I_{other}$) | |

**Step 2:**
- Enter the air infiltration rate of the space. This can be determined by measurement or estimation. See Supplement 5 for an estimation method.
- If the air infiltration rate exceeds 0.6, use 0.6.

Air infiltration rate:________________ ACH (max. 0.6)

**Step 3:**
- Calculate the required volume for non-fan-assisted appliances using the equation.
- Enter the required volume here and in Step 6.

Required volume:

$$\frac{21\ \text{ft}^3}{\text{ACH}}\left(\frac{I_{other}}{100{,}000\ \text{Btu/hr}}\right) = \frac{21\ \text{ft}^3}{\_\_\_\_}\left(\frac{\_\_\_\_\_\_\ \text{Btu/hr}}{100{,}000\ \text{Btu/hr}}\right)$$

$$\_\_\_\_\ (\_\_\_\_) = \_\_\_\_\_\_\ \text{ft}^3$$

**Step 4:**
- Enter the input ratings of all fan-assisted appliances in Table 2.
- Total the column.

**Table 2 Ratings for Fan-Assisted Appliances**

| Appliance | Input rating (Btu/hr) |
|---|---|
| Furnace | |
| Furnace | |
| Space heater | |
| Other | |
| Other | |
| Other | |
| Total | |

**Step 5:**
- Calculate the required volume for fan-assisted appliances using the equation.
- Enter the required volume here and below.

Required volume:

$$\frac{15\ \text{ft}^3}{\text{ACH}}\left(\frac{I_{fan}}{100{,}000\ \text{Btu/hr}}\right) = \frac{15\ \text{ft}^3}{\_\_\_\_}\left(\frac{\_\_\_\_\_\_\ \text{Btu/hr}}{100{,}000\ \text{Btu/hr}}\right)$$

$$\_\_\_\_\ (\_\_\_\_) = \_\_\_\_\_\_\ \text{ft}^3$$

**Step 6:**
- Add the two required volumes calculated to find the total required volume for all appliances.

Required volume (Non-fan assisted) =________________ ft$^3$
Required volume (Fan assisted) = ________________ ft$^3$
Total required volume = ____________ ft$^3$

**Step 7:**
- Enter the available room volume.

Available volume = ____________
Room volume = ____________ ft$^3$

**Step 8:**
- If the total required volume is greater than the available room volume, air openings are required. Check Yes and determine the opening size and location. If the total required volume is less than the available room volume, no air openings are required. Check No.

Air openings required? (Check one)

Yes ❑ No ❑

**Job:** ____________________ **Prepared by:** ____________________ **Date:** ____________



***WORKSHEET S6.4*** *Sizing Combustion Air Openings Using the KAIR Method.*

## CALCULATION WORKSHEET: SIZING COMBUSTION AIR OPENINGS

**Step 1:**

- Enter the input ratings of all appliances in Table 1.
- Total the column.
- Proceed to either Part A, Part B, or Part C.

**Table 1 Ratings for Appliances**

| Appliance | Input rating (Btu/hr) |
|---|---|
| Furnace | |
| Water heater | |
| Space heater | |
| | |
| | |
| Total | |

### A. All air from outdoors via two permanent openings (or vertical ducts).

**Step 2:**

- Where all air is to be taken from the outdoors, divide the total input of all gas appliances in the space by 4000.

Total input:

__________ /4000 = __________ in.$^2$/opening

**Step 3:**

- Select a duct with the area needed.
- Use Table 2 to calculate for square or round ducts. For other sizes or rectangular shapes, the duct size can be calculated.

Square duct: ____ in. × ____ in. = ____ in.$^2$

Round duct: ____ in. round = ____ in.$^2$

**Table 2 Sizes for Square and Round Ducts**

| Area of Square Ducts | | Area of Round Ducts | |
|---|---|---|---|
| Side (in.) | Area (in.) | Diameter (in.) | Area (in.$^2$) |
| 3 | 9 | 3 | 7.1 |
| 4 | 16 | 4 | 12.6 |
| 5 | 25 | 5 | 19.6 |
| 6 | 36 | 6 | 28.3 |
| 7 | 49 | 7 | 38.5 |
| 8 | 64 | 8 | 50.2 |
| 10 | 100 | 10 | 78.5 |
| 12 | 144 | 12 | 113.0 |

### B. All air from the outdoors via two horizontal ducts.

**Step 2:**

- Where all air is to be taken from the outdoors, divide the total input of all gas appliances in the space by 2000.

Total input:

__________ /2000 = __________in.$^2$/opening

**Step 3:**

- Select a duct with the area needed.
- Use Table 2 to calculate for square or round ducts. For other sizes or rectangular shapes, the duct size can be calculated.

Square duct: ____ in. × ____ in. = ____ in.$^2$

Round duct: ____ in. round = ____ in.$^2$

### C. All air from the outdoors via one opening.

**Step 2:**

- Where all air is to be taken from the outdoors using one opening, divide the total input of all appliances in the space by 3000.

Total input:

____________ /3000 = ____________in.$^2$

Provide one ____ in. square duct or one ____ in. round duct.

**Step 3:**

- Check minimum clearances: side and back, 1 in.; front, 6 in.

Side and back: ____ Clearances OK? (Check one)

Front: ____ Yes ❑ No ❑

**Step 4:**

- Check that total vent connector area is less than or equal to the opening area (Step 2).

Furnace connector diameter = ____

Divide by 2 = ____

Squared = ____ × 3.14 = ____

Water heater connector diameter = ____

Divide by 2 = ____

Squared = ____ × 3.14 = ____

Total vent connector area = ____

**Job:** ______________________ **Prepared by:** ______________________ **Date:** ______________



***WORKSHEET S6.5*** *Calculating the Size of Combustion Air Openings.*

## CALCULATION WORKSHEET: SIZING REDUCED BUILDING AREA OPENINGS

**Step 1:**

- Calculate the room volume.

Room volume: Room length:______________
Room width: ______________
Room height:______________

Room volume: = Length × width × height
= _________

**Step 2:**

- Calculate the total input of all appliances in the room.
- Enter the input rating of all appliances in Table 1.
- Total the column.
- Divide the total (room volume) by 1000 (of Btu/hr).

**Table 1 Appliances Table**

| Appliance | Input rating (Btu/hr) |
|---|---|
| Furnace | |
| Water heater | |
| Space heater | |
| Range | |
| Other | |
| Total | |
| Total/1000 | |

**Step 3:**

- Calculate the required volume.
- Enter the available room volume (Step 1).

Required volume:
= 50 $ft^3$ per 1000 Btu/hr (total/1000)
= 50 × ______
= _________

Available room volume:
= _________ $ft^3$

**Step 4:**

- Calculate the percentage of required volume that is available in the room.

Percentage of volume available: $= \dfrac{\text{Available volume}}{\text{Required volume}} \times 100$
= _________ × 100
= _________ %

**Step 5:**

- Calculate the percentage of volume required.

Percentage of volume available: = 100% – _________ %
= _________ %

**Step 6:**

- Calculate the reduced opening area.

Reduced opening = Required area × percentage reduction
= _________ $in.^2$ × _________ %
= _________ $in.^2$
or two _____ $in.^2$ by ________ $in.^2$ openings.

**CAUTION: Do not apply to one opening.**

**Job:** ______________________________ **Prepared by:** ______________________ **Date:** _________________



***WORKSHEET S6.6*** *Calculating the Size of Reduced Building Area Openings.*

## CALCULATION WORKSHEET: VENT SIZING, SINGLE APPLIANCE

**Step 1:**

- Determine the type of chimney or vent used and enter it to the right.
- If used, determine the type of vent connector material and enter it to the right.
- Select the appropriate table and enter it to the right.

Type of vent: ______________________

Type of connector: __________________

Height: ___________________________

Lateral: ___________________________

Table used: ________________________

**Step 2:**

- Draw a sketch of the proposed design in the space to the right. Use the back of this page or a separate sheet if more space is needed.

Note: Factors such as combining connectors prior to entering the common vent will affect capacity.

Proposed design sketch

**Step 3:**

- Determine the common vent size using the single appliance vent table. Read across the row with the correct height and lateral offset. Use the NAT Min column for draft hood appliances and the FAN Min and Max columns for draft hood appliances.

Vent size: ____________________

**Step 4:**

- Check for vent downsizing.
- If the vent is smaller than the draft hood outlet or flue collar, see 13.1.2 for restrictions.

Draft hood outlet/
flue collar diameter: ________________

**Step 5:**

- Check for excessive elbows.
- If more than two elbows are used, derate the single appliance vent table values by 10% for each elbow above 2 (see 13.1.3).

Number of elbows: ____________________

❑ Derating not required

❑ Derating required. Show calculation on a separate page.

**Job:** ______________________ **Prepared by:** ____________________ **Date:** ______________



***WORKSHEET S6.7*** *Calculating Vent Size for Single Appliance.*

## CALCULATION WORKSHEET: VENT SIZING, TWO APPLIANCES

**Step 1:**

- Determine the type of chimney or vent used and enter it to the right.
- Determine the type of vent connector material to be used and enter it to the right.
- Determine the number of appliances and enter it to the right.
- Select the appropriate table and enter it to the right.

Type of vent: ____________________

Type of connector: ____________________

Number of appliances: ____________________

Table used: ____________________

**Step 2:**

- Draw a sketch of the proposed design in the space to the right. Use the back of this page or a separate sheet if more space is needed.

Note: Factors such as combining connectors prior to entering the common vent will affect capacity.

Proposed design sketch

**Step 3:**

- Determine the vent connector size using the top portion of the multiple appliance vent table, as shown in Table 1.

**Step 4:**

- Check for vent connector size limits, 13.2.21.

**Table 1 Top of Multiple Appliance Vent Table**

| Appliance | Connector rise | Vent height | Input rating | Fan or natural | Connector size | Table used |
|---|---|---|---|---|---|---|
| Furnace | | | | | | |
| Water heater | | | | | | |
| | | | | | | |
| | | | | | | |
| Total | | | | | | |

**Step 5:**

- Check for excessive connector horizontal length using Table 2.
- If any connector horizontal length in feet exceeds 1.5 times the diameter in inches, derate the table by 10% for each length unit and recalculate connector length (see 13.2.2).

**Table 2 Connector Horizontal Length**

| Connector | Diameter | Max. length | Length | OK? |
|---|---|---|---|---|
| | | | | |
| | | | | |
| | | | | |
| | | | | |

**Step 6:**

- Determine the common vent size using the bottom portion of the multiple appliance vent table, as shown in Table 3.

**Table 3 Bottom of Multiple Appliance Vent Table**

| Appliance types | Vent height | Combined input | Common vent size | Table used |
|---|---|---|---|---|
| | | | | |

**Job:** ____________________ **Prepared by:** ____________________ **Date:** ____________________



***WORKSHEET S6.8*** *Calculating Vent Size for Two Appliances.*

## CALCULATION WORKSHEET: AIR CHANGES PER HOUR

**Step 1:**

- Enter the building parameters in the spaces to the right.

Calculation parameters:

No. of stories: __________

Area of living space: __________

Ceiling height: __________

Interior design temperature: __________

Coldest 1-week outdoor temperature: __________

**Step 2:**

- Determine the effective air leakage rate, $A_L$, by entering the number, area, or length construction features in Table 1.
- Multiply by the factors in Table 1 and add to find the total $A_L$.

**Table 1 Effective Air Leakage Area**

| Area | Component | Description | Size/No. | $A_L$/Unit | $A_L$ |
|---|---|---|---|---|---|
| Ceiling | | General | | 0.026 in.$^2$/ft$^2$ | |
| Walls | | Cont. air infilt. barrier | | 0.0022 in.$^2$/ft$^2$ | |
| Joints | Top plate | Band joist | | 0.005 in.$^2$/lftc | |
| | Sole plate | Caulked | | 0.04 in.$^2$/lftc | |
| Windows | Exterior | Double hung with pressurized | | 0.023 in.$^2$/lftc | |
| | Framing | Wood, caulked | | 0.004 in.$^2$/ft$^2$ | |
| Doors | Front exterior | Double | | 0.12 in.$^2$/ft$^2$ | |
| | Patio | Sliding | | 0.079 in.$^2$/ft$^2$ | |
| | Framing | Doors, wood | | 0.004 in.$^2$/ft$^2$ | |
| | Attic | Fold down | | 3.4 in.$^2$ ea. | |
| Fireplace | | With damper closed | | 0.62 in.$^2$/ft$^2$ | |
| Exhaust vents | Kitchen | With damper closed | | 0.8 in.$^2$ ea. | |
| | Bathroom | With damper closed | | 1.6 in.$^2$ ea. | |
| | Dryer | With damper | | 0.46 in.$^2$ ea. | |
| Penetrations | Plumbing | Vent stacks | | 0.3 in.$^2$ ea. | |
| | Vent | Furnace/water heater | | 3.1 in.$^2$ ea. | |
| | Wiring | Caulked or gasketed | | 0.3 in.$^2$ ea. | |
| | Electrical | Interior outlets on outside walls | | 0.023 in.$^2$ ea. | |
| | Ceiling lights | Surface mounted | | 0.13 in.$^2$ ea. | |
| | | | | $A_L$ = | |

**Step 3:**

- Calculate the air changes per hour (ACH).
- Calculate the house volume. The 0.8 reduction factor corrects for interior spaces not available for combustion (i.e., closets, wall spaces, cabinets).
- Determine values for $C_S$ and $C_W$ (from Tables S5.2, S5.3, or S5.4 as appropriate), and $U$.

House volume = Area × ceiling height × 0.8 =

$C_S$ = (from Table S5.2)

$C_W$ = (from Tables S5.3 and S5.4)

$U$ =

**Step 4:**

- Calculate the air flow through the house, $Q$.
- Calculate the air changes per hour, $I$.

$Q = A_L/C_S\ t + C_W\ U^2$

= __________

= _____ cfm

$I = Q \times 60$ min/hr/V

= __________

= _____ ACH

**Job:** __________ **Prepared by:** __________ **Date:** __________



***WORKSHEET S6.9*** *Calculating Air Changes per Hour.*

# SUPPLEMENT 7

# Clearance Distance for Gas Appliance Sidewall Venting

***Editor's Note:*** *This supplement provides an overview of a research project conducted under the auspices of the Fire Protection Research Foundation and completed in December of 2007. It attempted to provide information on the effects of sidewall venting where buildings are close together (5 ft and 10 ft). It is a preliminary study that was not intended to be a thorough review of all possible cases and variables. It is provided here to make readers of the handbook aware of it. The final report of the study can be downloaded from:*

*http://www.nfpa.org/assets/files/PDF/Research/Gas_Appliance_Sidewall_Venting.pdf*

*As with all NFPA Research Foundation Reports, there is no charge for downloading the Final Report.*

## INTRODUCTION

The objective of this work was to assess the effects of sidewall venting of flue gas from a high-efficiency, natural gas appliance in the presence of a neighboring home. The National Fire Protection Association (NFPA) and the American Gas Association (AGA) cosponsored this study as an initial attempt to identify basic criteria related to the dispersion of sidewall-vented flue gas between neighboring houses. For this study, Battelle performed a review of the literature for information on issues related to sidewall venting and a computational fluid dynamics (CFD) modeling study to investigate the effects of important parameters, such as the spacing between houses, wind direction, and vent terminal design.

## APPROACH

A project technical panel was assembled to guide the project and evaluate the results in accordance with the policies of the Fire Protection Research Foundation. This panel was comprised of representatives from NFPA and AGA member organizations.

There are a vast number of parameters that could be important in determining how flue gas will disperse in the real world, so it was agreed to limit the work as follows:

- A single appliance input rating, excess air, and efficiency
- Three different vent terminal designs
- Two different spacings between the houses
- Two different wind speeds and the outdoor ambient temperatures

The fundamental objective of this study was to take a first step toward understanding the physics and to gather insight as to the factors that may be important in generating guidelines or codes for sidewall venting.

A typical Category IV appliance with a 100,000 Btu/hr input rating operating with 40 percent excess combustion air was chosen as the test subject. Such an appliance, assuming roughly 92 percent efficiency, would produce a flue gas with a flow rate of 23 acfm and a temperature of 115°F. Two identical houses, each 36 ft wide, 25 ft on the sides, 20 ft high to the bottom of the roof, and with a peak height of 28 ft were used. Two house separation distances, 5 ft and 10 ft, were used as well. The vent would be located 1 ft from the ground, centered horizontally and extend 1 ft from the face of the house with the vent. The vent would be 2 in. in diameter and have two terminal designs, one with no terminal (straight vent) and one with a 6 in. diameter plate facing the vent outlet to force a radial dispersion of flue gas as it exits the vent (radial vent). Winter (0°F) and summer (75°F) outdoor temperatures were investigated, with no wind and a 7 mph wind blowing perpendicular to the houses and parallel (between) the houses. This agreed-upon set of parameters resulted in the eleven initial CFD cases. The definition of two additional cases was held in reserve to allow for investigation of questions or opportunities that might arise from the results of the initially defined set of cases. Table S7.1 lists all of the CFD cases that were run in this study. Cases 12 and 13, which are highlighted in Table S7.1, are the two additional cases that were defined after completion of the first eleven.

## PREVIOUS STUDIES

To begin work on the report, previous studies were reviewed to determine if any of the past work on venting of gas appliances could be used to form the basis for decision-making regarding the separation distance between buildings in the presence of a sidewall-vented appliance. Only a few relevant papers were found and these are summarized below.

The Gas Research Institute (GRI) funded research from 1988 to 1994 on the venting of gas appliances [1]. The objective of this program was to develop guidelines for the practical and safe venting of flue gases from mid- and high-efficiency gas-fired appliances. In 1992, the GRI venting program conducted a study to examine means to sidewall vent gas appliances in multistory buildings [2]. As part of this work, the issue of sidewall venting between closely spaced buildings was identified. Several European vent configurations were studied that allowed sidewall-vented appliances to be discharged into a manifold that ran vertically up the side of the building and exited above the roofline. The simplest design incorporated a vertical manifold with two openings, one at the top and one at the bottom. Sidewall vents could be discharged into the manifold, which would cause an upflow due to the buoyancy of the warm vent gases, while maintaining nearly atmospheric pressure in the vent due to the bottom opening. Several variations of this concept are currently in use in Europe, including a U-shaped vent with both openings above the roof. Vent gases are discharged into one leg of the manifold, and the other leg is used to supply combustion air to the appliance. These configurations allow sidewall venting to take place between buildings that are right next to each other because the manifold discharges the vent gases above the roofline.

From 1999 to 2002, Battelle examined the minimum clearance distances between sidewall vented gas appliances and adjacent building openings along the same wall as the sidewall vent [3, 4]. Dr. James Reuther used analytical expressions for gas jet dispersion to estimate dilution factors for vent gases being discharged horizontally from the side of a building. The results show that for a straight-out discharge of a sidewall vent, the minimum separation distance could be less than one foot between the vent outlet and any of the same building openings, and still provide adequate dilution to the vent gases to alleviate safety concerns for the occupants of the building. However, the study did not include the effects of wind on the vent discharge, the effect of different types of vent terminals, or the influence of adjacent buildings or structures on plume dispersion.

*TABLE S7.1 Parameter Matrix for CFD Modeling Cases*

| Case Number | Vent Type | House Separation (ft) | Outdoor Temperature (°F) | Wind Direction | Wind Speed (mph) |
|---|---|---|---|---|---|
| 1 | Radial | 5 | 0 | NA | 0 |
| 2 | Radial | 5 | 75 | NA | 0 |
| 3 | Radial | 10 | 0 | NA | 0 |
| 4 | Radial | 10 | 75 | NA | 0 |
| 5 | 2" Straight | 5 | 0 | NA | 0 |
| 6 | 2" Straight | 5 | 75 | NA | 0 |
| 7 | 2" Straight | 10 | 0 | NA | 0 |
| 8 | 2" Straight | 10 | 75 | NA | 0 |
| 9 | 2" Straight | 5 | 75 | East | 7 |
| 10 | 2" Straight | 5 | 75 | North | 7 |
| 11 | 2" Straight | 5 | 75 | South | 7 |
| 12 | 45° Down | 5 | 0 | NA | 0 |
| 13 | 3" Straight | 5 | 0 | NA | 0 |

In 2006 and 2007, Battelle conducted a study for the Gas Appliance Manufacturers Association (GAMA) on the minimum separation distance between sidewall vents and adjacent building openings in which wind effects and different types of vent terminals were examined [5]. This study used CFD models to compute plume dispersion and dilution factors, which include the effects of different wind speeds and directions, different types of vent terminals, different input ratings of appliances, and vent terminal height above grade. The findings indicated that the minimum separation distances between the vent discharge and the same wall openings currently in NFPA 54, *National Fuel Gas Code,* could probably be reduced if the industry could agree on a reasonable dilution factor for the vent gases [6]. The GAMA study did not calculate dilution factors for adjacent buildings located in the vicinity of the sidewall vent.

## CFD ANALYSIS

CFD simulations were used in this project to estimate the dispersion of flue gas from high-efficiency, sidewall-vented appliances. In CFD, a graphical representation of a fluid-dynamics problem is generated, and the fluid space in the domain is discretized using a computational mesh. CFD software is used to numerically solve the differential equations of fluid flow, energy, and species transport, three-dimensionally on the computational mesh, given conditions specified by the user on the boundaries of the fluid domain. If done correctly, the result is a reasonably accurate calculation of fluid velocity, pressure, temperature, and concentration on each node of the mesh. The solution can be examined and analyzed either graphically or with numerical calculations to determine the conditions anywhere in the domain. A commercial CFD software package was used for these analyses. Note that in the discussion that follows, House 1 refers to the house with the vent and House 2 refers to the neighboring house.

### Model Setup and Boundary Conditions

Exhibit S7.1 contains a diagram of the model used in the simulations without wind. For the cases without wind, the two houses are enclosed in a hemispherical dome, centered on the ground halfway between the houses, with a radius of 400 ft. The surface of the dome represents the sky, far from the houses, and is set as an atmospheric pressure boundary in the model. A pressure boundary in CFD allows fluid to flow into and out of the domain such that the pressure everywhere on the boundary is held constant at a specified value (atmospheric pressure in this case). The bottom face of the dome represents the ground and is modeled as a wall boundary where the air velocity is fixed at zero. The two houses are visible in the center of the model. House 1 (with the vent) is on the left (south) side, and House 2 (the neighboring house) is on the right (north) side. The houses are separated by a distance of 5 ft in the diagram.

Exhibit S7.2 contains a diagram of the model used in the simulations with wind. For the cases with wind, the two houses are enclosed in a rectangular box that is 800 ft × 800 ft horizontally and 400 ft high. The bottom face of the box represents the ground and is modeled as a wall boundary where the air velocity is fixed at zero. The top face of the box is modeled as a free-stream boundary far from all walls and obstacles where it is assumed that the velocity is not changing (zero shear stress). There is no friction or loss of mechanical energy (pressure) on a free-stream boundary. The side walls are used to generate a wind with a specified velocity at the outdoor ambient temperature. To generate a south wind, for example, the left side of the box is assigned a wind velocity where air will enter the domain uniformly. The opposing right face is assigned atmospheric pressure. The two sides of the box that run parallel with the wind direction (east and west, or front and back faces) are specified as free-stream boundaries, similar to the top face of the box.

- Dome has radius of 400 ft.
- Dome surfaces are set as atmospheric pressure boundaries.
- Ground and all other surfaces are set as standard walls (fluid velocity is zero).

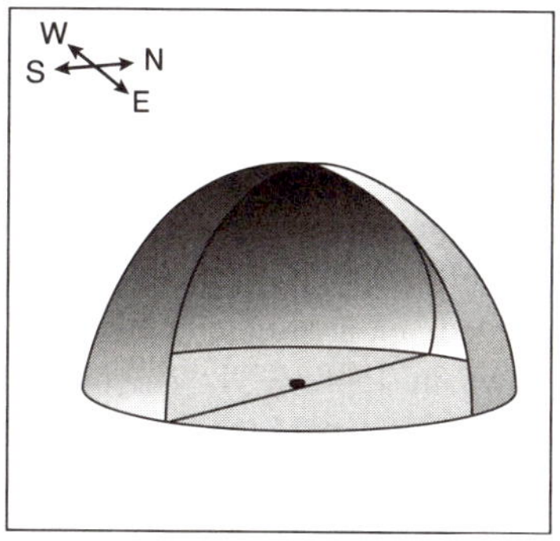


*EXHIBIT S7.1 Diagram of Model Used in Simulations Without Wind. (Courtesy of Battelle)*

- Rectangular box is 800 ft × 800 ft horizontally and 400 ft high.
- Top surface is set as a free stream boundary (zero shear stress).
- For a south wind, the left face is set as a velocity boundary and the right face is set as an atmospheric pressure boundary. The east and west (front and back) faces are set as free stream boundaries (zero shear stress).
- Similar procedure is used for an east/west wind.

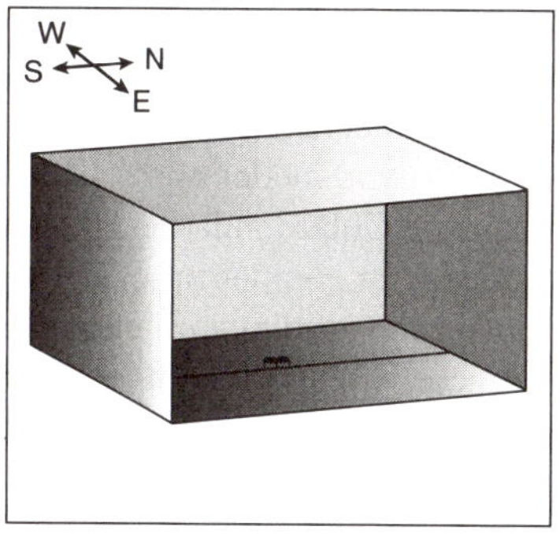


*EXHIBIT S7.2 Diagram of Model Used in Simulations with Wind. (Courtesy of Battelle)*

Exhibit S7.3 shows a diagram of four vent designs that were modeled in this study. In each diagram, the flow direction is from left to right. The straight vent (top left) and radial vent (top right) were the designs decided upon in the initial meetings on this project. The 45 degree down (bottom left) and three-inch straight (bottom right) vents were decided upon after completion of the first eleven cases shown in Table S7.1. The boundary condition set at the vent outlet, where the venting gas enters the fluid domain, depends on the appliance being modeled. The input rating, excess combustion air, and efficiency of the appliance determine the flow rate, composition, and temperature at the outlet of the vent. In this study, the appliance was fixed at 100,000 Btu/hr, 40 percent excess air, and 92 percent efficiency, resulting in a constant flow rate and temperature of 23 acfm and 115°F, respectively. The velocity of flue gas through a 2 in. vent, at a volumetric flow rate of 23 acfm, is 6.1 m/sec (20 ft/sec).

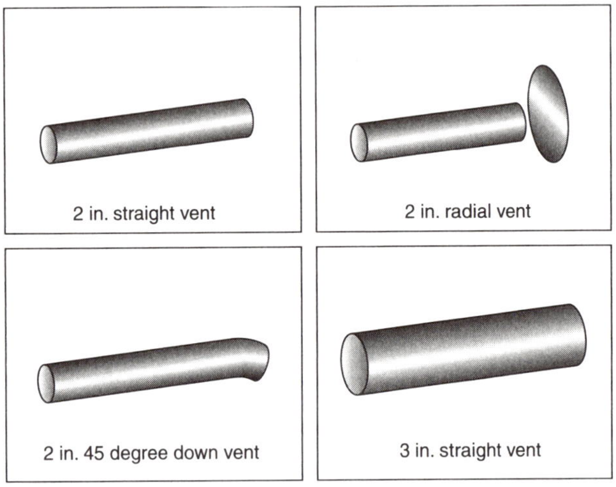


***EXHIBIT S7.3*** *Four Vent Designs Used in the Study. (Courtesy of Battelle)*

The concentration of carbon dioxide and water vapor in the flue gas are certainly important variables in these analyses and can be computed from combustion calculations using the fuel composition, excess air, and their associated temperatures. There is, however, a degree of complexity to modeling the flue gas as a multi-component gas in the CFD simulation. In order to track the concentration of each component of the flue gas through the fluid domain surrounding the house, the multi-component diffusion coefficient of each component would need to be known or calculated. The uncertainties in performing the multi-component calculations do not warrant the added complexity and computational cost. In similar analyses on past projects, we have chosen to model the flue gas as a single component, which we'll call "flue gas," with a representative diffusion coefficient in air and an ideal-gas density based on the temperature. The mass fraction of flue gas, by definition, has a value of one at the vent outlet. By tracking the concentration of flue gas in the domain surrounding the house, the model computes the degree to which the flue gas is diluted by the outside air at every point in the domain. The concentration of a given component, such as carbon dioxide, at every point in the domain can then be estimated with reasonable accuracy by assuming a concentration at the vent outlet and multiplying by the normalized concentration from the CFD calculation. Modeling the flue gas as a single-component gas with a maximum concentration at the vent outlet, which is defined as one, normalizes the concentration profile in the domain. It assumes that the concentration of any component of the flue gas is dispersed and diluted by the same amount as the flue gas itself.

## Method of Analyzing the Results

A parametric study was conducted using CFD simulations to determine the concentration profile of flue gas everywhere in the fluid domain, with emphasis on the region surrounding the vent and between the two houses. In each of the cases, the steady-state concentration profile of flue gas was computed on the face of House 1 (with the vent) and the face of House 2 (neighboring house) opposite the vent. The objective of the simulations was to determine the maximum concentration of flue gas (or carbon monoxide) on the surface of each house. The point of maximum concentration is also the point of minimum dilution, which is defined as the reciprocal of the maximum normalized concentration. It measures the degree to which the flue gas has been diluted in the worst case position on each house (e.g., a minimum dilution of 1000 means that the flue gas has been diluted by a factor of 1000 in the worst case location). The point of minimum dilution on House 1 was determined at distances greater than 4 ft from the vent because existing guidelines state that doors or windows should be located at least this distance from the vent location. The point of minimum dilution on House 2 could occur anywhere. No assumptions about the location of doors or windows on either house were made.

The contour plots in the presentation shown in Appendix A of the full report (*http://www.nfpa.org/assets/files//PDF/Research/Gas_Appliance_Sidewall_Venting.pdf*) show the normalized concentration of flue gas. The plots are scaled by color, from blue to red, with blue representing the lowest concentration and red representing the highest concentration. The scale is from a normalized concentration (or mole fraction) of 0.001 (dilution factor 1000 on the blue end) to 0.2 (dilution factor 5 on the red end), and is the same in all of the plots. Setting low and high limits for the contour plots allows for a view of the shape and trajectory of the plumes. There is no color in the plots where the concentration is outside the limits. In other words, at positions outside the edge of the plumes shown in the contour plots, the normalized concentration is below 0.001 and the dilution factor is greater than 1000.

## Results of Cases Initially Agreed Upon

The value and location of minimum dilution (maximum concentration) on the face of each house was recorded for each of the CFD cases run in this study. Table S7.2 lists the cases, their input parameters, and the results for Houses 1 and 2. The two cases that are highlighted at the bottom of the table are the two additional cases that were decided upon after completion of the initial eleven cases. The rationale for selecting those cases is discussed in the section titled Results for Two Final Cases of this report. The X and Y coordinates for the location of maximum concentration are given in feet, with the origin at the base of the house vertically and centered horizontally. The X coordinate measures how far horizontally the location of maximum concentration is from the center of the house, while the Y coordinate measures how far the location of maximum concentration is from the ground.

With the exception of the case with an east wind of 7 mph (Case 9), which is the direction that runs between the houses, the horizontal location of maximum concentration is within 4 ft from the center of either house. Unless deflected horizontally by wind, the plume generally rises directly in plane with the vent. The vertical location is affected most by the outdoor temperature (or more specifically, the difference in temperature between the flue gas and ambient air) and house separation distance. Temperature difference and separation distance influence the vertical location of maximum concentration because of the competing rates of flue gas convection and buoyancy. When the outdoor temperature is low, the flue gas will be more buoyant and rise more rapidly relative to the rate that it is forced horizontally by the vent. The opposite is true when the outdoor temperature is warmer. When the houses are separated by a greater distance, the plume has more time to rise before impacting the neighboring house.

The effects of parameters that varied in the first eleven cases are discussed in the sections that follow. These parameters include house separation distance, outdoor temperature, vent terminal design, and wind speed and direction.

**Effect of House Separation Distance.** Exhibit S7.4 shows very little effect of separation distance on the House 1 minimum dilution factor with a radial vent (dotted lines). The radial vent design traps the flue gas nearer to House 1 regardless of house separation distance. There is an effect of separation distance on the dilution factor for House 1 with the 2 in. straight vent (solid lines) because the straight vent design throws the flue gas away from House 1 toward House 2. When the separation distance is higher, the gas is allowed to disperse away from House 1 and is more diluted near House 1. Regardless of separation distance, however, the flue gas is diluted by more than a factor of 100 before reaching the surface of House 1.

Exhibit S7.5 shows a similar plot of minimum dilution factor on House 2, which is the neighboring house. The results are virtually opposite those in Exhibit S7.4 for House

*TABLE S7.2 Matrix of Results for All CFD Cases*

| | Parameters | | | | | Results for House 1 | | | | Results for House 2 | | | |
|---|---|---|---|---|---|---|---|---|---|---|---|---|---|
| Case | Vent | Bldg. Sep. (ft) | Wind Dir | Wind Speed (mph) | Amb Temp (°F) | Max Conc X Pos (ft) | Max Conc Y Pos (ft) | Max Conc | Min Dilution | Max Conc X Pos (ft) | Max Conc Y Pos (ft) | Max Conc | Min Dilution |
| 1 | Radial | 5 | NA | 0 | 0 | 0.15 | 7.25 | 8.00E-02 | 13 | −0.05 | 19.50 | 1.15E-03 | 869 |
| 2 | Radial | 5 | NA | 0 | 75 | −0.08 | 7.40 | 1.17E-01 | 9 | 0.45 | 19.73 | 1.62E-03 | 616 |
| 3 | Radial | 10 | NA | 0 | 0 | −0.15 | 7.25 | 7.55E-02 | 13 | −0.49 | 17.76 | 3.29E-06 | 304,300 |
| 4 | Radial | 10 | NA | 0 | 75 | −0.15 | 7.25 | 1.04E-01 | 10 | 0.03 | 14.99 | 6.40E-06 | 156,200 |
| 5 | 2" Straight | 5 | NA | 0 | 0 | 2.01 | 19.63 | 1.61E-03 | 622 | −0.46 | 1.53 | 8.46E-02 | 12 |
| 6 | 2" Straight | 5 | NA | 0 | 75 | 4.39 | 19.74 | 4.55E-03 | 220 | 0.32 | 1.07 | 9.91E-02 | 10 |
| 7 | 2" Straight | 10 | NA | 0 | 0 | −0.17 | 19.57 | 2.41E-04 | 4,145 | 0.15 | 10.14 | 2.05E-02 | 49 |
| 8 | 2" Straight | 10 | NA | 0 | 75 | 0.59 | 7.20 | 3.18E-05 | 31,420 | −0.06 | 2.30 | 5.93E-02 | 17 |
| 9 | 2" Straight | 5 | East | 7 | 75 | 17.85 | 0.82 | 6.68E-03 | 150 | 17.39 | 2.08 | 9.40E-04 | 1,064 |
| 10 | 2" Straight | 5 | North | 7 | 75 | −2.28 | 7.48 | 2.63E-02 | 38 | −2.38 | 1.51 | 6.60E-02 | 15 |
| 11 | 2" Straight | 5 | South | 7 | 75 | 3.66 | 17.48 | 8.57E-03 | 117 | 0.32 | 1.07 | 7.20E-02 | 14 |
| 12 | 45° Down | 5 | NA | 0 | 0 | −0.50 | 19.82 | 1.77E-03 | 564 | 0.49 | 0.28 | 1.20E-01 | 8 |
| 13 | 3" Straight | 5 | NA | 0 | 0 | 0.97 | 19.59 | 1.23E-03 | 813 | 0.31 | 3.53 | 8.48E-02 | 12 |

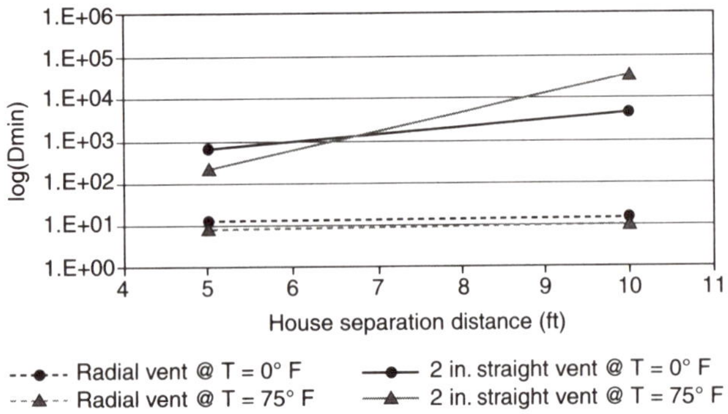


***EXHIBIT S7.4*** *Effect of House Separation Distance on the Minimum Dilution Factor for House 1.*

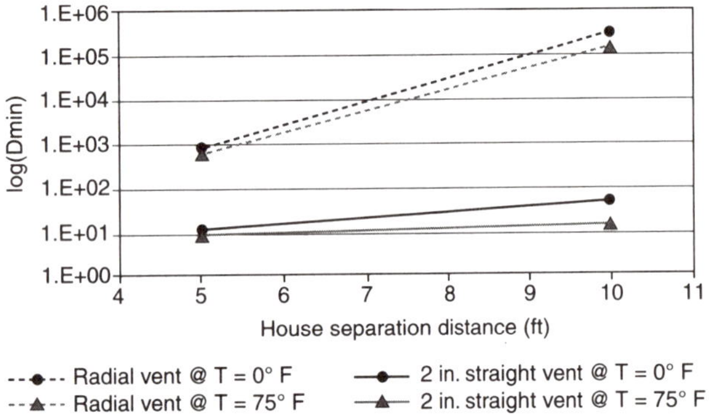


***EXHIBIT S7.5*** *Effect of House Separation Distance on the Minimum Dilution Factor for House 2.*

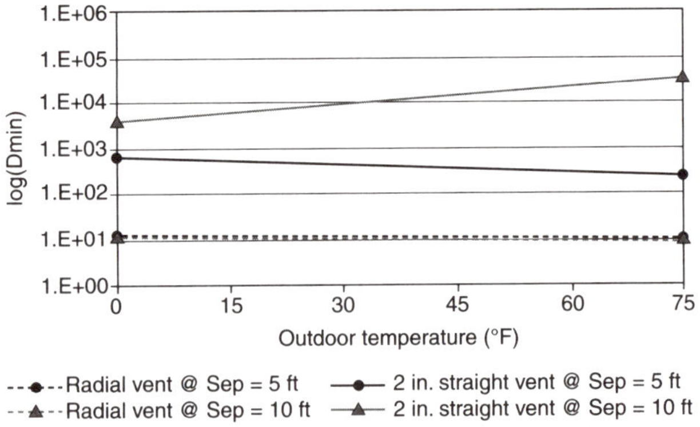


***EXHIBIT S7.6*** *Effect of Outdoor Temperature on the Minimum Dilution Factor for House 1.*

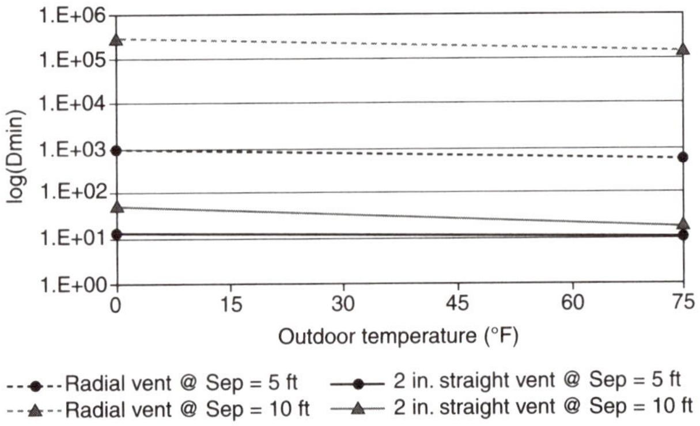


***EXHIBIT S7.7*** *Effect of Outdoor Temperature on the Minimum Dilution Factor for House 2.*

1. Because the radial vent traps the flue gas nearer to House 1, the dilution factors are much higher, and well above the threshold, on House 2. Because the straight vent throws the flue gas toward House 2, the dilution factors are much lower on House 2. The case with a 2 in. straight vent, an outdoor temperature of 0°F, and a house separation distance of 10 ft (Case 7) is the only straight vent case with no wind that meets the threshold requirement for the neighboring house. In that case, the rate of buoyancy (rise) is just sufficient to allow the flue gas to disperse and be diluted by a factor of more than 44 before reaching House 2.

**Effect of Outdoor Temperature.** Exhibits S7.6 and S7.7 show plots of minimum dilution factor versus outdoor temperature, without wind, for Houses 1 and 2, respectively. As discussed previously, outdoor temperature determines the rate of buoyancy or rise. When the outdoor temperature is cooler, the flue gas will rise more rapidly. Both plots show that temperature has a relatively small effect on dilution at the surface of either house. All of the plots thus far indicate that the dilution factors are governed more by vent design than by temperature. The effect of vent design is discussed in the next section.

**Effect of Vent Terminal Design.** Exhibits S7.8 and S7.9 show plots of minimum dilution factor versus vent terminal design, without wind, on Houses 1 and 2, respectively, for the two vent designs considered in the initial set of cases (radial vent and 2 in. straight vent). In these plots, the dotted lines are for cases at a house separation distance of 5 ft and the solid lines represent cases with a separation

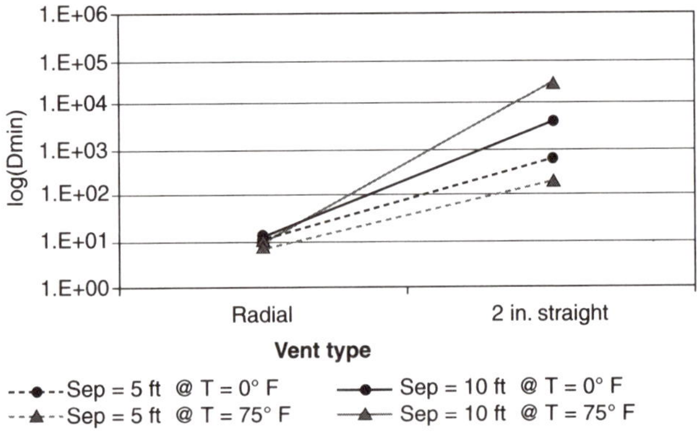


***EXHIBIT S7.8*** *Effect of Vent Terminal Design on the Minimum Dilution Factor for House 1.*

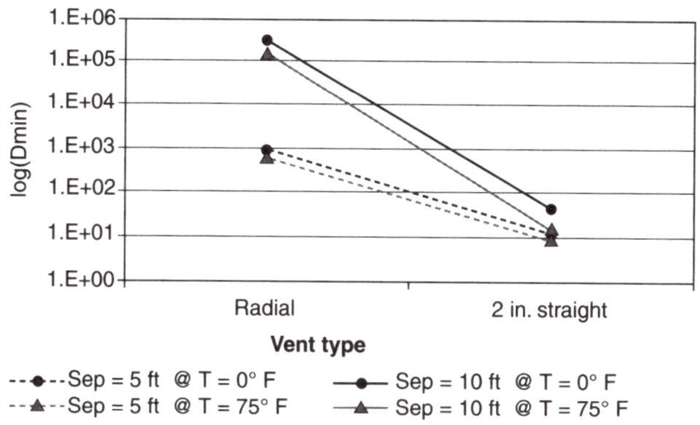


***EXHIBIT S7.9*** *Effect of Vent Terminal Design on the Minimum Dilution Factor for House 2.*

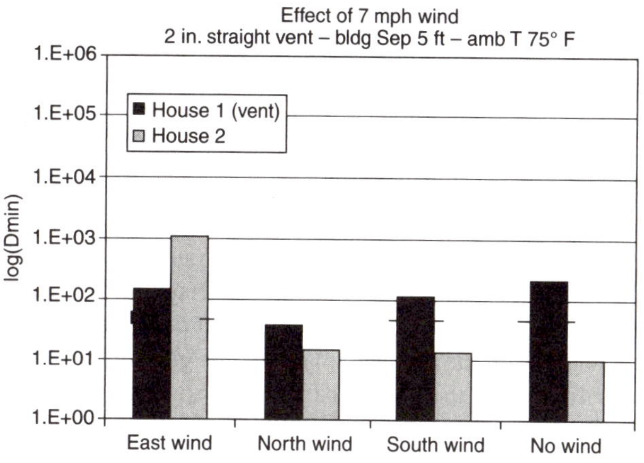


***EXHIBIT S7.10*** *Effect of Wind on the Minimum Dilution Factor for Both Houses.*

distance of 10 ft. These plots clearly show that in the absence of wind, the dilution factor is most dependent on vent terminal design. The radial vent clearly traps the flue gas nearer to House 1 (low dilution factor) and protects House 2 (high dilution factor). The 2 in. straight vent produces the opposite effect. It throws the flue gas toward House 2 (low dilution factor) and protects House 1 (high dilution factor).

**Effect of Wind Speed and Direction.** Two wind cases, both at a wind speed of 7 mph, were initially selected for analysis in this study, one perpendicular to the houses and one parallel, or between the two houses. After some discussion among members of the technical review committee, it was decided to run the perpendicular wind cases in both directions (i.e., wind from behind House 1 and wind from behind House 2) because it was not clear the effects would be the same. When the wind blows over a building, there is a zone of recirculation (recirculation shadow) on the downwind side of the building. There is turbulence on the downwind side that forces the air downward in a circular fashion and back toward the downstream face of the building. It was not clear that the effect of a shadow cast by House 1 in the presence of a south wind (from behind House 1) would be the same as a shadow cast by House 2 in the presence of a north wind (from behind House 2).

Exhibit S7.10 shows a bar chart that shows the dilution factor on both House 1 (blue bars) and House 2 (red bars) in each of the three wind cases. In each of the cases, the vent is a 2 in. straight vent, the house separation distance is 5 ft, and the outdoor temperature is 75°F. Case 6, which is the similar case with no wind, is included in the plot for comparison. The east wind, which is the direction between the houses and parallel to them, carries the flue gas away and dilutes the space between the houses considerably.

The effects of wind perpendicular to the houses are similar in each direction. As seen in Exhibit S7.10, the straight vent tends to carry the flue gas away from House 1 and toward House 2, resulting in higher dilution factors on House 1 (blue bar) and lower dilution factors on House 2 (red bar). The bars for the north and south wind cases are similar, but the dilution factor on House 1 is decreased compared to the case with no wind, and the dilution factor on House 2 is slightly increased compared to the case with no wind. Generally speaking, the turbulence that is created between the houses as a result of wind perpendicular to the houses tends to mix and disperse the flue gas. As a result, it reaches the surface of House 1 at a slightly higher concentration, and House 2 at a lower concentration, than it would without wind.

## Results for Two Final Cases

The definition of two final cases was held in reserve pending the insight that might be gained from the analysis of the results of the eleven cases initially defined. From those initial cases, it was determined that wind certainly can have a profound effect on, and in most cases be helpful with, flue gas dispersion. Wind helps disperse and sweep the flue gas away, particularly in this application where the vent is located between two buildings and the wind is forced to approach the vent at an angle. Outdoor temperature had little effect, and no real determination could be made as to a limit for house separation distance for safe venting. Vent design emerged as the dominant factor. It is also a factor over which code bodies, manufacturers, and installers have some control. Regardless of house separation distance, the radial vent design protects the neighboring house and traps the flue gas near the house with the vent, while the 2 in. straight vent sends the flue gas away from House 1 toward House 2. The two final cases were designed to begin answering whether it is possible to achieve sufficient dilution on both houses at a separation distance of 5 ft with different vent terminal designs. It was further decided to run the two cases with no wind and at a winter outdoor temperature of 0°F.

The two vent terminal designs selected for the final cases are shown in Exhibit S7.3. The 45-degree down vent has a nominal diameter of 2 in. and directs the flue gas downward at a 45-degree angle. It was selected because the interaction of the flue gas with the ground could create turbulence and allow the gas to disperse more before reaching House 2. The 3 in. straight vent design was chosen to decrease the momentum of the flue gas as it exits the vent. Increasing the diameter from 2 in. to 3 in. decreases the flue-gas velocity by a factor of 55 percent and could perhaps allow the gas to disperse more before reaching House 2.

Exhibit S7.11 shows a bar chart of minimum dilution factors on House 1 (blue bars) and House 2 (red bars) for the four vent designs modeled in this study. In all of the cases there is no wind, the house separation distance is 5 ft, and the outdoor temperature is 0°F. Both the 45-degree down vent and the 3 in. straight vent produce results similar to the 2 in. straight vent. The 45-degree down vent sends the flue gas along the ground, but it still reaches House 2 at a concentration similar to the 2 in. straight vent. Similarly, the 3 in. straight vent does not slow the flue gas down enough to lower the concentration as it reaches House 2.

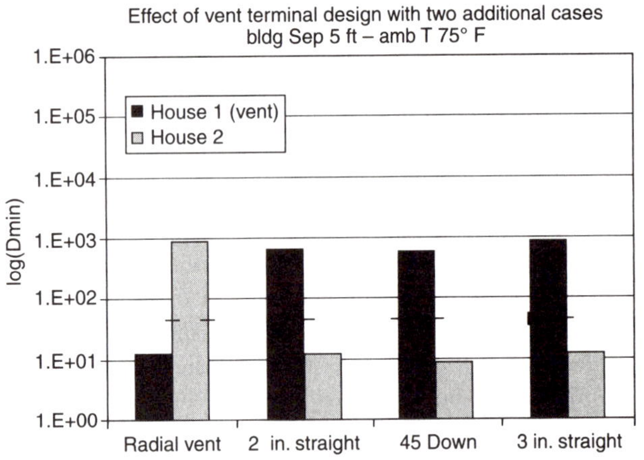


***EXHIBIT S7.11*** *Effect of Vent Terminal Design on the Minimum Dilution Factor for Both Houses.*

## SUMMARY

This study takes the first steps toward understanding the physics and identifying the important factors in sidewall venting of gas appliances between buildings. It provides an initial foundation for building a sound technical basis for establishing codes and guidelines. Fluid-dynamics modeling by itself is rarely sufficient for answering important real-world questions, particularly when those questions relate to safety, potential damage to property, or the success or failure of commercial products. To move forward, it will be important to establish the validity of these results, and to extend the modeling analyses to more real-world conditions.

There are a number of ways that one might consider to accomplish the necessary validation, extend the modeling, and move toward a more detailed, comprehensive, and technically sound justification for sidewall venting guidelines. It is possible to conduct experimental studies using a controlled environment and an appliance or set of appliances, and by constructing walls and/or inexpensive buildings. It would be possible then to visualize the plumes under various conditions and measure plume dispersion using tracer gas. CFD models of the experiments could be conducted in parallel to validate the models and refine the modeling procedures.

Another approach would be for the industry participants to note field cases where sidewall venting between houses resulted in problems such as ice buildup or infiltration of vent gases. If the industry participants were to note the conditions in the problem cases, such as appliance details, vent design, house separation, outdoor temperature, etc., then Battelle could use that information to construct and validate fluid-dynamics models. It would probably require a heating season to accumulate problem cases of interest, after which a set of them could be selected for analysis by Battelle. After validation, the modeling could then be used with confidence to answer questions related to conditions for which no direct data are available. While there would be lag time in generating the needed results, the cost of this approach is likely to be far lower than a detailed experimental study.

The results of this limited study do show that vent design is important for safely venting flue gas between buildings. This study showed that the radial vent design could possibly cause problems for the house with the vent, while straight vent designs could cause problems for a neighboring house that is less than 10 ft away. It seems reasonable that there are alternative vent-terminal designs between these limits that could prevent problems on both houses. There is an opportunity for industry to meet this need by designing new vent terminals that will disperse and dilute sidewall-vented flue gas between buildings.

## REFERENCES

1. Rutz, A. L., D. D. Paul, D. W. DeWerth, and R. A. Borgeson, "GRI's Venting Research Program: Activities at Battelle and A.G.A. Laboratories (1988–

1994)," Final Report No.: GRI-94/0371, Gas Research Institute, October 1994.

2. Barrett, R. E., "Venting Gas-Fired Appliances in Multistory Buildings: A Review of the State of the Art," Gas Research Institute Report No. GRI-92/0529, June 1992.
3. Reuther, J., N. Leslie, and R. Hemphill, "Justification for Sidewall Vent-Terminal Locations for Gas Appliances," 50th Proceeding of the International Appliance Technology Conference, pp. 231–242, IATC, 1999.
4. Reuther, J. J., and R. Hemphill, "Vent-terminal locations as related to air infiltration and indoor air quality," American Society of Heating, Refrigerating and Air-Conditioning, *ASHRAE Transactions*, **108**(1), pp. 563–571, 2002.
5. Paul, D., S. Ricci, S. Talbert, and J. Reuther, "Minimum Clearances for Sidewall Venting of Gas-Fired Appliances," Gas Appliance Manufacturers Association, Final Report prepared by Battelle, April 2007.
6. NFPA 54, *National Fuel Gas Code,* National Fire Protection Association, Quincy, MA, 2009 edition.

## SUPPLEMENT 8

# Technical/Substantive Changes from the 2006 Edition to the 2009 Edition of NFPA 54

***Editor's Note:*** *Supplement 8 is included in the handbook as a guide to the reader. The 2009 edition of the National Fuel Gas Code is compared here with the 2006 edition of the code to identify the major code changes. These major code changes are also explained to the reader in the Comments column.*

| *2006 Edition* | *2009 Edition* | *Comments* |
|---|---|---|
| The term *grade* is changed to *finished ground level* in 3.3.76.1, 7.1.6.1, 9.3.8.8, 12.4.3.6, and 12.9.2. | | To clarify the Code's intent that air intakes and sidewall vents be a distance above the finished ground level, not construction grade. The new term is consistent with other codes. |
| Not in 2006 edition. | **3.3.12 Bonding Jumper.** A reliable conductor to ensure the required electrical conductivity between metal parts required to be electrically connected. | Definitions for bonding jumper and grounding electrode added. Used in 7.13.4. Definitions extracted from *NFPA 70®, National Electrical Code®*. |
| Not in 2006 edition. | **3.3.56 Grounding Electrode.** A conducting object through which a direct connection to earth is established. [**70**, 2008] | |
| Not in 2006 edition. | **3.3.105.3 Excess Flow Valve (EFV).** A valve designed to activate when the fuel gas passing through it exceeds a prescribed flow rate. | Definition for excess flow valve was added. The term is used in new Section 5.13. |
| **5.4.1* General Considerations.** Gas piping systems shall be of such size and so installed as to provide a supply of gas sufficient to meet the maximum demand without undue loss of pressure between the point of delivery and the appliance. | **5.4.1* General Considerations.** Gas piping systems shall be of such size and so installed as to provide a supply of gas sufficient to meet the maximum demand and supply gas to each appliance inlet at not less than the minimum supply pressure required by the appliance. | Revised for clarity. The term *undue loss* is impossible to define and invites arbitrary enforcement of the code. The new text clearly states the intent. |

| *2006 Edition* | *2009 Edition* | *Comments* |
|---|---|---|
| **5.4.4 Allowable Pressure Drop.** The design pressure loss in any piping system under maximum probable flow conditions, from the point of delivery to the inlet connection of the appliance, shall be such that the supply pressure at the appliance is greater than the minimum pressure required for proper appliance operation. | **5.4.4 Allowable Pressure Drop.** The design pressure loss in any piping system under maximum probable flow conditions, from the point of delivery to the inlet connection of the appliance, shall be such that the supply pressure at the appliance is greater than or equal to the minimum pressure required by the appliance. | Revised to allow the design pressure to equal or exceed the "minimum pressure required by the appliance." Formerly the design pressure had to exceed the minimum required for "proper appliance operation," which was subjective and open to differing interpretations. |
| **5.5.1 Maximum Design Operating Pressure.** The maximum design operating pressure for piping systems located inside buildings shall not exceed 5 psi (34 kPa) unless one or more of the following conditions are met:<br>(1) through (4) Unchanged | **5.5.1 Maximum Design Operating Pressure.** The maximum design operating pressure for piping systems located inside buildings shall not exceed 5 psi (34 kPa) unless one or more of the following conditions are met:<br>(1) through (4) Unchanged<br>(5) The piping serves appliances or equipment used for agricultural purposes.<br>(6) The piping system is an LP-Gas piping system with a design operating pressure greater than 20 psi and complies with NFPA 58, *Liquefied Petroleum Gas Code*. | Added two items to the list of conditions under which pressure can exceed 5 psi (34 kPa).<br>The addition of "agricultural purposes" recognizes that higher pressures are typically used on farms for livestock facilities.<br>New item recognizes new coverage in NFPA 58 of piping systems between 20 and 50 psi (13 and 234 kPag), which is referenced. |
| **5.6.3.1 Steel.** Steel tubing shall comply with ASTM A 539, *Standard Specification for Electric Resistance-Welded Coiled Steel Tubing for Gas and Fuel Oil Lines*, or ASTM A 254, *Standard Specification for Copper Brazed Steel Tubing*. | **5.6.3.1 Steel.** Steel tubing shall comply with ASTM A 254, *Standard Specification for Copper Brazed Steel Tubing*. | Reference to ASTM 539 deleted because it has been withdrawn with no replacement. |
| **5.6.4.1** Plastic pipe, tubing, and fittings used to supply fuel gas shall be used outdoors underground only and shall conform to ASTM D 2513, *Standard Specification for Thermoplastic Gas Pressure Pipe, Tubing, and Fittings*. Pipe to be used shall be marked "gas" and "ASTM D 2513." | **5.6.4.1** Plastic pipe, tubing, and fittings used to supply fuel gas shall conform to ASTM D 2513, *Standard Specification for Thermoplastic Gas Pressure Pipe, Tubing, and Fittings*. Pipe to be used shall be marked "gas" and"ASTM D 2513." | The words "shall be used outdoors underground only" were deleted because installation is not covered by Chapter 5. This subject is adequately covered in Chapter 7, Gas Piping Installation, paragraph 7.1.7.1. |
| **5.6.8.2 Tubing Joints.** Tubing joints shall either be made with approved gas tubing fittings or be brazed with a material having a melting point in excess of 1000°F (538°C). Brazing alloys shall not contain more than 0.05 percent phosphorus. | **5.6.8.2 Tubing Joints.** Tubing joints shall either be made with approved gas tubing fittings, be brazed with a material having a melting point in excess of 1000°F (538°C), or made by press connect fittings complying with ANSI LC-4, *Press-Connect Copper and Copper Alloy. Fittings for Use in Fuel Gas Distribution Systems*. Brazing alloys shall not contain more than 0.05 percent phosphorus. | A new system of connecting copper tubing is added to the code, and the ANSI standard for the system is referenced.<br>The press-connect system uses special copper fittings that are swaged using a special tool. |
| **5.7.2.3** Gas meters shall be located at least 3 ft (0.9 m) from sources of ignition. | Not in 2009 edition. | The requirement was deleted because a gas meter is not a source of gas discharge and therefore should not have location restrictions from sources of ignition. |

| *2006 Edition* | *2009 Edition* | *Comments* |
|---|---|---|
| Not in 2006 edition. | **5.8.5.1 (3)** The regulator vent shall terminate at least 3 ft (0.9 m) from source of ignition. | New requirement for a minimum separation distance from a regulator vent termination to a source of ignition of at least 3 ft (0.9 m). A regulator vent can discharge gas and there is a possibility of an ignition. |
| Not in 2006 edition. | **5.13 Excess Flow Valve(s).** Where automatic excess flow valves are installed, they shall be listed and shall be sized and installed in accordance with the manufacturers' instructions. | Excess flow valves, while not required by NFPA 54, are being installed. If improperly sized they will not close to stop leakage; therefore, the committee added a requirement for proper sizing. |
| **7.1.5 Piping Through Foundation Wall.** Underground piping, where installed through the outer foundation or basement wall of a building, shall be encased in a protective pipe. The space between the gas piping and the building shall be sealed to prevent entry of gas or water. | **7.1.5 Piping Through Foundation Wall.** Underground piping installed through the outer foundation or basement wall of a building shall be encased in a protective sleeve or protected by an approved device or method. The space between the gas piping and the building or sleeve shall be sealed to prevent entry of gas and water. | Revised to allow specifically manufactured products that provide a seal and protection in addition to coring a concrete foundation wall and that provide no sleeve. |
| **7.1.7.2 Connections Between Metallic and Plastic Piping.** Connections made outdoors and underground between metallic and plastic piping shall be made with fittings conforming to either of the following:<br>(1) through (2) Unchanged | **7.1.7.2 Connections Between Metallic and Plastic Piping.** Connections made between metallic and plastic piping shall be made with fittings conforming to either one of the following:<br>(1) through (2) Unchanged<br>(3) ASTM F 2509, *Standard Specification for Field-assembled Anodeless Riser Kits for Use on Outside Diameter Controlled Polyethylene Gas Distribution Pipe and Tubing.* | A new reference standard ASTM F 2509 is added to the list of fittings that can be used to connect metallic and plastic pipe. It covers field-assembled anodeless riser kits. |
| **7.2.6.1** Piping shall be supported with pipe hooks, metal pipe straps, bands, brackets, or hangers suitable for the size of piping, of adequate strength and quality, and located at intervals so as to prevent or damp out excessive vibration. Piping shall be anchored to prevent undue strains on connected appliances and equipment and shall not be supported by other piping. Pipe hangers and supports shall conform to the requirements of ANSI/MSS SP-58, *Pipe Hangers and Supports — Materials, Design and Manufacture.* | **7.2.6.1** Piping shall be supported with metal pipe hooks, metal pipe straps, -metal bands, metal brackets, or metal hangers, or building structural components, suitable for the size of piping, of adequate strength and quality, and located at intervals so as to prevent or damp out excessive vibration. Piping shall be anchored to prevent undue strains on connected appliances and equipment and shall not be supported by other piping. Pipe hangers and supports shall conform to the requirements of ANSI/MSS SP-58, *Pipe Hangers and Supports — Materials, Design and Manufacture.* | Requirements for support of gas pipe are revised to be consistent with plumbing code requirements. Plastic materials are no longer allowed for straps, and hangers can creep (elongate, deform, or sag) with age, can have a short lifespan where exposed to sunlight, can lack sufficient strength to support heavy piping, are more susceptible to physical damage, can fail from stress cracking, and will certainly fail in a fire.<br><br>Gas piping could fail under its own weight if not properly supported and can also put stress on appliance controls and connections. |
| Not in 2006 edition. | **7.9.2.4 Shutoff Valve for Laboratories.** Each laboratory space containing two or more gas outlets installed on tables, benches, or in hoods in educational, research, commercial, and industrial occupancies shall have a single shutoff valve through which all such gas outlets | New requirement for a master shutoff valve in laboratories. This allows turning off the gas supply to all lab benches. Gas turret valves commonly used in laboratories can be opened inadvertently. |

| *2006 Edition* | *2009 Edition* | *Comments* |
|---|---|---|
| | are supplied. The shutoff valve shall be accessible and shall be located within the laboratory or located adjacent to the laboratory's egress door and shall be identified. | |
| **7.13 Electrical Bonding and Grounding.**<br>**7.13.1** Each aboveground portion of a gas piping system that is likely to become energized shall be electrically continuous and bonded to an effective ground-fault current path. Gas piping shall be considered to be bonded when it is connected to appliances that are connected to the appliance grounding conductor of the circuit supplying that appliance. | **7.13 Electrical Bonding and Grounding.**<br>**7.13.1 Pipe and Tubing Other than CSST.** Each aboveground portion of a gas piping system other than CSST that is likely to become energized shall be electrically continuous and bonded to an effective ground-fault current path. Gas piping, other than CSST, shall be considered to be bonded when it is connected to appliances that are connected to the appliance grounding conductor of the circuit supplying that appliance. | Revised bonding requirements for CSST, and reference to NFPA 780 is added. |
| | **7.13.2 CSST.** CSST gas piping systems shall be bonded to the electrical service grounding electrode system at the point where the gas service enters the building. The bonding jumper shall not be smaller than 6 AWG copper wire or equivalent | New requirement for bonding of CSST piping systems. This is intended to reduce punctures from lightning strikes. |
| **7.13.2** Gas piping shall not be used as a grounding conductor or electrode. | **7.13.3 Prohibited Use.** Gas piping shall not be used as a grounding conductor or electrode. This does not preclude the bonding of metallic piping to a grounding system. | Intent clarified. |
| | **7.13.4* Lightning Protection Systems** Where a lightning protection system is installed, the bonding of the gas piping shall be in accordance with NFPA 780, *Standard for the Installation of Lightning Protection Systems*. | Specific reference to NFPA 780 is added to clarify that 7.13.3 does not prevent bonding gas piping to lightning protection systems. |
| **8.2 Piping System, Appliance, and Equipment Leakage Check.** | **8.2 Piping System Leak Check.** | Title changed to Leak Check, which is used elsewhere in NFPA 54. |
| **9.3.1.5** Makeup air requirements for the operation of exhaust fans, kitchen ventilation systems, clothes dryers, and fireplaces shall be considered in determining the adequacy of a space to provide combustion air requirements. | **9.3.1.5** Where exhaust fans, clothes dryers, and kitchen ventilation systems interfere with the operation of appliances, makeup air shall be provided. | The requirement for makeup air is revised to be enforceable. Previously, the text only required "consideration." There is only one solution to the problem of negative pressure, and that is to provide makeup air. |
| **9.6.1.1 Commercial Cooking Appliances.** Commercial cooking appliances that are moved for cleaning and sanitation purposes shall be connected in accordance with the connector manufacturer's installation instructions using a listed appliance connector complying with ANSI Z21.69, *Connectors for Movable Gas Appliances*. | **9.6.1.1 Commercial Cooking Appliances.** Commercial cooking appliances that are moved for cleaning and sanitation purposes shall be connected in accordance with the connector manufacturer's installation instructions using a listed appliance connector complying with ANSI Z21.69/CSA 6.22, *Connectors for Movable Gas Appliances*. The commercial cooking appliance connector installation shall be configured in accordance with the manufacturer's installation instructions. | Provides additional guidance for the installation of commercial appliance connectors by referencing the connector manufacturer's installation instructions. |

| *2006 Edition* | *2009 Edition* | *Comments* |
|---|---|---|
| Not in 2006 edition. | **9.6.1.2 Commercial Cooking Appliances with Casters.** Movement of appliances with casters shall be limited by a restraining device installed in accordance with the connector and appliance manufacturer's installation instructions. | Requires commercial cooking appliances with casters to have a restraining device. |
| Not in 2006 edition. | **9.6.1.3* Suspended Low-Intensity Infrared Tube Heaters.** Suspended low-intensity infrared tube heaters shall be connected to the building piping system with a connector listed for the application in accordance with ANSI Z21.24/CGA 6.10, *Standard for Connectors for Gas Appliances*.<br>(A) The connector shall be installed in accordance with the tube heater installation instructions and shall be in the same room as the appliance.<br>(B) Only one connector shall be used per appliance. | Recently, gas leakage from fractured copper tubing has caused multiple major incidents. The cause was traced to the use of copper tubing as a gas connector on radiant tube heaters, which are not secure to the structure. Vibrations from the fan motor appear to be the source of vibration that caused the copper connector brittle failure. Copper tubing is now prohibited on such heaters by requiring only listed connectors in compliance with ANSI Z21.24 or UL 569. |
| **9.6.6 Gas Convenience Outlets.** Appliances shall be permitted to be connected to the building piping by means of a listed gas convenience outlet, in conjunction with a listed appliance connector, used in accordance with the manufacturer's installation instructions. | **9.6.6 Gas Convenience Outlets.** Gas convenience outlets shall be listed in accordance with ANSI Z21.90, *Gas Convenience Outlets and Optional Enclosures,* and installed in accordance with the manufacturer's installation instructions. | The standard now reflects that an ANSI standard for gas convenience outlets is available. |
| **9.6.7 Sediment Trap.** Where a sediment trap is not incorporated as a part of the appliance, a sediment trap shall be installed as close to the inlet of the appliance as practicable at the time of appliance installation. The sediment trap shall be either a tee fitting with a capped nipple in the bottom outlet as illustrated in Figure 9.6.7 or other device recognized as an effective sediment trap. Illuminating appliances, ranges, clothes dryers, decorative appliances for installation in vented fireplaces, gas fireplaces, and outdoor grills shall not be required to be so equipped. | **9.6.7 Sediment Trap.** Where a sediment trap is not incorporated as a part of the appliance, a sediment trap shall be installed downstream of the appliance shutoff valve as close to the inlet of the appliance as practical at the time of appliance installation. The sediment trap shall be either a tee fitting with a capped nipple in the bottom outlet as illustrated in Figure 9.6.7 or other device recognized as an effective sediment trap. Illuminating appliances, ranges, clothes dryers, decorative appliances for installation in vented fireplaces, gas fireplaces, and outdoor grills shall not be required to be so equipped. | Added "downstream of the appliance shutoff valve" to specify where the sediment trap must be installed. |
| **10.3.1 (1)** Central heating furnaces and low-pressure boilers shall be installed in a closet located in the bedroom or bathroom, the closet shall have a weather-stripped solid door with a self-closing device, and all combustion air shall be obtained from the outdoors. | **10.3.1 (1)** Central heating furnaces and low-pressure boilers shall be installed in a closet equipped with a weather-stripped door with no openings, and with a self-closing device. All combustion air shall be obtained from the outdoors in accordance with 9.3.3. | The word "solid" is deleted from subsection 10.3.1(1) to clarify the committee's intent that a solid core door is not required. The intent is that the door shall not have "openings" in the door such as louvers. |
| Not in 2006 edition. | **10.23.2** Unvented room heaters shall be listed in accordance with ANSI Z21.11.2, *Gas-Fired Room Heaters-Volume II, Unvented Room Heaters*, and shall be | Listing requirements for unvented room heaters were added. |

| *2006 Edition* | *2009 Edition* | *Comments* |
|---|---|---|
| | installed in accordance with the manufacturer's installation instructions. | |
| Not in 2006 edition. | **10.24.1** Stationary gas engines shall not be rigidly connected to the gas supply piping. | Added to recognize the need to isolate the piping system from vibrations from stationary gas engines. |
| **10.28.1 (1)** Water heater shall be installed in a closet equipped with a weather-stripped door with a self-closing device, and all combustion air shall be obtained from the outdoors in accordance with 9.3.3. | **10.28.1 (1)** Water heater shall be installed in a closet equipped with a weather-stripped door with no openings, and with a self-closing device. All combustion air shall be obtained from the outdoors in accordance with 9.3.3. | The committee's intent that that the door must not have "openings" in the door such as louvers is clarified. |
| Not in 2006 edition. | **10.32 Outdoor Open Flame Decorative Appliances.** Permanently fixed in place outdoor open flame decorative appliances shall be installed in accordance with the following.<br>**10.32.1 Listed Units.** Listed outdoor open flame decorative appliances shall be installed in accordance with the manufacturer's installation instructions.<br>**10.32.2 Unlisted Units.** Unlisted outdoor open flame decorative appliances shall be installed outdoors in accordance with the manufacturer's installation instructions and with clearances to combustible material of not less than 36 in. (910 mm) from the sides. In no case shall the appliance be located under overhead combustible construction.<br>**10.32.3 Connection to the Piping System.** The connection to the gas piping system shall be in accordance with 9.6.1 (1), (2), (4), or (5). | New section covering outdoor open flame decorative appliances in response to questions regarding their proper installation. |
| **12.4.4.2** Where automatically operated appliances are vented through a ventilating hood or exhaust system equipped with a damper or with a power means of exhaust, provisions shall be made to allow the flow of gas to the main burners only when the damper is open to a position to properly vent the appliance and when the power means of exhaust is in operation. | **12.4.4.2** Where automatically operated appliances, other than commercial cooking appliances, are vented through a ventilating hood or exhaust system equipped with a damper or with a power means of exhaust, provisions shall be made to allow the flow of gas to the main burners only when the damper is open to a position to properly vent the appliance and when the power means of exhaust is in operation. | This requirement was clarified to specifically exempt commercial cooking appliances vented using range hoods, in response to questions from enforcers. |
| Not in 2006 edition. | **12.5.3 Plastic Vent Joints.** Plastic pipe and fittings used to vent appliances shall be installed in accordance with the appliance manufacturer's installation instructions. Where primer is required, it shall be of a contrasting color. | This requirement easily allows inspectors to verify the use of primer, which is of utmost importance to proper joints in plastic pipe. Note that a specific color is not required, and manufacturers can use any contrasting color. |
| **12.6.3.2** Where an incinerator is vented by a chimney serving other appliances, the gas input to the incinerator shall not be included in calculating chimney size, | Not in 2009 edition. | Requirement deleted, as it was in conflict with 12.6.5.1. Solid fuel appliances and gas-fired appliances cannot share a chimney or flue. |

| *2006 Edition* | *2009 Edition* | *Comments* |
|---|---|---|
| provided the chimney flue diameter is not less than 1 in. (25 mm) larger in equivalent diameter than the diameter of the incinerator flue outlet. | | |
| **12.6.4.2** Chimneys shall be lined in accordance with NFPA 211, *Standard for Chimneys, Fireplaces, Vents, and Solid Fuel-Burning Appliances.*<br><br>*Exception: Existing chimneys shall be permitted to have their use continued when an appliance is replaced by an appliance of similar type, input rating, and efficiency.* | **12.6.4.2** Chimneys shall be lined in accordance with NFPA 211, *Standard for Chimneys, Fireplaces, Vents, and Solid Fuel–Burning Appliances.*<br><br>*Exception: Existing chimneys shall be permitted to have their use continued when an appliance is replaced by an appliance of similar type, input rating, and efficiency, where the chimney complies with 12.6.4, and the sizing of the chimney is in accordance with 12.6.3.* | References were added to the exception so it is understood that chimneys must be in good repair and sized properly. |
| **12.7.1 Application.** The installation of gas vents shall meet the following requirements:<br>(1) unchanged<br>(2) unchanged<br>(3) Vents serving Category I fan-assisted appliances shall be installed in accordance with the appliance manufacturer's instructions and Chapter 13 or other approved engineering methods.<br>(4) Gas vents installed within masonry chimneys shall be installed in accordance with the manufacturer's installation instructions. Gas vents installed within masonry chimneys shall be identified with a permanent label installed at the point where the vent enters the chimney. The label shall contain the following language: "This gas vent is for appliances that burn gas. Do not connect to solid or liquid fuel–burning appliances or incinerators." | **12.7.1 Installation.** The installation of gas vents shall meet the following requirements:<br>(1) unchanged<br>(2) unchanged<br>(3) Gas vents installed within masonry chimneys shall be installed in accordance with the manufacturer's installation instructions. Gas vents installed within masonry chimneys shall be identified with a permanent label installed at the point where the vent enters the chimney. The label shall contain the following language: "This gas vent is for appliances that burn gas. Do not connect to solid or liquid fuel–burning appliances or incinerators."<br>(4) Screws, rivets, and other fasteners shall not penetrate the inner wall of double wall gas vents, except at the transition from the appliance draft hood outlet, flue collar, or single wall metal connector to a double wall vent. | Former (3) is deleted as sizes of vents are covered in Section 12.7.3.<br><br>New (4) added to specifically prohibit penetration of the inner wall of double wall vents. This is a common violation of vent manufacturer's installation instructions, and the committee is advised that such instructions are rarely accessible on the job site. |
| **12.11.14.2** A vent connector made of a single-wall metal pipe shall not pass through a combustible exterior wall unless guarded at the point of passage by a ventilated metal thimble not smaller than the following:<br>(1), (2), (3), and Exception | **12.8.4.** Single-wall metal pipe shall not pass through a combustible exterior wall unless guarded at the point of passage by a ventilated metal thimble not smaller than the following:<br>(1), (2), (3), and Exception. | The phrase "vent connector" was replaced by "single-wall metal pipe," in 12.11.14.2 and its subparagraphs, in several locations in the paragraph. The section was relocated from vent connector section to metal piping section, which is a more appropriate location. |
| Not in 2006 edition. | **12.8.4.1 Prohibited use.** Single-wall metal pipe shall not be used as a vent in dwellings and residential occupancies. | Single-wall metal pipe is prohibited because it requires a large clearance to combustibles, is difficult to support, and can have a short life due to condensation and exposure to the elements. |

| *2006 Edition* | *2009 Edition* | *Comments* |
|---|---|---|
| **12.9.2** A mechanical draft venting system of other than direct-vent type shall terminate at least 4 ft (1.2 m) below, 4 ft (1.2 m) horizontally from, or 1 ft (300 mm) above any door, operable window, or gravity air inlet into any building. The bottom of the vent terminal shall be located at least 12 in. (300 mm) above grade. | **12.9.2** A mechanical draft venting system of other than direct-vent type shall terminate at least 4 ft (1.2 m) below, 4 ft (1.2 m) horizontally from, or 1 ft (300 mm) above any door, operable window, or gravity air inlet into any building. The bottom of the vent terminal shall be located at least 12 in. (300 mm) finished above ground level. | It is the committee's intent to measure from finished ground level, not construction grade. |
| **12.9.4** Through-the-wall vents for Category II and Category IV appliances and noncategorized condensing appliances shall not terminate over public walkways or over an area where condensate or vapor could create a nuisance or hazard or could be detrimental to the operation of regulators, relief valves, or other equipment. Where local experience indicates that condensate is a problem with Category I and Category III appliances, this provision shall also apply. | **12.9.4** Through-the-wall vents for Category II and Category IV appliances and noncategorized condensing appliances shall not terminate over public walkways or over an area where condensate or vapor could create a nuisance or hazard or could be detrimental to the operation of regulators, relief valves, or other equipment. Where local experience indicates that condensate is a problem with Category I and Category III appliances, this provision shall also apply. Drains for condensate shall be installed in accordance with the manufacturer's installation instructions. | Sentence was added to provide guidance on installation of condensate drains. |
| Not in 2006 edition. | **12.9.5** Where vents, including those for direct-vent appliances or combustion air intake pipes, penetrate outside walls of buildings, the annular spaces around such penetrations shall be permanently sealed using approved materials to prevent entry of combustion products into the building. | New requirement for sealing around vent penetration of outside walls. This is to prevent insects from entering the building and to prevent products of combustion from being drawn into the building, which can result in carbon monoxide production in combination with other factors. This was based on an incident. |
| Not in 2006 edition. | **12.11.4.1** Where two or more openings are provided into one chimney flue or vent, the openings shall either be at different levels, or the connectors shall be attached to the vertical portion of the chimney or vent at an angle of 45 degrees or less relative to the vertical. | Openings into common vents must be at different levels or if on the same level made at a 45-degree angle. |
| **12.11.9.2** The maximum horizontal length of a single-wall connector shall be 75 percent of the height of the chimney or vent, except for engineered systems. The maximum length of an individual connector for a chimney or vent system serving multiple appliances, from the appliance outlet to the junction with the common vent or another connector, shall be 100 percent of the height of the chimney or vent. | **12.11.9.2** The maximum horizontal length of a single-wall connector shall be 75 percent of the height of the chimney or vent, except for engineered systems. | This section was deleted to resolve a inconsistency between requirements added with the revised venting tables and older requirements. It limits the length of vent connectors to 75 percent of chimney height consistently throughout the code. |
| **13.1.3 Elbows.** Single-appliance venting configurations with zero (0) lateral lengths in Table 13.1(a), Table 13.1(b), and Table 13.1(e) shall not have elbows in the venting system. Single-appliance venting | **13.1.3 Elbows.** Single-appliance venting configurations with zero (0) lateral lengths in Tables 13.1(a), 13.1(b), and 13.1(e) shall not have elbows in the venting system. Single-appliance venting with | Revised to provide guidance on how to size vents using multiple offsets. |

| *2006 Edition* | *2009 Edition* | *Comments* |
|---|---|---|
| with lateral lengths include two 90 degree elbows. For each additional elbow up to and including 45 degrees, the maximum capacity listed in the venting tables shall be reduced by 5 percent. For each additional elbow greater than 45 degrees up to and including 90 degrees, the maximum capacity listed in the venting tables shall be reduced by 10 percent. | lateral lengths include two 90-degree elbows. For each additional elbow up to and including 45 degrees, the maximum capacity listed in the venting tables shall be reduced by 5 percent. For each additional elbow greater than 45 degrees up to and including 90 degrees, the maximum capacity listed in the venting tables shall be reduced by 10 percent. Where multiple offsets occur in a vent, the total lateral length of all offsets combined shall not exceed that specified in Tables 13.1(a) through (e). | |
| **13.1.11 Chimneys and Vent Locations.** Table 13.1(a) through Table 13.1(e) shall be used only for chimneys and vents not exposed to the outdoors below the roofline. A Type B vent or listed chimney lining system passing through an unused masonry chimney flue shall not be considered to be exposed to the outdoors. A Type B vent passing through an unventilated enclosure or chase insulated to a value of not less than R8 shall not be considered to be exposed to the outdoors. Table 13.1(c) in combination with Table 13.1(f) shall be used for clay-tile-lined exterior masonry chimneys, provided all of the following requirements are met:<br>(1) through (6) unchanged<br>(7) Where the conditions of 13.1.11(1) through (6) cannot be met, an alternative venting design shall be used, such as a listed chimney lining system.<br>*Exception: Vents serving listed appliances installed in accordance with the appliance manufacturer's installation instructions.* | **13.1.11 Chimneys and Vent Locations.** Table 13.1(a) through Table 13.1(e) shall be used only for chimneys and vents not exposed to the outdoors below the roofline. A Type B vent or listed chimney lining system passing through an unused masonry chimney flue shall not be considered to be exposed to the outdoors. A Type B vent passing through an unventilated enclosure or chase insulated to a value of not less than R8 shall not be considered to be exposed to the outdoors. Table 13.1(c) in combination with Table 13.1(f) shall be used for clay-tile-lined exterior masonry chimneys, provided all of the following requirements are met:<br>(1) through (6) unchanged | Item (7) and the Exception, essentially an exception to the listed requirements, is deleted as the subject is covered elsewhere in the code. |
| **13.2.5 Vent Offsets.** Where the common vertical vent is offset, the maximum capacity of the common vent shall be reduced in accordance with 13.2.6 and the horizontal length of the common vent offset shall not exceed 18 in./in. (18 mm/mm) of common vent diameter (D). | **13.2.5 Vent Offsets.** Where the common vertical vent is offset, the maximum capacity of the common vent shall be reduced in accordance with 13.2.6 and the horizontal length of the common vent offset shall not exceed 18 in./in. (18 mm/mm) of common vent diameter *(D)*. Where multiple offsets occur in a common vent, the total horizontal length of all offsets combined shall not exceed 18 in./in. (18 mm/mm) of the common vent diameter. | Openings into common vents must be at different levels or if on the same level made at 45-degree angle. |
| **13.2.22 Chimneys and Vent Locations.** Table 13.2(a) and Table 13.2(b) shall be used only for chimneys and vents not exposed to the outdoors below the roof line. A Type B vent passing through an unventilated enclosure or chase insulated | **13.2.22 Chimneys and Vent Locations.** Table 13.2(a) and Table 13.2(b) shall be used only for chimneys and vents not exposed to the outdoors below the roof line. A Type B vent passing through an unventilated enclosure or chase insulated | |

| *2006 Edition* | *2009 Edition* | *Comments* |
|---|---|---|
| to a value of not less than R8 shall not be considered to be exposed to the outdoors. Table 13.2(f), Table 13.2(g), Table 13.2(h), and Table 13.2(i) shall be used for clay-tile-lined exterior masonry chimneys, provided all the following conditions are met:<br>(1) through (5) unchanged<br>(6) Where the conditions of 13.2.22(1) through (5) cannot be met, an alternative venting design shall be used, such as a listed chimney lining system.<br>*Exception: Vents serving listed appliances installed in accordance with the appliance manufacturer's installation instructions.* | to a value of not less than R8 shall not be considered to be exposed to the outdoors. Table 13.2(f), Table 13.2(g), Table 13.2(h), and Table 13.2(i) shall be used for clay-tile-lined exterior masonry chimneys, provided all the following conditions are met:<br>(1) through (5) unchanged | Item (6) and the Exception, essentially an exception to the listed requirements, is deleted as the subject is covered elsewhere in the code. |
| Not in 2006 edition. | **A.5.6** Table A.5.6 is a list of piping materials and fittings that are allowed in the Code. | New Annex table summarizes the Code's piping and joining materials and methods in a table format. |
| Not in 2006 edition. | **A.9.6.1.3** The expansion and contraction of the heater and the vibration from the blower motor may lead to work hardening of the rigid pipe or semi-rigid metallic tubing, which may ultimately lead to fractures and leakage. Connectors for this type of heater should have adequate flexibility, temperature rating, and vibration resistance to accommodate the characteristics of the heater. Such flexible connectors for suspended heaters should meet the following criteria:<br>**(1)** Be determined to be appropriate for the application.<br>**(2)** Be specified by the heater manufacturer.<br>**(3)** Be installed in accordance with the manufacturer's installation instructions. | Explanatory information to provide guidance for new section 9.6.1.3. |
| Not in 2006 edition. | **A.12.8.4.1** The prohibition only applies to a vent entirely constructed of single-wall metal pipe located in a residential occupancy. The prohibition does not apply to single wall vent connectors used to connect an appliance to the vent as permitted in Section 12.11 and Chapter 13. | Explanatory material is provided to clarify the prohibition of single wall pipe in residential vents. While Section 12.8 is clearly the vent section, several knowledgeable users of the Code have misread the proposed prohibition to extend to vent connectors. |

# Code Index

**H**



**I**

P

# Commentary Index

## G

## J



## K



## L

## O

## Q



## R

# IMPORTANT NOTICES AND DISCLAIMERS CONCERNING AGA AND NFPA® DOCUMENTS

## NOTICE AND DISCLAIMER OF LIABILITY CONCERNING THE USE OF AGA AND NFPA DOCUMENTS

NFPA codes, standards, recommended practices, and guides, of which the document contained herein is one, and AGA's Z223.1 are developed through a consensus standards development process approved by the American National Standards Institute. This process brings together volunteers representing varied viewpoints and interests to achieve consensus on fire and other safety issues. While the AGA and the NFPA administer the process and establish rules to promote fairness in the development of consensus, they do not independently test, evaluate, or verify the accuracy of any information or the soundness of any judgments contained in their codes and standards.

The AGA and NFPA disclaim liability for any personal injury, property or other damages of any nature whatsoever, whether special, indirect, consequential or compensatory, directly or indirectly resulting from the publication, use of, or reliance on this document. The AGA and the NFPA also make no guaranty or warranty as to the accuracy or completeness of any information published herein.

In issuing and making this document available, the AGA and the NFPA are not undertaking to render professional or other services for or on behalf of any person or entity. Nor are the AGA and the NFPA undertaking to perform any duty owed by any person or entity to someone else. Anyone using this document should rely on his or her own independent judgment or, as appropriate, seek the advice of a competent professional in determining the exercise of reasonable care in any given circumstances.

The AGA and the NFPA have no power, nor do they undertake, to police or enforce compliance with the contents of this document. Nor do the AGA and the NFPA list, certify, test or inspect products, designs, or installations for compliance with this document. Any certification or other statement of compliance with the requirements of this document shall not be attributable to the AGA and the NFPA and is solely the responsibility of the certifier or maker of the statement.

## ADDITIONAL NOTICES AND DISCLAIMERS

### Updating of AGA and NFPA Documents

Users of NFPA codes, standards, recommended practices, and guides and AGA Z223.1 should be aware that these documents may be superseded at any time by the issuance of new editions or may be amended from time to time through the issuance of Tentative Interim Amendments. An official AGA or NFPA document at any point in time consists of the current edition of the document together with any Tentative Interim Amendments and any Errata then in effect. In order to determine whether a given document is the current edition and whether it has been amended through the issuance of Tentative Interim Amendments or corrected through the issuance of Errata, consult appropriate NFPA publications such as the National Fire Codes(r) Subscription Service, visit the NFPA website at www.nfpa.org, or contact the AGA or NFPA at the address listed below.

### Interpretations of AGA and NFPA Documents

A statement, written or oral, that is not processed in accordance with the interpretation procedures of the Z223 Committee or Section 6 of the NFPA Regulations Governing Committee Projects shall not be considered the official position of the Z223 Committee or the NFPA or any of its Committees and shall not be considered to be, nor be relied upon as, a Formal Interpretation.

### Patents

The AGA and the NFPA do not take any position with respect to the validity of any patent rights asserted in connection with any items which are mentioned in or are the subject of AGA and NFPA codes, standards, recommended practices, and guides, and the AGA and the NFPA disclaim liability for the infringement of any patent resulting from the use of or reliance on these documents. Users of these documents are expressly advised that determination of the validity of any such patent rights, and the risk of infringement of such rights, is entirely their own responsibility. AGA and NFPA adhere to applicable policies of the American National Standards Institute with respect to patents. For further information contact the AGA or NFPA at the address listed below.

### Law and Regulations

Users of these documents should consult applicable federal, state, and local laws and regulations. The AGA and NFPA do not, by the publication of their codes, standards, recommended practices, and guides, intend to urge action that is not in compliance with applicable laws, and these documents may not be construed as doing so.

### Copyrights

This document is copyrighted by the AGA and the NFPA. It is made available for a wide variety of both public and private uses. These include both use, by reference, in laws and regulations, and use in private self-regulation, standardization, and the promotion of safe practices and methods. By making this document available for use and adoption by public authorities and private users, the AGA and the NFPA do not waive any rights in copyright to this document.

Use of AGA and NFPA documents for regulatory purposes should be accomplished through adoption by reference. The term "adoption by reference" means the citing of title, edition, and publishing information only. Any deletions, additions, and changes desired by the adopting authority should be noted separately in the adopting instrument. In order to assist NFPA in following the uses made of its documents, adopting authorities are requested to notify the NFPA (Attention: Secretary, Standards Council) in writing of such use. For technical assistance and questions concerning adoption of AGA and NFPA documents, contact the AGA or the NFPA at the address below.

### For Further Information

All questions or other communications relating to AGA and NFPA codes, standards, recommended practices, and guides and all requests for information on AGA and NFPA procedures governing their codes and standards development process, including information on the procedures for requesting Formal Interpretations, for proposing Tentative Interim Amendments, and for proposing revisions to AGA or NFPA documents during regular revision cycles, should be sent to AGA headquarters, addressed to the attention of the Secretary, Accredited Standards Committee Z223, 400 N. Capitol Street, N.W., Washington, DC 20001 and to NFPA headquarters, addressed to the attention of the Secretary, Standards Council, NFPA, 1 Batterymarch Park, P.O. Box 9101, Quincy, MA 02269-9101. For more information about NFPA, visit the NFPA website at www.nfpa.org. For more information about AGA, visit the AGA website at www.aga.org.

# A Guide to Using the National Fuel Code Handbook

This seventh edition of the *National Fuel Gas Code Handbook* contains the complete text of the 2009 edition of NFPA® 54, *National Fuel Gas Code.*

The commentary text contains Frequently Asked Questions shown in the margin as FAQs. The FAQs answer the most commonly asked questions throughout the handbook.

222 Chapter 10 • Installation of Specific Appliances

FAQ ► Can a refrigeration coil be installed in a duct downstream of a forced air furnace?

A refrigeration coil can be installed with a forced air furnace only if the furnace blower can provide sufficient airflow.

The general requirement assumes that the furnace blower can deliver sufficient airflow at least 0.5 in. w.c. (120 Pa) external static pressure resistance to overcome the resistance imposed by the supply and return air duct systems plus the addition of a refrigeration coil. The requirement in 10.3.8(1) provides enough blower capacity for a nominal 0.15 in. w.c. (37 Pa) external static pressure for the supply duct system, 0.30 in. w.c. (75 Pa) for the cooling coil, and 0.05 in. w.c. (12.5 Pa) for the return air duct system and filters. Failure to provide sufficient airflow to overcome the external static pressure can lead to problems that include the following:

- Condensate freezing on the refrigeration coil during air-conditioning mode, which can block airflow through the system
- Overheating of the furnace heating element, which can cause premature failure of the furnace

Consult the furnace manufacturer's specifications to determine available airflow rates and external static pressures and recommendations for refrigeration coil applications. Consult the air conditioner manufacturer's specifications to determine the required airflow rate and external static pressure resistance of the coil.

(2) Furnaces shall not be located upstream from cooling units, unless the cooling unit is designed or equipped so as not to develop excessive temperature or pressure.

Direct-expansion cooling coils with a specified condensing unit are listed as part of the air-conditioning unit assembly. The requirements for listing generally include the protective devices on the system necessary to prevent the development of excessive temperatures and pressures in the cooling coil, as stated in 10.3.8(2), when the coil is installed downstream from the furnace. In direct expansion systems, the vapor in the cooling coil expands into the

*EXHIBIT 10.11 Typical Installation of a Horizontal Furnace Showing Plenum and Ducts. (Courtesy of*

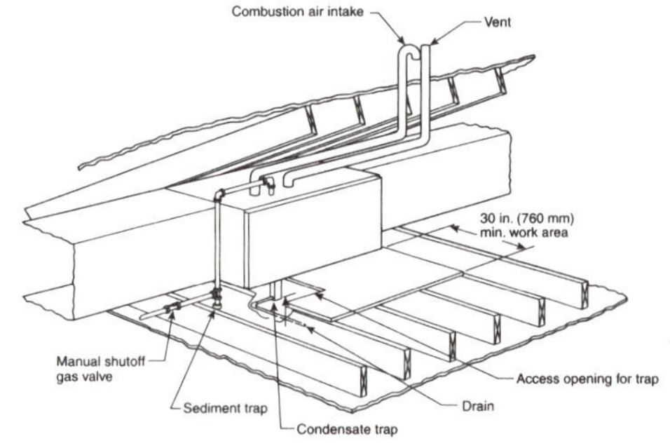


Note: Local codes may require a drain pan under the furnace and condensate trap when a condensing furnace is installed above finished ceilings.

*2009 National Fuel Gas Code Handbook*

SUPPLEMENT 8

## Technical/Substantive Changes from the 2006 Edition to the 2009 Edition of NFPA 54

***Editor's Note:*** *Supplement 8 is included in the handbook as a guide to the reader. The 2009 edition of the National Fuel Gas Code is compared here with the 2006 edition of the code to identify the major code changes. These major code changes are also explained to the reader in the Comments column.*

| *2006 Edition* | *2009 Edition* | *Comments* |
|---|---|---|
| The term *grade* is changed to *finished ground level* in 3.3.76.1, 7.1.6.1, 9.3.8.8, 12.4.3.6, and 12.9.2. | | To clarify the Code's intent that air intakes and sidewall vents be a distance above the finished ground level, not construction grade. The new term is consistent with other codes. |
| Not in 2006 edition. | **3.3.12 Bonding Jumper.** A reliable conductor to ensure the required electrical conductivity between metal parts required to be electrically connected. | Definitions for bonding jumper and grounding electrode added. Used in 7.13.4. Definitions extracted from *NFPA 70®, National Electrical Code®*. |
| Not in 2006 edition. | **3.3.56 Grounding Electrode.** A conducting object through which a direct connection to earth is established. [**70**, 2008] | |
| Not in 2006 edition. | **3.3.105.3 Excess Flow Valve (EFV).** A valve designed to activate when the fuel gas passing through it exceeds a prescribed flow rate. | Definition for excess flow valve was added. The term is used in new Section 5.13. |
| **5.4.1* General Considerations.** Gas piping systems shall be of such size and so installed as to provide a supply of gas sufficient to meet the maximum demand without undue loss of pressure between the point of delivery and the appliance. | **5.4.1* General Considerations.** Gas piping systems shall be of such size and so installed as to provide a supply of gas sufficient to meet the maximum demand and supply gas to each appliance inlet at not less than the minimum supply pressure required by the appliance. | Revised for clarity. The term *undue loss* is impossible to define and invites arbitrary enforcement of the code. The new text clearly states the intent. |

541

Eight supplements appear at the end of the handbook and are not a part of the Code or the commentary. They provide additional, useful information for Code users and include drawings, photographs, tables, and extended text. The supplements appear in black type inside a blue box.